P9-AQP-892

Methods in Enzymology

Volume 394
NUCLEAR MAGNETIC RESONANCE OF
BIOLOGICAL MACROMOLECULES
Part C

METHODS IN ENZYMOLOGY

EDITORS-IN-CHIEF

John N. Abelson   Melvin I. Simon

DIVISION OF BIOLOGY
CALIFORNIA INSTITUTE OF TECHNOLOGY
PASADENA, CALIFORNIA

FOUNDING EDITORS

Sidney P. Colowick and Nathan O. Kaplan

QP601
M49
v. 394

Methods in Enzymology

Volume 394

Nuclear Magnetic Resonance of Biological Macromolecules

Part C

EDITED BY

Thomas L. James

DEPARTMENT OF PHARMACEUTICAL CHEMISTRY
UNIVERSITY OF CALIFORNIA
SAN FRANCISCO, CALIFORNIA 94143

ELSEVIER
ACADEMIC
PRESS

AMSTERDAM • BOSTON • HEIDELBERG • LONDON
NEW YORK • OXFORD • PARIS • SAN DIEGO
SAN FRANCISCO • SINGAPORE • SYDNEY • TOKYO
Academic Press is an imprint of Elsevier

NO LONGER THE PROPERTY
OF THE
UNIVERSITY OF R. I. LIBRARY

Elsevier Academic Press
525 B Street, Suite 1900, San Diego, California 92101-4495, USA
84 Theobald's Road, London WC1X 8RR, UK

This book is printed on acid-free paper.

Copyright © 2005, Elsevier Inc. All Rights Reserved.

No part of this publication may be reproduced or transmitted in any form or by any
means, electronic or mechanical, including photocopy, recording, or any information
storage and retrieval system, without permission in writing from the Publisher.

The appearance of the code at the bottom of the first page of a chapter in this book
indicates the Publisher's consent that copies of the chapter may be made for
personal or internal use of specific clients. This consent is given on the condition,
however, that the copier pay the stated per copy fee through the Copyright Clearance
Center, Inc. (www.copyright.com), for copying beyond that permitted by
Sections 107 or 108 of the U.S. Copyright Law. This consent does not extend to
other kinds of copying, such as copying for general distribution, for advertising
or promotional purposes, for creating new collective works, or for resale.
Copy fees for pre-2005 chapters are as shown on the title pages. If no fee code
appears on the title page, the copy fee is the same as for current chapters.
0076-6879/2005 $35.00

Permissions may be sought directly from Elsevier's Science & Technology Right
Department in Oxford, UK: phone: (+44) 1865 843830, fax: (+44) 1865 853333,
E-mail: permissions@elsevier.com.uk. You may also complete your request on-line
via the Elsevier homepage (http://elsevier.com), by selecting
"Customer Support" and then "Obtaining Permissions."

For all information on all Academic Press publications
visit our Web site at www.books.elsevier.com

ISBN: 0-12-182799-2

PRINTED IN THE UNITED STATES OF AMERICA
05 06 07 08 9 8 7 6 5 4 3 2 1

Working together to grow
libraries in developing countries

www.elsevier.com | www.bookaid.org | www.sabre.org

ELSEVIER   BOOK AID   Sabre Foundation
              International

Table of Contents

Section I. Techniques: Spectral, Experimental, and Analytical

Section II. Proteomics

Section III. Challenging Systems for NMR

Section IV. Macromolecular Dynamics

Section V. Macromolecular Complexes

Section VI. Ligand Discovery

Contributors to Volume 394

Article numbers are in parentheses and following the names of contributors. Affiliations listed are current.

THOMAS B. ACTON (8), *Center for Advanced Biotechnology and Medicine, Department of Molecular Biology and Biochemistry, Rutgers University, Piscataway, New Jersey 08854*

JAMES ARAMINI (8), *Center for Advanced Biotechnology and Medicine, Department of Molecular Biology and Biochemistry, Rutgers University, Piscataway, New Jersey 08854*

CHERYL ARROWSMITH (5), *Ontario Cancer Institute and Department of Medical Biophysics, University of Toronto, Toronto, Ontario, M5G 2M9 Canada*

HANUDATTA S. ATREYA (4), *Department of Chemistry, University at Buffalo, The State University of New York, Buffalo, New York 14260*

MICHAEL C. BARAN (5, 8), *Center for Advanced Biotechnology and Medicine, Department of Molecular Biology and Biochemistry, Rutgers University, Piscataway, New Jersey 08854*

AD BAX (3), *National Institute of Diabetes and Digestive and Kidney Diseases, National Institutes of Health, Bethesda, Maryland 20892*

G. C. BENISON (7), *Department of Biochemistry and Molecular Biology, University of Georgia, Athens, Georgia 30602*

P. BERNADÓ (17), *Institut de Biologie Structurale, Jean Pierre Ebel, 38027 Grenoble, France*

PAUL B. CARD (1), *Departments of Biochemistry and Pharmacology, University of Texas Southwestern Medical Center, Dallas, Texas 75390*

YI-WEN CHIANG (8), *Center for Advanced Biotechnology and Medicine, Department of Molecular Biology and Biochemistry, Rutgers University, Piscataway, New Jersey 08854*

TERESA CLIMENT (8), *Center for Advanced Biotechnology and Medicine, Department of Molecular Biology and Biochemistry, Rutgers University, Piscataway, New Jersey 08854*

BONNIE COOPER (8), *Center for Advanced Biotechnology and Medicine, Department of Molecular Biology and Biochemistry, Rutgers University, Piscataway, New Jersey 08854*

LORENZO CORSINI (2), *Institute for Biophysical Chemistry, University of Frankfurt, 60439 Frankfurt, Germany*

L. DAVID FINGER (22), *Department of Chemistry and Biochemistry, University of California, Los Angeles, California 90095*

A. A. DE ANGELIS (14), *Department of Chemistry and Biochemistry, University of California, San Diego, La Jolla, California 92093*

FRANK DELAGLIO (3), *National Institute of Diabetes and Digestive and Kidney Diseases, National Institutes of Health, Bethesda, Maryland 20892*

NATALIA G. DENISSOVA (8), *Center for Advanced Biotechnology and Medicine, Department of Molecular Biology and Biochemistry, Rutgers University, Piscataway, New Jersey 08854*

SHAWN M. DOUGLAS (8), *Department of Molecular Biophysics and Biochemistry and Department of Computer Science, Yale University, New Haven, Connecticut 06520*

VOLKER DÖTSCH (2), *Institute for Biophysical Chemistry, University of Frankfurt, 60439 Frankfurt, Germany*

FLORIAN DURST (2), *Institute for Biophysical Chemistry, University of Frankfurt, 60439 Frankfurt, Germany*

H. JANE DYSON (11), *Department of Molecular Biology and Skaggs Institute for Chemical Biology, Scripps Research Institute, La Jolla, California 92037*

CHRISTIAN EICHMÜLLER (6), *Institute of Organic Chemistry, University of Innsbruck, A-6020 Innsbruck, Austria*

ELAN Z. EISENMESSER (21), *Department of Biochemistry, Brandeis University, Waltham, Massachusetts 02454*

CHARLES D. ELLIS (12), *Department of Biochemistry, Vanderbilt University School of Medicine, Nashville, Tennessee 37232*

JOHN K. EVERETT (8), *Center for Advanced Biotechnology and Medicine, Department of Molecular Biology and Biochemistry, Rutgers University, Piscataway, New Jersey 08854*

JULI FEIGON (22), *Department of Chemistry and Biochemistry, University of California, Los Angeles, California 90095*

J. GARCÍA DE LA TORRE (17), *Departamento de Química Física, Universidad de Murcia, 30071 Murcia, Spain*

KEVIN H. GARDNER (1), *Departments of Biochemistry and Pharmacology, University of Texas Southwestern Medical Center, Dallas, Texas 75390*

MARK GERSTEIN (8), *Department of Molecular Biophysics and Biochemistry and Department of Computer Science, Yale University, New Haven, Connecticut 06520*

C. V. GRANT (14), *Department of Chemistry and Biochemistry, University of California, San Diego, La Jolla, California 92093*

MICHAEL J. GREY (18), *Department of Biochemistry and Molecular Biophysics, Columbia University, New York, New York 10032*

ALEXANDER GRISHAEV (10), *National Institute of Diabetes and Digestive and Kidney Diseases, National Institutes of Health, Bethesda, Maryland 20892*

KRISTIN C. GUNSALUS (8), *Department of Biology, Center for Comparative Functional Genomics, New York Univeristy, New York, New York 10003*

PHILIP J. HAJDUK (23), *Abbott Laboratories, Abbott Park, Illinois 60064*

KATHLEEN B. HALL (19), *Department of Biochemistry and Molecular Biophysics, Washington University School of Medicine, St. Louis, Missouri 63110*

CHI KENT HO (8), *Center for Advanced Biotechnology and Medicine, Department of Molecular Biology and Biochemistry, Rutgers University, Piscataway, New Jersey 08854*

BERND HOFFMANN (6), *Institute of Theoretical Chemistry and Molecular Structural Biology, University of Vienna, A-1030 Vienna, Austria*

YUANPENG JANET HUANG (5), *Center for Advanced Biotechnology and Medicine, Department of Molecular Biology and Biochemistry, Rutgers University, Piscataway, New Jersey 08854*

JOHN F. HUNT (8), *Department of Biological Sciences, Columbia University, New York, New York 10027*

JEFFREY R. HUTH (23), *Abbott Laboratories, Abbott Park, Illinois 60064*

PETER M. HWANG (13), *Departments of Medical Genetics, Biochemistry, and Chemistry, University of Toronto, Toronto, Ontario, M5S 1A1 Canada*

MASAYORI INOUYE (8), *Department of Biochemistry, Robert Wood Johnson Medical School, University of Medicine and Dentistry of New Jersey, Piscataway, New Jersey 08854*

JAISON JACOB (12), *Department of Biochemistry, Vanderbilt University School of Medicine, Nashville, Tennessee 37232*

THOMAS L. JAMES (24), *Department of Pharmaceutical Chemistry, University of California, San Francisco, California 94143*

D. H. JONES (14), *Department of Chemistry and Biochemistry, University of California, San Diego, La Jolla, California 92093*

MURTHY D. KARRA (12), *Department of Biochemistry, Vanderbilt University School of Medicine, Nashville, Tennessee 37232*

LEWIS E. KAY (13), *Departments of Medical Genetics, Biochemistry, and Chemistry, University of Toronto, Toronto, Ontario, M5S 1A1 Canada*

DOROTHEE KERN (21), *Department of Biochemistry, Brandeis University, Waltham, Massachusetts 02454*

ROBERT KONRAT (6), *Institute of Theoretical Chemistry and Molecular Structural Biology, University of Vienna, A-1030 Vienna, Austria*

GEORG KONTAXIS (3), *Department of Theoretical Chemistry and Molecular Structural Biology, University of Vienna, A-1030 Vienna, Austria*

MIGUEL LLINÁS (10), *Department of Chemistry, Carnegie Mellon University, Pittsburgh, Pennsylvania 15213*

PETER J. LUKAVSKY (16), *MRC Laboratory of Molecular Biology, Cambridge, CB2 2QH, England*

LI CHUNG MA (8), *Center for Advanced Biotechnology and Medicine, Department of Molecular Biology and Biochemistry, Rutgers University, Piscataway, New Jersey 08854*

DAPHNE MACAPAGAL (8), *Center for Advanced Biotechnology and Medicine, Department of Molecular Biology and Biochemistry, Rutgers University, Piscataway, New Jersey 08854*

K. L. MAYER (7), *Department of Biochemistry and Molecular Biology, University of Georgia, Athens, Georgia 30602*

MORIZ MAYER (24), *Biovertis AG, 1080 Vienna, Austria*

M. F. MESLEH (14), *Department of Chemistry and Biochemistry, University of California, San Diego, La Jolla, California 92093*

GAETANO T. MONTELIONE (5, 8), *Center for Advanced Biotechnology and Medicine, Department of Molecular Biology and Biochemistry, Rutgers University and Robert Wood Johnson Medical School, Piscataway, New Jersey 08854*

HUNTER N. B. MOSELEY (5), *Center for Advanced Biotechnology and Medicine, Department of Molecular Biology and Biochemistry, Rutgers University, Piscataway, New Jersey 08854*

S. J. OPELLA (14), *Department of Chemistry and Biochemistry, University of California, San Diego, La Jolla, California 92093*

KIRILL OXENOID (12), *Department of Biochemistry, Vanderbilt University School of Medicine, Nashville, Tennessee 37232*

ARTHUR G. PALMER, III (18), *Department of Biochemistry and Molecular Biophysics, Columbia University, New York, New York 10032*

S. H. PARK (14), *Department of Chemistry and Biochemistry, University of California, San Diego, La Jolla, California 92093*

M. PONS (17), *Department de Química Orgànica, Universitat de Barcelona, 08028 Barcelona, Spain, and Laboratori de RMN de Biomolècules, Parc Cientific de Barcelona, 08028 Barcelona, Spain*

ROBERT POWERS (5), *Department of Chemistry, University of Nebraska—Lincoln, Lincoln, Nebraska 68588*

J. H. PRESTEGARD (7), *Complex Carbohydrate Research Center, University of Georgia, Athens, Georgia 30602*

JOSEPH D. PUGLISI (16), *Department of Structural Biology, Stanford University School of Medicine, Stanford, California 94305*

PARANJI K. RAJAN (8), *Center for Advanced Biotechnology and Medicine, Department of Molecular Biology and Biochemistry, Rutgers University, Piscataway, New Jersey 08854*

CAROL A. ROHL (9), *Department of Biomolecular Engineering, University of California, Santa Cruz, California 95064*

CHARLES R. SANDERS (12), *Department of Biochemistry and Center for Structural Biology, Vanderbilt University School of Medicine, Nashville, Tennessee 37232*

DARYL R. SAUER (23), *Abbott Laboratories, Abbott Park, Illinois 60064*

ZACH SERBER (2), *Department of Molecular Pharmacology, Stanford University School of Medicine, Stanford, California 94305*

RITU SHASTRY (8), *Center for Advanced Biotechnology and Medicine, Department of Molecular Biology and Biochemistry, Rutgers University, Piscataway, New Jersey 08854*

LIANG-YU SHIH (8), *Center for Advanced Biotechnology and Medicine, Department of Molecular Biology and Biochemistry, Rutgers University, Piscataway, New Jersey 08854*

ICHIO SHIMADA (20), *Division of Physical Chemistry, Graduate School of Pharmaceutical Sciences, The University of Tokyo, Hongo, Tokyo 113-0033, Japan*

SCOTT A. SHOWALTER (19), *Department of Biochemistry and Molecular Biophysics, Washington University School of Medicine, St. Louis, Missouri 63110*

FRANK SÖNNICHSEN (12), *Cleveland Center for Structural Biology and Department of Physiology and Biophysics, Case Western Reserve University, Cleveland, Ohio 44106*

OTHMAR STEINHAUSER (6), *Institute of Theoretical Chemistry and Molecular Structural Biology, University of Vienna, A-1030 Vienna, Austria*

CHAOHONG SUN (23), *Abbott Laboratories, Abbott Park, Illinois 60064*

G.V.T. SWAPNA (8), *Center for Advanced Biotechnology and Medicine, Department of Molecular Biology and Biochemistry, Rutgers University, Piscataway, New Jersey 08854*

THOMAS SZYPERSKI (4, 5), *Department of Chemistry, University at Buffalo, The State University of New York, Buffalo, New York 14260*

ROBERTO TEJERO (5), *Center for Advanced Biotechnology and Medicine, Department of Molecular Biology and Biochemistry, Rutgers University, Piscataway, New Jersey 08854*

CHANGLIN TIAN (12), *Department of Biochemistry, Vanderbilt University School of Medicine, Nashville, Tennessee 37232*

H. VALAFAR (7), *Department of Computer Science, University of South Carolina, Columbia, South Carolina 29208*

CHUNYU WANG (18), *Department of Biochemistry and Molecular Biophysics, Columbia University, New York, New York 10032*

GERHARD WIDER (15), *Institute for Molecular Biology and Biophysics, ETH Zurich, CH-8093 Zurich, Switzerland*

MICHAEL WILSON (8), *Department of Molecular Biophysics and Biochemistry and Department of Computer Science, Yale University, New Haven, Connecticut 06520*

PETER E. WRIGHT (11), *Department of Molecular Biology and Skaggs Institute for Chemical Biology, Scripps Research Institute, La Jolla, California 92037*

MAGNUS WOLF-WATZ (21), *Department of Biochemistry, Brandeis University, Waltham, Massachusetts 02454*

HAIHONG WU (22), *Department of Chemistry and Biochemistry, University of California, Los Angeles, California 90095*

MARGARET WU (8), *Center for Advanced Biotechnology and Medicine, Department of Molecular Biology and Biochemistry, Rutgers University, Piscataway, New Jersey 08854*

RONG XIAO (8), *Center for Advanced Biotechnology and Medicine, Department of Molecular Biology and Biochemistry, Rutgers University, Piscataway, New Jersey 08854*

METHODS IN ENZYMOLOGY

VOLUME 90. Carbohydrate Metabolism (Part E)
Edited by WILLIS A. WOOD

VOLUME 91. Enzyme Structure (Part I)
Edited by C. H. W. HIRS AND SERGE N. TIMASHEFF

VOLUME 92. Immunochemical Techniques (Part E: Monoclonal Antibodies and General Immunoassay Methods)
Edited by JOHN J. LANGONE AND HELEN VAN VUNAKIS

VOLUME 93. Immunochemical Techniques (Part F: Conventional Antibodies, Fc Receptors, and Cytotoxicity)
Edited by JOHN J. LANGONE AND HELEN VAN VUNAKIS

VOLUME 94. Polyamines
Edited by HERBERT TABOR AND CELIA WHITE TABOR

VOLUME 95. Cumulative Subject Index Volumes 61–74, 76–80
Edited by EDWARD A. DENNIS AND MARTHA G. DENNIS

VOLUME 96. Biomembranes [Part J: Membrane Biogenesis: Assembly and Targeting (General Methods; Eukaryotes)]
Edited by SIDNEY FLEISCHER AND BECCA FLEISCHER

VOLUME 97. Biomembranes [Part K: Membrane Biogenesis: Assembly and Targeting (Prokaryotes, Mitochondria, and Chloroplasts)]
Edited by SIDNEY FLEISCHER AND BECCA FLEISCHER

VOLUME 98. Biomembranes (Part L: Membrane Biogenesis: Processing and Recycling)
Edited by SIDNEY FLEISCHER AND BECCA FLEISCHER

VOLUME 99. Hormone Action (Part F: Protein Kinases)
Edited by JACKIE D. CORBIN AND JOEL G. HARDMAN

VOLUME 100. Recombinant DNA (Part B)
Edited by RAY WU, LAWRENCE GROSSMAN, AND KIVIE MOLDAVE

VOLUME 101. Recombinant DNA (Part C)
Edited by RAY WU, LAWRENCE GROSSMAN, AND KIVIE MOLDAVE

VOLUME 102. Hormone Action (Part G: Calmodulin and Calcium-Binding Proteins)
Edited by ANTHONY R. MEANS AND BERT W. O'MALLEY

VOLUME 103. Hormone Action (Part H: Neuroendocrine Peptides)
Edited by P. MICHAEL CONN

VOLUME 104. Enzyme Purification and Related Techniques (Part C)
Edited by WILLIAM B. JAKOBY

VOLUME 105. Oxygen Radicals in Biological Systems
Edited by LESTER PACKER

VOLUME 106. Posttranslational Modifications (Part A)
Edited by FINN WOLD AND KIVIE MOLDAVE

Section I

Techniques: Spectral, Experimental, and Analytical

[1] Identification and Optimization of Protein Domains for NMR Studies

By Paul B. Card and Kevin H. Gardner

Abstract

The success of genomic sequencing projects in recent years has presented protein scientists with a formidable challenge in characterizing the vast number of gene products that have subsequently been identified. NMR has proven to be a valuable tool in the elucidation of various properties for many of these proteins, allowing versatile studies of structure, dynamics, and interactions in the solution state. But the characteristics needed for proteins amenable to this kind of study, such as folding capability, long-term stability, and high solubility, require robust and expeditious methods for the identification and optimization of target protein domains. Here we present a variety of computational and experimental methods developed for these purposes and show that great care must often be taken in the design of constructs intended for NMR-based investigations.

Introduction

Although there have recently been major advances in the sequencing and annotation of genomic information from a wide variety of organisms, a large percentage of identified genes have no known function [including >40% of human genes as initially sequenced (Venter *et al.*, 2001)]. Structural biology has played an important role in addressing this shortcoming, as structural homology between proteins can provide valuable information about possible protein functions (Cort *et al.*, 1999; Zarembinski *et al.*, 1998). Whether practiced on individual systems or on a proteomic scale, a common challenge to structural biologists is the need to rapidly identify, express, and purify proteins that are amenable to nuclear magnetic resonance (NMR) spectroscopy, X-ray crystallography, and other biophysical techniques.

Solution NMR has played an integral role in these studies for a variety of reasons. These include the abilities to investigate protein structure in a solution state more closely resembling physiological conditions, monitor binding processes and interfaces, extract dynamic properties from relaxation and exchange parameters, and provide important avenues to study crystallization-resistant targets. However, the inherently low sensitivity of

Copyright 2005, Elsevier Inc.
All rights reserved.
0076-6879/05 $35.00

NMR requires that proteins be soluble to high concentrations and relatively stable over long periods of time. In addition, although many advances have been made for NMR studies of larger proteins and complexes using deuterium labeling (Gardner and Kay, 1998), TROSY-based pulse sequences (Pervushin, 2000), and higher magnetic field strengths, NMR methods have been most routinely applied to proteins under approximately 20 kDa to avoid problems associated with spectral overlap and broad line widths.

The method most often used to overcome this high-molecular-weight problem is the "divide-and-conquer" approach, where constituent domains of the full-length protein are individually cloned, expressed, and purified. Although this strategy helps keep resonance signals sharp by decreasing rotational correlation times and greatly simplifies the interpretation of NMR spectra by reducing the number of peaks, it requires the identification of well-folded, soluble fragments that retain both their native structure and function when removed from the context of full-length protein. Here we review a variety of computational and experimental approaches that have been successfully used to identify well-behaved protein domains for NMR studies.

Computational Methods

Computational algorithms have proven invaluable to molecular biologists by helping to identify open reading frames and homologous proteins and to classify gene families. In a similar manner, they are often used for the prediction and annotation of protein and domain families based on statistical models derived from multiple sequence alignments that are generated from the constantly increasing storehouse of genomic data. Identification of these relationships is particularly important for structural biologists, because many structurally similar members of a particular family of proteins can exhibit relatively low degrees of sequence homology. For example, PAS (Per-ARNT-Sim) protein–protein interaction domains have a high degree of structural conservation despite notoriously low similarities between their primary sequences (Hefti *et al.*, 2004; Taylor and Zhulin, 1999). For this reason, robust algorithms designed for the identification and annotation of protein domains and secondary structural elements are often used to locate independently folded, stable protein constructs that can be cloned and expressed for structural studies.

Domain Prediction Methods

An important development in computational tools designed for these purposes was the incorporation of hidden Markov models (HMMs) (Krogh *et al.*, 1994). Statistical methods for modeling, database searching, and generating multiple alignments are based on the assumption that even when structurally similar protein domains have widely divergent sequences, there should be conserved elements within the primary sequence that define the folding, structure, and function of the target. This information is encoded into an HMM with a training process involving representative members of a domain family, generating a matrix that describes the likelihood of finding any given amino acid (or insertions or deletions) at each location throughout the domain. These matrices can be used to generate a probability distribution for all possible sequences, allowing one to score the probability that a given query sequence is a member of the family described by the HMM. As such, these are powerful tools to search sequence databases for previously unidentified representatives of HMM-encoded domains, and they can be used to define common aspects of various members, including domain boundaries. One particularly successful implementation of HMM-based methodologies to define protein members, sequence alignments, and domain boundaries is Pfam (http://pfam.wustl.edu) (Sonnhammer *et al.*, 1998). Unlike many contemporary techniques, Pfam uses the entire span of the domain to search sequence databases for new members. Previous strategies typically included only well-conserved motifs, but HMM-profile methods allow less conserved regions, as well as insertions and deletions, to be dealt with and annotated in a more accurate manner.

Another important HMM-based tool, particularly for researchers interested in the study of domains used in signaling processes, is SMART (Simple Modular Architecture Research Tool, http://smart.embl-heidelberg. de) (Schultz *et al.*, 1998). Currently, SMART is based on multiple sequence alignments of 667 domain families that have been updated to include additional homologues predicted by computational methods and verified by their experimentally determined biological context. Taken together, these sequences were then used to make HMMs for the identification and annotation of novel or uncharacterized domains. Although Pfam demonstrated an increased ability to find incomplete domain sequences relative to SMART, the latter was designed to identify full-length domains, which is clearly important for structural biologists intending to study fully functional constructs. Another advantage of SMART is the inclusion of integral HMM-based tools to identify compositionally biased regions such as transmembrane segments, which can aid the rational design of protein fragments that are less likely to aggregate under the high concentrations necessary for NMR studies.

Secondary Structure Prediction Methods

There have been many techniques developed that utilize different heuristics for the prediction of protein secondary structure. For example, the PHD method (Rost and Sander, 1993) uses a multiple sequence alignment of the query sequence as input for a neural network initially trained on a nonredundant set of 130 protein chains, integrating evolutionary information into the prediction process. Alternatively, the PREDATOR algorithm (Frishman and Argos, 1996) is based on the recognition of potential hydrogen bonding between residues in a query amino acid sequence, using database-derived statistics on residue type occurrences in α-helical and β-strand secondary structural elements. Studies have shown that the shortcomings of these and other computational methods can be addressed by combining multiple methods to improve the accuracy of *de novo* predictions (Livingstone and Barton, 1996). This approach is well implemented by the JPred secondary structure prediction server (http://www.compbio.dundee.ac.uk/~www-jpred) (Cuff *et al.*, 1998). JPred further improves its accuracy by generating secondary structure predictions not only for the initial query sequence but also for a series of nonredundant homologues found with BLAST searches. In addition, a consensus is derived with scoring functions for each residue to give a weighted average prediction among the different methods and sequences that provides a more accurate identification of secondary structure and domain boundaries.

Experimental Methods

Parallel Fragment Cloning and Expression

Although computational methods have often demonstrated their ability to identify the locations of domains and secondary structure elements within protein sequences, this still leaves a range of challenges for an NMR spectroscopist intending to translate these predictions into large quantities of soluble domains for study. These complications stem from several sources, including poor sequence conservation and incomplete HMM-based domain models, as well as the many difficulties with predicting complex solution behavior directly from sequence.

To address these problems, methods have been developed to experimentally identify protein domains from within larger sequences. One of the most common approaches is based on limited proteolysis (Cohen *et al.*, 1995), where the domain-containing sequence and flanking regions are treated with small quantities of proteases. Protease-resistant fragments, typically identified and mapped within the protein sequence by mass spectrometry, are interpreted to be well-folded domains. Although this approach has often been successfully used, it is potentially susceptible to artifacts from proteolysis of solvent-exposed loops and other extended segments located within domains of interest.

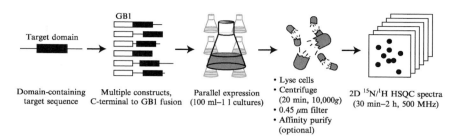

FIG. 1. Overview of parallel expression approach.

As an alternative strategy, we have relied on an NMR-based screen of related protein constructs with different N- and C-termini flanking a domain of interest identified by computational approaches (Fig. 1). Typically, six to eight constructs are designed with unique sets of termini based upon comparisons between orthologous proteins and secondary structure predictions. In parallel, these fragments are cloned as fusions to the C-terminal side of small, highly soluble auxiliary proteins, expressed in *Escherichia coli* cultures and used for two-dimensional (2D) ^{15}N/^{1}H heteronuclear single-quantum correlation (HSQC) experiments as detailed below. Obtaining such spectra at an early stage provides qualitative assessments of protein folding from amide proton chemical shift dispersion. In addition, these data provide early indicators of spectral quality that might reveal issues with peak doubling or heterogeneous line widths that might complicate experiments relying on amide proton detection.

Choice of Fusion Protein

Fusion proteins can aid structural studies of native protein domains by increasing expression levels, improving solubility, and providing opportunities to use affinity purification methods. Although these goals are accomplished to varying degrees depending on both the choice of fusion and the target (Hammarstrom *et al.*, 2002), there are several options widely available including maltose binding protein, glutathione *S*-transferase, and a variety of multihistidine tags.

We have routinely used a point mutant of the *Streptococcus* protein G beta 1 domain (mutant T2Q, here referred to as GB1) because it meets several of these goals and provides additional benefits for cell lysate screening (see the following section). Foremost, GB1 fusions typically express to very high levels and are quite soluble (Hammarstrom *et al.*, 2002; Huth *et al.*, 1997; Zhou *et al.*, 2001). GB1-containing proteins can be affinity purified on IgG resins, although the relatively harsh conditions needed to elute such proteins (pH < 4 or mild denaturant) have led us to generate His_6-tagged versions of GB1 for Ni(II) affinity purification (Harper *et al.*, 2003). GB1 is also very small (56 amino acid residues) and separated from the target protein by a flexible linker of approximately 12 residues in our expression vectors (Amezcua *et al.*, 2002; Harper *et al.*, 2003). As such, GB1 produces clear, well-resolved peaks in $^{15}N/^1H$ HSQC spectra, conveniently providing a positive control for establishing that fusions are present at high concentration and not grossly aggregated.

E. coli *Cell Lysate Screening by NMR*

To confirm the folded state of proteins up to approximately 20 kDa in molecular mass, we have found it useful to obtain 2D $^{15}N/^1H$ HSQC spectra directly on *E. coli* cell lysates from cultures overexpressing the proteins of interest (Gronenborn and Clore, 1996). For this, we use 100-ml to 1-liter cultures of *E. coli*, grown in uniformly ^{15}N-labeled media, containing plasmids overexpressing the target fusion protein under the control of a T7-based promoter. After overnight expression at 20°, cell pellets from these growths are lysed, clarified with high-speed centrifugation (20 min, 10,000g), and filtered through a 0.45-μm filter. Lysates are then concentrated to approximately 1 ml or can alternatively be subjected to a rapid affinity-based purification step (Woestenenk *et al.*, 2003). With or without initial purification, this general approach can be used in a straightforward manner to manually screen six to eight samples, or potentially larger numbers, with the aid of automation.

Lysates prepared in this manner typically generate 2D $^{15}N/^1H$ HSQC spectra with reasonable signal-to-noise in a fairly short amount of time (30 min to 2 h) on commonly available 500-MHz spectrometers. It has been shown that spectra recorded from *E. coli* lysates or intact cells overexpressing small proteins are dominated by signals originating from the protein of interest (Gronenborn and Clore, 1996; Serber *et al.*, 2004). This stems from the ability of common bacterial expression systems to often produce target proteins as the single most abundant protein in *E. coli*. In contrast, many host proteins are either present in low abundance or participate in large complexes, hampering their detection by NMR. Signals from a number of small nitrogen-containing metabolites, such as cyclic enterobacterial common antigen (Erbel *et al.*, 2003, 2004), can often be seen in these spectra, although the small number of these peaks typically do not interfere with analysis.

To demonstrate this approach, Fig. 2 shows a comparison of $^{15}N/^1H$ HSQC spectra of lysates from cells expressing GB1 fusions of several constructs for the N-terminal PAS domain of human PAS kinase (Amezcua *et al.*, 2002). Spectra from each of the fusion proteins are superimposed with a spectrum of free GB1 recorded under comparable conditions. Although some GB1-derived crosspeaks show chemical shift changes due to the presence of the fusion partner, it is easy to distinguish peaks from GB1 and the PAS fragment. Examination of spectra obtained from GB1 fused to residues 131–208 of human PAS kinase (Fig. 2A) reveals that signals originating from the PAS domain have poor amide proton chemical shift dispersion (7.8 ppm $< \delta <$ 8.4 ppm) and heterogeneous line widths. These results are consistent with this fragment not adopting a well-ordered tertiary structure. Addition of 29 residues to the C-terminus shows a marked improvement in both chemical shift dispersion and line width (Fig. 2B), indicating that these residues are required for folding. Further addition of residues on the C-terminal side simply increases the crowding in the center of the spectrum (Fig. 2C), suggesting that they remain unfolded and do not stably associate with the PAS domain.

Deuterium Exchange Mass Spectrometry

A second class of experimental approaches to rapidly identify protein domains are mass spectrometry–based measurements of deuterium exchange. As is widely appreciated in the protein NMR community, labile hydrogens can readily exchange with deuterium when proteins are placed into D_2O-containing solvents. Although this exchange is quite rapid for solvent-exposed sites, it is significantly slowed by the hydrogen bonding and burial that accompany tertiary structure formation. These exchange rates can be quantified using a variety of NMR-based methods, most of which entail the successive acquisition of spectra during incubation in deuterated solvents. Although such approaches provide site-specific rate measurements, their requirement for milligram quantities of purified protein precludes their use for high-throughput stability screening methods.

For such applications, a more viable alternative is provided by mass spectrometry methods, which sacrifice the ability to measure ^2H exchange rates at specific amides in favor of higher sensitivity. Among these techniques, matrix-assisted laser desorption/ionization time-of-flight (MALDI-TOF) mass spectrometry has been particularly useful for monitoring deuterium exchange in a wide range of samples ranging from purified proteins (Mandell *et al.*, 1998) to *E. coli* lysates. This has been utilized by Oas and co-workers in their SUPREX (*s*tability of *u*npurified *p*roteins from *r*ates of H/D *ex*change) method (Ghaemmaghami *et al.*, 2000), in which crude extracts from recombinant *E. coli* cultures are diluted into a D_2O-exchange buffer containing several concentrations of denaturant. After allowing incubation for a set period of time, exchange is stopped

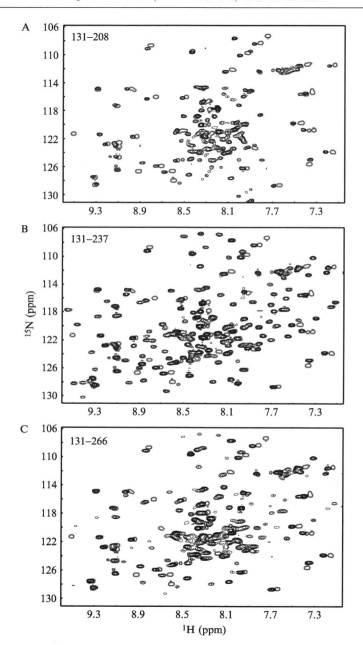

by adding a cold, acidic buffer containing the MALDI matrix material. By measuring the increase in mass relative to a fully protonated sample, exchange rates can be obtained that can be further analyzed to calculate the free energy of protein folding (ΔG_f). Initially demonstrated on several proteins in lysates (Ghaemmaghami *et al.*, 2000), this has since been extended to determine the stabilization provided by protein-binding ligands (Powell *et al.*, 2002) and to measure protein stability in living *E. coli* cells (Ghaemmaghami and Oas, 2001).

Solubility Screening

After a construct is identified that is well folded and free from aggregation, solvent screening is often required to find conditions that maximize long-term solubility. Ionic strength, temperature, amphipathicity, reducing capacity, and pH are all important parameters that can greatly affect the behavior of proteins in solution. There is a wide range of buffers, stabilizers, and affinity tags that can be tested for their ability to keep proteins soluble for long periods of time. Several methods have been used to identify optimal solvent conditions for NMR studies of particular proteins; two of the best documented are the microdialysis button test (Bagby *et al.*, 1997) and the microdrop screen (Lepre and Moore, 1998). Both of these approaches require a small quantity of protein (~0.1–0.2 mg/sample) that is distributed into different solvent conditions and visually assessed for precipitation over time. In the microdialysis button technique, very small aliquots of protein are placed in specially designed reservoirs covered by dialysis membranes, allowing the gradual exchange into new solvent conditions while ensuring that protein concentration remains unchanged. The microdrop screening method instead uses a setup common to many crystallization trials, where small drops of protein solution on slides are suspended over many different types of buffer in sealed chambers. Vapor diffusion brings each sample to equilibrium, providing an effective screen for conditions while allowing the protein concentration to change during the screening process. An extensive review of useful solvent conditions and the screening methods mentioned here is presented elsewhere (Bagby *et al.*, 2001).

FIG. 2. Utility of *E. coli* cell lysate screening for determining domain boundaries. Fragments of human PAS kinase were expressed in *E. coli* BL21(DE3) as C-terminal fusions to GB1 using a T7 RNA polymerase expression vector (Amezcua *et al.*, 2002). Each spectrum was generated from a lysate generated by a 1-liter culture grown in uniformly ^{15}N-labeled M9 media, which was concentrated to 1 ml before acquiring 2D ^{15}N/^{1}H HSQC spectra (40 min, 37°). (A–C) An overlay of spectra from GB1–PAS kinase fusions (black contours; PAS kinase residues are listed on each panel) with a reference from an isolated GB1 sample (red). (See color insert.)

Higher-Throughput, Library-Based Approaches

The methods detailed above are well suited for application to a small number of constructs at one time. Practically, this imposes only minor constraints on a standard search for well-behaved domains within a protein target. However, certain applications may necessitate screening much larger libraries than are feasible with these approaches. For cases in which little computationally derived information is available regarding domain structure, it may be necessary to resort to generating large libraries of random fragments to screen larger regions of protein for folded domains. Alternatively, to optimize the expression and solubility of single proteins, a combination of DNA shuffling and random mutagenesis can be used to produce libraries encoding diverse variants of the target (Pedelacq *et al.*, 2002). These collections of randomly generated proteins can then be fused N-terminally to a variety of reporter proteins, including green fluorescent protein, chloramphenicol acetyltransferase, and various β-galactosidase fragments, to permit large-scale solubility screening (e.g., Maxwell *et al.*, 1999; Pedelacq *et al.*, 2002; Waldo *et al.*, 1999; Wigley *et al.*, 2001). Generated in this way, the functional activities of these reporters (fluorescence, antibiotic resistance, and enzymatic activity) are modulated by the solubility of candidate fragments. These reporter functions can easily be surveyed in bacterial colonies grown on solid media, allowing the solubility of thousands of protein constructs to be rapidly assessed. Although these approaches are not in widespread use at this time, they represent a potentially powerful tool that should be relatively straightforward to implement in a wide variety of laboratory settings.

Caveats to the Divide-and-Conquer Approach

Despite the success of the reductionist strategy to produce individual domains for biochemical and structural studies, it is important to keep in mind that this is not a universally applicable approach. Many cases have been described in which a domain, homologues of which can often be well behaved as isolated from different proteins, adopts a series of obligate intra- or intermolecular interactions to promote its folding or function.

For example, such interactions may arise from the incorporation of additional secondary structure elements into a domain as we have found with a photosensory PAS domain from oat phototropin 1 (Harper *et al.*, 2003). While evaluating several PAS-containing fragments for NMR studies, we observed significant chemical shift changes in $^{15}N/^{1}H$ HSQC spectra for constructs that extended past residue 540 (Fig. 3). These chemical shift changes provided evidence for intramolecular interactions that we

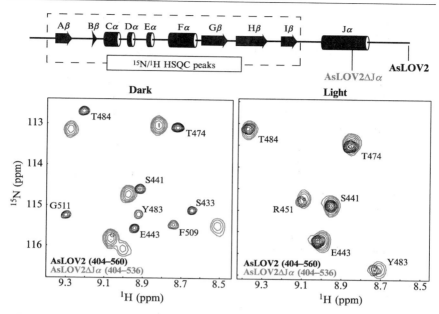

FIG. 3. Discovery of novel secondary structural elements by comparison of multiple constructs in the photosensory LOV2 domain of *Avena sativa* phototropin 1 (AsLOV2). Comparisons of ^{15}N/^{1}H HSQC peaks originating from residues within the predicted PAS domain showed significant chemical shift changes among constructs with differing C-termini, as demonstrated in spectra from AsLOV2 (black) and the shorter AsLOV2ΔJα (red). The data strongly suggested that the C-terminal region interacted with the PAS domain in the dark. When these experiments were repeated with blue light illumination, these ^{15}N/^{1}H HSQC peaks became independent of the construct length, providing evidence that light-induced conformational changes led to the displacement of the C-terminal Jα helix. Adapted with permission from Harper *et al.* (2003). (See color insert.)

subsequently established to be caused by a novel helix (Jα) binding across an exposed β-sheet on the PAS domain surface. This interaction is broken in the lit-state structure, suggesting that is it is critically involved with a light-dependent conformational change.

Intramolecular stabilizing interactions can also be found between domains, as noted in studies of the fourth and fifth PDZ domains of rat GRIP1 (Feng *et al.*, 2003; Zhang *et al.*, 2001). PDZ domains normally fold into self-contained stable mixed α/β domain structures, but attempts to express PDZ4 or PDZ5 from GRIP1 failed to yield stably folded domains. However, when a tandem construct of PDZ4-5 was expressed with the native nine-residue linker coupling the two, both domains were well folded. In addition, *in trans* mixing experiments failed to fold either domain, suggesting that this apparent chaperoning function worked only if the two domains were covalently linked.

Although it is clear that reductionist strategies have been very successful in many cases at solving individual domain structures to high resolution, it is clear from these and other examples that the techniques we have reviewed here should initially be applied on fragments containing significant amounts of sequence flanking the domain(s) of interest to survey for possible intramolecular interactions. In cases in which such interactions are found, it should be emphasized that chemical shifts, structures, and other data provided by studying individual domains are still extremely useful as they provide an important foundation for further investigations of larger complexes and fragments. In particular, the development of intein-based protein splicing methods holds much promise in this area, as they allow the comparison of NMR signals from domains in isolation to signals in multi-domain constructs (Otomo *et al.*, 1999; Muir, 2003). The practicality of such studies has also been improved by the development of deuterium labeling (Gardner and Kay, 1998) and TROSY (Pervushin, 2000) methods for studying larger systems in general. These methods suggest an exciting future of "mixed resolution" NMR approaches that integrate high-resolution structural data on individual domains with lower resolution information on intra- and intermolecular interactions, providing further insights into the structure and function of a wider range of biological systems.

Acknowledgments

We thank all of the members of the Gardner laboratory for contributing to the development and refinement of these methods within the group. This work has been supported by grants from the National Institutes of Health (CA90601 and CA95471) and the Robert A. Welch Foundation (I-1424) to K.H.G.

References

Amezcua, C. A., Harper, S. M., Rutter, J., and Gardner, K. H. (2002). Structure and interactions of PAS kinase N-terminal PAS domain: Model for intramolecular kinase regulation. *Structure (Camb.)* **10,** 1349–1361.

Bagby, S., Tong, K. I., Liu, D., Alattia, J. R., and Ikura, M. (1997). The button test: A small scale method using microdialysis cells for assessing protein solubility at concentrations suitable for NMR. *J. Biomol. NMR* **10,** 279–282.

Bagby, S., Tong, K. I., and Ikura, M. (2001). Optimization of protein solubility and stability for protein nuclear magnetic resonance. *Methods Enzymol.* **339,** 20–41.

Cohen, S. L., Ferre-D'Amare, A. R., Burley, S. K., and Chait, B. T. (1995). Probing the solution structure of the DNA-binding protein Max by a combination of proteolysis and mass spectrometry. *Protein Sci.* **4,** 1088–1099.

Cort, J. R., Koonin, E. V., Bash, P. A., and Kennedy, M. A. (1999). A phylogenetic approach to target selection for structural genomics: Solution structure of YciH. *Nucleic Acids Res.* **27**, 4018–4027.

Cuff, J. A., Clamp, M. E., Siddiqui, A. S., Finlay, M., and Barton, G. J. (1998). JPred: A consensus secondary structure prediction server. *Bioinformatics* **14**, 892–893.

Erbel, P. J., Barr, K., Gao, N., Gerwig, G. J., Rick, P. D., and Gardner, K. H. (2003). Identification and biosynthesis of cyclic enterobacterial common antigen in *Escherichia coli. J. Bacteriol.* **185**, 1995–2004.

Erbel, P. J., Seidel, R., Macintosh, S. E., Gentile, L. N., Amor, J. C., Kahn, R. A., Prestegard, J. H., McIntosh, L. P., and Gardner, K. H. (2004). Cyclic enterobacterial common antigen: Potential contaminant of bacterially expressed protein preparations. *J. Biomol. NMR* **29**, 199–204.

Feng, W., Shi, Y., Li, M., and Zhang, M. (2003). Tandem PDZ repeats in glutamate receptor-interacting proteins have a novel mode of PDZ domain-mediated target binding. *Nat. Struct. Biol.* **10**, 972–978.

Frishman, D., and Argos, P. (1996). Incorporation of non-local interactions in protein secondary structure prediction from the amino acid sequence. *Protein Eng.* **9**, 133–142.

Gardner, K. H., and Kay, L. E. (1998). The use of ^{2}H, ^{13}C, ^{15}N multidimensional NMR to study the structure and dynamics of proteins. *Annu. Rev. Biophys. Biomol. Struct.* **27**, 357–406.

Ghaemmaghami, S., and Oas, T. G. (2001). Quantitative protein stability measurement *in vivo. Nat. Struct. Biol.* **8**, 879–882.

Ghaemmaghami, S., Fitzgerald, M. C., and Oas, T. G. (2000). A quantitative, high-throughput screen for protein stability. *Proc. Natl. Acad. Sci. USA* **97**, 8296–8301.

Gronenborn, A. M., and Clore, G. M. (1996). Rapid screening for structural integrity of expressed proteins by heteronuclear NMR spectroscopy. *Protein Sci.* **5**, 174–177.

Hammarstrom, M., Hellgren, N., van Den Berg, S., Berglund, H., and Hard, T. (2002). Rapid screening for improved solubility of small human proteins produced as fusion proteins in *Escherichia coli. Protein Sci.* **11**, 313–321.

Harper, S. M., Neil, L. C., and Gardner, K. H. (2003). Structural basis of a phototropin light switch. *Science* **301**, 1541–1544.

Hefti, M. H., Francoijs, K. J., de Vries, S. C., Dixon, R., and Vervoort, J. (2004). The PAS fold: A redefination of the PAS domain based upon structural prediction. *Eur. J. Biochem.* **271**, 1198–1208.

Huth, J. R., Bewley, C. A., Jackson, B. M., Hinnebusch, A. G., Clore, G. M., and Gronenborn, A. M. (1997). Design of an expression system for detecting folded protein domains and mapping macromolecular interactions by NMR. *Protein Sci.* **6**, 2359–2364.

Krogh, A., Brown, M., Mian, I. S., Sjolander, K., and Haussler, D. (1994). Hidden Markov models in computational biology. Applications to protein modeling. *J. Mol. Biol.* **235**, 1501–1531.

Lepre, C. A., and Moore, J. M. (1998). Microdrop screening: A rapid method to optimize solvent conditions for NMR spectroscopy of proteins. *J. Biomol. NMR* **12**, 493–499.

Livingstone, C. D., and Barton, G. J. (1996). Identification of functional residues and secondary structure from protein multiple sequence alignment. *Methods Enzymol.* **266**, 497–512.

Mandell, J. G., Falick, A. M., and Komives, E. A. (1998). Measurement of amide hydrogen exchange by MALDI-TOF mass spectrometry. *Anal. Chem.* **70**, 3987–3995.

Maxwell, K. L., Mittermaier, A. K., Forman-Kay, J. D., and Davidson, A. R. (1999). A simple *in vivo* assay for increased protein solubility. *Protein Sci.* **8**, 1908–1911.

Muir, T. W. (2003). Semisynthesis of proteins by expressed protein ligation. *Annu. Rev. Biochem.* **72**, 249–289.

Otomo, T., Teruya, K., Uegaki, K., Yamazaki, T., and Kyogoku, Y. (1999). Improved segmental isotope labeling of proteins and application to a larger protein. *J. Biomol. NMR* **14,** 105–114.

Pedelacq, J. D., Piltch, E., Liong, E. C., Berendzen, J., Kim, C. Y., Rho, B. S., Park, M. S., Terwilliger, T. C., and Waldo, G. S. (2002). Engineering soluble proteins for structural genomics. *Nat. Biotechnol.* **20,** 927–932.

Pervushin, K. (2000). Impact of transverse relaxation optimized spectroscopy (TROSY) on NMR as a technique in structural biology. *Q. Rev. Biophys.* **33,** 161–197.

Powell, K. D., Ghaemmaghami, S., Wang, M. Z., Ma, L., Oas, T. G., and Fitzgerald, M. C. (2002). A general mass spectrometry-based assay for the quantitation of protein-ligand binding interactions in solution. *J. Am. Chem. Soc.* **124,** 10256–10257.

Rost, B., and Sander, C. (1993). Prediction of protein secondary structure at better than 70% accuracy. *J. Mol. Biol.* **232,** 584–599.

Schultz, J., Milpetz, F., Bork, P., and Ponting, C. P. (1998). SMART, a simple modular architecture research tool: Identification of signaling domains. *Proc. Natl. Acad. Sci. USA* **95,** 5857–5864.

Serber, Z., Corsini, L., Durst, F., and Doetsch, V. (2004). In-cell NMR spectroscopy. *Methods Enzymol.,* this volume, Chapter 2.

Sonnhammer, E. L., Eddy, S. R., Birney, E., Bateman, A., and Durbin, R. (1998). Pfam: Multiple sequence alignments and HMM-profiles of protein domains. *Nucleic Acids Res.* **26,** 320–322.

Taylor, B. L., and Zhulin, I. B. (1999). PAS domains: Internal sensors of oxygen, redox potential, and light. *Microbiol. Mol. Biol. Rev.* **63,** 479–506.

Venter, J. C., Adams, M. D., Myers, E. W., Li, P. W., Mural, R. J., Sutton, G. G., Smith, H. O., Yandell, M., Evans, C. A., Holt, R. A., *et al.* (2001). The sequence of the human genome. *Science* **291,** 1304–1351.

Waldo, G. S., Standish, B. M., Berendzen, J., and Terwilliger, T. C. (1999). Rapid protein-folding assay using green fluorescent protein. *Nat. Biotechnol.* **17,** 691–695.

Wigley, W. C., Stidham, R. D., Smith, N. M., Hunt, J. F., and Thomas, P. J. (2001). Protein solubility and folding monitored *in vivo* by structural complementation of a genetic marker protein. *Nat. Biotechnol.* **19,** 131–136.

Woestenenk, E. A., Hammarstrom, M., Hard, T., and Berglund, H. (2003). Screening methods to determine biophysical properties of proteins in structural genomics. *Anal. Biochem.* **318,** 71–79.

Zarembinski, T. I., Hung, L. W., Mueller-Dieckmann, H. J., Kim, K. K., Yokota, H., Kim, R., and Kim, S. H. (1998). Structure-based assignment of the biochemical function of a hypothetical protein: A test case of structural genomics. *Proc. Natl. Acad. Sci. USA* **95,** 15189–15193.

Zhang, Q., Fan, J. S., and Zhang, M. (2001). Interdomain chaperoning between PSD-95, Dlg, and Zo-1 (PDZ) domains of glutamate receptor-interacting proteins. *J. Biol. Chem.* **276,** 43216–43220.

Zhou, P., Lugovskoy, A. A., and Wagner, G. (2001). A solubility-enhancement tag (SET) for NMR studies of poorly behaving proteins. *J. Biomol. NMR* **20,** 11–14.

[2] In-Cell NMR Spectroscopy

By Zach Serber, Lorenzo Corsini, Florian Durst, and
Volker Dötsch

Abstract

The role of a protein inside a cell is determined by both its location and its conformational state. Although fluorescence techniques are widely used to determine the cellular localization of proteins *in vivo*, these approaches cannot provide detailed information about a protein's three-dimensional state. This gap, however, can be filled by NMR spectroscopy, which can be used to investigate both the conformation as well as the dynamics of proteins inside living cells. In this chapter we describe technical aspects of these "in-cell NMR" experiments. In particular, we show that in the case of ^{15}N-labeling schemes the background caused by labeling all cellular components is negligible, while ^{13}C-based experiments suffer from high background levels and require selective labeling schemes. A correlation between the signal-to-noise ratio of in-cell NMR experiments with the overexpression level of the protein shows that the current detection limit is 150–200 μM (intracellular concentration). We also discuss experiments that demonstrate that the intracellular viscosity is not a limiting factor since the intracellular rotational correlation time is only approximately two times longer than the correlation time in water. Furthermore, we describe applications of the technique and discuss its limitations.

Introduction

The noninvasive nature of magnetic resonance techniques makes them ideal tools not only for *in vitro* applications to study solutions of molecules but also to investigate living cells and tissues (Bachert, 1998; Cohen *et al.*, 1989; Gillies, 1994; Kanamori and Ross, 1997; Li *et al.*, 1996; Spindler *et al.*, 1999). During the past three decades, *in vivo* nuclear magnetic resonance (NMR) spectroscopy has become a well-established technique for the observation of metabolites and metal ions and the investigation of metabolic fluxes in systems ranging from suspensions of bacteria and other cells to entire perfused organs (Degani *et al.*, 1994). In addition, magnetic resonance imaging has developed into a very important diagnostic tool in clinical applications by providing images of internal organs. Together magnetic resonance imaging and *in vivo* magnetic resonance spectroscopy

Copyright 2005, Elsevier Inc.
All rights reserved.
0076-6879/05 $35.00

are able to provide information ranging from the metabolic state of living cells to the shape and function of entire organs, which cannot be achieved by any other technique. So far, one gap in this range has been the investigation of macromolecules *in vivo*, which has been possible only in a few very special cases (Brown *et al.*, 1977; Jue and Anderson, 1990; Kreutzer *et al.*, 1992; Tran *et al.*, 1998). Recently, however, we and others have demonstrated that magnetic resonance spectroscopy can be used to study proteins and other biological macromolecules inside living cells (Dedmon *et al.*, 2002; Hubbard *et al.*, 2003; Lippens and Bohin, 1999; Serber and Dötsch, 2001; Serber *et al.*, 2001a,b; Wieruszeski *et al.*, 2001). The main difference between these in-cell NMR techniques used to study macromolecules and the classic *in vivo* NMR spectroscopy used to investigate metabolites, ions, and other small molecules is how one distinguishes the resonance lines of the molecule of interest from the background of all other molecules that are present in a cell. In the case of small molecule *in vivo* NMR, this distinction is often achieved because the molecule of interest is highly abundant and has the most prominent resonances of the entire spectrum. Alternatively, the resonances can be made visible by adding molecules labeled with NMR-active isotopes (mainly ^{13}C) to the cell suspension where they either diffuse through the cellular membrane or are actively transported into the interior of the cells. Due to the small number of resonance lines of the labeled molecules, one-dimensional spectra are normally sufficient to investigate their interaction with other cellular components. In contrast, in-cell NMR spectroscopy of macromolecules relies on expressing these molecules and labeling them with NMR active isotopes directly in the cells. One exception is the injection of purified and labeled proteins into large cells like *Xenopus* oocytes or their introduction into cell extracts. The high number of resonance lines in these macromolecules almost always requires the measurement of two-dimensional NMR spectra. In this chapter, we will concentrate only on the application of NMR spectroscopy to macromolecules in living cells and will not focus on the classical *in vivo* NMR spectroscopy of small molecules.

What Is the Aim of In-Cell NMR Spectroscopy?

The aim of in-cell NMR spectroscopy is to investigate the conformation and dynamics of macromolecules in their natural environment. The power of in-cell NMR spectroscopy lies in observing changes in the structures of biological macromolecules and in their interaction with other cellular components. The exact position of a resonance line of an atomic nucleus (the chemical shift) is a sensitive function of its magnetic environment. Changes in this environment, caused by posttranslational modifications, conformational changes, or binding events, result in changes in the positions of the resonance line (Fig. 1). If differences between the in-cell spectra and the *in vitro* spectra exist, the cause for these differences can be investigated by attempting to simulate the *in vivo* conditions *in vitro*. For example, the laboratory of Gary Pielak has shown that the bacterial protein FlgM is completely unfolded in its purified *in vitro* state but gains structure inside living *Escherichia coli* cells (Dedmon *et al.*, 2002). By adding either 400 g/liter of bovine serum albumin (BSA), 450 g/liter of ovalbumin, or 450 g/liter of glucose to a purified sample of FlgM, they could obtain the same spectral features that were observed in the in-cell spectrum, suggesting that the structural differences between the *in vitro* and the *in vivo* forms of the protein are induced by molecular crowding (Berg *et al.*, 2000; Minton, 2000, 2001).

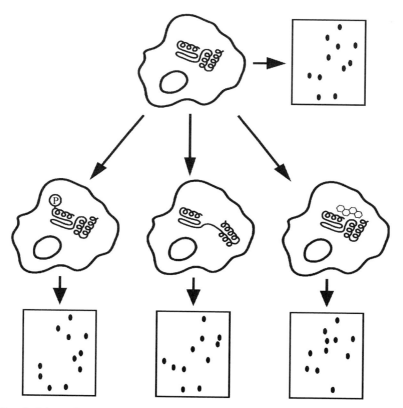

FIG. 1. Schematic representation of potential applications of in-cell NMR experiments. Changes in the chemical shifts can be used to monitor posttranslational modifications, conformational changes, or protein–drug interactions inside living cells. Reproduced from Serber and Dötsch (2001), with permission.

In principle, in-cell NMR could also allow researchers to eliminate the often time-consuming purification procedures in structure determination projects by directly measuring all necessary NMR spectra inside the overexpressing cells. However, it is unlikely that in-cell NMR will be used to determine full three-dimensional structures *in vivo*, since traditional *in vitro* methods offer several key advantages, such as sharper line width and more stable conditions over the course of a several-day multidimensional experiment, that outweigh the potential time savings from skipping the purification.

In the following sections we will discuss some important aspects of in-cell NMR spectroscopy and compare the method to standard high-resolution NMR experiments with purified *in vitro* samples of macromolecules.

Labeling Techniques

The main prerequisite for observing a specific protein species inside living cells by NMR spectroscopy is that the NMR signals of this protein species of interest must be distinguishable from the NMR resonances of all other proteins, nucleic acids, and other cellular components. To achieve this goal, different labeling techniques can be employed as discussed below in detail. The specific labeling scheme will depend on the kind of macromolecule that is to be investigated as well as on the types of cells that are used. Because *E. coli* is so far the most often used organism, we will focus this discussion first on bacterial cells and then provide some remarks on labeling schemes in other cell types. We will discuss both labeling with ^{15}N and with ^{13}C. In addition, ^{19}F can be employed for labeling purposes (Brindle *et al.*, 1989, 1994). Although ^{19}F is the only naturally occurring fluorine isotope, its low abundance in living organisms creates a virtually zero background. However, labeling with fluorine requires chemical modification of the building blocks of biological macromolecules by replacing a hydrogen atom with a fluorine atom, thus creating unnatural systems with potentially different chemical behavior. For this reason fluorine labeling will not be discussed in this chapter. Most of the techniques that are described in this chapter were developed with four different proteins: NmerA, calmodulin, B1 domain of protein G, and FKBP. All of them were cloned into a pET-11a or a pET-9a vector (Stratagene), and the bacterial strain used was *E. coli* BL21 (DE3).

^{15}N Labeling

Background Signals. Originally we were very concerned about growing the bacterial cells on ^{15}N-labeled medium, since this would not only label the protein of interest but every single nitrogen in the cell with ^{15}N. We therefore used a two-step protocol aimed at reducing the level of potential background signals (Serber *et al.*, 2001a,b). First, we grew *E. coli* cells harboring an overexpression plasmid in unlabeled LB medium. After reaching the desired optical density, we harvested the cells by centrifugation and resuspended them in ^{15}N-labeled minimal M9 medium. Ten minutes after resuspension, the cells were induced with isopropylthiogalactoside (IPTG). As a second step for reducing the background level, we added the drug rifampicin (Campbell *et al.*, 2001; Richardson and Greenblatt, 1996; Sippel and Hartmann, 1968) to the bacterial culture to a concentration of 35 μM 40 min after adding IPTG. Rifampicin inhibits the bacterial RNA polymerase but not the polymerase of the bacteriophage T7. Proteins under the control of a T7 promoter can, therefore, be selectively expressed in *E. coli*. The 40-min delay time between the induction with IPTG and the adding of rifampicin is necessary to allow the bacteria to produce enough T7 polymerase. However, comparison of in-cell NMR spectra of samples that were produced in the presence or in the absence of rifampicin was virtually identical and did not show any sign of an increased level of background signals in the absence of the drug (Serber *et al.*, 2001b) (Fig. 2). In each spectrum, a small number of very sharp signals is detectable in the center of the spectrum. The sharpness of these lines suggests that they do not originate from protein signals but from the incorporation of ^{15}N into small molecules like amino acids.

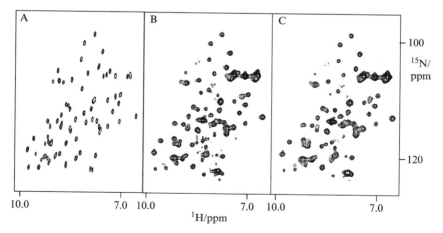

FIG. 2. Comparison of HSQC spectra of a purified *in vitro* sample of NmerA (A) with two in-cell spectra of the same protein. The spectrum in (B) was measured with a sample containing rifampicin and the spectrum in (C) with a sample without the drug. No difference in the level of background signals can be detected between the spectra (B) and (C).

We also investigated the influence of switching the medium from unlabeled LB medium to ^{15}N-labeled minimal medium prior to induction (Serber et al., 2001b). Three different protocols were used to produce in-cell NMR samples of NmerA. First, we grew the bacteria in ^{15}N-labeled minimal medium to an optical density of 0.8 and induced the expression of NmerA by addition of IPTG in the same medium. Second, we grew the bacteria in ^{15}N-labeled minimal medium to an optical density of 0.8, harvested them by centrifugation, and resuspended them in fresh ^{15}N-labeled minimal medium before induction with IPTG. Finally, we grew the bacteria in LB medium, harvesting them by centrifugation, and resuspended them in ^{15}N-labeled minimal medium to the same optical density as the previous sample. All three spectra showed a very similar level of background signals, suggesting that switching the type of medium prior to induction has a negligible effect on the suppression of these signals. The three spectra did show, however, large differences in the intensity of the protein peaks. The sample obtained by growing and expressing the protein in the same minimal medium exhibited the lowest sensitivity. Switching the medium to fresh ^{15}N-labeled minimal medium prior to induction increased the spectral quality several fold. The type of medium used to grow the bacteria in the first phase before induction seems to have only a very small influence on the resulting spectrum, with the sample that was initially grown in LB medium showing a slightly higher sensitivity than the spectrum that was grown in minimal medium.

Investigation of the Influence of the Overexpression Level. The results described above suggest that the amount of background signals that arises from ^{15}N incorporation into other cellular components is small and furthermore insensitive to the specific growth and induction protocol used. This leaves the overexpression level as well as the rotational correlation time of the protein in the cytoplasm as crucial parameters for in-cell NMR. We tested the lower limit for the observation of overexpressed proteins inside living bacteria by inducing NmerA expression for varying amounts of time (Serber et al., 2001b). The results are summarized in Fig. 3, which compares in-cell spectra measured after different induction times with the corresponding overexpression level. Ten minutes after induction, the in-cell heteronuclear single-quantum coherence (HSQC) spectrum shows only some background signals, and NmerA cannot be detected on a polyacrylamide gel. After 30 min, some weak protein resonances become visible in the HSQC spectrum and a faint band of NmerA appears on the gel. One hour postinduction all resonances seen in in-cell NMR experiments of NmerA are visible, and after 2 h the signals become even stronger. The corresponding gel lanes show a prominent NmerA band. These results demonstrate that the detection limit for proteins that tumble freely in the

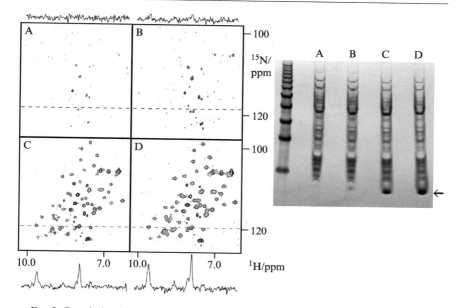

Fig. 3. Correlation of the sensitivity of in-cell NMR spectra of NmerA with the expression level of the protein. The spectrum in (A) was recorded 10 min after induction, the spectrum in (B) 30 min after induction, and the spectra in (C) and in (D) 1 h and 2 h after induction, respectively. The letters on top of each lane of the SDS–polyacrylamide gel correspond to the individual HSQC spectra. In the cross section of spectrum (B) the peaks are right at the detection limit. Each spectrum was measured with four scans per increment in less than 10 min on a 500-MHz instrument equipped with a cryoprobe. Reproduced from Serber *et al.* (2001b), with permission.

bacterial cytoplasm is just 1–2% of the total soluble protein and that an overexpression level of 5% produces in-cell NMR spectra of good quality. This detection limit of 1–2% of the total soluble protein corresponds to an intracellular concentration of NmerA of approximately 150–200 μM (Serber and Dötsch, 2001).

Improvement of Spectral Quality by Expression in Labeled Rich Media. The experiments described above showed that the overexpression level is one of the most important factors influencing the spectral quality of in-cell NMR experiments and suggested that the type of expression medium used could have a strong influence. We investigated this question by comparing in-cell NMR spectra of NmerA grown on minimal medium with spectra of samples grown on labeled rich medium (Serber *et al.*, 2001b). The rich media for these experiments were created by diluting a 10 times concentrated stock solution of ^{15}N-labeled and deuterated medium with H_2O.

This labeling scheme with a deuterated carbon source dissolved in H_2O has the advantage that it produces a high level of side chain deuteration (approximately 80% deuteration on methyl groups and 50% deuteration on the α-carbons, depending on the amino acid type) but keeps the amide nitrogens protonated (Markus *et al.*, 1994). Separate experiments have shown that a 50% deuteration level of the α-carbons leads to a 2-fold reduction of the amide proton R_2 relaxation rate, which significantly reduces the line width (Löhr *et al.*, 2003; Markus *et al.*, 1994). Comparison of the in-cell NMR spectra of the two NmerA samples expressed on ^{15}N-labeled minimal medium and on the ^{15}N-labeled and deuterated rich medium clearly showed a 2- to 3-fold higher sensitivity of the sample expressed on rich medium (Fig. 4). This higher sensitivity can be attributed both to the higher protein expression level in the rich medium as well as to the effect of the deuteration. The comparison of one-dimensional cross sections through peaks of the HSQC spectra showed a reduction in the amide proton line width from an average of 55 Hz in the nondeuterated sample to 40 Hz in the partially deuterated sample.

Selective Amino Acid Labeling. The larger line width of the in-cell NMR spectra causes greater peak overlap relative to *in vitro* spectra (Serber *et al.*, 2001a). One potential method to overcome this problem is selective ^{15}N labeling of only certain types of amino acids. This method is

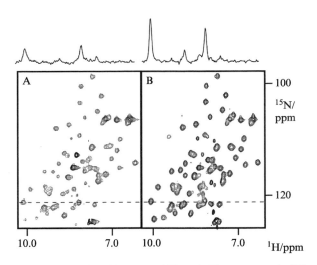

FIG. 4. Comparison of two in-cell spectra of NmerA. The spectrum in (A) was measured with a sample grown on ^{15}N-labeled M9 minimal medium and the spectrum in (B) with a sample grown on ^{15}N-labeled and deuterated medium. Reproduced from Serber *et al.* (2001b), with permission.

particularly powerful if only a certain type of amino acid is of interest, e.g., a residue in the active site of an enzyme. However, not all amino acids are equally suited for amino acid type selective labeling. Depending on their role in the metabolic pathways in the cell and on the presence of transamidases, adding ^{15}N-labeled amino acids to the growth media results in significant levels of cross-labeling of other amino acid types (Lee *et al.*, 1995; McIntosh and Dahlquist, 1990; Muchmore *et al.*, 1989; Ou *et al.*, 2001; Waugh, 1996). Amino acids that are useful for selective labeling are the ones that are located at the end of biosynthetic pathways, such as lysine, arginine, or histidine (Waugh, 1996). For other amino acids, special auxotrophic strains or different media can be used to suppress cross-labeling (McIntosh and Dahlquist, 1990). Many different auxotrophic *E. coli* strains are known that can be used for selective labeling of any of the naturally occurring 20 amino acids (McIntosh and Dahlquist, 1990; Waugh, 1996). In addition, many yeast auxotrophic strains have been created, albeit not for NMR labeling purposes. One drawback, at least of the auxotrophic bacterial strains, is that they often show a reduced expression rate, which reduces the quality of the in-cell NMR spectra. Alternatively, cross-labeling can be suppressed in BL21(DE3) bacteria by adding unlabeled amino acids (Cheng *et al.*, 1995; Reese and Dötsch, 2003). Selective labeling of leucines can, for example, be achieved by adding unlabeled Val and Ile, selective labeling of Ile by adding unlabeled Leu and Val, and selective labeling of Phe by adding unlabeled Tyr (and vice versa). For other amino acids, even more complicated media have to be used (Cheng *et al.*, 1995).

One advantage of amino acid type selective labeling is that the level of background signals can be minimal and, for some types of amino acid, virtually background-free in-cell NMR spectra can be obtained (Serber *et al.*, 2001b). Furthermore, labeling with certain amino acids is of particular interest for labeling in other cell types, for example, eukaryotic cells. Media for uniform labeling of proteins in some eukaryotic cells are commercially available, for example, for yeast and for insect cells. However, these media are very expensive. An interesting alterative to these expensive media is to supplement unlabeled growth media with labeled amino acids that are incorporated into the overexpressed protein and to use these labeled amino acids as probes in in-cell NMR experiments.

^{13}C Labeling. The labeling schemes described above allow researchers to obtain conformational and dynamic information mainly about the protein's backbone and on a few side chains containing nitrogen. To investigate side chains in general, carbon-based labeling schemes have to be used. ^{13}C-based in-cell NMR experiments provide several advantages over ^{15}N-based experiments. First, the sensitivity of detecting methylene and methyl groups is higher due to the larger number of protons directly attached to

the heteronucleus as compared to the single amide proton. Second, carbon-bound protons do not chemically exchange with protons of the bulk water as amide protons do. This exchange can significantly reduce the signal intensity and even broaden the resonance beyond detection. Finally, methyl groups belong to the slowest relaxing spins based on their fast internal rotation, which further increases the sensitivity of methyl group detection (Hajduk *et al.*, 2000; Kay and Torchia, 1991; Pellecchia *et al.*, 2002). Because methyl groups also show the best proton-to-heteronucleus ratio, they are the most attractive side chain probes for in-cell NMR experiments. However, carbon-based labeling schemes also have disadvantages, including higher costs and a significantly higher background level. Comparison of in-cell NMR spectra of calmodulin expressed in *E. coli* grown on M9 minimal medium containing fully ^{13}C-labeled glucose with the corresponding uninduced sample showed that only the high field shifted methyl groups of the protein could be unambiguously assigned as calmodulin resonances (Serber *et al.*, 2004). For all other peaks the high background level excluded such an assignment. Similar results were also obtained from in-cell NMR experiments with cyclic osmoregulated periplasmic glucan expressed in *Ralstonia solanacearum* (Wieruszeski *et al.*, 2001). Full ^{13}C labeling with glucose resulted in many peaks, making the assignment very difficult except for the characteristic region of the anomeric resonances. The high background level observed in these experiments starkly contrasts with the low background level previously observed in ^{15}N,^{1}H-HSQC–based in-cell NMR applications. The greater abundance of carbon than nitrogen in small molecules is the most likely reason for the excessive ^{13}C background level. In addition, many nitrogen-bound protons in these small molecules will exchange very fast with the bulk water, thus effectively broadening their resonances beyond detection. These results have demonstrated that full ^{13}C labeling is less useful than ^{15}N labeling for in-cell NMR studies unless resonances with unique chemical shifts (like high field-shifted methyl groups or anomeric carbons) are used as probes for the macromolecules.

The solution to this high background problem is more selective labeling procedures. In previous experiments we were able to obtain background-free NMR spectra when we switched from full ^{15}N labeling to amino acid-type selective labeling, for example, with ^{15}N-labeled lysine (Serber *et al.*, 2001b). Because methyl groups are the most attractive labeling target for selective labeling procedures, we have focused on developing labeling schemes solely for methyl groups. Several labeling schemes for methyl groups have been developed that mainly focus on reintroducing selectively protonated and ^{13}C-labeled methyl groups in otherwise deuterated proteins (Gardner *et al.*, 1997; Goto *et al.*, 1999; Kay, 2001; Rosen *et al.*, 1996).

Growing *E. coli* on M9 minimal medium in D_2O with 3 g/liter of proto-nated and ^{13}C-labeled pyruvate as the only carbon source significantly reduced the amount of background signals (Serber *et al.*, 2004). However, a substantial amount still remained, which, in particular for quantitative applications, would make this method problematic.

Virtually background-free in-cell NMR spectra can be obtained with even more selective labeling schemes. In-cell NMR spectra of a sample expressed on M9 minimal medium supplemented with ^{13}C-methyl group– labeled methionine at a concentration of 250 mg/liter showed in addition to the methionine methyl group peaks of the protein of interest (calmodulin) only the methyl group of the free methionine and some metabolic products with almost identical chemical shifts (Serber *et al.*, 2004) (Fig. 5). The very intense peak of the free methionine could be reduced by resuspending the bacteria in 50 ml of methionine-free medium followed immediately by centrifugation and final sample preparation. Through this washing step ~90% of the free methionine (and the other products) was removed. Longer washing times did not result in significantly better suppression of

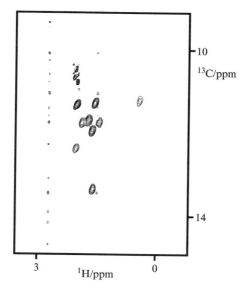

FIG. 5. Spectrum of the methionine methyl group region of calmodulin in *E. coli*. The sample was grown on M9 minimal medium supplemented with ^{13}C-methyl group–labeled methionine. Only the free methionine (the sharp signal at 10.5 ppm in the ^{13}C dimension) and the nine methionine methyl groups of calmodulin are visible.

the free methionine peak, suggesting that most of the free methionine is actually in the extracellular milieu and not contained in the cells.

Although methionine methyl group labeling results in high–quality spectra with high sensitivity and low background level, this particular labeling scheme suffers from the low abundance of methionine in proteins, which reaches only 2.4% of all amino acids. Another attractive amino acid for methyl group labeling is alanine, since it reaches 8.3% abundance, is commercially available in a methyl group only ^{13}C-labeled form, and is quite inexpensive. Its main disadvantage is that ^{13}C-methyl group–labeled alanine produces a significant level of background signals in in-cell NMR experiments (Cheng et al., 1995; Serber et al., 2004). This level can, however, be reduced if, instead of standard M9 minimal medium, a supplemented minimal medium is used to which other unlabeled amino acids and nucleotides have been added (Cheng et al., 1995; Serber et al., 2004). Another attractive probe for in-cell NMR experiments is the δ-methyl group of isoleucine, since it shows very slow relaxation. Isoleucine is commercially available only in a fully ^{13}C-labeled form, which produces very low background levels in in-cell NMR experiments. Alternatively, methyl-group–labeled isoleucine can also be prepared biosynthetically from selectively labeled α-ketobutyrate (Goto et al., 1999). However, the costs for this labeling strategy are high and the risk for background peaks is higher than for isolated isoleucine.

Selective labeling is not only crucial for protein in-cell NMR but also for the detection of other macromolecules in vivo. Wieruszeski and colleagues (2001) had shown that detection of resonances of the cyclic osmoregulated periplasmic glucan in the periplasm of Ralstonia solanacearum is hindered by a high level of background signals if the cells are grown on fully ^{13}C-labeled glucose. The level of background signals could be significantly reduced when the cells were grown on glucose selectively labeled on the C_1 position with [13]C. In contrast, growing the cells on glucose labeled on the C_2 position led to a higher background level. These examples show that in the case of ^{13}C-based labeling, selective labeling schemes are necessary, and the amount of background signals depends crucially on the labeling strategy.

The Rotational Correlation Time of Proteins Inside Cells

A prerequisite for liquid state NMR spectroscopy is that the molecules of interest tumble freely in solution with a rotational correlation time that is not longer than a couple of tens of nanoseconds. Slow tumbling of a molecule causes broadening of the resonance lines, which can lead to their complete disappearance. Because the rotational tumbling rate is a function of the viscosity of the medium in which the macromolecule is dissolved, the cellular viscosity becomes a crucial parameter for in-cell NMR experiments. So far, several techniques have been used to study the intracellular viscosity of different cell types. In yeast cells an intracellular viscosity twice that of water was observed through measurements of ^{19}F relaxation times of selectively tryptophan-fluorinated enzymes (Williams *et al.*, 1997). In fibroblasts, viscosities as low as 1.2–1.4 times that of water were determined by fluorescence polarization experiments (Bicknese *et al.*, 1993; Dayel *et al.*, 1999; Fushimi and Verkman, 1991; Luby-Phelps *et al.*, 1993). Other measurements in higher cells using ^{13}C relaxation (Endre *et al.*, 1983), ^{1}H line width (Livingston *et al.*, 1983), and electron spin resonance (Mastro *et al.*, 1984) also showed an upper limit of the intracellular viscosity for most cell types of twice that of water. To further investigate the intracellular rotational correlation time of proteins in the bacterial cytoplasm, we have performed R_1 and R_2 relaxation measurements of the ^{15}N backbone nuclei of the protein NmerA and have compared them to the relaxation measurements of a purified *in vitro* sample (L. Corsini and V. Dötsch, unpublished results). This analysis showed that the rotational correlation time of NmerA inside the cytoplasm is approximately twice as long as *in vitro*. Due to the linear relationship between the viscosity, the rotational correlation time, and the molecular mass of a protein, this 2-fold increase in the viscosity leads to a 2-fold increase in the apparent molecular mass of a macromolecule. The recent introduction of TROSY (Pervushin *et al.*, 1997) and similar techniques (Riek *et al.*, 1999) into the field of high-resolution NMR spectroscopy has significantly increased the molecular weight range of proteins that are amenable to NMR spectroscopy. In particular, the TROSY version of ^{15}N, ^{1}H-HSQC experiment can be applied to macromolecules above 100 kDa. The combination of TROSY and the relatively low viscosity of the cellular medium suggests that for most proteins investigated so far by traditional NMR methods, the intracellular viscosity will not be a limiting factor for in-cell NMR experiments.

The situation changes, however, if interaction between the macromolecule of interest and cellular components occur. In particular, binding to large cellular components like membranes or DNA increases the rotational correlation time, causing line broadening and potential loss of signals. As an example, we were unable to observe in-cell NMR signals from FKBP (Michnick *et al.*, 1991; Moore *et al.*, 1991) even with the use of high levels of deuteration and TROSY-based pulse sequences despite the fact that this small, 10-kDa protein is highly expressed in *E. coli* and behaves well *in vitro*. Disrupting the bacterial membrane by adding lysozyme to the bacterial slurry in the NMR tube, however, allowed us to observe the normal ^{15}N, ^{1}H-HSQC of FKBP. This result suggested that FKBP is involved in larger complexes inside the bacterial cell that reduce its tumbling rate sufficiently to broaden its resonance lines beyond the detection limit. Disrupting the cellular membrane releases FKBP from these complexes and makes it observable. In contrast to backbone labeling with ^{15}N, labeling of the methionine methyl groups with ^{13}C made two of the three methionine methyl groups detectable in in-cell NMR experiments (Serber et al., 2004). These results are in agreement with the finding that FKBP is never free inside the bacterial cytoplasm but instead is constantly interacting with other proteins, trying to act as a peptidyl-prolyl isomerase. In the case of very large complexes, detection of flexible methyl groups seems to be superior to TROSY and similar backbone-oriented techniques. This is also supported by in vitro NMR applications with very large protein complexes (Kreishman-Deitrick et al., 2003).

If even larger protein complexes with concomitant longer rotational correlation times should be investigated by in-cell NMR experiments, solid-state NMR techniques have to be employed. Preliminary experiments that we have conducted on proteins deposited in inclusion bodies in E. coli suggest that in-cell solid-state NMR experiments are technically feasible.

NMR Techniques

So far both traditional high-resolution liquid-state NMR spectroscopy (Dedmon et al., 2002; Hubbard et al., 2003; Serber et al., 2001a,b, 2004; Shimba et al., 2003) as well as high-resolution magic angle spinning techniques (Lippens and Bohin, 1999; Wieruszeski et al., 2001) have been used and, very recently, even solid-state methods have been employed. The two main challenges for preparing samples for in-cell NMR experiments are to produce a sample of high homogeneity and to keep the cells alive during the measurements. Both parameters, the homogeneity and the cell survival, increase with decreasing cell concentration, albeit at the expense of lower sensitivity. In our experience, samples for high-resolution NMR experiments containing a bacterial slurry with a 30–50% cell content are optimal. Although the rotational correlation time of the cellular proteins does not depend on the macroscopic viscosity of the entire sample and a denser packing of cells should, in principle, be feasible, the quality of the NMR spectra of tightly packed cells decreases (Serber et al., 2001b). We attribute this result to nonuniform cell distributions in the more densely packed samples that increase the inhomogeneity of the entire sample and contribute to line broadening. Similar results have been observed by the laboratory of Guy Lippens (Wieruszeski et al., 2001).

The second method to obtain in-cell NMR spectra is high-resolution magic angle spinning. In heterogeneous systems, magnetic susceptibility gradients are generated at the interfaces of regions with different magnetic properties. These gradients lead to significant line broadening that can render obtaining high-resolution spectra impossible. Fortunately, the magnetic susceptibility of water and cells is very similar. However, the effect of susceptibility gradients as well as macroscopic inhomogeneities is evident from a relatively large line width of approximately 40 Hz for a small 7-kDa protein like NmerA at a temperature of 37°. Furthermore, R_2 relaxation experiments have demonstrated a fast ^{15}N relaxation rate of 30–35 s^{-1}. The line width of proteins in in-cell NMR experiments can be reduced by high-resolution magic angle spinning. By spinning a sample of living *R. solanacearum* at rates of 6 kHz, the research group of Guy Lippens was able to obtain high-resolution ^{13}C, ^1H-HSQC spectra of a periplasmic glucan and of proteins in the cytoplasm. The major drawback of the magic angle spinning method is that only a small amount of cells fits into the rotor. Combined with the fact that current high-resolution magic angle spinning probe heads are not cryogenic probes, the sensitivity of this method is lower than the sensitivity of liquid-state NMR experiments.

The second crucial parameter that limits in-cell NMR experiments is the survival rate of the cells. The survival rate depends critically on the type of cell used as well as on the technique and the medium used to produce the sample. In general, bacteria survive longer when grown and resuspended in labeled full medium than in minimal medium. However, not every commercially available full medium is equally well suited for in-cell experiments. In our experience, deuterated and ^{15}N-labeled medium combines the benefits of long cell survival and high sensitivity due to partial deuteration and high expression level. In these media, survival rates of several hours are possible. However, even during the first few hours of NMR measurement, differences in the bacteria occur that change the state of the intracellular milieu. In particular, pH changes occur that can be monitored through the protonation states of suitable histidines or other pH-sensitive resonances (see below). These changes have to be taken into account during the planning and execution of the experiments. Series of NMR experiments, such as relaxation series, are therefore best carried out in an interleaved mode instead of in a traditional sequential series.

For most higher order cells, the problems of keeping them alive for longer times in the NMR tube increase. Insect cells, while having the advantage of potential high expression rates and growing in suspension, are, for example, very delicate compared to bacterial cells. Exceptions are yeast, which possess a very robust cell wall, and *Xenopus* oocytes, which also have a tough outer sphere. However, if necessary, cells can be kept alive for longer times in the NMR tubes if these tubes are modified to allow for the constant exchange of the medium to supply fresh nutrients as well as oxygen (Degani *et al.*, 1994; McGovern, 1994). Classic *in vivo* NMR experiments of small molecules have, for example, been carried out even with mammalian cells in special bioreactors that contain lines for the constant exchange of the medium. For these experiments, cells are often embedded into a gel matrix and can be kept alive for several days (McGovern, 1994). Unfortunately, most of these modifications are based on 10-mm NMR tubes, and the larger diameter creates additional problems for high-resolution NMR spectroscopy, in particular for cryogenic probes. Further complications arise from the fact that the medium to be exchanged has to be kept at 37° (or the temperature that the cells require). This means that the medium transfer lines also have to be insulated.

Practical Aspects

In-cell NMR experiments differ from traditional high-resolution liquid-state NMR experiments in many ways ranging from sample preparation to the actual measurement. One of the biggest differences is that for each in-cell NMR experiment a new sample has to be made that cannot be stored and used later for additional experiments (one exception are *Xenopus* oocytes). On the other hand, each sample preparation involves a much smaller volume than for a typical *in vitro* NMR experiment. To prepare a typical 500-μl NMR sample with an $\sim$30% cell slurry, only about 50 ml of culture is required. As described in the section on "labeling techniques," labeling costs can be reduced by first growing the cells on unlabeled rich media to a high optical density. Typically, we harvest bacteria grown on 70 ml LB medium at an optical density of 1.4–1.6 and resuspend them in 50 ml labeled medium, either minimal medium or labeled rich medium. Approximately 5–10 min after resuspension, protein overexpression is started by adding IPTG. Three to four hours after induction, cells are harvested by centrifugation (170g for 25 min) and resuspended in a small volume of medium. Rich protonated media contain many small molecules such as amino acids and sugars, which can produce spectral artifacts (T_1 noise) during the measurement. The best resuspension media are, therefore, deuterated full media, since they keep the cells alive longer than minimal medium and minimize the amount of spectral artifacts. To obtain high quality in-cell NMR spectra it is necessary to take great care during the last sample preparation step, the resupending of the cells after the final centrifugation step. During this resuspending step, which must be carried out carefully so as not to lyse the bacteria, the slurry has to be made as homogeneous as possible and macroscopic clumps of cells have to be removed. A bacterial sample prepared in this way shows only little sedimentation tendency over a period of several hours.

If the overexpression media contain labeled amino acids (for example, methyl group-labeled methionine), which create very prominent signals that can obscure the smaller protein signals, the concentration of these amino acids in the final sample can be reduced by an additional washing step. For this purpose, the cells are resuspended again in 50 ml of fresh medium, not supplemented with the particular amino acid, and immediately harvested again by centrifugation.

The final step of sample preparation involves placing the bacterial slurry into the NMR tube. We usually add 40 μl of D_2O to the empty NMR tube and slowly add the bacterial slurry to the bottom of the tube with the help of a long glass pipette. If the sample has been grown on deuterated medium, no further addition of D_2O is necessary.

The NMR measurement itself also differs in some aspects from the traditional experiment with a purified *in vitro* sample. In particular, shimming is not possible on the bacterial slurry. In our laboratory, we prepare an extra shim sample that contains the same medium as the in-cell sample and has the exact same volume. Shimming is performed only with this sample. Other differences include the temperature of the measurement, which is dictated by the optimal temperature of the cells used during the experiment. For bacteria we have used 37°, for yeast 30°, for insect cells 27°, and for oocytes 16°. Finally, the time of the actual measurement is governed by the survival time of the cells in the NMR tube and should be kept as short as possible. The acquisition of a series of experiments such as relaxation studies is best performed by measuring the spectra in an interleaved mode as a pseudo-three-dimensional experiment instead of the typical sequential mode. In this way, changes in the state of the cells have the same effect on all spectra of the series.

The procedures described here reflect only our experience with a limited number of cells and proteins. In-cell NMR experiments with other cell types might require different strategies and optimizations.

Applications

Investigation of Conformation and Dynamics

The dependence of the chemical shift on chemical modification, conformational changes, or binding events has been used in countless studies of proteins or other macromolecules in *in vitro* experiments. The same effect can also be used to investigate the conformation and state of posttranslational modification in in-cell NMR experiments. Hubbard and co-workers (2003) have used in-cell NMR spectroscopy to investigate the ion binding status of the bacterial two-component signal transduction protein CheY in its natural environment, the bacterial cytoplasm. Through comparison of the in-cell spectrum with *in vitro* spectra of the protein complexed with different ions, they could show that CheY preferentially binds Mg^{2+} ions in the *E. coli* cytoplasm. Small additional changes in the chemical shifts might indicate further interactions with other components; however, that could not be conclusively determined so far. During our investigation of calmodulin in bacterial cells, we observed chemical shifts characteristic of the apo form of the protein, showing that the intracellular Ca^{2+} concentration is not high enough to make the calcium-bound form the major conformation in the bacterial cytoplasm (Serber and Dötsch, 2001). Furthermore, additional peaks could be observed suggesting that more than one conformation is present under these *in vivo* conditions (Fig. 6).

Other in-cell NMR applications focus on the *in vivo* conformation of a protein, for example, the already mentioned investigation of the conformation of the bacterial protein FlgM, which is completely unfolded in the purified *in vitro* state but adopts some secondary structure elements in the bacterial cytoplasm (Dedmon *et al.*, 2002). Dedmon and co-workers could show that this induction of structure is caused by the high concentration of other (macro)molecules inside the cell, known as molecular crowding.

NMR spectroscopy is also a very well-established technique to investigate the dynamics of proteins, and many different techniques have been developed to study dynamics ranging from picosecond motions to millisecond motions (Palmer *et al.*, 2001). We have used R_1 and R_2 relaxation experiments to investigate the dynamics of the bacterial protein NmerA

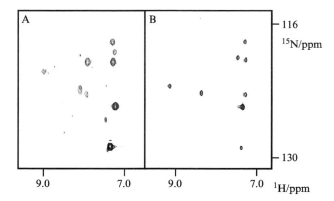

FIG. 6. Comparison of an in-cell NMR spectrum of calmodulin (A), selectively ^{15}N labeled on lysine with an identically labeled *in vitro* sample (B). The strong peak at 129 ppm in the ^{15}N dimension represents a metabolic product of lysine. All other peaks are calmodulin resonances. Reproduced from Serber and Dötsch (2001), with permission.

in its natural environment. These experiments have indicated that the relaxation rate of the metal binding loop of the protein is considerably faster *in vivo* than *in vitro* (L. Corsini and V. Dötsch, unpublished results). Although the cause of this result is still subject to an ongoing investigation, the example shows that dynamic differences between the *in vitro* and the *in vivo* forms of a protein can exist.

Drug Binding

High-resolution liquid-state NMR spectroscopy is widely used in the pharmaceutical industry as a screening tool for protein–drug interaction. As mentioned above, the basis for this application of NMR spectroscopy is the fact that binding events lead to changes in the magnetic environment of the nuclei in the binding site resulting in changes in the chemical shifts. However, *in vitro* NMR screening has—like all *in vitro* screening tools— the disadvantage that an interaction that is observed might not occur in the same way *in vivo*. Potential problems include the inability of a drug molecule to cross the cellular membrane, its fast metabolization, its binding to other cellular components with higher affinity than to its intended target, or differences in the target protein conformation between its *in vitro* and *in vivo* states. These disadvantages of *in vitro* screens can, in principle, be overcome by using in-cell NMR experiments for screening. Chemical shift changes observed in in-cell NMR experiments indicate that the drug molecules can cross the cellular membrane and interact with the target protein in its natural environment. Using in-cell NMR experiments, Hubbard and co-workers (2003) could show that the drug BRL-16492PA that binds to the bacterial two-component signal transduction protein CheY *in vitro* also binds to the same protein inside the bacterial cytoplasm. They concluded this from observing virtually identical chemical shift changes in the $^{15}N,^{1}H$-HSQC spectrum of CheY upon adding the drug either to a purified *in vitro* sample or to a slurry of *E. coli* overexpressing the protein.

In addition to amide protons, which are normally used to monitor drug–protein interactions due to their high chemical shift dispersion, methyl groups are excellent indicators for interaction. A recent investigation has found that within a set of 191 crystal structures of protein–ligand complexes, 92% of the ligands had a heavy atom within 6 Å of a methyl group while only 82% had a heavy atom within the same distance of an amide proton (Stockman and Dalvit, 2002). In addition, as described above, methyl groups provide a significantly higher sensitivity than amide protons, thus making methyl groups very attractive alternatives to amide protons for monitoring protein–drug interactions by in-cell NMR experiments. As an example, we investigated the interaction of calmodulin with the known drug phenoxybenzamine hydrochloride (Serber *et al.*, 2004). Calmodulin binds to several drugs that mimic its interaction with peptides. The main site of interaction is a hydrophobic pocket that is lined with methionines, which makes the methionine-methyl group an excellent indicator for drug–calmodulin interactions. We added the drug to the expressing *E. coli* culture half an hour prior to centrifugation. Although no differences in chemical shift between an *in vitro* sample and the in-cell sample could be detected, some of the peaks in the in-cell spectrum showed increased line broadening, indicating a weak interaction with the drug. This result is in agreement with reports that phenoxybenzamine interacts with calmodulin only in its calcium-bound form and our own results that had demonstrated that calmodulin in the bacterial cytoplasm exists mainly in the calcium-free apo form, suggesting that at most a weak interaction should be observable. To investigate if and how much of the drug had been taken up by the bacteria, we harvested the cells by centrifugation and measured a ^{1}H-1D spectrum of the supernatant. No sign of the drug could be detected. In contrast, the resuspended bacteria pellet showed strong signals from the drug. Further investigations showed that after cell lysis almost all of the drug was still associated with the cell debris, suggesting that phenoxybenzamine is mainly associated with the bacterial membrane. The high local concentration of phenoxybenzamine near the bacterial membrane is most likely responsible for the observed weak interaction between the calcium-free protein and the drug. Although the example reported here—calmodulin expressed in *E. coli*—is not a biologically relevant system, it demonstrates the advantages of in-cell NMR experiments, which are able to detect both the protein resonances as well as the drug.

pH Determination in Living Cells

In-cell NMR spectroscopy can also be used to determine the tautomerization state of histidines in the cellular environment (Schmidt *et al.*, 1991; Shimba *et al.*, 2003; Sudmeier *et al.*, 2003). Histidines are frequently found in the active site of proteins, where their exact protonation state and tautomeric form determine their participation in enzymatic activity. In the past, NMR spectroscopy has been used to investigate the tautomerization and protonation state of histidines *in vitro*. Recently, we have extended these investigations to in-cell NMR experiments based on measuring the values of the C–N coupling constants of the $C^{\varepsilon 1}$ and $C^{\delta 2}$ (Shimba *et al.*, 2003). The values of these coupling constants depend on the protonation and tautomeric state of the imidazole ring and can be determined from the ratio of the $C^{\varepsilon 1}$–H and $C^{\delta 2}$–H cross-peaks in constant time HSQC experiments measured with and without amplitude modulation by the C–N coupling.

Alternatively, if the pK_a value of the histidine side chain is known, measurement of these coupling constants can be used to determine the intracellular pH. Using this method, we determined the pH in the bacterial cytoplasm under the conditions of our NMR experiments to be 7.1 ± 0.1.

Challenges and Future Directions

The goal of in-cell NMR experiments is to investigate the conformation and dynamics of proteins in their natural environment. The relative insensitivity of NMR spectroscopy relative to other spectroscopic methods, however, currently requires the overexpression of the investigated protein to levels of at least 2–5% of total soluble protein. For some proteins like NmerA, which accumulates in the bacterial cytoplasm to levels of up to 6% of total soluble protein in response to mercurials (Fox and Walsh, 1982; Miller, 1999; Misra *et al.*, 1985), this high overexpression level is close to their natural cytoplasmic concentrations. For most other proteins, however, these overexpression levels exceed their natural concentrations and can potentially also influence the behavior of a protein. In particular, if binding events with cellular components are the focus of the in-cell NMR experiments, the relative concentrations of all binding partners have to be considered. On the other hand, in-cell NMR spectroscopy is an ideal tool for investigations of nonspecific interactions like the interaction of the peptidyl-prolyl isomerase FKBP with proteins or the effect of molecular crowding.

Another factor that can change the cytoplasmic environment is the high cellular density in the NMR tube. This high density leads to oxygen starvation for the bacteria, switching them to an anaerobic state, which changes the metabolism of the bacteria and influences the intracellular pH. The influence of the high cell density is, for example, evident from the fact that no detectable protein expression can be induced in a bacterial sample placed inside an NMR tube. These problems, however, can be solved by employing modified NMR tubes or bioreactors for the NMR experiments that can be used to exchange media and provide the bacteria with oxygen (Degani *et al.*, 1994; McGovern, 1994). Several different designs for these bioreactors have already been used for *in vivo* spectroscopy with small molecules. If no bioreactors are available, in-cell NMR experiments should be kept short to minimize the influence of changes in the cellular state.

The biggest challenge for in-cell NMR spectroscopy is to increase the sensitivity and spectral quality of the experiments. Currently, the resolution is compromised by the large line width of the peaks. In the case of NmerA, relaxation measurements have demonstrated that the rotational correlation time has increased only by a factor of 2 relative to an *in vitro* sample. Consequently, the increase in line width is not caused by the higher viscosity of the bacterial cytoplasm but by other factors, linked to the inhomogeneity of the sample, which can at least partially be averaged out by magic angle spinning.

Obviously, increasing the sensitivity of NMR experiments is of paramount importance for in-cell NMR. Higher sensitivity will reduce the necessary intracellular concentration that will better approximate the natural environment of the macromolecule investigated. Fortunately, the introduction of cryoprobes has dramatically increased the sensitivity of NMR instruments in recent years, and further sensitivity improvements are expected. The sensitivity of in-cell NMR experiments can also be increased by detecting methyl groups instead of amide protons (Serber et al., 2004). The higher number of protons attached to the heteronucleus combined with the fast internal rotation and the lack of line broadening due to chemical exchange with the bulk water makes methyl group detection approximately three times more sensitive than amide proton detection, thus reducing the minimum intracellular concentration from roughly 150–200 to 50–70 μM. This reduction is particularly interesting for in-cell NMR experiments in eukaryotic cells where overexpression levels are often lower than in E. coli. In addition to yeast, Xenopus oocytes, and insect cells, mammalian cells transfected with lentiviruses could become an interesting system for studying the behavior of proteins in their intracellular environment.

References

Bachert, P. (1998). Pharmacokinetics using fluorine NMR in vivo. Prog. Nucl. Magn. Reson. Spectrosc. **33**, 1–56.

Berg, B., Wain, R., Dobson, C. M., and Ellis, R. J. (2000). Macromolecular crowding perturbs protein refolding kinetics. EMBO J. **19**, 3870–3875.

Bicknese, S., Periasamy, N., Shohet, S. B., and Verkman, A. S. (1993). Cytoplasmic viscosity near the cell plasma membrane: Measurement by evanescent field frequency-domain microfluorimetry. Biophys. J. **65**, 1272–1282.

Brindle, K. M., Fulton, A. M., and Williams, S. P. (1994). Combined NMR and molecular genetics approach to studying enzymes in vivo. In "NMR in Physiology and Biomedicine" (R. J. Gillies, ed.), pp. 237–261. Academic Press, San Diego, CA.

Brindle, K. M., Williams, S. P., and Boulton, M. (1989). 19F NMR detection of a fluorine-labeled enzyme in vivo. FEBS Lett. **255**, 121–124.

Brown, F. F., Campbell, I. D., Kuchel, P. W., and Rabenstein, D. C. (1977). Human erythrocyte metabolism studies by 1H spin echo NMR. FEBS Lett. **82**, 12–16.

Campbell, E. A., Korzheva, N., Mustaev, A., Murakami, K., Nair, S., Goldfarb, A., and Darst, S. A. (2001). Structural mechanism of rifampicin inhibition of bacterial RNA polymerase. Cell **104**, 901–912.

Cheng, H., Westler, W. M., Xia, B., Oh, B. H., and Markley, J. L. (1995). Protein expression, selective isotope labeling, and analysis of hyperfine-shifted NMR signals of anabaena 7120 vegetative [2Fe-2S]ferredoxin. Arch. Biochem. Biophys. **316**, 619–634.

Cohen, J. S., Lyon, R. C., and Daly, P. F. (1989). Monitoring intracellular metabolism by nuclear magnetic resonance. *Methods Enzymol.* **177,** 435–452.

Dayel, M. J., Hom, E. F., and Verkman, A. S. (1999). Diffusion of green fluorescent protein in the aqueous-phase lumen of endoplasmic reticulum. *Biophys. J.* **76,** 2843–2851.

Dedmon, M. M., Patel, C. N., Young, G. B., and Pielak, G. J. (2002). FlgM gains structure in living cells. *Proc. Natl. Acad. Sci. USA* **99,** 12681–12684.

Degani, H., Ronen, S. M., and Furman-Haran, E. (1994). Breast cancer: Spectroscopy and imaging of cells and tumors. *In* "NMR in Physiology and Biomedicine" (R. J. Gillies, ed.), pp. 329–351. Academic Press, San Diego, CA.

Endre, Z. H., Chapman, B. E., and Kuchel, P. W. (1983). Intra-erythrocyte microviscosity and diffusion of specifically labelled [glycyl-alpha-13C]glutathione by using 13C NMR. *Biochem. J.* **216,** 655–660.

Fox, B., and Walsh, C. T. (1982). Mercuric reductase. Purification and characterization of a transposon-encoded flavoprotein containing an oxidation-reduction-active disulfide. *J. Biol. Chem.* **257,** 2498–2503.

Fushimi, K., and Verkman, A. S. (1991). Low viscosity in the aqueous domain of cell cytoplasm measured by picosecond polarization microfluorimetry. *J. Cell. Biol.* **112,** 719–725.

Gardner, K. H., Rosen, M. K., and Kay, L. E. (1997). Global folds of highly deuterated, methyl-protonated proteins by multidimensional NMR. *Biochemistry* **36,** 1389–1401.

Gillies, R. J. (1994). "NMR in Physiology and Biomedicine." Academic Press, San Diego, CA.

Goto, N. K., Gardner, K. H., Mueller, G. A., Willis, R. C., and Kay, L. E. (1999). A robust and cost-effective method for the production of Val, Leu, Ile (delta 1) methyl-protonated 15N-, 13C-, 2H-labeled proteins. *J. Biomol. NMR* **13,** 369–374.

Hajduk, P. J., Augeri, D. J., Mack, J., Mendoza, R., Yang, J., Betz, S. F., and Fesik, S. W. (2000). NMR-based screening of proteins containing 13C-labeled methyl groups. *J. Am. Chem. Soc.* **122,** 7898–7904.

Hubbard, J. A., MacLachlan, L. K., King, G. W., Jones, J. J., and Fosberry, A. P. (2003). Nuclear magnetic resonance spectroscopy reveals the functional state of the signalling protein CheY *in vivo* in *Escherichia coli. Mol. Microbiol.* **49,** 1191–1200.

Jue, T., and Anderson, S. (1990). 1H NMR observation of tissue myoglobin: An indicator of cellular oxygenation *in vivo. Magn. Reson. Med.* **13,** 524–528.

Kanamori, K., and Ross, B. D. (1997). Glial alkalinization detected *in vivo* by 1H-15N heteronuclear multiple quantum coherence transfer NMR in severely hyperammonemic rat. *J. Neurochem.* **68,** 1209–1220.

Kay, L. (2001). Nuclear magnetic resonance methods for high molecular weight proteins. *Methods Enzymol.* **339,** 174–203.

Kay, L. E., and Torchia, D. A. (1991). The effect of dipolar cross correlation on 13C methyl-carbon T1, T2 and NOE measurements in macromolecules. *J. Magn. Reson.* **95,** 536–547.

Kreishman-Deitrick, M., Egile, C., Hoyt, D. W., Ford, J. J., Li, R., and Rosen, M. K. (2003). NMR analysis of methyl groups at 100–500 kDa: Model systems and Arp2/3 complex. *Biochemistry* **42,** 8579–8586.

Kreutzer, U., Wang, D. S., and Jue, T. (1992). Observing the 1H NMR signal of the myoglobin Val-E11 in myocardium: An index of cellular oxygenation. *Proc. Natl. Acad. Sci. USA* **89,** 4731–4733.

Lee, K. M., Androphy, E. J., and Baleja, J. D. (1995). A novel method for selective isotope labeling of bacterially expressed proteins. *J. Biomol. NMR* **5,** 93–96.

Li, C. W., Negendank, W. G., Murphy-Boesch, J., Padavic-Shaller, K., and Brown, T. R. (1996). Molar quantitation of hepatic metabolites *in vivo* in proton-decoupled, nuclear

Overhauser effect enhanced 31P NMR spectra localized by three-dimensional chemical shift imaging. *NMR Biomed.* **9,** 141–155.

Lippens, G., and Bohin, J.-P. (1999). Structural diversity of the osmoregulated periplasmic glucans of gram-negative bacteria by a combined genetics and nuclear magnetic resonance approach. *In* "NMR in Supramolecular Chemistry" (M. Pons, ed.), Vol. 191, pp. 191–226. NATO Advanced Research Series, Kluwer Academic Publishers, Dordrecht.

Livingston, D. J., La Mar, G. N., and Brown, W. D. (1983). Myoglobin diffusion in bovine heart muscle. *Science* **220,** 71–73.

Löhr, F., Katsemi, V., Hartleib, J., Günther, U., and Rüterjans, H. (2003). A strategy to obtain backbone resonance assignments of deuterated proteins in the presence of incomplete amide 2H/1H back-exchange. *J. Biomol. NMR* **25,** 291–311.

Luby-Phelps, K., Mujumdar, S., Mujumdar, R. B., Ernst, L. A., Galbraith, W., and Waggoner, A. S. (1993). A novel fluorescence ratiometric method confirms the low solvent viscosity of the cytoplasm. *Biophys. J.* **65,** 236–242.

Markus, M. A., Dayie, K. T., Matsudaira, P., and Wagner, G. (1994). Effect of deuteration on the amide proton relaxation rates in proteins. Heteronuclear NMR experiments on Villin 14T. *J. Magn. Reson. B* **105,** 192–195.

Mastro, A. M., Babich, M. A., Taylor, W. D., and Keith, A. D. (1984). Diffusion of a small molecule in the cytoplasm of mammalian cells. *Proc. Natl. Acad. Sci. USA* **81,** 3414–3418.

McGovern, K. A. (1994). Bioreactors. *In* "NMR in Physiology and Biomedicine" (R. J. Gillies, ed.), pp. 279–293. Academic Press, San Diego, CA.

McIntosh, L. P., and Dahlquist, F. W. (1990). Biosynthetic incorporation of 15N and 13C for assignment and interpretation of nuclear magnetic resonance spectra of proteins. *Q. Rev. Biophys.* **23,** 1–38.

Michnick, S. W., Rosen, M. K., Wandless, T. J., Karplus, M., and Schreiber, S. L. (1991). Solution structure of FKBP, a rotamase enzyme and receptor for FK506 and rapamycin. *Science* **252,** 836–839.

Miller, S. M. (1999). Bacterial detoxification of Hg(II) and organomercurials. *Essays Biochem.* **34,** 17–30.

Minton, A. P. (2000). Implication of macromolecular crowding for protein assembly. *Curr. Opin. Struct. Biol.* **10,** 34–39.

Minton, A. P. (2001). The influence of macromolecular crowding and macromolecular confinement on biochemical reactions in physiological media. *J. Biol. Chem.* **276,** 10577–10580.

Misra, T. K., Brown, N. L., Haberstroh, L., Schmidt, A., Goddette, D., and Silver, S. (1985). Mercuric reductase structural genes from plasmid R100 and transposon Tn501: Functional domains of the enzyme. *Gene* **34,** 253–262.

Moore, J. M., Peattie, D. A., Fitzgibbon, M. J., and Thomson, J. A. (1991). Solution structure of the major binding protein for the immunosuppressant FK506. *Nature* **351,** 248–250.

Muchmore, D. C., McIntosh, L. P., Russell, C. B., Anderson, D. E., and Dahlquist, F. W. (1989). Expression and nitrogen-15 labeling of proteins for proton and nitrogen-15 nuclear magnetic resonance. *Methods Enzymol.* **177,** 44–73.

Ou, H. D., Lai, H. C., Serber, Z., and Dötsch, V. (2001). Efficient identification of amino acid types for fast protein backbone assignments. *J. Biomol. NMR* **21,** 269–273.

Palmer, A. G., Kroenke, C. D., and Loria, J. P. (2001). Nuclear magnetic resonance methods for quantifying microsecond-to-millisecond motions in biological macromolecules. *Methods Enzymol.* **339,** 204–238.

Pellecchia, M., Meininger, D., Dong, Q., Chang, E., Jack, R., and Sem, D. S. (2002). NMR-based structural characterization of large protein-ligand interactions. *J. Biomol. NMR* **22,** 165–173.

Pervushin, K., Riek, R., Wider, G., and Wüthrich, K. (1997). Attenuated T2 relaxation by mutual cancellation of dipole–dipole coupling and chemical shift anisotropy indicates an avenue to NMR structures of very large biological macromolecules in solution. *Proc. Natl. Acad. Sci. USA* **94,** 12366–12371.

Reese, M. L., and Dötsch, V. (2003). Fast mapping of protein–protein interfaces by NMR spectroscopy. *J. Am. Chem. Soc.* **125,** 14250–14251.

Richardson, J. P., and Greenblatt, J. (1996). Control of RNA chain elongation and termination. In *"Escherichia coli* and Salmonella" (F. C. Neidhardt, ed.), Vol. 1, pp. 822–848. ASM Press, Washington, D.C.

Riek, R., Wider, G., Pervushin, K., and Wüthrich, K. (1999). Polarization transfer by cross-correlated relaxation in solution NMR with very large molecules. *Proc. Natl. Acad. Sci. USA* **96,** 4918–4923.

Rosen, M. K., Gardner, K. H., Willis, R. C., Parris, W. E., Pawson, T., and Kay, L. E. (1996). Selective methyl group protonation of perdeuterated proteins. *J. Mol. Biol.* **263,** 627–636.

Schmidt, J. M., Thuring, H., Werner, A., Rüterjans, H., Quaas, R., and Hahn, U. (1991). Two-dimensional 1H, 15N-NMR investigation of uniformly 15N-labeled ribonuclease T1. Complete assignment of 15N resonances. *Eur. J. Biochem.* **197,** 643–653.

Serber, Z., and Dötsch, V. (2001). In-cell NMR spectroscopy. *Biochemistry* **40,** 14317–14323.

Serber, Z., Keatinge-Clay, A. T., Ledwidge, R., Kelly, A. E., Miller, S. M., and Dötsch, V. (2001a). High-resolution macromolecular NMR spectroscopy inside living cells. *J. Am. Chem. Soc.* **123,** 2446–2447.

Serber, Z., Ledwidge, R., Miller, S. M., and Dötsch, V. (2001b). Evaluation of parameters critical to observing proteins inside living *Escherichia coli* by in-cell NMR spectroscopy. *J. Am. Chem. Soc.* **123,** 8895–8901.

Serber, Z., Straub, W., Corsini, L., Nomura, A. M., Shimba, N., Craik, C. S., Ortiz de Montellano, P., and Dötsch, V. (2004). Methyl groups as probes for proteins and complexes in in-cell NMR experiments. *J. Am. Chem. Soc.* **126,** 7119–7125.

Shimba, N., Serber, Z., Ledwidge, R., Miller, S. M., Craik, C. S., and Dötsch, V. (2003). Quantitative identification of the protonation state of histidines *in vitro* and *in vivo*. *Biochemistry* **42,** 9227–9234.

Sippel, A., and Hartmann, G. (1968). Mode of action of rifamycin on the RNA polymerase reaction. *Biochim. Biophys. Acta* **157,** 218–219.

Spindler, M., Saupe, K. W., Tian, R., Ahmed, S., Matlib, M. A., and Ingwall, J. S. (1999). Altered creatine kinase enzyme kinetics in diabetic cardiomyopathy. A (31P) NMR magnetization transfer study of the intact beating rat heart. *J. Mol. Cell. Cardiol.* **31,** 2175–2189.

Stockman, B. J., and Dalvit, C. (2002). NMR screening techniques in drug discovery and drug design. *Prog. Nucl. Magn. Reson. Spectrosc.* **41,** 187–231.

Sudmeier, J. L., Bradshaw, E. M., Haddad, K. E., Day, R. M., Thalhauser, C. J., Bullock, P. A., and Bachovchin, W. W. (2003). Identification of histidine tautomers in proteins by 2D 1H/13C(delta2) one-bond correlated NMR. *J. Am. Chem. Soc.* **125,** 8430–8431.

Tran, T. K., Kreutzer, U., and Jue, T. (1998). Observing the deoxy myoglobin and hemoglobin signals from rat myocardium *in situ*. *FEBS Lett.* **434,** 309–312.

Waugh, D. S. (1996). Genetic tools for selective labeling of proteins with α-^{15}N-aminoacids. *J. Biomol. NMR* **8,** 184–192.

Wieruszeski, J.-M., Bohin, A., Bohin, J.-P., and Lippens, G. (2001). *In vivo* detection of the cyclic osmoregulated oeriplasmic glucan of Ralstonia solanacearum by high resolution magic angle spinning NMR. *J. Magn. Reson.* **151,** 118–123.

Williams, S. P., Haggle, P. M., and Brindle, K. M. (1997). F-19 NMR measurements of the rotational mobility of proteins *in vivo*. *Biophys. J.* **72,** 490–498.

[3] Molecular Fragment Replacement Approach to Protein Structure Determination by Chemical Shift and Dipolar Homology Database Mining

By GEORG KONTAXIS, FRANK DELAGLIO, and AD BAX

Abstract

A novel approach is described for determining backbone structures of proteins that is based on finding fragments in the protein data bank (PDB). For each fragment in the target protein, usually chosen to be 7–10 residues in length, PDB fragments are selected that best fit to experimentally determined one-bond heteronuclear dipolar couplings and that show agreement between chemical shifts predicted for the PDB fragment and experimental values for the target fragment. These fragments are subsequently refined by simulated annealing to improve agreement with the experimental data. If the lowest-energy refined fragments form a unique structural cluster, this structure is accepted and side chains are added on the basis of a conformational database potential. The sequential backbone assembly process extends the chain by translating an accepted fragment onto it. For several small proteins, with extensive sets of dipolar couplings measured in two alignment media, a unique final structure is obtained that agrees well with structures previously solved by conventional methods. With less dipolar input data, large, oriented fragments of each protein are obtained, but their relative positioning requires either a small set of translationally restraining nuclear Overhauser enhancements (NOEs) or a protocol that optimizes burial of hydrophobic groups and pairing of β-strands.

Introduction

With the completion of the sequencing of the human and many other genomes and the availability of an abundance of protein sequence data, there is a strong demand for rapid determination of tertiary protein structures. There are two main experimental avenues toward obtaining atomic resolution protein structures: X-ray crystallography and solution state nuclear magnetic resonance (NMR) spectroscopy. The process of structure determination by X-ray crystallography is already quite streamlined due to the availability of robotics for optimizing crystallization conditions, high-intensity synchrotron radiation sources, and standardized, semiautomated

Copyright 2005, Elsevier Inc.
All rights reserved.
0076-6879/05 $35.00

analysis software. Structure determination by NMR spectroscopy on the other hand is still a time-consuming and labor-intensive process with a turnaround time typically on the order of several months, which additionally requires ^{15}N, ^{13}C, and, for larger proteins, ^{2}H isotopic enrichment. Usually, an NMR structure determination project proceeds in several stages: assignment of backbone resonances using pairs of now standard triple-resonance experiments, assignment of side chain resonances using ^{13}C-, ^{1}H- or ^{15}N-mediated TOCSY- and COSY-type experiments, followed by assignment of NOE cross-peaks, and structure calculation.

The introduction of facile methods for weakly aligning proteins relative to the magnetic field now also allows measurement of residual dipolar couplings (RDCs). In favorable cases, the alignment necessary for a non-vanishing dipolar interaction can be imposed on the solute macromolecules directly by the magnetic field (Bothner-by *et al.*, 1985; Kung *et al.*, 1995; Tjandra *et al.*, 1996; Tolman *et al.*, 1995), but more commonly an anisotropic aqueous medium is used. Many such media are now available, including lyotropic liquid crystalline solutions of phospholipid bicelles (Tjandra and Bax, 1997), Pf1, *fd*, or TM phage particles (Clore *et al.*, 1998b; Hansen *et al.*, 1998), cellulose crystallites (Fleming *et al.*, 2000), and polyethylene glycol (Ruckert and Otting, 2000) or cetylpyridinium halide-based bilayers (Barrientos *et al.*, 2000; Prosser *et al.*, 1998). Anisotropically compressed, low-density polyacrylamide gels (Chou *et al.*, 2001a; Ishii *et al.*, 2001; Meier *et al.*, 2002; Sass *et al.*, 2000; Tycko *et al.*, 2000; Ulmer *et al.*, 2003) and suspensions of magnetically oriented purple membrane fragments (Koenig *et al.*, 1999; Sass *et al.*, 1999) also have proven useful for this purpose.

RDCs are global parameters in the sense that they restrain the orientations of the corresponding dipolar interaction vectors all relative to a single reference frame, often referred to as the principal axis frame of the alignment tensor. In this respect, they differ in nature from NOEs and dihedral restraints derived from J couplings, which report on atomic positions relative to one another. Besides improving local geometry (Chou *et al.*, 2001b; Tjandra *et al.*, 1997), RDCs have been shown to be highly useful for determining the relative orientation of individual domains in multisubunit proteins, nucleic acids, and their complexes (Braddock *et al.*, 2001; Clore, 2000; Lukavsky *et al.*, 2003).

In the principal frame of the alignment tensor, the dipolar coupling is given by

$$D_{ij}(\theta,\phi) = D_{\mathrm{a}}[(3\cos^2\theta_{ij} - 1) + 3/2\,R\sin^2\theta_{ij}\cos^2\phi_{ij}] \qquad (1)$$

where θ and ϕ are the polar angles of the dipolar interaction vector, r_{ij}, in the alignment frame; D_{a} is the magnitude of the alignment tensor, which includes constants related to the magnetogyric ratio and internuclear distance of nuclei i and j, and R is the rhombicity of the alignment tensor (Bax *et al.*, 2001). Clearly, with a single experimental $D_{ij}(\theta,\phi)$ value, and two variable parameters, in general an infinite number of (θ,ϕ) solutions exist. This degeneracy may be partly lifted if RDCs in a different alignment medium and with a different, independent alignment tensor are available (Ramirez and Bax, 1998). However, even in this case, a vector orientation can never be distinguished from its inverse because both orientations lead to the same dipolar coupling. Only once the "handedness" of a local element of structure that involves several dipolar interactions in at least two alignment frames is known, can the absolute orientation of such a fragment and thereby of its vectors be determined (Al-Hashimi *et al.*, 2000). As a consequence, when attempting to build a full protein structure that simultaneously satisfies Eq. (1) for all dipolar interactions, the number of false minima scales exponentially with this number of couplings. Solving this problem by means of a "brute force" simulated annealing or Monte Carlo program on a full protein has proven very difficult. However, when first assembling local substructures, this problem is no longer intractible, and a number of recently proposed approaches rely on this principle (Andrec *et al.*, 2001; Delaglio *et al.*, 2000; Hus *et al.*, 2000; Rohl and Baker, 2002).

Our present approach represents a much improved and more stable version of the molecular fragment replacement (MFR) method described earlier, which derived backbone torsion angles from searching the PDB for seven-residue peptide fragments that fit experimental dipolar couplings in a fragment of the target protein (Delaglio *et al.*, 2000). In the original procedure, a starting model was first built using these backbone torsion angles and subsequently refined by optimizing agreement between this model and the full set of dipolar couplings. Although the method results in reasonable structures, provided that a nearly complete set of dipolar coupling is available, convergence to a satisfactory final structure and accuracy of its local details remain limited by the quality of the fragments of the original search. However, in favorable cases, substantial regions of a protein can be assembled from such data, even in cases in which all dipolar couplings and assignments are derived from a single experiment (Zweckstetter and Bax, 2001).

Our MFR approach is related to work by Annila *et al.* (1999), who proposed to use dipolar couplings for finding structurally homologous proteins in the PDB. Instead of searching for complete proteins, the MFR program searches the PDB for structural homology for only 7–10 residues at a time. The idea of using small substructures from a database of representative protein structures as "templates" for building a structure was pioneered by Jones and co-workers and has been very successful in

X-ray crystallography (Jones and Thirup, 1986). The approach has also been applied to solving structures on the basis of NOEs, where it searches a database for substructures compatible with the experimental NOEs (Kraulis and Jones, 1987). However, because only short and medium range NOEs can be used in the search process, obtaining the correct tertiary fold remains very difficult with such a method. Other approaches relying on database substructures have also been described in recent years. Work by the Baker group (Rohl and Baker, 2002) relies on selecting a large number of database fragments that are roughly compatible with the experimental parameters measured for the corresponding target fragment and then using efficient Monte Carlo methods to assemble these fragment into a common structure with reasonable packing properties, where the fragments retain an orientation needed to satisfy dipolar coupling restraints. A method proposed by Andrec et al. (2001) is similar in spirit to our own MFR method but uses "postprocessing" to distinguish correct from incorrect fragments by comparing them with the overlapping region of an adjacent fragment. Using a so-called bounded-tree search, self-consistent sets of overlapping fragments can be identified relatively rapidly, resulting in a backbone structure.

A different approach to building complete protein backbone structures from dipolar couplings, which does not rely on a database for finding suitable substructures, has been proposed by Hus et al. (2000) and Giesen et al. (2003). It is conceptually somewhat similar to approaches pursued in determining polypeptide structure on the basis of ^{15}N–^1H dipolar couplings derived from solid-state NMR measurements (Brenneman and Cross, 1990; Marassi and Opella, 1998; Nishimura et al., 2002; Wu et al., 1995) and conducts a systematic search in Cartesian space when adding a peptide plane to the chain. The approach requires a very complete set of dipolar couplings when applied de novo, but it has other applications too. For example, it was shown to be particularly powerful for pinpointing the precise structural differences in the backbone at the active site of a 27-kDa enzyme [methionine sulfoxide reductase (MsrA) from Erwinia chrysanthemi] and its Escherichia coli homologue, for which an X-ray structure was available (Beraud et al., 2002). Conceptually, Hus' method shares features with a method devised by Mueller et al. (2000), which determines the (usually 4-fold degenerate) peptide plane orientations compatible with experimental dipolar couplings prior to finding a chain compatible with these orientations. Finally, Fowler et al. (2000) demonstrated it is feasible to get information on the fold from ^{15}N–$^1H^N$, $^1H^N$–$^1H^N$, and $^1H^N$–$^1H^\alpha$ couplings without the need for ^{13}C enrichment.

Our improved MFR approach, which we refer to as MFR+, represents a much more versatile and stable version of the original MFR method. The method differs from all previous database substructure methods by intro-duction of an intermediate step where the fragments are refined with respect to the experimental observables (shifts, couplings, and possibly short and medium range NOEs or torsion angle restraints) prior to their final selection and incorporation into a structure. It also utilizes the unique advantage of dipolar tensor parameters to maintain reasonable orientations for each fragment at all stages of the substructure assembly. The user has complete freedom to specify weighting factors used and to define the minimal criteria for deeming a selection to be "reliable." With the default settings, and with dipolar couplings available from two different media, the program can rapidly generate backbone structures for the proteins ubiquitin and GB3 that are considerably less than 1 Å from their true structure. With less experimental data, accurate partial structures can be obtained, which subse-quently can be used to assemble the structure either manually or by using docking algorithms (Clore, 2000; Clore and Schwieters, 2003).

Description of the MFR+ Method

The routines to conduct the homology search, visualize the results, and the assembly of a structure were written in the Tcl/Tk language and use "NMRWish," an in-house version of the Tcl/Tk interpreter "wish," which has been customized by the addition of routines to handle and manipulate tables, databases, and PDB format files. It can perform chemical shift (CS) and dipolar coupling (DC) simulations, carry out coordinate align-ments, handle restraints (NOE, dihedral, CS, DC, J), and also includes facilities for structure calculation and a molecular dynamics engine, named DYNAMO (available through http://spin.niddk.nih.gov/bax).

The MFR+ method uses chemical shifts and dipolar couplings as input for database mining and subsequent model building. As mentioned earlier, MFR is closely related to a method for protein fold recognition based on "dipolar homology," where a database is searched for either a complete protein or a protein domain, compatible with a given set of dipolar cou-plings (Annila *et al.*, 1999). This "protein fold recognition" procedure typically requires that some degree of sequence homology is available, so that a sequence alignment can be performed, and it is known which residue in the database protein corresponds to which residue in the target protein. The MFR approach is not subject to this requirement and simply searches for all substructures in the PDB that are compatible with dipolar couplings and chemical shifts measured for a given fragment of the target protein. Provided the fragment is chosen to be small enough in size, invariably there are a large number of fragments in the database that exhibit a reasonable fit to the experimental parameters (Du *et al.*, 2003), but frequently they do not represent a unique structural cluster. For such small fragment sizes, the experimental information therefore is less discriminating in defining the correct database substructures than when searching for full proteins. However, when choosing too large a fragment length, it is possible that no adequate hits may be found in the database. In practice, we find a fragment size in the 7–10 residue range to be optimal for MFR+, with larger fragments preferred if couplings are available in only a single medium, or if only relatively few couplings (two or less) are available per residue.

The MFR approach is based on the premise that if two fragments can be fit to the same set of dipolar couplings and exhibit similar secondary chemical shift (i.e., deviation from random coil shifts) patterns, they are likely to have similar structures. By comparing with one another the best "hits" found in the database for a given set of experimental dipolar couplings and secondary chemical shifts, a degree of certainty is obtained about how well the database substructures represent the corresponding fragment in the query protein.

Several stages can be distinguished in the MFR+ method:

1. Search of the database for fragments with dipolar, chemical shift, and sequence homology; optionally, other NMR observables based on local structure such as J couplings or NOEs can also be used.
2. Low temperature simulated annealing refinement of an initial set of best matches found in the search, followed by automated selection of a subset of these refined structures.
3. Assembly of a tertiary structure.

A flow diagram of this procedure is presented in Fig. 1, and a detailed discussion of the steps involved will be presented below. As mentioned above, for finding the structurally feasible peptide conformations, we rely on the PDB. Although for very short fragments, a small number of proteins is sufficient to represent all possible conformations found in nature (Jones and Thirup, 1986; Jones et al., 1991), for larger peptide fragments such as those used in our MFR+ approach this is no longer the case, and a database as large as possible is desirable. At the same time, structures with high sequence homology, or low-resolution structures with considerable uncertainty in their coordinates, carry little useful information. As a compromise, we have filtered the PDB to retain only X-ray–derived structures, solved at a resolution of ≤ 2.2 Å, and to discard structures that are more than $\sim 90\%$ identical to an already selected protein. This final database then contains 893 entries (distributed as part of the MFR+ package). For testing the MFR+ program, which is carried out on proteins for which the structure already is available in this database (e.g., ubiquitin or GB3),

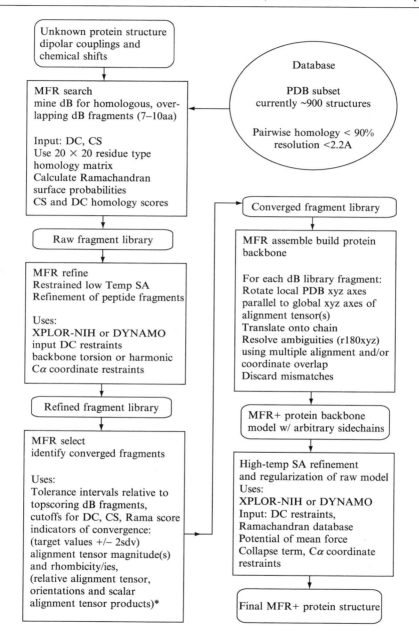

the relevant identical and/or highly homologous structures are excluded from the database search.

Search Procedure

As a first step, the protein under study (target protein) is broken up into small, overlapping peptide fragments, typically 7–10 residues in length. So, for a 100-residue protein, 91–94 such query fragments are generated. In a first, preliminary evaluation, singular value decomposition (SVD) (Losonczi *et al.*, 1999; Sass *et al.*, 1999) is used to determine how well these couplings fit to any of the roughly 180,000 substructures in the database. In practice, for reliable evaluation at least a dozen or more couplings per fragment are required, as there are five degrees of freedom (i.e., five independent elements of the alignment tensor) in the fitting process. In its simplest and fastest mode of operation, only the 10 or 20 best-fitting fragments are retained.

The SVD fitting procedure, while fast, does not take advantage of information that frequently is known prior to the start of the search, such as the magnitude and rhombicity of the alignment tensor, which can be obtained by inspection of the "powder pattern" (Bryce and Bax, 2004; Clore *et al.*, 1998a). Alternatively, these two parameters frequently can be derived at good accuracy by fitting dipolar couplings, measured for a stretch of residues that carries a clear α-helical chemical shift signature, to an idealized model α-helix. Such information can be incorporated into nonlinear least-squares optimization techniques, but carrying out such an optimization many thousands of times per query fragment is exceedingly slow. As a compromise, an alternate strategy is to retain a larger subset of the best SVD results, e.g., 1000, and subject this subset to restrained least-squares minimization, using prior knowledge of magnitude and rhombicity. As an interesting side note, the MFR SVD search procedure itself provides an automated method for estimating tensor magnitude and rhombicty, before the structure is known, since the collection of best-fitting fragments found by SVD will have magnitudes and rhombicities that cluster around the true values for the intact target protein (Fig. 2). In practice, average values over the collection of fragments are weighted to give the highest

---

FIG. 1. Flow diagram of the MFR+ protein structure determination protocol. aa, amino acid; CS, chemical shift; dB, database; DC, dipolar coupling; PDB, Protein Data Bank; Rama, Ramachandran map; sdv, standard deviation; SA, simulated annealing. The selection criteria involving relative alignment tensor orientation, marked with an asterisk, applies only to the case in which multiple alignment media are used.

importance to regions where the fragments show the greatest structural consensus. Results of this method are shown in Fig. 2, which indicates that tensor parameters for 11 different samples could be predicted with a root mean square (RMS) error of less than 5%. For the case in which multiple media are used for a given protein, the same procedure can also be used for determining the relative orientation of the alignment tensors.

During the database search, a dipolar coupling score F_{DC} is used to measure the goodness of fit between a target fragment's set of N observed dipolar couplings and the values calculated for a database substructure. The score is a simple weighted root-mean-square deviation (RMSD):

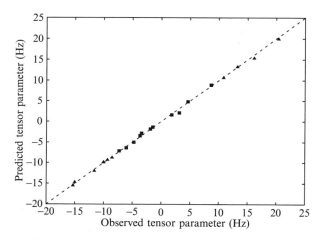

FIG. 2. Plot of dipolar coupling tensor parameters obtained from SVD fits of experimental dipolar couplings to known structures vs. values predicted automatically by the MFR search procedure, without prior knowledge of the structure. (▲) tensor magnitude D_a; (■) rhombicity, R, times tensor magnitude D_a. Values are shown for 11 different samples, comprising the proteins ubiquitin (two media), DinI (two media), Gb3 (five media), and γ-crystallin (two media; J. Wu, personal communication). All values are scaled relative to $^1D_{HN}$ couplings. The MFR tensor estimates are weighted averages of the SVD-derived tensor parameters obtained for database fragments selected by the MFR search. For calculating this weighted average, first for each set of selected database fragments, corresponding to target protein fragment j, linear averages, $<D_a(j)>$ and $<R(j)>$, are calculated. Then, the weighted average over the chain is calculated as $<D_a> = \sum_{j=1,...,N} w_j <D_a(j)>/\sum_{j=1,...,N} w_j$, and similarly for $<R>$, where the summation extends over all N fragments. The weighting factor is given by $w_j = \exp(-\alpha q_j)$, where $\alpha = 3$ Å^{-1}, and q_j is the backbone RMSD over the collection of selected database segments for fragment j. Typical values found for q are 0.1–0.3 Å for helical regions, 0.2–0.5 Å for well-defined β-sheet regions, and 0.8–2.0 Å for regions where MFR results are ambiguous. In all cases, tensor parameters are predicted to an accuracy of better than 0.5 Hz.

$$F_{\mathrm{DC}} = \left\{ (1/N) \sum_{ij} [w_{ij}(D_{ij}^{\mathrm{obs}} - D_{ij}^{\mathrm{calc}})]^2 \right\}^{1/2} \tag{2}$$

In practice, the weighting factor w_{ij} is selected to scale all types of couplings into a similar range, commonly as $(\gamma_N \gamma_H / r_{\mathrm{NH}}^3)/(\gamma_i \gamma_j / r_{ij}^3)$. The weighting factor w_{ij} can also be scaled to include adjustment for the estimated uncertainties in the measured couplings. In general, no such adjustment is needed if the estimated random error in a given type of coupling is less than about $\pm 15\%$ of the applicable D_a value.

The chemical shift score for each fragment is calculated by comparing observed values with those expected for the database substructure. These latter values are derived using (ϕ, ψ)-dependent chemical shift surfaces, obtained from the TALOS chemical shift database (Cornilescu *et al.*, 1999).

For each (ϕ, ψ), the program looks up the average secondary chemical shift on this (ϕ, ψ) surface and its rms spread, $\sigma(\phi, \psi)$. The chemical shift score is then defined as

$$F_{\mathrm{CS}} = \left\{ \sum_i [(\mathrm{CS}_i^{\mathrm{obs}} - \mathrm{CS}_i^{\mathrm{calc}})/\sigma_i(\phi, \psi)]^2 / N \right\}^{1/2} \tag{3}$$

Other terms can be used too in the search for suitable database fragments. Considering that the search does not account for residue type, but that certain residues in the database (e.g., Gly) have very different (ϕ, ψ) distributions from those seen for other residues, one such additional term is referred to as "Ramachandran surface quality," or "ϕ, ψ surface quality." It describes how well the trial fragment falls into the most favored region of the Ramachandran plot of the residue(s) in the target protein, onto which it is mapped. It is calculated as the normalized probability for a particular residue to assume a particular ϕ, ψ combination, and it is therefore a measure of how likely the target sequence can assume the conformation of the database substructure. For an N-residue fragment this surface score is defined as

$$F_{\mathrm{SURF}} = \sum_i - \ln[p_i(\phi_i, \psi_i)/p_{i,\mathrm{max}}]/N \tag{4}$$

where $p_i(\phi_i, \psi_i)$ represents the database population of the (ϕ_i, ψ_i) conformation for the target residue type at position i, and $p_{i,\mathrm{max}}$ is the population of the most favored conformation of that particular residue type.

Sequence homology information can also be used in the fragment scoring process. A homology score is generated to penalize for "mutations" between the target sequence and the sequence of a database substructure, according to

$$F_{\text{HOMO}} = \left(\sum_i h_{jk}(i)^2 / N \right)^{1/2} \tag{5}$$

where $h_{jk}(i)$ are the elements of a residue-type similarity matrix between residue type j at position i in the target protein fragment and residue type k in the corresponding position in the database substructure. There are several possible schemes for homology scoring, and in practice a given scheme is selected by substituting the appropriate table corresponding to the desired homology matrix. In the present work, the matrix used is derived from the similarity in the residue-specific Ramachandran map distributions, $p(\phi,\psi)$:

$$h_{jk} = A \left\{ \sum_{\phi,\psi=0°,\ldots,359°} ([p_k(\phi,\psi) - p_j(\phi,\psi)] p_j(\phi,\psi))^2 \right\}^{1/2} \tag{6}$$

where A is an arbitrary factor, used to scale the results to a convenient range, and each $p(\phi,\psi)$ distribution is normalized such that its maximum value is unity. Equation (6) results in an asymmetric score matrix (especially for substitutions involving Gly residues) reflecting the fact that, for example, for a Gly residue in the target fragment, the presence of a non-Gly residue in the database fragment biases the selected fragment toward negative ϕ angles; this is more of a concern than the inverse scenario. (Note that a Gly residue in the trial fragment, with a bias toward positive ϕ, usually is already penalized by an increased Ramachandran surface quality score.) To illustrate the weight factors for residue similarity identified in this manner, Table I provides a condensed form of the 20×20 matrix. However, it is the full matrix that is used by the software.

Additional information, such as sequential or medium range backbone NOEs, or dihedral restraints also can be introduced at this stage of the fragment search, but this latter information was not needed in the application to small proteins with relatively complete sets of dipolar couplings, discussed in this chapter.

The final score for a particular fragment is a weighted sum of the individual terms:

$$F_{\text{TOTAL}} = \sum_i c_i F_i \tag{7}$$

where i = CS, DC, surface (SURF), homology (HOMO). Empirically, a set with $c_{\text{DC}} = 1.0$, $c_{\text{CS}} = 0.2$, $c_{\text{SURF}} = 0.2$, $c_{\text{HOMO}} = 0.2$ was found to give close to optimal results. These coefficients were chosen such that the dipolar term dominates, but the other terms remain significant. It can be useful to also evaluate which fragments are selected if the DC term is scaled down, and the

TABLE I
HOMOLOGY FACTORS, h_{jk}, DERIVED FROM EQ. (6)[a]

Residue	A	RK	N	D	CST	QE	HFWY	G	IV	LM	P
A	0.0	0.6	1.8	0.9	2.1	0.4	1.5	8.7	2.8	0.6	5.1
RK	0.5	0.0	1.6	0.6	1.7	0.4	1.1	8.6	2.3	0.5	5.0
N	0.7	0.7	0.0	0.4	1.4	0.5	1.4	8.2	2.4	0.7	5.0
D	0.5	0.4	1.1	0.0	1.7	0.3	1.4	8.5	2.6	0.6	4.9
CST	0.9	0.9	1.5	0.8	0.0	0.8	1.3	8.7	1.7	0.9	4.8
QE	0.4	0.4	1.6	0.7	1.8	0.0	1.3	8.6	2.5	0.4	5.2
HFWY	0.9	0.9	1.9	1.1	1.7	1.0	0.0	8.6	1.9	0.8	5.0
G	0.4	0.4	1.5	0.7	1.9	0.5	1.2	0.0	2.8	0.5	4.9
IV	0.8	1.1	2.0	1.1	1.8	0.8	1.3	8.7	0.0	0.8	5.3
LM	0.5	0.6	1.8	0.8	1.8	0.4	1.2	8.7	2.2	0.0	5.2
P	4.0	4.3	4.9	4.6	5.0	4.4	4.8	9.7	5.5	4.3	0.0

[a] The full matrix used by MFR+ consists of 20 × 20 elements. The grouping above reflects the high degree of similarity of the coefficients pertaining to any given group.

CS term is given a high weight. A set of coefficients we commonly use for this purpose is $c_{DC} = 0.1$, $c_{CS} = 1.0$, $c_{SURF} = 0.1$, $c_{HOMO} = 0.1$. This alternate weighting scheme is particularly useful when in a given region of the polypeptide backbone the number of observed dipolar couplings is low. If both searches are conducted, the results are pooled together, for example, by retaining the 10 best fragments of each search. In a subsequent refinement (see below), database fragments that are close to their true structure will better converge to the correct solution, and the presence of lower quality fragments, usually selected on the basis of their chemical shifts, therefore does not pose a problem. Typical search parameters are summarized in Table II.

A convenient way to visualize the search results displays the collection of fragments as a "Ramachandran flight path" of the peptide fragments, which connects the (ϕ_i, ψ_i) position in the Ramachandran map of residue i to the (ϕ_{i+1}, ψ_{i+1}) and (ϕ_{i-1}, ψ_{i-1}) position of residue $i + 1$ and $i - 1$ (Fig. 3). The "true" structure needs to be represented by an unbroken path, and outliers can be identified very easily this way. Figure 3A shows that at this stage the ϕ, ψ angles of the selected substructures still exhibit a considerable spread. However, as described below, the quality of these fragments and the width of their ϕ, ψ distribution can be vastly improved by refining these fragments with respect to their dipolar couplings.

Fragment Refinement

Next, the backbone angles from the substructures resulting from the search are used to build fragments of the target protein, with the correct residue types, but with initial side chain orientations being random. As the

TABLE II
FILES AND PARAMETERS USED DURING THE MFR+ DATABASE HOMOLOGY SEARCH AND
THEIR TYPICAL NAMES OR VALUES

Input file or parameter	Name or typical value
Dipolar coupling table(s)[a]	dObs[*].tab
Backbone chemical shift table[a]	csObs.tab
Reference structure[b]	ref.pdb
Name of output fragment table	mfr.tab
Location of PDB files	$PDBH_DIR[c]
List of PDB files to search	$PDBH_TAB[c]
csW/dcW/surfW/homoW[d]	0.2/1.0/0.2/0.2 or 1.0/0.1/0.1/0.1
segLength[e]	7 (default)–10
scoreCount[f]	10 (default)–20
csThresh[g]	2.2 (default)
undefFrac[h]	0.9 (default)
Additional parameters, calculated for each saved fragment in the output fragment table	
da[i]	From fit
dr[i]	From fit
scalarProduct[j]	−1 to 1
cosXX, cosYY, cosZZ[k]	0 to 1

[a] Dipolar coupling and backbone chemical shift input tables are PALES and TALOS input formats, respectively.

[b] Covalent template of the target protein, either in extended or randomized conformation, or as a model derived from other sources, in PDB format. The program can display agreement between selected fragments and the template (to be used for test purposes, when the true structure is known).

[c] UNIX environment variables.

[d] Weighting factors for F_{DC} [Eq. (2)], F_{CS} [Eq. (3)], F_{SURF} [Eq. (4)], and F_{HOMO} [Eq. (5)] in the total fragment score [Eq. (7)].

[e] Number of residues per segment.

[f] Number of database fragments retained after first round of MFR+ search.

[g] Chemical shift cut-off value. No fragment with F_{CS} greater than this value is retained.

[h] If a fraction smaller than undefFrac of the experimental data can be mapped onto the database fragment (e.g., due to a missing N–H vector for Pro in a database fragment), the database fragment is not considered.

[i] Axial and rhombic components of local best fit alignment tensor (obtained from SVD fit), calculated for each alignment medium used.

[j] Normalized generalized scalar product of local best fit (SVD) alignment tensors; this is a number in the [−1.0 to 1.0] range.

[k] cosXX, cosYY, and cosZZ specify the relative orientation of two alignment tensor axis frames (OXYZ and O′X′Y′Z′). cosXX = | cos[angle (OX O′X′)] |, etc.

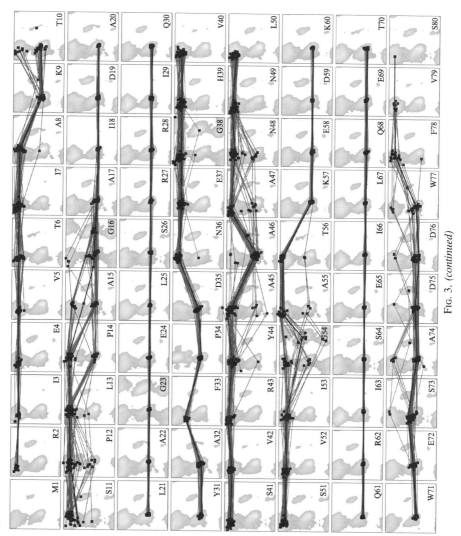

FIG. 3. *(continued)*

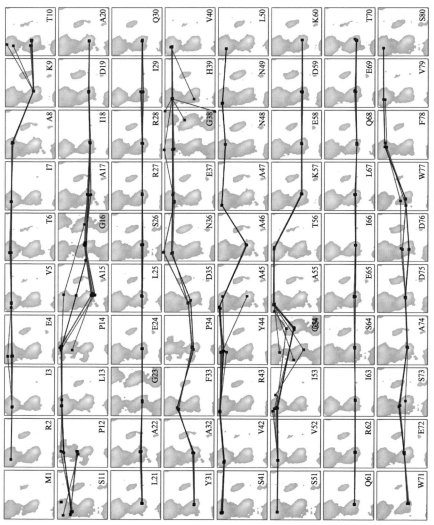

B

dipolar couplings are extremely sensitive to internuclear vector orientations, even relatively small variations in these orientations can result in a large dipolar residual and a wrong (usually too small) magnitude and orientation of the best-fit alignment tensor (Zweckstetter and Bax, 2002). This can render the long-range information content of the residual dipolar couplings difficult to utilize and also can make it difficult to distinguish good database substructures from false hits, which accidentally yield some above average agreement with the dipolar couplings.

Both these problems can be addressed by refining the selected database fragments by means of a gentle, low-temperature simulated annealing protocol against the experimental RDCs (and other restraints, if available), using either an in-house modified version of X-PLOR (Brunger, 1993; Schwieters et al., 2003) (referred to as XPLOR-NIH, available through http://nmr.cit.nih.gov) or the program DYNAMO (available through http://spin.niddk.nih.gov/bax). Parameters used during this refinement are listed in Table III. Refinement in the presence of dipolar couplings is carried out using an artificial, harmonic coordinate restraint term (similar to a non-crystallographic symmetry term), which keeps backbone C^α atoms of the refined fragment reasonably close ($< \sim 1$ Å) relative to the initial database substructure but allows small adjustments in the orientations of the individual peptide planes, while usually increasing the agreement with the dipolar couplings considerably.

Incorporation of dipolar coupling restraints into X-PLOR structure calculation and refinement has been described previously (Schwieters et al., 2003; Tjandra et al., 1997), and relies on the use of a tetraatomic orthonormal "pseudomolecule," OXYZ, to represent the principal axis system of the alignment tensor, whose orientation is allowed to float. Both XPLOR-NIH and DYNAMO software packages now allow the magnitude and rhombicity of the alignment tensor to float during refinement, but this option is not used in the current evaluation of the performance of the

---

FIG. 3. Summary of MFR search results for DinI, (A) before and (B) after refinement of the fragments with respect to dipolar restraints. Each subpanel displays the (ϕ,ψ) backbone angles found for a given residue in the MFR search, with gray regions marking the most occupied region in the database for the particular residue type. Solid lines connect (ϕ,ψ) pairs found for adjacent residues in the same fragment, and "dead ends" such as observed for P12 in (A) correspond to the fragment where this residue is the last in the selected stretch. In (A), blue (dark gray) lines correspond to MFR search results heavily weighted toward dipolar couplings (using parameters of Table II); red (light gray) lines correspond to MFR results when emphasizing chemical shifts. After refinement, using parameters of Table III and convergence criteria of Table IV, the majority of residues display unique (ϕ,ψ) backbone angles (B). The black line connects the (ϕ,ψ) pairs seen in the previously determined NMR structure (PDB entry 1GHH). (See color insert.)

TABLE III
TYPICAL PARAMETERS FOR LOW-TEMPERATURE SIMULATED ANNEALING REFINEMENT OF
DATABASE FRAGMENTS USING EITHER XPLOR-NIH OR DYNAMO

Parameter	Simulated annealing protocol	
	Fragment refinement	Structure regularization
Temperature (K)[a]	$500 \rightarrow 1$	$1000 \rightarrow 1$
Temperature step (K)	10	10
Number of steps	3000	20000
Timestep (fs)	3	3
Masses (amu)	100	100
Force parameters[b]		
k_{bond} (kcal mol^{-1} Å^{-2})	1000	1000
k_{angle} (kcal mol^{-1} rad^{-2})	$400 \rightarrow 1000$	$400 \rightarrow 1000$
$k_{improper}$ (kcal mol^{-1} rad^{-2})	$100 \rightarrow 1000$	$100 \rightarrow 1000$
k_{vdw} (kcal mol^{-1} Å^{-4})	$0.01 \rightarrow 1$	$0.01 \rightarrow 1$
Repel[c]	0.8	0.8
k_{rama} (kcal mol^{-1})	$0.002 \rightarrow 1$	$0.002 \rightarrow 1$
k_{bor} (kcal mol^{-1} rad^{-2})[d]	$1 \rightarrow 40$	$1 \rightarrow 40$
k_{cen} (kcal mol^{-1} rad^{-2})[d]	$1 \rightarrow 10$	$1 \rightarrow 10$
k_{dipo} (kcal mol^{-1} Hz^{-2})	$0.01 \rightarrow 1$	$0.01 \rightarrow 1$
$k_{collapse}$ (kcal mol^{-1} Å^{-2})	—	50
k_{harm} (kcal mol^{-1} Å^{-2})[e]	$10 \rightarrow 0.1$	$10 \rightarrow 0.1$
k_{cdih} (kcal mol^{-1} rad^{-2})[f]	—	$300 \rightarrow 1$

[a] Temperature control is achieved by coupling to a heat bath with a coupling constant of 10 fs.
[b] Weighting factors for the energy terms (if any) are included in the force constants.
[c] Scale factor for van der Waals radius.
[d] Force constants for vector angle restraint energy term (border and center exclusion) as defined in Meiler et al. (2000).
[e] Harmonic coordinate restraint, applied to C$^\alpha$ only.
[f] Force constant for dihedral restraints (if available).

MFR+ approach. We find that whenever reasonably reliable values for the alignment parameters can be extracted, either from the histogram of observed dipolar couplings (Bryce and Bax, 2004; Clore et al., 1998a), or from the first round of the database fragment search (Fig. 2), it is better to leave these values fixed during refinement.

The dipolar energy term is of the form $E_{dipo} = k_{dipo} (D^{obs} - D^{calc})^2$. To avoid large initial erratic forces, the program starts with the alignment frame in an optimal orientation, determined by best fitting the experimental dipolar couplings to the coordinates of the database peptide. Using additional a priori information about the relative orientations of the alignment tensors, applicable in cases in which measurements have been carried out in multiple media, is advantageous at this point. As shown for the magnitude and rhombicity of the alignment tensors (Fig. 2), the relative tensor orientations can also be obtained from a weighted average of the results of the first round of the MFR database search. In practice, we define the relative orientation of the alignment tensors by so-called "vector angle restraints" (Meiler *et al.*, 2000) between the axes of the two pseudomolecules (OXYZ and O'X'Y'Z'), representing the two alignment tensors.

The refinement protocol also takes advantage of a very weak, database-derived dihedral energy term or potential of mean force, which is commonly referred to as a "Ramachandran term." This term disfavors conformations that are not or very sparsely represented in the PDB and can improve the local quality of structures (Kuszewski *et al.*, 1997). Another important benefit of this potential energy term is that it positions the side chains in the orientations found to be most likely for the corresponding backbone torsion angles. The parameters for the refinement protocol are summarized in Table III.

After refinement, it is usually fairly straightforward to separate the correctly refined substructures from those that involve "false positives" in the initial database search. In this process, the refined structures are ranked according to their DC residual, CS agreement, as well as their "surface quality," calculated using Eqs. (2)–(4). Fragments beyond an adjustable cut-off threshold for F_{DC}, F_{CS}, and F_{SURF} are discarded.

Even though the best-fitting substructures were selected from the database and subsequently refined, if even these best fits are relatively poor, these substructures frequently are structurally quite diverse and the precise value of the dipolar residual becomes a less discriminating factor. This is illustrated by a plot of backbone coordinate RMSD vs. normalized dipolar score, using $[(0.8D_a)^2 + (0.6D_r)^2]^{-1/2}$ for normalization (Clore and Garrett, 1999) (Fig. 4). For fragments for which the RMSD between best-fit and observed dipolar couplings exceeds ca. $0.2D_a$, the spread relative to the true structure rapidly increases, and above $0.3D_a$ essentially no correlation between the conformation of the database substructure and the target fragment remains. Therefore, such fragments are discarded at this stage.

For the converged fragments, the "energies" typically fall within a small margin (typically 10–20%) of the lowest value observed for that fragment, and only those fragments that are within this margin are accepted. Typical tolerance values, tolDC, tolCS, and tolSurf, for the selection of converged fragments are listed in Table IV.

Additional criteria for validation include the requirement that the magnitude and rhombicity of the local "best fit" alignment tensor, as determined by SVD, falls within an adjustable fraction (typically ±20%) of the values used during refinement (Table IV). This latter selection can eliminate "false positives" that did not converge properly in the course of

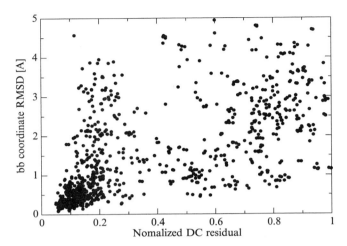

FIG. 4. Plot of backbone coordinate RMSD when best fitting the refined nine-residue fragment library to the X-ray structure (PDB entry 1UBQ) vs. normalized dipolar coupling RMSD. Prior to calculating the dipolar coupling RMSD, couplings are normalized by dividing each difference between observed and best-fitted coupling by $[(0.8 \, Da)^2 + (0.6 \, Dr)^2]^{1/2}$. The normalized dipolar residual used for making the plot corresponds to the average of the two different data sets available from the two alignment media.

the dipolar coupling refinement. Similarly, if data from more than one alignment medium are available, a requirement is that the relative orientation and normalized scalar product (Sass *et al.*, 1999) of the tensors fall within a specified margin, typically two standard deviations from what is observed for the full set of fragments.

The results of the refinement and the selection of accepted fragments again are inspected by viewing the Ramachandran flight path (Fig. 3B). As expected, the refinement of the individual substructures yields considerably smaller spreads in ϕ, ψ angles and improved agreement relative to the reference structure. Typically, when inspecting the full protein in this manner, distinctly different solutions are observed at several locations, as seen, for example, for residues T10, P14, and G54 in DinI (Fig. 3B). However, as discussed below, many of these ambiguities are resolved in an automated manner at the assembly stage.

Assembly of Structure from Fragments

All the manipulations for model building are carried out in the Tcl/Tk script language, using the DYNAMO class of routines within "NMRWish." A key routine in this process, dynAlign, is used to best-fit

TABLE IV

SELECTION CRITERIA AND THEIR TYPICAL SETTINGS FOR MFR+
SELECTION OF CONVERGED FRAGMENTS AFTER REFINEMENT[a]

Parameter	Typical value
tolDC[b]	10–33%
tolCS[b]	10–33%
tolSurf[b]	10–33%
daTol[c]	2 SD[d]
rTol[c]	2 SD[d]
scalarTol[e]	2 SD[d]
tolXX/YY/ZZ[e]	2 SD[d]

[a] Various "quality checks" are applied to remove fragments that result from accidental "false positives" in the MFR+ search stage. Knowledge of alignment tensor magnitude(s) and rhombicity(ies) (cf. Fig. 2) and relative orientations (if more than one alignment medium is employed) is used to define tolerance limits. Peptide fragments with deviations greater than the specified tolerance margins are discarded at the "quality check" stage.

[b] Maximum allowed deviation of F_{DC} [Eq. (2)], F_{CS} [Eq. (3)], and F_{Surf} [Eq. (4)] relative to the best of the refined database fragment in that particular residue range.

[c] Maximum allowed deviation relative to the average da and dr values (see Table II), when considering the entire ensemble of selected database fragments.

[d] SD, standard deviation.

[e] Maximum deviation of generalized scalar product and cosXX, cosYY, and cosZZ (see Table II) relative to the averaged value, when considering the entire ensemble of selected database fragments.

a new fragment onto the growing chain, by minimizing the backbone coordinate RMSD for the overlapping residues. In principle, there are three translational and three rotational degrees of freedom in this process. However, as the orientation of the fragment relative to the alignment frame is already known uniquely at this stage (or with 4-fold ambiguity in case of only a single alignment tensor), dynAlign has the ability to freeze the rotational degrees of freedom while optimizing the translational parameters only. In case of a single, axially symmetric or nearly axially symmetric alignment tensor, rotation about the z-axis is also allowed as an adjustable parameter.

The procedure for the assembly of the structure is visualized in Fig. 5. In a first step, the refined fragments are rotated such that their coordinate frames coincide with the principal frame(s) of their local alignment

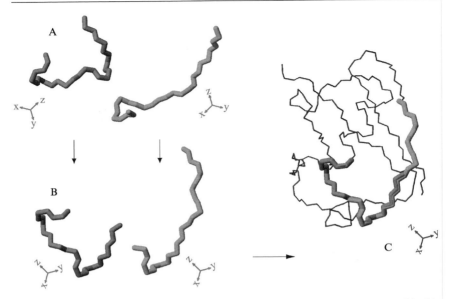

FIG. 5. Pictorial representation of the MFR+ assembly process, illustrated for ubiquitin. (A) Backbone representations of MFR+-derived fragments for residues 13–21 (red, light gray) and 17–25 (green, dark gray) fragments in arbitrary relative orientation, together with their corresponding alignment tensor frames. (B) After rotation such that their respective local alignment tensors have identical orientations. (C) Fragments are translated (with fixed orientation) such that the coordinate RMSD relative to the previously assembled chain is minimized. The figure was generated using the program MOLMOL (Koradi et al., 1996). (See color insert.)

tensor(s). When dipolar couplings from only a single medium are available, there is a 4-fold ambiguity in the orientation of the fragment. If dipolar couplings from multiple alignment media are available, the 4-fold degeneracy can easily be resolved (Al-Hashimi et al., 2000) prior to assembly, but this is by no means a prerequisite. In practice, a computationally convenient solution generates all four orientations for each alignment frame and compares at the assembly stage, discussed below, which of the possible pairwise combinations yields the correct, compatible orientation for the fragment relative to the prebuilt fraction of the chain, as judged from the lowest coordinate RMSD, when optimizing their relative translation while maintaining a consistent relative alignment tensor orientation (Al-Hashimi et al., 2000).

In the assembly step, the N- and C-terminal residue of each fragment are discarded, as they typically are less well defined by the data. Subsequently, each of the accepted, shortened fragments is translated onto the growing chain by minimizing the coordinate RMSD relative to the previously built fraction of the chain (Fig. 5). All fragments that fit within an adjustable threshold (typically requiring a coordinate RMSD, max RMSDpRes $\leq 0.2 \times N$ Å, where N is the number of overlapping residues) are retained for generating the final structure. As mentioned above, when data from only a single medium are available, four distinctly different fragment orientations agree equally well with the experimental data (Al-Hashimi *et al.*, 2000), but usually at most one of these will yield a reasonable coordinate fit. This chain extension process is repeated until the assembly of the protein structure is complete. Parameters for the assembly process are presented in Table V. It is worth noting that in the case of

TABLE V
PARAMETERS AND TYPICAL VALUES FOR MFR+ CHAIN ASSEMBLY

Parameter	Typical value
Table with converged fragments	frag.tab
Dipolar coupling table(s)	dObs[*].tab
dadrFlag[a]	0 or 1
da, dr (for each tensor)[b]	From Table II
MaxRMSD[c]	[0.0–1.0 Å]
maxRMSDpRes[d]	[0.1–0.3 Å/residue]
minOVLPCount[e]	[1 – fragLength] = 2 (default)
skipFirst[f]	0 or 1

[a] Flag to select whether SVD (0) or nonlinear Powell minimization is used to fit a database fragment to experimental dipolar couplings.

[b] Determined from the fragment search (Fig. 2). Parameters da and dr are used only if dadrFlag $\neq$ 0.

[c] Maximum acceptable backbone (N, Cα, C^1) coordinate RMSD between a new refined database fragment and the overlapping region of the previously assembled protein backbone. If the current backbone coordinate RMSD is smaller, the fragment is accepted; if it is larger, it not used in model building. When maxRMSD is set to zero, the parameter maxRMSDpRes is used instead to decide if a fragment should be accepted or rejected.

[d] Used only if maxRMSD = 0. maxRMSDpRes is the maximum acceptable coordinate RMSD per residue overlap between a new refined database fragment and the previously assembled protein backbone. Above this threshold the fragment will be discarded.

[e] Minimum number of residues required to overlap between a new fragment and a previously assembled protein chain.

[f] Flag to decide whether to discard (>0) or retain (0) N- and C-terminal residues of a database fragment in the assembly of the MFR model.

axially symmetric alignment, particularly when only data in a single alignment medium are available, the rotation about the z-axis of the alignment tensor is undefined and in that case it must be treated as an additional degree of freedom when building a fragment onto the chain. This additional degeneracy can pose significant problems in the assembly process and may require that additional data, such as sparse NOEs, are available.

In the final step, coordinates of all accepted fragments are averaged, and the resulting structure, which may have local nonphysical geometry, is regularized by simulated annealing using a protocol similar to the one used for refinement of individual fragments (Table III), but using 20,000 instead of 3000 steps, and a 2-fold (1000 K vs. 500 K) higher starting temperature.

Application to Model Proteins

Application of the MFR+ method is demonstrated for three proteins for which extensive sets of experimental backbone dipolar couplings were available, ubiquitin, GB3, and DinI. Crystallographically determined structures are available for ubiquitin and GB3, and NMR structures are available for all three. The method has also been applied to several slightly larger proteins for which dipolar couplings were simulated, including thioredoxin, profilin, and interleukin-1β. In all these applications, standard parameters (Table II) were used, with a query fragment length of nine residues, but nearly identical results were obtained using seven- or eight-residue fragments. For ubiquitin, GB3, and DinI, relatively complete sets of $^1D_{NH}$, $^1D_{NC'}$, $^1D_{C\alpha H\alpha}$, $^1D_{C\alpha C'}$, and $^2D_{C'HN}$, reported previously, were used (Ottiger and Bax, 1998; Ramirez $et\ al.$, 2000; Ulmer $et\ al.$, 2003). For the other three proteins, these couplings were generated with the program PALES, for alignment tensors predicted by PALES for media of bicelles and Pf1 (Zweckstetter and Bax, 2000; Zweckstetter $et\ al.$, 2004). The number of dipolar coupling and chemical shift restraints for each protein is listed in Table VI.

Ubiquitin

Ubiquitin has served as a test case for several programs that determine the NMR structure by nonconventional methods. Using a method similar to MFR+, but using only consistent sets of overlapping fragments and best fitting the coordinates of these, Andrec $et\ al.$ (2001) obtained a model with a backbone coordinate RMSD of 2.4 Å relative to the X-ray structure. Using the same input data, the ROSETTA method of the Baker group (Rohl and Baker, 2002), operating in torsion angle space, resulted in models that deviate from the X-ray reference structure by only 1.03–1.17 Å.

TABLE VI
RESULTS OF THE MFR+ METHOD APPLIED TO MODEL PROTEINS[a]

Protein	Number of residues	PDB entry	N_{CS}[b]	N_{DC} medium A[b]	N_{DC} medium B[b]	Backbone RMSD to PDB (Å)	All atom RMSD to PDB (Å)	Angular ϕ/ψ RMSD (°)[c]
Ubiquitin	76	1UBQ	378	333	325	0.70 (0.81)[d]	1.66 (1.79)[d]	7.5/9.1
DinI	81	1GHH	389	343	209	1.51 (0.79)[e]	2.34 (1.92)[e]	9.6/7.6
GB3	56	2IGD	273	211	231	0.68 (1.12)[d]	1.49 (2.01)[d]	11.7/11.7
Thioredoxin	105	1ERT	416	501	501	0.83	1.67	8.0/7.7
Profilin	125	1ACF	598	588	588	0.44	0.85	6.0/8.9
Interleukin-1β	153	4ILB	574	707	707	1.95 (1.41)[f]	2.74 (2.41)[f]	10.4/9.9

[a] Experimental data are used for ubiquitin, DinI, and GB3; input RDC data for thioredoxin, profilin, and interleukin-1β are simulated using PALES software, but chemical shifts are experimental. Unless otherwise noted, reported coordinate RMSD values refer to residues 2–72 (ubiquitin), 2–77 (DinI), 2–55 (GB3), 2–104 (thioredoxin), 2–124 (profilin), and 2–152 (interleukin-1β).

[b] N_{CS} is the number of chemical shifts; N_{DC} (A) and N_{DC} (B) are the numbers of dipolar couplings available in media A (bicelles) and B (charged bicelles for ubiquitin; Pf1 for all others).

[c] Excludes regions where large crankshaft errors have occurred in model building and refinement (N52/G53 in ubiquitin, G58/G59 in profilin, S21/G22 in interleukin-1β).

[d] Using dipolar data from a single medium (bicelles) only.

[e] For residues 2–53.

[f] For residues 4–134.

ROSETTA proved particularly tolerant to incomplete RDC data. For example, using only D_{NH} input values, the fold could still be predicted reliably (backbone RMSD 2.75 Å) The original version of our MFR program yielded a backbone structure that differed by 0.88 Å from the X-ray structure (Delaglio *et al.*, 2000). The sequential chain building method of Hus *et al.* (2000) yielded comparable results (RMSD 1.0 Å), using the same sets of residual dipolar couplings. The present version of MFR+ yields a backbone structure that differs by 0.70 Å from the crystal structure (PDB entry 1UBQ) (Vijay-Kumar *et al.*, 1987) shown in Fig. 6A, 0.72 Å from the NMR structure (PDB entry 1D3Z) (Cornilescu *et al.*, 1998). This latter number increases to 0.81 Å, if data from only one alignment medium are used. The RMSD for all nonhydrogen atoms relative to the X-ray (1.66 Å) or lowest energy NMR structure (1.68 Å) is considerably larger, resulting from the lack of experimental data restraining the side chain orientations.

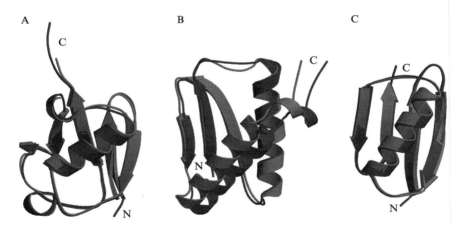

FIG. 6. Comparison of MFR+-derived structures and previously solved X-ray (A and C) and NMR (B) structures of (A) ubiquitin, (B) DinI, and (C) GB3. Blue (light gray) structures correspond to the PDB reference coordinates; red (dark gray) represents the MFR+-derived ribbon. For ubiquitin (A), the disordered C-terminus (residues 74–76) could not be built by the MFR+ method and is not shown. For DinI (B), the structure could not be assembled uniquely from dipolar couplings acquired in a single medium. Even with data from two media, some ambiguity remains (Fig. 3B) and is responsible for the erroneous lateral displacement of helix α2 by ca. 3.7 Å relative to the reference structure. For clarity, superposition of the MFR+ and reference structure is optimized for residues 2–53, highlighting the displacement of helix 2. Reference structures correspond to PDB entries 1UBQ, 1GHH, and 1IGD. Figures were generated using the programs Molscript (Kraulis, 1991) and Raster 3D (Merritt and Murphy, 1994). (See color insert.)

For the four C-terminal residues, the MFR+ program was unable to find a unique solution. This is not surprising, considering that these residues are dynamically highly disordered (Tjandra *et al.*, 1995; Wand *et al.*, 1996). The inability to define a unique structure in the presence of such extensive motion suggests that the MFR+ automatically recognizes such regions. However, for other regions in the protein where increased dynamics is known to take place but is less extreme (e.g., around residues G10, K11, I23, E24, K48, and G53), and where several amide resonances and their corresponding dipolar couplings are unobservable as a result of conformational exchange, MFR+ can faithfully define the backbone structure. For these cases, the increased internal dynamics affects at most only a few sequential residues, which does not significantly perturb the search for optimally fitting nine-residue substructures. Clearly, the final NMR model calculated with MFR+ no longer carries a signature of this increased internal dynamics. However, the increased dynamics will generally be evident from the raw spectra, which exhibit either exchange-broadened weak resonances or motionally narrowed, intense resonances.

DinI

DinI is a small globular protein of 81 residues, implicated in DNA repair, for which an NMR structure has been obtained (PDB entry 1GHH) from both NOE and extensive dipolar coupling restraints (Ramirez *et al.*, 2000). Compared to ubiquitin, this proved to be a more challenging test case. NMR data were collected at lower concentrations, resulting in lower signal-to-noise ratios and a less complete set of dipolar couplings. Two sets of RDCs were available, one nearly complete set, acquired in bicelles, and a somewhat less complete set, due to stronger homonuclear $^1H-^1H$ dipolar broadening resulting from overalignment, obtained in a Pf1 solution.

For an extended loop region, K9 to G16, several of the fragments did not find satisfactory substructures in the database that met the standard cut-off criteria. In fact, comparing the NMR structure for this region with all fragments in our database indicates that although the search indeed selects the best fitting fragment, none of the database fragments found by the MFR+ search agrees to better than 0.8 Å. The dipolar search is affected significantly by structural differences of this magnitude. A subsequent evaluation, searching simply for database fragments with the closest backbone RMSD relative to this fragment of the previously determined DinI structure, confirms that no fragments that are closer than 0.8 Å in backbone structure are present in the database.

A number of false positives in the MFR+ search, which differed sub-stantially in structure but accidentally yielded better than random RMSDs between observed and best-fitted dipolar couplings, was also obtained in the search for the K9–G16 fragment. This problem was compounded by the presence of two Pro residues at positions 12 and 14, which resulted in far fewer dipolar couplings for the fragments that include these residues. As illustrated in Fig. 3, it was not possible to unambiguously extract reliable fragments with the standard protocol. However, because at the chain building stage none of the erroneously identified fragments yielded a suitable match to the previously built chain, even the large degeneracy encountered in this loop region did not prevent successful chain extension.

For DinI, the chemical shifts and dipolar coupling fragment searches are clearly indicative of two long α-helices, stretching from G16 to A32 and from K57 to W77 (with a kink at S73). Such stretches of helix, which typically result in very good hits when searching the database, are very helpful in accurately defining the relative orientations as well as the mag-nitudes and rhombicities of the two alignment tensors. Using this addition-al information, the search hits were then screened for the correct magnitude and relative orientation of the alignment tensors. Several frag-ments failed to converge to a unique structure in the course of their refinement, despite having almost indistinguishable scores and energies, e.g., fragments around T10–A15, A45–N48, and G54. Nevertheless, there was sufficient overlap between converged fragments that the chain could be built. As mentioned above, the requirement that a fragment must give a reasonable backbone coordinate match to the previous one is key in resolving such remaining ambiguities.

The final backbone differs rather substantially, by 1.51 Å, from the previously determined NMR structure (PDB entry 1GHH). As illustrated in Fig. 6B, a comparison of the two structures reveals that this high RMSD results from a lateral translational error in the position of the second long helix: If residues 2–53 are superimposed, helix $\alpha2$ is shifted by 3.7 Å from that seen in 1 GHH, apparently caused by an incorrect formation of the reverse turn centered at G54. For residues R2–I53, the backbone RMSD is only 0.79 Å, and the C-terminal kinked helix, K57–W77, agrees with the previously determined solution structure to within 0.44 Å.

GB3

A very high-resolution X-ray structure, solved at 1.1 Å resolution (PDB entry 1IGD) (Derrick and Wigley, 1994), is available for this protein, and the solution structure (PDB entries 1P7E/1P7F) (Ulmer *et al.*, 2003) agrees with this structure to within 0.3 Å.

Using dipolar couplings from only one alignment medium (bicelles), the standard protocol resulted in a model that differed by a backbone RMSD of 1.12 Å from the X-ray structure (1IGD). When using dipolar couplings from both bicelle and phage media, this RMSD decreased to 0.68 Å. The recently reported solution structure and the two structures derived with MFR+ are superimposed on one another in Fig. 6C. The backbone RMSD of the MFR+ model relative to the NMR structure (1P7E) is smaller than found relative to the X-ray structure, both when using data from only one alignment medium (bicelles; 0.89 Å) and when using data measured in bicelles and phage media (0.49 Å). This smaller difference reflects a previously noted small change in the twist of the β-sheet, which is constrained by intermolecular hydrogen bonds in the crystalline lattice (Derrick and Wigley, 1994).

Tests Using Simulated Data

To further test the MFR method on larger systems, residual dipolar couplings were simulated for three proteins, using alignment tensors predicted on the basis of their three-dimensional structure. The targets chosen were proteins from the TALOS database (Cornilescu *et al.*, 1999), for which nearly complete backbone chemical shift assignments and high-resolution X-ray crystal structures were available. Data for thioredoxin, profilin, and interleukin-1β were simulated using a version of the program PALES (Zweckstetter and Bax, 2000), which has been modified to include the effects of electrostatic alignment as appropriate for phage (Zweckstetter *et al.*, 2004). Two different sets of residual dipolar couplings, corresponding to neutral bicelles and Pf1 phage media, were simulated for each protein. Couplings that are generally not measured with the standard methods (residues preceding Pro and residues with missing shifts) were removed from the coupling tables. Noise was added to the simulated, normalized couplings (1.0 Hz for thioredoxin and profilin; 1.5 Hz for interleukin-1β). Increasing the rms error in the simulated data up to 30% of the applicable D_a value has little effect on the regions of the target protein that yield unique fragments in the database search at low noise levels. However, it tends to increase the width of the selected fragment distribution for regions that exhibit ambiguity when smaller errors in the simulated data are used.

For thioredoxin [105 residues, X-ray crystal structure PDB entry 1ERT (Weichsel et al., 1996)] the chemical shifts were taken from Qin *et al.* (1996). For residues 2–104, the resulting MFR+ model exhibits a backbone RMSD of 0.83 Å relative to the X-ray structure. This difference (Fig. 7A) results mainly from small translational displacements of secondary structure

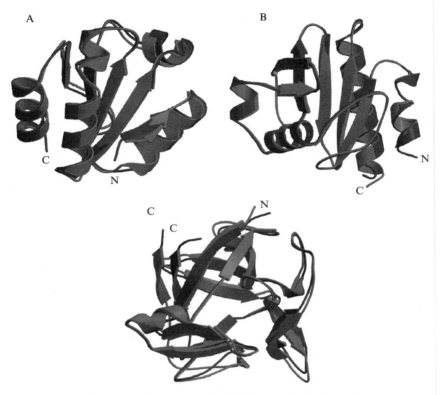

FIG. 7. Comparison of MFR+ structures [red (dark gray)] and X-ray reference structures [blue (light gray)] for (A) reduced human thioredoxin (PDB entry 1ERT), (B) profilin (PDB entry 1ACF), and (C) interleukin-1β (PDB entry 4ILB). MFR+ structures were derived from experimental chemical shifts and using two sets of dipolar couplings, simulated for the X-ray structures, for media containing 50 mg/ml bicelles and 15 mg/ml Pf1. Figures were generated using the programs Molscript (Kraulis, 1991) and Raster 3D (Merritt and Murphy, 1994). (See color insert.)

elements, which presumably could be corrected by the use of a very minimal set of NOEs. Thioredoxin had one problematic loop region around M74–P75. At M74, in the region connecting the C-cap of α3 to β4, the backbone adopts an unusual conformation ($\phi = -80°$, $\psi = -107°$), slightly outside the allowed region of the Ramachandran plot. The presence of the Pro residue reduces the number of dipolar couplings and resulted in a "crankshaft" difference in this critical region, while retaining the correct orientation for adjacent residues. As a consequence, the whole region comprising strands β4, β5, and α4 shows a small lateral displacement relative to the X-ray structure. Other interesting features, such as a kink in helix $\alpha 2$ at position I38 and a twist of strand $\beta 5$ at residue V86, are correctly recognized by MFR+.

Application to profilin [125 residues, X-ray crystal structure PDB entry 1ACF (Fedorov *et al.*, 1994); NMR structure PDB entry 2PRF (Archer *et al.*, 1994; Vinson *et al.*, 1993)] presented an interesting case due to its relatively high Gly contents (17 Gly residues). Gly residues can adopt unusual geometries that are less well represented in the PDB. However, with a backbone RMSD of only 0.44 Å, the final MFR+ model agrees remarkably well with the 1ACF structure that was used to generate the dipolar couplings (Fig. 7B).

For interleukin-1β [150 residues, X-ray crystal structure PDB entry 4ILB (Veerapandian *et al.*, 1992)] the chemical shifts were taken from Clore *et al.* (1990). Due to its complicated topology, this protein proved to be the most challenging test case for MFR+, largely because it consists almost exclusively of β-sheet. Although for conventional NOE-based structure determination this is generally beneficial, the large number of reverse turns and the possibility for the accumulation of error when the chain gets longer pose significant challenges to the MFR+ procedure. Moreover, due to the absence of a clear α-helical segment in interleukin-1β, it is less straightforward to extract accurate alignment tensor parameters. The inherently higher structural variability of β-strands makes these less suitable for such a purpose. Therefore, the relative orientation of the two alignment tensors could not be established a priori and could not be used as a restraint during the initial MFR search.

The presence of two adjacent Gly residues in the last loop (G139 and G140), which adopt an unusual conformation in a sparsely occupied region of the Ramachandran map, resulted in the lack of properly matching fragments in the database, and no prediction could be made for this region. Because the relative orientation of the last β-strand relative to the rest of the protein could be inferred from the two sets of RDCs, coordinates for G139 and G140 were added to the model by adding an extended Gly–Gly dipeptide to bridge the gap prior to subjecting the full protein to another cycle of regularization and refinement. However, because no NOEs were used in this process, the local geometry resulting from this procedure is only very approximate, resulting in a displacement by over 4 Å of the last β-strand relative to the crystal structure (Fig. 7C). This misplacement increased the overall backbone RMSD to 1.95 Å, but a considerably smaller difference (RMSD 1.41 Å) is obtained when considering only residues 5–134, excluding the last β-strand and its preceding loop. Other turns for which MFR+ did not yield accurate conformations include those between $\beta 2$ and $\beta 3$ and between $\beta 6$ and $\beta 7$.

Concluding Remarks

The MFR+ approach provides a remarkably direct way to determine solution NMR structures from protein backbone RDC data, either without or with inclusion of a small set of local backbone NOE data. The approach utilizes only protein backbone data and thereby bypasses the side chain and NOE assignment step. However, resulting structures have limitations that are distinct from those encountered in conventional, NOE-based structural studies. The most significant limitation of the MFR+ method in its application to full backbone structures is its requirement for relatively complete sets of residual dipolar couplings, preferably in two media. Short gaps of one residue at a time, which may result from exchange broadening, rapid solvent exchange, or the presence of a Pro residue, are typically easily bridged. Gaps longer than two residues at a time, especially in loop regions, often make it impossible to define the local structure uniquely. In such cases, only pieces or subdomains of a protein can be built reliably, and additional information such as NOE contacts or hydrophobic packing-based modeling is required to assemble these pieces correctly. Regions of the protein for which no well-matching substructures can be found in the database tend to be more problematic when applying the MFR+ procedure. This problem is compounded by the fact that the absence of tightly fitting fragments in the database occurs almost exclusively outside regions of well-defined secondary structure. If the best database hits are still relatively poor and structurally diverse, this can result in additional complications in the assembly process.

Owing to the symmetry of the dipolar interaction, if a backbone bond is parallel to any of the three principal axes of the alignment tensor, all couplings will be invariant to a 180° rotation about this bond, and data from a second alignment medium are required to resolve such an ambiguity. If the dataset for the other alignment medium happens to be incomplete for this region, it may be impossible to distinguish the two cases. In practice, it is usually the torsion angle ψ that causes such problems, as for many values of ψ, $\psi + 180°$ also falls in the allowed region of the Ramachandran map. With the exception of Gly residues, most 180° changes in ϕ result in severe steric clashes and then are easily filtered out.

As mentioned earlier, problems also arise when attempting to build extended loop regions that are not very well represented in the PDB. Particularly if the density of RDCs is low in such a region, it may become impossible to uniquely define matching substructures in the database. Such problematic areas usually can be spotted at an early stage of the fragment search by divergence in the "Ramachandran flight map patterns," or at a later stage by inspection of the corresponding flight maps for the selected, refined fragments (Fig. 3).

Although our data demonstrate that reliable backbone models for small and medium sized proteins can be built on the basis of quite complete dipolar coupling and chemical shift data, application of the MFR+ program is not limited to these cases. Zweckstetter previously has shown that in favorable cases, fragments of a structure can be derived from chemical shifts together with as few as two dipolar couplings per residue (Zweckstetter and Bax, 2001). It is primarily at the assembly phase that additional data are needed. In its simplest form, tight backbone torsion angle restraints of the uniquely defined fragments could be used in conjunction with a limited number of backbone–backbone NOEs as restraints in a regular simulated annealing protocol to determine full structures in these cases. Alternatively, more sophisticated "docking" procedures based on rigid body refinement (Clore and Schwieters, 2002; Schwieters and Clore, 2001; Schwieters *et al.*, 2003) or Monte Carlo–based "shuffling" approaches (Rohl and Baker, 2002) may be used for assembling these fragments into a final structure. Note that the known orientations of each fragment (except for a 4-fold degeneracy) provide important additional restraints during such an assembly procedure.

Axially symmetric alignment tensors yield data that are less discriminating when building a structure from dipolar couplings than highly asymmetric (rhombic) alignment tensors (Delaglio *et al.*, 2000). In the axially symmetric case, only the unique axis (z-axis) of the local alignment frame is defined, and a rotational degree of freedom around this axis remains in the placement of fragments.

Our results demonstrate that it is possible for relatively small proteins to completely determine the backbone structures from backbone dipolar couplings. However, the MFR+ allows for convenient use of local NOEs too, when searching the database. The sequential $d_{H\alpha HN}(i, i + 1)$ connectivity can be particularly useful for excluding fragments with the wrong ψ angle. Long-range NOEs are incorporated most easily after the initial model has been built, using a simulated annealing refinement protocol, where the backbone torsion angles are restrained relatively tightly to the values obtained from the initial model (excepting residues that proved uncertain after the fragment refinement procedure), and the dipolar coupling and NOE restraints are incorporated in the usual manner.

In its present implementation, the MFR+ method focuses only on building protein backbone structures, and side chains are essentially positioned according to their most likely conformation for the corresponding ϕ, ψ angles, using a database-derived empirical energy term (Dunbrack and Karplus, 1994; Kuszewski *et al.*, 1997). Our results indicate that this rather crude approach to side chain modeling yields reasonable results, with total increases in RMSD between the all-heavy-atom model and the corresponding reference structure typically being less than 1 Å. This is not much worse than found for many medium resolution NMR structures in the PDB, if corresponding X-ray structures are taken as the reference. Not surprisingly, however, a substantial subset of side chains is poorly positioned with such an approach, but these tend to be easily identified by their very poor van der Waals contacts, using programs such as AQUA or PROCHECK (Laskowski *et al.*, 1996).

Although the MFR+ method here has been presented as a method for building protein structures without recourse to NOEs, it is likely that it will become most valuable in a hybrid approach where it is used to define structures of smaller fragments that subsequently require few NOEs for assembling them into a structure. Such an approach will be much less demanding in terms of completeness of the dipolar coupling data and will take advantage of the subset of NOEs that is frequently identified very easily, including H^N-H^N interactions.

References

Al-Hashimi, H. M., Valafar, H., Terrell, M., Zartler, E. R., Eidsness, M. K., and Prestegard, J. H. (2000). Variation of molecular alignment as a means of resolving orientational ambiguities in protein structures from dipolar couplings. *J. Magn. Reson.* **143**, 402–406.

Andrec, M., Du, P. C., and Levy, R. M. (2001). Protein backbone structure determination using only residual dipolar couplings from one ordering medium. *J. Biomol. NMR* **21**, 335–347.

Annila, A., Aitio, H., Thulin, E., and Drakenberg, T. (1999). Recognition of protein folds via dipolar couplings. *J. Biomol. NMR* **14**, 223–230.

Archer, S. J., Vinson, V. K., Pollard, T. D., and Torchia, D. A. (1994). Elucidation of the poly-L-proline binding-site in Acanthamoeba profilin-I by NMR-spectroscopy. *FEBS Lett.* **337**, 145–151.

Barrientos, L. G., Dolan, C., and Gronenborn, A. M. (2000). Characterization of surfactant liquid crystal phases suitable for molecular alignment and measurement of dipolar couplings. *J. Biomol. NMR* **16**, 329–337.

Bax, A., Kontaxis, G., and Tjandra, N. (2001). Dipolar couplings in macromolecular structure determination. *Methods Enzymol.* **339**, 127–174.

Beraud, S., Bersch, B., Brutscher, B., Gans, P., Barras, F., and Blackledge, M. (2002). Direct structure determination using residual dipolar couplings: Reaction-site conformation of methionine sulfoxide reductase in solution. *J. Am. Chem. Soc.* **124**, 13709–13715.

Bothner-by, A. A., Gayathri, C., Vanzijl, P. C. M., Maclean, C., Lai, J. J., and Smith, K. M. (1985). High-field orientation effects in the high-resolution proton NMR-spectra of diverse porphyrins. *Magn. Reson. Chem.* **23**, 935–938.

Braddock, D. T., Cai, M. L., Baber, J. L., Huang, Y., and Clore, G. M. (2001). Rapid identification of medium- to large-scale interdomain motion in modular proteins using dipolar couplings. *J. Am. Chem. Soc.* **123**, 8634–8635.

Brenneman, M. T., and Cross, T. A. (1990). A method for the analytic determination of polypeptide structure using solid-state nuclear magnetic-resonance—the metric method. *J. Chem. Phys.* **92**, 1483–1494.

Brunger, A. T. (1993). "XPLOR: A System for X-ray Crystallography and NMR, 3.1 Ed." Yale University Press, New Haven, CT.

Bryce, D. L., and Bax, A. (2004). Application of correlated residual dipolar couplings to the determination of the molecular alignment tensor magnitude of oriented proteins and nucleic acids. *J. Biomol. NMR* **28,** 273–287.

Chou, J. J., Gaemers, S., Howder, B., Louis, J. M., and Bax, A. (2001a). A simple apparatus for generating stretched polyacrylamide gels, yielding uniform alignment of proteins and detergent micelles. *J. Biomol. NMR* **21,** 377–382.

Chou, J. J., Li, S. P., Klee, C. B., and Bax, A. (2001b). Solution structure of Ca2+-calmodulin reveals flexible hand-like properties of its domains. *Nat. Struct. Biol.* **8,** 990–997.

Clore, G. M. (2000). Accurate and rapid docking of protein-protein complexes on the basis of intermolecular nuclear Overhauser enhancement data and dipolar couplings by rigid body minimization. *Proc. Natl. Acad. Sci. USA* **97,** 9021–9025.

Clore, G. M., and Garrett, D. S. (1999). R-factor, free R, and complete cross-validation for dipolar coupling refinement of NMR structures. *J. Am. Chem. Soc.* **121,** 9008–9012.

Clore, G. M., and Schwieters, C. D. (2002). Theoretical and computational advances in biomolecular NMR spectroscopy. *Curr. Opin. Struct. Biol.* **12,** 146–153.

Clore, G. M., and Schwieters, C. D. (2003). Docking of protein-protein complexes on the basis of highly ambiguous intermolecular distance restraints derived from H-1(N)/N-15 chemical shift mapping and backbone N-15-H-1 residual dipolar couplings using conjoined rigid body/torsion angle dynamics. *J. Am. Chem. Soc.* **125,** 2902–2912.

Clore, G. M., Bax, A., Driscoll, P. C., Wingfield, P. T., and Gronenborn, A. M. (1990). Assignment of the side-chain H-1 and C-13 resonances of interleukin-1-beta using double-resonance and triple-resonance heteronuclear 3-dimensional NMR-spectroscopy. *Biochemistry* **29,** 8172–8184.

Clore, G. M., Gronenborn, A. M., and Bax, A. (1998a). A robust method for determining the magnitude of the fully asymmetric alignment tensor of oriented macromolecules in the absence of structural information. *J. Magn. Reson.* **133,** 216–221.

Clore, G. M., Starich, M. R., and Gronenborn, A. M. (1998b). Measurement of residual dipolar couplings of macromolecules aligned in the nematic phase of a colloidal suspension of rod-shaped viruses. *J. Am. Chem. Soc.* **120,** 10571–10572.

Cornilescu, G., Marquardt, J. L., Ottiger, M., and Bax, A. (1998). Validation of protein structure from anisotropic carbonyl chemical shifts in a dilute liquid crystalline phase. *J. Am. Chem. Soc.* **120,** 6836–6837.

Cornilescu, G., Delaglio, F., and Bax, A. (1999). Protein backbone angle restraints from searching a database for chemical shift and sequence homology. *J. Biomol. NMR* **13,** 289–302.

Delaglio, F., Kontaxis, G., and Bax, A. (2000). Protein structure determination using molecular fragment replacement and NMR dipolar couplings. *J. Am. Chem. Soc.* **122,** 2142–2143.

Derrick, J. P., and Wigley, D. B. (1994). The 3rd igg-binding domain from streptococcal protein-G—an analysis by X-ray crystallography of the structure alone and in a complex with Fab. *J. Mol. Biol.* **243,** 906–918.

Du, P. C., Andrec, M., and Levy, R. M. (2003). Have we seen all structures corresponding to short protein fragments in the Protein Data Bank? An update. *Protein Eng.* **16,** 407–414.

Dunbrack, R. L., and Karplus, M. (1994). Conformational-analysis of the backbone-dependent rotamer preferences of protein side-chains. *Nat. Struct. Biol.* **1,** 334–340.

Fedorov, A. A., Magnus, K. A., Graupe, M. H., Lattman, E. E., Pollard, T. D., and Almo, S. C. (1994). X-ray structures of isoforms of the actin-binding protein profilin that differ in their affinity for phosphatidylinositol phosphates. *Proc. Natl. Acad. Sci. USA* **91,** 8636–8640.

Fleming, K., Gray, D., Prasannan, S., and Matthews, S. (2000). Cellulose crystallites: A new and robust liquid crystalline medium for the measurement of residual dipolar couplings. *J. Am. Chem. Soc.* **122,** 5224–5225.

Fowler, C. A., Tian, F., and Prestegard, J. H. (2000). An NMR method for the rapid determination of protein folds using dipolar couplings. *Biophys. J.* **78**, 2827.

Giesen, A. W., Homans, S. W., and Brown, J. M. (2003). Determination of protein global folds using backbone residual dipolar coupling and long-range NOE restraints. *J. Biomol. NMR* **25**, 63–71.

Hansen, M. R., Mueller, L., and Pardi, A. (1998). Tunable alignment of macromolecules by filamentous phage yields dipolar coupling interactions. *Nat. Struct. Biol.* **5**, 1065–1074.

Hus, J. C., Marion, D., and Blackledge, M. (2000). *De novo* determination of protein structure by NMR using orientational and long-range order restraints. *J. Mol. Biol.* **298**, 927–936.

Ishii, Y., Markus, M. A., and Tycko, R. (2001). Controlling residual dipolar couplings in high-resolution NMR of proteins by strain induced alignment in a gel. *J. Biomol. NMR* **21**, 141–151.

Jones, T. A., and Thirup, S. (1986). Using known substructures in protein model-building and crystallography. *EMBO J.* **5**, 819–822.

Jones, T. A., Zou, J., Cowan, S. W., and Kjeldgaard, M. (1991). Improved methods for building protein models in electron density maps and location of errors in these models. *Acta Crystallogr. A.* **47**, 110–119.

Koenig, B. W., Hu, J. S., Ottiger, M., Bose, S., Hendler, R. W., and Bax, A. (1999). NMR measurement of dipolar couplings in proteins aligned by transient binding to purple membrane fragments. *J. Am. Chem. Soc.* **121**, 1385–1386.

Koradi, R., Billeter, M., and Wuthrich, K. (1996). MOLMOL: A program for display and analysis of macromolecular structures. *J. Mol. Graph* **14**, 51–55.

Kraulis, P. J. (1991). MOLSCRIPT: A program to produce both detailed and schematic plots of protein structures. *J. Appl. Crystallogr.* **24**, 946–950.

Kraulis, P. J., and Jones, T. A. (1987). Determination of 3-dimensional protein structures from nuclear magnetic-resonance data using fragments of known structures. *Proteins* **2**, 188–201.

Kung, H. C., Wang, K. Y., Goljer, I., and Bolton, P. H. (1995). Magnetic alignment of duplex and quadruplex DNAs. *J. Magn. Reson. Ser. B* **109**, 323–325.

Kuszewski, J., Gronenborn, A. M., and Clore, G. M. (1997). Improvements and extensions in the conformational database potential for the refinement of NMR and X-ray structures of proteins and nucleic acids. *J. Magn. Reson.* **125**, 171–177.

Laskowski, R. A., Rullmann, J. A. C., MacArthur, M. W., Kaptein, R., and Thornton, J. M. (1996). AQUA and PROCHECK-NMR: Programs for checking the quality of protein structures solved by NMR. *J. Biomol. NMR* **8**, 477–486.

Losonczi, J. A., Andrec, M., Fischer, M. W. F., and Prestegard, J. H. (1999). Order matrix analysis of residual dipolar couplings using singular value decomposition. *J. Magn. Reson.* **138**, 334–342.

Lukavsky, P. J., Kim, I., Otto, G. A., and Puglisi, J. D. (2003). Structure of HCVIRES domain II determined by NMR. *Nat. Struct. Biol.* **10**, 1033–1038.

Marassi, F. M., and Opella, S. J. (1998). NMR structural studies of membrane proteins. *Curr. Opin. Struct. Biol.* **8**, 640–648.

Meier, S., Haussinger, D., and Grzesiek, S. (2002). Charged acrylamide copolymer gels as media for weak alignment. *J. Biomol. NMR* **24**, 351–356.

Meiler, J., Blomberg, N., Nilges, M., and Griesinger, C. (2000). A new approach for applying residual dipolar couplings as restraints in structure elucidation. *J. Biomol. NMR* **16**, 245–252.

Merritt, E. A., and Murphy, M. E. P. (1994). Raster3d Version-2.0—a program for photorealistic molecular graphics. *Acta Crystallogr. Sect. D-Biol. Crystallogr.* **50**, 869–873.

Mueller, G. A., Choy, W. Y., Skrynnikov, N. R., and Kay, L. E. (2000). A method for incorporating dipolar couplings into structure calculations in cases of (near) axial symmetry of alignment. *J. Biomol. NMR* **18**, 183–188.

Nishimura, K., Kim, S. G., Zhang, L., and Cross, T. A. (2002). The closed state of a H+ channel helical bundle combining precise orientational and distance restraints from solid state NMR-1. *Biochemistry* **41**, 13170–13177.

Ottiger, M., and Bax, A. (1998). Determination of relative N-H-N N-C', C-alpha-C', and C(alpha)-H-alpha effective bond lengths in a protein by NMR in a dilute liquid crystalline phase. *J. Am. Chem. Soc.* **120**, 12334–12341.

Prosser, R. S., Losonczi, J. A., and Shiyanovskaya, I. V. (1998). Use of a novel aqueous liquid crystalline medium for high-resolution NMR of macromolecules in solution. *J. Am. Chem. Soc.* **120**, 11010–11011.

Qin, J., Clore, G. M., and Gronenborn, A. M. (1996). Ionization equilibria for side-chain carboxyl groups in oxidized and reduced human thioredoxin and in the complex with its target peptide from the transcription factor NF kappa B. *Biochemistry* **35**, 7–13.

Ramirez, B. E., and Bax, A. (1998). Modulation of the alignment tensor of macromolecules dissolved in a dilute liquid crystalline medium. *J. Am. Chem. Soc.* **120**, 9106–9107.

Ramirez, B. E., Voloshin, O. N., Camerini-Otero, R. D., and Bax, A. (2000). Solution structure of DinI provides insight into its mode of RecA inactivation. *Protein Sci.* **9**, 2161–2169.

Rohl, C. A., and Baker, D. (2002). *De novo* determination of protein backbone structure from residual dipolar couplings using rosetta. *J. Am. Chem. Soc.* **124**, 2723–2729.

Ruckert, M., and Otting, G. (2000). Alignment of biological macromolecules in novel nonionic liquid crystalline media for NMR experiments. *J. Am. Chem. Soc.* **122**, 7793–7797.

Sass, H. J., Musco, G., Stahl, S. J., Wingfield, P. T., and Grzesiek, S. (2000). Solution NMR of proteins within polyacrylamide gels: Diffusional properties and residual alignment by mechanical stress or embedding of oriented purple membranes. *J. Biomol. NMR* **18**, 303–309.

Sass, J., Cordier, F., Hoffmann, A., Rogowski, M., Cousin, A., Omichinski, J. G., Lowen, H., and Grzesiek, S. (1999). Purple membrane induced alignment of biological macromolecules in the magnetic field. *J. Am. Chem. Soc.* **121**, 2047–2055.

Schwieters, C. D., and Clore, G. M. (2001). Internal coordinates for molecular dynamics and minimization in structure determination and refinement. *J. Magn. Reson.* **152**, 288–302.

Schwieters, C. D., Kuszewski, J. J., Tjandra, N., and Clore, G. M. (2003). The Xplor-NIH NMR molecular structure determination package. *J. Magn. Reson.* **160**, 65–73.

Tjandra, N., and Bax, A. (1997). Direct measurement of distances and angles in biomolecules by NMR in a dilute liquid crystalline medium [see comments]. *Science* **278**, 1111–1114.

Tjandra, N., Feller, S. E., Pastor, R. W., and Bax, A. (1995). Rotational diffusion anisotropy of human ubiquitin from N-15 NMR relaxation. *J. Am. Chem. Soc.* **117**, 12562–12566.

Tjandra, N., Grzesiek, S., and Bax, A. (1996). Magnetic field dependence of nitrogen-proton J splittings in N-15-enriched human ubiquitin resulting from relaxation interference and residual dipolar coupling. *J. Am. Chem. Soc.* **118**, 6264–6272.

Tjandra, N., Omichinski, J. G., Gronenborn, A. M., Clore, G. M., and Bax, A. (1997). Use of dipolar H1-N15 and H1-C13 couplings in the structure determination of magnetically oriented macromolecules in solution. *Nat. Struct. Biol.* **4**, 732–738.

Tolman, J. R., Flanagan, J. M., Kennedy, M. A., and Prestegard, J. H. (1995). Nuclear magnetic dipole interactions in field-oriented proteins—information for structure determination in solution. *Proc. Natl. Acad. Sci. USA* **92**, 9279–9283.

Tycko, R., Blanco, F. J., and Ishii, Y. (2000). Alignment of biopolymers in strained gels: A new way to create detectable dipole-dipole couplings in high-resolution biomolecular NMR. *J. Am. Chem. Soc.* **122**, 9340–9341.

Ulmer, T. S., Ramirez, B. E., Delaglio, F., and Bax, A. (2003). Evaluation of backbone proton positions and dynamics in a small protein by liquid crystal NMR spectroscopy. *J. Am. Chem. Soc.* **125,** 9179–9191.

Veerapandian, B., Gilliland, G. L., Raag, R., Svensson, A. L., Masui, Y., Hirai, Y., and Poulos, T. L. (1992). Functional implications of interleukin-1-beta based on the 3-dimensional structure. *Proteins* **12,** 10–23.

Vijay-Kumar, S., Bugg, C. E., and Cook, W. J. (1987). Structure of ubiquitin refined at 1.8 A resolution. *J. Mol. Biol.* **194,** 531–544.

Vinson, V. K., Archer, S. J., Lattman, E. E., Pollard, T. D., and Torchia, D. A. (1993). 3-Dimensional solution structure of Acanthamoeba profilin-I. *J. Cell Biol.* **122,** 1277–1283.

Wand, A. J., Urbauer, J. L., McEvoy, R. P., and Bieber, R. J. (1996). Internal dynamics of human ubiquitin revealed by C-13-relaxation studies of randomly fractionally labeled protein. *Biochemistry* **35,** 6116–6125.

Weichsel, A., Gasdaska, J. R., Powis, G., and Montfort, W. R. (1996). Crystal structures of reduced, oxidized, and mutated human thioredoxins: Evidence for a regulatory homodimer. *Structure* **4,** 735–751.

Wu, C. H., Ramamoorthy, A., Gierasch, L. M., and Opella, S. J. (1995). Simultaneous characterization of the amide H-1 chemical shift, H-1-N-15 dipolar, and N-15 chemical-shift interaction tensors in a peptide-bond by 3-dimensional solid-state NMR-spectroscopy. *J. Am. Chem. Soc.* **117,** 6148–6149.

Zweckstetter, M., and Bax, A. (2000). Prediction of sterically induced alignment in a dilute liquid crystalline phase: Aid to protein structure determination by NMR. *J. Am. Chem. Soc.* **122,** 3791–3792.

Zweckstetter, M., and Bax, A. (2001). Single-step determination of protein substructures using dipolar couplings: Aid to structural genomics. *J. Am. Chem. Soc.* **123,** 9490–9491.

Zweckstetter, M., and Bax, A. (2002). Evaluation of uncertainty in alignment tensors obtained from dipolar couplings. *J. Biomol. NMR* **23,** 127–137.

Zweckstetter, M., Hummer, G., and Bax, A. (2004). Prediction of charge-induced molecular alignment of biomolecules dissolved in dilute liquid-crystalline phases. *Biophys. J.* **86,** 3444–3460.

[4]   Rapid NMR Data Collection

By Hanudatta S. Atreya and Thomas Szyperski

Abstract

Rapid data collection is an area of intense research in biomolecular NMR spectroscopy, in particular for high-throughput structure determination in structural genomics. NMR data acquisition and processing protocols for rapidly obtaining high-dimensional spectral information aim at avoiding sampling limited data collection and are reviewed here with emphasis on G-matrix Fourier transform NMR spectroscopy.

Copyright 2005, Elsevier Inc.
All rights reserved.
0076-6879/05 $35.00

Introduction

Multidimensional nuclear magnetic resonance (NMR) spectroscopy is pivotal for pursuing NMR-based structural biology (Cavanagh *et al.*, 1996; Wüthrich, 1986). In many instances, it is desirable to obtain multidimensional spectral information as rapidly as possible. First, the costs related to spectrometer usage are reduced, and the throughput of samples per NMR spectrometer can be increased. Second, the requirement for longevity of NMR samples is alleviated. Third, a higher time resolution can be achieved to study dynamic processes by multidimensional spectra. The first two objectives are at the heart of NMR-based structural genomics, which aims at establishing NMR spectroscopy as a powerful tool for exploring protein "fold space" and yielding at least one experimental structure for each family of protein sequence homologues (Montelione *et al.*, 2000).

Fast acquisition of multidimensional spectra is, however, limited by the need to sample (several) indirect dimensions. This restriction can be coined the "NMR sampling problem": above a threshold at which the measurement time is long enough to ensure a workable signal-to-noise ratio, the sampling of indirect dimensions determines the requirement for instrument time. In this "sampling-limited" data collection regime (Szyperski *et al.*, 2002), valuable instrument time is invested to meet the sampling demand rather than to achieve sufficient "signal averaging." Hence, techniques to speed up NMR data collection focus on *avoiding* this regime, that is, they are devised to push data collection into the "sensitivity-limited" regime in order to properly adjust NMR measurement time to sensitivity requirements. In view of the well-known fact that NMR measurement times tend to increase with molecular weight (Fig. 1), rapid sampling approaches for accurate adjustment of measurement times on the one hand and methodology developed to study large systems on the other [e.g., transverse relaxation optimized spectroscopy (Pervushin *et al.*, 1997) or protein deuteration (Gardner and Kay, 1998)] are complementary.

The implementation of rapid data collection protocols avoiding sampling limitations requires that the number of acquired free induction decays (FIDs), i.e., the number of data points sampled in the indirect dimensions, is reduced. Notably, phase-sensitive acquisition of an ND Fourier transform (FT) NMR experiment requires sampling of $N-1$ indirect dimensions with $n_1 \times n_2 \times \cdots \times n_{N-1}$ complex points, representing $2^{N-1} \times (n_1 \times n_2 \times \cdots \times n_{N-1})$ FIDs. A steep increase of the minimal measurement time, T_m, with dimensionality results: acquiring 16 complex points in each indirect dimension (with one scan per FID each second) yields $T_m(3D) = 0.5$ h, $T_m(4D) = 9.1$ h, $T_m(5D) = 12$ days, and $T_m(6D) = 1.1$ years.

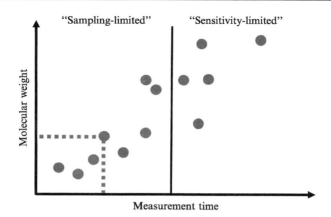

FIG. 1. Sketch of system size *vs.* required NMR measurement time with "sampling-limited" and "sensitivity-limited" data collection regimes. The dashed lines indicate that methodology for rapid NMR data collection allows the measurement time to be accurately adjusted to sensitivity requirements.

When reducing the number of acquired FIDs, the key challenge is to preserve the multidimensional spectral information that can be obtained by conventional linear sampling with appropriately long maximal evolution times in all indirect dimensions. Moreover, trimming the number of sampled data points may in turn require processing techniques that complement, or replace, widely used Fourier transformation of time domain data. Hence, we shall review approaches to reduce the sampling demand as well as associated processing techniques. To document the impact of rapid NMR data sampling for high-throughput structure determination, we shall present a brief survey of the application of a specific fast data acquisition technique, that is, reduced-dimensionality NMR spectroscopy, in the Northeast Structural Genomics Consortium (http://www.nesg.org).

Rapid NMR Data Collection

Currently available approaches to accelerate NMR data collection in biological NMR spectroscopy are summarized in Table I. One could classify as "basic" those that require only adjustment of acquisition parameter(s), while modification of the radiofrequency (rf) pulse scheme and/or more sophisticated data processing protocols are required for "advanced" approaches.

TABLE I
APPROACHES TO ACCELERATE NMR DATA COLLECTION

Approach	Acceleration	Data processing[a]
Basic		
Reduction[b] of t_{max}	$<\sim$1.5 for each dimension	FT
Reduction[c] of t_{rel}	$<\sim$2[d]	FT
Aliasing	$<\sim$3	FT
Advanced		
Reduction of t_{max}/sparse sampling	$<\sim$3 for each dimension	MER or TWD/FT
Reduction of t_{rel}: L-optimization[e]	$\sim$2–3	FT
Simultaneous acquisition	$<\sim$2	FT
Hadamard spectroscopy	$<\sim$2	HT
Single scan ND NMR	$\sim$10–50 for each dimension	FT
RD NMR spectroscopy	$\sim$10	FT
GFT NMR spectroscopy	$\sim$10[k] for $(N,N-K)$D	GFT

[a] FT, Fourier transformation; HT, Hadamard transformation; MER, maximum entropy reconstruction; TWD, three-way decomposition; GFT, G-matrix Fourier transformation. Note that FT is quite often employed in conjunction with linear prediction (Ernst *et al.*, 1987; Stephenson, 1988) to reduce truncation artifacts.

[b] t_{max} represents the maximal evolution time in an indirect dimension of a multidimensional NMR experiment.

[c] t_{rel} indicates the relaxation delay between the FIDs acquired for a multidimensional NMR experiment.

[d] Assuming that t_{rel} in studies of biological macromolecules is usually set between $\sim T_1(^1H)$ and $\sim 2T_1(^1H)$.

[e] L-optimization indicates longitudinal ^1H relaxation optimization (Pervushin *et al.*, 2002).

Basic Approaches

First, maximum evolution times, t_{max}, may be reduced in the indirect dimensions. Provided that truncation artifacts are kept at an acceptable level by using linear prediction (Ernst *et al.*, 1987; Stephenson, 1988), it is often possible to reduce t_{max} up to about a factor of 2. The resulting increase in line widths, however, implies that spectral resolution and signal-to-noise ratios are compromised as a result of the acceleration of data collection. Second, the relaxation delay between scans, t_{rel}, may be reduced, as was recognized for the rapid acquisition of NOESY data for smaller molecules (Köck and Griesinger, 1994). When reducing t_{rel}, sensitivity is traded off for increased acquisition speed. This approach is limited by (1) the duty cycle of spectrometers and (2) the quite rapid loss of sensitivity when relaxation delays become shorter than the longitudinal relaxation time T_1 of the excited nucleus (Ernst *et al.*, 1987). At best, acceleration by a factor of about two is thus achieved in routine applications in biological NMR spectroscopy (Table I). Third, the spectral width can be reduced, so that NMR signals are aliased. In protein NMR spectroscopy, significant signal aliasing is feasible in the aliphatic ^{13}C dimension, since ^{13}C^{ali} and ^{1}H^{ali} chemical shift values are strongly correlated (Cavanagh et al., 1996). As a result, the spectral width in the ^{13}C^{ali} dimension can be reduced up to about 3-fold in ^{13}C^{ali}-resolved multidimensional experiments. In most other cases, however, the acceleration that can be achieved by folding peaks is rather moderate. Taken together, the "basic" approaches (Table I) for accelerating data acquisition may help reduce the measurements times but are insufficient to compensate for increases in measurement time by orders of magnitude resulting from sampling of several indirect dimensions.

Advanced Approaches

To efficiently overcome sampling limitations, more sophisticated or "advanced" strategies for NMR data collection and processing have been devised (Table I). We shall first address techniques that make it possible to avoid the disadvantages arising from shortened maximal evolution times, t_{max}, and shortened relaxation delay between scans, t_{rel}, as described in the previous section. Then, we shall outline approaches that require additional efforts to reprogram rf pulse schemes in separate subsections.

A loss in spectral resolution arising from shortened maximal evolution periods in indirect dimensions may be avoided by using maximum entropy reconstruction (MER) (Hoch, 1985, 1989; Hoch and Stern, 1996, 2001; Hoch et al., 1990; Hore, 1985; Laue et al., 1985; Sibisi, 1983; Sibisi et al., 1984; Stern et al., 2002). Essentially, such processing techniques promise to reduce truncation artifacts without losing resolution. However, given the use of MER, it turns out that "sparse sampling" approaches (see below) are superior when compared to a direct truncation of time domain data.

Avoiding the sensitivity loss associated with rapid pulsing represents a particular challenge. Recently, Pervushin et al. (2002) developed the concept of longitudinal ^{1}H relaxation optimization ("L-optimization") to enhance the sensitivity of "out-and-back" triple resonance experiments. In out-and-back experiments, amide proton polarization is transferred to α- and/or β-carbons for frequency labeling and subsequently transferred back to the amide proton for signal detection. L-optimization is achieved by keeping the polarization of the aliphatic protons along the z-axis, which enhances longitudinal relaxation of amide protons during the relaxation delay. Very recently, L-optimization has been suggested for rapid data sampling, since sensitivity reduction at very short t_{rel} can be avoided

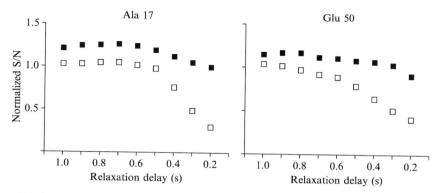

FIG. 2. The "normalized" S/N of two representative cross-peaks detected for the 17-kDa NESGC target protein ER75 (Liu *et al.*, 2004) in the 2D ^{15}N,^{1}H plane of RD 3D HNN<u>CAHA</u> (Szyperski *et al.*, 1998) (open squares) and L-optimized (Atreya and Szyperski, 2004; Pervushin *et al.*, 2002) RD 3D HNN<u>CAHA</u> (black squares) as a function of the relaxation delay between scans. The resonance assignment for the two peaks is indicated.

(Atreya and Szyperski, 2004). Figure 2 displays the relative sensitivity associated with two representative cross-peaks detected for a 17-kDa protein as a function of t_{rel} in the first two-dimensional (2D) [^{15}N–^{1}H] plane [t_1(^{13}C,^{1}H) = 0] of L-optimized and conventional three-dimensional (3D) HNN<u>CAHA</u> (Szyperski *et al.*, 1998) spectra. The data show that a rapid loss of sensitivity at $t_{rel} < 0.6$ s can be avoided, thus establishing L-optimization as a valuable approach supporting rapid data collection.

Sparse Sampling. In conventional multidimensional NMR spectroscopy, indirect dimensions are sampled *linearly*, i.e., indirect evolution periods are incremented by using a fixed delay ("increment") Δt. The length of the increment defines the spectral width according to the Nyquist theorem as being $1/\Delta t$ (Ernst *et al.*, 1987). Either MER or the combined use of three-way decomposition (TWD) and FT (Gutamas *et al.*, 2002; Korzhnev *et al.*, 2001; Orekhov *et al.*, 2001, 2003) allows *nonlinearly* sampled time domains to be converted into frequency domain data (Table I). In principle, it is possible to implement nonlinear sampling by probing an indirect dimension at arbitrary time points. In practice, however, a spacing of multiples of Δt is chosen between points, that is, the linear sampling protocol is being "diluted out" by randomly omitting a fraction of the time points of linear sampling. Hence, nonlinear sampling schemes are also referred to as "sparse sampling" protocols.

To sample at "higher density" where the signal is stronger, it is advantageous to randomly select data points for sparse sampling so that the probability of selection matches the signal envelope. Accordingly, random sampling for nondecaying signals (Hoch and Stern, 1996), e.g., those resulting from constant-time evolution periods, results in exponentially weighted random sampling for exponentially decaying signals (Barna and Laue, 1987; Barna et al., 1987; Cieslar et al., 1993; Schmieder et al., 1993, 1994) or sine-modulated exponentially weighted random sampling for a signal building up according to a sine function (e.g., antiphase signals in COSY) (Schmieder et al., 1993, 1994). Several applications using MER (Barna and Laue, 1987; Hoch, 1989; Schmieder et al., 1993, 1994) or TWD (Orekhov et al., 2003) have thus far demonstrated that between 50% and 75% of the data points can be omitted from a conventional spectrum acquired with linear sampling, without necessarily sacrificing spectral resolution.

One-dimensional nonlinear sampling schemes can, in principle, be combined to sample two- or possibly even higher-dimensional subspaces of a multidimensional NMR experiment in a nonlinear fashion (Hoch and Stern, 2001). This would lead to correspondingly increased acquisition speed. For example, nonlinear sampling of an entire 2D subspace might reduce the minimal measurement time by about an order of magnitude. Future research needs to determine if multiple sparse sampling is routinely feasible.

Simultaneous Acquisition of Multidimensional NMR Spectra. The simultaneous acquisition of two multidimensional NMR spectra represents a straightforward concept to speed up data acquisition. This approach has been pioneered by Falmer for 2D $^{13}C, ^1H/^{15}N, ^1H$-heteronuclear multiple-quantum correlation (HMQC) (Farmer, 1991), followed by Boelens et al. (1994) for 3D $^1H/^{13}C/^{15}N$ triple resonance spectra and by Farmer and Mueller (1994) and Pascal et al. (1994) for simultaneous acquisition of 3D ^{15}N- and $^{13}C^{ali}$-resolved $^1H, ^1H$-NOESY. Subsequently, additional schemes for simultaneous acquisition of 3D triple resonance experiments (Hu et al., 2001; Mariani et al., 1994; Pang et al., 1998; Xia et al., 2002) as well as an improved implementation for heteronuclear-resolved NOESY (Xia et al., 2003) have been reported. The primary challenge when implementing two different magnetization transfer pathways within the same rf pulse scheme is evidently due to the need to compromise on delays tuned for polarization transfer. In turn, this leads to some intrinsic loss in sensitivity for both experiments and prevents taking full advantage of the theoretically possible acceleration (Table I). Moreover, the concept of simultaneous data acquisition cannot be generalized for arbitrary pairs of NMR experiments.

Simultaneous acquisition of 3D ^{15}N- and $^{13}C^{ali}$-resolved $^1H, ^1H$-NOESY (Pascal et al., 1994; Xia et al., 2003), however, appears to be advantageous due to the neat spectral separation of amide and aliphatic proton chemical shifts (Cavanagh et al., 1996; Wüthrich, 1986). In addition, data analysis is facilitated for two heteronuclear NOESY experiments that are acquired in a single data set. This promises to compensate for (1) some loss of intrinsic sensitivity arising from simultaneous acquisition and (2) the drawback that sensitivity enhancement schemes (Kay *et al.*, 1992) cannot be employed simultaneously for $^{15}N,^1H$- and $^{13}C,^1H$-heteronuclear single-quantum correlation (HSQC).

Simultaneous acquisition of triple-resonance spectra can be implemented (Xia *et al.*, 2002) to constructively use axial peak magnetization (Szyperski *et al.*, 1996), which is usually discarded in conventional NMR spectroscopy. Then, additional correlations can be detected "for free." For example, when recording an "out-and-back" 3D HNN(CA)HA experiment, the delay for polarization transfer from $^{13}C^\alpha$ to $^1H^\alpha$ is set to a compromise value to enable detection of peaks arising from glycyl residues. As a result, an incomplete "INEPT-step" leaves a considerable fraction of the magnetization on $^{13}C^\alpha$. Thus, in 3D HNN[CAHA] (Xia *et al.*, 2002), an 3D HNNCA spectrum can simultaneously be acquired without additional investment of spectrometer time.

Hadamard NMR Spectroscopy. Multidimensional NMR data acquisition can be accelerated if the spectral range that is sampled is sparsely covered with resonance lines. In this case, excitation with selective rf pulses (Kessler *et al.*, 1991) employed in conjunction with Hadamard encoding allows a focus on the resonances for which multidimensional shift correlations shall be detected (Bircher *et al.*, 1990, 1991; Blechta and Freman, 1993; Bolinger and Leigh, 1988; Brutscher, 2004; Goelman, 1994; Goelman *et al.*, 1990; Kupče and Freeman, 1993, 2003a,b). Here, we shall limit the discussion to salient features of importance for biological NMR spectroscopy. First, the accurate positioning of selective pulses, which are mandatory to replace conventional sampling and to effectively exclude empty spectral regions, requires a priori knowledge of chemical shifts. Second, spectral ranges in protein NMR are usually covered rather well with resonance lines (Wüthrich, 1986). Hence, it can be anticipated that the achievable acceleration is, in practice, less than roughly a factor of two (provided that chemical shifts shall be measured with acceptable precision in the "Hadamard dimension"). Third, nonideality of selective rf pulses may complicate proper dissection of signals that are not well separated, i.e., signals arising from the same resonance are observed in more than a single subspectrum (Brutscher, 2004). However, Hadamard spectroscopy appears to be a highly valuable tool for multidimensional studies of smaller organic molecules exhibiting a wide dispersion of resonances.

Single-Scan Acquisition of Two-Dimensional NMR Spectra. Very recently, Frydman *et al.* (2002, 2003) introduced a remarkable experimental scheme for the acquisition of 2D NMR spectra with a single scan. In this approach, which constitutes an interface between high-resolution NMR and magnetic resonance imaging, the indirect chemical shift evolution is spatially encoded by use of pulse field gradients and then "read out" in conjunction with the chemical shift evolution in the direct dimension within a single scan. This leads to distinct line shape features in the indirect dimension (Shapira *et al.*, 2004b). In principle, this acquisition scheme can be extended to an arbitrary number of dimensions, that is, the implementation of 3D and 4D single scan acquisition (Shrot and Frydman, 2003). Reduced sensitivity, however, currently limits use to 2D single-scan acquisition for biological macromolecules. Notably, 2D single-scan data acquisition appears to be of high interest for real-time 2D NMR identification of chromatographic separations (Shapira *et al.*, 2004a).

Reduced Dimensionality (RD) NMR Spectroscopy. In view of the staggering increase of measurement time with increasing dimensionality, the joint sampling of two (or more) dimensions appears to be an attractive approach to tackle the NMR sampling problem. The first implementation of joint sampling of two chemical shift evolution periods dates back to the introduction of two-spin coherence (TSC) NMR spectroscopy (Szyperski *et al.*, 1993a). In TSC NMR, one of the transverse components of a two-spin coherence is detected in quadrature, while the chemical shift evolution of the second transverse component gives rise to a cosine modulation of the detected signal. This results in an in-phase splitting encoding the second shift, thereby effectively projecting the ND onto an $(N-1)$D spectrum. [For the sake of clarity it shall be mentioned that TSC NMR has also been named MQ NMR (Simorre *et al.*, 1994) or ZQ/DQ NMR (Rexroth *et al.*, 1995).] Subsequently, the TSC NMR concept was generalized, which led to the introduction of reduced dimensionality (RD) NMR spectroscopy (Szyperski *et al.*, 1993b). Starting from arbitrary rf pulse schemes, quadrature detection of one chemical shift evolution period is omitted, and the sampling of this period is achieved in conjunction with another period that is phase-sensitively sampled. The resulting peak pattern is as described for TSC NMR. Figure 3 shows a 2D HNNCA spectrum that was acquired for an 11-kDa protein (Szyperski *et al.*, 1993b).

RD NMR was further refined (Figs. 4, 5, and 6) by introducing (1) the relative scaling of the two jointly sampled indirect evolution times (Brutscher *et al.*, 1994; Szyperski *et al.*, 1993b, 1994) and (2) the use of time proportional phase incrementation (TPPI) (Marion and Wüthrich, 1983) to position the carrier of the projected dimension at the edge of the spectral range (Brutscher *et al.*, 1995b; Szyperski *et al.*, 1995). As a key step toward preserving the full potentialities of the parent experiment upon projection, the detection of central peaks defining the center of the RD NMR peak pairs was introduced (Fig. 5). This can be achieved either by incomplete INEPT (Ding and Gronenborn, 2002; Szyperski *et al.*, 1995) or by using ^{13}C

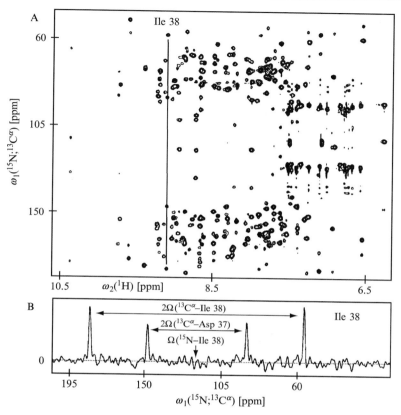

FIG. 3. (A) Contour plot of a 2D HN<u>N</u>CA (Szyperski *et al.*, 1993b) spectrum recorded for the 11-kDa mixed disulfide of *Escherichia coli* glutaredoxin (C14S) and glutathione (600 MHz ^{1}H resonance frequency; pH 6.5). $140(t_1/^{15}N;^{13}C^\alpha) \times 512(t_2/^1H)$ complex points were acquired. The ^{15}N chemical shifts were detected in quadrature, so that $\Omega(^{15}N) \pm \Omega(^{13}C')$ is observed along ω_1. (B) Cross section along ω_1 showing the RD NMR peak pair. The assignments of the intraresidue and sequential peak pairs are indicated. Reproduced with permission from Szyperski *et al.* (1993b). Copyright 1993 *Journal of the American Chemical Society.*

steady-state magnetization (Szyperski *et al.*, 1996). Central peaks serve to (1) identify peak pairs if chemical shifts in all but the RD dimension are degenerate (Szyperski *et al.*, 1995, 1996) and (2) recover sensitivity by symmetrization about the position of the central peaks (Brutscher *et al.*, 1995b; Szyperski *et al.*, 1995). In many cases, the inspection of peak intensities allows unambiguous identification and grouping of peak pairs (Fig. 5A), which is due to "spin relaxation time labeling" (Szyperski *et al.*,

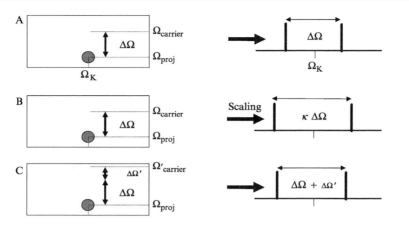

Fig. 4. Sketch demonstrating that (A) an RD NMR spectrum can alternatively be recorded by scaling chemical shift evolution periods (B) (Brutscher *et al.*, 1995a,b; Szyperski *et al.*, 1994, 1995) and/or in a manner placing central peaks and peak pairs in distinct spectral regions by employing TPPI (Marion and Wüthrich, 1983) (C). This ensures that spectral crowding remains as in conventional multidimensional NMR.

2002). If peak pairs with identical intensities are observed (Fig. 5B), central peak detection (Fig. 5D) is required. Alternatively, peak pairs can be identified by recording a second spectrum in which the projected shift evolution period is scaled with a factor close to 1.0 (Fig. 5C). This ensures that the maximal evolution times, and thus line shapes, do not change. The resulting small changes in the chemical shifts of peaks can be readily identified, and we name this approach "perturbation scaling" of shift evolution periods.

Löhr and Rüterjans (1995) introduced the 2-fold application of the RD NMR approach, in which the transfer amplitude is cosine modulated twice with the chemical shifts of two different nuclei. This results in a 2-fold reduction of the dimensionality. Recently, Ding and Gronenborn (2002) developed valuable experiments in which the transfer amplitude is cosine modulated twice within the same dimension.

An important innovation was published by Brutscher *et al.* (1995a), who developed the phase-sensitive joint sampling of two chemical shifts. Their approach could be named "edited phase-sensitive" RD NMR, since the salient new feature was the ability to separate the RD NMR peak pair into two subspectra. Very recently, this approach was extended to the joint sampling of two shift evolution periods (Kozminski and Zhukov, 2003). As an alternative to such spectral editing, scaling of shift evolution periods employed in conjunction with TPPI was introduced (Szyperski *et al.*, 1995).

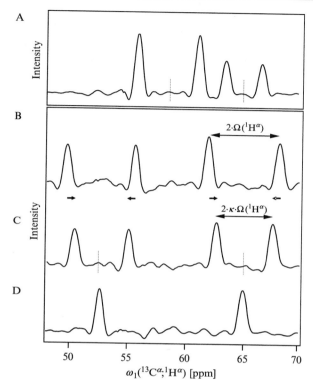

FIG. 5. Unambiguous identification of peak pairs registered in RD and/or GFT NMR. (A) The peak intensities allow two peak pairs that exhibit degenerate chemical shifts to be grouped in other, conventionally sampled, dimensions. Such "spin relaxation time labeling" (Szyperski *et al.*, 2002) quite often provides unambiguous assignments even in the absence of central peaks. (B) Peak pairs arising from two different spin systems have nearly the same intensities, and additional information is required to group them. (C) Rerecording of the same spectrum with slightly changed scaling of the projected chemical shift evolution period ($\kappa \sim 0.8$–0.9). Such "perturbation scaling" (see text) allows pairs of peaks belonging to the same spin system to be identified, since doublets are "compressed" as a result of the scaling (or expanded if $\kappa \sim 1.1$–1.2). (D) The detection of central peaks allows straightforward identification of peak pairs (Szyperski *et al.*, 1995, 1996). All cross sections are taken along $\omega_1(^{13}C^\alpha;^1H^\alpha)$ from L-optimized (Atreya and Szyperski, 2004; Pervushin *et al.*, 2002) 3D HNN<u>CAHA</u> (Szyperski *et al.*, 1998) recorded within 4 h on a 1.0-mM Z-domain sample (Tashiro *et al.*, 1997; Zheng *et al.*, 2004) at 25° using a Varian INOVA spectrometer operating at a ^{1}H resonance frequency of 750 MHz.

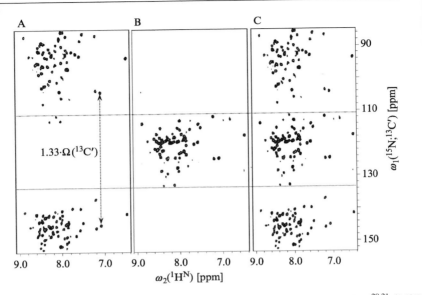

FIG. 6. Contour plots of peak pairs (A) and central peaks (B) of L-optimized[20,21] (3,2)D HN<u>N</u>CO recorded within 120 s for an 8-kDa protein Z-domain (Tashiro *et al.*, 1997; Zheng *et al.*, 2004) at 25° on a Varian INOVA spectrometer operating at a ^{1}H resonance frequency of 750 MHz. The spectra illustrate the use of TPPI and chemical shift scaling to place central peaks and peak pairs in separate spectral regions (Szyperski *et al.*, 1995) (see also Fig. 4). As an example, the ^{13}C' shift evolution was scaled by $\kappa = 0.66$ relative to the ^{15}N shift evolution. This leads, as indicated, to a peak pair separation of $1.33 \cdot \Omega(^{13}C')$ when compared to $2 \cdot \Omega(^{13}C')$ in the "nonscaled" congener. To compensate for the reduction in separation of the two peaks, the phase of the first 90° pulse exciting the ^{13}C' spin is subject to TPPI (Marion and Wüthrich, 1983). For the spectrum shown here, the apparent carrier position was moved by 1700 Hz, so that peak pairs at $\Omega(^{15}N) \pm 2\kappa \cdot \Omega(^{13}C')$ and central peaks at $\Omega(^{15}N)$ are placed into two distinct spectral regions (C).

This allows central peaks and components of the peak pairs to be placed into distinct spectral regions. As an example, Fig. 6 shows an L-optimized (Atreya and Szyperski; 2004; Pervushin *et al.*, 2002) 2D HN<u>N</u>CO spectrum recorded with scaling of the ^{13}C' chemical shift evolution and subsequent adjustment of the ^{13}C' carrier position by employing time-proportional phase incrementation. The spectrum was recorded in 120 s for an 8-kDa protein on a 750-MHz spectrometer equipped with a conventional ^{1}H/^{15}N/^{13}C probe.

The coincrementation of two evolution periods is a common feature of RD and accordion NMR spectroscopy. However, accordion spectroscopy serves to cosample real observables with the indirect shift evolution period.

Accordion spectroscopy was introduced by Ernst and co-workers in the early 1980s; it is based on joint sampling of a chemical shift evolution period and a mixing period of exchange spectroscopy (Bodenhausen and Ernst, 1981). Subsequently, accordion spectroscopy has been used for measurement of nuclear spin relaxation times (Kay and Prestegard, 1988; Mandel and Palmer, 1994), TOCSY mixing (Kontaxis and Keeler, 1995), or scalar coupling constants (Tolman and Prestegard, 1996). In contrast, RD NMR focuses on the joint sampling of a second chemical shift, which is a complex observable (i.e., the jointly sampled observable possesses phase). This requires the distinct experimental design of RD NMR.

G-Matrix Fourier Transform (GFT) NMR Spectroscopy. GFT NMR spectroscopy (Kim and Szyperski, 2003) represents a generalization of RD NMR and aims at providing high-dimensional spectral information with both accuracy and speed. GFT NMR spectroscopy results from "modules" derived for RD NMR and combines (1) multiple phase-sensitive RD NMR, (2) multiple "bottom-up" central peak detection, and (3) (time domain) editing of the components of the chemical shift multiplets. This resulting data acquisition scheme requires additional processing of time domain data, the so called "G-matrix" transformation (Table I). Hence, the acronym GFT indicates a combined G-matrix and Fourier transformation.

The phase-sensitive joint sampling of *several* indirect dimensions of a high-dimensional NMR experiment (Fig. 7) requires that the spectral width, SW_{GFT}, in the resulting combined "GFT dimension" is set to $SW_{GFT} = \sum \kappa_j SW_j$, where SW_j and κ_j represent, respectively, the jth spectral width and the factor to scale the sampling increments of the jth dimension, which enable adjustment for maximal evolution times (Kim and Szyperski, 2003). As a result, the "sampling demand" increases only *linearly* when dimensions are added for joint sampling, that is, the minimal measurement time of a GFT NMR experiment scales with the sum of the number of complex points required to sample the individual dimensions. In sharp contrast, the minimal measurement time of a conventional multidimensional NMR scales with the *product* of the number of complex points (see example in the Introduction). Hence, employment of GFT NMR makes it possible to reduce measurement times by about an order of magnitude *for each dimension* that is being added to the joint sampling scheme (Table I).

As described, RD NMR yields doublets ("peak pairs") that arise from the joint sampling of two chemical shift evolution periods (Figs. 5 and 6). In GFT NMR, the joint sampling of several shift evolution periods generates more complicated multiplet structures, which were named "chemical shift multiplets" (Figs. 7 and 8) (Kim and Szyperski, 2003). If all projected shifts are measured in a cosine-modulated fashion, the components of the chemical shift multiplet are all inphase. Depending on which and how many

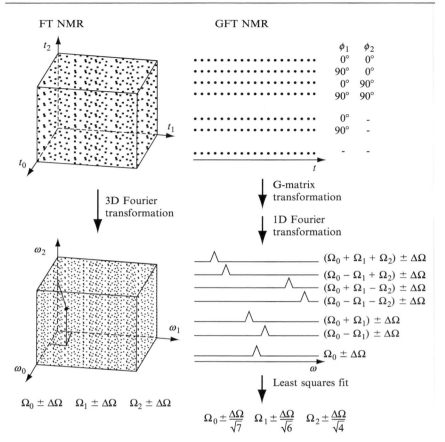

FIG. 7. Comparison of the conventional sampling of a 3D time domain subspace of an ND FT NMR experiment (on the left) with the phase sensitive joint sampling of the three dimensions in an $(N, N-2)$D GFT NMR (on the right), i.e., with $K = 2$. Processing of the FT NMR experiment requires a 3D FT of the subspace, while the GFT NMR experiment requires time domain editing of chemical shift multiplet components by application of the G-matrix and 1D FT of the resulting data sets. For the GFT NMR experiment, the phase settings of ϕ_1 and ϕ_2 of the rf pulses creating transverse magnetization for frequency labeling with Ω_1 and Ω_2 are indicated on the right. Instead of a single peak in FT NMR, which encodes three chemical shifts, one an overdetermined system of equations is obtained. A least-squares fit calculation yields the three shifts from the position of seven peaks. In a GFT NMR experiment with constant time chemical shift evolution periods, the lines forming the chemical shift multiplets have the same width as the resonances in FT NMR. This yields the same standard deviation $\Delta\Omega$ for the identification of peak positions in the two experiments. Hence, the precision of the chemical shift measurements obtained after the least-squares fit is increased in GFT NMR (Atreya and Szyperski, 2004; Kim and Szyperski, 2003, 2004). Reproduced with permission from Kim and Szyperski (2003). Copyright 2003 *Journal of the American Chemical Society.*

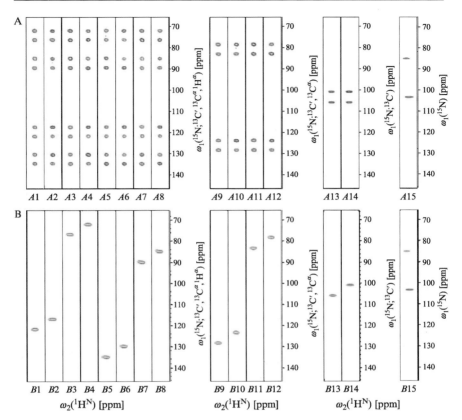

FIG. 8. $\omega_1[(^{15}N;^{13}C',^{13}C^\alpha,^1H^\alpha)$, $\omega_2(^1H^N)]$, $[\omega_1(^{15}N;^{13}C',^{13}C^\alpha)$, $\omega_2(^1H^N)]$, $[\omega_1(^{15}N;^{13}C')$, $\omega_2(^1H^N)]$, and $[\omega_1(^{15}N)$, $\omega_2(^1H^N)]$ strips taken from the (5,2)D HACACONHN GFT NMR experiment (see Fig. 9). Positive and negative contour levels are shown in red and blue, respectively. (A) Spectra $A1 \ldots A15$ containing the chemical shift multiplets obtained after FT. (B) Spectra $B1 \ldots B15$ containing the individual edited chemical shift multiplet components obtained after G-matrix FT. Reproduced with permission from Kim and Szyperski (2003). Copyright 2003 *Journal of the American Chemical Society.*

shifts are measured in a sine-modulated manner, various components become antiphase. Recording of all combinations of cosine and sine modulations then allows the components of the shift multiplet to be edited into subspectra (Fig. 7). In particular, G-matrix transformation enables this editing to be performed in the time domain. This is advantageous when linear prediction of time domain data is applied, because the S/N for each

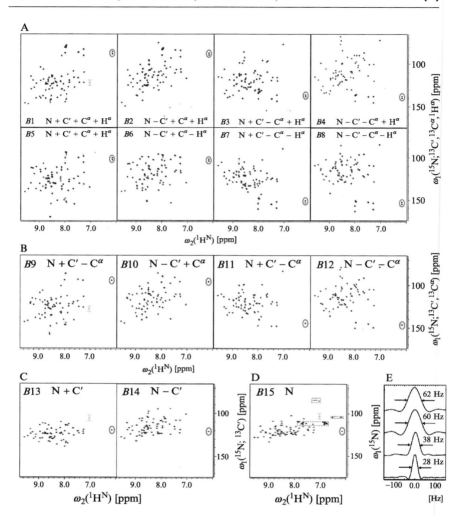

FIG. 9. The 15 2D planes constituting the (5,2)D HACACONHN GFT NMR experiment ($K = 3$) recorded for the 8.6-kDa protein ubiquitin. The linear combination of chemical shifts detected in a given plane is indicated. (A) The basic spectra $B1$ to $B8$. (B) First-order central peak spectra $B9$ to $B12$. (C) Second-order central peak spectra $B13$ and $B14$. (D) Third-order central peak spectrum $B15$. Signals arising from side chain moieties are in dashed boxes. (E) Cross sections taken along $\omega_1(^{15}N;^{13}C',^{13}C^\alpha,^1H^\alpha)$ at the peak of Ser-20 in $B1$ (at the top), along $\omega_1(^{15}N;^{13}C',^{13}C^\alpha)$ in $B9$ (second from top), along $\omega_1(^{15}N;^{13}C')$ in $B13$ (third from top), and along $\omega_1(^{15}N)$ in $B15$ (at the bottom). The sections are indicated in green in the corresponding panel. Comparison of sections from $B1$ and $B9$ shows that signals do not broaden with increasing K, while the smaller line widths observed in spectra $B13$ to $B15$ result from longer t_{max} values. The 15 signals detected on the backbone amide proton of Ile-36 are circled in red. Doublets are observed in $B1 \ldots B8$ since Gly-35 exhibits nondegenerate $^1H^\alpha$ chemical shifts, yielding the correlation of six shifts: $\delta(^1H^{\alpha 2}) = 4.135 \pm$

multiplet component is increased while a single component remains after editing for each subspectrum (Fig. 7).

The GFT NMR formalism embodies a generally applicable NMR data acquisition scheme (Kim and Szyperski, 2003). If $m = K + 1$ chemical shift evolution periods of an ND experiment are jointly sampled in a single indirect GFT dimension, $2^m - 1$ different $(N-K)$D spectra represent the GFT NMR experiment containing the information of the parent ND experiment. Hence, such a set of $2^m - 1$ subspectra was named an $(N, N-K)$D GFT NMR experiment. As an example, Fig. 9 shows the $2^4 - 1 = 15$ planes of a (5,2)D HACACONHN GFT NMR experiment recorded for a 8.6-kDa protein with four scans per real increment in 138 min, i.e., the minimal measurement time with a single scan per increment amounts to 33 min. In contrast, a conventional 5D HACACONHN sampled with $10(t_1/{}^1H^\alpha) \times 11(t_2/{}^{13}C^\alpha) \times 22(t_3/{}^{13}C') \times 13(t_4/{}^{15}N) \times 512(t_5/{}^1HN)$ complex data points would have required 5.8 days of spectrometer time with a single scan per real data point, i.e., a 250-fold reduction in minimal measurement time could be achieved. Moreover, the processed (5,2)D HACACONHN frequency domain data have a total size of 16 MByte, while a hypothetical 5D spectrum with the same digital resolution would represent a file of 618 GByte. Hence, employment of GFT NMR allows accurate adaptation of measurement times without sacrificing digital resolution. It was recently proposed that the projected GFT subspectra (recorded with different scaling of the chemical shift evolution periods) be used to "reconstruct" the parent ND spectrum (Coggins et al., 2004; Kupče and Freeman, 2003c,d, 2004). Such "projection-reconstruction" efforts might be of value whenever manual analysis of the reconstructed ND spectrum itself is more straightforward than the analysis of the set of projections.

Chemical shifts are multiply encoded in the shift multiplets registered in GFT NMR experiments. This corresponds to performing statistically independent multiple measurements, so that the chemical shifts can be obtained with high precision (Fig. 7) (Kim and Szyperski, 2003, 2004). Moreover, the well-defined peak pattern of the shift multiplets allows implementation of robust algorithms for peak picking (Moseley et al.,

---

0.006 ppm, $\delta({}^1H^{\alpha 1}) = 3.929 \pm 0.006$ ppm, $\delta({}^{13}C^\alpha) = 46.10 \pm 0.019$ ppm, $\delta({}^{13}C') = 173.911 \pm 0.017$ ppm for Gly-35, and $\delta({}^{15}N) = 120.295 \pm 0.043$ ppm and $\delta({}^1H^N) = 6.174 \pm 0.005$ ppm for Ile-36. The standard deviations of the indirectly detected chemical shifts were estimated from a Monte Carlo simulation (Kim and Szyperski, 2003). Notably, phase-sensitive editing of the chemical shift multiplets yields increasing peak dispersion (and thus resolution) in each of the constituent spectra compared to 2D ${}^{15}N,{}^1H$-HSQC (panel B15). Reproduced with permission from Kim and Szyperski (2003). Copyright 2003 Journal of the American Chemical Society.

2004). Both features make GFT NMR highly amenable to automated analysis.

Processing of Rapidly Sampled NMR Data Sets

In this section, we shall briefly survey processing protocols that are used in conjunction with the rapid sampling protocols (Table I) described in the previous section.

Maximum Entropy Reconstruction

MER of NMR spectra is based on the maximum entropy principle, which itself can be viewed as a generalization of the "principle of insufficient reason": maximizing the entropy of a probability distribution yields the most uniform distribution, given a set of constraints reflecting our knowledge about the probabilities (Jaynes, 1979). The goal of MER is to calculate the frequency domain spectrum with maximal entropy, which shall represent the most "uniform" spectrum being consistent with the experimental *time* domain data (Sibisi, 1983; Sibisi *et al.*, 1984). Hence, any deviation (a "peak") from a uniform distribution (the baseline and its offset) must originate from the experimental data serving as constraints. Mathematically, this can be achieved by maximizing the entropy and establishing consistency with experimental data, which corresponds to maximizing the target function, *TF* (Hoch and Stern, 1996):

$$TF(\mathbf{f}) = S(\mathbf{f}) - \lambda C(\mathbf{f}) \tag{1}$$

where $\mathbf{f}$ represents the data points of the reconstructed spectrum, $S(\mathbf{f})$ is a measure for its entropy, $C(\mathbf{f})$ reflects the consistency of the reconstructed spectrum with the experimental data, and λ is a Lagrange multiplier.

The appropriate form of the entropy measure has been a subject of debate (Hoch, 1989). One currently chosen definition for $C(\mathbf{f})$ is (Hoch and Stern, 1996):

$$C(\mathbf{f}) = \frac{1}{2} \sum_{k=0}^{M-1} |m_k - d_k|^2 \tag{2}$$

where M indicates the number of acquired data points, and m_k and d_k represent, the time domain data points obtained by inverse FT of the reconstructed spectrum and the experimental data, respectively. As implicated in Eq. (1), the entropy $S(\mathbf{f})$ is maximized given that $C(\mathbf{f}) \leq C_0$. Importantly, Eq. (2) also allows nonlinearly (or sparsely) sampled time domain data to be used for reconstruction of the frequency domain spectrum.

It has been shown (Hoch *et al.*, 1990) that

$$S(\mathbf{f}) = \sum_{n=0}^{N-1} R(f_n) \qquad (3)$$

represents a suitable measure for the entropy provided that phase-sensitively detected time domain data are available. $R(f_n)$ is the contribution of data point f_n to the overall entropy, with

$$R(f_n) = -\frac{f_n}{\alpha} \log \left(\frac{\left| f_n/\alpha + \sqrt{4 + |f_n|^2/\alpha^2} \right|}{2} \right) - \sqrt{4 + |f_n|^2/\alpha^2} \qquad (4)$$

where α is a scaling factor (Hoch *et al.*, 1990). Maximizing *TF* [Eq. (1)] for usually rather large NMR data sets requires the use of computational "first-order methods" such as steepest descent or conjugate gradients, that is, algorithms utilizing the Hessian matrix of second derivatives are most often not feasible (Hoch, 1989; Hoch and Stern, 2001).

Importantly, MER does not require a priori information about line shapes. However, if such information is available, it enhances the performance of MER. The same holds if uniform scalar couplings are manifested in the spectrum. Those can be considered when maximizing the entropy, which yields a decoupled spectrum (e.g., see Shimba *et al.*, 2003). The computational time required to recreate the spectrum by MER depends both on the number of data points and the spectral width. In the case of multidimensional NMR data, MER can be applied either sequentially or simultaneously to several dimensions. In practice, sequential application of MER is preferred due to computational requirements associated with such processing (Hoch, 1989; Hoch and Stern, 2001).

Three-Way Decomposition Analysis of Multidimensional NMR Data

It was recently proposed that a linearly sampled time-domain data set be reconstructed from a sparsely sampled set (Gutamas *et al.*, 2002; Korzhnev *et al.*, 2001; Orekhov *et al.*, 2001, 2003) by using TWD techniques (Carroll and Chang, 1970; Ibraghimov, 2002). The reconstructed time domain data set is subsequently processed by FT. The "missing" time domain points are predicted from the available nonlinearly sampled data set using least-squares fit. For a sparsely sampled 3D data, the target function (*TF*)

$$TF = \sum_{ijk} K_{ijk} \left| S_{ijk} - \sum_{m=1}^{M} a^m \cdot F1_i^m \cdot F2_j^m \cdot F3_k^m \right|^2 \qquad (5)$$

is minimized (Orekhov *et al.*, 2003). The sparse sampling scheme is encoded in the matrix K, where $K_{ijk} = 0$ or $K_{ijk} = 1$ for omitted or recorded data points, respectively (i.e., only recorded data points contribute to TF). S_{ijk} are the experimental data points (the indices i, j, and k enumerate the time domain data points of the 3D spectrum), $F1_i^m$, $F2_j^m$, and $F3_k^m$ are one-dimensional vectors referred to as "shapes" along each of the dimensions, and the a^m-represent the amplitudes of the M components (e.g., frequencies) that characterize the final 3D spectrum. The total number of measured data points is given by $I \times J \times K$, where I, J, and K correspond to the number of sampled data points in each of three dimensions. On the other hand, the number of fitted parameters is equal to $M(I + J + K - 2)$. Hence, the fit is largely "overdetermined." This justifies the use of a least-squares approach [Eq. (5)] for reconstructing the full-time domain data set (Orekhov *et al.*, 2003).

Hadamard NMR Spectroscopy

Hadamard NMR spectroscopy is based on selective excitation of NMR lines and their encoding in the form of a "Hadamard matrix" (Bolinger and Leigh, 1988; Goelman *et al.*, 1990). The experiment is carried out using an array of N selective pulses irradiating N different regions of the spectrum simultaneously. The phases of the individual selective pulses are then varied in a series of scans, that is, they are encoded based on a Hadamard matrix of order N in each of the successive scans. Linearly combining the responses from all N scans ("Hadamard transformation") enables the N individual responses to be separated. The selective irradiation scheme with Hadamard encoding can be applied to either the excitation pulses or to spin inversion pulses.

As an example, consider a spectrum that is divided into four distinct regions. The individual regions are excited simultaneously using a "multiplex irradiation sequence" delivering selective rf pulses in all four regions simultaneously. The phases of the rf pulses are varied according to the following table, which represents a Hadamard matrix of order 4 (an "$H-4$ matrix"):

Number of scans	Spectral component			
	1	2	3	4
1	+	+	+	+
2	+	−	+	−
3	+	+	−	−
4	+	−	−	+

In the first scan, all rf pulse phases have the same sign. In the subsequent scans, the phases are modulated following the signs given in rows 2 to 4 of the Hadamard matrix. The decoding, that is, the separation of the individual components, is achieved by forming linear combinations according to the same matrix (Hadamard transformation). For example, simply adding scans 1 through 4 (first row of matrix) "isolates" the first component since all others cancel. A major drawback of Hadamard spectroscopy is evidently due to the fact that precise measurement of chemical shifts in the "Hadamard dimension" would require an unrealistically large number of highly selective pulses.

To achieve rapid data sampling, it is desirable to reduce the number of employed rf pulses, M, and thus the number of scans. For improved artifact suppression, however, the order, N, of the Hadamard matrix (which corresponds to the actual number of scans) is often chosen to be greater than M. For example, the first scan (corresponding to the first row of the Hadamard matrix) represents contributions with the same sign from all pulses. Hence, this scan is usually discarded, because it is more sensitive to spectrometer imperfections than the rows with equal number of plus and minus signs.

G-Matrix Transformation

The joint sampling of chemical shift evolution periods in GFT NMR generates chemical shift multiplets (Fig. 7). In turn, G-matrix transformation of time domain data allows the components of the multiplets to be edited into different subspectra (Kim and Szyperski, 2003). If an ND FT NMR experiment is acquired as an $(N,N-K)$D GFT NMR experiment, $K + 1$ chemical shifts are measured in a single "GFT dimension" in which *linear combinations* of the jointly sampled shifts are detected phase sensitively. The remaining frequency axes in the resulting $(N-K)$D subspectra are sampled as in conventional NMR (Cavanagh *et al.*, 1996). To indicate which chemical shifts are jointly sampled, the corresponding nuclei are underlined in the name of the experiments, as was proposed for RD NMR (Szyperski *et al.*, 1993a,b). For example, a 5D HACACONHN experiment can be acquired as (5,2)D HACACONHN ($K = 3$), wherein the GFT dimension encodes the chemical shifts of $^1H^\alpha$, $^{13}C^\alpha$, $^{13}C'$, and ^{15}N (Fig. 8). If a particular chemical shift denoted Ω_0 is chosen out of the set of $K + 1$ jointly sampled shifts as the phase sensitively detected "center shift" (e.g., ^{15}N in Fig. 8), sampling of the remaining K shifts, Ω_1, Ω_2, ... , Ω_k, generates chemical shift multiplets centered about Ω_0. Thus, the linear combinations $\Omega_0 \pm \Omega_1 \ldots \pm \Omega_k$ are measured in the GFT dimension.

Notably, a particular choice of Ω_0 determines the strategy for central peak detection (see below). In addition to jointly incrementing $K + 1$ shifts, the phases Φ_j of the rf pulses exciting spins of type j ($j = 1 \ldots K$) are systematically varied between $0°$ and $90°$ to register both cosine- and sine-modulated data sets (Fig. 7). This results in 2^k "basic spectra," and G-matrix transformation of these subspectra affords editing of the shift components.

Considering that each of the 2^k subspectra contains a real and an imaginary part, a total of 2^{k+1} data sets is obtained. These can we written as a 2^{k+1}-dimensional vector:

$$\mathbf{S}(K) = \begin{bmatrix} C_K \\ S_K \end{bmatrix} \otimes \cdots \otimes \begin{bmatrix} C_1 \\ S_1 \end{bmatrix} \otimes \begin{bmatrix} C_0 \\ S_0 \end{bmatrix} \tag{6}$$

where $c_j = \cos(\Omega_j t)$ and $s_j = \sin(\Omega_j t)$ and t define the evolution time in the indirect GFT dimension. Multiplication of $\mathbf{S}(K)$ with the G-matrix according to

$$\mathbf{T}(K) = \mathbf{G}(K) \cdot \mathbf{S}(K) \tag{7}$$

yields the desired vector $\mathbf{T}(K)$, which comprises the edited subspectra. It is convenient to write $\mathbf{T}(K)$ in complex notation since, after FT, the dispersive imaginary parts of the edited spectra are discarded. Then, the G-matrix represents a $2^K \times 2^{K+1}$ complex matrix (Kim and Szyperski, 2003):

$$\mathbf{G}(K) = \begin{bmatrix} 1 & i \\ 1 & -i \end{bmatrix}_1 \otimes \cdots \otimes \begin{bmatrix} 1 & i \\ 1 & -i \end{bmatrix}_K \otimes \begin{bmatrix} 1 & i \end{bmatrix} \tag{8}$$

As an example for time domain editing, the G-matrix transformation of the basic spectra of (4,2)D GFT NMR ($N = 4$; $K = 2$) shall be illustrated, with Ω_0, Ω_1, and Ω_2 being the three jointly sampled shifts. The real and imaginary parts of the $2^K = 4$ spectra are denoted S_{jr} ($j = 1 \ldots 4$) and S_{ji} ($j = 1 \ldots 4$) and are proportional to

$$
\begin{aligned}
S_{1r} &\propto \cos(\Omega_0 t)\cos(\Omega_1 t)\cos(\Omega_2 t) \\
S_{1i} &\propto \cos(\Omega_0 t)\sin(\Omega_1 t)\cos(\Omega_2 t) \\
S_{2r} &\propto \cos(\Omega_0 t)\cos(\Omega_1 t)\sin(\Omega_2 t) \\
S_{2i} &\propto \cos(\Omega_0 t)\sin(\Omega_1 t)\sin(\Omega_2 t) \\
S_{3r} &\propto \sin(\Omega_0 t)\cos(\Omega_1 t)\cos(\Omega_2 t) \\
S_{3i} &\propto \sin(\Omega_0 t)\sin(\Omega_1 t)\cos(\Omega_2 t) \\
S_{4r} &\propto \sin(\Omega_0 t)\cos(\Omega_1 t)\sin(\Omega_2 t) \\
S_{4i} &\propto \sin(\Omega_0 t)\sin(\Omega_1 t)\sin(\Omega_2 t)
\end{aligned}
\Rightarrow \mathbf{S}(2) = \begin{bmatrix} S_{1r} \\ S_{1i} \\ S_{2r} \\ S_{2i} \\ S_{3r} \\ S_{3i} \\ S_{4r} \\ S_{4i} \end{bmatrix} \tag{9}
$$

The G-matrix [$K = 2$; Eq. (8)] is given by

$$\mathbf{G}(2) = \begin{bmatrix} 1 & 0 & 0 & -1 & 0 & -1 & -1 & 0 \\ 0 & 1 & 1 & 0 & 1 & 0 & 0 & -1 \\ 1 & 0 & 0 & 1 & 0 & -1 & 1 & 0 \\ 0 & 1 & -1 & 0 & 1 & 0 & 0 & 1 \\ 1 & 0 & 0 & -1 & 0 & 1 & 1 & 0 \\ 0 & 1 & 1 & 0 & -1 & 0 & 0 & 1 \\ 1 & 0 & 0 & 1 & 0 & 1 & -1 & 0 \\ 0 & 1 & -1 & 0 & -1 & 0 & 0 & -1 \end{bmatrix} \tag{10}$$

This results in vector $\mathbf{T}(2) = \mathbf{G}(2) \cdot \mathbf{S}(2)$ [Eq. (7)], which, in complex notation, is given by

$$\mathbf{T}(2) = \begin{bmatrix} e^{i(\Omega_0 + \Omega_1 + \Omega_2)t} \\ e^{i(\Omega_0 + \Omega_1 - \Omega_2)t} \\ e^{i(\Omega_0 - \Omega_1 + \Omega_2)t} \\ e^{i(\Omega_0 - \Omega_1 - \Omega_2)t} \end{bmatrix} \tag{11}$$

As a result, the four linear combinations $\Omega_0 \pm \Omega_1 \pm \Omega_2$ are measured in one of each of the four subspectra [given by $\mathbf{T}(2)$].

For frequency domain editing, each of the 2^K basic subspectra [i.e., $\mathbf{S}(K)$; see Eqs. (6) and (9)] constituting the $(N, N-K)$D GFT NMR experiment are first Fourier transformed separately. To obtain absorptive peaks, a zero-order phase correction of $n \times 90°$ is applied along the GFT dimension with n denoting the number of sine modulations. Subsequently, a linear combination of these in the frequency domain yields the edited 2^K subspectra. If the 2^K-dimensional vector $\mathbf{A}(K)$ represents the appropriately phased spectra containing the chemical shift multiplets (Fig. 7, upper panel) and $\mathbf{B}(K)$ contains the resulting 2^K edited spectra, we have

$$\mathbf{B}(K) = \mathbf{F}(K) \cdot \mathbf{A}(K) \tag{12}$$

where $\mathbf{F}(K)$ represents a $2^K \times 2^K$ matrix that can be readily derived from $\mathbf{F}(K-1)$ by tensor product formation:

$$\mathbf{F}(K) = \mathbf{F}(K-1) \otimes \mathbf{F}(1), \qquad \text{with } \mathbf{F}(1) = \begin{bmatrix} 1 & 1 \\ 1 & -1 \end{bmatrix} \tag{13}$$

The G- and F-matrices for time and frequency domain editing are related to each other as described by Kim and Szyperski (2003).

To restore the full potentialities of the parent ND experiment, peaks defining the centers of the components forming the chemical shift multiplets need to be recorded (Kim and Szyperski, 2003). Thus, for $(N, N-K)$D

GFT NMR it is also necessary to measure $\Omega_0 \pm \Omega_1 \ldots \pm \Omega_{K-1}, \Omega_0 \pm \Omega_1 \ldots$ $\pm \Omega_{K-2}, \ldots, \Omega_0 \pm \Omega_1$ and Ω_0, where spectra obtained by omission of m chemical shifts are denoted to be of the mth order. This implies that 2^{K-1} additional spectra for central peak detection are recorded. Central peak acquisition can be achieved in three different ways (Atreya and Szyperski, 2004; Kim and Szyperski 2003; Szyperski et al., 1995, 1996, 2002; Xia et al., 2002):

1. All subspectra constituting the $(N,N-K)$D GFT NMR experiment are acquired by successive omission of shift evolution periods from the ND FT NMR rf pulse scheme affording the basic spectra. This strategy requires that in total $2^{K+1}-1$ subspectra are recorded.

2. Central peaks can be obtained from incomplete polarization transfer. The exclusive use of this approach corresponds to their simultaneous acquisition in the 2^K basic spectra, that is, the total number of subspectra remains identical to the number of basic spectra.

3. Heteronuclear steady-state magnetization can be used, and two subspectra are required for each order of the central peaks obtained in such a way. This increases the number of subspectra to be recorded 2-fold for each projected dimension, so that a total of 4^K subspectra are recorded.

Depending on the experiment, it is possible to combine the three options above. The second and the third option allow magnetization otherwise yielding unwanted "axial peaks" to be used (Szyperski et al., 1996), and central peaks to be registered even if the resonances in the higher-order spectra are broadened. The use of option (3) requires additional "preprocessing," as described in the supplementary information of Kim and Szyperski (2003).

Application of Reduced-Dimensionality NMR Spectroscopy in Structural Genomics

Rapid NMR data acquisition is critical in structural genomics projects in which 3D structures of proteins are to be solved in a high-throughput manner. It has been proposed at the outset of establishing structural genomics consortia that RD NMR will play an important role in NMR-based structural genomics (Montelione et al., 2000). In fact, solution of the NMR sampling problem is a prerequisite to optimally adjust measurement times (Fig. 1). RD and GFT NMR spectroscopy are particularly attractive because both kinds of spectroscopy (1) are easy to implement and can readily be combined with transverse relaxation optimization (Pervushin *et al.*, 1997), (2) do not require additional hardware, (3) do not require prior chemical shift information, and (4) allow implementation of robust algorithms for peak picking and automated assignments (Moseley *et al.*, 2004). In our laboratory, RD NMR has been applied in the framework of the Northeast Structural Genomics consortium (NESGC) for proteins ranging in molecular mass from 7 to 22 kDa. Figure 10 shows a plot of the total measurement time required for a defined set of 3D RD NMR experiments (which provide 4D spectral information) *vs.* the molecular mass. All data were recorded by using conventional ^{1}H/^{13}C/^{15}N triple resonance probes at 600 and 750 MHz ^{1}H resonance frequency. The graph indicates that RD NMR effectively allows NMR measurement time to be adjusted while obtaining the desired 4D spectral information (note that the minimal measurement time for acquiring a single 4D spectrum is about 2 days). Moreover, when considering the use of cryogenic probes, which reduce measurement times by a factor of 10 or more (Monleón *et al.*, 2002), it becomes apparent that enhanced sampling strategies beyond RD NMR are required to take full advantage of the newest generation of high-field NMR hardware. We expect that GFT NMR, or variants thereof, that are employed in conjunction with supporting techniques such as L-optimization (Table I) will turn out to be of critical value in the future.

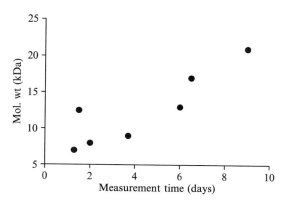

FIG. 10. Plot of the molecular weight of NESGC protein targets *vs.* the total measurement time of RD NMR spectra required to obtain (nearly) complete resonance assignments. All NMR data were collected by using a conventional ^{1}H/^{15}N/^{13}C triple resonance probe on 600- or 750-MHz spectrometers.

Conclusions

The development of techniques for rapid acquisition of multidimensional NMR data is a rapidly growing field. In view of the fact that increasing spectrometer sensitivity can be used to approach larger system and/or to speed up data collection, such methodology is pivotal to take best advantage of costly high-field NMR equipment. Moreover, the design of rapid sampling protocols shall allow integration with technology for automated data analysis. It currently appears that the use of RD/GFT NMR spectroscopy in conjunction with other approaches such as simultaneous acquisition, L-optimization, or its combination with sparse sampling will develop into a most powerful arsenal of techniques to effectively solve the NMR sampling problem.

Acknowledgments

Our research was supported by the Protein Structure Initiative of the National Institutes of Health (PM GM62413-01) and the National Science Foundation (MCB 0075773). We thank our colleagues of the Northeast Structural Genomics Consortium (NESGC), in particular Gaetano Montelione, for fruitful discussions.

References

Atreya, H. S., and Szyperski, T. (2004). G-matrix Fourier transform NMR spectroscopy for complete protein resonance assignment. *Proc. Natl. Acad. Sci. USA* **101,** 9642–9647.

Barna, J. C. J., and Laue, E. D. (1987). Conventional and exponential sampling for 2D NMR experiments with application to a 2D NMR spectrum of a protein. *J. Magn. Reson.* **75,** 384–389.

Barna, J. C. J., Laue, E. D., Mayger, M. R., Skilling, J., and Worall, S. J. P. (1987). Exponential sampling, an alternate method for sampling in two-dimensional NMR experiments. *J. Magn. Reson.* **73,** 69–77.

Bircher, H. R., Müller, C., and Bigler, P. (1990). Improvements for the 1D HOHAHA experiment. *J. Magn. Reson.* **89,** 146–152.

Bircher, H. R., Müller, C., and Bigler, P. (1991). Improved sensitivity for the 2D analogue of the 3D HOHAHA-COSY experiment. *Magn. Reson. Chem.* **29,** 726–729.

Blechta, V., and Freeman, R. (1993). Multisite Hadamard NMR spectroscopy. *Chem. Phys. Lett.* **215,** 341–346.

Bodenhausen, G., and Ernst, R. R. (1981). The accordion experiment, a simple approach to 3-dimensional NMR-spectroscopy. *J. Magn. Reson.* **45,** 367–373.

Boelens, R., Burgering, M., Fogh, R. H., and Kaptein, R. (1994). Time-saving methods for heteronuclear multidimensional NMR of (C-13, N-15) doubly labeled protein. *J. Biomol. NMR* **4,** 201–213.

Bolinger, L., and Leigh, J. S. (1988). Hadamard spectroscopic imaging (HSI) for multivolume localization. *J. Magn. Reson.* **80,** 162–167.

Brutscher, B. (2004). Combined frequency- and time-domain NMR spectroscopy. Application to fast protein resonance assignment. *J. Biomol. NMR* **29,** 57–64.

Brutscher, B., Morelle, N., Cordier, F., and Marion, D. (1995a). Determination of an initial set of NOE-derived distance contraints for the structure determination of N-15/C-13-labeled proteins. *J. Magn. Reson.* **B109**, 238–242.

Brutscher, B., Cordier, F., Simorre, J.-P., Caffrey, M. S., and Marion, D. (1995b). High-resolution 3D HNCOCA experiment applied to a 28-kDa paramagnetic protein. *J. Biomol. NMR* **5**, 202–206.

Brutscher, B., Simorre, J.-P., Caffrey, M. S., and Marion, D. (1994). Design of a complete set of 2-dimensional triple resonance experiments for assigning labeled proteins. *J. Magn. Reson.* **B105**, 77–82.

Carroll, J. D., and Chang, J. (1970). Analysis of individual differences in multidimensional scaling via an N-way generalization of the 'Eckart-Young' decomposition. *Psychometrica* **35**, 283–319.

Cavanagh, C., Fairbrother, W. J., Palmer, A. G., and Skelton, N. J. (1996). "Protein NMR Spectroscopy." Academic Press, San Diego, CA.

Cieslar, C., Ross, A., Zink, T., and Holak, T. A. (1993). Efficiency in multidimensional NMR by optimized recording of time point-phase pairs in evolution periods and their selective linear transformation. *J. Magn. Reson.* **B101**, 97–101.

Coggins, B. E., Venters, R. A., and Zhou, P. (2004). Generalized reconstruction of *n*-D NMR spectra from multiple projections: Application to the 5-D HACACONH spectrum of protein G B1 domain. *J. Am. Chem. Soc.* **126**, 1000–1001.

Ding, K., and Gronenborn, A. M. (2002). Novel 2D triple-resonance NMR experiments for sequential resonance assignments of proteins. *J. Magn. Reson.* **156**, 262–268.

Ernst, R. R., Bodenhausen, G., and Wokaun, A. (1987). "Principles of Nuclear Magnetic Resonance in One and Two Dimensions." Oxford University Press (Clarendon), London.

Farmer, B. T. (1991). Simultaneous [^{13}C-^{15}N]-HMQC, a pseudo-triple-resonance experiment. *J. Magn. Reson.* **93**, 635–641.

Farmer, B. T., and Mueller, L. (1994). Simultaneous acquisition of [13C,15N]- and [15N,15N]-separated 4D gradient-enhanced NOESY spectra in proteins. *J. Biomol. NMR* **4**, 673–687.

Frydman, L., Lupulescu, A., and Scherf, T. (2003). Principles and features of single scan two-dimensional NMR spectroscopy. *J. Am. Chem. Soc.* **125**, 9204–9217.

Frydman, L., Scherf, T., and Lupulescu, A. (2002). The acquisition of multidimensional NMR spectra within a single scan. *Proc. Natl. Acad. Sci. USA* **99**, 15858–15862.

Gardner, K. H., and Kay, L. E. (1998). The use of ^{2}H, ^{13}C, ^{15}N multidimensional NMR to study the structure and dynamics of proteins. *Annu. Rev. Biophys. Biomol. Struct* **27**, 357–406.

Goelman, G. (1994). Fast Handamard spectroscopic imaging techniques. *J. Magn. Reson.* **B104**, 212.

Goelman, G., Subramanium, V. H., and Leigh, J. S. (1990). Transverse Hadamard spectroscopic imaging technique. *J. Magn. Reson.* **89**, 437–454.

Gutamas, A., Jarvoll, P., Orekhov, V. Y., and Billeter, M. (2002). Three-way decomposition of a complete 3D ^{15}N-NOESY-HSQC. *J. Biomol. NMR* **24**, 191–201.

Hoch, J. C. (1985). Maximum entropy signal processing of two-dimensional NMR data. *J. Magn. Reson.* **64**, 436–440.

Hoch, J. C. (1989). Modern spectrum analysis in nuclear magnetic resonance—alteratives to the Fourier transform. *Methods Enzymol.* **176**, 216–241.

Hoch, J. C., and Stern, A. S. (1996). "NMR Data Processing." Wiley-Liss, New York.

Hoch, J. C., and Stern, A. S. (2001). Maximum entropy reconstruction, analysis and deconvolution in multidimensional nuclear magnetic resonance. *Methods Enzymol.* **338**, 159–178.

Hoch, J. C., Stern, A. S., Donoho, D. L., and Johnstone, I. M. (1990). Maximum entropy reconstruction of complex (phase-sensitive) spectra. *J. Magn. Reson.* **86,** 236–246.

Hore, P. J. (1985). NMR data processing using the maximum entropy method. *J. Magn. Reson.* **62,** 561–567.

Hu, W. D., Gosser, Y. Q., Xu, W. J., and Patel, D. J. (2001). Novel 2D and 3D multiple-quantum bi-directional HCNCH experiments for the correlation of ribose and base protons/carbons in 13C/15N labeled RNA. *J. Biomol. NMR* **20,** 167–172.

Ibraghimov, I. (2002). Application of the three-way decomposition for matrix compression. *Numer. Linear Algebra Appl.* **9,** 551–565.

Jaynes, E. T. (1979). Where do we stand on maximum entropy? *In* "The Maximum Entropy Formalism" (R. D. Levine and M. Tribus, eds.), pp. 15–118. MIT Press, Cambridge, MA.

Kay, L. E., and Prestegard, J. H. (1988). Spin-lattice relaxation of coupled spins from 2D accordion spectroscopy. *J. Magn. Reson.* **77,** 599–605.

Kay, L. E., Keifer, P., and Saarinen, T. (1992). Pure absorption gradient enhanced heteronuclear single quantum coherence spectroscopy with increased sensitivity. *J. Am. Chem. Soc.* **114,** 10663–10665.

Kessler, H., Mronga, S., and Gemmecker, G. (1991). Multi-dimensional NMR experiments using selective pulses. *Magn. Reson. Chem.* **29,** 527–557.

Kim, S., and Szyperski, T. (2003). GFT NMR, a new approach to rapidly obtain precise high-dimensional spectral information. *J. Am. Chem. Soc.* **125,** 1385–1393.

Kim, S., and Szyperski, T. (2004). GFT NMR experiment for polypeptide backbone and $^{13}C^{\beta}$ chemical shift assignment. *J. Biomol. NMR* **28,** 117–130.

Köck, M., and Griesinger, C. (1994). FAST NOESY experiments—An approach for fast structure determination. *Angew. Chem. Int. Ed. Engl.* **33,** 332–334.

Kontaxis, G., and Keeler, J. (1995). The accordion approach for "Taylored" TOCSY. *J Magn. Reson.* **115,** 35–41.

Korzhnev, D. M., Ibraghimov, I. V., Billeter, M., and Orehov, V. Y. (2001). MUNIN: Application of three-way decomposition to the analysis of heteronuclear NMR relaxation data. *J. Biomol. NMR* **21,** 263–268.

Kozminski, W., and Zhukov, I. (2003). Multiple quadrature detection in reduced dimensionality experiments. *J. Biomol. NMR* **26,** 157–166.

Kupče, E., and Freeman, R. (1993). Multisite correlation spectoscopy with soft pulses. A new phase-encoding scheme. *J. Magn. Reson.* **A105,** 310–315.

Kupče, E., and Freeman, R. (2003a). Frequency-domain Hadamard spectroscopy. *J. Magn. Reson.* **162,** 158–165.

Kupče, E., and Freeman, R. (2003b). Two-dimensional Hadamard spectroscopy. *J. Magn. Reson.* **162,** 300–310.

Kupče, E., and Freeman, R. (2003c). Reconstruction of the three-dimensional NMR spectrum of a protein from a set of plane projections. *J. Biomol. NMR* **27,** 383–387.

Kupče, E., and Freeman, R. (2003d). Projection-reconstruction of three-dimensional NMR spectra. *J. Am. Chem. Soc.* **125,** 13958–13959.

Kupče, E., and Freeman, R. (2004). Fast reconstruction of four-dimensional NMR spectra from plane projections. *J. Biomol. NMR* **28,** 391–395.

Laue, E. D., Skilling, J., Staunton, J., Sibisi, S., and Brereton, R. G. (1985). Maximum entropy method in nuclear magnetic resonance spectroscopy. *J. Magn. Reson.* **62,** 437–452.

Liu G., Li, Z., Chang, Y., Acton, T., Montelione, G. T., Murray, D., and Szyperski, T. High-quality structural models for a large family of homologous proteins derived from structures of *E. coli* proteins YgdK and SufE, enhancers of cystein desulfurase activity. Submitted.

Löhr, F., and Rüterjans, H. (1995). A new experiment for the sequential assignment of backbone resonances in proteins. *J. Biomol. NMR* **6,** 189–197.

Mandel, A. M., and Palmer, A. G. (1994). Measurement of relaxation-rate constants using constant-time accordion NMR spectroscopy. *J. Magn. Reson.* **110,** 62–72.

Mariani, M., Tessari, M., Boelens, R., Vis, H., and Kaptein, R. (1994). Assignment of the protein backbone from a single 3D, N-15, C-13, time-shared HXYH experiment. *J. Magn. Reson.* **B104,** 294–297.

Marion, D., and Wüthrich, K. (1983). Application of phase sensitive two-dimensional correlated spectroscopy (COSY) for measurements of ^{1}H–^{1}H spin–spin coupling-constants in proteins. *Biochem. Biophys. Res. Commun.* **113,** 967–974.

Monleón, D., Colson, K., Moseley, H. N. B., Anklin, C., Oswald, R., Szyperski, T., and Montelione, G. T. (2002). Rapid analysis of protein backbone resonance assignments using cryogenic probes, a distributed Linux-based computing architecture, and an integrated set of spectral analysis tools. *J. Struct. Funct. Genom.* **2,** 93–101.

Montelione, G. T., Zheng, D., Huang, Y., Gunsalus, C., and Szyperski, T. (2000). Protein NMR spectroscopy for structural genomics. *Nat. Struct. Biol.* **7,** 982–984.

Moseley, H. N. B., Riaz, N., Aramini, J. M., Szyperski, T., and Montelione, G. T. (2004). A generalized approach to automated NMR peak list editing: Application to reduced dimensionality triple resonance spectra. *J. Magn. Reson.* **170,** 263–277.

Orekhov, V. Y., Ibraghimov, I., and Billeter, M. (2001). MUNIN: A new approach to multi-dimensional NMR spectra interpretation. *J. Biomol. NMR* **20,** 49–60.

Orekhov, V. Y., Ibraghimov, I., and Billeter, M. (2003). Optimizing resolution in multidimensional NMR by three-way decomposition. *J. Biomol. NMR* **27,** 165–173.

Pang, Y. X., Zeng, L., Kurochkin, A. V., and Zuiderweg, E. R. P. (1998). High-resolution detection of five frequencies in a single 3D spectrum: HNHCACO—a bidirectional coherence transfer experiment. *J. Biomol. NMR* **11,** 185–190.

Pascal, S. M., Muhandiram, D. R., Yamazaki, T., Forman-Kay, J. D., and Kay, L. E. (1994). Simultaneous acquisition of N-15 edited and C-13 edited NOE spectra of proteins dissolved in H$_2$O. *J. Magn. Reson.* **B103,** 197–201.

Pervushin, K., Riek, R., Wider, G., and Wüthrich, K. (1997). Attenuated T-2 relaxation by mutual cancellation of dipole–dipole coupling and chemical shift anisotropy indicates an avenue to NMR structures of very large biological macromolecules in solution. *Proc. Natl. Acad. Sci. USA* **94,** 12366–12371.

Pervushin, K., Vogeli, B., and Eletsky, A. (2002). Longitudinal ^{1}H relaxation optimization in TROSY NMR spectroscopy. *J. Am. Chem. Soc.* **124,** 12898–12902.

Rexroth, A., Schmidt, P., Szalma, S., Geppert, T., Schwalbe, H., and Griesinger, C. (1995). New principle for the determination of coupling constants that largely suppresses differential relaxation effects. *J. Am. Chem. Soc.* **117,** 10389–10390.

Schmieder, P., Stern, A. S., Wagner, G., and Hoch, J. C. (1993). Application of nonlinear sampling schemes to COSY-type spectra. *J. Biomol. NMR* **3,** 569–576.

Schmieder, P., Stern, A. S., Wagner, G., and Hoch, J. C. (1994). Improved resolution in triple-resonance spectra by nonlinear sampling in the constant-time domain. *J. Biomol. NMR* **4,** 483–490.

Shapira, B., Karton, A., Aronzon, D., and Frydman, L. (2004a). Real-time 2D NMR identification of analytes undergoing continuous chromatographic separation. *J. Am. Chem. Soc.* **126,** 1262–1265.

Shapira, B., Lupulescu, A., Shrot, Y., and Frydman, L. (2004b). Line shape considerations in ultrafast 2D NMR. *J. Magn. Reson.* **166,** 152–163.

Shimba, N., Stern, A. S., Craik, C. S., Hoch, J. C., and Dotsch, V. (2003). Elimination of ^{13}C$^\alpha$ splitting in protein NMR spectra by deconvolution with maximum entropy reconstruction. *J. Am. Chem. Soc.* **125,** 2382–2383.

Shrot, Y., and Frydman, L. (2003). Single-scan NMR spectroscopy in arbitrary dimensions. *J. Am. Chem. Soc.* **125,** 11385–11396.

Sibisi, S. (1983). Two-dimensional reconstructions from one-dimensional data by maximum entropy. *Nature* **301,** 134–136.

Sibisi, S., Skilling, J., Brereton, R. G., Laue, E. D., and Staunton, J. (1984). Maximum entropy signal processing in practical NMR spectroscopy. *Nature* **311,** 446–447.

Simorre, J.-P., Brutscher, B., Caffrey, M. S., and Marion, D. (1994). Asignment of NMR-spectra of proteins using triple-resonance 2-dimensional experiments. *J. Biomol. NMR* **4,** 325–333.

Stephenson, D. S. (1988). Linear prediction and maximum entropy methods in NMR spectroscopy. *Prog. NMR Spectrosc.* **20,** 515–626.

Stern, A. S., Li, K. B., and Hoch, J. C. (2002). Modern spectrum analysis in multidimensional NMR spectroscopy: Comparison of linear-prediction extrapolation and maximum-entropy reconstruction. *J. Am. Chem. Soc.* **124,** 1982–1993.

Szyperski, T., Wider, G., Bushweller, J. H., and Wüthrich, K. (1993a). 3D $^{13}C-^{15}N$ heteronuclear two-spin coherence spectroscopy for polypeptide backbone assignments in $^{13}C-^{15}N$-double labeled proteins. *J. Biomol. NMR* **3,** 127–132.

Szyperski, T., Wider, G., Bushweller, J. H., and Wüthrich, K. (1993b). Reduced dimensionality in triple resonance NMR experiments. *J. Am. Chem. Soc.* **115,** 9307.

Szyperski, T., Pellecchia, M., and Wüthrich, K. (1994). 3D $H^{\alpha/\beta}C^{\alpha/\beta}(CO)NHN$, a projected 4D NMR experiment for the sequential correlation of polypeptide $^1H^{\alpha/\beta}$, $^{13}C^{\alpha/\beta}$, and backbone ^{15}N and $^1H^N$ chemical shifts. *J. Magn. Reson.* **B105,** 188–191.

Szyperski, T., Braun, D., Fernandez, C., Bartels, C., and Wüthrich, K. (1995). A novel reduced-dimensionality triple resonance experiment for efficient polypeptide backbone assignment, 3D C*O*HNN*C*A. *J. Magn. Reson.* **B108,** 197–203.

Szyperski, T., Braun, D., Banecki, B., and Wüthrich, K. (1996). Useful information from axial peak magnetization in projected NMR experiments. *J. Am. Chem. Soc.* **118,** 8147–8148.

Szyperski, T., Banecki, B., Braun, D., and Glaser, R. W. (1998). Sequential assignment of medium-sized $^{15}N/^{13}C$-labeled proteins with projected 4D triple resonance NMR experiments. *J. Biomol. NMR* **11,** 387–405.

Szyperski, T., Yeh, D. C., Sukumaran, D. K., Moseley, H. N. B., and Montelione, G. T. (2002). Reduced-dimensionality NMR spectroscopy for high-throughput protein resonance assignment. *Proc. Natl. Acad. Sci. USA* **99,** 8009–8014.

Tashiro, M., Tejero, R., Zimmerman, D. E., Celda, B., Nilsson, B., and Montelione, G. T. (1997). High-resolution solution NMR structure of the Z domain of staphylococcal protein A. *J. Mol. Biol.* **272,** 573–590.

Tolman, J. R., and Prestegard, J. H. (1996). Measurement of amide $^{15}N-^1H$ one-bond couplings in proteins using accordion heteronuclear shift correlation experiments. *J. Magn. Reson.* **B112,** 269–274.

Wüthrich, K. (1986). "NMR of Proteins and Nucleic Acids." Wiley, New York.

Xia, Y., Arrowsmith, C., and Szyperski, T. (2002). Novel projected 4D triple resonance experiments for polypeptide chemical shift assignment. *J. Biomol. NMR* **24,** 41–51.

Xia, Y., Yee, A., Arrowsmith, C. H., and Gao, X. (2003). $^1H^C$ and $^1H^N$ total NOE correlations in a single 3D experiment. ^{15}N and ^{13}C time-sharing in t_1 and t_2 dimensions for simultaneous data acquisition. *J. Biomol. NMR* **27,** 193–203.

Zheng, D., Aramini, J. M., and Montelione, G. T. (2004). Validation of helical tilt angles in the solution NMR structure of the Z domain of staphylococcal protein A by combined analysis of residual dipolar coupling and NOE data. *Protein Sci.* **13,** 549–554.

Section II

Proteomics

[5] An Integrated Platform for Automated Analysis of Protein NMR Structures

By Yuanpeng Janet Huang, Hunter N. B. Moseley, Michael C. Baran, Cheryl Arrowsmith, Robert Powers, Roberto Tejero, Thomas Szyperski, and Gaetano T. Montelione

Abstract

Recent developments provide automated analysis of NMR assignments and three-dimensional (3D) structures of proteins. These approaches are generally applicable to proteins ranging from about 50 to 150 amino acids. In this chapter, we summarize progress by the Northeast Structural Genomics Consortium in standardizing the NMR data collection process for protein structure determination and in building an integrated platform for automated protein NMR structure analysis. Our integrated platform includes the following principal steps: (1) standardized NMR data collection, (2) standardized data processing (including spectral referencing and Fourier transformation), (3) automated peak picking and peak list editing, (4) automated analysis of resonance assignments, (5) automated analysis of NOESY data together with 3D structure determination, and (6) methods for protein structure validation. In particular, the software AutoStructure for automated NOESY data analysis is described in this chapter, together with a discussion of practical considerations for its use in high-throughput structure production efforts. The critical area of data quality assessment has evolved significantly over the past few years and involves evaluation of both intermediate and final peak lists, resonance assignments, and structural information derived from the NMR data. Methods for quality control of each of the major automated analysis steps in our platform are also discussed. Despite significant remaining challenges, when good quality data are available, automated analysis of protein NMR assignments and structures with this platform is both fast and reliable.

Introduction

With the advent of multidimensional and triple-resonance strategies for determining resonance assignments and three-dimensional (3D) structures, it has become increasingly clear that protein NMR spectra have the quality and information content to allow largely automated and standardized analyses of assignments and structures for small proteins. This has been realized over the past few years in the development of automated methods

Copyright 2005, Elsevier Inc.
All rights reserved.
0076-6879/05 $35.00

for many of the steps in production nuclear magnetic resonance (NMR) protein structure analysis. These advances are significant demonstrations of NMR as a powerful and accessible tool for biophysical chemistry, drug design, and functional genomics. In this chapter, we summarize our efforts in standardizing the NMR data collection process, building an integrated platform for automated NMR structure analysis, and demonstrating its impact for the Northeast Structural Genomics (NESG) Consortium.

Overview of the Automated Protein Structure Analysis Process

The principal steps of automated NMR protein structure analysis are outlined in Fig. 1. These include (1) standardized data collection and organization, (2) processing (including spectral referencing and Fourier

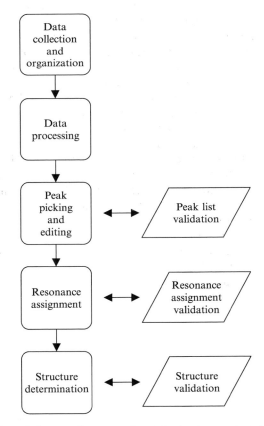

FIG. 1. Flowchart of the overall process of protein structure analysis from NMR data.

transformation), (3) peak picking and peak list editing, (4) resonance assignment, and (5) structure determination (including analysis of conformational constraints, NOESY assignment, residual dipolar coupling (RDC) data analysis, and 3D structure generation). In building an automated data analysis platform, the input and output of each of these steps must be organized in a self-consistent way, ideally using a relational database (Baran *et al.*, 2002; Zolnai *et al.*, 2003). A key issue for automated analysis is validation of the completeness, quality, and consistency of data generated in each of these principal steps. Recent efforts have focused on peak list validation, resonance assignment validation, and structure validation. A critical issue for automation is data quality. These validation steps, and estimates in uncertainties in the derived information, are critical both for defining a robust and reliable automation process and for interpreting the resulting resonance assignments and 3D structures.

Standardized Data Collection and Organization

The Organizational Challenge

The process of NMR-based protein structure analysis is challenged by requirements for properly executing, processing, and analyzing many separate NMR experiments. Unlike biomolecular crystallography, which generally involves a single type of data collection experiment, an NMR protein structure determination may require proper collection and analysis of 10–20 individual two-, three-, and four-dimensional (2D, 3D, and 4D) NMR spectra. These data must be highly self-consistent, as the input to the structure calculations is a composite generated from across these many data sets.

Standardized Data Collection

The challenges of organization for automated data analysis begin with data collection. As protein structure analysis relies on data from many different NMR experiments, it is critical that these data be self-consistent and fairly complete. Self-consistency can be particularly problematic when mixing data collected on different NMR spectrometers and/or using different samples of the protein under investigation. Efforts must be made to minimize spectrum-to-spectrum variability. In our laboratories, we generally collect all the data needed for a protein structure analysis back-to-back on the same sample and usually with the same NMR instrument. However, it is not always possible to collect in this manner, and even this strategy does not ensure consistency across spectra since sample heating effects can depend on decoupler duty cycles, which are different across NMR experiments. Fortunately, the latest generation NMR probes, and particularly cryogenic probes, exhibit less sample heating from decoupling than previous generation probes.

Another critical organizational issue for automated data analysis is the use of a standardized set of NMR pulse sequences for data collection. Each implementation of a sophisticated NMR experiment involves data collection and processing parameters that are unique to that implementation. It is very difficult to construct an analysis platform that is completely flexible with respect to all possible permutations. Well-defined sets of NMR data collection strategies create the basis for a robust analysis platform, providing consistent types of input data and guiding users to a better understanding of which NMR experiments are essential, optional (but useful), or superfluous. In general, different protein classes (e.g., small $^{15}N,^{13}C$-enriched proteins vs. larger perdeuterated $^{15}N,^{13}C$ proteins) require different data collection strategies, but a standardized set of experiments for each of these general classes can be defined. Within our "standard data collection sets," some experiments are defined as "required" while others are labeled "optional." Typically, "optional" experiments are carried out only when the quality evaluation of the "required" set deems it necessary.

It is also valuable to define the adjustable (sample dependent) and fixed parameters of data collection and processing for each NMR experiment in each "standard set." For example, in generating triple-resonance spectra for automated analysis of resonance assignments, we constrain the digital resolution in "matching dimensions of complementary spectra" (e.g., the ^{13}C dimensions of HNCA and HNcoCA spectra) to be identical to maximize accuracy in matching intraresidue and sequential cross peaks between these spectra. In the activities of our structural genomics project (www.nesg.org), one of the most critical innovations providing high-efficiency NMR structure generation has been the establishment of standardized data collection strategies and carefully considered default data collection and processing parameters.

The development of a package for employing reduced-dimensionality (RD) NMR spectroscopy (Szyperski *et al.*, 1993) for complete protein resonance assignment (Szyperski *et al.*, 2002) exemplifies this point. The "RDpack" (Y. Xia, D. K. Sukumaran, C. Arrowsmith, and T. Szyperski, unpublished) comprises pulse sequences, parameter sets, scripts, and macros for efficient *de novo* implementation of RD NMR experiments as well as rapid adjustment of parameter sets when using VARIAN INOVA spectrometers (Fig. 2). The RDpack is freely available for academic users and contains 11 experiments. 3D $\underline{H}^{\alpha\beta}C^{\alpha\beta}$coNH, 3D $\underline{HACA}$coNH, and 3D

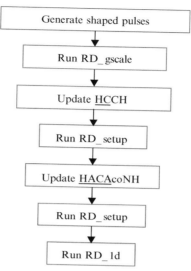

FIG. 2. Flowchart outlining the use of the RDpack. Steps required solely for the *de novo* implementation of RD NMR experiments are shown in red boxes, and steps required for rapid adjustment of parameter sets are displayed in green boxes. First, shaped pulses are generated by use of shell scripts, and the power levels for pulsed field gradients are adjusted to the available hardware configuration (using the macro RD_gscale). Second, the 3D HCCH parameter set is updated by providing proton and carbon high-power pulse widths, power levels, and carrier positions. Execution of the macro "RD_setup" transfers these parameters to the entire suite of RD experiments. Third, the 3D HACAcoNH parameter set is updated by providing nitrogen high-power pulse width, power level, and carrier position. These parameters are then transferred to nitrogen-resolved RD experiments by use of RD_setup. Finally, the macro RD_1d starts the acquisition of the first FID of all 11 parameter sets, while also allowing rapid assessment of the relative sensitivity of the various experiments. (See color insert.)

HCccoNH-TOCSY sequentially correlate proton and carbon shifts of residue $i-1$ with the amide proton and nitrogen shifts of residue i, 3D $\underline{H}^{\alpha\beta}\underline{C}^{\alpha\beta}NH$, 3D HN$\underline{CAHA}$, and 3D $\underline{H}^{\alpha\beta}\underline{C}^{\alpha\beta}$coHA provide complementary intraresidue connectivities, and 3D HN<$\underline{CO}$, $\underline{CA}$> affords both sequential and intraresidue connectivities. Aliphatic side chains are assigned by use of 3D H$\underline{C}$CH COSY and TOCSY, while aromatic spin system identification relies on 2D HBCBcgcdHD and 2D ^{1}H-TOCSY H$\underline{C}$H COSY. All parameter sets offer flags to conveniently select (1) central peak acquisition from ^{13}C steady-state magnetization (Szyperski *et al.*, 1996), (2) transverse relaxation optimized spectroscopy (TROSY) (Pervushin *et al.*, 1997) type data acquisition, and (3) ^{2}H decoupling.

Local Data Organization and Archiving

Biomolecular NMR research groups require efficient and simple access to archival NMR data, both for routine storage purposes and for the development and testing of novel computational methods for data analysis. Common methods of archiving raw NMR data [usually in the form of time domain free-induction decay (FID) data] in use in most biomolecular NMR laboratories are often inefficient, outdated, and error prone, leading to frequent loss of valuable data that are both hard and expensive to obtain. The growing demands on data organization and formatting in submitting NMR data and structures to public databases like the BMRB (Seavey et al., 1991) and the PDB (Berman et al., 2000) also require simple methods of harvesting NMR data and moving this information from the NMR laboratory into appropriate archival formats. This is particularly challenging for the several pilot projects in structural proteomics (Chance et al., 2002; Gong et al., 2003; Heinemann et al., 2000; Kennedy et al., 2002; Terwilliger, 2000; Yokoyama et al., 2000), which are being encouraged to submit into the public domain many more data items than have been traditionally expected from a conventional structural biology project. The goal of a standardized archive is not only to increase laboratory productivity through organization but also to support future NMR methods development by organizing laboratory data into a format that can easily be retrieved, reproduced, and shared across the community. If properly organized and archived, these data will be invaluable to the NMR community in efforts to develop new data collection and analysis technologies.

Examples of recently described NMR Laboratory Information Management System (LIMS) solutions are the Sesame (Zolnai et al., 2003) and SPINS (Baran et al., 2002) databases. SPINS (*S*tandardized *P*rote*I*n *N*MR *S*torage) (Baran et al., 2002) is an object-oriented relational database and data model that provides facilities for high-volume NMR data archival, data organization, and dissemination of raw NMR FID data to the public domain by automatic preparation of the header files needed for simple submission to the BMRB (Seavey et al., 1991).

NMR Spectral Processing

Several NMR spectral processing issues need to be carefully considered for successful automated data analysis. Particularly important is accurate and precise chemical shift referencing in the direct and indirect dimensions using IUPAC-defined referencing methods (Wishart et al., 1995), with dimethylsilapentane-5-sulfonic acid (DSS) as the reference compound. Accurate ^{13}C, ^{15}N, and ^{1}H referencing is essential for ensuring the development of an accurate database of chemical shift values (Zhang *et al.*, 2003). Proper chemical shift referencing for aliphatic ^{13}C and ^{1}H resonances is also critical for accurate amino acid typing (Grzesiek and Bax, 1993; Moseley *et al.*, 2001; Zimmerman *et al.*, 1997) and secondary structure analysis (Wishart and Sykes, 1994), generating information that is used in most automated assignment and structure analysis programs. In our laboratories, we externally calibrate the synthesizer offsets on each NMR spectrometer with a sample of 1 mM DSS in $^{2}H_2O$ at neutral pH and at multiple temperatures and then use these calibrations to define the corresponding chemical shift value of the carrier offset in each dimension of each NMR spectrum (Monleon *et al.*, 2002).

As with NMR data collection, similar amounts of zero-filling and/or linear prediction and similar window functions should be applied to matching dimensions across spectra to provide comparable final digital resolutions (Montelione *et al.*, 1999; Moseley *et al.*, 2001). This allows the use of the tightest possible "match tolerances" in later steps of automated analysis. We typically zero-fill the direct H^N dimension to 1024 complex points and 2-fold linear predict and zero-fill each indirect dimension to 256 or 512 complex points. Even though this copious increase in digital resolution goes beyond the usual theoretical recommendations, such processing can aid peak picking software that does not interpolate peak centers well. The use of linear prediction also suppresses severe Fourier truncation artifacts (e.g., sinc wiggles) and reduces line broadening effects of window functions (Koehl, 1999). This can have a significant impact in crowded regions of a spectrum. Linear prediction generally produces cleaner spectra and better shaped peaks, thus improving the performance of the peak picking algorithms, providing higher-quality peak lists, and ultimately improving the performance of later automated analysis steps (Moseley *et al.*, 2001). It is also critical to apply ridge-suppression and baseline correction in each spectral dimension to improve their quality, which can be very important for restrictive peak picking steps that arise later (Monleon *et al.*, 2002).

Several high-quality NMR processing programs have been developed over the past several years, including Felix (Molecular Simulations, Inc., San Diego, CA), NMRPipe (Delaglio *et al.*, 1995), PROSA (Güntert *et al.*, 1992), VNMR (Varian, Inc., Palo Alto, CA), and XWinNMR (Bruker Analytik GmbH, Karlsruhe, Germany). NMR data processing requires expert knowledge of many technical concepts and terms, presenting barriers to scientists not familiar with the deeper details of NMR spectroscopy. However, many of the parameters associated with the referencing and processing of NMR data, though specific to the pulse sequence program and particular spectrometer used to record the data, are relatively sample independent. Given the constraints of the data collection process as defined by the NMR pulse sequence, only a few adjustable parameters need to be considered by a user, and most of these can be set to usable default values based on general laboratory experience. Accordingly, there are several steps in the analysis of NMR data that may be viewed as routine tasks but often demand nontrivial amounts of time, knowledge of NMR theory, and familiarity with technical features of the specific data collection methods and/or processing software.

To address these data organization issues, we have developed AutoProc (Monleon *et al.*, 2002), a data dictionary together with a set of software tools designed to allow a nonexpert in NMR spectroscopy to accurately reference multidimensional NMR spectra, generate and run appropriate conversion scripts, and process NMR data using the software package NMRPipe (Delaglio *et al.*, 1995). AutoProc takes as input FID files along with libraries of spectrometer and pulse-sequence-specific description (table) files. It converts the data into a processing format, references the data in the direct and indirect dimensions using spectrometer-specific calibrations, and creates processing scripts suitable for running NMRPipe. It is straightforward to modify AutoProc to work with other script-based processing software like Felix (Molecular Simulations, Inc., San Diego, CA) or PROSA (Güntert *et al.*, 1992).

Peak Picking

Peak picking represents one of the crucial steps of NMR data analysis that has resisted successful automation for the purpose of automated resonance assignment and structure determination. This is due largely to cross-peak overlap and artifacts associated with large peaks, especially solvent and diagonal peaks. Multidimensional NMR spectra often exhibit artifacts of baseline distortions, intense solvent lines, ridges, and/or sinc wiggles. These problems are sometimes exacerbated by different processing methods that can dramatically affect line shape, intensity, and resolution of peaks as well as the severity of spectral artifacts.

Most automated peak pickers (Eccles *et al.*, 1991; Garrett *et al.*, 1991; Goddard and Kneller, 2000; Herrmann *et al.*, 2002; Koradi *et al.*, 1998; Orekhov *et al.*, 2001) rely on properties of an individual peak along with a model of the noise generated in the spectrum to determine whether a peak is valid or not, though one approach has looked at comparative properties of doublets (Andrec and Prestegard, 1998). Many programs perform restricted peak picking or filtered peak picking, which is a form of peak list editing in which one peak list is filtered against another in comparable dimensions (Goddard and Kneller, 2000; Monleon *et al.*, 2002; Zimmerman *et al.*, 1997). ATNOS (Herrmann *et al.*, 2002a) is software for automated

NOESY peak picking. It uses NOESY symmetry relationships along with restrictive peak picking against an assigned resonance list to guide the automated peak picking while using a ridge detection method to minimize peak picking along ridges. ANTOS has been used together with NOESY assignment and the structure determination software CANDID (Herrmann et al., 2002b) and DYANA (Güntert et al., 1997) to iteratively identify and assign NOESY cross peaks.

In our laboratories, peak picking is usually done using the restrictive peak picking and peak editing facilities in the program Sparky (Goddard and Kneller, 2000) or XEasy. Additional software, AutoPeak (Monleon et al., 2002), uses peak lists generated from manually peak picked 2D ^{15}N–^{1}H-heteronuclear multiple-quantum correlation (HSQC), ^{13}C–^{1}H-HSQC spectra as frequency filters across raw peak lists from 3D spectra. For the peaks that pass these filters, Sparky reports line width, root mean square fits to Lorenzian line shape, and peak intensity data can be used to further filter artifactual entries in the initial peak list table. Despite the sophistication of these automatic peak picking and editing methods, it is generally necessary to follow up with further editing (inclusion and exclusion) of peak lists by manual inspection of the spectra. This manual editing is guided by a data completeness quality report generated from initial analysis of data [i.e., the examine_spin_systems.pl (ESS) report from the AutoPeak software]. For an experienced spectroscopist, peak list editing for a typical set of NMR spectra used for backbone resonance assignments is completed in about 1 day and can be streamlined by doing some of the peak list editing while some data collection is still in progress (Moseley et al., 2001).

Interspectral Registration and Quality Assessment of Peak Lists

Quality assessment of input peak lists for further steps in the automated NMR analysis is crucial for the success of automation. We use several quality assessments of peak lists when judging if the peak lists are good enough for the later steps of automation. These include (1) peak list registration, (2) the examine_expected_peaks.pl (EEP), and (3) the ESS reports of the AutoPeak software suite (Moseley et al., 2001). The first quality assessment is the ability to register peak lists to each other in their comparable dimensions. Registration is an often overlooked step that is absolutely required for good performance in automated resonance assignment and NOESY assignment steps. In our current platform, a distance matrix approach [calculate_registration (Monleon et al., 2002)] is used to register peak lists from different spectra using resonance frequencies common to pairs of spectra. This approach has the added benefit of providing standard deviations of matching frequencies that can be used to derive appropriate tolerances for later steps in the automated NMR data analysis. These standard deviations, along with a count of the peaks that contributed to their calculation, provide scores that can be used to assess the quality of the peak lists. Interspectral registration data, and other (EEP and ESS) spin system quality reports provided by the AutoPeak software suite, are used to determine if a set of peak lists is of good enough quality for automated NMR analysis and to identify problematic or incomplete peak lists.

AutoAssign: Automated Analysis of Backbone Resonance Assignments

Significant progress has been made recently in automated analysis of resonance assignments, particularly using triple-resonance NMR data. Several laboratories are developing programs that automate either backbone or complete resonance assignments [reviewed in Baran et al. (2004), Moseley and Montelione (1999), and Zimmerman and Montelione (1995)]. Most automated programs use the same general analysis scheme that originates from the classical strategy developed by Wüthrich and co-workers (Billeter et al., 1982; Wagner and Wüthrich, 1982; Wüthrich, 1986).

Most commonly used algorithms for automated analysis of resonance assignments include the following steps (Moseley and Montelione, 1999): (1) register peak lists in comparable dimensions (registering/aligning), (2) group resonances into spin systems (grouping), (3) identify amino acid type of spin systems (typing), (4) find and link sequential spin systems into segments (linking), and (5) map spin system segments onto the primary sequence (mapping). Different automation programs implement each step with varying degrees of success; however, overall robustness is often dictated by the performance of the weakest step. The different automated resonance assignment programs are typically categorized by the methods they use in the mapping step. These methods include simulated annealing/Monte Carlo algorithms (Buchler et al., 1997; Leutner et al., 1998; Lukin et al., 1997), genetic algorithms (Bartels et al., 1996, 1997), exhaustive search algorithms (Andrec and Levy, 2002; Atreya et al., 2000; Coggins and Zhou, 2003; Güntert et al., 2000), heuristic comparison to predicted chemical shifts derived from homologous proteins (Gronwald et al., 1998), and heuristic best-first algorithms (Hyberts and Wagner, 2003; Li and Sanctuary, 1997; Zimmerman et al., 1994, 1997).

We develop and use the automated backbone resonance assignment program AutoAssign (Moseley et al., 2001; Zimmerman et al., 1997). AutoAssign is a constraint-based expert system (heuristic best first mapping algorithm) designed to determine backbone H^N, H^α, $^{13}C'$, $^{13}C^\alpha$, ^{15}N, and $^{13}C^\beta$ resonance assignments from peak lists derived from a set of triple resonance spectra with common H^N–^{15}N resonance correlations. The original implementation of AutoAssign was written in the programming language LISP with a Tcl/Tk-based graphical user interface (GUI) (Zimmerman et al., 1997). The current version of AutoAssign is written in C++ with a Java-based GUI (Moseley et al., 2001). The program can handle data obtained on uniformly ^{15}N–^{13}C doubly labeled, uniformly or partially deuterated, ^2H–^{15}N–^{13}C triply labeled, and selectively methyl-protonated, uniformly, or partially deuterated, ^2H–^{15}N–^{13}C triply labeled protein samples.

AutoAssign requires five different types of peak lists but may use up to nine different types of peak lists representing data obtained from a variety of triple resonance experiments, and a ^{15}N–H^N-HSQC spectrum. These nine types of peak lists represent information from the following nine types of experiments: HSQC*, HNCO, HNCACB*, HNcoCACB*, HNCA*, HNcoCA*, HNcaCO, HNcaHA, and HNcocaHA. Those peak lists marked by an asterisk are required by the program; however, using all nine types of data obtains the best performance (Moseley et al., 2001).

Key components of the processing, peak picking, and automated assignment software, AutoProc (Monleon et al., 2002), NMRPipe (Delaglio et al., 1995), AutoPeak (Monleon et al., 2002), Sparky (Goddard and Kneller, 2000), and AutoAssign (Moseley et al., 2001; Zimmerman et al., 1997), have been integrated together to provide a platform for rapid analysis of resonance assignments from triple resonance data. This prototype "integrated backbone resonance assignment platform" (Monleon et al., 2002) was applied to data collected from the small protein bovine pancreatic trypsin inhibitor (BPTI) using a first-generation high-sensitivity triple resonance NMR cryoprobe. Six NMR spectra were recorded in each of two sessions on a 500-MHz NMR system, requiring 28 and 4 h of data collection time, respectively. Fourier transforms were carried out using a cluster of Linux-based computers, and complete analysis of the seven spectra collected in each session was carried out in about 2 h. Nearly complete backbone resonance assignments and secondary structures (based on chemical shift data) for a 58-residue protein were determined in less than 30 h, including data collection, processing, and analysis time. In this optimum case of this small well-behaved protein providing excellent spectra, extensive backbone resonance assignments could also be obtained using less than 6 h of data collection and processing time. These results demonstrate the feasibility of high-throughput triple resonance NMR for determining resonance assignments and secondary structures of small proteins.

Automated Analysis of Side-Chain Resonance Assignments

Although several approaches have been found to provide robust automation of backbone resonance assignments, a robust approach to automated side-chain assignments is not yet generally available. The program GARANT (Bartels *et al.*, 1996) supports automated backbone and side-chain assignments. Recently, a combined approach of using GARANT and AUTOPSY (Koradi *et al.*, 1998) together demonstrates promising results in automating both peak picking and resonance assignments, including many side-chain aromatic ^{1}H resonance assignments (Malmodin *et al.*, 2003).

The principal challenge in automated analysis of side-chain resonances is incompleteness in experimental peak lists generally available for this task. Most published efforts in automating side-chain resonance assignments (Bartels *et al.*, 1997; Coggins and Zhou, 2003; Hyberts and Wagner, 2003) focus on HCCcoNH-TOCSY (Grzesiek *et al.*, 1993; Logan *et al.*, 1992; Montelione *et al.*, 1992) and use statistical comparisons to ^{13}C side chain resonance values of amino acid residues to assign the chemical shifts. These H^N-detected ^{13}C–^{13}C-TOCSY spectra are simple to interpret, but are often quite incomplete. Generally, no single spectrum has all side chain carbon resonances due to differences in TOCSY transfer efficiencies for short-chain and long-chain amino acids, although more complete data can sometimes be obtained by coadding spectra recorded with different isotropic mixing times (Celda and Montelione, 1993). Although fairly complete HCCcoNH-TOCSY data can sometimes be obtained for proteins of <10 kDa, and analyzed automatically with published methods, relaxation effects generally prevent the experiment from working well with larger proteins unless they are partially deuterated (Farmer and Venters, 1995; Gschwind *et al.*, 1998; Lin and Wagner, 1999). For these reasons, a robust approach for automated side-chain assignments should utilize HCCcoNH-TOCSY recorded with multiple mixing times, as well as other data such as HCCH-COSY (Bax *et al.*, 1990a; Ikura *et al.*, 1990; Kay *et al.*, 1990) and/or HCCH-TOCSY (Bax *et al.*, 1990b; Fesik *et al.*, 1990).

Resonance Assignment Validation Software

As with peak picking, quality assessment of resonance assignments is crucial for robustness in later steps of the automated NMR analysis. For this purpose, we have developed a set of computer utilities called the Assignment Validation Software (AVS) suite (Moseley *et al.*, 2004) for rigorously evaluating and validating a set of protein resonance assignments before submission to the BMRB and/or use in subsequent structure and/or functional analysis, without the need of a 3D structure. They serve the purpose of providing strict consistency checks for detecting possible errors and identifying "suspicious" assignments that deserve closer scrutiny prior to NOESY spectral analysis and 3D structure generation.

AutoStructure: Automated Analysis of NOESY Data

One of the principal goals of automated structure determination programs involves iterative analysis of multidimensional NOESY data. Several fully automated heuristic approaches for NOESY interpretation and structure calculation have been developed, including NOAH (Mumenthaler and Braun, 1995; Mumenthaler et al., 1997), ARIA (Nilges, 1995; Nilges et al., 1997), CANDID (Herrmann et al., 2002b), AutoStructure (Huang et al., 2003), a simulated annealing assignment approach implemented in XPLOR (Kuszewski et al., 2004), and other generally less developed programs (Adler, 2000; Grishaev and Llinas, 2002; Gronwald et al., 2002). The NOAH, ARIA, and CANDID programs utilize an iterative *top-down* data interpretation approach, having the following steps in common: (1) ambiguous proton–proton interactions from unassigned NOESY cross peaks, together with unambiguously assigned proton–proton interactions, are incorporated into structure calculations and generate a new set of model structures; (2) ambiguous proton–proton interactions are iteratively trimmed using the resulting model structures if they are far apart in the intermediate model structures. One key difference between NOAH and ARIA/CANDID is how ambiguous peaks are converted into distance constraints: NOAH creates an unambiguous constraint for each ambiguous proton–proton interaction while ARIA/CANDID uses an ambiguous constraint strategy (Nilges, 1995; Nilges et al., 1997) that generates only one ambiguous distance constraint for each ambiguous peak.

AutoStructure (Huang et al., 2003) uses an iterative *bottom-up topology-constrained approach* to analyze NOE peak lists and generate protein structures. AutoStructure first builds an initial fold based on intraresidue and sequential NOESY data, together with characteristic NOE patterns of secondary structures, including helical medium-range NOE interactions and interstrand β-sheet NOE interactions, and unique long-range packing NOE interactions based on chemical shift matching and symmetry considerations. Unassigned NOESY cross peaks are not used in structure calculations. Additional NOESY cross peaks are iteratively assigned using intermediate structures and the knowledge of high-order topology constraints of α-helix and β-sheet packing geometries. This protocol, in principle, resembles the methodology that an expert would utilize in manually solving a protein structure by NMR. The program AutoStructure has been combined with the structure generation programs DYANA (Güntert *et al.*, 1997) or XPLOR/CNS (Brünger, 1992; Brünger *et al.*, 1998).

The Control Flow of AutoStructure

The first step of AutoStructure (Fig. 3) is to match the chemical shifts from the NOESY peak list with the chemical shifts from the resonance assignment table using a loose match tolerance Δ_1 (typical values are 0.05 ppm for ^{1}H and 0.5 ppm for ^{13}C or ^{15}N). Aliased peaks can be directly matched to unaliased chemical shifts; there is no need for manually

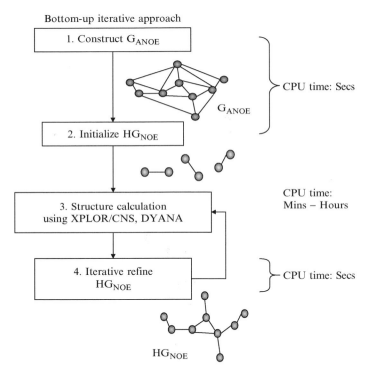

FIG. 3. The control flow of AutoStructure. AutoStructure uses a bottom-up iterative approach. It has four major steps. Step 1 constructs an ambiguous distance network G_{ANOE}, in which all vertices represent protons and proton pairs are connected when their chemical shift values are matched with an NOE peak's chemical shift values within a loose match tolerance. A heuristic HG_{NOE} is initialized from G_{ANOE} at Step 2. After HG_{NOE} is initialized, an initial fold is generated at Step 3. Step 4 iteratively refines HG_{NOE} from the structures generated from Step 3. (See color insert.)

unfolding aliased peaks or generating aliased chemical shifts. AutoStructure builds an ambiguous distance network (G_{ANOE}) from the chemical shift matching, in which nodes represent protons from the resonance assignment table, and edges represent NOE cross peaks linking all possible matched proton pairs. The rest of the steps of AutoStructure involve building a heuristic subgraph (HG_{NOE}) from G_{ANOE}, which is as close to the true distance network (representing the true 3D structures) as possible.

In Step 2, HG_{NOE} is initialized using all well-matched (within a tighter tolerance Δ_2) NOE-linked proton pairs that are connected by only two, three, or four covalent bonds (Wüthrich *et al.*, 1983) or belong to one of the $H^\alpha H^N(i, i + 1)$, $H^\beta H^N(i, i + 1)$, or $H^N H^N(i, i + 1)$ sequential NOE connections commonly observed in protein NOESY spectra (Billeter *et al.*, 1982). These close proton pair connections are anticipated from the amino acid sequence of the protein. A similar approach of reliably identifying intraresidue and sequential NOESY peaks is often used by experts in the process of manual analysis of NOESY data. At this step, Auto Structure also attempts to minimize site-specific chemical shift differences between the resonance assignment table and the NOESY peak list, due to interspectral variations of temperature and sample conditions. If proton h_i is involved in at least three NOE interactions (degree of vertex $h_i \geq 3$), its resonance frequency $\delta(h_i)$ in the refined resonance assignment list R' is updated with the median value derived from these linked NOE cross peaks. Match tolerances (Δ_1) for those protons with refined chemical shifts are then set to a narrower tolerance, and linking edges with large mismatches resulting from these protons with updated chemical shift values are removed from G_{ANOE}. This step simulates the expert analysis process of refining chemical shift values to be used in NOESY analysis from the frequencies of interpreted NOESY cross peaks.

After refining the resonance assignment table with intraresidue NOESY data, AutoStructure identifies helices and β-sheets, including interstrand alignments, by discovering patterns of NMR data that characterize secondary structures. This part of the algorithm uses chemical shift index (CSI) values (Wishart and Sykes, 1994), $^3J(H^N-H^\alpha)$ scalar coupling data (Wüthrich, 1986), and characteristic NOE contact patterns. These NOE contact patterns, characteristic of canonical secondary structures, are identified in G_{ANOE} and then added into the HG_{NOE} heuristic distance network using constraints implied by unique features of these secondary and tertiary structures already identified by the NMR data. At the same time, edges that represent linked proton pairs that are inconsistent with the geometries of identified secondary structures are removed from G_{ANOE}. In these ways, both local and long-range constraints indicated by the secondary structure topology are used to further build HG_{NOE} from G_{ANOE} prior to the actual structure generation process.

At the end of Step 2, AutoStructure identifies unique NOE connections (h1, h2, p) with frq(p) = 1 from G_{ANOE} and selectively adds into HG_{NOE} those that are supported by a large number of potential interresidue contacts in a contact map generated from the G_{ANOE} network that has been interpreted to this point. A well-matched NOE-linked proton pair (h1, h2, p) is identified as a unique connection if the number of possible proton–proton interactions linked to the peak is unique [frq(p) = 1]. At this point, symmetry features of multidimensional NOESY spectra are also considered to resolve ambiguities due to chemical shift degeneracy for peaks with frq(p) > 1.

In Step 3, AutoStructure constructs protein model structures. The program generates distance constraints directly from HG_{NOE} by calibrating the peak's intensities assuming a simple two-spin approximation and binning them into upper-bound distance classes as described by Wüthrich and co-workers (Mumenthaler et al., 1997; Wüthrich, 1986; Wüthrich et al., 1983). Dihedral angle constraints are generated from local NOE and scalar coupling data using the conformational grid search program HYPER (Tejero et al., 1999). Hydrogen bond distance constraints are identified based on the observation of helix and β-sheet NOE contact patterns, together with analysis of amide hydrogen exchange data and 3D structures when available (Wüthrich, 1986). Potential cis-peptide bonds and disulfide bonds are identified and reported to the user for expert validation. After validation, these special structural features are manually added into the constraint list. AutoStructure generates input constraint lists suitable for either XPLOR/CNS or DYANA for protein structure calculations. Structures are usually generated using a coarse-grain parallel calculation strategy on a Linux cluster, although the program can also be run on a single processor system, such as a Linux-based laptop computer.

In Step 4, a set of N model structures that best satisfy the resulting constraints is used to evaluate and refine the self-consistency of HG_{NOE}. First, distances (of the sum of inverse sixth powers of individual degenerate proton–proton distances) between all NOE-linked proton pairs of HG_{NOE} are calculated. Proton pairs with internuclear distances that violate the corresponding constraints by greater than $dvio_{min}$ in all of these N initial structures are removed from the HG_{NOE} distance network. The resulting HG_{NOE} is then used to regenerate another set of 3D model structures, which are again used for self-consistency analysis. This process of identifying inconsistent constraints within HG_{NOE} by 3D structure generation and analysis of constraint violations is repeated until no more inconsistent proton pair interactions remain in HG_{NOE}.

The resulting HG_{NOE} distance network and its corresponding model structures are considered to be self-consistent and are subsequently used as templates to refine and expand HG_{NOE}. First, AutoStructure analyzes the topology of the initial or intermediate structures, and trims G_{ANOE} down based on *topology constraints* implied by helical-packing and β-sheet packing geometries based on the "ridges into grooves model" (Chothia, 1984; Chothia *et al.*, 1981; Cohen *et al.*, 1982; Janin and Chothia, 1980). Next, AutoStructure further expands HG_{NOE} by adding NOE-linked proton pairs from G_{ANOE} that are well supported by the intermediate 3D structures. During this process, HG_{NOE} is further refined by removing any NOE assignments to long-range interactions associated with "orphan contacts" that may have evolved in the structure evolution process. Steps 3 and 4 are repeated several times (typically nine times) to iteratively refine the resulting structures. During this process, AutoStructure continues to refine the resonance assignment table using the resulting self-consistent HG_{NOE}.

Description of Input Data for AutoStructure

AutoStructure uses the following input data: (1) protein amino acid sequence and a list of resonance assignments (set R), (2) a list of the multidimensional (i.e., 2D, 3D, or 4D) NOESY cross-peak frequencies (which may be aliased) and intensities (set NOE), (3) a list of scalar coupling constant data (optional), (4) a list of slow amide ^{1}H exchange data (optional), and (5) other manually analyzed constraints when available, such as RDC (Tjandra and Bax, 1997), disulfide bond, and dihedral angle (Cornilescu *et al.*, 1999) constraint data. NOESY peak lists are generated using third-party automatic spectrum peak picking programs, usually followed by some manual editing. Dimeric proteins can also be analyzed when interchain NOESY cross-peak data are available from X-filtered NOESY experiments (Clore *et al.*, 1994), as demonstrated for coil–coil helix dimers (Greenfield *et al.*, 2001, 2003).

Quality Control Issues of Input Data for Autostructure

1. *Requirements for resonance assignment table.* AutoStructure uses a chemical shift index method (Wishart and Sykes, 1994) for secondary structure analysis and therefore requires accurate chemical shift referencing for C^α, C^β, and H^α resonances. This chemical shift index method relies on the use of the recommended IUPAC chemical shift referencing method with DSS as the reference compound. High-quality AutoStructure calculations require the input resonance assignment table to be more than 85% complete. For each aromatic residue, at least one

aromatic side chain proton should be assigned for AutoStructure to define its ring packing.

2. *Requirements for NOE peak lists.* Peak lists do not have to be perfect. AutoStructure can handle the presence of artifactual peaks and incompleteness; however, inaccurate or imprecise peak picking can considerably limit the performance of the program. Intense solvent lines, ridges, and/or sinc wiggles should be manually inspected and removed from the peak lists. Many NOE peaks may overlap with solvent lines and become difficult to peak pick. However, collecting $3D^{13}C$-NOESY in D_2O can minimize such problems. AutoStructure can handle aliased/folded peaks. High-quality AutoStructure calculations require the input peak list (set NOE) to contain at least 90% real cross peaks.

3. *Requirements for matching the NOE peak lists and resonance assignments.* AutoStructure calculates an M score that estimates the percent of predicted conformation-independent two- and three-bond connected NOE-linked proton pairs that are missing from the NOE peak lists. Four factors can contribute to high M scores: (1) misalignment between chemical shifts from NOE peak lists and the resonance assignment table, (2) significant differences in the digital resolutions between chemical shifts from NOE peak lists and the resonance assignment table, (3) poor quality of NOE peak lists, and (4) incorrect resonance assignments. A high M score (i.e., >25%) suggests that at least one of the input data sets (R and/or NOE) is of inadequate quality and needs to be improved. Those predicted two- and three-bond connected NOE-linked proton pairs missing from the NOE peak lists are reported to aid the user in improving the corresponding chemical shift assignments and/or identifying the expected NOESY cross peaks in the corresponding NOESY spectrum.

AutoStructure requires that all NOESY spectra be accurately referenced relative to the values of chemical shifts reported in the resonance assignment table. For each frequency dimension, the software computes the overall average chemical shift match difference from these predicted NOE-linked proton pairs. Consistent spectral referencing is achieved using these differences as global reference correction factors for the target spectrum, providing a tighter match between NOE peak lists and resonance assignment table, and allowing the use of smaller matching tolerances for further NOESY interpretation.

Using AutoStructure

AutoStructure is implemented in a combination of C/C++ programs, Perl programs, and shell scripts. It can be run in batch model or using the GUI. AutoStructure distribution on the Linux platform is freely available to academic users at http://www-nmr.cabm.rutgers.edu. AutoStructure analyzes NOEs and generates constraints for structure calculations. At least one of the structure calculation programs XPLOR/CNS, DYANA is required to be installed before running AutoStructure for iterative NOESY data analysis.

AutoStructure can automatically generate constraints for XPLOR/CNS or DYANA structure calculations. Manual constraints, including RDCs, can also be used in structure calculations and the resulting structures used for iterative analysis of AutoStructure. An initial structure model or homology model can also be used as input for AutoStructure analysis.

AutoStructure can also be used at varies stages of the resonance assignment procedure for validation. For example, given backbone resonance assignments and 3D ^{15}N-NOESY peak lists, AutoStructure can assign all backbone intraresidue, sequential, and medium-range NOEs, and identify all secondary structure elements. These NOE connectivities are commonly used for cross-validation of backbone sequential connectivity derived from triple resonance methods. Given nearly complete backbone and side-chain resonance assignments and 3D HCCH-COSY peak lists, AutoStructure can also assign all peaks in the 3D HCCH-COSY peak list for validation of the two-bond and three-bond connectivities of the side-chain resonances.

Testing AutoStructure

AutoStructure was developed and tested using several different experimental input data sets. For all test proteins, low atomic root-mean-square deviations (RMSDs) were obtained across the final structures, which, by conventional criteria, are indicative of high-quality structure determinations. The AutoStructure program has been used in over a dozen protein structure determinations (e.g., Aramini *et al.*, 2003; Greenfield *et al.*, 2001, 2003; Huang *et al.*, 2003; Makokha *et al.*, 2004; Ramelot *et al.*, 2003). Figure 4 shows AutoStructure results for the human basic fibroblast growth factor (154 amino acid residues), together with a comparison to the structure obtained by manual analysis of the same NMR data (Moy *et al.*, 1996) and by X-ray crystallography (Zhu *et al.*, 1991). Figure 4 also presents a *de novo* structure determination for a homodimeric 33-residue-per-chain, coiled-coil protein using AutoStructure (Greenfield *et al.*, 2001).

Minimal Constraint Approaches to Rapid Automated
  Fold Determination

Medium-accuracy fold information can often provide key clues about protein evolution and biochemical function(s). Extending ideas originally proposed by Kay and co-workers for determining low-resolution structures

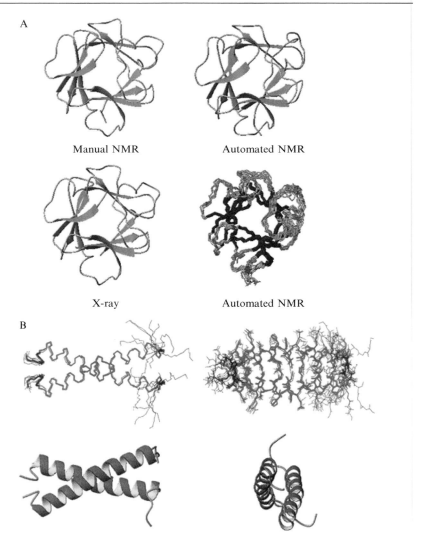

FIG. 4. Results of automatic analysis of protein structures from NMR data. (A) Comparison of backbone structures of human basic fibroblast growth factor (FGF) determined by manual analysis of NMR data (PDB code 1 bld), by automated analysis of the same NMR data using AutoStructure/XPLOR, or by X-ray crystallography (PDB code 1bas). The superposition of 10 NMR structures of human basic FGF computed by AutoStructure with XPLOR is also shown. Backbone conformations are shown only for residues 29 to 155, since the N-terminal polypeptide segment is not well defined in either the automated or manual analyses. For this portion of the structure, the backbone RMSD values within the families of structures determined by AutoStructure are ~0.7 Å, and the backbone

of larger proteins (Gardner *et al.*, 1997), a largely automatic strategy has been developed for rapid determination of medium-accuracy protein backbone structures using deuterated, $^{13}C,^{15}N$-enriched protein samples with selective protonation of side-chain methyl groups ($^{13}CH_3$) (Zheng *et al.*, 2003). Data collection includes acquiring NMR spectra for automatically determining assignments of backbone and side-chain ^{15}N, H^N resonances, and side-chain $^{13}CH_3$ methyl resonances. Conformational constraints are automatically derived using these chemical shifts, amide $^1H/^2H$ exchange, NOESY spectra, and RDC data. The total time required for collecting and analyzing such NMR data and generating medium-resolution but accurate protein folds can potentially be as short as a few days (Zheng *et al.*, 2003).

Structure Quality Assessment Tools

One of the most important challenges in modern protein NMR is to develop a fast and sensitive structure quality assessment measure that can evaluate the "goodness-of-fit" of a 3D structure compared with its NOESY peak lists and indicate the correctness of its fold. This is especially critical for automated NOESY interpretation and structure determination approaches. One approach uses an NMR R factor similar to that used in X-ray crystallography, which often requires computationally intensive, complete relaxation matrix calculations (Gonzalez *et al.*, 1991; Gronwald *et al.*, 2000; Zhu *et al.*, 1998). We have developed a set of quality Recall, Precision, and F-measure (NMR RPF) scores from information retrieval to assess the global "goodness-of-fit" (Huang *et al.*, 2005). These statistical RPF scores are quite rapid to compute, since NOE assignments and complete relaxation matrix calculations are not required, and are valuable in assessing protein NMR structure accuracy.

The quality of an NMR structure is also defined by a number of structural parameters including fold and packing quality, deviations of bond lengths and bond angles from standard values, backbone and side-chain dihedral angle distributions, hydrogen bond geometry, and close contacts between atoms. Currently there is no single comprehensive structure validation program that takes all these structural parameters into

---

RMSD between the AutoStructure and the X-ray crystal structure or manually determined NMR structures is ~0.8 Å. (B) Solution NMR structure of TM1bZip N-terminal segment of human α-tropomyosin determined by AutoStructure with DYANA (Greenfield *et al.*, 2001). The top panels show superpositions of backbone (*left*) and all heavy (*right*) atoms, respectively. Secondary structures are colored in red. The bottom panel shows ribbon diagrams of one representative structure. (See color insert.)

account to evaluate the overall quality of the structure. However, a number of different individual structure quality software packages exist that report scores quantifying some key structural parameters, such as ProCheck_nmr (Laskowski *et al.*, 1996), WHAT IF (Vriend, 1990), Verify 3D (Eisenberg *et al.*, 1997), PDB Validation Software (Westbrook *et al.*, 2003), and MAGE (Word *et al.*, 2000). In the NESG Consortium, we have developed an overall structure quality report that takes into account output from all of the programs mentioned above, and others, and evaluates their output based on a Z-score that normalizes each of these scores against a set of high-resolution X-ray crystal structures. This tool handles all data format conversions required to run the software mentioned above and presents the output as a series of easy-to-read reports and graphs for one-step structure quality evaluation.

An Integrated Platform for Automated NMR Structure Analysis

Protein NMR spectroscopists depend on a number of software packages to facilitate the analysis of data. For this reason, the process of solving a protein structure by NMR presents a formidable technical challenge to scientists. Although a number of software packages have been developed for the analysis of NMR data, a comprehensive solution for the complete automated analysis of NMR data from FIDs to three-dimensional structures is not yet available. Users choose between a number of different software programs each specialized in a certain step of the structure determination process. As a result, a dramatic learning curve exists for scientists to become proficient enough with all the necessary software to do their job. Furthermore, invaluable time is often wasted on trivial tasks such as preparing the output of one program to be usable for the next. Also, interlaboratory and in some cases even intralaboratory data exchange becomes extremely difficult when people are using a number of different formats required by the various pieces of software available. To add to this complexity, with data passing between so many sources, organization quickly becomes a problem. Precious data are often lost due to disorganization. This can lead to irreproducible results and curb the development of future technologies.

The CCPN effort (Fogh *et al.*, 2002) (http://www.bio.cam.ac.uk/nmr/ccp/) is attempting to address these problems in data organization and pipelining by developing a detailed data model to capture the complete NMR structure determination process. The data model is not only a standard solution for NMR databases to be implemented under, but also an application programming interface (API) to unify the development of future NMR software. The ANSIGv3.3 (Helgstrand *et al.*, 2000) spectral visualization software is an example of software developed over the CCPN data model.

The SPINS (Baran *et al.*, 2002) software provides an alternative solution to the integration problem. The SPINS data model is designed to easily accommodate any software available to the community. Rather than designing a data model for the world to adopt, the SPINS data model is intended for internal use by SPINS as a means to easily integrate any software. The SPINS data model was designed to be compatible with the BMRB NMRStar format, thus ensuring compatibility with other public domain efforts.

The current implementation of SPINS integrates several pieces of third-party software (Fig. 5), presenting them as a single application to the user. The SPINS software makes use of the following programs: (1) the SPINS (Baran *et al.*, 2002) database for storage and organization of raw FIDs, peak lists, chemical shift lists, constraint lists, 3D structures, and other intermediate results; (2) AutoProc (Monleon *et al.*, 2002), a spectral referencing and processing script-generating program; (3) NMRPipe (Delaglio *et al.*, 1995) for executing multidimensional Fourier transformations using scripts generated by AutoProc; (4) NMRDraw (Delaglio *et al.*, 1995) spectral visualization software for evaluating spectral quality; (5) SPARKY (Goddard and Kneller, 2000) spectral visualization software, launched out of SPINS, for peak picking and interactive peak list editing; (6) AutoPeak software (Monleon *et al.*, 2002; Moseley *et al.*, 2001) for interspectral registration, automated peak list editing, and peak data validation; (7) AutoAssign (Moseley *et al.*, 2001; Zimmerman *et al.*, 1997) automated backbone assignment software; (8) Assignment Validation Suite (AVS) software (Moseley *et al.*, 2004), providing statistical and graphical tools for validating the quality of the assignments; and (9) Auto-Structure, along with DYANA (Güntert *et al.*, 1997), XPLOR-nih (Schwieters *et al.*, 2003), or CNS (Brünger *et al.*, 1998) to iteratively assign NOESY peak lists and generate 3D structures.

The SPINS software provides an integrated process and user interface for using the software packages described above without having to worry about the numerous I/O complexities associated with data analysis using multiple software packages. Furthermore, the process is warehoused by the underlying SPINS database, making it completely reproducible. The completed process can be automatically exported in a standard format (NMRStar 3.1) for submission to the BMRB (Seavey *et al.*, 1991).

Conclusions

Recent developments provide automated analysis of NMR assignments and 3D structures. These approaches are generally applicable to proteins ranging from about 50 to 150 amino acids. Although progress over the past

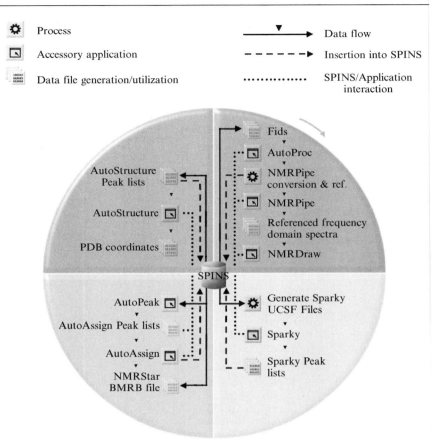

Process

Accessory application

Data file generation/utilization

Data flow

Insertion into SPINS

SPINS/Application
interaction

Fids

AutoProc

NMRPipe
conversion & ref.

AutoStructure
Peak lists

NMRPipe

AutoStructure

Referenced frequency
domain spectra

PDB coordinates

NMRDraw

SPINS

AutoPeak

Generate Sparky
UCSF Files

AutoAssign Peak lists

Sparky

AutoAssign

NMRStar
BMRB file

Sparky Peak
lists

FIG. 5. The integrated SPINS platform for automated analysis of NMR data. This figure depicts the flow of data through the SPINS software from raw FIDs to backbone assignments. (1) The raw FID data are housed in the SPINS database. (2) AutoProc queries the SPINS database for autoreferencing and processing of experimental data using NMRPipe. (3) Sparky software is used for manual peak picking and peak list editing. (4) AutoPeak software is used to validate peak lists as well as prepare AutoAssign input. (5) AutoAssign software is used for automated backbone resonance assignments. The SPINS platform also integrates Auto-Structure software for NOESY data analysis, together with DYANA/CNS/XPLOR software for 3D structure generation and software providing estimates of structure quality scores. (See color insert.)

few years is encouraging, even for small proteins more work is required before automated structure analysis is routine. In particular, general methods for automated analysis of side-chain resonance assignments are not yet well developed, though current efforts in this area are quite promising.

Moreover, little work has focused on the specific problems associated with nucleic acid structures. The critical area of quality assessment has evolved significantly over the past few years and involves evaluation of both intermediate and final peak lists, resonance assignments, and structural information derived from the NMR data. However, although various resonance assignment and 3D structure "R factors" are beginning to be used, no community-wide consensus has been reached on how to evaluate the accuracy and precision of a protein NMR structure. Despite these significant challenges, when good quality data are available, automated analysis of protein NMR assignments and structures is both fast and reliable. Moreover, automation methods are beginning to have a broad impact on the structural NMR community.

Acknowledgments

We thank J. Aramini, A. Bhattacharya, G. Sahota, D. Snyder, G. V. T. Swapna, and D. Zheng for useful discussions and for their efforts over the past several years in developing automated NMR data analysis algorithms and software. The authors' recent work on automated NMR data analysis has been supported by the NIH Protein Structure Initiative (P50-GM62413).

References

Adler, M. (2000). Modified genetic algorithm resolves ambiguous NOE restraints and reduces unsightly NOE violations. *Proteins* **39,** 385–392.

Andrec, M., and Levy, R. M. (2002). Protein sequential resonance assignments by combinatorial enumeration using 13C alpha chemical shifts and their (i, i−1) sequential connectivities. *J. Biomol. NMR* **23,** 263–270.

Andrec, M., and Prestegard, J. H. (1998). A Metropolis Monte Carlo implementation of bayesian time-domain parameter estimation: Application to coupling constant estimation from antiphase multiplets. *J. Magn. Reson.* **130,** 217–232.

Aramini, J. M., Huang, Y. J., Cort, J. R., Goldsmith-Fischman, S., Xiao, R., Shih, L. Y., Ho, C. K., Liu, J., Rost, B., Honig, B., Kennedy, M. A., Acton, T. B., and Montelione, G. T. (2003). Solution NMR structure of the 30S ribosomal protein S28E from *Pyrococcus horikoshii*. *Protein Sci.* **12,** 2823–2830.

Atreya, H. S., Sahu, S. C., Chary, K. V., and Govil, G. (2000). A tracked approach for automated NMR assignments in proteins (TATAPRO). *J. Biomol. NMR* **17,** 125–136.

Baran, M. C., Moseley, H. N., Sahota, G., and Montelione, G. T. (2002). SPINS: Standardized protein NMR storage. A data dictionary and object-oriented relational database for archiving protein NMR spectra. *J. Biomol. NMR* **24,** 113–121.

Baran, M. C., Huang, Y. J., Moseley, H., and Montelione, G. T. (2004). Automated analysis of protein NMR assignments and structures. *Chem. Rev.* **104,** 3541–3556.

Bartels, C., Billeter, M., Güntert, P., and Wüthrich, K. (1996). Automated sequence-specific NMR assignment of homologous proteins using the program GARANT. *J. Biomol. NMR* **7,** 207–213.

Bartels, C., Güntert, P., Billeter, M., and Wüthrich, K. (1997). GARANT—A general algorithm for resonance assignment of multidimensional nuclear magnetic resonance spectra. *J. Comput. Chem.* **18,** 139–149.

Bax, A., Clore, G. M., Driscoll, P. C., Gronenborn, A. M., Ikura, M., and Kay, L. E. (1990a). Practical aspect of proton–carbon–carbon–proton three-dimensional correlation spectroscopy of labeled proteins. *J. Magn. Reson.* **87,** 620–627.

Bax, A., Clore, G. M., and Gronenborn, A. M. (1990b). 1H-1H correlation via isotropic mixing of 13C magnetization, a new three-dimensional approach for assigning 1H and 13C spectra of 13C-enriched proteins. *J. Magn. Reson.* **88,** 425–431.

Berman, H. M., Westbrook, J., Feng, Z., Gilliland, G., Bhat, T. N., Weissig, H., Shindyalov, I. N., and Bourne, P. E. (2000). The Protein Data Bank. *Nucleic Acids Res.* **28,** 235–242.

Billeter, M., Braun, W., and Wüthrich, K. (1982). Sequential resonance assignments in protein 1H nuclear magnetic resonance spectra. Computation of sterically allowed proton-proton distances and statistical analysis of proton-proton distances in single crystal protein conformations. *J. Mol. Biol.* **155,** 321–346.

Brünger, A. T. (1992). X-PLOR, Version 3.1: "A System for X-ray Crystallography and NMR." Yale University Press, New Haven, CT.

Brünger, A. T., Adams, P. D., Clore, G. M., DeLano, W. L., Gros, P., Grosse-Kunstleve, R. W., Jiang, J. S., Kuszewski, J., Nilges, M., Pannu, N. S., Read, R. J., Rice, L. M., Simonson, T., and Warren, G. L. (1998). Crystallography & NMR system: A new software suite for macromolecular structure determination. *Acta Crystallogr. D Biol. Crystallogr.* **54,** 905–921.

Buchler, N. E., Zuiderweg, E. R., Wang, H., and Goldstein, R. A. (1997). Protein heteronuclear NMR assignments using mean-field simulated annealing. *J. Magn. Reson.* **125,** 34–42.

Celda, B., and Montelione, G. T. (1993). Total correlation spectroscopy (TOCSY) of proteins using co-addition of spectra recorded with several mixing times. *J. Magn. Reson.* **B101,** 189–193.

Chance, M. R., Bresnick, A. R., Burley, S. K., Jiang, J. S., Lima, C. D., Sali, A., Almo, S. C., Bonanno, J. B., Buglino, J. A., Boulton, S., Chen, H., Eswar, N., He, G., Huang, R., Ilyin, V., McMahan, L., Pieper, U., Ray, S., Vidal, M., and Wang, L. K. (2002). Structural genomics: A pipeline for providing structures for the biologist. *Protein Sci.* **11,** 723–738.

Chothia, C. (1984). Principles that determine the structure of proteins. *Annu. Rev. Biochem.* **53,** 537–572.

Chothia, C., Levitt, M., and Richardson, D. (1981). Helix to helix packing in proteins. *J. Mol. Biol.* **145,** 215–250.

Clore, G. M., Omichinski, J. G., Sakaguchi, K., Zambrano, N., Sakamoto, H., Appella, E., and Gronenborn, A. M. (1994). High-resolution structure of the oligomerization domain of p53 by multidimensional NMR. *Science* **265,** 386–391.

Coggins, B. E., and Zhou, P. (2003). PACES: Protein sequential assignment by computer-assisted exhaustive search. *J. Biomol. NMR* **26,** 93–111.

Cohen, F. E., Sternberg, M. J., and Taylor, W. R. (1982). Analysis and prediction of the packing of alpha-helices against a beta-sheet in the tertiary structure of globular proteins. *J. Mol. Biol.* **156,** 821–862.

Cornilescu, G., Delaglio, F., and Bax, A. (1999). Protein backbone angle restraints from searching a database for chemical shift and sequence homology. *J. Biomol. NMR* **13,** 289–302.

Delaglio, F., Grzesiek, S., Vuister, G. W., Zhu, G., Pfeifer, J., and Bax, A. (1995). NMRPipe: A multidimensional spectral processing system based on UNIX pipes. *J. Biomol. NMR* **6,** 277–293.

Eccles, C., Güntert, P., Billeter, M., and Wüthrich, K. (1991). Efficient analysis of protein 2D NMR spectra using the software package EASY. *J. Biomol. NMR* **1,** 111–130.

Eisenberg, D., Luthy, R., and Bowie, J. U. (1997). VERIFY3D: Assessment of protein models with three-dimensional profiles. *Methods Enzymol.* **277,** 396–404.

Farmer, B. T., and Venters, R. A. (1995). Assignment of side-chain 13C resonances in perdeuterated proteins. *J. Am. Chem. Soc.* **117,** 4187–4188.

Fesik, S. W., Eaton, H. L., Olejniczak, E. T., Zuiderweg, E. R., McIntosh, L. P., and Dahlquist, F. W. (1990). 2D and 3D NMR spectroscopy employing 13C-13C magnetization transfer by isotropic mixing. Spins system identification in large proteins. *J. Am. Chem. Soc.* **112,** 886–888.

Fogh, R., Ionides, J., Ulrich, E., Boucher, W., Vranken, W., Linge, J. P., Habeck, M., Rieping, W., Bhat, T. N., Westbrook, J., Henrick, K., Gilliland, G., Berman, H., Thornton, J., Nilges, M., Markley, J., and Laue, E. (2002). The CCPN project: An interim report on a data model for the NMR community. *Nat. Struct. Biol.* **9,** 416–418.

Gardner, K. H., Rosen, M. K., and Kay, L. E. (1997). Global folds of highly deuterated, methyl-protonated proteins by multidimensional NMR. *Biochemistry* **36,** 1389–1401.

Garrett, D. S., Powers, R., Gronenborn, A. M., and Clore, G. M. (1991). A common sense approach to peak picking in two-, three-, and four-dimensional spectra using automatic computer analysis of contour diagrams. *J. Magn. Reson.* **95,** 214–230.

Goddard, T. D., and Kneller, D. G. (2000). "SPARKY 3." University of California, San Francisco, CA.

Gong, W. M., Liu, H. Y., Niu, L. W., Shi, Y. Y., Tang, Y. J., Teng, M. K., Wu, J. H., Liang, D. C., Wang, D. C., Wang, J. F., Ding, J. P., Hu, H. Y., Huang, Q. H., Zhang, Q. H., Lu, S. Y., An, J. L., Liang, Y. H., Zheng, X. F., Gu, X. C., and Su, X. D. (2003). Structural genomics efforts at the Chinese Academy of Sciences and Peking University. *J. Struct. Funct. Genomics* **4,** 137–139.

Gonzalez, C., Rullmann, J. A. C., Bonvin, A. M. J. J., Boelens, R., and Kaptein, R. (1991). Toward an NMR R factor. *J. Magn. Reson.* **91,** 659–664.

Greenfield, N. J., Huang, Y. J., Palm, T., Swapna, G. V., Monleon, D., Montelione, G. T., and Hitchcock-DeGregori, S. E. (2001). Solution NMR structure and folding dynamics of the N terminus of a rat non-muscle alpha-tropomyosin in an engineered chimeric protein. *J. Mol. Biol.* **312,** 833–847.

Greenfield, N. J., Swapna, G. V., Huang, Y., Palm, T., Graboski, S., Montelione, G. T., and Hitchcock-DeGregori, S. E. (2003). The structure of the carboxyl terminus of striated alpha-tropomyosin in solution reveals an unusual parallel arrangement of interacting alpha-helices. *Biochemistry* **42,** 614–619.

Grishaev, A., and Llinas, M. (2002). CLOUDS, a protocol for deriving a molecular proton density via NMR. *Proc. Natl. Acad. Sci. USA* **99,** 6707–6712.

Gronwald, W., Willard, L., Jellard, T., Boyko, R. F., Rajarathnam, K., Wishart, D. S., Sonnichsen, F. D., and Sykes, B. D. (1998). CAMRA: Chemical shift based computer aided protein NMR assignments. *J. Biomol. NMR* **12,** 395–405.

Gronwald, W., Kirchhofer, R., Gorler, A., Kremer, W., Ganslmeier, B., Neidig, K. P., and Kalbitzer, H. R. (2000). RFAC, a program for automated NMR R-factor estimation. *J. Biomol. NMR* **17,** 137–151.

Gronwald, W., Moussa, S., Elsner, R., Jung, A., Ganslmeier, B., Trenner, J., Kremer, W., Neidig, K. P., and Kalbitzer, H. R. (2002). Automated assignment of NOESY NMR spectra using a knowledge based method (KNOWNOE). *J. Biomol. NMR* **23,** 271–287.

Grzesiek, S., and Bax, A. (1993). Amino acid type determination in the sequential assignment procedure of uniformly 13C/15N-enriched proteins. *J. Biomol. NMR* **3,** 185–204.

Grzesiek, S., Anglister, J., and Bax, A. (1993). Correlation of backbone amide and aliphatic side-chain resonances in 13C/15N-enriched proteins by isotropic mixing of 13C magnetization. *J. Magn. Reson.* **101,** 114–119.

Gschwind, R. M., Gemmecker, G., and Kessler, H. (1998). A spin system labeled and highly resolved ed-H(CCO)NH-TOSCY experiment for the facilitated assignment of proton side chains in partially deuterated samples. *J. Biomol. NMR* **11,** 191–198.

Güntert, P., Dötsch, V., Wider, G., and Wüthrich, K. (1992). Processing of multi-dimensional NMR data with the new software PROSA. *J. Biomol. NMR* **2,** 619–629.

Güntert, P., Mumenthaler, C., and Wüthrich, K. (1997). Torsion angle dynamics for NMR structure calculation with the new program DYANA. *J. Mol. Biol.* **273,** 283–298.

Güntert, P., Salzmann, M., Braun, D., and Wüthrich, K. (2000). Sequence-specific NMR assignment of proteins by global fragment mapping with the program MAPPER. *J. Biomol. NMR* **18,** 129–137.

Heinemann, U., Frevert, J., Hofmann, K., Illing, G., Maurer, C., Oschkinat, H., and Saenger, W. (2000). An integrated approach to structural genomics. *Prog. Biophys. Mol. Biol.* **73,** 347–362.

Helgstrand, M., Kraulis, P., Allard, P., and Hard, T. (2000). Ansig for Windows: An interactive computer program for semiautomatic assignment of protein NMR spectra. *J. Biomol. NMR* **18,** 329–336.

Herrmann, T., Güntert, P., and Wüthrich, K. (2002a). Protein NMR structure determination with automated NOE-identification in the NOESY spectra using the new software ATNOS. *J. Biomol. NMR* **24,** 171–189.

Herrmann, T., Güntert, P., and Wüthrich, K. (2002b). Protein NMR structure determination with automated NOE assignment using the new software CANDID and the torsion angle dynamics algorithm DYANA. *J. Mol. Biol.* **319,** 209–227.

Huang, Y. J., Powers, R., and Montelione, G. T. (2005). Protein NMR recall, precision, and F-measure scores (RPF scores): Structure quality assessment measures based on information retrieval statistics. *J. Am. Chem. Sec.* In press.

Huang, Y. J., Swapna, G. V., Rajan, P. K., Ke, H., Xia, B., Shukla, K., Inouye, M., and Montelione, G. T. (2003). Solution NMR structure of ribosome-binding factor A (RbfA), a cold-shock adaptation protein from *Escherichia coli. J. Mol. Biol.* **327,** 521–536.

Hyberts, S. G., and Wagner, G. (2003). IBIS–a tool for automated sequential assignment of protein spectra from triple resonance experiments. *J. Biomol. NMR* **26,** 335–344.

Ikura, M., Kay, L. E., and Bax, A. (1990). A novel approach for sequential assignment of 1H, 13C, and 15N spectra of proteins: Heteronuclear triple-resonance three-dimensional NMR spectroscopy. Application to calmodulin. *Biochemistry* **29,** 4659–4667.

Janin, J., and Chothia, C. (1980). Packing of alpha-helices onto beta-pleated sheets and the anatomy of alpha/beta proteins. *J. Mol. Biol.* **143,** 95–128.

Kay, L. E., Ikura, M., and Bax, A. (1990). Proton–proton correlation via carbon–carbon coupling: A three-dimensional NMR approach for the assignment of aliphatic resonances in proteins labeled with carbon-13. *J. Am. Chem. Soc.* **112,** 888–889.

Kennedy, M. A., Montelione, G. T., Arrowsmith, C. H., and Markley, J. L. (2002). Role for NMR in structural genomics. *J. Struct. Funct. Genomics* **2**, 155–169.

Koehl, P. (1999). Linear prediction spectral analysis of NMR data. *Prog. NMR Spectrosc.* **34**, 257.

Koradi, R., Billeter, M., Engeli, M., Güntert, P., and Wüthrich, K. (1998). Automated peak picking and peak integration in macromolecular NMR spectra using AUTOPSY. *J. Magn. Reson.* **135**, 288–297.

Kuszewski, J., Schwieters, C. D., Garrett, D. S., Byrd, R. A., Tjandra, N., and Clore, G. M. (2004). Completely automated, highly error-tolerant macromolecular structure determination from multidimensional nuclear overhauser enhancement spectra and chemical shift assignments. *J. Am. Chem. Soc.* **26**, 6258–6273.

Laskowski, R. A., Rullmann, J. A., MacArthur, M. W., Kaptein, R., and Thornton, J. M. (1996). AQUA and PROCHECK-NMR: Programs for checking the quality of protein structures solved by NMR. *J. Biomol. NMR* **8**, 477–486.

Leutner, M., Gschwind, R. M., Liermann, J., Schwarz, C., Gemmecker, G., and Kessler, H. (1998). Automated backbone assignment of labeled proteins using the threshold accepting algorithm. *J. Biomol. NMR* **11**, 31–43.

Li, K. B., and Sanctuary, B. C. (1997). Automated resonance assignment of proteins using heteronuclear 3D NMR. 2. Side chain and sequence-specific assignment. *J. Chem. Inf. Comput. Sci.* **37**, 467–477.

Lin, Y., and Wagner, G. (1999). Efficient side-chain and backbone assignment in large proteins: Application to tGCN5. *J. Biomol. NMR* **15**, 227–239.

Logan, T. M., Olejniczak, E. T., Xu, R. X., and Fesik, S. W. (1992). Side chain and backbone assignments in isotopically labeled proteins from two heteronuclear triple resonance experiments. *FEBS Lett.* **314**, 413–418.

Lukin, J. A., Gove, A. P., Talukdar, S. N., and Ho, C. (1997). Automated probabilistic method for assigning backbone resonances of (13C, 15N)-labeled proteins. *J. Biomol. NMR* **9**, 151–166.

Makokha, M., Huang, Y. J., Montelione, G. T., Edison, A. S., and Barbar, E. (2004). The solution structure of the pH-induced monomeric dyein light chain LC8 from Drosophilla. *Protein Sci.* **13**, 727–734.

Malmodin, D., Papavoine, C. H., and Billeter, M. (2003). Fully automated sequence-specific resonance assignments of hetero-nuclear protein spectra. *J. Biomol. NMR* **27**, 69–79.

Monleon, D., Colson, K., Moseley, H. N., Anklin, C., Oswald, R., Szyperski, T., and Montelione, G. T. (2002). Rapid analysis of protein backbone resonance assignments using cryogenic probes, a distributed Linux-based computing architecture, and an integrated set of spectral analysis tools. *J. Struct. Funct. Genomics* **2**, 93–101.

Montelione, G. T., Lyons, B. A., Emerson, S. D., and Tashiro, M. J. (1992). An efficient triple resonance experiment using carbon-13 isotropic mixing for determining sequence-specific resonance assignments of isotopically enriched proteins. *J. Am. Chem. Soc.* **114**, 10974–10975.

Montelione, G. T., Rios, C. B., Swapna, G. V. T., and Zimmerman, D. E. (1999). Biological magnetic resonance. *In* "NMR Pulse Sequences and Computational Approaches for Automated Analysis of Sequence-Specific Backbone Resonance Assignments in Proteins" (E. Berliner and N. R. Krishna, eds.), Vol. 17, pp. 81–130. Plenum, New York.

Moseley, H. N., and Montelione, G. T. (1999). Automated analysis of NMR assignments and structures for proteins. *Curr. Opin. Struct. Biol.* **9**, 635–642.

Moseley, H. N., Monleon, D., and Montelione, G. T. (2001). Automatic determination of protein backbone resonance assignments from triple resonance nuclear magnetic resonance data. *Methods Enzymol.* **339,** 91–108.

Moseley, H. N., Sahota, G., and Montelione, G. T. (2004). Assignment validation software suite for the evaluation and presentation of protein resonance assignment data. *J. Biomol. NMR* **28,** 341–355.

Moy, F. J., Seddon, A. P., Bohlen, P., and Powers, R. (1996). High-resolution solution structure of basic fibroblast growth factor determined by multidimensional heteronuclear magnetic resonance spectroscopy. *Biochemistry* **35,** 13552–13561.

Mumenthaler, C., and Braun, W. (1995). Automated assignment of simulated and experimental NOESY spectra of proteins by feedback filtering and self-correcting distance geometry. *J. Mol. Biol.* **254,** 465–480.

Mumenthaler, C., Güntert, P., Braun, W., and Wüthrich, K. (1997). Automated combined assignment of NOESY spectra and three-dimensional protein structure determination. *J. Biomol. NMR* **10,** 351–362.

Nilges, M. (1995). Calculation of protein structures with ambiguous distance restraints. Automated assignment of ambiguous NOE crosspeaks and disulphide connectivities. *J. Mol. Biol.* **245,** 645–660.

Nilges, M., Macias, M. J., O'Donoghue, S. I., and Oschkinat, H. (1997). Automated NOESY interpretation with ambiguous distance restraints: The refined NMR solution structure of the pleckstrin homology domain from beta-spectrin. *J. Mol. Biol.* **269,** 408–422.

Orekhov, V. Y., Ibraghimov, I. V., and Billeter, M. (2001). MUNIN: A new approach to multi-dimensional NMR spectra interpretation. *J. Biomol. NMR* **20,** 49–60.

Pervushin, K., Riek, R., Wider, G., and Wüthrich, K. (1997). Attenuated T2 relaxation by mutual cancellation of dipole-dipole coupling and chemical shift anisotropy indicates an avenue to NMR structures of very large biological macromolecules in solution. *Proc. Natl. Acad. Sci. USA* **94,** 12366–12371.

Ramelot, T. A., Ni, S., Goldsmith-Fischman, S., Cort, J. R., Honig, B., and Kennedy, M. A. (2003). Solution structure of Vibrio cholerae protein VC0424: A variation of the ferredoxin-like fold. *Protein Sci.* **12,** 1556–1561.

Schwieters, C. D., Kuszewski, J. J., Tjandra, N., and Clore, M. G. (2003). The Xplor-NIH NMR molecular structure determination package. *J. Magn. Reson.* **160,** 65–73.

Seavey, B. R., Farr, E. A., Westler, W. M., and Markley, J. L. (1991). A relational database for sequence-specific protein NMR data. *J. Biomol. NMR* **1,** 217–236.

Szyperski, T., Wider, G., Bushweller, J. H., and Wüthrich, K. (1993). Reduced dimensionality in triple resonance experiments. *J. Am. Chem. Soc.* **115,** 9307–9308.

Szyperski, T., Braun, D., Banecki, B., and Wüthrich, K. (1996). Useful information from axial peak magnetization in projected NMR experiments. *J. Am. Chem. Soc.* **118,** 8147–8148.

Szyperski, T., Yeh, D. C., Sukumaran, D. K., Moseley, H. N., and Montelione, G. T. (2002). Reduced-dimensionality NMR spectroscopy for high-throughput protein resonance assignment. *Proc. Natl. Acad. Sci. USA* **99,** 8009–8014.

Tejero, R., Monleon, D., Celda, B., Powers, R., and Montelione, G. T. (1999). HYPER: A hierarchical algorithm for automatic determination of protein dihedral-angle constraints and stereospecific C beta H2 resonance assignments from NMR data. *J. Biomol. NMR* **15,** 251–264.

Terwilliger, T. C. (2000). Structural genomics in North America. *Nat. Struct. Biol.* **7,** 935–939.

Tjandra, N., and Bax, A. (1997). Direct measurement of distances and angles in biomolecules by NMR in a dilute liquid crystalline medium. *Science* **278,** 1111–1114.

Vriend, G. (1990). WHAT IF: A molecular modeling and drug design program. *J. Mol. Graph.* **8,** 52–56.

Wagner, G., and Wüthrich, K. (1982). Sequential resonance assignments in protein 1H nuclear magnetic resonance spectra. Basic pancreatic trypsin inhibitor. *J. Mol. Biol.* **155,** 347–366.

Westbrook, J., Feng, Z., Burkhardt, K., and Berman, H. M. (2003). Validation of protein structures for the protein data bank. *Methods Enzymol.* **374,** 370–385.

Wishart, D. S., and Sykes, B. D. (1994). The 13C chemical-shift index: A simple method for the identification of protein secondary structure using 13C chemical-shift data. *J. Biomol. NMR* **4,** 171–180.

Wishart, D. S., Bigam, C. G., Yao, J., Abildgaard, F., Dyson, H. J., Oldfield, E., Markley, J. L., and Sykes, B. D. (1995). 1H, 13C and 15N chemical shift referencing in biomolecular NMR. *J. Biomol. NMR* **6,** 135–140.

Word, J. M., Bateman, R. C., Jr., Presley, B. K., Lovell, S. C., and Richardson, D. C. (2000). Exploring steric constraints on protein mutations using MAGE/PROBE. *Protein Sci.* **9,** 2251–2259.

Wüthrich, K. (1986). "NMR of Proteins and Nucleic Acids." John Wiley, New York.

Wüthrich, K., Billeter, M., and Braun, W. (1983). Pseudo-structures for the 20 common amino acids for use in studies of protein conformations by measurements of intramolecular proton-proton distance constraints with nuclear magnetic resonance. *J. Mol. Biol.* **169,** 949–961.

Yokoyama, S., Hirota, H., Kigawa, T., Yabuki, T., Shirouzu, M., Terada, T., Ito, Y., Matsuo, Y., Kuroda, Y., Nishimura, Y., Kyogoku, Y., Miki, K., Masui, R., and Kuramitsu, S. (2000). Structural genomics projects in Japan. *Nat. Struct. Biol.* **7,** 943–945.

Zhang, H., Neal, S., and Wishart, D. S. (2003). RefDB: A database of uniformly referenced protein chemical shifts. *J. Biomol. NMR* **25,** 173–195.

Zheng, D., Huang, Y. J., Moseley, H. N., Xiao, R., Aramini, J., Swapna, G. V., and Montelione, G. T. (2003). Automated protein fold determination using a minimal NMR constraint strategy. *Protein Sci.* **12,** 1232–1246.

Zhu, L., Dyson, H. J., and Wright, P. E. (1998). A NOESY-HSQC simulation program, SPIRIT. *J. Biomol. NMR* **11,** 17–29.

Zhu, X., Komiya, H., Chirino, A., Faham, S., Fox, G. M., Arakawar, T., Hsu, B. T., and Rees, D. C. (1991). Three-dimensional structures of acidic and basic fibroblast growth factors. *Science* **251,** 90–93.

Zimmerman, D. E., and Montelione, G. T. (1995). Automated analysis of nuclear magnetic resonance assignments for proteins. *Curr. Opin. Struct. Biol.* **5,** 664–673.

Zimmerman, D., Kulikowski, C., Wang, L., Lyons, B., and Montelione, G. T. (1994). Automated sequencing of amino acid spin systems in proteins using multidimensional HCC(CO)NH-TOCSY spectroscopy and constraint propagation methods from artificial intelligence. *J. Biomol. NMR* **4,** 241–256.

Zimmerman, D. E., Kulikowski, C. A., Huang, Y., Feng, W., Tashiro, M., Shimotakahara, S., Chien, C., Powers, R., and Montelione, G. T. (1997). Automated analysis of protein NMR assignments using methods from artificial intelligence. *J. Mol. Biol.* **269,** 592–610.

Zolnai, Z., Lee, P. T., Li, J., Chapman, M. R., Newman, C. S., Phillips, G. N., Jr., Rayment, I., Ulrich, E. L., Volkman, B. F., and Markley, J. L. (2003). Project management system for structural and functional proteomics: Sesame. *J. Struct. Funct. Genomics* **4,** 11–23.

[6] Rapid Assessment of Protein Structural Stability and Fold Validation via NMR

By Bernd Hoffmann, Christian Eichmüller,
Othmar Steinhauser, and Robert Konrat

Abstract

In structural proteomics, it is necessary to efficiently screen in a high-throughput manner for the presence of stable structures in proteins that can be subjected to subsequent structure determination by X-ray or NMR spectroscopy. Here we illustrate that the ^{1}H chemical distribution in a protein as detected by ^{1}H NMR spectroscopy can be used to probe protein structural stability (e.g., the presence of stable protein structures) of proteins in solution. Based on experimental data obtained on well-structured proteins and proteins that exist in a molten globule state or a partially folded α-helical state, a well-defined threshold exists that can be used as a quantitative benchmark for protein structural stability (e.g., foldedness) in solution. Additionally, in this chapter we describe a largely automated strategy for rapid fold validation and structure-based backbone signal assignment. Our methodology is based on a limited number of NMR experiments (e.g., HNCA and 3D NOESY-HSQC) and performs a Monte Carlo–type optimization. The novel feature of the method is the opportunity to screen for structural fragments (e.g., template scanning). The performance of this new validation tool is demonstrated with applications to a diverse set of proteins.

Introduction

The genome sequencing projects are delivering vast amounts of protein sequences encoding functionally important proteins, which are putative protein therapeutics and/or targets for the pharmaceutical industry. The concept of "structural proteomics" or "structural genomics" [e.g., the elucidation of the three-dimensional (3D) structures of the encoded proteins] is based on the empirical finding that protein function cannot always be deduced from the primary sequence but is coded in its 3D shape (Jones and Thornton, 1997; Kasuya and Thornton, 1999; Russell, 1998; Russel et al., 1998; Thornton et al., 1991). Beyond that, structural proteomics efforts will also enlarge the database of known protein structures and provide a sufficiently large basis set of structures to allow for an efficient

Copyright 2005, Elsevier Inc.
All rights reserved.
0076-6879/05 $35.00

determination of structure based on homology modeling techniques (Karplus *et al.*, 1999; Koppensteiner *et al.*, 2000; Ota *et al.*, 1999; Sander and Schneider, 1991; Sippl and Weitckus, 1992). To date protein structures are determined either by X-ray crystallography or nuclear magnetic resonance (NMR) spectroscopy. One important issue in large-scale structural proteomics is target selection or the identification of promising proteins suitable for determination of structure.

The ^{1}H NMR chemical shifts are governed by the details of the 3D solution structures of proteins. Although an enormous amount of ^{1}H experimental data exist that underscores this relationship (Seavey *et al.*, 1991), ^{1}H chemical shift information was mainly used as a prerequisite for assignment of relevant structural constraints [e.g., distance-dependent nuclear Overhauser enhancement (NOE) or dihedral angle constraints] (Wüthrich, 1986). Here we use the statistics of the ^{1}H chemical shift distribution to probe protein structural stability in solution. The method uses the autocorrelation function of the ^{1}H spectra of proteins, which are easily obtained and do not require isotope labeling of the proteins. We demonstrate that a significant correlation exists between the autocorrelation function, the topological complexity (expressed as the relative contact order), and protein structural stability of the protein. Data obtained on a diverse set of folded proteins with native structures and partially folded proteins [e.g., the molten globule state of α-lactalbumin (Kuwajima, 1996) and the partially folded oncogenic transcription factor v-Myc] (Fieber *et al.*, 2001) demonstrate that the method can be used to efficiently screen for protein structural stability in a high-throughput manner, with possible beneficial applications to large-scale structural genomics efforts (Kim, 1998) currently underway in the United States (Terwilliger, 2000), Europe (Heinemann, 2000), and Japan (Yokoyama *et al.*, 2000). Additionally, the statistical significance of the observed empirical correlation can also be used to study tertiary structural features of proteins without the need of tedious ^{1}H signal assignment. As a first example, the Ca^{2+}-induced fold stabilization in α-lactalbumin is discussed. The sensitive dependence of ^{1}H chemical shift distribution in the low contact order regime (e.g., partially folded and/or unfolded proteins) also suggests fruitful applications to protein folding studies.

NMR spectroscopy continues to make significant contributions in the challenging area of structural genomics (Prestegard *et al.*, 2001; Staunton *et al.*, 2003), and even high-throughput applications are becoming feasible. This growing impact is due to recent advances in protein preparations, spectrometer hardware, data analysis, and pulse sequence developments. One of the most time-consuming bottlenecks in the process of structure elucidation by NMR is the signal assignment of backbone and side chain

^{1}H, ^{13}C, and ^{15}N resonances, which is a prerequisite for the subsequent gathering of information about protein structure, dynamics, and intermolecular interactions from NMR spectra. Ongoing progress in the development of more powerful spectrometer equipment and pulse sequences has been accompanied by increasing efforts to partly or fully automate the signal assignment procedure. In recent years, numerous research groups reported the development of assignment programs or software packages. A detailed description of these methods is beyond the scope of this chapter. Instead, we refer to an exhaustive review by Gronwald and Kalbitzer (2004) and references therein. The signal assignment process can be subdivided into several steps: (1) grouping of resonances from one or more spectra to spin systems, (2) association of spin systems with amino acid types, (3) linking of spin systems to shorter or longer fragments, and (4) mapping of fragments to the primary sequence. Although some of the reported programs concentrate on one of these steps, others tackle several steps at once. The underlying tools and procedures to accomplish these tasks include tree search algorithms, best-first deterministic approaches, exhaustive searches, genetic algorithms, threshold accepting, Monte Carlo simulations coupled with energy minimization algorithms, neural networks, and others. Most of the programs, in particular those for assignment of larger proteins, rely on a specific set of (numerous) two-dimensional (2D) and 3D NMR spectra or a considerable minimum amount of NMR data to produce reliable results. Therefore, assignment programs are often quite demanding in terms of spectrometer time necessary to acquire sufficient input data. In addition, existing assignment programs are sensitive to missing or incorrect input data (resulting from signal overlap, relaxation processes, noise, and artifacts) and fail to find the correct assignment under nonideal conditions.

In this chapter, we present a new tool for structure-based signal assignment and protein fold validation. It requires minimal NMR data input and the existence of a structure homologue. Although it is reminiscent of existing NMR software packages (Hitchens *et al.*, 2003), it is novel as it also allows for screening for structural fragments (e.g., template scanning).

Materials and Methods

Fold Stability Analysis

The following protein samples were used in this study: α-lactalbumin (Acharya *et al.*, 1991), lysozyme (Diamond, 1974), MutS (Tollinger *et al.*, 1998), creatine kinase (Rao *et al.*, 1998), ubiquitin (Vijay-Kumar *et al.*, 1987), bovine pancreatic trypsin inhibitor (BPTI) (Parkin *et al.*, 1996), myoglobin

(Maurus *et al.*, 1998), v-Myc (Fieber *et al.*, 2001), and bovine serum albumin (BSA) (Janatova *et al.*, 1968). Lysozyme, α-lactalbumin, creatine kinase, BPTI, myoglobin, and BSA were purchased from Sigma (St. Louis, MO) and used without further purification, while v-Myc (Fieber *et al.*, 2001) and MutS (Tollinger *et al.*, 1998) were prepared as described previously. Ca^{2+}-depleted α-lactalbumin was prepared by overnight dialysis using a buffer at pH 1.5 and subsequently refolded by adjusting the pH to 6.5. The molten globule state of α-lactalbumin was prepared by adjusting the pH of the protein solution to 2.5. The pH of the protein solution was carefully controlled with a pH meter. All NMR experiments were performed on a Varian UNITY*Plus* 500-MHz spectrometer equipped with a pulse field gradient unit and a triple resonance probe with actively shielded z gradients. All spectra were recorded at 26°. Water suppression was achieved with a presaturation and WATERGATE detection scheme. For the $^{13}C,^{15}N$-labeled proteins ubiquitin and MutS the first trace [omitting the nuclear Overhauser enhancement spectroscopy (NOESY) mixing period and with ^{13}C, ^{15}N-decoupling during acquisition] of a $^{13}C,^{15}N$-NOESY-hetero nuclear single-quantum correlation (HSQC) (Pascal *et al.*, 1994) spectrum was used.

In contrast to the previously published application of random matrix theory to the statistical analysis of protein ^1H chemical shifts (Lacelle, 1984), the applied statistical analysis used the autocorrelation function of protein one-dimensional (1D) ^1H spectra. NMR spectra were processed and analyzed using NMRPipe (Delaglio *et al.*, 1995) software. Acquisition parameters were as follows: spectral width, 12,000 Hz; number of spectral points, 11,392 for 1D ^1H spectra and 1536 for spectra acquired using a $^{13}C,^{15}N$-NOESY-HSQC, respectively; zero filling, 24 K. However, we have demonstrated that the exact number of spectral points does not influence the outcome of the statistical analysis (data not shown). Residual water was eliminated by deleting the spectral region 4.90–4.55 ppm. To eliminate possible errors introduced by the elimination of the spectral region around the water resonance, the autocorrelation function was calculated for several spectra in which different spectral regions (around the water resonance frequency) were eliminated. No changes in the autocorrelation function $C(\omega)$ were observed. Intensities were extracted from the 1D ^1H spectra with a perl script using the function nLinLS provided with NMRPipe (Delaglio *et al.*, 1995) (and calculated as integrals over 10 data points). The ^1H spectrum for the theoretical random coil peptide was calculated using the sequence of α-lactalbumin and the published random coil shifts for short peptides (Wishart *et al.*, 1995). Shifts for each proton were additionally randomized within ± 0.02 ppm. The resulting data files were used to calculate the autocorrelation functions. The obtained autocorrelation functions,

$C(\omega)$, were normalized to the value at the smallest available frequency difference (0.01 ppm). The raw data were incorporated into the program package xmgr and numerically averaged (averaging window, 50 data points). The values of the autocorrelation function at frequency 0.5 ppm, $C(0.5)$ were used as measures of protein structural stability.

Contact orders were determined from structural coordinates in the Protein Data Bank (Berman *et al.*, 2000). Relative contact orders were calculated according to the published procedure by Baker and co-workers (Plaxco *et al.*, 1998) (see also http://depts.washington.edu/bakerpg). The contact order for the partially folded v-Myc protein was calculated based on the solution structure, which revealed an α-helical conformation for the leucine zipper region comprising residues 384–411 (Fieber *et al.*, 2001). The unfolded segments of v-Myc were taken as random coils and thus neglected for the calculation of the relative contact order.

Fold Validation

Cross-peaks are automatically picked in the HNCA and ^{15}N-NOESY-HSQC (Cavanagh *et al.*, 1996) spectra employing Nmrview software (Johnson and Blevins, 1994). The peak picking in the ^{15}N-NOESY-HSQC is restricted to the H^N–H^N NOEs in the ^{1}H spectral window from 6 to 12 ppm. Artifacts and noise peaks are deleted manually. In addition, an in-house written software tool is used to filter out cross-peaks arising from J-coupled asparagine and glutamine side chain amide resonances.

HNCA cross-peaks are then grouped into individual spin systems (i), with (i) being an arbitrary reference number. Cross-peaks that are separated by less than the digital resolution ($\sim$0.2 ppm in the ^{15}N dimension and less than $\sim$0.02 ppm in the direct ^{1}H dimension) are assumed to belong to the same spin system. The more intense cross-peak is assigned to the $C\alpha(i)$ nucleus, whereas the less intense signal is attributed to the $C\alpha(i-1)$ nucleus. The observation of more than two aligned cross-peaks is indicative for overlapping residues with degenerate ^{1}H and ^{15}N backbone amide frequencies. In these cases $C\alpha(i)$ and $C\alpha(i-1)$ resonances cannot be distinguished and hence these chemical shifts are not included in the input shift table. If only one (^{1}H, ^{15}N, ^{13}C) correlation is observed within boundaries of digital resolution, the $C\alpha(i-1)$ chemical shift is assumed to coincide with the $C\alpha(i)$ chemical shift. The collection of a supplementary HN(CO)CA (Cavanagh *et al.*, 1996) dataset is recommended to obtain a complete and correct input shift list with clear discrimination of $C\alpha(i)$ and $C\alpha(i-1)$ resonances.

The arbitrary reference numbers attributed to spin systems detected in the HNCA experiment are transferred to residues observed in the

^{15}N-NOESY-HSQC spectrum and a list including all potential H^N–H^N NOEs is generated. The identification of the dipolar coupling partner of a specific H^N–H^N NOE (preliminary assignment in F2) is achieved in the following fully automated manner: (1) It is checked for each individual H^N–H^N NOE arising from dipolar interaction between residues i and j and with chemical shift coordinates (^1H_i/^1H_j/$^{15}N_i$) whether a symmetric NOE exists at the position ($^1H_j \pm 0.03$/$^1H_i \pm 0.03$/$^{15}N_j$). If only one symmetric NOE partner is found in the NOE list, the residue j is in all likelihood the dipolar coupling partner of residue i. (2) If multiple symmetric NOEs are found, no clear assignment is feasible in the indirect dimension F2; all residues giving rise to the symmetric NOE represent potential dipolar coupling partners. The intensity of the symmetric NOE (j/i) with respect to the NOE (i/j) is neglected in this analysis for the sake of simplicity. The NOE (i/j) is then duplicated according to the number of potential coupling partners j, whose preliminary reference numbers are assigned to the F2 dimension of each of these duplicated NOEs. Although, with this procedure, wrong NOEs are included in the NOE input list, at least one of the "cloned" NOEs will have the correct assignment. (3) If for a specific NOE at the position (^1H_i/^1H_j/$^{15}N_i$) no symmetric NOE (^1H_j/^1H_i/$^{15}N_j$) is found, potential coupling partners can be unraveled by inspecting the $^1H^N$ chemical shifts of all experimentally observed amino acids. The NOE (i/j) is again multiplied in the NOE input list according to the number of residues j having an H^N chemical shift of $^1H_j \pm 0.03$ ppm and the reference numbers of these residues are assigned to the F2 dimension of the NOE (i/j). (4) If no potential dipolar coupling partner is found in steps (1) to (3) (i.e., NOE between a backbone amide proton and an aromatic side chain proton), the NOE (i/j) is omitted from the H^N–H^N NOE input list.

As a result of this procedure, two input files are obtained. The first one contains all experimentally observed residues with their arbitrary reference numbers as well as their backbone $C\alpha(i)$ and $C\alpha(i-1)$ chemical shifts. The second input file lists all potential H^N–H^N NOEs that are observed among the query protein residues.

Prediction of Query Protein Chemical Shifts and NOEs

Chemical shifts of the homology model were either obtained by taking chemical shift values deposited in the BMRB database (Doreleijers *et al.*, 2003) or by shift prediction employing ShiftX software (Neal *et al.*, 2003) (see Table II). Homology model secondary chemical shifts are calculated by subtracting random coil shifts from H, N, and $C\alpha$ chemical shifts. These secondary shifts are subsequently added sequence specifically to the query protein random coil shifts according to the sequence alignment between homology model and query protein. This yields chemical shift predictions for the query protein. Query protein H^N–H^N NOEs are predicted by computing all pairs of backbone amide protons with distances shorter than 5 Å from the homology model atom coordinates.

Monte Carlo Simulation

The Monte Carlo simulation (Metropolis *et al.*, 1953) attempts to find the best global mapping of experimentally observed spin systems onto the query protein primary sequence. A start configuration is generated by randomly assigning experimentally observed residues to residue positions in the primary sequence. The program is able to handle unoccupied sequence positions that occur when the number of experimentally observed residues is smaller than the total number of query protein residues. Multiple random changes are generated by choosing two query protein sequence positions A and B and by exchanging the experimentally observed residues characterized by their chemical shift values and H^N–H^N NOEs between both positions. After each Monte Carlo step the objective function E (analog of energy) is evaluated with respect to its value before the rearrangement. The random change proposed to the system is accepted or rejected according to the Metropolis criterion, i.e., if $E_2 \leq E_1$, the step is necessarily accepted; if $E_2 > E_1$, the step is accepted with the probability of $p = \exp(E_1 - E_2)/kT$, with $k = 1$. The start temperature T is set to a value that is considerably larger than the largest ΔE normally encountered. The temperature is held constant for several thousand Monte Carlo steps and is then lowered in multiplicative steps, each amounting to a 1–5% decrease in T with respect to the previous temperature value. When T has reached a value where further efforts to reduce the objective function E become inconclusive, the first cycle of the Monte Carlo simulation is stopped. The uniqueness of assignment is assessed by running 10–20 independent Monte Carlo assignment cycles.

In our approach the objective function E is defined as $E = -\log P$, with P being an overall probability scoring value:

$$P = \text{TAN} \cdot \exp(-f_1 \cdot \phi\text{RMSD}_1) \cdot \exp(-f_2 \cdot \phi\text{RMSD}_2) \cdot \text{CA} \qquad (1)$$

The Tanimoto coefficient TAN is a measure of the number of experimentally observed NOEs that coincide with predicted NOEs for a given tentative assignment. It is defined as $c \cdot w/(a + b - c)$, where a and b are the number of experimentally observed and predicted NOEs, respectively, and c is the number of matching NOEs in both input lists A and B. The weighing factor $w = b/a$ ensures that TAN can reach its maximum value of 1 even if a does not equal b. Although the Tanimoto coefficient forces the system into configurations with a maximum number of coinciding experimental and predicted H^N–H^N NOEs, the second term $\exp(-f_1 \cdot \phi RMSD_1)$ with

$$\phi RMSD_1 = \left\{ \sum (\Delta C\alpha_{i,k}^2 + \Delta C\alpha_{i-1,k-1}^2 + \Delta C\alpha_{j,i}^2 + \Delta C\alpha_{j-1,l-1}^2) \right\}^{1/2} / c \quad (2)$$

ensures that the average root mean square deviation (RMSD) of the four query protein $C\alpha$ chemical shifts $C\alpha_i$, $C\alpha_{i-1}$, $C\alpha_j$, and $C\alpha_{j-1}$ and the corresponding predicted shifts $C\alpha_k$, $C\alpha_{k-1}$, $C\alpha_l$, and $C\alpha_{l-1}$ is minimized for all c coinciding pairs of experimental and predicted NOEs, with i/j and k/l being dipolar-coupled partners in the query protein and homology model, respectively. Query protein residues whose H^N–H^N NOEs do not coincide with predicted NOEs at a specific Monte Carlo step or with no detectable H^N–H^N NOEs do not contribute to the term $\phi RMSD_1$. To account for these residues, the third factor $\exp(-f_2 \cdot \phi RMSD_2)$ with

$$\phi RMSD_2 = \left\{ \sum (\Delta C\alpha_{m,n}^2 + \Delta C\alpha_{m-1,n-1}^2) \right\}^{1/2} / z \quad (3)$$

is introduced into the probability scoring function [Eq. (1)]. This term is a measure for the overall matching of experimental shifts $C\alpha_m$ and $C\alpha_{m-1}$ with the corresponding predicted shifts $C\alpha_n$ and $C\alpha_{n-1}$ in a specific configuration. The expression $\phi RMSD_2$ forces the z experimentally observed spin systems to move toward configurations with an overall good match of experimental and predicted $C\alpha$ chemical shifts. The factors f_1 and f_2 in Eq. (1) represent empirically determined weighing factors and were set to $f_1 = f_2$ in our test runs. The factor CA in Eq. (1) represents the percentage of residues whose $C\alpha_{i-1}$ chemical shifts match the $C\alpha_l$ shifts of the predecessor within a user-defined tolerance value (0.15 ppm). If the total number of experimentally observed residues is smaller than the number of residues in the query protein sequence, a constant number of residue positions remain "unoccupied" throughout the Monte Carlo simulation. Spin systems with no predecessor or successor are treated as adjacent residues with matching sequential $C\alpha$ chemical shifts.

Our assignment and structure validation software is written in the programming language "C." The program is streamlined with respect to CPU time requirements. This is achieved by avoiding noninteger arithmetic and by outsourcing the most time-consuming computational steps from the actual Monte Carlo/simulated annealing procedure. To this end, lookup values contributing to the $\phi RMSD_1$, $\phi RMSD_2$, and CA terms of the probability scoring function [Eq. (1)] are computed in advance for all combinations of scalar- and/or dipolar-coupled residue pairs that may be encountered at any of the query protein sequence positions in the course of the subsequent Monte Carlo run. In addition to these measures, after each Monte Carlo step, the scoring function is not evaluated from scratch, i.e., by summing up the contributions of all residues in that particular tentative assignment. Instead, only changes induced by those residues subjected to the random change are calculated. Our test calculations were executed on a Pentium-grade Linux PC performing $\sim$30,000 Monte Carlo steps per second. The required CPU time for 20 independent assignment runs ranges from $\sim$30 min for medium sized proteins ($\sim$150 residues) to 2 h for MBP.

Results

Fold Stability

An outline of the method for rapid assessment of protein stability is illustrated in Fig. 1. The starting point is a conventional protein ^{1}H 1D spectrum in which the residual water is eliminated by simply deleting the spectral region 4.90–4.55 ppm. This data file is used to calculate the auto-correlation function $C(\omega)$. The autocorrelation function is the Fourier transform (FT) of the product between the free induction decay (FID) and its complex conjugate and is thus related to the distribution function of the frequency and relaxation rate differences, respectively. It is important to realize that the lack of specific long-range contacts in unfolded states compared to well-structured proteins leads to a significant narrowing of the distribution function. The obtained autocorrelation functions $C(\omega)$ are normalized to the value at the smallest available frequency difference (0.01 ppm) and numerically smoothed. In Fig. 2 typical (smoothed) auto-correlation functions $C(\omega)$ are shown. From inspection of Fig. 2, it is obvious that there is a clear distinction between a well-folded protein with pronounced structural stability and partially folded or unfolded states. The α-lactalbumin molten globule state at pH 2.5 was chosen as an example of a partially folded state (Fig. 2, blue line), and the dashed black line indicates a theoretical autocorrelation function $C(\omega)$ assuming ^{1}H random coil shifts for the protons of α-lactalbumin. We have found that the primary sequence of the protein does not significantly influence $C(\omega)$; thus the random coil data presented in Fig. 2 can be regarded as representative for a completely unfolded protein in solution. It is evident from Fig. 2 that partially folded proteins as evidenced by the α-lactalbumin molten globule at pH 2.5 or v-Myc (Fieber et al., 2001) are remarkably different from native proteins and display a significant reduction of the autocorrelation function $C(\omega)$.

A closer inspection of Fig. 2 reveals that although the overall appearances of the various $C(\omega)$ for folded proteins (Fig. 2, black, red, and green

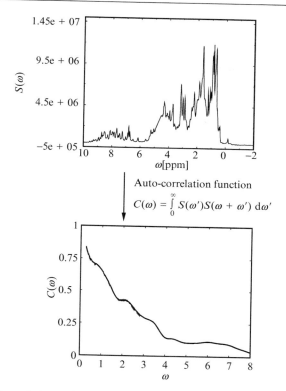

Auto-correlation function

$$C(\omega) = \int\limits_0^\infty S(\omega')S(\omega + \omega')\, d\omega'$$

FIG. 1. Outline of the statistical analysis. The starting point of the method is the experimental protein 1D ^{1}H spectra (consisting typically of 2400 data points or 0.01 spectral resolution), in which a residual water signal is eliminated by discarding the spectral region between 4.55 and 4.90 ppm. The obtained autocorrelation function, $C(\omega)$, is normalized to the value at the smallest available energy difference (0.01 ppm), numerically smoothed (typically by averaging over 50 data points). The value of the autocorrelation function at a frequency difference of 0.5 ppm, $C(0.5)$, is taken as a measure for cooperative structural properties of the proteins and can be used as a quantitative measure of protein structural stability.

lines) are remarkably similar, there are noticeable differences. Specifically, the slight additional maxima for $C(\omega)$ at larger frequencies (around 6 ppm) suggest the possibility to extract structural features of proteins from the autocorrelation function, which is reminiscent of CD spectroscopy. It is interesting to note that the α-lactalbumin molten globule (Fig. 2, blue line) displays this slight additional maximum in the autocorrelation function $C(\omega)$, suggesting that the partly folded molten globule state comprises polypeptide fragments with extended chain conformations. We have made

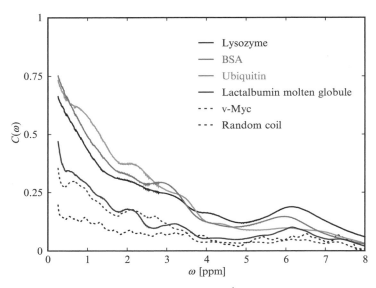

FIG. 2. Autocorrelation functions of protein 1D ^{1}H spectra. The following proteins are shown: lysozyme (black), BSA (red), ubiquitin (green), the molten globule state of α-lactalbumin (blue), the partially folded protein v-Myc (blue, dashed line), and a theoretical random coil polypeptide assuming random coil ^{1}H chemical shifts (black, dashed line). Only energy difference data with $\Delta\omega > 0.25$ ppm are shown (see text). (See color insert.)

similar observations (e.g., additional maxima at larger frequencies) for the β-catenin binding fragment of the T-cell factor-4 (TCF4) for which an extended conformation (in additional to a C-terminal α-helix) was observed in the crystal structure. CD spectroscopy of apo-TCF4, however, indicated a random coil in solution. Preliminary NMR data obtained for ^{13}C,^{15}N-labeled apo-TCF4 also provided evidence for the prevalence of extended local structure elements in solution (data not shown). It thus may be feasible to study partially folded protein states by means of the proposed autocorrelation function analysis. For a completely unfolded state, however (Fig. 2, dashed black line), no additional maxima are observed, which again results from the significantly reduced dispersion observed in 1D ^{1}H spectra of unfolded proteins (Wishart *et al.*, 1995).

To derive an unbiased measure for protein structural stability in solution and given the fact that the autocorrelation function is unknown, we propose the autocorrelation function value $C(\omega)$ at 0.5 ppm, $C(0.5)$, as a benchmark of fold stability. Although we have also tested alternative measures such as information theory, methods of moments, and nonlinear curve-fitting, we prefer to use the $C(0.5)$ value, partly because the $C(0.5)$

value can be related to the heterogeneity of the individual protein ^{1}H resonances. Table I lists $C(0.5)$ values obtained for the various proteins. It can be seen that there is a significant difference between proteins that exhibit a well-defined solution structure (e.g., lysozyme, myoglobin, ubiquitin, creatine kinase, MutS) and proteins that exist in partly folded states (e.g., the α-lactalbumin molten globule at pH 2.5, the oncogenic transcription factor v-Myc). Whereas natively folded proteins display $C(0.5)$ values >0.5, partially folded or unfolded proteins have values of <0.4. A $C(0.5)$ threshold value of 0.4–0.5 thus significantly discriminates between these two regimes.

We then explored whether there is a quantifiable relationship between the native state topology of a protein and the statistics of the ^{1}H chemical shift distribution obtained from the autocorrelation analysis of protein 1D ^{1}H spectra. The topological complexity was specified numerically according to a procedure proposed by Plaxco et al. (1998). We have used the relative contact order, which reflects the relative importance of local and nonlocal residue contacts to the global fold of a protein. The relative contact order, CO, can be interpreted as the average primary sequence

TABLE I

STATISTICAL ANALYSIS OF PROTEIN ^{1}H CHEMICAL SHIFT DISTRIBUTIONS[a]

Protein (PDB code)	$C(0.5)$	CO (%)
Lysozyme (6LYZ)	0.58	11.1
Creatine kinase (2CRK)	0.54	7.5
α-Lactalbumin (1A4V)	0.62	9.7
BPTI (1BPI)	0.60	15.9
Myoglobin (1AZI)	0.72	7.9
MutS (1BE1)	0.57	9.1
Ubiquitin (1UBQ)	0.65	14.9
BSA	0.66	—[b]
v-Myc	0.28	2.0[c]
α-Lactalbumin molten globule	0.34	—[b]
Ubiquitin in 10 M urea	0.32	—[b]
Random coil	0.14	—[b]

[a] The decay of the autocorrelation function $C(\omega)$ is described by its value at a frequency difference of 0.5 ppm (see Materials and Methods). The relative contact order (Plaxco et al., 1998) (CO) is taken as a measure of the topological complexity of proteins.
[b] No structure/contact order available.
[c] Calculated based on the solution structure of v-Myc (Fieber et al., 2001). Only the C-terminal α-helix comprising residues 384–411 were considered.

distance between all pairs of contacting residues along the polypeptide chain (normalized by the total number of residues in the protein). Figure 3 shows the relationship between the topological complexity of the proteins (described by the relative contact order) and the ^{1}H chemical shift distribution described by the $C(0.5)$ value. For example, proteins with compact 3D structures (large relative contact order) display $C(0.5)$ values between about 0.54 and 0.72, respectively. From Fig. 3 it can be seen that the $C(0.5)$ values for the various natively folded proteins are below 0.75, the only exception being myoglobin, which has an attached heme moiety and is thus different compared to the other unligated proteins. Additionally, in the 1D ^{1}H spectra of myoglobin, signals from the bound heme moiety

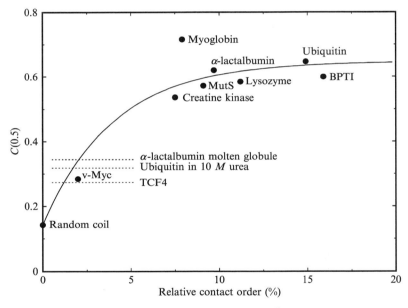

FIG. 3. The relationship between the statistics of protein ^{1}H chemical shifts and protein topology. The autocorrelation function $C(\omega)$ (Fig. 2) was approximated by its value at a frequency of 0.5 ppm, $C(0.5)$. The correlation between the relative contact order of proteins and $C(0.5)$ defines a criterion for the identification of a folded protein. Black symbols depict experimental values; the gray circle denotes the calculated value for a random coil peptide; and the dotted horizontal lines indicate experimental $C(0.5)$ values for the α-lactalbumin molten globule (pH 2.5) and ubiquitin denatured in 10 M urea. For these conformationally flexible proteins no contact order could be calculated. Proteins with $C(0.5) > 0.5$ exhibit a well-defined global fold and exist in a well-structured form in solution. The solid line represents a fit to the experimental data using the analytical function $C(0.5) = A_0 + A_\infty[1 - \exp(-A_1 \cdot \text{CO})]$.

are not suppressed and thus contribute to the observed ^{1}H autocorrelation function. The relationship $C(0.5)$ vs. relative contact order, CO (Fig. 3), presumably reflects the common folding principles of proteins that are based on the chemical similarities of the amino acid building blocks. It also indicates the existence of a threshold value for $C(0.5)$. Most likely, this reflects the upper limit of structural or topological complexity in proteins, which is due to the avoidance of steric clashes of amino acid side chains upon contraction of the polypeptide chain. A detailed understanding of the true relationship between the correlation of energy levels in proteins, as reflected in $C(\omega)$, and the topological complexity of proteins (relative contact order) will provide some (qualitative) insight into the cooperativity of protein structures but is beyond the scope (and not of particular relevance to the proposed applications) of this chapter.

Partially folded proteins or proteins with molten globule-like behavior, however, display significantly smaller $C(0.5)$ values (<0.4). The larger $C(0.5)$ value (0.34) observed for the α-lactalbumin molten globule compared to the partially folded oncogenic transcription factor v-Myc (0.28) suggests more cooperative long-range interactions in this dynamic protein state. Indeed, there is evidence that the molten globule of α-lactalbumin has a native-like overall fold with weak but reasonably well-defined tertiary interactions (Alexandrescu et al., 1993; Baum et al., 1989; Chakraborty et al., 2001; Chyan et al., 1993; Dobson, 1994; Peng and Kim, 1994; Peng et al., 1995; Redfield et al., 1999; Wu et al., 1995). In contrast, v-Myc exists as a partially folded protein displaying a well-defined C-terminal α-helix and a "nascent" helix in the N-terminal basic domain with no evidence for significant long-range order (Fieber et al., 2001).

It is also illuminating to compare our findings on the molten globule state of α-lactalbumin with data obtained using NMR spin diffusion as a probe for protein compactness and residual structure in molten globule states (Griko and Kutyshenko, 1994; Kutyshenko and Cortijo, 2000). The rigidity parameter (G) was introduced as a measure for residual structure in proteins subjected to denaturing conditions, such as temperature, denaturing agents, and changes in pH. G is defined as the intensity ratio between conventional 1D ^{1}H spectra and spin diffusion spectra for certain spectral regions of proteins (e.g., amide, aromatic, and or aliphatic). G values of $\sim$0.1 were obtained for denatured (unfolded) proteins, whereas values of $\sim$0.5 have been found for native and, surprisingly, molten globule states (Griko and Kutyshenko, 1994; Kutyshenko and Cortijo, 2000). This was suggestive of the existence of native-like tertiary structures in molten globules. NOEs and other data also supported the notion that molten globules exist in significantly compact structural ensembles (Balbach et al., 1997; Choy et al., 2001). However, the fact that a native-like spectral appearance is observed does not imply that a molten globule exists as a compact, impermeable sphere (Griko and Kutyshenko, 1994; Kutyshenko and Cortijo, 2000). Our finding that the α-lactalbumin molten globule is significantly less compact and less ordered than well-structured native proteins emphasizes the notion that a molten globule is best described as a native-like but noncooperative assembly of the constituent core regions of the polypeptide chain (Schulman and Kim, 1996; Schulman et al., 1997). The lack of cooperativity in molten globules is observed as a significantly faster decay of the autocorrelation function, described with $C(0.5)$, compared to native proteins (see Fig. 2), which exist as densely packed polypeptide chains of a highly cooperative nature. Interestingly, our findings are also consistent with recent NMR experiments that also demonstrated a noncooperative unfolding of the α-lactalbumin molten globule by probing unfolding events at individual residues (Schulman and Kim, 1996; Schulman et al., 1997). Finally, the dynamic nature of the transiently formed structural ensemble of a molten globule is indicated by effective transverse spin relaxation (e.g., extreme line-broadening due to motional dynamics in the millisecond to microsecond time scale), which typically precludes direct NMR studies of molten globules (Last et al., 2001).

Interestingly, ubiquitin denatured in 10 M urea displays a $C(0.5)$ value of 0.32 similar to the values of the molten globule of α-lactalbumin and v-Myc. The observation of small $C(0.5)$ values is consistent with the notion that the auto correlation function predominantly probes cooperative long-range interactions in well-defined protein folds. It also suggests, however, that urea-denatured ubiquitin exhibits some residual structure. Similar observations (e.g., the prevalence of residual structure in denatured proteins) have been made for the denatured forms of 434-repressor (Neri et al., 1992) and the fragment $\Delta131\Delta$ of staphylococcal nuclease (Shortle and Ackerman, 2001).

Encouraged by the quality of the data, we investigated the possibility of using this analysis to probe protein stability in general and to determine whether the accuracy of the method is sufficiently high to monitor subtle changes of protein structural stability (foldedness) upon, for example, ligand binding. As a first example, we present data obtained on monitoring stability changes of α-lactalbumin upon Ca^{2+} binding. α-Lactalbumin is the regulatory component of the lactose synthase complex that catalyzes the biosynthesis of lactose. It has a bipartite structure and consists of two lobes. The α-domain is composed of four α-helices (and two short 3_{10} helices), whereas the smaller β-domain consists of a triple-stranded antiparallel β-sheet and a 3_{10} helix, linked by a series of loops (Acharya et al., 1991; Calderone et al., 1996; Pike et al., 1996). All known α-lactalbumin crystal structures revealed a conserved Ca^{2+}-binding site, formed by the side chain

β-carboxylate groups of three aspartic acid residues, two backbone carbonyl oxygens, and two bound water molecules (contributing two oxygens to the metal coordination site), which are arranged in a distorted pentagonal bipyramidal coordination sphere (Acharya et al., 1991; Anderson et al., 1997; Calderone et al., 1996; Pike et al., 1996). The apparent K_{Ca} of α-lactalbumin (Wijesinha-Bettoni et al., 2001) is of the order of 10^6–10^7 M^{-1} at physiological pH levels. The impact of Ca^{2+} on α-lactalbumin protein folding has been investigated (Anderson et al., 1997; Troullier et al., 2000; Wijesinha-Bettoni et al., 2001). Upon formation of a loosely defined protein state, Ca^{2+} binding drives the formation of the α-lactalbumin native state, presumably in a cooperative manner (Forge et al., 1999; Kuwajima et al., 1989; Troullier et al., 2000). The structural role of Ca^{2+} and its influence on the stability of α-lactalbumin were also demonstrated by means of hydrogen exchange protection (Wijesinha-Bettoni et al., 2001). It was observed that Ca^{2+} binding stabilizes the structure of native bovine α-lactalbumin; at pH 8 the Ca^{2+}-depleted (apo) form of has a melting point T_m of 34°, compared to 64° for the Ca^{2+}-loaded (holo) form. Although apo α-lactalbumin displays a native-like structure, as inferred from CD, fluorescence, and low-resolution NMR data, and the helical content of apo α-lactalbumin is equal to (or even slightly greater than) holo α-lactalbumin, the hydrogen-exchange results indicated that the Ca^{2+}-binding loop and the C-helix are stabilized in the holo form. Recently, the crystal structure of apo α-lactalbumin was solved, and the X-ray data additionally corroborated the previous finding that Ca^{2+} causes an increase in stability but little structural change (Chrysina et al., 2000; Wijesinha-Bettoni et al., 2001).

The Ca^{2+}-depleted α-lactalbumin was titrated with a concentrated stock solution of $CaCl_2$ until a 10-fold molar excess of Ca^{2+} over α-lactalbumin was reached. Each solution of varying Ca^{2+}/α-lactalbumin concentration ratio was subjected to the analysis of the ^1H chemical shift distribution. The titration curve that is obtained is shown in Fig. 4. It can be seen that the elimination of Ca^{2+} resulted in a significant reduction in the $C(0.5)$ value. The addition of Ca^{2+} leads to an increase in the $C(0.5)$ value. Given the structural similarities of apo and holo α-lactalbumin, the significant increase of $C(0.5)$ reflects the increased protein stability of the Ca^{2+}-loaded form compared to the Ca^{2+}-depleted form of α-lactalbumin. It should be noted that this change in protein stability is not obvious from a simple inspection of the 1D ^1H spectra. The dashed line in Fig. 4 was calculated using the well-known Ca^{2+} association constant of α-lactalbumin, $K_{Ca} = 10^6$ M^{-1} (literature value of $K_{Ca} = 10^6$–10^7 M^{-1} at physiological pH) (Wijesinha-Bettoni et al., 2001). The agreement between the theoretical curve [by using the experimentally obtained $C(0.5)$ values for the Ca^{2+}-depleted and for the Ca^{2+}-saturated form, respectively] and the

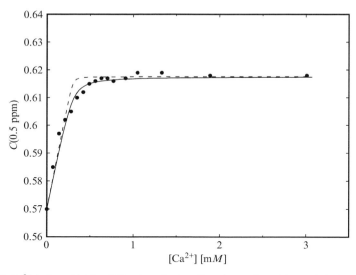

FIG. 4. Ca^{2+}-induced fold stabilization of α-lactalbumin monitored by statistical analysis of ^1H chemical shift distribution. The value of the autocorrelation function at 0.5 ppm, $C(0.5)$, was taken as a measure for protein structural stability (e.g., foldedness, see text). The black line represents the titration curve fitted to the experimental values ($K_d = 1.2 \times 10^{-5}$ M), the dashed black line the theoretical titration curve using the published Ca^{2+} dissociation constant K_d of α-lactalbumin (Wijesinha-Bettoni et al., 2001), $K_a = 10^{-6}$ M.

experimental $C(0.5)$ values vs. Ca^{2+} concentration is remarkably good. The experimentally determined dissociation constant (1.2×10^{-5} M, Fig. 4 solid line) at pH 6.5 convincingly demonstrates that the proposed method can be used to study subtle changes of protein structural stability caused by binding of metals and/or small-molecular-weight ligands. The small reduction of the observed K_d is presumably due to the lower pH used in the present study.

Fold Validation

We present here a Monte Carlo/Simulated Annealing (MC/SA) program for automated backbone H^N, N, and $C\alpha$ chemical shift assignment and structure validation. The program requires minimal NMR data input and the existence of a 3D structure of a homology model. Arbitrary reference numbers are attributed to experimentally observed residues in 3D HNCA and 3D ^{15}N-NOESY-HSQC spectra. Thus, each residue of the protein is represented by four resonance frequencies [^{15}N, $^1H^N$, $^{13}C\alpha(i)$, and $^{13}C\alpha(i-1)$, respectively]. Input lists containing query protein

C$\alpha(i)$ and C$\alpha(i - 1)$ chemical shifts of these residues as well as backbone amide H^N–H^N NOEs are generated. Based on a precise sequence alignment between query protein and homology model, the homology model is used to predict Cα chemical shifts and H^N–H^N NOEs for stretches of structurally equivalent query protein residues. Starting from an arbitrary start configuration, the Monte Carlo algorithm picks randomly pairs of sequence positions and swaps the residues tentatively assigned to these positions. An overall scoring value is computed after each of the multiple Monte Carlo steps to determine whether the proposed random change is a step toward the correct assignment or not. The random changes proposed to the system are accepted or rejected according to the Metropolis criterion. The Monte Carlo algorithm is coupled with a Simulated Annealing protocol that forces the system into a low-energy configuration in which experimentally observed and predicted NMR shifts and NOEs match best. An in-depth description of the objective function is provided in Materials and Methods. The objective function includes mathematical terms accounting for matching of experimentally observed and predicted NMR parameters such as Cα shifts and H^N–H^N NOEs as well as sequential C$\alpha(i)$/C$\alpha(i - 1)$ shift matching along the query protein sequence.

We have checked the performance of our Monte Carlo–based assignment and structure validation program with five query proteins (calmodulin, 150 residues; MBP, 370 residues; Q83, 150 residues; ICln, 168 residues; and CypD, 165 residues) differing in size and tertiary structure in a series of 15 test runs (Table II, Fig. 5). Whereas calmodulin is a purely α-helical protein, both Q83 and ICln feature β-barrel structures surrounded by a varying number of helices. MBP is a large two-domain protein with each domain being made up of numerous strands and helices. The CypD structure is so far unknown. However, CypD shares a high degree of sequence similarity with its homologue CypA, which is made up of an eight-stranded barrel surrounded by three helices and various extended loop segments.

In the case of MBP and calmodulin, we have chosen their X-ray structures as a homology model. Under these idealized conditions (1) the homology model sequence covers the entire query protein sequence, (2) shift and NOE predictions are available for all query protein residue positions, and (3) sets of "experimental" and predicted H^N–H^N NOEs, which were both calculated from the MBP and calmodulin atom coordinates assuming a 5 Å distance cutoff, are identical for NMR-observable residues with available chemical shifts in the BMRB database. Atom coordinate–derived "experimental" NOEs for NMR-unobservable residues (i.e., with no chemical shifts reported in the BMRB data bank) were omitted from the input file. Therefore, the input list of "experimental"

TABLE II
DATA INPUT AND ASSIGNMENT ACCURACY OF 15 MONTE CARLO-BASED ASSIGNMENT TEST RUNS[a]

Test run	Query protein (QP)/homology model (HM)	Non-Pro residues in QP sequence (n)	Non-Pro residues in alignment QP/HM (n)	Experimentally observed QP residues within alignment (n)	HM-based NOE predictions (+sequential NOEs) (n)	Experimentally observed QP NOEs (n)	Correct/erroneous assignment (n) (% correct assignments)
1	Calmodulin/calmodulin	146	146	139 BMRB 4284	323 PDB 1CFF	300 PDB 1CFF	145/1(99%)
2	MBP/MBP	349	349	330 BMRB 4354	697 PDB 1EZO	656 PDB 1EZO	334/15 (96%)
3	Q83/NGL	151	102	99 BMRB 4664	203 (+43) PDB 1NGL	243 PDB 1JZU	96/6 (94%)
4	Q83/NGL	151	85	82 BMRB 4664	171 (+53) PDB 1NGL	243 PDB 1JZU	81/4 (95%)
5	Q83/NGL	151	33	33 BMRB 4664	59 (+109) PDB 1NGL	243 PDB 1JZU	33/0 (100%)
6	Q83/NGL	151	35	32 BMRB 4664	68 (+103) PDB 1NGL	243 PDB 1JZU	33/2 (94%)
7	Q83/NGL	151	27	24 BMRB 4664	48 (+109) PDB 1NGL	243 PDB 1JZU	24/3 (89%)
8	Q83/NGL	151	26	26 BMRB 4664	51 (+113) PDB 1NGL	243 PDB 1JZU	23/3 (88%)
9	Q83/NGL	151	26	26 BMRB 4664	41 (+114) PDB 1NGL	243 PDB 1JZU	26/0 (100%)
10	ICln/UNC-89	158	74	73	170 (+87) PDB 1FHO	163[b]	69/5 (93%)
11	ICln/UNC-89	158	38	34	87 (+117) PDB 1FHO	163[b]	34/4 (89%)
12	ICln/UNC-89	158	36	36	67 (+119) PDB 1FHO	163[b]	35/1 (97%)
13	CypD/CypA	158	158	153	321 PDB 1CWB	791[c]	152/1 (99%)
14	CypD/CypA	158	54	54	101 (+104) PDB 1CWB	791[c]	49/5 (91%)
15	CypD/CypA	158	55	55	108 (+101) PDB 1CWB	791[c]	55/0 (100%)

[a] Column 1, number of test run. Column 2, query protein and structure homologue. Column 3, number of nonproline residues in the query protein sequence. Column 4, number of query protein residues that are aligned with and structurally similar to the homology model. Column 5, number of NMR-detectable residues of column 4; if chemical shift information was obtained from the BMRB database, the corresponding query protein entry code is provided. Column 6, number of homology model-based H^N–H^N NOE predictions as derived from homology model atom coordinates; homology model PDB entry codes are provided; the numbers in parentheses refer to additional sequential H^N–H^N NOE predictions that were introduced for query protein residues with no structure similarity to the homology model. Column 7, number of experimentally observed query protein H^N–H^N NOEs; the query protein PDB entry codes is provided, if "experimental" NOEs were calculated from query protein atom coordinates. Column 8, number of residues in column 4 that were correctly and erroneously assigned and percentage of correctly assigned residues.

[b] Synthetic H^N–H^N NOEs derived from query protein atom coordinates.

[c] H^N–H^N NOE input list results from a manual signal assignment performed with a ^{15}N-NOESY-HSQC and was generated as outlined in Materials and Methods.

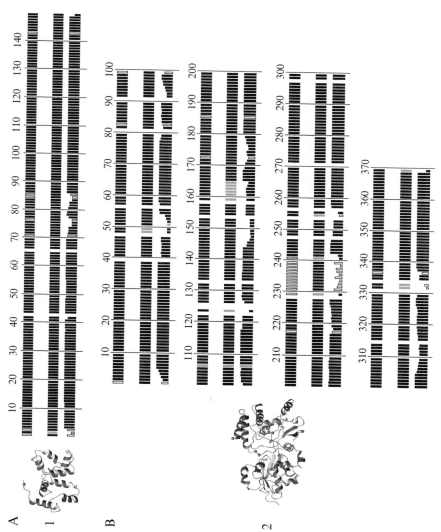

Fig. 5. (continued)

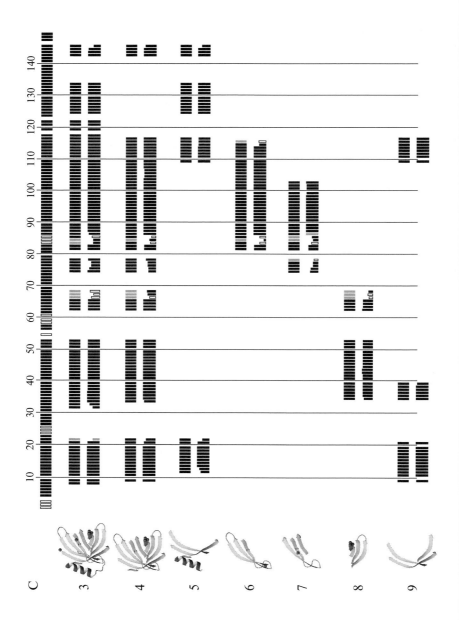

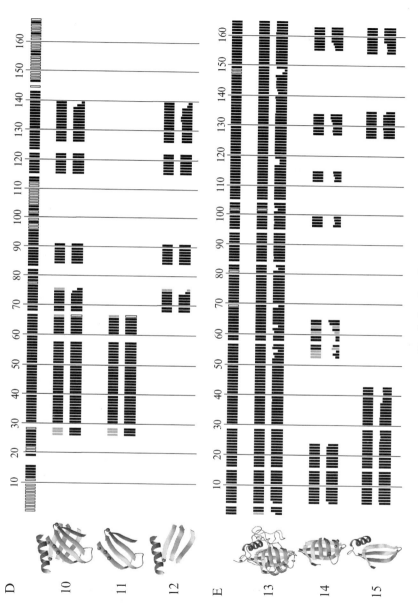

FIG. 5. (continued)

NOEs comprises 93% (calmodulin) and 94% (MBP) of the predicted NOEs. Homology model shift estimates were generated by using ShiftX (Neal *et al.*, 2003) software and thus differ from experimental chemical shifts obtained from the BMRB data bank.

In the case of the query proteins ICln and Q83, test conditions were much closer to real situations. ICln and Q83 structure homologues were identified through knowledge-based potential methodology (Domingues *et al.*, 1999; Sippl, 1993, 1995). Segments of the PH domain from the *Caenorhabditis elegans* muscle protein UNC-89 (PDB-ID, 1FHO: BMRB accession number, 4373) and of human neutrophil gelatinase-associated lipocalin NGL (PDB-ID, 1NGL; BMRB accession number, 4267) were revealed to share strong structural similarities with 47% of the ICln and 71% of the Q83 primary sequence, respectively. Therefore, these homology segments were used for shift and NOE predictions for structurally equivalent query protein segments. For the remaining Q83 and ICln residues with no structure similarity with NGL and UNC-89, respectively, random coil shifts and sequential H^N-H^N NOEs were added to the input lists. A synthetic set of Q83 H^N-H^N NOEs computed from Q83 atom coordinates represents again a complete and unambiguous input list. The ratio of Q83 "experimental" NOEs and predicted NOEs based on the structural model (1NGL) is close to 1. In contrast to Q83, the list of experimental ICln NOEs was obtained from a manually picked and edited ^{15}N-NOESY-HSQC spectrum of the protein. This results in a decreased ratio in experimental to predicted NOEs of $\sim$0.7, since complete observation of dipolar-coupled

---

FIG. 5. Graphic representation of assignment results obtained with our Monte Carlo–based approach for calmodulin (A), maltose-binding protein (B), Q83 (C), Icln (D), and CypD (E). The upper row in (A–E) represents the entire query protein primary sequence with each bar symbolizing one residue. Missing bars indicate residue positions occupied by prolines. NMR-detectable residues are shown as black bars and NMR-unobservable residues as unfilled black bars. Each individual test run is numbered as in Table I. The result of each test run is summarized by two rows of bars. Those query protein residues that are not part of homology segments (which have no structural equivalent in the homology model) as well as prolines were omitted from both rows. Upper row: correct assignment, black; erroneous assignment, gray. Lower row: residue with $C\alpha(i - 1)$ chemical shift matching/mismatching with $C\alpha(i)$ chemical shifts of its predecessor, black/gray, respectively. Filled bars represent NMR-observable residues; unfilled bars indicate NMR-unobservable residues. Note that no statement about interresidue $C\alpha$ chemical shift matching/mismatching can be made for residues adjacent (C-terminal) to NMR-unobservable residues. Maximal bar height symbolizes 100% assignment reproducibility after 20 independent assignment cycles. Reduced reproducibility is accordingly indicated by lower bar heights. The ribbon drawings display those parts of the query proteins that are made of the residues shown in the upper and lower rows. The positions of erroneously assigned residues are shown as small spheres. All ribbon drawings were generated with MOLMOL (Koradi *et al.*, 1996). (See color insert.)

backbone amide protons is hampered by shift degeneracies and relaxation processes.

Cyclophilin P chemical shift and NOE input lists originated exclusively from NMR spectra and, therefore, test conditions were most demanding in calculations performed with these data sets. CypD shares ~90% sequence identity with human cyclophilin A (PDB ID, 1cwb, X-ray structure, complexed with cyclosporin; BMRB entry code, 2208). However, no structural information is yet available for CypD. Input lists containing experimental as well as predicted $C\alpha$ chemical shifts and $H^N–H^N$ NOEs were generated as described in Material and Methods. A total of 412 $H^N–H^N$ NOEs were peak picked in the ^{15}N-NOESY-HSQC spectrum of CypD in the spectral window from 6 to 12 ppm. For 183 NOEs, a single dipolar coupling partner was identified. For the remaining 229 NOEs either more than one or no symmetric cross-peak was detected. By taking into consideration all potential dipolar coupling partners, these 229 NOEs were duplicated to a total of 776 NOEs (see Materials and Methods). Under the assumption that the correct dipolar coupling partner is assigned to at least one of the duplicated NOEs, this resulted in adding 547 wrong NOEs to the input file list. Redundant NOEs (i/j j/i) were filtered out, leaving 791 experimentally observed dipolar coupling interactions in the NOE input file list. Using a 5 Å distance cutoff, 321 $H^N–H^N$ NOEs were predicted for CypD based on the analysis of the atom coordinates of the CypA structure homologue. Thus, the ratio of experimental to predicted NOEs amounts to 2.5 and is considerably higher than for the previously mentioned test query proteins.

In our test examples, 72% (ICln) to 97% (CypD) of the nonproline residues were NMR detectable. The large majority of query residues within homology segments were NMR detectable; most of the undetected residues fall into sequence regions having no sequence alignment with the homology model and for which no homology model–based shift and NOE predictions are available (Fig. 5).

A first series of test runs (Table II and Fig. 5) was performed for all query proteins with the entirety of available input data (test runs 1, 2, 3, 10, and 13). In subsequent test calculations (4–9, 11–12, and 14–15) test conditions were artificially rendered more demanding. The goal of subsequent test runs was to check whether our Monte Carlo–based assignment procedure produces satisfying results if shift and NOE predictions are available only for smaller structural elements, i.e., if the homology model covers only smaller building blocks of the query protein. To this end, the homology model–based shift and NOE predictions for the query proteins ICln, Q83, and CypD were deleted for varying homology segments. Predicted NMR data were retained for smaller structural subunits comprising only 17–59% of the query protein primary sequence. Shift and NOE predictions that were deleted for certain query protein segments were replaced by random coil chemical shifts and sequential H^N–H^N NOEs. This results in a considerable decrease in the total number of homology model–based NOE predictions and in a change in the ratio of experimental to predicted NOEs.

In each test run, the program was allowed to assign all experimentally observed query protein spin systems to all nonproline residues of the query protein primary sequence, whether homology model–derived shift and NOE predictions were available for a specific sequence position or not. To assess the reproducibility of the assignment, each test run was performed 20 times. If the assignment for a specific query protein residue position was ambiguous, i.e., if after 20 assignment cycles more than one spin system was attributed to that position, the spin system that occurred most often was chosen for the final assignment. (This method does not necessarily result in a unique assignment, since a specific spin system might be retrieved at more than one residue position of the query protein.)

In spite of the diversity of test conditions, the assignment accuracy is satisfying in all cases. In the first series of test runs performed with the entirety of available homology model–based shift and NOE predictions, the percentage of correctly assigned residues within homology segments ranges from 93% (ICln) to 99% (calmodulin, CypD). The slightly lowered value of successfully assigned ICln residues in test run 10 may be due to the fact that an increased number of ICln residues (28%) was not detected by NMR and that homology model–derived shift and NOE predictions were available for only 47% of the ICln primary sequence. Surprisingly, our Monte Carlo assignment algorithm performed equally well in the second series of test calculations in which test conditions were rendered more demanding by retaining homology model–based shift and NOE predictions for smaller structural motifs. In these test runs, the percentage of correctly assigned residues ranges from 88% to 100% within these smaller building blocks.

In our test runs 75–100% of residues are assigned with a reproducibility of 75% or higher. The vast majority of these residues are correctly assigned. Based on our results, we can define the rule of thumb that the assignment of a specific residue is correct if it is part of a stretch of four or more consecutive residues that do not have any $C\alpha$ chemical shift mismatches and display an assignment reproducibility of >75% for each residue. Surprisingly, a clear majority of assignments with considerably higher uncertainties are still correct. Erroneous assignments may become manifested in $C\alpha$ chemical shift mismatches between adjacent residues. In addition, wrong assignments are evident if a certain experimentally observed residue appears (in rare cases) more than once in the final assignment list as a result of the final selection procedure performed after multiple independent Monte Carlo cycles as described above. (These are residues with an assignment reproducibility <50%.) Certain query protein residue positions appear to be more prone to erroneous assignments than others, and special care should be taken in the evaluation of these positions. Within homology segments, most of the erroneous assignments occur (1) at residue positions adjacent to prolines (e.g., MBP residues Pro-48-49-50), (2) at the N- or C-terminal ends of homology segments (e.g., ICln residues 27–29), (3) at sequence positions whose corresponding residues are not detected (e.g., Q83 residue position 85), or (4) in combinations of these situations.

Discussion

The statistical interpretation of chemical shifts was pioneered in the late 1960s by Schaefer and Yaris (1969) when they demonstrated that the complicated ^{13}C and ^1H NMR spectra of the cyclic tetramer of polypropylene oxide can be interpreted by an analysis of the spin hamiltonian in terms of the statistical theory of energy levels. Later, their suggestions were taken up by Lacelle (1984), who applied the approach to a vitamin (vitamin B_{12}), an antibiotic (alamethicin), and a protein (trypsin inhibitor homologue K) and showed that the method indeed provides, at least, a qualitative estimation of the degree of correlation between energy levels via a characterization of the spacing distribution of energy levels.

Here we systematically studied a diverse selection of proteins, comprising pure α-helical as well as α/β proteins, the relationship between the ^1H chemical shift distribution and protein structural stability in solution (e.g., foldedness and/or topological complexity of protein). The strategy was initiated by the idea of developing a robust, straightforward method to analyze protein spectra and to investigate the possibility of probing protein structural stability by a general method without the need of time-consuming assignment strategies, as this would be of significance to ongoing large-scale structural genomics efforts devoted to structural characterization of a vast number of proteins. In the analysis of the ^1H chemical shift distribution, we calculate the autocorrelation function $C(\omega)$ of the 1D ^1H spectra and take the value of $C(\omega)$ at 0.5 ppm as a quantitative benchmark to discriminate between folded, partially folded, and random-coil proteins. We do not attempt to physically interpret this parameter but rather use it as a quantitative means to probe fold stability. The analysis of the protein set convincingly demonstrated that it is indeed possible to probe protein structural stability through this simple analysis of 1D ^1H protein NMR spectra. There is a significant difference between proteins exhibiting a well-defined 3D structure and proteins that appear to be unfolded, partially folded, or that exist in a molten globule state.

The particular merits of the method are the ease of implementation, small amount of material (given the advent of more sensitive NMR detection schemes, e.g., cryoprobes), the high-throughput capability, and the fact that no isotope labeling is necessary. We thus foresee several obvious applications. First, recent genomic sequencing efforts have provided the coding DNA sequences of a large number of unknown genes and structural genomics or structural proteomics (Prestegard et al., 2001) attempts to provide 3D structural information of proteins encoded by the sequenced genes. Irrespective of the method of structure determination (X-ray or NMR spectroscopy), NMR is expected to play a significant role in structural genomics activities (Prestegard et al., 2001), as, for example, ^{15}N-filtered H/D exchange-based NMR experiments (Prestegard et al., 2001) (e.g., the identification of rapidly exchanging amide protons) and simple 1D experiments (Rehm et al., 2002) have already been demonstrated to be very effective to screen expressed and purified proteins for stability, structural disorder, and/or sample conditions that are favorable for crystallization. The data presented in this chapter suggest that this spectral autocorrelation method will be very valuable for this purpose, as the method does not require isotope labeling and also provides a means to identify metals and/or small ligands as well as macromolecular interactions that may be relevant for fold stabilization and function.

In contrast to structural genomics efforts that aim at characterizing folded proteins, a recently proposed target selection strategy focuses on unusual and uncharacterized soluble proteins in Mycoplasma genitalium, the smallest autonomously replicating organism (Balasubramanian et al., 2000). The aim of this approach was to identify proteins that show atypical behavior in terms of structural stability (foldedness), for example, proteins that are "unstructured" in the absence of a binding partner or that exhibit unusual thermodynamic properties. In this study, CD spectroscopy was used to probe the integrity of folding and to investigate the thermodynamic stability. As an alternative to optical methods, a mass spectrometry–based approach for protein stability screening was recently designed, which can even be extended to in vivo studies (Ghaemmaghami and Oas, 2001; Ghaemmaghami et al., 2000). With its ease of implementation, numerical analysis, and high-throughput capability, the proposed method should prove to be an additional important element of modern proteomic technology.

Second, the results obtained on the titration of α-lactalbumin with Ca^{2+} show that the proposed method can detect binding through changes of the ^{1}H chemical shift distribution, which in turn reflect protein stability changes. Given the well-established link between thermodynamic protein stability and ligand binding (Pace and McGrath, 1980), it may also be possible to use the high-throughput capability of the proposed method to screen large ligand libraries (Diercks *et al.*, 2001; Moore, 1999). If ^{13}C, ^{15}N-labeling of the protein is available, the method can be applied equally to protein–protein and protein–nucleic acid complexes. In particular, the approach can be applied to identify proteins that are only loosely defined structurally and undergo conformational restructuring or even adopt a well-defined native structure only upon binding to their authentic binding partners (for a review, see Wright and Dyson, 1999), a phenomenon that remarkably and unexpectedly is even more pronounced in higher organisms (Dunker and Obradovic, 2001).

Finally, data obtained on partially folded proteins (the native-like α-lactalbumin molten globule and the partially folded oncogenic transcription factor v-Myc) suggest fruitful applications of the proposed method to studies of molten globules and protein folding (Dolgikh *et al.*, 1981; Kuwajima, 1989; Ptitsyn, 1995). For example, site-directed mutagenesis has been successfully applied to obtain a quantitative measurement of the contributions of individual residues to the stability of molten globules (Hughson *et al.*, 1991). Additionally, studying the contribution of individual residues to the protein structural stability of molten globules may be valuable for understanding this important protein state.

The tremendous advance in the large-scale gene sequencing of whole genomes poses an enormous challenge to NMR spectroscopy. New integrated approaches are necessary to enable NMR spectroscopy to solve protein structures in a high-throughput manner and thus to keep pace with the generation of huge amounts of sequence information. In this context many research groups, including ours, have focused their efforts on the development of new programs devised to speed up the process of NMR structure elucidation. The program presented here is a powerful new tool for rapid sequence-specific assignment of backbone resonances of uniformly ^{13}C- and ^{15}N-labeled globular proteins and structure validation. Our approach requires minimal NMR data input from two 3D spectra and therefore a reduced amount of spectrometer time. The need for more extensive data collection is circumvented by using chemical shift and NOE predictions derived from a 3D structure of a query protein homologue. Although additional data input is not mandatory for obtaining correct assignments, further chemical shift and interresidue connectivity information can easily be included for Hα, Cβ, and C' nuclei. It is, however, important to note that the performance of our program in its present form depends on the sequence alignment accuracy of structurally equivalent blocks of the query protein and its homologue. We have observed that the assignment accuracy deteriorates in particular if query protein segments predicted to form a β-sheet structure are inaccurately aligned with the homology model sequence (data not shown). A modification of the current form of the objective function, in particular the replacement of the Tanimoto coefficient by a more sophisticated expression, might help to eliminate the pitfall of improper sequence alignment. As test runs with CypD input data have clearly demonstrated, our program is robust enough to tolerate numerous ambiguous H^N–H^N NOEs resulting from the inability to clearly identify the majority of dipolar-coupled pairs of backbone amide protons on the sole basis of a 3D ^{15}N-NOESY-HSQC spectrum. In addition, even if homology model–based shift and NOE predictions are missing for certain residue stretches, the algorithm is still able to find the correct assignments for the remaining protein segments. The latter feature represents a distinct advantage of the program described in this chapter over existing assignment and structure validation programs and suggests fruitful applications in the scanning of query proteins for the presence of structure templates. Template scanning and motif recognition are useful when complete homology model covering the entire query protein sequence is not available and/or to study protein modules in the context of multidomain proteins.

Conclusions

We have demonstrated that protein structural stability is reflected in the distribution of protein ^{1}H chemical shifts. A method was proposed that does not require isotope labeling but instead uses easily obtainable 1D ^{1}H spectra, from which the spectral autocorrelation function is calculated. The method allows a significant and reliable distinction between unfolded or partially folded proteins and proteins with well-defined global folds. Additionally, the precision of the method is sufficient to discern subtle differences in protein structural stability between, for example, the molten globule state of α-lactalbumin with a native-like overall fold and the partially folded (displaying a single α-helix and lacking long-range tertiary interactions) oncogenic transcription factor v-Myc with possible applications to protein folding studies. Data obtained on the Ca^{2+}-depleted apo and the Ca^{2+}-loaded holo form of α-lactalbumin additionally suggest that the method is able to detect subtle changes in protein stability caused by ligand binding. The method can easily be adjusted for screening purposes using NMR flow probes and micromanipulator robots and should consequently prove useful for target selection in high-throughput structural genomics and the identification of experimental conditions to optimize protein stability and crystal formation.

As the number of experimental protein structures is expected to significantly increase in the foreseeable future, comparative structure prediction will become an essential tool in structural genomics. Although the reliability of structural modeling approaches is well documented and the precision (and accuracy) of predicted structures is sufficiently high to draw conclusions about putative biochemical functionality, there is still a demand for experimental verification and/or subsequent structural refinement. Given the robustness and reliability of our proposed strategy, we anticipate fruitful applications of the methodology in ongoing structural genomics efforts.

References

Acharya, K. R., Ren, J. S., Stuart, D. I., Phillips, D. C., and Fenna, R. E. (1991). Crystal structure of human alpha-lactalbumin at 1.7 Å resolution. *J. Mol. Biol.* **221,** 571–581.

Alexandrescu, A. T., Evans, P. A., Pitkeathly, M., Baum, J., and Dobson, C. M. (1993). Structure and dynamics of the acid-denatured molten globule state of alpha-lactalbumin: A two-dimensional NMR study. *Biochemistry* **32,** 1707–1718.

Anderson, P. J., Brooks, C. L., and Berliner, L. J. (1997). Functional identification of calcium binding residues in bovine alpha-lactalbumin. *Biochemistry* **36,** 11648–11654.

Balasubramanian, S., Schneider, T., Gerstein, M., and Regan, L. (2000). Proteomics of *Mycoplasma genitalium*: Identification and characterization of unannotated and atypical proteins in a small model genome. *Nucleic Acids Res.* **28,** 3075–3082.

Balbach, J., Forge, V., Lau, W. S., Jones, J. A., van Nuland, N. A., and Dobson, C. M. (1997). Detection of residue contacts in a protein folding intermediate. *Proc. Natl. Acad. Sci. USA* **94,** 7182–7185.

Baum, J., Dobson, C. M., Evans, P. A., and Hanley, C. (1989). Characterization of a partly folded protein by NMR methods: Studies on the molten globule state of guinea pig alpha-lactalbumin. *Biochemistry* **28,** 7–13.

Berman, H. M., Westbrook, J., Feng, Z., Gilliland, G., Bhat, T. N., Weissig, H., Shindyalov, I. N., and Bourne, P. E. (2000). The Protein Data Bank. *Nucleic Acids Res.* **28,** 235–242.

Calderone, V., Giuffrida, M. G., Viterbo, D., Napolitano, L., Fortunato, D., Conti, A., and Acharya, K. R. (1996). Amino acid sequence and crystal structure of buffalo alpha-lactalbumin. *FEBS Lett.* **394,** 91–95.

Cavanagh, J., Fairbrother, W. J., Palmer, A. G., and Skelton, N. G. (1996). "Protein NMR Spectroscopy: Principles and Practice." Academic Press, San Diego, CA.

Chakraborty, S., Ittah, V., Bai, P., Luo, L., Haas, E., and Peng, Z. (2001). Structure and dynamics of the alpha-lactalbumin molten globule: Fluorescence studies using proteins containing a single tryptophan residue. *Biochemistry* **40,** 7228–7238.

Chrysina, E. D., Brew, K., and Acharya, K. R. (2000). Crystal structures of apo- and holo-bovine alpha-lactalbumin at 2.2-Å resolution reveal an effect of calcium on inter-lobe interactions. *J. Biol. Chem.* **275,** 37021–37029.

Choy, W. Y., and Forman-Kay, J. D. (2001). Calculation of ensembles of structures representing the unfolded state of an SH3 domain. *J. Mol. Biol.* **308,** 1011–1032.

Chyan, C. L., Wormald, C., Dobson, C. M., Evans, P. A., and Baum, J. (1993). Structure and stability of the molten globule state of guinea-pig alpha-lactalbumin: A hydrogen exchange study. *Biochemistry* **32,** 5681–5691.

Delaglio, F., Grzesiek, S., Vuister, G. W., Zhu, G., Pfeifer, J., and Bax, A. (1995). NMRPipe: A multidimensional spectral processing system based on UNIX pipes. *J. Biomol. NMR* **6,** 277–293.

Diamond, R. (1974). Real-space refinement of the structure of hen egg-white lysozyme. *J. Mol. Biol.* **82,** 371–391.

Diercks, T., Coles, M., and Kessler, H. (2001). Applications of NMR in drug discovery. *Curr. Opin. Chem. Biol.* **5,** 285–291.

Dobson, C. M. (1994). Protein folding. Solid evidence for molten globules. *Curr. Biol.* **4,** 636–640.

Dolgikh, D. A., Gilmanshin, R. I., Brazhnikov, E. V., Bychkova, V. E., Semisotnov, G. V., Venyaminov, S., and Ptitsyn, O. B. (1981). Alpha-Lactalbumin: Compact state with fluctuating tertiary structure? *FEBS Lett.* **136,** 311–315.

Domingues, F., Koppensteiner, W. A., Jaritz, M., Prlic, A., Weichenberger, C., Wiederstein, M., Wiederstein, M., Flöckner, H., Lackner, P., and Sippl, M. J. (1999). Sustained performance of knowledge-based potentials in fold recognition. *Proteins* **3,** 112–120.

Doreleijers, J. F., Mading, S., Maziuk, D., Sojourner, K., Yin, L., Zhu, J., Markley, J. L., and Ulrich, E. L. (2003). BioMagResBank database with sets of experimental NMR constraints corresponding to the structures of over 1400 biomolecules deposited in the Protein Data Bank. *J. Biomol. NMR* **26,** 139–146.

Dunker, A. K., and Obradovic, Z. (2001). The protein trinity—linking function and disorder. *Nat. Biotechnol.* **19,** 805–806.

Fieber, W., Schneider, M. L., Matt, T., Krautler, B., Konrat, R., and Bister, K. (2001). Structure, function, and dynamics of the dimerization and DNA-binding domain of oncogenic transcription factor v-Myc. *J. Mol. Biol.* **307,** 1395–1410.

Forge, V., Wijesinha, R. T., Balbach, J., Brew, K., Robinson, C. V., Redfield, C., and Dobson, C. M. (1999). Rapid collapse and slow structural reorganisation during the refolding of bovine alpha-lactalbumin. *J. Mol. Biol.* **288,** 673–688.

Ghaemmaghami, S., and Oas, T. G. (2001). Quantitative protein stability measurement *in vivo*. *Nat. Struct. Biol.* **8,** 879–882.

Ghaemmaghami, S., Fitzgerald, M. C., and Oas, T. G. (2000). A quantitative, high-throughput screen for protein stability. *Proc. Natl. Acad. Sci. USA* **97,** 8296–8301.

Griko, Y. V., and Kutyshenko, V. P. (1994). Differences in the processes of beta-lactoglobulin cold and heat denaturations. *Biophys. J.* **67,** 356–363.

Gronwald, W., and Kalbitzer, H. R. (2004). Automated structure determination of proteins by NMR spectroscopy. *Prog. Nucl. Magn. Reson. Spectrosc.* **44,** 33–96.

Heinemann, U. (2000). Structural genomics in Europe: Slow start, strong finish? *Nat. Struct. Biol.* **7**(Suppl.), 940–942.

Hitchens, T. K., Lukin, J. A., Zhan, Y. P., McCallum, S. A., and Rule, G. S. (2003). MONTE: An automated Monte Carlo based approach to nuclear magnetic resonance assignment of proteins. *J. Biomol. NMR* **25,** 1–9.

Hughson, F. M., Barrick, D., and Baldwin, R. L. (1991). Probing the stability of a partly folded apomyoglobin intermediate by site-directed mutagenesis. *Biochemistry* **30,** 4113–4118.

Janatova, J., Fuller, J. K., and Hunter, M. J. (1968). The heterogeneity of bovine albumin with respect to sulfhydryl and dimer content. *J. Biol. Chem.* **243,** 3612–3622.

Johnson, B. A., and Blevins, R. A. (1994). NMRView: A computer program for the visualization and analysis of NMR data. *J. Biomol. NMR* **4,** 603–614.

Jones, S., and Thornton, J. M. (1997). Prediction of protein–protein interaction sites using patch analysis. *J. Mol. Biol.* **272,** 133–143.

Karplus, K., Barrett, C., Cline, M., Diekhans, M., Grate, L., and Hughey, R. (1999). Predicting protein structure using only sequence information. *Proteins* **3**(Suppl.), 121–125.

Kasuya, A., and Thornton, J. M. (1999). Three-dimensional structure analysis of PROSITE patterns. *J. Mol. Biol.* **286,** 1673–1691.

Kim, S. H. (1998). Shining a light on structural genomics. *Nat. Struct. Biol.* **5**(Suppl.), 643–645.

Koppensteiner, W. A., Lackner, P., Wiederstein, M., and Sippl, M. J. (2000). Characterization of novel proteins based on known protein structures. *J. Mol. Biol.* **296,** 1139–1152.

Koradi, R., Billeter, M., and Wuthrich, K. (1996). MOLMOL: A program for display and analysis of macromolecular structures. *J. Mol. Graph.* **14,** 29–32.

Kutyshenko, V. P., and Cortijo, M. (2000). Water-protein interactions in the molten-globule state of carbonic anhydrase b: An NMR spin-diffusion study. *Protein Sci.* **9,** 1540–1547.

Kuwajima, K. (1989). The molten globule state as a clue for understanding the folding and cooperativity of globular-protein structure. *Proteins* **6,** 87–103.

Kuwajima, K. (1996). The molten globule state of alpha-lactalbumin. *FASEB J.* **10,** 102–109.

Kuwajima, K., Mitani, M., and Sugai, S. (1989). Characterization of the critical state in protein folding. Effects of guanidine hydrochloride and specific Ca2+ binding on the folding kinetics of alpha-lactalbumin. *J. Mol. Biol.* **206,** 547–561.

Lacelle, S. (1984). Random matrix theory in biological nuclear magnetic resonance. *Biophys. J.* **46,** 181–186.

Last, A. M., Schulman, B. A., Robinson, C. V., and Redfield, C. (2001). Probing subtle differences in the hydrogen exchange behavior of variants of the human alpha-lactalbumin molten globule using mass spectrometry. *J. Mol. Biol.* **311,** 909–919.

Maurus, R., Bogumil, R., Nguyen, N. T., Mauk, A. G., and Brayer, G. (1998). Structural and spectroscopic studies of azide complexes of horse heart myoglobin and the His-64 → Thr variant. *Biochem. J.* **332**(Pt. 1), 67–74.

Metropolis, N., Rosenbluth, A. W., Rosenbluth, M. N., Teller, A. H., and Teller, E. (1953). Equation of state calculations by fast computing machines. *J. Chem. Phys.* **21,** 1087–1092.

Moore, J. M. (1999). NMR screening in drug discovery. *Curr. Opin. Biotechnol.* **10,** 54–58.

Neal, S., Nip, A. M., Zhang, H., and Wishart, D. S. (2003). Rapid and accurate calculation of protein 1H, 13C, and 15N chemical shifts. *J. Biomol. NMR* **26,** 215–240.

Neri, D., Billeter, M., Wider, G., and Wüthrich, K. (1992). NMR determination of residual structure in a urea-denatured protein, the 434-respressor. *Science* **257,** 1559–1563.

Ota, M., Kawabata, T., Kinjo, A. R., and Nishikawa, K. (1999). Cooperative approach for the protein fold recognition. *Proteins* **3**(Suppl.), 126–132.

Pace, C. N., and McGrath, T. (1980). Substrate stabilization of lysozyme to thermal and guanidine hydrochloride denaturation. *J. Biol. Chem.* **255,** 3862–3865.

Parkin, S., Rupp, B., and Hope, H. (1996). Structure of bovine pancreatic trypsin inhibitor at 125K: Definition of carboxyl-terminal residues Gly57 and Ala58. *Acta Crystallogr. D Biol. Cryst.* **52,** 18–29.

Pascal, S. M., Muhandiram, D. R., Yamazaki, T., Forman-Kay, J. D., and Kay, L. E. (1994). Simultaneous acquisition of ^{15}N- and ^{13}C-edited NOE spectra of proteins dissolved in H2O. *J. Magn. Reson.* **103B,** 197–201.

Peng, Z. Y., and Kim, P. S. (1994). A protein dissection study of a molten globule. *Biochemistry* **33,** 2136–2141.

Peng, Z. Y., Wu, L. C., and Kim, P. S. (1995). Local structural preferences in the alpha-lactalbumin molten globule. *Biochemistry* **34,** 3248–3252.

Plaxco, K. W., Simons, K. T., and Baker, D. (1998). Contact order, transition state placement and the refolding rates of single domain proteins. *J. Mol. Biol.* **277,** 985–994.

Pike, A. C., Brew, K., and Acharya, K. R. (1996). Crystal structures of guinea-pig, goat and bovine alpha-lactalbumin highlight the enhanced conformational flexibility of regions that are significant for its action in lactose synthase. *Structure* **4,** 691–703.

Prestegard, J. H., Valafar, H., Glushka, J., and Tian, F. (2001). Nuclear magnetic resonance in the era of structural genomics. *Biochemistry* **40,** 8677–8685.

Ptitsyn, O. B. (1995). Molten globule and protein folding. *Adv. Protein Chem.* **47,** 83–229.

Rao, J. K., Bujacz, G., and Wlodawer, A. (1998). Crystal structure of rabbit muscle creatine kinase. *FEBS Lett.* **439,** 133–137.

Redfield, C., Schulman, B. A., Milhollen, M. A., Kim, P. S., and Dobson, C. M. (1999). Alpha-lactalbumin forms a compact molten globule in the absence of disulfide bonds. *Nat. Struct. Biol.* **6,** 948–952.

Rehm, T., Huber, R., and Holak, T. A. (2002). Application of NMR in structural proteomics: Screening for proteins amenable to structural analysis. *Structure* **10,** 1613–1618.

Russell, R. B. (1998). Detection of protein three-dimensional side-chain patterns: New examples of convergent evolution. *J. Mol. Biol.* **279,** 1211–1227.

Russell, R. B., Sasieni, P. D., and Sternberg, M. J. (1998). Supersites within superfolds. Binding site similarity in the absence of homology. *J. Mol. Biol.* **282,** 903–918.

Sander, C., and Schneider, R. (1991). Database of homology-derived protein structures and the structural meaning of sequence alignment. *Proteins* **9,** 56–68.

Schaefer, J., and Yaris, R. (1969). Random matrix theory and nuclear magnetic resonace spectral distributions. *J. Chem. Phys.* **51,** 4469–4474.

Schulman, B. A., and Kim, P. S. (1996). Proline scanning mutagenesis of a molten globule reveals non-cooperative formation of a protein's overall topology. *Nat. Struct. Biol.* **3,** 682–687.

Schulman, B. A., Kim, P. S., Dobson, C. M., and Redfield, C. (1997). A residue-specific NMR view of the non-cooperative unfolding of a molten globule. *Nat. Struct. Biol.* **4,** 630–634.

Seavey, B. R., Farr, E. A., Westler, W. M., and Markley, J. L. (1991). A relational database for sequence-specific protein NMR data. *J. Biomol. NMR* **1,** 217–236.

Shindyalov, I. N., and Bourne, P. E. (2000). The Protein Data Bank. *Nucleic Acids Res.* **28,** 235–242.

Shortle, D., and Ackerman, M. S. (2001). Persistence of native-like topology in a denatured protein in 8M urea. *Science* **293,** 487–489.

Sippl, M. J. (1993). Recognition of errors in three-dimensional structures of proteins. *Proteins.* **17,** 355–362.

Sippl, M. J. (1995). Knowledge based potentials for proteins. *Curr. Opin. Struct. Biol.* **5,** 229–235.

Sippl, M. J., and Weitckus, S. (1992). Detection of native-like models for amino acid sequences of unknown three-dimensional structure in a data base of known protein conformations. *Proteins* **13,** 258–271.

Staunton, D., Owen, J., and Campbell, I. D. (2003). NMR and structural genomics. *Acc. Chem. Res.* **36,** 207–214.

Terwilliger, T. C. (2000). Structural genomics in North America. *Nat. Struct. Biol.* **7**(Suppl.), 935–939.

Thornton, J. M., Flores, T. P., Jones, D. T., and Swindells, M. B. (1991). Protein structure. Prediction of progress at last. *Nature* **354,** 105–106.

Tollinger, M., Konrat, R., Hilbert, B. H., Marsh, E. N., and Krautler, B. (1998). How a protein prepares for B12 binding: Structure and dynamics of the B12-binding subunit of glutamate mutase from *Clostridium tetanomorphum.* *Structure* **6,** 1021–1033.

Troullier, A., Reinstadler, D., Dupont, Y., Naumann, D., and Forge, V. (2000). Transient non-native secondary structures during the refolding of alpha-lactalbumin detected by infrared spectroscopy. *Nat. Struct. Biol.* **7,** 78–86.

Vijay-Kumar, S., Bugg, C. E., and Cook, W. J. (1987). Structure of ubiquitin refined at 1.8 Å resolution. *J. Mol. Biol.* **194,** 531–544.

Wijesinha-Bettoni, R., Dobson, C. M., and Redfield, C. (2001). Comparison of the structural and dynamical properties of holo and apo bovine alpha-lactalbumin by NMR spectroscopy. *J. Mol. Biol.* **307,** 885–898.

Wishart, D. S., Bigam, C. G., Holm, A., Hodges, R. S., and Sykes, B. D. (1995). 1H, 13C, and 15N random coil NMR chemical shifts of the common amino acids. I. Investigations of nearest-neighbor effects. *J. Biomol. NMR* **5**, 67–81.

Wright, P. E., and Dyson, H. J. (1999). Intrinsically unstructured proteins: Re-assessing the protein structure-function paradigm. *J. Mol. Biol.* **293**, 321–331.

Wu, L. C., Peng, Z. Y., and Kim, P. S. (1995). Bipartite structure of the alpha-lactalbumin molten globule. *Nat. Struct. Biol.* **2**, 281–286.

Wüthrich, K. (1986). "NMR of Proteins and Nucleic Acids." Wiley, New York.

Yokoyama, S., Matsuo, Y., Hirota, H., Kigawa, T., Shirouzu, M., Kuroda, Y., Kurumizaka, H., Kawaguchi, S., Ito, Y., Shibata, T., Kainosho, M., Nishimura, Y., Inoue, Y., and Kuramitsu, S. (2000). Structural genomics projects in Japan. *Prog. Biophys. Mol. Biol.* **73**, 363–376.

[7] Determination of Protein Backbone Structures from Residual Dipolar Couplings

By J. H. Prestegard, K. L. Mayer, H. Valafar, and G. C. Benison

Abstract

There are a number of circumstances in which a focus on determination of the backbone structure of a protein, as opposed to a complete all-atom structure, may be appropriate. This is particularly the case for structures determined as a part of a structural genomics initiative in which computational modeling of many sequentially related structures from the backbone of a single family representative is anticipated. It is, however, also the case when the backbone may be a stepping-stone to more targeted studies of ligand interaction or protein–protein interaction. Here an NMR protocol is described that can produce a backbone structure of a protein without the need for extensive experiments directed at side chain resonance assignment or the collection of structural information on side chains. The procedure relies primarily on orientational constraints from residual dipolar couplings as opposed to distance constraints from NOEs. Procedures for sample preparation, data acquisition, and data analysis are described, along with examples from application to small target proteins of a structural genomics project.

Introduction

Residual dipolar couplings (RDCs) are now widely used as a source of constraints in the determination of the structure of biomolecules. Several reviews on the subject have appeared (Al-Hashimi and Patel, 2002; Bax *et al.*, 2001; de Alba and Tjandra, 2002; Prestegard *et al.*, 2000; Tolman,

METHODS IN ENZYMOLOGY, VOL. 394

Copyright 2005, Elsevier Inc.
All rights reserved.
0076-6879/05 $35.00

2001; Zhou *et al.*, 1999), and there are numerous examples of application to proteins in the more recent literature (Alexandrescu and Kammerer, 2003; Assfalg *et al.*, 2003; Beraud *et al.*, 2002; Lin *et al.*, 2004; Nair *et al.*, 2003; Ohnishi *et al.*, 2004; Tossavainen *et al.*, 2003; Zheng *et al.*, 2004). However, in the majority of cases, use has been as a supplement to other structural information, rather than a primary source of information. Although this is often appropriate, there are cases in which use as a primary source of structural data should be considered. One case arises in the context of the structural genomics initiative (Adams *et al.*, 2003; Chance *et al.*, 2002; Montelione *et al.*, 2000). This initiative set as its goal the production of massive numbers of three-dimensional protein structures in an effort to leverage the information flowing from whole genome sequencing efforts of the previous decade (Burley *et al.*, 1999; Norvell and Machalek, 2000). Production of experimental structures for each gene sequenced would obviously be impossible, but production of a sufficient number of structures to populate "fold space" (10,000 structures in 10 years) might be possible with adequate attention to automation of existing methodology and development of new methodology for determination of structure. With representative structures in each "fold family," computer modeling methods would then be able to build structures for most sequences. However, even with the reduced number of structures to be determined, the methodology developed would have to be efficient.

For nuclear magnetic resonance (NMR), substantial efficiency might be gained by focusing on backbone structures as opposed to structures complete with all side chain atoms. Given the intent to model most structures starting from representative structures that may have as little as 30% sequence identity, a backbone structure for a fold family representative should be adequate. As much as 70% of the side chains would, after all, be replaced in the course of modeling a new protein. Although structural information is obviously limited, backbone structures may have direct application in other areas as well. Drug discovery programs often rely on perturbations of resonances from just backbone atoms to identify ligand-binding sites (Fesik, 1999, 2001; Hajduk *et al.*, 1999, 2002). Orientational data collected on ^{15}N–^{1}H pairs of atoms along the backbone is often enough to align elements of protein complexes (Clore and Schwieters, 2003; Dosset *et al.*, 2001; Weaver and Prestegard, 1998). Furthermore, a backbone structure, with assignments of resonances from backbone atoms, is potentially a good starting point for more complete structure determination. These considerations make presentation of backbone structure determination methodology particularly appropriate in this volume.

Focusing on backbone atoms is, unfortunately, not entirely compatible with a traditional nuclear Overhauser enhancement (NOE)-based approach to protein structure. NOEs stem from very short-range interactions; we can easily measure an NOE for a pair of protons at 3 Å separation, but at 6 Å, the steep $1/r^6$ distance dependence reduces signal by a factor of 64 and nearly eliminates the possibility of observation. Except for β-sheets, atoms in remote parts of a protein backbone seldom come within 6 Å of one another. Reliance on RDC data enters at this point. RDCs show orientational dependence in addition to distance dependence, and these orientational dependencies give rise to constraints that are effective in relating remote parts of the backbone, no matter what their separation in space.

In what follows we present a particular strategy that evolved from an attempt to produce protein structures efficiently as a part of a structural genomics project (Adams *et al.*, 2003). This is, of course, not the only such strategy; there are others that use different sets of RDC data and ones that use fundamentally different algorithms for the determination of structure (Andrec *et al.*, 2001a,b; Delaglio *et al.*, 2000; Haliloglu *et al.*, 2003; Rohl and Baker, 2002). The strategy presented here shares some of the philosophy of these other methods, but also has some unique characteristics. One is that it was designed to work with lower levels of ^{13}C enrichment; this can lead to significant cost savings in some situations, but it also avoids some of the complexities associated with ^{13}C decoupling and the loss of signal due to multiple magnetization transfer pathways. The ability to use simpler pulse sequences and exploit selective pathways compensates for some of the loss in sensitivity with lower levels of labeling. In addition, prior assignment of resonances is not essential; this avoids the need for collection of separate resonance assignment experiments.

In the following we present the basic rationale for converting RDCs into a three-dimensional backbone structure, a set of experiments that has proven useful in acquiring the required data, and step-by-step examples of applications to structural genomics targets. Although the presentation centers on efficiency of structural genomics applications (data acquisition takes about one-third the time of a complete NOE-based structure), many aspects of the procedure should be of more general utility. Efficient collection of the same subsets of RDC data used in structural genomics applications may find application in refinement of structures based on NOE data, and the procedures described for orienting backbones may have application to assembly of subunits into multiprotein complexes.

Origin of RDCs

RDCs arise from the same basic interaction that gives rise to the NOE, namely a through-space dipole–dipole interaction between nuclei that possess magnetic moments. For a pair of weakly coupled spin one-half nuclei this interaction can be represented as in Eq. (1). Here r is the distance between nuclei i and j, γ_{ij} are the magnetogyric ratios for the nuclei, μ_0 is the permeability of space, h is Planck's constant, and θ is the angle between the internuclear vector and the magnetic field. For a directly bonded pair of nuclei, r is fixed and θ becomes the primary source of structural information. Clearly, a constraint on θ for every bond vector along a protein backbone would be a powerful determinant of molecular geometry.

$$D_{ij} = -\frac{\mu_0 \gamma_i \gamma_j h}{(2\pi r)^3} \left\langle \frac{3\cos^2\theta - 1}{2} \right\rangle \tag{1}$$

The brackets in Eq. (1) denote averaging over the rapid molecular tumbling that occurs in solution. When this tumbling makes the interaction vector sample directions in space isotropically, the interaction averages to zero, and no dipole–dipole contributions to splittings are observed. This is why NMR spectroscopists working in solution seldom worry about direct dipole–dipole contributions to their spectra and rely on indirect spin relaxation contributions such as the NOE for structural information. However, the average can be made nonzero by inducing partial alignment of the molecules studied. As long as molecular tumbling remains rapid, a single average interaction that adds to multiplet splittings in coupled spectra results. Alignment is commonly accomplished with the use of aqueous liquid crystal media such as bicelles or filamentous bacteriophage that interact weakly with the molecule of interest to give departures from isotropic sampling by one part in 10^3 or 10^4. A large number of media for inducing alignment now exist, as do paramagnetic tags that aid self-alignment of molecules in high magnetic fields (Barbieri et al., 2002; Wohnert et al., 2003). These are described in recent reviews (Bax et al., 2001; Prestegard and Kishore, 2001; Prestegard et al., 2004).

Partial alignment, unfortunately, introduces hidden unknowns into Eq. (1). It is usually necessary to characterize the level of alignment, the asymmetry of alignment, and the direction of the principal alignment axes as seen from the point of the molecule under study. The five additional parameters needed can be considered to be three Euler angles to define the alignment axes (α, β, and γ), a principal order parameter (S_{zz}), and an asymmetry parameter (η). Determining these parameters requires that a significant number of RDCs be measured for each molecular fragment. However, the extra information obtained is also valuable. If fragments are part of the same rigid structure, they must share the same S_{zz} and η. If they do not, the existence of internal motions is suggested. Also, the alignment axes must coincide if the fragments are oriented so that they properly represent parts of the same rigid molecule. This will be the basis of our strategy for determining structure.

Fortunately, RDCs are easily measured. When all units in Eq. (1) are in SI units, D_{ij} is given in Hertz and corresponds to the dipole–dipole contribution to the splitting of the doublets that would be observed for each member of a weakly coupled pair of spins. When the spins are directly bonded, substantial through-bond (scalar, J_{ij}) couplings also exist, and the observed splitting would be the sum of scalar and dipole–dipole couplings. Hence, RDCs are measured as differences in splittings of partially aligned $(J_{ij} + D_{ij})$ and isotropic (J_{ij}) spectra.

Although there are a number of ways to extract alignment parameters and evaluate fragment geometry (Delaglio et al., 2000; Hus et al., 2000; Schwieters et al., 2003), the procedure used in the protocol presented here rests on recasting Eq. (1) in terms of elements of a 3×3 order matrix, S_{kl} [see Eq. (2) and Saupe (1968)]. $D_{\max ij}$ is the coupling constant for a pair of nuclei at a 1.0 Å separation with their internuclear vector along the magnetic field, and $\cos(\theta_{k,l})$ are the direction cosines relating the internuclear vector to the axes (x, y, and z) of an arbitrarily chosen molecular fragment frame.

$$D_{ij} = \frac{D_{\max ij}}{r^3} \sum_{k,l} S_{kl} \cos(\theta_k) \cos(\theta_l) \tag{2}$$

Because the order matrix is traceless and symmetric, only five order matrix elements are independent in the expression for each RDC. For a given trial geometry of a fragment, the $\cos(\theta_{k,l})$ are also known, making a set of equations of the form of Eq. (2) solvable for any five or more independent RDC measurements. Singular value decomposition can be used to give a best least-squares solution for the order parameters (Losonczi et al., 1999). The order parameters can then be used to backcalculate RDCs for comparison to experiment and evaluation of trial geometries for the fragment (Valafar and Prestegard, 2004).

Conversion of RDCs to Structure

The general procedure for converting RDCs to a protein backbone structure begins with the determination of fragment geometries using the program REDCRAFT (REsidual Dipolar Coupling Residue Assembly and Filter Tool) (H. Valafar and J. H. Prestegard, unpublished observations; available at http://secnmr.org). This uses the order matrix evaluation described above and backcalculation of RDCs to select proper fragment geometries. It also uses a number of other filters for allowed fragment geometry including a Ramachandran space filter and a torsion angle filter that is based on the Karplus equation for three-bond scalar couplings. The program begins by first evaluating geometry solutions using incremented sets of ϕ, ψ values for pairs of peptide planes connected by a common α carbon. These geometries are ranked based on agreement with RDCs and other filters; then the top choices from the ranked list for one unit are combined with the top choices from a sequentially connected second unit by overlaying the C- and N-terminal peptide planes. Selection of this second unit is usually based on the overlap of the intraresidue C_α chemical shift seen for the first residue with the interresidue C_α chemical shift seen for the second residue in HNCA-style experiments. The proposed geometries for the new fragment, now containing three peptide planes and two sets of ϕ, ψ angles, are evaluated and ranked for a next round of extension. Ambiguities in selection of a next residue do, of course, arise. These can be resolved to some extent by excluding those connections that produce no acceptable matches to experimental RDCs. However, at some point the process terminates because ambiguities are unresolvable or data needed for connection are missing due to the occurrence of a proline or just lack of observable data. Multiple fragments of five or six C_α carbons are usually sufficient to proceed with a first round of structure determination.

Coordinates for the structurally defined fragments are next transformed to a common principal alignment frame (PAF). The Euler angles needed to accomplish this can be found from the transformations that diagonalize the order matrix for each fragment. The software package REDCAT provides a means of extracting Euler angles (Valafar and Prestegard, 2004). When all fragments are in their common PAF, assembly into a complete structure remains a problem of translation only. There is one caveat; the insensitivity of Eqs. (1) and (2) to a 180° rotation about any principal axis leads to a 4-fold degeneracy in possible fragment orientations. This problem is sometimes obviated when fragments are sequentially separated by only one or two residues, but the problem can be more generally solved by considering alignments from RDCs collected in a second medium (Al-Hashimi et al., 2000). When alignment frames for different media differ in nontrivial ways, only one of the four possibilities for relative alignment of fragments will appear in both sets, allowing a definitive choice of orientations.

Translation of fragments to produce a fully assembled structure is accomplished under distance constraints from a small number of interfragment NOEs observed in ^{15}N-edited NOE data sets or from expected covalent connections of fragments once they are placed in sequence. This is done manually using common molecular graphics programs to monitor constraint distances while fragments are translated. The number of observable long-range backbone-to-backbone proton NOEs is obviously small due to the relatively long distances found between backbone atoms, but combining NOE constraints with limitations imposed by covalent connection through the small numbers of residues between fragments, and limitations imposed by van der Waals contact, produces adequate numbers of constraints. Minimally, three translational constraints per fragment are needed.

Obtaining constraints from a covalent connection requires placement of fragments in proper sequential positions. Data used to do this can come primarily from correlating C_α chemical shifts with amino acid type. The use of C_α shift deviations from random coil values for each amino acid is commonly used to evaluate secondary structure after resonance assignment (Cornilescu *et al.*, 1999; Wishart and Sykes, 1994; Wishart *et al.*, 1992), but, in our case, local structure, as opposed to resonance assignment, is known for each fragment; here, the process can be reversed to use shift departures from those found for secondary structure (ϕ, ψ) types to identify specific amino acids. This cannot be done with certainty site by site, but connected sets of shifts in fragments containing five or more C_α carbons give high probabilities of placement at unique positions in a known protein sequence. This process has been automated in a program called SEASCAPE (SEquential Assignment by Structure and Chemical shift Aided Probability Estimation) (Morris *et al.*, 2004). Definitive assignment and further fragment extension can be aided by ^{15}N-edited TOCSY data. These data are also useful in assignment of H_α chemical shifts for identification of backbone-to-backbone NOE constraints involving H_α protons.

Final assembled structures can be refined by minimization in programs such as XPLOR-NIH using a combination of NOE and RDC constraints (Schwieters *et al.*, 2003). Use of this refinement step is facilitated by having the assembled fragments in their principal alignment frame. Entry of coordinates for pseudoatoms defining an alignment frame then involves just simple displacements of pseudoatoms along *x*, *y*, and *z* axes. The axial alignment and rhombicity parameters that this program normally uses can be entered from known relationships to order parameters provided by the REDCAT program; the axial alignment parameter, D_a, and the rhombicity parameter, R, are given as $1/2(D_{max} \cdot S_{zz})$ and $2/3\eta$, respectively.

Hence, we have a complete protocol for structure determination of protein backbones based primarily on RDC data. Five or more pieces of RDC data are collected about a C_α carbon connecting two peptide planes and allowed ϕ and ψ angles are identified based on best fits to the RDC data (this is usually done for data from two different alignment media). Other C_α carbons are connected based on chemical shift overlap in HNCA-style spectra, eventually generating fragments of known geometry of five or more such carbons. These fragments are placed in sequence using backbone chemical shift data, and the orientations in space of the fragments are determined. NOEs and connectivity restraints allow translation of fragments to form a crude structure, and this structure is minimized under a combination of RDC and NOE constraints. These procedures are illustrated in what follows with specific examples taken from application to structural genomics targets.

Experiments for Obtaining RDCs

For the procedure described here, the collection of adequate numbers of RDCs to define local geometry (ϕ, ψ) and orientation of a peptide fragment is based on just three experiments, an IPAP ^{1}H–^{15}N heteronuclear single-quantum correlation (HSQC) (Ottiger et al., 1998), an HNCA-ECOSY (Weisemann et al., 1994), and an IPAP-HNCO (Tian et al., 2001). These experiments were selected not only for their efficiency but for their applicability to samples having lower levels of ^{13}C enrichment. The use of lower levels of ^{13}C enrichment was undertaken in our case to explore potential cost savings, but there are also protein expression systems, particularly ones for eukaryotic proteins, in which contributions of cell mass and materials from conditioning phases to carbon sources make it difficult to attain a high level of enrichment (Wood and Komives, 1999). The couplings returned in our experiments are illustrated in Fig. 1. These include ^{1}H$_{N}$–^{15}N, ^{1}H$_{N}$–^{13}CO, ^{15}N–^{13}CO, ^{13}C$_{\alpha}$–^{1}H$_{\alpha}$, ^{1}H$_{\alpha(i)}$–^{1}H$_{N}$, and ^{1}H$_{\alpha(i-1)}$–^{1}H$_{N}$ couplings. For a segment of two-peptide planes, nine couplings are potentially available; this number (assuming the absence of accidental degeneracies due to parallel orientation of interaction vectors) is adequate to determine the seven parameters needed to define local geometry and fragment orientation.

The ^{1}H$_{N}$–^{15}N residual dipolar coupling is most easily measured in a variation of the two-dimensional (2D) ^{1}H–^{15}N HSQC experiment in which

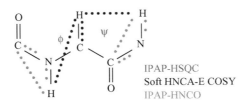

Fig. 1. Illustration of the nine residual dipolar couplings that can be measured for a dipeptide unit using the low percentage ^{13}C labeling scheme. Data from two consecutive residues are combined to determine the dihedral angles ϕ and ψ around the central C$_{\alpha}$ carbon. (See color insert.)

the ^1H_N–^{15}N coupling is allowed to evolve along with the ^{15}N chemical shift. Due to the doubling in the number of peaks as compared to a decoupled HSQC experiment, peak overlap can become a hindrance to the measurement of D_{HN-N} for proteins longer than approximately 80 residues. In these cases, it is helpful to record the upfield and downfield multiplet components in separate spectra using the IPAP-HSQC experiment (Fig. 2) (Ottiger *et al.*, 1998). This experiment is similar to an HSQC but optionally includes a spin-echo element on ^{15}N preceding the t_1 evolution period. In the presence of the spin-echo element, magnetization is labeled in t_1 according to $\sin(\omega t_1)\sin(\pi J_{NH})$, and the resulting doublets are antiphase in t_1. In the absence of the spin-echo element, magnetization is labeled according to $\cos(\omega t_1)\cos(\pi J_{NH})$ in t_1, resulting in in-phase doublets. The in-phase and antiphase spectra are added and subtracted to give the upfield and downfield multiplet components, respectively.

$D_{H\alpha HN}$ and $D_{C\alpha H\alpha}$ are measured with an HNCA-ECOSY pulse sequence (Fig. 3) (Tian *et al.*, 2001). In this experiment, magnetization originating on H_N is transferred through an INEPT element to ^{15}N. ^{15}N is then allowed to evolve during the constant-time t_1 period, with WALTZ decoupling of protons. From there, another INEPT element is used to transfer magnetization to C_α. The C_α–H_α coupling and the C_α chemical shift are then allowed to evolve together during t_2. This allows the observation of the $D_{C\alpha H\alpha}$ residual dipolar coupling in the ^{13}C dimension. Magnetization is

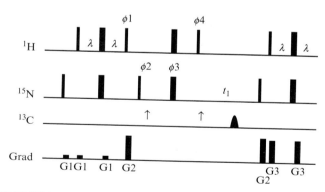

Fig. 2. IPAP-HSQC pulse sequence (Ottiger, 1998). The phase cycle is $\phi1 = -y,y$; $\phi2 = x,x,-x,-x$ (in-phase spectrum) and $\phi2 = -y,-y,y,y$ (antiphase spectrum); $\phi3 = 4(x),4(y),4(-x),4(-y)$; $\phi4 = 8(x),8(-x)$; reciever $= x,-x,-x,x$ (in-phase spectrum) and $x,-x,-x,x,-x,x,x,-x$ (antiphase spectrum). To achieve quadrature detection in t_1, States-TPPI incrementation is applied to $\phi2$ and $\phi3$. The portion of the pulse sequence between the arrows ($\uparrow$) is ommitted for the in-phase spectrum. The final pulse on ^1H during the INEPT transfer is a 3919 compound pulse, used for water suppression. Gradients employed are G1 = (4 G/cm, 1 ms), G2 = (26 G/cm, 0.25 ms), and G3 = (24 G/cm, 1 ms).

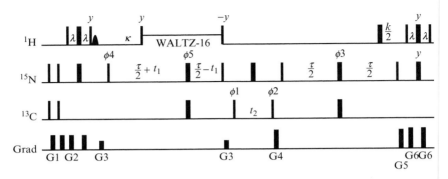

Fig. 3. The HNCA-ECOSY pulse sequence. The phase cycle is $\phi1 = x,-x$; $\phi2 = x$, $x,-x,-x$; $\phi3 = 8(x),8(y)$; $\phi4 = x$; $\phi5 = x,x,x,x,- x,-x,-x,-x$; receiver $= x,-x,-x,x,x,-x,-x,x$, $-x,x,x,-x,-x,x,x,-x$. Quadrature detection in t_2 is achieved by cycling $\phi1$ in States-TPPI fashion. Quadrature detection in t_1 is achieved by States-TPPI cycling of $\phi4$ and $\phi5$. Gradients employed are G1 = (20 G/cm, 0.5 ms), G2 = (20 G/cm, 1 ms), G3 = (12 G/cm, 1 ms), G4 = (26 G/cm, 1 ms), G5 = (26 G/cm, 1 ms), G6 = (28 G/cm, 1 ms).

then transferred through ^{15}N to ^{1}H$_N$ for detection. The ECOSY principle is used to transfer coherence from each doublet component seen in the ^{13}C dimension selectively to a single doublet component of the ^{1}H$_N$ doublet seen in the ^{1}H dimension. This means that the H$_\alpha$ spin (the source of coupling for both doublets) must experience an even number of π rotations after t_2 so that the α and β spin states are not mixed. The elimination of two of the four peaks that would normally be seen in a COSY spectrum (or fully coupled HSQC) allows the observation of the relatively small ^{1}H$_N$–H$_\alpha$ coupling as horizontal (ν_3) displacements of the remaining diagonally displaced peaks.

When the experiment is performed using a high-Q probe such as a cold probe, radiation damping of the water signal becomes an important consideration. The best water suppression is achieved through empirical adjustment of the soft pulse following the initial ^{1}H–^{15}N transfer. Both the phase and the amplitude of this pulse can be adjusted to give the smallest remaining water signal. Improvement in water suppression can also be obtained by replacing the final 180° pulse on ^{1}H with a 3-9-1-9 selective pulse (Sklenar et al., 1993). The version shown is not a sensitivity-enhanced version, but design of such a version is likely to be possible.

The HNCA-ECOSY pulse sequence, in particular, illustrates the point that there are advantages to low levels of ^{13}C enrichment other than savings in materials costs. One such advantage is due to the absence of $J_{C\alpha C\beta}$ and $J_{C\alpha C'}$ couplings in natural abundance or partially enriched samples. These couplings are large enough that if various coupling elimination techniques are not used with fully enriched samples, they will dominate the ^{13}C line width. This is a problem for both experiments designed to obtain sequential connectivities and experiments designed to measure $^1H_\alpha$–$^{13}C_\alpha$ couplings. In standard HNCA experiments used for sequential connectivity, attempts to remove couplings between C_α and C_β carbons are seldom made. This limits resolution and makes correlation of $C_{\alpha i}$ and $C_{\alpha(i-1)}$ chemical shifts of limited utility for sequential connection of residues, hence, the popularity of CACBNH experiments in which both C_α and C_β interresidue correlations are made. With low enrichment ^{13}C–^{13}C couplings do not occur with significant probability and chemical shift correlations become much more valuable. In fact, the combination of $C_{\alpha i}$ – $C_{\alpha(i-1)}$ connectivities and $D_{C\alpha H\alpha}$ couplings measured from the i and $i + 1$ residues (to be described below) can yield sequential assignments comparable in quality to what can be obtained from a combination of C_α and C_β connectivities. Another advantage comes from the fact that connectivities from H_N to the $C_{\alpha i}$ and $C_{\alpha(i-1)}$ carbons seldom occur in the same molecule when ^{13}C enrichment is low. This means that magnetization is not divided between two pathways, and some of the signal loss from low enrichment is regained.

For measurement of $^{13}C_\alpha$–$^1H_\alpha$ couplings in highly enriched samples, the effects of ^{13}C–^{13}C couplings can be removed by constant-time techniques (Bax *et al.*, 2001). Commonly used is the CT-(HA)CA(CO)NH experiment. Here a relatively long CT period is used to improve resolution of peaks, but this usually comes at some cost in sensitivity and in restrictions on the choice of evolution times. The passage of magnetization through the carbonyl carbon is used to eliminate the transfer of magnetization from both intra- and interresidue C_α carbons. This simplifies spectra, but requires that other experiments be used to establish residue connectivities.

The smallest backbone residual dipolar coupling considered here is $D_{NC'}$. It is also the most difficult to measure. It is useful to measure $D_{NC'}$ as a splitting in the ^{15}N dimension due to the favorable relaxation properties of ^{15}N. The most important consideration in this type of experiment, when done with partially ^{13}C-enriched samples, is that ^{15}N magnetization arising from non-^{13}C-labeled molecules must be filtered out by the pulse sequence. In a sample at natural abundance in ^{13}C or enriched at the 16% level commonly used in our studies, the majority of the ^{15}N magnetization arises from non-^{13}C-labeled molecules, so the filtering must be efficient. In particular, it is not sufficient to run an ordinary ^1H–^{15}N correlation experiment in the absence of ^{13}C decoupling (Wang *et al.*, 1998). One successful pulse sequence for the measurement of $D_{NC'}$ has been the 2D IPAP-HNCO sequence (Fig. 4). This also allows the measurement of $D_{HN-C'}$. This pulse sequence begins with an INEPT-type transfer of magnetization

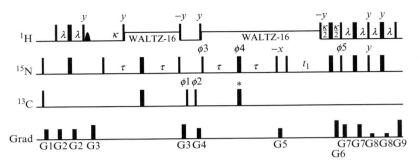

FIG. 4. The IPAP-HNCO pulse sequence. The phase cycle is $\phi1 = x,-x$; $\phi2 = x,x,-x,-x$; $\phi4 = x,x, \; x,x,-x,-x,-x,-x$; $\phi5 = x$; receiver $= x,-x,-x,x$. The antiphase spectrum is recorded by setting $\phi3 = y$ and excluding the pulse marked with an asterisk (*). The in-phase spectrum is recorded by setting the phase $\phi3 = x$ and including the pulse marked with an asterisk. Sensitivity enhancement is achieved by inverting the sign of gradient G6 for the second component of each t_1 increment. Gradients employed are G1 = (20 G/cm, 0.5 ms), G2 = (20 G/cm, 1 ms), G3 = (12 G/cm, 1 ms), G4 = (26 G/cm, 1 ms), G5 = (15 G/cm, 1.5 ms), G6 = (26 G/cm, 1 ms), G7 = (28 G/cm, 1 ms), G8 = (6 G/cm, 1 ms), G9 = (26 G/cm, 0.15 ms).

from ^1H_N to ^{15}N. Filtering for ^{13}C-labeled molecules is achieved by using the $J_{NC'}$ coupling to generate $N_zC'_z$ magnetization, while ^{15}N magnetization arising from non-^{13}C-labeled molecules remains transverse. Pulsed field gradients then dephase the transverse magnetization. Finally, the remaining ^{15}N magnetization is allowed to evolve and is then transferred back to H_N for detection. The result is efficient selection for magnetization residing on ^{15}N bound to ^{13}C. The sequence also uses the ECOSY principle; in this case the common source of couplings is the carbonyl carbon, $^{13}C'$. The $^{15}N-C'$ coupling is measured in t_1, and the H_N-C' coupling is measured in the directly detected dimension. The ^{13}C nucleus experiences no pulses between the end of the t_1 evolution period and the end of the pulse sequence; therefore the α and β spin states are not mixed, allowing ECOSY-type separation. The sequence reported here is similar to that previously described (Tian *et al.*, 2001), except that in the current sequence, the upfield and downfield components of each multiplet appear in separate spectra. This separation is very helpful in relieving the overlap due to the small size of $J_{NC'}$ and $J_{HN,C'}$. This becomes more important as the size of the protein under study increases, resulting in broader lines and more numerous peaks.

Measuring RDCs in the above spectra can be a formidable task. Some of this is due to the sheer volume of data and the need to make proper correlations of peaks between data sets. Here automatic peak picking routines available in programs such as NMRPipe and NMRDraw are useful (Delaglio *et al.*, 1995). Scripts can be designed to transfer default assignments from one set to another and automatic deposition to databases can facilitate subsequent analysis of data. In many cases precision of measurement with these tools is adequate. We estimate that agreement of RDCs with structures produced is inherently limited to about 10% of the total range of couplings by the inaccuracy of the peptide geometries used to build our models. Natural out-of-plane distortions for the N–H bond vector are, for example, estimated to be as much as 6° (MacArthur and Thornton, 1996). If this variation occurs near the magic angle, the corresponding change in a measured RDC is 10% of the range. Hence, measurement with a precision of better than 10% is not useful in our application. For ^1H–^{15}N couplings of $+20$ Hz to -10 Hz, the required 3 Hz precision is for most data sets attainable with standard peak picking routines. Where needed, better precision can be obtained by manually picking peak centers in displayed columns or fitting peaks to Lorenzian or other line shapes.

There are some special problems that occur in HNCO-ECOSY spectra and in the measurement of H_N–H_α couplings from HNCA-ECOSY spectra. These couplings can be small, and line shapes can be distorted by various cross-correlation effects and differential relaxation of in-phase and anti-phase components. Here we have found simulation of combined peak shapes useful as well as empirical offsets of measured couplings to compensate for relaxation. We nevertheless typically raise error estimates for these couplings to approximately half the relevant line width.

Sample Preparation

Isotope Labeling

Samples are prepared to have a high level of ^{15}N enrichment but more modest levels of ^{13}C enrichment. In our case ^{13}C incorporation is achieved through the use of a mixture of C1-^{13}C-glucose and C2-^{13}C-glucose as a carbon source in minimal media designed for use with an *Escherichia coli* host. This results in a 15–20% ^{13}C-labeled sample with nearly random distribution of labeled sites. As outlined above, this system has a number of advantages over those that provide full ^{13}C isotope labeling: the isotope costs for the sample preparation are somewhat lower than those for a fully labeled protein (this could be lower if demand for C1-^{13}C-glucose and C2-^{13}C-glucose were higher). In addition, spectra of aligned samples are simplified through the reduction in the number of long-range couplings.

To accomplish protein expression, a pET plasmid containing the clone for the protein of interest is transformed into *E. coli* BL21(DE3) and plated onto M9 plates (Sambrook *et al.*, 1989) containing the appropriate antibiotic. The cells are grown for around 20 h at 37°, since they will grow more slowly and require more time to produce colonies than on a rich medium plate. This transformation step is important to select for colonies that grow in M9; however, isotope labels are not needed in this step. The following day a colony is picked and used to inoculate 50 ml of M9-containing antibiotic and 2 g ^{13}C-1 glucose, 1 g ^{13}C-2 glucose and 1 g ^{15}N ammonium chloride as the sole ^{13}C and ^{15}N sources, in a 250-ml flask with baffles. The cells are incubated for about 20 h at 37° with shaking. The next morning, 1 liter of M9 with antibiotics is inoculated with a 20-ml aliquot of the overnight culture. The culture is induced with IPTG (0.5–1.0 mM) when the OD$_{600}$ is ∼0.7 (usually about 4 h after inoculation). After induction, the temperature of the growth can remain at 37° or be lowered (25° or 30°, etc.) depending on the expression level of the particular protein. The cells are harvested 6–8 h after induction, depending on the temperature being used (longer times for lower temperatures).

Sample Preparation and Alignment

As outlined in the introduction, measurement of residual dipolar couplings requires that the protein be aligned in a liquid crystalline medium. It is also useful to have RDC data collected under multiple alignment conditions (usually two), as additional alignment sets can help resolve the 4-fold degeneracy that results from one medium (see above). Although it is ideal to collect a full RDC data set under each condition, it may be sufficient to collect one full set and a partial set (e.g., ^{1}HN–^{15}N) for a second medium. There are many types of alignment being utilized for these purposes (Prestegard and Kishore, 2001), but we typically rely on Pf1 filamentous phage (Hansen *et al.*, 1998) and polyethylene glycol-alkylether (PEG) bicelles (Ruckert and Otting, 2000). The PEG bicelles, in turn, can be doped with negative (SOS, SDS) or positive (CTAB) agents to provide yet another set of alignment conditions.

Preparation of a sample aligned with Pf1 phage is fairly straightforward. Because the protein will be diluted by the alignment medium, a concentrated stock solution of protein is used. Pf1 phage (Hansen *et al.*, 1998) is usually provided at a concentration of 50 mg/ml. The working sample is prepared at concentrations of about 1 mM protein and 10 mg/ml phage by making the appropriate dilutions. D$_2$O is added to a final concentration of 10%. The ^{2}H splitting is measured (see below); if it is too low, the concentration of phage can be increased to 15 or 20 mg/ml as needed.

Preparation of a sample aligned with polyethylene glycol-alkylether (PEG) bicelles is slightly more involved, but still achievable. Both C8E5 (pentaethylene glycol octyl ether) and C12E5 (pentaethylene glycol dode-cyl ether) can be used; C12E5 usually provides a proper level of alignment at room temperature. A stock solution of 8% (w/v) PEG is prepared and diluted to a working concentration of about 4%. As with the phage sample, a concentrated stock solution of protein is used; it contains 10% D_2O and is in buffer at the correct pH. The 8% PEG stock solution consists of 50 μl C12E5, 16 μl hexanol, and 250 μl buffer containing 10% D_2O at the correct pH. The C12E5 and buffer are mixed well with vortexing. The hexanol is added in 4-μl increments, with vortexing after each addition. The solution goes from clear to milky, turbid, and then to transluscent and viscous with lots of bubbles. Hexanol is added until the solution goes clear again. If it becomes milky/turbid again, the solution has gone past the nematic phase. The working sample of 4% PEG is prepared by diluting the protein and PEG 1:1 with vortexing. After incubation at room temperature overnight, the measured ^2H splitting (see below) should be about 15 Hz. PEG can be doped with CTAB or SOS to provide a second alignment tensor. The ratio of PEG:CTAB is typically 27:1; the ratio of PEG:SOS is typically 30:1.

It is not always guaranteed that the liquid crystalline medium will align in the magnetic field, or that if it does, the protein will also align. There is also the possibility that the protein will interact with the alignment medi-um, which may result in a much higher degree of alignment than is desired. Hence, the strength of alignment in a particular liquid crystalline medium needs to be assessed. To determine the level of alignment of the liquid crystals themselves, a simple one-dimensional ^2H spectrum is obtained. This is often done on a sample without protein added, to check the stability of the medium alone. Alignment results in splitting of the ^2H water resonance from the 10% D_2O present in the sample. When the medium is homogeneous, a symmetrical doublet is produced (Fig. 5). If the medium is not fully aligned, a third peak corresponding to the isotropic signal is commonly seen (it may take an hour or more for the sample to equilibrate to a uniform oriented medium). The separation of the peaks is measured in Hz. Although this splitting does not directly relate to the level of protein alignment, a splitting of 15 Hz often results in magnitudes of ^1H_N–^{15}N couplings in the 15–20 Hz range for the protein. The quality of both the medium and the data that will be produced can be further assessed from an IPAP ^{15}N–^1H HSQC experiment before a full data set is collected. This can be done on a sample labeled only with ^{15}N if conservation of the ^{13}C protein is desired, but for the final data set, these RDCs must be recollected on the ^{13}C sample so the entire data set is self-consistent.

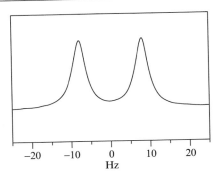

Fig. 5. ^{2}H spectrum of an aligned protein sample. The sample was 0.5 mM PF0255 aligned in 50 mM sodium phosphate, 100 mM KCl, 90% H_2O, and 4% (w/v) C12E5-hexanol bicelles. The ^{2}H splitting measured was 16 Hz, which corresponded to magnitudes of ^1H_N–^{15}N couplings in the range of −9 to 15 Hz.

Examples of Spectra

$D_{HN,N}$

^1H_N–^{15}N RDCs can be measured using simple versions of a 2D coupled H–N HSQC. For larger proteins, however, an IPAP version of this sequence (Ottiger et al., 1998) that reduces spectral overlap is preferred. The pulse sequence for this experiment was shown in Fig. 2, and examples of this spectrum under isotropic and aligned conditions are shown in Fig. 6 for a 13.8-kDa target protein (PF0385). This protein from *Pyrococcus furiosus* is annotated only as a conserved hypothetical protein and has no significant sequence identity to anything in the Protein Data Bank. The spectrum is plotted with peaks color coded for the sum and difference of the in-phase and antiphase components. These are normally observed and analyzed in separate spectral panels. The separation of the two peaks in the vertical dimension is the sum of the RDC and the scalar coupling ($D_{HN,N}$ + $^1J_{HN,N}$) in the aligned spectrum and the scalar coupling alone in the isotropic spectrum ($^1J_{HN,N}$). Note that on alignment the magnitudes of some couplings decrease and the magnitudes of some couplings increase, corresponding to preferred angles near zero and near 90°, respectively. Also note that an increase in the magnitude of a coupling for ^1H_N–^{15}N actually corresponds to a negative RDC; the scalar one bond coupling is negative for ^1H_N–^{15}N because of the negative magnetogyric ratio of ^{15}N. Keeping signs straight is important when multiple types of RDCs are to be combined in a structural analysis.

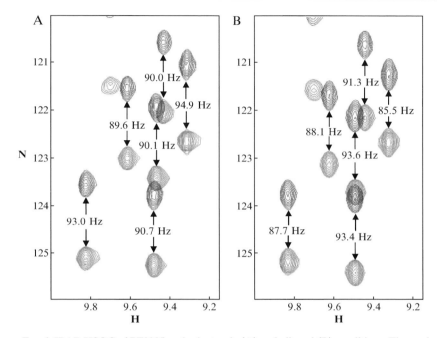

FIG. 6. IPAP-HSQC of PF0385 under isotropic (A) and aligned (B) conditions. The peaks corresponding to the sum and difference spectra are colored red and black, respectively. $^{15}N-^{1}H$ splittings are indicated. The isotropic sample contained 1 mM PF0385 in 20 mM sodium phosphate, 15 mM KCl, 90% H_2O. The aligned sample contained 0.5 mM PF0385 in 20 mM sodium phosphate, 85 mM NaCl, 10 mg/ml Pf1 phage, 90% H_2O. (See color insert.)

A large number of points in the t_1 dimension has been collected to provide resolution sufficient for accurate measurement of the coupling (a 100 ms acquisition time is usually sufficient). This experiment is generally very sensitive and produces highly accurate coupling values. As an example, data collection for the ^{15}N-coupled HSQC includes 256 complex t_1 points and 2048 t_2 points collected over 2 h using a cryogenic probe. Prior to Fourier transformation, data in the direct dimension were corrected for solvent, multiplied by a squared sine bell shifted by 90°, and zero-filled to 4096 points. Data in the indirect dimension were multiplied by a squared sine bell shifted by 90°, linear predicted from 256 to 512, and zero-filled to 4096 points. Peaks were picked using the automated peak picking function in NMRPipe (Delaglio *et al.*, 1995) and inspected manually for accuracy. Peaks were classified as good, slightly overlapped/distorted, and severely overlapped/distorted, so that more weight could be given to the most reliable data. Arbitrary peak labels were transferred in automatically from an isotropic HSQC spectrum. J or $J + D$ values were automatically calculated as the difference in Hertz between the coupled peaks in the ^{15}N dimension, and the data were stored in a database for automated recovery and analysis.

$D_{C\alpha H\alpha}$, $D_{H\alpha(t-1)HN}$, $D_{H\alpha(t)HN}$

Couplings involving C_α and H_α were collected using a three-dimensional soft HNCA-ECOSY (Weisemann *et al.*, 1994). The pulse sequence was shown in Fig. 3, and typical spectra are shown in Figs. 7 and 8 for a 7.8-kDa target protein (PF0255) under isotropic and aligned conditions. PF0255 is a *Pyrococcus furiosus* protein that was annotated as a DNA-directed RNA polymerase subunit. However, it had less than 23% sequence identity to its nearest neighbor in the Protein Data Bank. The HNCA-ECOSY experiment provides C_α–H_α couplings in the indirect ^{13}C dimension and H_α–H_N

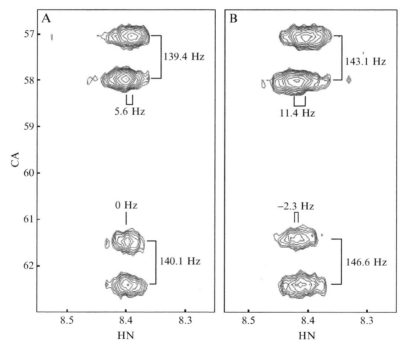

Fig. 7. HNCA-ECOSY of PF0255 under isotropic (A) and aligned (B) conditions. $^{1}H_\alpha$-$^{1}H^N$ and $^{1}H_{\alpha(i-1)}$ -$^{1}H^N$ splittings are indicated. The samples contained 1.0 mM PF0255 (isotropic) or 0.5 mM PF0255 (aligned) in 50 mM sodium phosphate, 100 mM KCl, 90%H2O. The aligned sample contained 4% (w/v) C12E5-hexanol bicelles.

couplings in the direct dimension for both intra- and interresidue peaks. Identification of the interresidue resonances is aided by the observation that the $H_\alpha(i-1)$–H_N coupling is zero under isotropic conditions. Examples of this are shown in Fig. 7. The pair of peaks at 62 ppm and the pair at 57.5 ppm in the carbon dimension correspond to an interresidue pair and an intraresidue pair, respectively. Note that the interresidue pair at 62 ppm in the aligned case shows a small negative splitting. This corresponds to the through space RDC observable for this pair with a four-bond covalent separation. The relative intensity of the peaks can also be used as a guide to assigning intra- and interresidue status (intraresidue is usually more intense). Even using a sample with a low level of isotopic labeling, the sensitivity of this experiment is adequate for the study of proteins with molecular mass <20 kDa. Despite this being a three-dimensional experiment, there are sometimes issues of spectral overlap that compromise accuracy for the couplings measured. For this reason, we classify the peaks as good, overlapped, or questionable, and we use larger errors for the less accurate peaks.

The HNCA-ECOSY experiment also provides our primary way of connecting residues that are adjacent in sequence. Because both intraresidue and (i)–$(i-1)$ connectivities occur in a single H_N column, an intraresidue C_α shift in one column can be matched with an interresidue C_α shift in another to make the connection. Splittings for a given C_α–H_α pair are also measurable from both sets of cross peaks and the (i)–$(i-1)$ splitting from one column must match the (i) splitting from the second column for the correct connection. Since these splittings vary widely under aligned conditions, this is extremely useful additional connectivity information. The expanded view of the HNCA-ECOSY shown in Fig. 8 illustrates the utility of this information.

Fragment connection is performed in a three-step process that is designed to be as efficient and automated as possible. It begins with an automated filter step that eliminates any residue that does not match both peaks in the coupled pair within a given chemical shift cutoff. A generous cutoff of 0.2 ppm in the ^{13}C dimension is used to avoid discounting a possible match between weak, overlapped, or distorted peaks. The remaining small number of possible matches is inspected manually using a script written in NMR View (Johnson and Blevins, 1994). This allows classification of possible matches from all spectra, eliminating poor candidates. The final list of possible matches is then run through an algorithm that chains pairs together into possible fragments. All possibilities are saved for later structural analysis when some inconsistent assemblies will be eliminated. The entire process is extremely efficient, taking on the order of 1 h for a 100-residue protein.

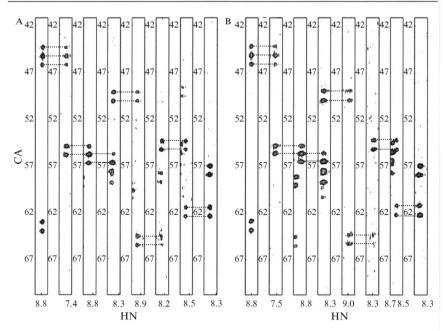

Fig. 8. HNCA-ECOSY strip plot of a fragment of PF0255 connected using C_α chemical shifts and RDCs under isotropic (A) and aligned (B) conditions. The samples contained 1.0 mM PF0255 (isotropic) or 0.5 mM PF0255 (aligned) in 50 mM sodium phosphate, 100 mM KCl, 90% H_2O. The aligned sample contained 4% (w/v) C12E5-hexanol bicelles. Identification of the first residue as a glycine (a triplet owing to the two H_α protons) and the downfield shifting of the C_α of residues 5 and 7 led to the assignment of this fragment to the sequence GKYAIRVR.

Data collection for the soft HNCA-ECOSY used as an example (0.5 mM protein in aligned medium) included 72 t_1 (^{13}C) points for an acquisition time of about 20 ms, 16 (^{15}N) t_2 points, and 2048 t_3 points collected over 37 h using a cryogenic probe on a 600-MHz spectrometer. Prior to Fourier transformation, data in the direct dimension were apodized as described above and zero-filled to 4096 points. Data in the ^{13}C dimension were linear predicted from 72 to 128 points, apodized, and zero-filled to 256 points. Data in the ^{15}N dimension were linear predicted from 16 to 32, apodized, and zero-filled to 64 points. J or $J + D$ values were automatically calculated as the difference in Hertz between the coupled peaks in the ^{1}H and ^{13}C dimensions and stored in a database for automated recovery and analysis.^{13}C$_\alpha$ chemical shifts were also stored for use in fragment identification (see below).

$D_{HN\text{-}C'}, D_{NC'}$

Couplings involving the carbonyl carbon are collected using a 2D modified HNCO experiment (Tian *et al.*, 2001) in a manner very similar to that of the coupled HSQC. As seen in the example spectra in Fig. 9, the $D_{NC'}$ and $D_{HN\text{-}C'}$ couplings are quite small (1–8 Hz) and are often difficult to measure. Like the HNCA-ECOSY, the multiple ECOSY peaks that appear for each H_N lead to overlap. For this reason, an IPAP version (F. Tian, unpublished results) (Fig. 4) was utilized for the larger target protein (PF0385, 13.8 kDa). This experiment may also provide more accurate measurement of the couplings in smaller proteins. The sum and difference spectra (red and black, respectively) have been overlaid to highlight the offsets in both the direct (^1H) and indirect (^{15}N) dimensions. Note that peaks are displaced along the positive diagonal in the isotropic spectrum indicating that the couplings in the two directions have the opposite sign.

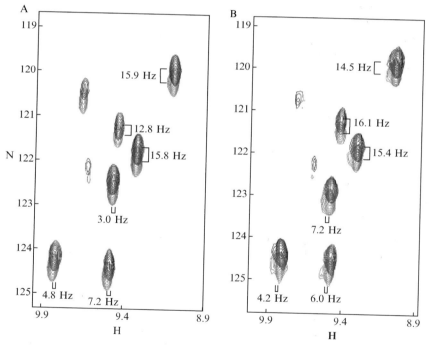

FIG. 9. IPAP-HNCO of PF0385 under isotropic (A) and aligned (B) conditions. The samples were 1 m*M* PF0385 in 20 m*M* sodium phosphate, 15 m*M* KCl, 90% H_2O for the isotropic and the same buffer, 85 m*M* NaCl, 10 mg/ml Pf1 phage for aligned. The peaks corresponding to the sum and difference spectra are colored red and black, respectively. (See color insert.)

However, there are sometimes exceptions in the aligned case. Because the RDC contribution to the CN splitting is never larger than the scalar coupling, we know this coupling to be negative, and the direction of the offset allows determination of the absolute sign of the RDC.

Data collection for the IPAP-HNCO shown included 256 t_1 points and 2048 t_2 points over 16 h at 600 MHz using a cryogenic probe. Prior to Fourier transformation, data in the direct dimension were apodized and zero-filled to 4096 points. Data in the ^{15}N dimension were linear predicted from 256 to 512, apodized, and zero-filled to 4096 points. The data were processed using IPAP scripts that combined the in-phase and antiphase signals to give the sum and difference spectra. Peaks were picked from both spectra, and the peak lists combined to allow for calculation of the RDCs. J or $J + D$ values were automatically calculated as the difference in Hertz between the coupled peaks in the ^{15}N or ^1H dimension, and the data were stored in a database for automated recovery and analysis.

An ^{15}N-edited NOESY-HSQC was also collected to aid in fragment assembly of the two proteins discussed. Identification of a small number (typically about a dozen) of interfragment NOEs provides information on the spatial arrangement of the fragments. The NOESY-HSQC is typically collected with 128 t_1 points, 16 t_2 points, and 2048 t_3 points over 16 h. Also a TOCSY-HSQC was collected in parallel with the NOESY-HSQC. This provides a convenient way of correlating H_α resonance position to HSQC cross-peaks for use in assignment of NOE constraints. In addition, information about amino acid type can be obtained and utilized for sequence-specific assignment of fragments.

Data Analysis

Data Assessment from Powder Patterns

Before embarking on the task of structure determination, it is advantageous to make a general assessment of the quality of various types of data and identify any anomalous points that may contaminate analysis. One simple tool is a direct comparison of distributions RDCs from different nuclear pairs (histograms showing the number of couplings measured at each possible coupling value). In principle, with a sufficiently large number of couplings, all space should be sampled and powder patterns should result that differ only by scaling factors related to the sizes of magnetogyric ratios for the coupled nuclei, the signs of these ratios, and distances between nuclei. Any inexplicable differences between two different sources of data will highlight systematic errors in data treatment (such as incorrect assumptions about signs of couplings); the patterns will also help to identify outliers that need to be manually reexamined, and they will give estimates of order parameters that can be used as filters for proper geometric solutions.

The first step in performing a powder pattern analysis is the conversion of all RDC values from units of Hertz to unitless measurements. This conversion can be performed using Eq. (3).

$$s = \frac{D \times r^3}{D_{max}} \qquad (3)$$

Here D denotes the experimental data, r is the length of the internuclear vector that joins the two nuclei of interest, and D_{max} is the maximum observable RDC for two nuclei at 1 Å distance under perfect alignment. Values for D_{max} and r used in our calculations are shown in Table I. Figure 10 illustrates a comparison of the distribution of RDCs for ^1H–^{15}N and ^1H–$^{13}C_\alpha$ couplings. Rather than plotting simple histograms, smoothed curves have been plotted utilizing Parsen density estimation (Fukunaga, 1990). The data are for PF0255, the same protein used to illustrate the HNCA-ECOSY data set described above.

TABLE I
MAXIMUM RDC VALUES AND BOND DISTANCES

	D_{max}	$r(Å)$
C–N	6125	1.335
N–HN	24350	1.010
C–H	−60400	2.035
C_α–H_α	−60400	1.090

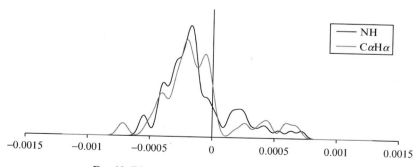

FIG. 10. Distributions of N–H_N and C_α–H_α RDCs.

The agreement shown in Fig. 10 is fairly good. There is a possible outlier at a value of −0.0007 in the C_α–H_α set. This data point was checked and appeared to be a valid measurement. Outlying data may not in all cases be due to error but may be due to sparse sampling of vectors in space. Also, in the case of C_α–H_α data, the outliers can originate from glycines, since the coupling value is often reported as the sum of both C_α–$H_{\alpha 1}$ and C_α–$H_{\alpha 2}$ residual dipolar couplings.

The fact that the central maximum lies at a negative value for both distributions in Fig. 10 suggests that the relative signs of couplings have been assigned properly (largest couplings negative for C_α–H_α and positive for H–N). A simplistic analysis of the shape of the distribution would assign extreme values to S_{zz} and S_{yy} in accord with the convention $|S_{zz}| \geq |S_{yy}|$. The third-order parameter (S_{xx}) can be calculated by utilizing the traceless property of the order tensor ($S_{xx} = -S_{zz} - S_{yy}$). Data for H–N measurements in Fig. 10 would set $S_{zz} = 0.00075$, $S_{yy} = -0.0006$, and $S_{xx} = -0.00015$. It can theoretically be proven that the most frequently observed value of RDC for a uniformly distributed set of vectors in space is the value corresponding to S_{xx} (Valafar and Prestegard, 2003; Varner et al., 1996). This would imply that the highest peak in the powder pattern should correspond to S_{xx}. This is again consistent with the numbers suggested above. Other more sophisticated methods such as maximum likelihood (Warren and Moore, 2001) can be employed to provide a better estimate of the order parameters by fitting the entire powder pattern rather than just the extrema.

The above order parameter estimates can be used as additional filters for fragment geometry solutions coming from the program to be described in the next section. The number of data points used to find allowed geometries and order matrix elements will be small, and false geometries can easily arise in combination with principal order parameters that are inconsistent with the distributions described above. Once several fragments with reliable geometries and consistent order parameters are found, it is often better to substitute principal order parameters from fragment solutions for the powder pattern estimates described above.

Structure Determination Using REDCRAFT

The actual search for fragment geometries consistent with RDC data is performed by a program named REDCRAFT (H. Valafar, unpublished observations). This program runs as either a single processor version or a Linux cluster version. As described in the introduction, it constructs backbone geometries for peptide fragments in a manner consistent with allowed Ramachandran space, measured $^3J_{\text{HN-H}\alpha}$ scalar couplings, and an experimentally collected set of RDCs.

Although the program itself is very general, the input for the current version has been tailored to the particular set of couplings produced by the experiments described above. This input consists of repeating blocks of data in the format described in Table II, one block for every residue of the fragment. The first line of a block begins with designation of the amino acid type for the residue. Because we work only with backbone data, and side chains are irrelevant, types are designated as either GLY (signifying glycine) or ALA (signifying alanine). The line continues with $^3J_{HN-H\alpha}$ for the residue and concludes with a comment. Comments can be any string data, but we find this field a convenient place to store residue numbers or peak numbers that correlate data with peaks in a reference $^{15}N-^1H$ HSQC data set.

The remaining six lines begin with one of the six RDCs that can be detected through a particular $^{15}N-^1H_N$ pair (in Hertz with the proper sign). Note that in the case of glycines, the $D_{C\alpha-H\alpha}$ is actually $D_{C\alpha1-H\alpha1} + D_{C\alpha2-H\alpha2}$. Similarly, the value of the $^3J_{HN-H\alpha}$ is reported as the sum of both couplings, and for a residue following glycine, the $D_{H\alpha(i-1)-HN}$ value is reported as a sum value. Any missing RDCs are indicated as 999. Each coupling is followed by a scaling factor that is used to account for the information content of the various couplings and concludes with another comment field. The scaling factors are based on the ratio of $D_{max,N-HN}$ to the maximum coupling for a given entry (Table I) and the ratio of

TABLE II
REPEATING BLOCKS OF DATA FOR REDCRAFT INPUT
(FORMAT LEFT, EXAMPLE RIGHT)

AA type	$^3J_{HN-H\alpha}$	Comments	GLY	15.792	Peak 52
D_{C-N}	ε_{C-N}		1.745	2	
D_{N-HN}	ε_{N-HN}		3.715	1	
D_{C-H}	ε_{C-H}		−8.034	1.333	
$D_{C\alpha-H\alpha}$	$\varepsilon_{C\alpha-H\alpha}$		−30.35	1	Sum of Ds
$D_{H\alpha-HN}$	$\varepsilon_{H\alpha-HN}$		5.985	0.888	Sum of Ds
$D_{H\alpha(i-1)-HN}$	$\varepsilon_{H\alpha(i-1)-HN}$		−1.075	1.333	
AA type	$^3J_{HN-H\alpha}$	Comments	ALA	4.2581	Peak 20
D_{C-N}	ε_{C-N}		0.316	1.333	
D_{N-HN}	ε_{N-HN}		−9.039	1	
D_{C-H}	ε_{C-H}		−0.209	0.888	Weak
$D_{C\alpha-H\alpha}$	$\varepsilon_{C\alpha-H\alpha}$		−14.938	1	
$D_{H\alpha-HN}$	$\varepsilon_{H\alpha-HN}$		2.528	1.333	
$D_{H\alpha(i-1)-HN}$	$\varepsilon_{H\alpha(i-1)-HN}$		−0.68	1.333	Sum of Ds

estimated percentage experimental error for D_{HN-N} to that for a given entry relative to a standard error. The result is a simple ratio of the estimated average error of H_N couplings relative to error of the coupling in question. Because of structural noise, errors for our more precise measurements are often set to 10% of the range of couplings observed. Others reflect experimental precision of measurement with reasons for deviation often included in the comment field. REDCRAFT can take multiple input files, with each one containing data from a different orientation medium. In this case a common geometry satisfying all data is found.

The time required for geometry searches depends on the depth of search into the ranked lists of allowed ϕ and ψ angles for each dipeptide fragment to be joined. Fragments as large as 20 residues can be processed using the parallel version of REDCRAFT with a reasonable search depth (1000) in less than a few hours. An entire protein of size 70 residues can be processed in less than a day. Under the conditions where data are numerous and of sufficiently high quality (such as for rubredoxin; BMRB accession number 5926), the search depth can be quite shallow, and a 25-residue fragment can be successfully processed under an hour on a typical desktop computer.

The final outcome of this analysis consists of a single file containing a ranked list of all examined structures. These are described by departures in ϕ and ψ from the extended starting conformation (IUPAC angles can be obtained by subtracting 180°). The top entry will be the geometry that best fits the experimental RDCs. Table III shows every tenth entry for the top 100 solutions produced by REDCRAFT for a seven-residue fragment from the C-terminus of the PF0255 protein (K46–V52). The last column indicates the root-mean-square deviation (RMSD) between the backcalculated scaled RDCs (using the best solution order tensor for the given geometry) and the measured RDCs. The data used in this particular calculation included 38 RDCs from alignment in a bacteriophage medium and 24 RDCs from alignment in a C12E5/hexanol bicelle medium. The derived conformation obviously agrees well with the data as indicated by the RMSD values and clustering of the ϕ and ψ values. The particular segment examined terminates because of a proline in the next position. A second segment can be examined independently beginning with the residue after proline (G54–R61).

In the case presented, we were able to sequentially place the fragments as being before and after proline 53. REDCRAFT is able to accommodate missing data, including that for an entire residue such as proline. Therefore, an entire 16-residue segment was also assembled using 67 RDCs for the first medium and 57 RDCs from the second medium. As a final option in the REDCRAFT determination protocol, angles determined for each

TABLE III
SAMPLING OF TOP 100 SOLUTIONS FOR A FRAGMENT OF SEVEN RESIDUES[a]

$\Delta\phi_1$	$\Delta\psi_1$	$\Delta\phi_2$	$\Delta\psi_2$	$\Delta\phi_3$	$\Delta\psi_3$	$\Delta\phi_4$	$\Delta\psi_4$	$\Delta\phi_5$	$\Delta\psi_5$	$\Delta\phi_6$	$\Delta\psi_6$	RMSD
−70	−40	−80	−60	−90	170	90	30	−70	100	−90	150	1.843248
−70	−40	−80	−50	−90	170	90	30	−70	100	−90	150	1.888763
−70	−40	−70	−60	−90	170	90	30	−70	100	−90	140	1.914637
−70	−30	−80	−50	−90	170	90	30	−70	100	−90	150	1.92957
−70	−40	−70	−60	−90	170	100	30	−70	100	−100	150	1.93773
−70	−30	−80	−60	−90	170	100	40	−80	100	−100	150	1.94264
−80	−30	−80	−60	−80	170	90	40	−80	100	−90	150	1.949104
−70	−30	−80	−60	−90	170	100	30	−80	100	−90	150	1.955443
−70	−30	−80	−60	−80	170	90	30	−70	100	−90	150	1.959502
−70	−40	−70	−60	−80	170	90	30	−70	100	−90	150	1.965129

[a] Results in six sets of torsion angles are being determined.

fragment can be locally refined to obtain a better match with the RDCs. This is done by a simple implementation of Monte Carlo sampling of a window size indicated by the user (usually ±5° for a 10° grid size). At the end of this procedure, a perl script (Mol_Scr.prl) is used to generate a script for MolMol (Koradi et al., 1996) to construct the atomic coordinates of the fragments. The predicted structure coming from this procedure applied to the entire 16-residue C-terminal fragment of PF0255 will be described more completely below.

Fragment Validation Using REDCAT

Upon the successful determination of fragment geometries, it is important to perform a more detailed error analysis. This can be done with the program REDCAT (Valafar and Prestegard, 2004). The atomic coordinates produced from the previous section, along with the RDCs used to determine the coordinates, can serve as input for the program. Using the "Prepare Input File" and "Import RDC" functions of this program, REDCAT input files can be conveniently produced. The first step for confirmation of a structure is to perform an error analysis and inspect the individual contributions of each piece of RDC data to the overall RMSD reported by REDCRAFT. Large errors should originate only from entries that have been identified as inaccurate measurements. Other large errors can result if there are segments with substantial degrees of internal motion (spin relaxation measurements or amide proton exchange experiments can be used to independently identify these segments) or from simple misassignments. In either case, this may be cause for a redetermination of fragment geometry with the problematic data corrected or eliminated.

REDCAT also provides solutions for principal order parameters. In cases of multiple fragments, it is important to verify that these are consistent between fragments, usually within 15% for S_{zz} and 30% for S_{xx} and S_{yy}.

Fragment Orientation Using REDCAT

In cases in which multiple fragments, as opposed to one continuous fragment, are obtained from REDCRAFT, it is important to put each fragment in its PAF for assembly into a complete molecule. Once each fragment has been described in its PAF, RDCs from a second alignment medium can be used to resolve inversion degeneracies (Al-Hashimi et al., 2000; Valafar et al., 2004), and then fragments can be translated under NOE and covalent constraints to produce a final structure. REDCAT offers tools to solve for the angles that transform a fragment to its PAF and to rotate initial fragment coordinates into that PAF. This rotation is an important step in assembling an intact structure when connectivities between residues terminate periodically throughout the length of the protein sequence. Even without direct connection, the proper relative orientations of separate fragments can be determined, because fragments from the same structure must share a common alignment frame. The frame produced by REDCAT is actually arbitrarily chosen from a 4-fold degenerate set that differs by rotations of 180° about each of the three Cartesian axes. In assembling fragments, the degeneracy can be resolved by comparing the possible relative orientations of two fragments as determined in two different alignment media. In general, relative orientations will appear the same for only one choice within the degenerate sets. The process is illustrated in Fig. 11 for the pre- and postproline pieces in the C-terminal segment of PF0255. In this illustration the pieces have been produced by dividing the 16-residue structure produced by REDCAT at the proline, but the process would be the same had the proper sequential connection through proline not been found. At the left of each line in the figure is the first half of the segment. This is followed by the second half in each of the four degenerate PAF frames. The first line is from a sample aligned in bacteriophage. The second line is from a sample aligned in C5E12/hexanol bicelles. All structures in the second line have been rotated by transformations that superimpose the first half fragment of the second line with the first half fragment in the first line. Note that only relative orientations involving the 53–58ref structure in line one and the 53–58y structure in line two appear the same. This points to the proper relative orientation for assembly. It is also reassuring that this puts the C-terminus of the first half in proximity to the N-terminus of the second half for easy covalent connection through a proline.

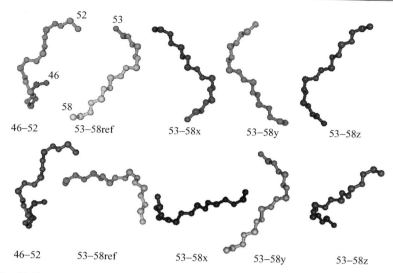

FIG. 11. Fragment alignment using RDC data for PF0255 from two different media. At the left of each line is the piece of the fragment before proline 53. The remaining four depictions of the piece after proline have been produced by rotating the reference structure by 180° about the x, y, and z axes of the principal alignment frame. The structures in the second line have been rotated to overlay the first piece in both lines using the program chimera (Huang et al., 1996). (See color insert.)

In general, when fragment geometries and alignments have been determined separately as described above, we will not have placed these connected sets of residues into a position in the overall sequence prior to fragment structure determination. In cases in which we have to rely primarily on data from the three RDC experiments described above, we can use a program named SEASCAPE to aid in this placement. This program relies on the information content of C_α chemical shifts of connected sets of residues once the secondary structure (ϕ, ψ) dependence is removed (Morris et al., 2004). Essentially the connected fragment, with its associated ϕ, ψ angles and C_α shift, is moved along the sequence, calculating a probability score for each placement. This is depicted in Fig. 12 for the two halves of the C-terminal fragment of PF0255. Fragment 1 (minus the first residue) has only one position in the sequence where the probability score is significant (height of bar), namely beginning at residue 47. Fragment 2 (using only data for residues 54–58) has two positions with significant probability for placement, but the most probable begins at residue 54. It is clear that the two fragments could have been correctly placed on either

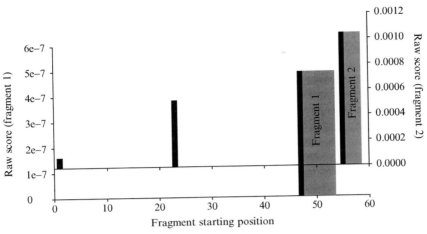

Fig. 12. Sequential placement of fragments using local structure and C_α chemical shifts. Fragment 1, based on highest probability, is positioned beginning with K47 and ending with V52. The position of the first residue in the fragment is indicated with a black bar while placement of the subsequent residues is shown in gray. No other placement has significant probability. Placement of fragment 2, again based on hightest probability, begins with G54 and ends with A57. Although other positions are possible, beginning with G1 or G23, the relative probabilities are half or less than that of beginning position G54.

side of the central proline based on just C_α shifts and local structure information.

Translation to connect fragments in the example shown here will be done only under constraints from the allowed geometry of the connecting proline. In general, small numbers of backbone-to-backbone NOE constraints would be added. For a trans proline, the appropriate distance between CO in the first half and N of the residue following proline in the second half is 3.2–4.2 Å. The terminal ϕ and ψ angles of each fragment are not well defined, and free rotation of these is allowed in making connections. This final structure can be refined with programs such as XPLOR-NIH (Schwieters *et al.*, 2003) to add missing residues, optimize bond geometry, and translate fragments.

Structure Quality Assessment

For the C-terminus of PF0255 used in the illustrations above there is now an X-ray structure (PDB number 1RYQ) that allows evaluation of the final structure produced by REDCRAFT. This structure was determined by the X-ray component of our pilot center during the course of our investigations (B. C. Wang *et al.*, unpublished observations). We will discuss only the NMR results from the fragment produced by running the intact 16-residue segment discussed above. A least-squares overlay of the backbone atoms from the central part of the segment (residues 45–57) produces an RMSD from the X-ray structure of 1.1 Å. This overlay is depicted in Fig. 13. The entire X-ray structure is shown for illustration. Additional fragments were in fact determined by the NMR approach, but agreement diverges at the amino-terminus. This is largely because the NMR structure was pursued on an apo form of this Zn/Fe protein, and without metal the amino-terminus is partially disordered. However, the excellent agreement for the C-terminal segment demonstrates the feasibility of constructing accurate backbone structures using primarily RDC information.

To meet efficiency objectives of the structural genomics initiative, the procedure we have outlined was designed to use a small subset of RDC data acquisition experiments and to proceed independently of separate

FIG. 13. Backbone superimposition of the RDC-based structure for residues 46–58 of PF0255 on the corresponding segment from the X-ray structure: 1RYQ. (See color insert.)

resonance assignment experiments. For other applications, there will certainly be other useful experiments for the measurement of RDCs, and there may be sound reasons to conduct experiments directed at independent sequential assignments. However, the data analysis tools we have described are quite generally applicable and should be useful in analyzing these expanded sets of data. For the future, application to larger proteins is clearly an important goal. Although we expect applications to proteins of 20 kDa to be possible with the procedure described, applications have so far been only to proteins under 15 kDa. Moving beyond this limit will certainly require alteration in the types of experiment used to acquire RDCs, the use of more extensive experiments for independent assignment, and the incorporation of more complementary data, such as that from NOEs, or relaxation enhancement by paramagnetic sites.

Acknowledgments

We would like to thank Peter LeBlond and Laura Morris for their assistance in preparing figures and tables for this chapter. We also thank all members of the Southeast Collaboratory for Structural Genomics for making the samples and data discussed in this chapter available to us. This work was supported by a grant from the National Institute of General Medical Sciences, GM062407.

References

Adams, M. W. W., Dailey, H. A., Delucas, L. J., Luo, M., Prestegard, J. H., Rose, J. P., and Wang, B. C. (2003). The Southeast Collaboratory for Structural Genomics: A high-throughput gene to structure factory. *Acc. Chem. Res.* **36,** 191–198.

Alexandrescu, A. T., and Kammerer, R. A. (2003). Structure and disorder in the ribonuclease S-peptide probed by NMR residual dipolar couplings. *Protein Sci.* **12,** 2132–2140.

Al-Hashimi, H. M., and Patel, D. J. (2002). Residual dipolar couplings: Synergy between NMR and structural genomics. *J. Biomol. NMR* **22,** 1–8.

Al-Hashimi, H. M., Valafar, H., Terrell, M., Zartler, E. R., Eidsness, M. K., and Prestegard, J. H. (2000). Variation of molecular alignment as a means of resolving orientational ambiguities in protein structures from dipolar couplings. *J. Magn. Reson* **143,** 402–406.

Andrec, M., Du, P., and Levy, R. M. (2001a). Protein backbone structure determination using only residual dipolar couplings from one ordering medium. *J. Biomol. NMR* **21,** 335–347.

Andrec, M., Du, P., and Levy, R. M. (2001b). Protein structural motif recognition via NMR residual dipolar couplings. *J. Am. Chem. Soc.* **123,** 1222–1229.

Assfalg, M., Bertini, I., Turano, P., Mauk, A. G., Winkler, J. R., and Gray, H. B. (2003). N-15-H-1 residual dipolar coupling analysis of native and alkaline-K79A *Saccharomyces cerevisiae* cytochrome c. *Biophys. J.* **84,** 3917–3923.

Barbieri, R., Bertini, I., Cavallaro, G., Lee, Y. M., Luchinat, C., and Rosato, A. (2002). Paramagnetically induced residual dipolar couplings for solution structure determination of lanthanide binding proteins. *J. Am. Chem. Soc.* **124,** 5581–5587.

Bax, A., Kontaxis, G., and Tjandra, N. (2001). Dipolar couplings in macromolecular structure determination. *Methods Enzymol.* **339,** 127–174.

Beraud, S., Bersch, B., Brutscher, B., Gans, P., Barras, F., and Blackledge, M. (2002). Direct structure determination using residual dipolar couplings: Reaction-site conformation of methionine sulfoxide reductase in solution. *J. Am. Chem. Soc.* **124,** 13709–13715.

Burley, S. K., Almo, S. C., Bonanno, J. B., Capel, M., Chance, M. R., Gaasterland, T., Lin, D. W., Sali, A., Studier, F. W., and Swaminathan, S. (1999). Structural genomics: Beyond the Human Genome Project. *Nat. Genet.* **23,** 151–157.

Chance, M. R., Bresnick, A. R., Burley, S. K., Jiang, J. S., Lima, C. D., Sali, A., Almo, S. C., Bonanno, J. B., Buglino, J. A., Boulton, S., *et al.* (2002). Structural genomics: A pipeline for providing structures for the biologist. *Protein Sci.* **11,** 723–738.

Clore, G. M., and Schwieters, C. D. (2003). Docking of protein-protein complexes on the basis of highly ambiguous intermolecular distance restraints derived from 1H/15N chemical shift mapping and backbone 15N-1H residual dipolar couplings using conjoined rigid body/torsion angle dynamics. *J. Am. Chem. Soc.* **125,** 2902–2912.

Cornilescu, G., Delaglio, F., and Bax, A. (1999). Protein backbone angle restraints from searching a database for chemical shift and sequence homology. *J. Biomol. NMR* **13,** 289–302.

de Alba, E., and Tjandra, N. (2002). NMR dipolar couplings for the structure determination of biopolymers in solution. *Prog. Nucl. Magn. Reson. Spectrosc.* **40,** 175–197.

Delaglio, F., Grzesiek, S., Vuister, G. W., Zhu, G., Pfeifer, J., and Bax, A. (1995). Nmrpipe—a multidimensional spectral processing b on Unix pipes. *J. Biomol. NMR* **6,** 277–293.

Delaglio, F., Kontaxis, G., and Bax, A. (2000). Protein structure determination using molecular fragment replacement and NMR dipolar couplings. *J. Am. Chem. Soc.* **122,** 2142–2143.

Dosset, P., Hus, J. C., Marion, D., and Blackledge, M. (2001). A novel interactive tool for rigid-body modeling of multi-domain macromolecules using residual dipolar couplings. *J. Biomol. NMR* **20,** 223–231.

Fesik, S. W. (1999). NMR as a tool in drug research. *FASEB J.* **13,** A1422.

Fesik, S. W. (2001). The use of NMR in cancer drug discovery. *Clin. Cancer Res.* **7,** 3827s.

Fukunaga, K. (1990). "Introduction to Statistical Pattern Recognition," 2nd ed. Academic Press, New York.

Hajduk, P. J., Meadows, R. P., and Fesik, S. W. (1999). NMR-based screening in drug discovery. *Q. Rev. Biophys* **32,** 211–240.

Hajduk, P. J., Betz, S. F., Mack, J., Ruan, X. A., Towne, D. L., Lerner, C. G., Beutel, B. A., and Fesik, S. W. (2002). A strategy for high-throughput assay development using leads derived from nuclear magnetic resonance-based screening. *J. Biomol. Screen* **7,** 429–432.

Haliloglu, T., Kolinski, A., and Skolnick, J. (2003). Use of residual dipolar couplings as restraints in *ab initio* protein structure prediction. *Biopolymers* **70,** 548–562.

Hansen, M. R., Mueller, L., and Pardi, A. (1998). Tunable alignment of macromolecules by filamentous phage yields dipolar coupling interactions. *Nat. Struct. Biol.* **5,** 1065–1074.

Huang, C. C., Couch, G. S., *et al.* (1996). Chimera: An extensible molecular modeling application constructed using standard components. Paper presented at Pacific Symposium on Biocomputing, January 3–6, Hawaii.

Hus, J. C., Marion, D., and Blackledge, M. (2000). *De novo* determination of protein structure by NMR using orientational and long-range order restraints. *J. Mol. Biol.* **298,** 927–936.

Johnson, B. A., and Blevins, R. A. (1994). NMR view—a computer-program for the visualization and analysis of NMR data. *J. Biomol. NMR* **4,** 603–614.

Koradi, R., Billeter, M., and Wuthrich, K. (1996). MOLMOL: A program for display and analysis of macromolecular structures. *J. Mol. Graph* **14,** 51–55.

Lin, Y. J., Dancea, F., Lohr, F., Klimmek, O., Pfeiffer-Marek, S., Nilges, M., Wienk, H., Kroger, A., and Ruterjans, H. (2004). Solution structure of the 30 kDa polysulfide-sulfur transferase homodimer from Wolinella succinogenes. *Biochemistry* **43,** 1418–1424.

Losonczi, J. A., Andrec, M., Fischer, M. W. F., and Prestegard, J. H. (1999). Order matrix analysis of residual dipolar couplings using singular value decomposition. *J. Magn. Reson* **138,** 334–342.

MacArthur, M. W., and Thornton, J. M. (1996). Deviations from planarity of the peptide bond in peptides and proteins. *J. Mol. Biol.* **264,** 1180–1195.

Montelione, G. T., Zheng, D. Y., Huang, Y. P. J., Gunsalus, K. C., and Szyperski, T. (2000). Protein NMR spectroscopy in structural genomics. *Nat. Struct. Biol.* **7,** 982–985.

Morris, L. C., Valafar, H., and Prestegard, J. H. (2004). Assignment of protein backbone resonances using connectivity, torsion angles and C-13(alpha) chemical shifts. *J. Biomol. NMR* **29,** 1–9.

Nair, M., McIntosh, P. B., Frenkiel, T. A., Kelly, G., Taylor, I. A., Smerdon, S. J., and Lane, A. N. (2003). NMR structure of the DNA-binding domain of the cell cycle protein Mbp1 from *Saccharomyces cerevisiae. Biochemistry* **42,** 1266–1273.

Norvell, J. C., and Machalek, A. Z. (2000). Structural genomics programs at the US National Institute of General Medical Sciences—Foreword. *Nat. Struct. Biol.* **7,** 931.

Ohnishi, S., Lee, A. L., Edgell, M. H., and Shortle, D. (2004). Direct demonstration of structural similarity between native and denatured eglin C. *Biochemistry* **43,** 4064–4070.

Ottiger, M., Delaglio, F., and Bax, A. (1998). Measurement of J and dipolar couplings from simplified two-dimensional NMR spectra. *J. Magn. Reson* **131,** 373–378.

Prestegard, J. H., and Kishore, A. I. (2001). Partial alignment of biomolecules: An aid to NMR characterization. *Curr. Opin. Chem. Biol.* **5,** 584–590.

Prestegard, J. H., Al-Hashimi, H. M., and Tolman, J. R. (2000). NMR structures of biomolecules using field oriented media and residual dipolar couplings. *Q. Rev. Biophys.* **33,** 371–424.

Prestegard, J. H., Bougault, C. M., and Kishore, A. I. (2004). Residual dipolar coupling in structure determination of biomolecules. *Chem. Rev.* **104,** 3519–3540.

Rohl, C. A., and Baker, D. (2002). *De novo* determination of protein backbone structure from residual dipolar couplings using rosetta. *J. Am. Chem. Soc.* **124,** 2723–2729.

Ruckert, M., and Otting, G. (2000). Alignment of biological macromolecules in novel nonionic liquid crystalline media for NMR experiments. *J. Am. Chem. Soc.* **122,** 7793–7797.

Sambrook, J., Fritsch, E. F., and Maniatis, T. (1989). "Molecular Cloning: A Laboratory Manual," 3rd Ed. Cold Spring Harbor Laboratory Press, Cold Spring Harbor, NY.

Saupe, A. (1968). Recent results in the field of liquid crystals. *Angew. Chem. Int. Ed. Engl.* **7,** 97.

Schwieters, C. D., Kuszewski, J. J., Tjandra, N., and Clore, G. M. (2003). The Xplor-NIH NMR molecular structure determination package. *J. Magn. Reson* **160,** 65–73.

Sklenar, V., Piotto, M., Leppik, R., and Saudek, V. (1993). Gradient-tailored water suppression for H-1-N-15 Hsqc experiments optimized to retain full sensitivity. *J. Magn. Reson. Ser. A* **102,** 241–245.

Tian, F., Valafar, H., and Prestegard, J. H. (2001). A dipolar coupling based strategy for simultaneous resonance assignment and structure determination of protein backbones. *J. Am. Chem. Soc.* **123,** 11791–11796.

Tolman, J. R. (2001). Dipolar couplings as a probe of molecular dynamics and structure in solution. *Curr. Opin. Struct. Biol.* **11,** 532–539.

Tossavainen, H., Permi, P., Annila, A., Kilpelainen, I., and Drakenberg, T. (2003). NMR solution structure of calerythrin, an EF-hand calcium-binding protein from *Saccharopolyspora erythraea. Eur. J. Biochem.* **270,** 2505–2512.

Valafar, H., and Prestegard, J. H. (2003). Rapid classification to a protein fold family using a statistical analysis of dipolar couplings. *Bioinformatics* **19**, 1–8.

Valafar, H., and Prestegard, J. H. (2004). REDCAT: A residual dipolar coupling analysis tool. *J. Magn. Reson.* **167**, 228–241.

Valafar, H., Mayer, K. L., Bougault, C. M., Leblond, P., Jenney, F. E., Brereton, P. S., Adams, M. W. W., and Prestegard, J. H. (2004). Backbone solution structures of proteins using residual dipolar couplings. *J. Struct. Funct. Genomics*, In press.

Varner, S. J., Vold, R. L., and Hoatson, G. L. (1996). An efficient method for calculating powder patterns. *J. Magn. Reson. Ser. A* **123**, 72–80.

Wang, Y. X., Marquardt, J. L., Wingfield, P., Stahl, S. J., Lee-Huang, S., Torchia, D., and Bax, A. (1998). Simultaneous measurement of H-1-N-15, H-1-C-13′, and N-15-C-13′ dipolar couplings in a perdeuterated 30 kDa protein dissolved in a dilute liquid crystalline phase. *J. Am. Chem. Soc.* **120**, 7385–7386.

Warren, J. J., and Moore, P. B. (2001). A maximum likelihood method for determining D-a(PQ) and R for sets of dipolar coupling data. *J. Magn. Reson* **149**, 271–275.

Weaver, J. L., and Prestegard, J. H. (1998). Nuclear magnetic resonance structural and ligand binding studies of BLBC, a two-domain fragment of barley lectin. *Biochemistry* **37**, 116–128.

Weisemann, R., Ruterjans, H., Schwalbe, H., Schleucher, J., Bermel, W., and Griesinger, C. (1994). Determination of H(N),H-alpha and H(N), C′ coupling-constants in C-13,N-15-labeled proteins. *J. Biomol. NMR* **4**, 231–240.

Wishart, D. S., and Sykes, B. D. (1994). The C-13 chemical-shift index—a simple method for the identification of protein secondary structure using C-13 chemical-shift data. *J. Biomol. NMR* **4**, 171–180.

Wishart, D. S., Sykes, B. D., and Richards, F. M. (1992). The chemical-shift index—a fast and simple method for the assignment of protein secondary structure through NMR-spectroscopy. *Biochemistry* **31**, 1647–1651.

Wohnert, J., Franz, K. J., Nitz, M., Imperiali, B., and Schwalbe, H. (2003). Protein alignment by a coexpressed lanthanide-binding tag for the measurement of residual dipolar couplings. *J. Am. Chem. Soc.* **125**, 13338–13339.

Wood, M. J., and Komives, E. A. (1999). Production of large quantities of isotopically labeled protein in Pichia pastoris by fermentation. *J. Biomol. NMR* **13**, 149–159.

Zheng, D. Y., Aramini, J. M., and Montelione, G. T. (2004). Validation of helical tilt angles in the solution NMR structure of the Z domain of staphylococcal protein A by combined analysis of residual dipolar coupling and NOE data. *Protein Sci.* **13**, 549–554.

Zhou, H. J., Vermeulen, A., Jucker, F. M., and Pardi, A. (1999). Incorporating residual dipolar couplings into the NMR solution structure determination of nucleic acids. *Biopolymers* **52**, 168–180.

[8] Robotic Cloning and Protein Production Platform of the Northeast Structural Genomics Consortium

By Thomas B. Acton, Kristin C. Gunsalus, Rong Xiao,
Li Chung Ma, James Aramini, Michael C. Baran, Yi-Wen Chiang,
Teresa Climent, Bonnie Cooper, Natalia G. Denissova,
Shawn M. Douglas, John K. Everett, Chi Kent Ho,
Daphne Macapagal, Paranji K. Rajan, Ritu Shastry,
Liang-yu Shih, G.V.T. Swapna, Michael Wilson,
Margaret Wu, Mark Gerstein, Masayori Inouye,
John F. Hunt, and Gaetano T. Montelione

Abstract

In this chapter we describe the core Protein Production Platform of the Northeast Structural Genomics Consortium (NESG) and outline the strategies used for producing high-quality protein samples using *Escherichia coli* host vectors. The platform is centered on 6X-His affinity-tagged protein constructs, allowing for a similar purification procedure for most targets, and the implementation of high-throughput parallel methods. In most cases, these affinity-purified proteins are sufficiently homogeneous that a single subsequent gel filtration chromatography step is adequate to produce protein preparations that are greater than 98% pure. Using this platform, over 1000 different proteins have been cloned, expressed, and purified in tens of milligram quantities over the last 36-month period (see Summary Statistics for All Targets, http://www.nmr.cabm.rutgers.edu/bioinformatics/ZebaView/). Our experience using a hierarchical multiplex expression and purification strategy, also described in this chapter, has allowed us to achieve success in producing not only protein samples but also many three-dimensional structures. As of December 2004, the NESG Consortium has deposited over 145 new protein structures to the Protein Data Bank (PDB); about two-thirds of these protein samples were produced by the NESG Protein Production Facility described here. The methods described here have proven effective in producing quality samples of both eukaryotic and prokaryotic proteins. These improved robotic and/or parallel cloning, expression, protein production, and biophysical screening technologies will be of broad value to the structural biology, functional proteomics, and structural genomics communities.

Copyright 2005, Elsevier Inc.
All rights reserved.
0076-6879/05 $35.00

Introduction

The Northeast Structural Genomics Consortium (NESG) is a pilot project designed to evaluate the feasibility and value of structural genomics. Its primary goals are to develop and refine new technologies for high-throughput protein production and structure determination by both NMR and X-ray crystallography and to apply these technologies in determining representative structures of the domain sequence families that constitute eukaryotic proteomes. The project (http://www.nesg.org), one of 11 pilot projects supported by the United States National Institutes of Health Protein Structure Initiative (http://www.nigms.nih.gov/psi/), is developing technology aimed at optimizing each stage of the structure determination pipeline.

One of the most important challenges to the emerging field of structural genomics is the preparation of protein samples suitable for the determination of three-dimensional structures. This sample preparation challenge is different from those encountered in most previous genome-wide initiatives, such as the Human Genome Sequencing Project or microarray gene expression studies, which focus on preparing nucleic acid samples (Lander, 1999; Lander et al., 2001; Winzeler et al., 1999), or involve production of only small quantities of proteins for functional studies (Ito et al., 2001; Uetz et al., 2000). Nucleic acids all have generally similar biophysical properties, allowing similar and well-defined purification and preparation techniques to be employed in high-throughput processes. Other genome-wide studies that focus on proteins such as yeast two-hybrid screens (Giot et al., 2003; Ito et al., 2001; Li et al., 2004; Rain et al., 2001; Uetz et al., 2000) require relatively small amounts of proteins, often expressed in a eukaryotic organism (yeast), and do not usually require protein purification to derive experimental information. Structural genomics projects require the production of tens of milligram quantities of soluble, highly purified protein samples. These proteins often have diverse biophysical properties, making the preparation of suitable samples more difficult, especially when considering high-throughput methods. The target proteins of the NESG (http://www-nmr.cabm.rutgers.edu/bioinformatics/ZebaView/index) are composed of protein domain families sharing structure and sequence similarity selected from the proteomes of archaea, eubacteria, and eukaryotic organisms, many of which are difficult to express in Escherichia coli expression systems. In addition, the NESG Consortium utilizes both nuclear magnetic resonance (NMR) and X-ray crystallographic methods of protein structure determination (Montelione and Anderson, 1999). Protein samples suitable for rapid three-dimensional (3D) structure determination by NMR and X-ray crystallography generally require $^{13}C, ^{15}N$ isotope enrichment or selenomethionine labeling. This necessitates that our protein production platform not only has high throughput but also is flexible enough to handle preparation of both protein sample types. Considering these challenges, one of the major contributions structural genomics will have on science is the development of new technologies that enhance our capabilities in the areas of protein expression and purification, and improve our abilities to deliver protein samples suitable for NMR, X-ray crystallography, and diverse biological studies.

In this chapter, we describe the high-throughput cloning and protein production platform we have developed at Rutgers University for the preparation and screening of protein samples amenable to structural determination by X-ray crystallography and/or NMR spectroscopy. The laboratory of Cheryl Arrowsmith at the Ontario Cancer Institute and the University of Toronto also produces proteins for structure studies by the NESG. This related, though distinct, platform has been described elsewhere (Yee *et al.*, 2002, 2003). The process (and most statistics) described in this chapter are specifically for the Rutgers component of the NESG protein production effort.

Although the Rutgers protein production effort for the NESG is currently limited to protein production in *E. coli*, this platform is quite flexible, providing for cloning and expression of a wide range of proteins from archaea, eubacteria, and eukaryotic organisms. The robotic platform is highly efficient, currently providing the capacity to clone and evaluate expression and solubility of 100 proteins per week and to produce tens of milligram quantities of 15–20 purified proteins per week for both NMR and crystallization screening. In addition to its central role in driving our structural genomics effort, the platform is a prototype of protein production technologies that will soon become commonplace in traditional structural biology, biochemistry, and proteomics projects.

Protein Production Platform

Targets and Bioinformatics Infrastructure

Most of the current NESG target proteins are full-length polypeptide chains shorter than 340 amino acids, selected from domain sequence clusters (Liu and Rost, 2004; Liu *et al.*, 2004), which are organized in the PEP/CLUP (http://cubic.bioc.columbia.edu/pep/) domain cluster database (Carter *et al.*, 2003). Each of these protein sequence clusters consists of three or more proteins (or protein fragments) corresponding to putative structural domains whose 3D structure is not known experimentally and cannot be accurately modeled through homology. The NESG focuses on

TABLE I
EUKARYOTIC TARGET GENOMES OF THE NESG CONSORTIUM

Organism	Number of targets
Arabidopsis thaliana	2242
Caenorhabditis elegans	340
Drosophila melanogaster	263
Homo sapiens	2857
Saccharomyces cerevisiae	584

domain families that include at least one representative from a set of five eukaryotic target organisms (Table I). These correspond to domain families constituting the eukaryotic proteome.

Zeba View (http://www-nmr.cabm.rutgers.edu/bioinformatics/Zeba View/), a web-based interactive summary of key NESG target information, functions as the "Official Target List" of the NESG project (Wunderlich *et al.*, 2004), and SPINE (Structural Proteomics in the Northeast; http://SPINE.nesg.org/), a web-based project database, organizes and coordinates detailed information about the protein production and structure analysis processes carried out in the multiple sites of the NESG Consortium (Bertone *et al.*, 2001; Goh *et al.*, 2003). SPINE is a laboratory information management system (LIMS) for most of the steps of the protein production process, as well as a data warehouse of information collected from other laboratory information management systems used by the NESG Consortium through an XML-based data exchange language (Wunderlich *et al.*, 2004). Each NESG protein target is assigned a NESG id code, the first letter(s) of which indicate the organism from which the target is cloned, the last letter the institute at which the protein is produced, followed by a serial number (e.g., HR32, human, Rutgers, target number 32).

Multiplex Expression Vector System

Highly homogeneous protein samples with minimal numbers of disordered nonnative residues are generally required for successful protein crystallization and for structure determination by X-ray crystallography or NMR. Protein samples for crystallization should ideally exhibit >98% homogeneity on sodium dodecyl sulfate (SDS) polyacrylamide gels. Moreover, whereas affinity tags are generally required for high-throughput purification protocols (Crowe *et al.*, 1994; Sheibani, 1999), large disordered tags can frustrate crystallization efforts and often exhibit strong sharp peaks and associated artifacts of Fourier transform processing in NMR spectra. Protein samples must be produced in soluble form and at high yield, as tens of milligram quantities are needed for crystallization and NMR experiments. For NMR studies, the high cost of uniform enrichment with ^{13}C isotopes generally demands high efficiency isotope incorporation. For example, in *E. coli* expression systems, we typically aim for production yields of 10–50 mg of purified protein per liter of fermentation using defined minimal media (MJ9) optimized for producing isotopically enriched proteins (Acton *et al.*, 2005; Jansson *et al.*, 1996). These constraints define the primary design features of vectors expressing protein open reading frames (ORFs) for use in structural genomics projects.

The advent of genomic studies has led to the introduction of several new cloning technologies, many of which are optimized for high throughput, including various systems of ligase-independent cloning. These cloning strategies, such as the Gateway (Invitrogen), TOPO (Invitrogen), and Creator (Clontech) systems, exploit various forms of recombinational cloning (Abremski and Hoess, 1984; Hartley *et al.*, 2000; Sauer, 1994; Shuman, 1994). These systems generally exhibit high cloning efficiency and significantly fewer cloning steps, both of which are advantageous for high-throughput procedures. However, as a consequence of the mechanisms of recombinatorial cloning, these strategies generally result in protein products with a significant number of nonnative amino-acid residues attached to one or both ends of the protein molecule. For example, an N-terminal His-tag fusion in the Gateway system results in the addition of 22 extraneous amino-acid residues. These nonnative residues can interfere with crystallization and other structural studies. It is possible to introduce a protease cleavage site downstream of N-terminal tags and thereby cleave off the extraneous residues from the recombination site (Yee *et al.*, 2002, 2003). However, there are often problems with such systems, including protease specificity and/or contaminating proteases that lead to unwanted cleavage(s), incomplete cleavage leading to nonhomogeneous samples or low yields, and the overall cost and complexity that this step adds to the high-throughput pipeline. In addition, many of the most useful sources of cDNA libraries for eukaryotic organisms in Gateway libraries do not include a stop codon allowing both N- and C-terminal fusions to be produced from a single entry clone. This unfortunately adds residues from both recombination sites, producing a protein with a very significant number of nonnative residues. These large numbers of extraneous residues may contribute to the limited success of structural genomics projects using such systems. Moreover, the cleavage pattern of commonly used site-specific proteases, such as the TEV protease (Kapust *et al.*, 2002), leave four to six residues on the N-terminal side of the recognition site, limiting their usefulness with C-terminal fusion tags.

With these issues in mind, we chose to create an expression vector set that would utilize a classic restriction endonuclease-ligase-dependent mechanism of cloning that could allow the generation of constructs with a minimum number of extraneous residues, while avoiding the requirement for protease cleavage and subsequent purification. Additionally, although a number of different systems for recombinant protein production in bacteria or eukaryotic cells are currently available (Geisse et al., 1996; Makrides, 1996), we opted to base our effort on isopropylthiogalactoside (IPTG)-inducible systems already in use for high-level expression in bacteria (Bujard et al., 1987; Studier et al., 1990), which readily allow isotope and L-selenomethionine (SeMet) enrichment. These pET vector systems also allow use of autoinduction media (Studier, 2004). Our focus was to create a flexible system that could efficiently generate an array of combinations for rapid screening of optimal expression conditions. Because every protein has different properties, which currently cannot be predicted in advance, we wanted the ability to produce the same protein as an N- or C-terminal hexa-histidine (6X-His) fusion (for rapid affinity purification and expression/solubility/NMR screening) as well as a nontagged version (for use in structure determination, if preferable). We also wanted to produce each of these protein variants under a number of different expression conditions (by varying promoters, bacterial strains, etc.), as optimal expression conditions generally vary from one protein to another.

To meet these criteria, we created a "Multiplex Vector Kit" consisting of a set of nine compatible expression vectors. The essential features and the minimal polylinker sequences of this vector set are shown in Fig. 1. As a starting point, we used commercially available E. coli expression vectors differing in the choice of promoter (T7, T7 lac, or T5 lac lac) and placement of a 6X-His tag at the N- or C-terminus, which we modified to suit our needs. As some of these commercial vectors have very limited polylinkers, we have engineered into these vectors an expanded "minimal common polylinker" (MCP) containing a set of restriction endonuclease (RE) sites shared by vectors with more extensive polylinkers. We placed the MCP in all three reading frames (1, 2, and 3) with respect to the 6X-His tag, allowing us to minimize the nonnative residues added to an ORF and to control the identity of amino acid residues that are added between the native sequence and the 6X-His tag. This generated a set of three vectors from each starting vector and also created two new sets of vector cognates that allow the choice of NcoI in place of NdeI as an option for in-frame ATG cloning (Fig. 1).

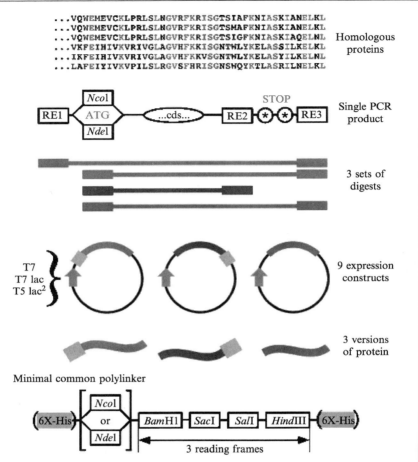

```
...VQWEMEVCKLPRLSLNGVRFKRISGTSIAFKNIASKIANELKL
...VQWEMEVCKLPRLSLNGVRFKRISGTSMAFKNIASKIANELKL
...VQWEMEVCKLPRLSLNGVRFKRISGTSIGFKNIASKIAQELNL    Homologous
...VKFEIHIVKVRIVGLAGVHFKKISGNTWLYKELASSILKELKL      proteins
...IKFEIHIVKVRIVGLAGVHFKKVSGNTWLYKELASYILKELNL
...LAFEIYIVKVPILSLRGVSFHRISGNSWQYKTLASRILNELKL
```

FIG. 1. Strategy for multiplexed protein expression. One or more representatives from a family of homologous proteins or protein domains are chosen. For each domain to be expressed, a PCR product is designed to contain either an *Nco*I or an *Nde*I site at the initiator ATG for the coding sequence along with three additional restriction sites (RE1, RE2, and RE3) from the MCP for cloning different versions of the protein into the various expression vectors. RE1 and either *Nco*I or *Nde*I are included in the 5' PCR primer, while the 3' primer includes RE2 followed by one or two stop codons and RE3. Using three different combinations of digests, nine different expression variants can be generated from the same PCR primers, differing in the promoter driving expression, the placement of an affinity tag (if any), and the identity of any nonnative amino-acid residues that result from the cloning strategy. The minimum common polylinker found in each of the nine custom expression vectors is shown at the bottom. (See color insert.)

A key goal of our design was to develop a vector set with the flexibility to create the maximum number of different expression constructs using a minimum number of different polymerase chain reaction (PCR) primers, considering that one of the greatest expenses in high-throughput cloning is the cost of oligonucleotide primers and high-fidelity Taq polymerases. With the vector modifications we introduced, a single PCR product can be designed so that only three different double restriction endonuclease digestions are sufficient to generate up to nine different expression constructs in parallel: three protein variants (N- or C-tagged 6X-His fusions, plus the nonfusion) each driven by any of the three promoters (T7, T7 lac, or T5 lac lac). A schematic of this strategy is shown in Fig. 1 (see legend for details). To accommodate all cloning options, PCR primers must be designed to introduce a specific arrangement of restriction sites into the resulting PCR product, which contains the protein coding sequence of interest flanked on either end by restriction sites compatible with the vector polylinkers (RE1, NdeI/NcoI, RE2, and RE3). A detailed discussion of the essential issues that must be considered in designing the PCR product and cloning with this expression kit is presented elsewhere (Acton et al., 2005; Everett et al., 2004).

We designed the "Multiplex Vector Kit" to allow cloning into several (at least nine) different vectors from a single PCR product. However, in the course of our work we have identified a second strategy for which the resulting common polylinker can be used to implement a significant cost-saving advantage. In the first stage of this procedure, we design primer pairs for cloning each of a large set of ORFs into one of the C-terminal 6X-His vectors, which allows a decreased number of nucleotides per primer as only one six-base restriction site is added. These PCR products are then cloned into one of the C-terminal fusion vectors from the Multiplex Vector Kit. In the second stage, each of the resulting C-terminal fusion constructs can then be used as a PCR template for amplification using a set of primers that anneals to the vector sequence flanking the ORF while introducing a stop codon and a restriction site at the 3' end of the gene (Fig. 2). This is accomplished using a 3' primer that anneals to the 6X-His coding sequence, bubbling off a restriction site and a stop codon, and then annealing back to the 3' restriction site into which the gene was originally cloned. The resulting PCR product is then cloned into an N-terminal 6X-His or nontagged vector of the Multiplex Vector Kit using the original 5' RE site together with the newly added 3' RE. This adds only two nonnative residues derived from the original 3' RE site that is now directly followed by the introduced stop codon. In this manner, all nine different expression constructs can be derived with a minimal number of initial primers and the cost-effective common primers.

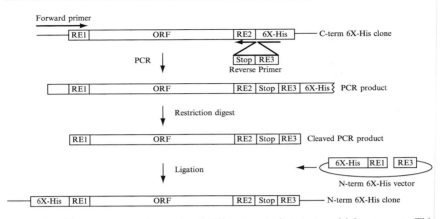

Fig. 2. Using common primers for shuffling targets between multiplex vectors. This schematic illustrates a strategy for producing all other construct variants in the "Multiplex Vector Kit" from a single C-terminal 6X-His construct using primers that are not gene specific. The forward primer anneals to the common vector sequence upstream of the coding sequence and will incorporate the initial RE1 site. The reverse primer anneals to the 6X-His coding sequence and includes a new restriction site (RE3), present in the other vectors, as well as a TAG stop codon. The primer then anneals to the original RE2 site, directing elongation from this point. The resulting PCR product is cleaved with RE1 and RE3, removing the common vector sequence and the original 6X-His coding sequence, and then ligated into a similarly cleaved vector. In this illustration, the ORF is then cloned into an N-terminal 6X-His fusion vector; this strategy can also be used to produce nontagged variants.

RT-PCR for High-Throughput Cloning of Eukaryotic ORFs

Using the strategy described above, high-throughput cloning of structural genomics targets from prokaryotic genomes is relatively straightforward. In particular, the absence of introns allows for the direct use of genomic DNA as a PCR template. As a consequence, although the primer sets for each ORF differ, the DNA template is common to all of the PCR reactions. This allows for simple manual or robotic manipulation using a common PCR cocktail containing all of the required buffers, enzymes, and the genomic template. However, a bottleneck emerges when cloning from most eukaryotic target genomes, since many of these genes contain introns. This necessitates using cDNA as a PCR template, resulting in several complications. Most importantly, adequate cDNA libraries must be obtained. The highest quality full-length libraries will be those arising from large-scale projects such as the Drosophila Gene Collection, I.M.A.G.E., or the *Caenorhabditis elegans* ORFeome project (Reboul *et al.*, 2003; Rubin *et al.*, 2000; Stapleton *et al.*, 2002; Strausberg *et al.*, 1999, 2002). A structural genomics project focusing on thousands of eukaryotic targets requires thousands of cDNA clones to serve as PCR templates; handling this set of individual cDNA clones incurs significant logistical complications and additional costs. For example, in genomic-scale cloning it is most practical to acquire the entire gene set, which is not only costly but also requires sufficient resources for archiving and retrieving the reagents. In addition, the use of target-specific cDNA templates complicates robotic automation, since each individual template must be transferred to an appropriate well in a PCR plate, presenting additional bioinformatics, robotic programming, and material costs. The resulting increased complexity also lengthens the time required for setting up the PCR reactions and generally has a negative effect on the outcome of the amplification.

To circumvent the problems associated with using target-specific cDNA templates and to increase throughput, we instituted a reverse transcriptase strategy to produce a common cDNA pool for use as a PCR template. In this strategy, we use polyadenylated mRNA or total RNA from various tissues, cell types, and developmental stages together with oligo(dT) primers to carry out reverse transcriptase reactions. Briefly, oligo(dT$_{12-18}$) (Invitrogen) is annealed to 5 μg of RNA in a volume of 275 μl by heating to 70° for 10 min followed by incubation on ice for 15 min. The volume is raised to 500 μl with the addition of Powerscript Reverse Transcriptase (Clontech), the corresponding first strand synthesis buffer, free dNTPs, and RNase-free water. The reaction is incubated for 60 min at 42° allowing first strand cDNA synthesis to occur, followed by digestion with RNase H (New England Biolabs) ensuring the removal of RNA that might interfere with PCR amplification of our target sequences (for greater detail see Acton et al., 2005). For each organism, cDNAs from several tissues, cell types, and/or developmental stages are then mixed to form a common cDNA pool. This cDNA pool is then added to the PCR cocktail mix, much like adding bacterial genomic DNA, and used as a common cDNA template in PCR reactions.

Robotic Vector Construction with the Biorobot 8000

To clone in a high-throughput manner we have automated each step of our restriction endonuclease-ligase-dependent cloning strategy using a Biorobot 8000 (Qiagen). Figure 3 outlines each of these steps of vector construction; steps shown in blue typeface are completely automated by the robot, while those in red are semiautomated, requiring some manual manipulations. A detailed description of the entire process is provided in Acton et al. (2005). Briefly, 96 protein targets are chosen for cloning and the primer pair for each ORF is determined using the Primer Prim'er oligonucleotide design program (Everett et al., 2004). The

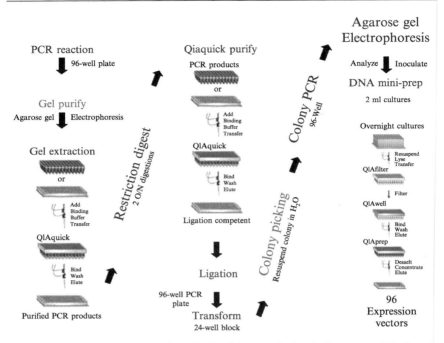

FIG. 3. Biorobot 8000 cloning schematic. Each key step in the cloning strategy is indicated; blue type denotes those steps that are completely automated, and red type indicates those steps that require some manual input. Roughly 1 week of one full-time equivalent is needed to complete all of the cloning steps for 96 target proteins. Several of the procedures are modifications of Qiagen-based protocols, such as the Qiaquick Purification and the DNA Mini-Prep protocols. However, most have been completely created in the Rutgers NESG Protein Production laboratory. A more detailed description of the cloning procedure, as well as the automated protocols, are provided elsewhere (Acton *et al.*, 2005). (See color insert.)

Primer Prim'er program (available on-line at http://www-nmr.cabm.rutgers. edu/bioinformatics/) generates order forms for the primer sets, which are transmitted directly to the primer vendor, typically Qiagen. Forward and reverse primers are grouped together, with the forward and reverse primers for each specific ORF in identical well positions on two separate order forms, synthesized by Qiagen, and provided in 96-well format with the concentration of each primer normalized to 50 μM in deionized water. The two 96-well primer blocks containing the reverse or forward primers are placed on the robot deck, and a Qiasoft 4.0 program written to automate the PCR setup is run. In this program, a PCR reaction mix, containing all necessary components for PCR amplification, is added to each well in a 96-well PCR plate. This includes the dNTPs, Advantage HF2 high-fidelity polymerase (Clontech), and its corresponding buffer and TaqStart antibody (Clontech). The latter sequesters the polymerase prior to thermo-cycling, which we have found to greatly decrease background amplification products caused from mispriming during the low temperature PCR setup, while also increasing the yield of correctly amplified products. The program then commands the pipette head to transfer 100 pmol of the appropriate forward and reverse primers from the primer blocks into the corresponding well for each target in the PCR plate. An Applied Biosystems 9700 thermocycler is used for the amplification with 35 total cycles. Each cycle contains a 10-s 90° melting step, a 30-s annealing step (50–55°), and a 1-min 68° elongation step. An annealing temperature step increase after 10 rounds of amplification is included to take into account the contribution of the extra bases added for the restriction sites (for greater detail, see Acton *et al.*, 2005).

Following PCR amplification, the products are separated on a 2% agarose gel, and the DNA bands are visualized using a low-energy ultraviolet (UV) lightwand. The correct-size fragments are easily identified, since the primer design program organizes the ORFs in the plate by increasing size (Everett *et al.*, 2004). The proper DNA fragments are then manually excised from the gel using a scalpel and relocated into the appropriate well of a 96-well S-Block (Qiagen). A completely automated 96-well gel extraction is carried out using reagents from the Qiagen Gel Extraction Kit and a QIAquick 96-well column PCR Cleanup plate. The resulting purified PCR products are then subjected to two restriction endonuclease digestions to allow for directional cloning, generally using *Nde*I and *Xho*I at the 5′ and 3′ ends, respectively. Following the second restriction digestion, an automated 96-well Mini-Elute DNA purification and elution into water is performed. Ligation into an appropriate precut expression vector is then carried out. Briefly, a 96-well PCR plate is chilled on the robot deck and a reaction mixture containing 100 ng of a similarly digested vector, ligase buffer, ligase (100 U, New England Biolabs), and water is trans erred to each well. Three- to 6- fold-molar excess (generally 1 or 2 μl) of the highly purified and cleaved DNA PCR product is added to the appropriate well for a 20-μl final volume, mixed, and incubated overnight at 16°.

Having completed vector construction, the next step of the process involves robotic transformation into *E. coli* cells in 24-well format. A 1-μl aliquot of each overnight ligation well is pipetted by the Biorobot 8000 into a corresponding well in another 96-well PCR plate prechilled at 0° on the robot deck. Each well of this plate contains 15 μl of XL-1 ultracompetent cells (Stratagene). A transformation procedure is then carried out on the robot deck keeping the PCR plate at 0° until a manual heat shock. SOC (100 μl) is added to each well, and the plate is incubated at 37° for 1 h.

The transformation is completed by pipetting the entire contents of each well into the corresponding wells of four 24-well blocks. Each block well contains 2 ml of Luria broth (LB) medium/Agar with ampicillin and 5–10 (3-mm-diameter) glass beads. The contents are dispersed using the robot's platform shaker and the glass beads, the latter of which are then poured off the plate. Following overnight incubation at 37°, two colonies per ORF are harvested and resuspended in 50 μl of sterile water. Colony-picking is the most labor-intensive step in the process outlined in Fig. 3. Colony PCR, using primers flanking the MCS, is set up robotically in 96-well format, and the results are visualized by agarose gel electrophoresis, identifying clones with correct-size inserts. These clones are then subcultured overnight, and plasmid DNA is isolated using a completely automated Qiagen 96-well DNA mini-prep procedure.

Archiving Expression Vectors

Considerable time, effort, and funds have gone into making each expression constructs, and during the structural determination process it is often necessary to produce multiple large-scale protein preparations. It is of the utmost importance that each expression construct is archived in a manner sufficient to allow easy retrieval and secure storage. Before each construct is miniprepped, two glycerol stocks in 96-well format are produced by aliquoting from the overnight culture and adding glycerol to a final concentration of 20% followed by flash freezing in dry ice. In addition, after the DNA miniprep is completed, an aliquot of each new construct is added to the appropriate well of two new 96-well plates that are then lyophilized to dryness, while the original is kept in liquid form as a working stock. Both the glycerol stocks and the DNA plates are then stored at −80°, with duplicates residing in separate freezers. The position of each plate and the contents of each well are then uploaded into the SPINE database such that each construct record in SPINE has associated DNA and glycerol stock locations. In this manner, the location of each clone in either form can be quickly located, using the Web-based SPINE LIMS, and subsequently retrieved.

Robotic Protein Expression Screening with the Biorobot 8000

The large number of expression constructs created by the automation of cloning also necessitates a large capacity screening process to evaluate the efficiency of protein expression and protein solubility in a high-throughput manner. Although all of the expression constructs could potentially be screened on a preparative scale, this would be costly and inefficient, since a large fraction of targeted proteins is observed to be either insoluble or not expressed in *E. coli*. The goal of small-scale expression is to predict the expression and solubility of each construct on the preparative scale. It should therefore be as representative of the large-scale conditions as possible. Moreover, to the degree that the analytical scale screening results correlate with large scale expression results, the smaller scale experiments can be used to explore different expression conditions, such as alternate bacterial strains, since different conditions sometimes produce significantly different expression or solubility results.

The scheme in Fig. 4 outlines our robotic 96-well expression and solubility screening process. Similar to the cloning schematic, completely automated steps are shown in blue and the partially automated steps are in red. Briefly, the starting material for the expression screening is the miniprep DNA derived from the cloning steps. Although it is possible to clone directly into an expression strain, we prefer to initially transform into a more stable cloning strain for archival purposes and then perform fresh transformations as needed in appropriate expression strains. We generally use the codon-enhanced BL21 (DE3)pMgK strain, containing plasmid-derived genes for arginine and isoleucine tRNA, since the codon usage in bacteria can be quite different than in eukaryotic organisms resulting in poor translation (Chen and Inouye, 1990; Ikemura, 1985; Sorensen *et al.*, 1989). After the completely automated transformation, individual colonies are picked from the four 24-well plates and inoculated into the corresponding well of a 96-well block (2.2 ml) containing 0.5 ml of LB medium per well. This initial culture is grown for 6 h at 37°, and a small aliquot from each well is added by the Biorobot 8000 to the corresponding well of a fresh 96-well block containing 0.5 ml of MJ9 minimal media (Jansson *et al.*, 1996) for overnight growth. Following saturated growth, the robot performs a 1:20 dilution into the corresponding well of one of four 24-square-well blocks (10 ml maximum volume/well) containing 2 ml of MJ9 media (Jansson *et al.*, 1996), covered with Airpore tape (Qiagen), and grown to mid-log phase (2–3 h growth) with vigorous shaking at 37°. The small volume of media in conjunction with the gas-permeable tape allows for excellent aeration, similar to the baffled Furnbach flasks used for large-scale protein synthesis, allowing the results of our analytical expression testing to more accurately mirror the results of subsequent preparative-scale fermentations. Once mid-log phase (0.5–1.0 OD_{600} units) has been reached, determined by sampling several wells in each plate, expression is induced with IPTG, the temperature is shifted to 17°, and the cultures are grown overnight with vigorous shaking. It has been previously reported, and we have also observed, that low temperature induction is often helpful in aiding solubility (Shirano and Shibata, 1990). Cells are harvested by centrifugation, the pellets are resuspended in lysis buffer

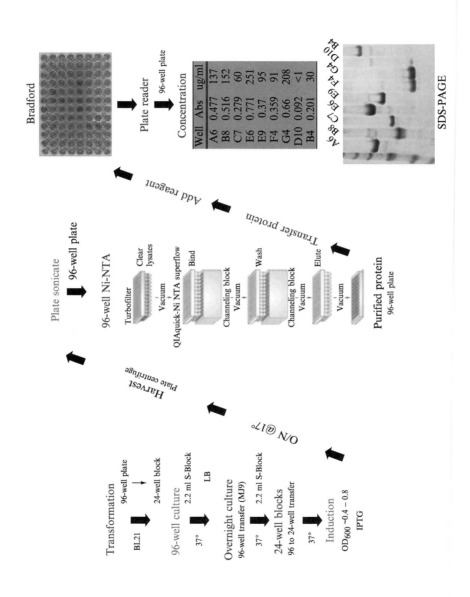

(50 mM NaH$_2$PO$_4$, 300 mM NaCl, 10 mM 2-mercaptoethanol) and robotically transferred to a 96-well PCR plate. A 96-well sonicator is used to
break open the cells, the lysates are added directly to the robot platform,
and the His-tagged recombinant proteins are purified using a modified 96-
well Ni-NTA purification protocol (Qiagen).

An aliquot of each well is next transferred to a fresh 96-well plate
containing Bradford reagent (Bradford, 1976) and the absorbance
measured using a plate reader (see Fig. 4). The concentration of soluble,
expressed protein competent for Ni binding is automatically calculated from
the absorbance, and constructs returning greater than a calculated 5 mg of
protein per liter of culture are marked for large-scale expression and purification. The SDS–polyacrylamide gel electrophoresis (PAGE) gel in Fig. 4
shows a representative sample of proteins purified in this manner, together
with data demonstrating the good correlation between protein concentration estimates by automated Bradford and Comassie Blue band stain intensity. These data are archived in the SPINE database and used to identify
constructs providing good protein expression and solubility for scale up and
biophysical analysis.

Fermentation and Preparative-Scale Protein Expression

Although the analytical expression analysis is invaluable in ascertaining
the behavior of each target when expressed in bacteria, the amount of
protein provided (10–500 μg) is not sufficient for crystallization experiments or structure determination. Therefore, the expression process needs
to be scaled up such that 10–100 mg of purified protein can be produced,
necessitating larger culture volumes. Our process for preparative-scale
protein expression, shown in Fig. 5, has been designed to optimize conditions with respect to yield, cost, throughput, and the different structural
determination approaches. We opted against using 1 liter or small fed-
batch fermenters mainly for reasons of cost, both of equipment and

FIG. 4. High-throughput analytical scale protein expression screening using robotic methods.
This schematic shows the step-by-step procedure used for small-scale expression screening.
Completely automated steps are shown in blue, and red denotes steps that are partially
automated. The entire process is conducted in 96-well plates or a corresponding number of 24-
well blocks. The right top shows a modified 96-well Bradford assay (Bradford, 1976) with
aliquots from the 96-well Ni-NTA purification. The plate configuration is the normal 8 rows by
12 columns. More intense blue wells denote a higher concentration of purified protein and hence
constructs that express high levels of soluble proteins. These targets are slated for large-scale
production. The relative concentration is calculated by the 96-well plate reader and is reported
in spreadsheet format (see the blue box). An SDS–PAGE gel shows the results of the
purification and relative agreement with the calculated values. (See color insert.)

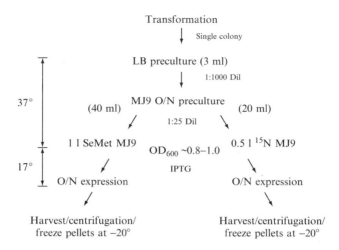

FIG. 5. Preparative-scale protein expression. Schematic of protein expression for NMR and X-ray crystallography samples. Each target is transformed into an appropriate BL21 (DE3) strain and subcultured into minimal media. Each target diverges into two pathways, for isotope enrichment and selenomethionine labeling, respectively. Preliminary growth occurs at 37° and the temperature is shifted to 17° upon IPTG induction. O/N, overnight.

reagents, as well as the fact that the prohibitive cost limits parallelization. Therefore, a strategy based on growth in 2-liter baffled Furnbach flasks was chosen, based on the simplicity of the technique, the low cost of the required equipment, and the ease of parallelization. Though not used in our platform, it is also possible to utilize disposable 2-liter plastic bottles in place of these Furnbach flasks (Millard et al., 2003). Having decided on this method, the growth conditions were optimized to provide high yields while maintaining ease and throughput.

The growth medium for protein production is MJ9, a modified minimal medium containing a stronger buffering system and supplemental vitamins and trace elements (Jansson et al., 1996), which has been optimized for efficient isotopic enrichment of proteins. We have found that MJ9 medium can support the same cell density and protein expression levels as rich media such as LB (data not shown), although not as high as superrich media such as Terrific Broth (Tartof and Hobbs, 1987). NMR studies of proteins generally require enrichment with ^{15}N, ^{13}C, and/or ^{2}H isotopes, using minimal media in which the sole sources of carbon (glucose) and nitrogen (ammonium ion) are uniformly enriched with ^{13}C and ^{15}N, respectively. In the absence of a structural model suitable for applying molecular replacement methods, high-throughput X-ray crystallography of protein structures is most efficient using single (SAD) and multiple anomalous diffraction (MAD) methods

(Dodson, 2003; Hendrickson and Ogata, 1997), which are generally readily carried out with SeMet substituted protein samples (Doublie *et al.*, 1996; Hendrickson and Ogata, 1997; Hendrickson *et al.*, 1990). Both isotopic ^{15}N, ^{13}C, and/or ^{2}H enrichment and SeMet labeling are carried out in our platform using MJ9 minimal media (Jansson *et al.*, 1996).

As shown in Fig. 5, fermentation for protein sample production is split into two branches, based on our need to produce proteins for NMR and X-ray analysis with their isotope or amino acid derivatives. The process begins with transformation of the target expression vector into the appropriate BL21(DE3) strain of *E. coli*, followed by an LB preculture. This preculture is then used to inoculate two overnight cultures (20 and 40 ml for ^{15}N and SeMet incorporation, respectively), which are grown to saturation. The entire volumes of each overnight culture are then used to inoculate each of two 2-liter baffled flasks per target, one containing 0.5 liter of MJ9 supplemented with uniformly (U)-$^{15}NH_4$ salts (1–2 g/liter) as the sole source of nitrogen and the other with 1 liter of MJ9 containing SeMet (L-selenomethionine at 60 mg/liter). When SeMet is included in the media, cells down-regulate the synthesis of methionine and incorporate the SeMet into nascent proteins (Doublie *et al.*, 1996). The cultures are incubated at 37° until OD_{600} ~0.8–1.0 units, equilibrated to 17°, and induced with IPTG (1 mM final concentration). Incubation with vigorous shaking in a 17° room continues overnight followed by harvesting through centrifugation. Aliquots of the induced cells are taken and SDS–PAGE analysis is performed on sonicated samples to assay for expression and solubility. The cell pellets, an isotope-enriched sample, and a SeMet-containing sample are generated for each target in this manner, and then stored at −20° until called for through the SPINE information management system by the protein purification team. To maintain cost-effectiveness (as ^{13}C enrichment is considerably more expensive than ^{15}N enrichment), the initial isotope-enriched sample is produced with ^{15}N enrichment only. If NMR screening results on this sample (described below) indicate that the protein is amenable to structural determination by NMR, additional protein samples are prepared with U-^{15}N,^{13}C enrichment (and sometimes also partial or complete ^{2}H enrichment) for 3D structure determinations.

Protein Purification

For crystallization and structural studies, it is imperative that the protein samples are highly homogeneous. The need to produce protein samples of sufficient purity while retaining high throughput is the primary challenge of this section of the pipeline. This is especially significant when considering the fact that proteins have such diverse biophysical characteristics.

The first step in allowing high-throughput handling is the addition of the 6X-His affinity tag to impart a similar chromatographic characteristic to all of the proteins, thus allowing a common purification technique [immobilized metal affinity chromatography (IMAC)] to be used for all samples, parallelization, and thus high-throughput handling (Crowe *et al.*, 1994; Sheibani, 1999). Like the fermentation pipeline, our protein purification strategy is also divided into two branches, producing samples for NMR (Fig. 6) and X-ray crystallography studies (Fig. 7). In both cases, cell pellets are resuspended in lysis buffer (defined above) with 10 mM imidazole (Sigma), lysed by sonication, and centrifuged to pellet the insoluble portion. The resultant supernatant is then applied to nickel-charged Hi-Trap fast protein liquid chromatography (FPLC) columns (Pharmacia) or nickel-nitrilotriacetic acid (Ni-NTA) agarose (Qiagen) open columns. The loaded columns are then washed with lysis buffer in two steps containing increasing amounts of imidazole, and finally eluted with lysis buffer containing 250 mM imidazole. Previously we utilized an AKTAexplorer 3D system running six HisTrap columns sequentially for automated purification of up to six targets. However, this was a timely process that lacked the robustness to handle our high-throughput needs. We have now incorporated a four-module AKTAxpress system and an automated two-step purification using 16 Hi-trap Ni columns and four HiPrep 26/10 desalting columns. This strategy allows for the purification and buffer exchange of 16 target proteins in less than 12 h. This system also has the ability to perform 16 Ni-affinity purifications and gel filtration chromatography steps in an automated fashion and has thus far proven extremely robust.

The path for NMR and X-ray samples diverge at this point in the purification scheme (cf. Figs. 5 and 6). The preparations slated for NMR are sufficiently pure (80–90%) at this step for NMR screening, as described below. IMAC purified SeMet-labeled proteins destined for crystallization screening are next concentrated by ultrafiltration to ~10 mg/ml, exchanged into storage buffer [50 mM Na$_2$HPO$_4$, 10 mM D,L-dithiothreitol (DTT), 300 mM NaCl, 100 mM arginine, 250 mM imidazole, 5 mM 2-mercaptoethanol, and 10% glycerol (pH 8.0)], flash-frozen in aliquots, and stored at −80° until prepared for aggregation screening or preparative gel filtration chromatography.

NMR Screening of Ni-NTA-Purified Samples

Although the spectra of many of the targets can be improved with further purification, overall the general amenability to structural determination by NMR can be ascertained with the IMAC purified preparations. Briefly, the protein preparation is divided into three fractions. Each

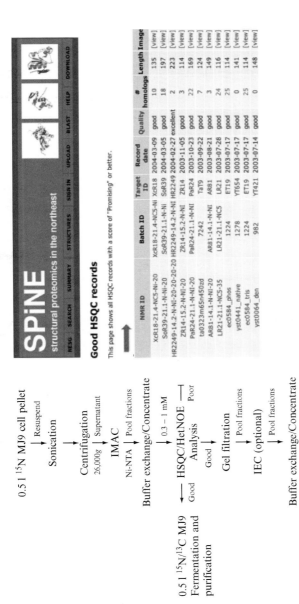

Fig. 6. (*Left*) Protein purification for NMR screening. Isotope-enriched cell pellets are resuspended in lysis buffer, sonicated, and cleared by centrifugation. Following Ni-NTA (Qiagen) IMAC purification, protein-containing fractions are pooled, concentrated, and exchanged into three buffers, which vary pH among other components. HSQC and HetNOE analysis is performed on these samples. If "good" spectra are obtained, further purification (gel filtration and optionally ion-exchange chromatography) is performed. (*Right*) View of "Good HSQC" Summary Page from the SPINE Database. This page lists those samples that are amenable to structural determination by NMR. Important aspects of the interface include the ability to view an image of the two-dimensional ^{15}N-^{1}H-HSQC for the listed target by selecting "[view]" under the image column. In addition, the number found in the "# homologs" column indicates how many additional protein targets from this Rost cluster family are in the NESG target list; selecting this link provides a list of these homologs and the progress by the consortium on each member of the family. (See color insert.)

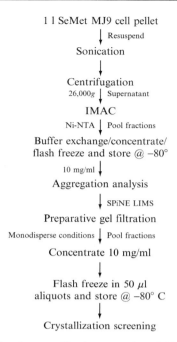

1 l SeMet MJ9 cell pellet

⏐ Resuspend

Sonication

Centrifugation

26,000g ⏐ Supernatant

IMAC

Ni-NTA ⏐ Pool fractions

Buffer exchange/concentrate/
flash freeze and store @ −80°

10 mg/ml ⏐

Aggregation analysis

⏐ SPiNE LIMS

Preparative gel filtration

Monodisperse conditions ⏐ Pool fractions

Concentrate 10 mg/ml

Flash freeze in 50 μl
aliquots and store @ −80° C

Crystallization screening

FIG. 7. Protein purification for crystallization screening. Cell pellets from selenomethio-nine-labeled protein fermentations are resuspended in lysis buffer, sonicated, and cleared by centrifugation. IMAC purification is performed and fractions containing the protein of interest are exchanged into storage buffer. Aliquots are then exchanged into a series of test buffers, and the aggregation state of the protein is assayed by analytical gel filtration and static light scattering (see Fig. 7). Once buffer conditions favoring a single stable species (e.g., monomer, dimer, etc.) are reported, preparative gel filtration in the corresponding buffer is performed. Protein samples are then concentrated to ∼10 mg/ml and the preparation is divided into 50-μl aliquots and flash frozen. These samples are then used for high-throughput crystallization screening.

fraction is exchanged into one of three NMR sample buffers (Table II) that differ in pH (4.5, 5.5, and 6.5, including only pH values different from the pI of the protein), using dialysis cassettes. This acts as the first step in sample optimization for NMR data collection. The dialyzed samples are concentrated to 0.3–1 mM in a final volume of 500 μl, transferred to 5-mm NMR tubes (Wilmad, 535PP), and stored at 4° until NMR data are collected. The sample description, including protein concentration and buffer conditions, is then entered into the SPINS NMR database (Baran et al., 2002), and a subset of this information is transferred automatically

TABLE II
NMR SCREENING BUFFERS

pH	Buffer
6.5 ± 0.1	20 mM MES, 100 mM NaCl, 5 mM CaCl$_2$, 10 mM DTT, 0.02% sodium azide, 5% D$_2$O
5.5 ± 0.1	20 mM NaOAc, 100 mM NaCl, 5 mM CaCl$_2$, 10 mM DTT, 0.02% sodium azide, 5% D$_2$O
4.5 ± 0.1	20 mM NaOAc, 100 mM NaCl, 5 mM CaCl$_2$, 10 mM DTT, 0.02% sodium azide, 5% D$_2$O

into the central SPINE database using an XML exchange language (see Wunderlich *et al.*, 2004, for a description of our basic XML exchange dictionary). SPINS also provides a Web-based list of all targets ready for NMR screening and archives key experimental data and data collection parameters from the NMR instrument.

NMR screening is performed using 500 or 600 MHz NMR spectrometers and is divided into two major components, with each target characterized in several (typically the three pHs described above) buffer conditions. Screening records two-dimensional ^{15}N–^{1}H heteronuclear single-quantum coherence (HSQC) and two-dimensional ^{15}N–^{1}H heteronuclear Overhauser effect (HetNOE) spectra, both usually performed at 20°. The spectral dispersion, together with the relative number of negative-valued peaks in the HetNOE spectrum, quickly indicates if the protein is largely folded; samples exhibiting minimal spectral dispersion and large numbers of negative HetNOE peaks are scored as "unfolded." Samples exhibiting very broad or relatively few peaks, and which are not characterized as "unfolded" by HetNOE data, are scored as "poor." A score of "unfolded" or "poor" indicates that these target samples are not amenable to structure determination by NMR. Samples providing well-resolved and fairly complete HSQC spectra are scored for their amenability to structural determination by NMR, subjectively rated as "excellent," "good," or "promising" (Yee *et al.*, 2002, 2003), based on the dispersion of resonances and the percentage of expected peaks (defined by the primary sequence) detected. This information, together with the raw free-induction decay (FID) data and a representative 2D plot of the spectrum, is archived into the Standardized Protein NMR Data Storage and Analysis System (SPINS) database (Baran *et al.*, 2004). These data are then transferred automatically from SPINS to the SPINE database, which is accessible over the internet to the entire NESG Consortium.

Protein samples with an NMR screening score of good or excellent are amenable to structural determination by NMR. NESG researchers are quickly informed of these targets through a "Good HSQC" table generated by SPINE (Fig. 6, right). Based on the results in SPINE, researchers then select targets with good or excellent HSQC spectra for pursuit. Following email notification that a target has been selected for NMR structure determination, the original ^{15}N-enriched protein sample is then further purified by gel filtration chromatography and ion-exchange chromatography (the latter only if the gel filtration–purified sample is not sufficiently pure) and concentrated to 0.3–1.0 mM in an optimized buffer (as indicated by the "button test," described in the section " 'Button Tests' to Optimize Condition for NMR Studies") using ultrafiltration, producing initial samples ready for production data collection. In addition, once selected for structural determination through the "Good HSQC" table, the protein target is also scheduled for refermentation in ^{15}N,^{13}C-enriched minimal media and production of a fully double-enriched sample.

Aggregation Screening of Ni-NTA-Purified Protein Samples

It is now well established that proteins that are monodisperse in solution are more likely to produce crystals during screening trials than polydisperse or aggregated samples (Ferre-D'Amare and Burley, 1994, 1997; Manor et al., 2005). In an effort to increase the number of samples that produce crystals, the NESG has developed a system that measures the aggregation state of protein samples following gel filtration FPLC, using a combination of static light scattering and refractive index (Manor et al., 2005). In this system, analytical gel filtration is carried out using an Agilent 1100 liquid chromatography system with a Shodex Protein KW-802.5 size-exclusion column. The effluent is detected using (1) static light scattering at three angles (45°, 90°, and 135°) measured with a miniDawn (or Dawn) static light-scattering system (Wyatt Technology), (2) absorbance at 280 nm, and (3) refractive index using an Optilab Interferometric Refractometer (Wyatt Technology). Analysis of these data provides estimates of shape-independent weight-average molecular mass (MW$_w$) and characteristics of the biopolymer mass distributions.

For polydisperse samples, a significant percentage of the mass injected into the system is distributed in multiple elution species, whereas for monodisperse samples the vast majority (>90%) of the mass injected elutes as a single species (e.g., all monomer, all dimer, etc.). Key data from these analyses are archived in the SPINE LIMS. Representative data from an aggregation screen analysis is shown from the corresponding SPINE view in Fig. 8. Using this system, buffer conditions (salt conditions and other

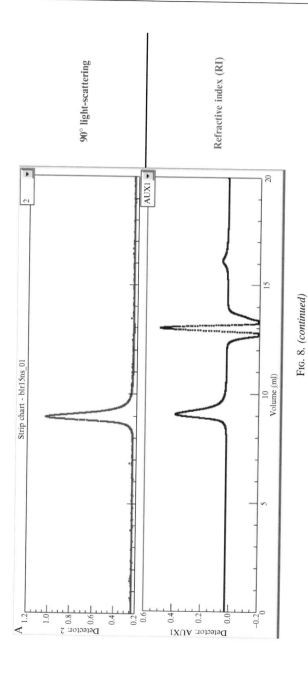

Fig. 8. *(continued)*

B

Aggregation Screening Record for blr15

⬤ **Recommended Buffers** :

- No Salt: 10mM Tris, 5mM DTT

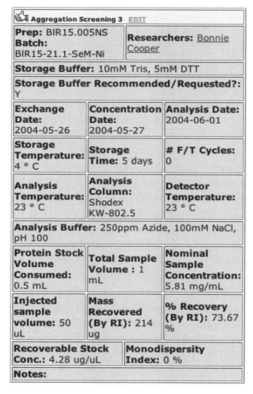

Aggregation Screening 3 _EDIT_	
Prep: BIR15.005NS **Batch:** BIR15-21.1-SeM-Ni	**Researchers:** Bonnie Cooper
Storage Buffer: 10mM Tris, 5mM DTT	
Storage Buffer Recommended/Requested?: Y	

Exchange Date: 2004-05-26	**Concentration Date:** 2004-05-27	**Analysis Date:** 2004-06-01
Storage Temperature: 4 ° C	**Storage Time:** 5 days	**# F/T Cycles:** 0
Analysis Temperature: 23 ° C	**Analysis Column:** Shodex KW-802.5	**Detector Temperature:** 23 ° C

Analysis Buffer: 250ppm Azide, 100mM NaCl, pH 100		
Protein Stock Volume Consumed: 0.5 mL	**Total Sample Volume :** 1 mL	**Nominal Sample Concentration:** 5.81 mg/mL
Injected sample volume: 50 uL	**Mass Recovered (By RI):** 214 ug	**% Recovery (By RI):** 73.67 %

Recoverable Stock Conc.: 4.28 ug/uL	**Monodispersity Index:** 0 %
Notes:	

FIG. 8. Aggregation screening. A combination of analytical gel filtration, static light scattering, and refractive index detects the volume and mass of each protein species in solution. (A) The static light scattering of NESG target protein BlR15 in a "no salt" buffer is shown in the top chromatogram and indicates a single peak under these buffer conditions. The bottom chromatogram traces the refractive index; the single peak in the lower molecular weight region (corresponding to the single peak in the light scattering) shows that most of the mass injected into the column is contained in this peak, indicating that the majority of the protein in this buffer is monomeric. (B) The Aggregation Screening results for NESG target BlR15 are summarized in this view from SPINE. (See color insert.)

additives) can be varied in a high-throughput manner and tested for their ability to produce an environment favorable for a monodisperse population. More specifically, IMAC-purified protein is exchanged into a series of several different buffers through an overnight dialysis at 4°. Each sample containing the protein exchanged into a specific buffer is then injected into the FPLC, and the species present in solution are separated based on size using a Shodex gel filtration column. The radius and mass of the species responsible for each peak are computed in real time. The aggregation state of the protein is thus characterized under different buffer conditions, and the buffer promoting the highest degree of monodispersity is chosen for the next step of preparative gel filtration purification.

Gel Filtration Chromatography

Preparative-scale gel filtration chromatography is generally performed under the buffer conditions that most favor monodispersity. The Rutgers Protein Production facility has a series of AKTA FPLC (Amersham Biosciences) chromatography systems including four Primes, one Purifier, one Explorer, and finally a four module AKTAxpress system. Each system is configured with a HiLoad 26/60 Superdex 75 gel filtration column(s) (Amersham Biosciences), with the capacity to safely load up to 10 ml of sample volume. Gel filtration columns are first equilibrated with buffers generally favoring monodispersity; for example, no salt (NS) buffer (10 mM Tris, 5 mM DTT, pH 7.5) promotes monodispersity for target BIR15 (Fig. 8). Protein samples are loaded onto the column, and the peak containing the monodisperse protein is collected. Gel filtration purified samples are then either prepared for NMR data collection or concentrated to ~10 mg/ml using an Ultrafree Centrifugal Filter Unit (Millipore), flash frozen in small (50-μl) aliquots to minimize the effects of protein dehydration upon freezing, and stored at −80°. These frozen samples are shipped to collaborators for crystallization screening.

"Button Tests" to Optimize Conditions for NMR Studies

One striking statistic from the HSQC screening is the fact that more than 25% of the samples produced have "good" (i.e., promising, good, or excellent) HSQC scores, indicating their structure is likely solvable by NMR. However, many of the targets that initially produce good spectra exhibit various forms of sample instability during data collection, including proteolysis, oxidation, deamidation, and slow precipitation between the time they are prepared and the completion of NMR data collection. Different temperatures and buffer conditions can produce significant differences in both spectral quality and sample stability. For example, in some cases cocktails of protease inhibitors are added to inhibit proteolysis during NMR data collection. Generally, even for a "high-throughput screening" platform, it is necessary to optimize buffer conditions before committing to significant NMR data collection time.

Currently, the sample stability issue that is most limiting to NMR structure production is the *slow precipitation* of subject proteins in NMR samples. Some 25% of our gel filtration–purified NMR samples exhibit slow precipitation over days or weeks after the sample is prepared, which can severely frustrate data collection efforts. To screen for conditions that avoid this *slow precipitation* behavior, we have implemented microscale buffer screening using microdialysis buttons (Bagby *et al.*, 1997) to identify conditions that stabilize the protein preparations. One advantage of this system is the small amount of protein sample needed for analysis (~50 μg for a dozen conditions), allowing a large range of conditions to be tested, including pH, salt, and other additives at varying concentrations. Table III lists 12 conditions, corresponding to three pH values, and the presence or absence of NaCl, L-arginine, and DTT. An individual microdialysis button of the protein sample is dialyzed against each of these buffers at 4°, and these "buttons" are then examined for signs of protein precipitation overnight using a dissecting microscope, and again after 1 week; the button is then moved to the same buffer at 20° and observed for another 2 weeks. We have often found that 100 mM L-arginine is useful for stabilizing protein samples against slow precipitation. Buffer conditions promoting sample stability identified from this assay are combined with information

TABLE III
BUFFER CONDITIONS OF INITIAL TESTS FOR STABILITY WITH RESPECT TO
SLOW PROTEIN PRECIPITATION

Buffer	NaCl	DTT	Arginine
50 mM ammonium acetate, pH 5.0	0	0	0
50 mM ammonium acetate, pH 5.0	0	10 mM	0
50 mM ammonium acetate, pH 5.0	0.1 M	10 mM	0
50 mM ammonium acetate, pH 5.0	0	10 mM	0.1 M
50 mM MES, pH 6.0	0	0	0
50 mM MES, pH 6.0	0	10 mM	0
50 mM MES, pH 6.0	0.1 M	10 mM	0
50 mM MES, pH 6.0	0	10 mM	0.1 M
50 mM Bis. Tris, pH 6.5	0	0	0
50 mM Bis. Tris, pH 6.5	0	10 mM	0
50 mM Bis. Tris, pH 6.5	0.1 M	10 mM	0
50 mM Bix. Tris, pH 6.5	0	10 mM	0.1 M

from the NMR screening data to ascertain optimal buffer conditions for NMR data collection.

Quality Control

Proteins prepared for NMR or X-ray crystallographic studies are all analyzed for homogeneity by SDS–PAGE and validated for molecular weight by matrix-assisted laser-desorption-induced time-of-flight (MALDI-TOF) mass spectrometry. When inconsistencies are observed, the expression constructs are validated by DNA sequencing. Data generated at each stage of the production pipeline, along with analytical results, spectra, comments, records of interlaboratory shipments, the names of the individuals involved in each production step, and other aspects of the production process, are archived in the SPINE database. Summaries of these data are available in public domain (http://nesg.org/).

Capacity of the Platform

Based on our current levels of success in producing diffraction quality crystals or samples amenable for NMR studies, we calculate that to reach our goal of determining 100–200 novel structures per year requires the capacity to produce 600 target proteins on the 10–50 mg scale per year, each with high (>98%) homogeneity. Our conservative estimate with current protein target list characteristics suggests that this requires producing roughly 2500 expression constructs per year. In the sections above, we outlined a scalable platform for high-throughput protein production of samples suitable for structural determination, including the needed technologies and infrastructure. The platform as it stands is producing these target numbers of expression constructs (2500 per year) and purified proteins (600 per year in 10–50 mg quantities) for both NMR and X-ray crystallization experiments.

Hierarchical Multiplex Expression

The ability to clone and analyze the expression of targets in a high-throughput manner has also allowed us to become more proficient and successful at rescuing targets that were not initially suitable for structural determination because of low expression, low solubility, poor NMR spectral quality, or poor quality crystals. The first layer of our hierarchical multiplex strategy is the use of multiple homologues from a particular target family (Fig. 1). This process is managed through the SPINE database. Next, the platform allows for cloning targets into a series of expression vectors with different placements of the affinity tag or promoters driving expression.

For example, a 96-well plate was recently assembled with targets ranging from bacteria to human proteins, all of which were previously cloned into a C-terminal 6X-His expression vector and were either not expressed or insoluble. These targets were then rapidly subcloned into an N-terminal expression vector (using a single universal primer and our robotic platform). Over 30% of the resulting N-terminal 6XHis-tagged proteins were expressed and soluble at levels amenable to preparative-scale expression and purification.

Several other new technologies are being explored for expanding our hierarchical-multiplex expression platform. *E. coli*–based cold-shock induction vectors (Qing *et al.*, 2004) have allowed production of many targets that are not expressed or insoluble in pET-derived expression vectors. High-throughput robotic-based expression technology also allows for varying other parameters of the expression conditions, such as the bacterial host strains, and efforts are in progress to explore and develop improved chaperone-supplemented bacterial host strains. Efforts are also in progress to develop robotic technologies for protein production in cell-free wheat germ (Ma *et al.*, 2005; Morita *et al.*, 2003; Sawasaki *et al.*, 2002), *Pichia pastoris* (Boettner *et al.*, 2002; Prinz *et al.*, 2004; Wood and Komives, 1999), and *Saccharomyces cervisiae* (Boettner *et al.*, 2002; Holz *et al.*, 2003; Prinz *et al.*, 2004) expression systems, particularly for eukaryotic protein targets. Accordingly, by hierarchical multiplexing of expression technologies, a significant number of targets that have not passed current analyses may eventually be produced in a form amenable to 3D structure determination.

Data Integration and Sharing

The SPINE (Bertone *et al.*, 2001; Goh *et al.*, 2003) and SPINS (Baran *et al.*, 2002) databases collect information from all steps in the protein production process, including information on the cloning and small-scale expression, large-scale fermentation and protein preparations, aggregation screening, and NMR screening. SPINE also tracks sample shipments between laboratories of the NESG Consortium. It is a central component of our protein production pipeline, acting to integrate the entire process of sample production and analysis and to organize data that will be invaluable for optimizing the sample production process and learning about physical and biochemical properties of proteins.

Summary

We have outlined our strategy and platform for producing high-quality protein samples using *E. coli* expression hosts. Our protein purification process is centered on 6X-His affinity tag-IMAC affinity purification, allowing all of our targets to have identical initial purification procedures and the implementation of high-throughput parallel methods. In most cases, these 6X-His-tagged proteins are sufficiently pure that a single ensuing gel filtration chromatography step is adequate to produce protein preparations that are greater than 98% homogeneous, a level that we have observed is sufficient for structural studies. Protein structures have generally been determined by NMR and X-ray crystallography with the small 6X-His tags on the protein targets, although in a few cases the tags have been removed by cloning into related vectors that provide tagless proteins. Our targets include primarily proteins that comprise the proteomes of the eukaryotic model organisms and their prokaryotic homologues (Fig. 9), and our current strategies have proven effective in producing samples containing structured eukaryotic and prokaryotic proteins.

The target list of the NESG Consortium (Liu and Rost, 2004; Liu *et al.*, 2004; Wunderlich *et al.*, 2004), roughly two-thirds of which are eukaryotic proteins (Fig. 9), is generally more challenging than those pursued by structural genomics projects focused exclusively on prokaryotic proteins, which tend to be easier to produce in bacterial host systems. Despite these challenges, over 1000 different protein targets have been cloned, expressed, and purified in tens of milligram quantities over the past 36-month period (see Summary Statistics for All Targets, http://www-nmr.cabm.rutgers.edu/bioinformatics/ZebaView/) in the Rutgers facility; current production rates

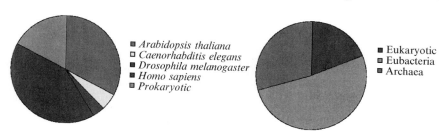

■ *Arabidopsis thaliana*
□ *Caenorhabditis elegans*
■ *Drosophila melanogaster*
■ *Homo sapiens*
□ *Prokaryotic*

■ Eukaryotic
■ Eubacteria
■ Archaea

Fig. 9. (*Left*) Phylogenetic distribution of NESG target proteins. Currently, 80% of the NESG targets are from the eukaryotic model organisms *Arabidopsis thaliana, Caenorhabditis elegans, Drosophila melanogaster*, and *Homo sapiens*. As illustrated in the pie chart, the majority of proteins are derived from the human and *Arabidopsis* genomes. The remaining 20% of NESG targets for Rutgers protein production (*S. cerevisiae* efforts are focused in Toronto) are of prokaryotic origin, including proteins from both archaea and eubacteria; see Table I for a complete listing of the eukaryotic organisms and the number of proteins targeted from each of these proteomes. (*Right*) Phylogenetic distribution of NESG protein structures deposited in the PDB. As indicated in the pie chart, ~20% of the NESG structures are of eukaryotic proteins, ~30% are of archaeal origin, and the remaining are structures of proteins from eubacteria. However, the majority of these prokaryotic proteins for which structures have been determined are members of protein domain families that also include eukaryotic members. (See color insert.)

are about 12 purified protein targets in tens of milligram quantities per week. Our experience using the hierarchical multiplex expression and purification strategy, also described in this chapter, has allowed us to achieve success in producing not only protein samples but also three-dimensional structures. As of December 2004, the NESG Consortium has deposited over 145 new protein structures to the PDB; about two-thirds of these protein samples were produced by the Rutgers NESG Protein Production Facility. Roughly 20% of the NESG protein structures are of eukaryotic proteins (Fig. 9), demonstrating the broad applicability of the platform. The sample production and screening technologies described here are scalable and, as demonstrated by several other chapters in this volume, efficiencies of sample production and structure determination methods in use by the NESG Consortium continue to improve. In addition to their role in our pilot structural genomics initiative, these improved robotic and/or parallel cloning, expression, protein production, and bio-physical screening technologies will be of broad value to the structural biology, functional proteomics, and structural genomics communities.

Acknowledgments

We thank Drs. S. Anderson, P. Manor, F. Piano, and A. Yee for helpful advice in developing this protein production platform. This work is supported by Grant P50-GM62413 from the Protein Structure Initiative of the National Institutes of Health, Institute of General Medical Sciences.

References

Abremski, K., and Hoess, R. (1984). Bacteriophage P1 site-specific recombination. Purification and properties of the Cre recombinase protein. *J. Biol. Chem.* **259,** 1509–1514.

Acton, T. B., Gunsalus, K., Xiao, R., Ma, L., Chiang, Y., Clement, T., Everett, J. K., Shastry, R., Denissova, N., Palacios, D., *et al.* (2005). The protein sample production platform of the Northeast Structural Genomics Consortium. *J. Struct. Funct. Genomics.* Submitted.

Bagby, S., Tong, K. I., Liu, D., Alattia, J. R., and Ikura, M. (1997). The button test: A small scale method using microdialysis cells for assessing protein solubility at concentrations suitable for NMR. *J. Biomol. NMR* **10,** 279–282.

Baran, M. C., Moseley, H. N., Sahota, G., and Montelione, G. T. (2002). SPINS: Standardized protein NMR storage. A data dictionary and object-oriented relational database for archiving protein NMR spectra. *J. Biomol. NMR* **24,** 113–121.

Baran, M. C., Haung, Y. J., Moseley, H. N., and Montelione, G. T. (2004). Automated analysis of protein NMR assignments and structures. *Chem. Rev.* **104,** 3541–3556.

Bertone, P., Kluger, Y., Lan, N., Zheng, D., Christendat, D., Yee, A., Edwards, A. M., Arrowsmith, C. H., Montelione, G. T., and Gerstein, M. (2001). SPINE: An integrated tracking database and data mining approach for identifying feasible targets in high-throughput structural proteomics. *Nucleic Acids Res.* **29,** 2884–2898.

Boettner, M., Prinz, B., Holz, C., Stahl, U., and Lang, C. (2002). High-throughput screening for expression of heterologous proteins in the yeast *Pichia pastoris*. *J. Biotechnol.* **99,** 51–62.

Bradford, M. M. (1976). A rapid and sensitive method for the quantitation of microgram quantities of protein utilizing the principle of protein dye binding. *Anal. Biochem.* **131,** 248–254.

Bujard, H., Gentz, R., Lanzer, M., Stuber, D., Muller, M., Ibrahimi, I., Hauptle, M. T., and Dobberstein, B. (1987). A T5 promotor based transcription-translation system for the analysis of proteins *in vivo* and *in vitro*. *Methods Enzymol.* **155,** 416–433.

Carter, P., Liu, J., and Rost, B. (2003). PEP: Predictions for entire proteomes. *Nucleic Acids Res.* **31,** 410–413.

Chen, G. F., and Inouye, M. (1990). Suppression of the negative effect of minor arginine codons on gene expression; preferential usage of minor codons within the first 25 codons of the *Escherichia coli* genes. *Nucleic Acids Res.* **18,** 1465–1473.

Crowe, J., Dobeli, H., Gentz, R., Hochuli, E., Stuber, D., and Henco, K. (1994). 6xHis-Ni-NTA chromatography as a superior technique in recombinant protein expression/purification. *Methods Mol. Biol.* **31,** 371–387.

Dodson, E. (2003). Is it jolly SAD? *Acta Crystallogr D Biol. Crystallogr.* **59,** 1958–1965.

Doublie, S., Kapp, U., Aberg, A., Brown, K., Strub, K., and Cusack, S. (1996). Crystallization and preliminary X-ray analysis of the 9 kDa protein of the mouse signal recognition particle and the selenomethionyl-SRP9. *FEBS Lett.* **384,** 219–221.

Everett, J. K., Acton, T. B., and Montelione, G. T. (2004). Primer Prim'r: A web based server for automated primer design. *J. Struct. Funct. Genomics* **5,** 13–21.

Ferre-D'Amare, A. R., and Burley, S. K. (1994). Use of dynamic light scattering to assess crystallizability of macromolecules and macromolecular assemblies. *Structure* **2,** 357–359.

Ferre-D'Amare, A. R., and Burley, S. K. (1997). Dynamic light scattering in evaluating crystallizability of macromolecule. *Methods Enzymol.* **276,** 157–166.

Geisse, S., Gram, H., Kleuser, B., and Kocher, H. P. (1996). Eukaryotic expression systems: A comparison. *Protein Expr. Purif.* **8,** 271–282.

Giot, L., Bader, J. S., Brouwer, C., Chaudhuri, A., Kuang, B., Li, Y., Hao, Y. L., Ooi, C. E., Godwin, B., Vitols, E., *et al.* (2003). A protein interaction map of *Drosophila melanogaster*. *Science* **302,** 1727–1736.

Goh, C. S., Lan, N., Echols, N., Douglas, S. M., Milburn, D., Bertone, P., Xiao, R., Ma, L. C., Zheng, D., Wunderlich, Z., *et al.* (2003). SPINE 2: A system for collaborative structural proteomics within a federated database framework. *Nucleic Acids Res.* **31,** 2833–2838.

Hartley, J. L., Temple, G. F., and Brasch, M. A. (2000). DNA cloning using *in vitro* site-specific recombination. *Genome Res.* **10,** 1788–1795.

Hendrickson, W. A., and Ogata, C. M. (1997). Phase determination from multiwavelength anomalous diffraction measurements. *Methods Enzymol.* **276,** 494–523.

Hendrickson, W. A., Horton, J. R., and LeMaster, D. M. (1990). Selenomethionyl proteins produced for analysis by multiwavelength anomalous diffraction (MAD): A vehicle for direct determination of three-dimensional structure. *EMBO J.* **9,** 1665–1672.

Holz, C., Prinz, B., Bolotina, N., Sievert, V., Bussow, K., Simon, B., Stahl, U., and Lang, C. (2003). Establishing the yeast *Saccharomyces cerevisiae* as a system for expression of human proteins on a proteome-scale. *J. Struct. Funct. Genomics* **4,** 97–108.

Ikemura, T. (1985). Codon usage and tRNA content in unicellular and multicellular organisms. *Mol. Biol. Evol.* **2,** 13–34.

Ito, T., Chiba, T., Ozawa, R., Yoshida, M., Hattori, M., and Sakaki, Y. (2001). A comprehensive two-hybrid analysis to explore the yeast protein interactome. *Proc. Natl. Acad. Sci. USA* **98,** 4569–4574.

Jansson, M., Li, Y.-C., Jendenberg, L., Anderson, S., and Montelione, G. T. (1996). High-level production of uniformly ^{15}N- and ^{13}C-enriched fusion proteins in *Escherichia coli*. *J. Biomol. NMR* **7,** 131–141.

Kapust, R. B., Tozser, J., Copeland, T. D., and Waugh, D. S. (2002). The P1′ specificity of tobacco etch virus protease. *Biochem. Biophys. Res. Commun.* **294,** 949–955.

Lander, E. S. (1999). Array of hope. *Nat. Genet.* **21,** 3–4.

Lander, E. S., Linton, L. M., Birren, B., Nusbaum, C., Zody, M. C., Baldwin, J., Devon, K., Dewar, K., Doyle, M., FitzHugh, W., *et al.* (2001). Initial sequencing and analysis of the human genome. *Nature* **409,** 860–921.

Li, S., Armstrong, C. M., Bertin, N., Ge, H., Milstein, S., Boxem, M., Vidalain, P. O., Han, J. D., Chesneau, A., Hao, T., *et al.* (2004). A map of the interactome network of the metazoan *C. elegans. Science* **303,** 540–543.

Liu, J., and Rost, B. (2004). CHOP proteins into structural domain-like fragments. *Proteins* **55,** 678–688.

Liu, J., Hegyi, H., Acton, T. B., Montelione, G. T., and Rost, B. (2004). Automatic target selection for structural genomics on eukaryotes. *Proteins* **56,** 188–200.

Ma, L. C., Sawasaki, T., Tsuchimochi, M., Mazda, S., Gunsalus, K. C., Macapagal, D., Shastry, R., Ho, C. K., Acton, T. B., Endo, Y., and Montelione, G. T. (2005). Evaluation of a wheat germ cell-free protein production system for expression and solubility screening of eukaryotic proteins. *J. Struct. Funct. Genomics.* Submitted.

Makrides, S. C. (1996). Strategies for achieving high-level expression of genes in *Escherichia coli. Microbiol. Rev.* **60,** 512–538.

Manor, P., Shen, J., Satterwhite, R., Kuzin, A., Forohar, F., Benach, J., Smith, P., Montelione, G. T., Acton, T. B., and Hunt, J. (2005). *Protein solution aggregation characteristics and crystallization.* In preparation.

Millard, C. S., Stols, L., Quartey, P., Kim, Y., Dementieva, I., and Donnelly, M. I. (2003). A less laborious approach to the high-throughput production of recombinant proteins in *Escherichia coli* using 2-liter plastic bottles. *Protein Expr. Purif.* **29,** 311–320.

Montelione, G. T., and Anderson, S. (1999). Structural genomics: Keystone for a Human Proteome Project. *Nat. Struct. Biol.* **6,** 11–12.

Morita, E. H., Sawasaki, T., Tanaka, R., Endo, Y., and Kohno, T. (2003). A wheat germ cell-free system is a novel way to screen protein folding and function. *Protein Sci.* **12,** 1216–1221.

Prinz, B., Schultchen, J., Rydzewski, R., Holz, C., Boettner, M., Stahl, U., and Lang, C. (2004). Establishing a versatile fermentation and purification procedure for human proteins expressed in the yeasts *Saccharomyces cerevisiae* and *Pichia pastoris* for structural genomics. *J. Struct. Funct. Genomics* **5,** 29–44.

Qing, G., Ma, L., Khorchid, A., Swapna, G. V. T., Mal, T. K., Takayama, M. M., Xia, B., Sangita Phadtare, S., Ke, H., Acton, T., *et al.* (2004). Cold-shock induced high-yield protein production in *Escherichia coli. Nat. Biotechnol.* **22,** 877–882.

Rain, J. C., Selig, L., De Reuse, H., Battaglia, V., Reverdy, C., Simon, S., Lenzen, G., Petel, F., Wojcik, J., Schachter, V., *et al.* (2001). The protein-protein interaction map of *Helicobacter pylori. Nature* **409,** 211–215.

Reboul, J., Vaglio, P., Rual, J. F., Lamesch, P., Martinez, M., Armstrong, C. M., Li, S., Jacotot, L., Bertin, N., Janky, R., *et al.* (2003). *C. elegans* ORFeome version 1.1: Experimental verification of the genome annotation and resource for proteome-scale protein expression. *Nat. Genet.* **34,** 35–41.

Rubin, G. M., Hong, L., Brokstein, P., Evans-Holm, M., Frise, E., Stapleton, M., and Harvey, D. A. (2000). A *Drosophila* complementary DNA resource. *Science* **287,** 2222–2224.

Sauer, B. (1994). Site-specific recombination: Developments and applications. *Curr. Opin. Biotechnol.* **5,** 521–527.

Sawasaki, T., Ogasawara, T., Morishita, R., and Endo, Y. (2002). A cell-free protein synthesis system for high-throughput proteomics. *Proc. Natl. Acad. Sci. USA* **99,** 14652–14657.

Sheibani, N. (1999). Prokaryotic gene fusion expression systems and their use in structural and functional studies of proteins. *Prep. Biochem. Biotechnol.* **29,** 77–90.

Shirano, Y., and Shibata, D. (1990). Low temperature cultivation of *Escherichia coli* carrying a rice lipoxygenase L-2 cDNA produces a soluble and active enzyme at a high level. *FEBS Lett.* **271,** 128–130.

Shuman, S. (1994). Novel approach to molecular cloning and polynucleotide synthesis using vaccinia DNA topoisomerase. *J. Biol. Chem.* **269,** 32678–32684.

Sorensen, M. A., Kurland, C. G., and Pedersen, S. (1989). Codon usage determines translation rate in *Escherichia coli. J. Mol. Biol.* **207,** 365–377.

Stapleton, M., Carlson, J., Brokstein, P., Yu, C., Champe, M., George, R., Guarin, H., Kronmiller, B., Pacleb, J., Park, S., *et al.* (2002). A *Drosophila* full-length cDNA resource. *Genome Biol.* **3,** 80–88.

Strausberg, R. L., Feingold, E. A., Klausner, R. D., and Collins, F. S. (1999). The mammalian gene collection. *Science* **286,** 455–457.

Strausberg, R. L., Feingold, E. A., Grouse, L. H., Derge, J. G., Klausner, R. D., Collins, F. S., Wagner, L., Shenmen, C. M., Schuler, G. D., Altschul, S. F., *et al.* (2002). Generation and initial analysis of more than 15,000 full-length human and mouse cDNA sequences. *Proc. Natl. Acad. Sci. USA* **99,** 16899–16903.

Studier, F. W. (2004). Personal communication.

Studier, F. W., Rosenberg, A. H., Dunn, J. J., and Dubendorff, J. W. (1990). Use of T7 RNA polymerase to direct expression of cloned genes. *Methods Enzymol.* **185,** 60–89.

Tartof, K. D., and Hobbs, C. A. (1987). Improved media for growing plasmid and cosmid clones. *Bethesda Res. Lab. Focus* **9,** 12.

Uetz, P., Giot, L., Cagney, G., Mansfield, T. A., Judson, R. S., Knight, J. R., Lockshon, D., Narayan, V., Srinivasan, M., Pochart, P., *et al.* (2000). A comprehensive analysis of protein-protein interactions in *Saccharomyces cerevisiae. Nature* **403,** 623–627.

Winzeler, E. A., Schena, M., and Davis, R. W. (1999). Fluorescence-based expression monitoring using microarrays. *Methods Enzymol.* **306,** 3–18.

Wood, M. J., and Komives, E. A. (1999). Production of large quantities of isotopically labeled protein in *Pichia pastoris* by fermentation. *J. Biomol. NMR* **13,** 149–159.

Wunderlich, Z., Acton, T. B., Liu, J., Kornhaber, G., Everett, J., Carter, P., Lan, N., Echols, N., Gerstein, M., Rost, B., and Montelione, G. T. (2004). The protein target list of the Northeast Structural Genomics Consortium. *Proteins* **56,** 181–187.

Yee, A., Chang, X., Pineda-Lucena, A., Wu, B., Semesi, A., Le, B., Ramelot, T., Lee, G. M., Bhattacharyya, S., Gutierrez, P., *et al.* (2002). An NMR approach to structural proteomics. *Proc. Natl. Acad. Sci. USA* **99,** 1825–1830.

Yee, A., Pardee, K., Christendat, D., Savchenko, A., Edwards, A. M., and Arrowsmith, C. H. (2003). Structural proteomics: Toward high-throughput structural biology as a tool in functional genomics. *Acc. Chem. Res.* **36,** 183–189.

[9] Protein Structure Estimation from Minimal Restraints Using Rosetta

By CAROL A. ROHL

Abstract

RosettaNMR combines the Rosetta *de novo* structure prediction method with limited NMR experimental data for rapid estimation of protein structure. The *de novo* Rosetta algorithm predicts protein three-dimensional structures using only sequence information by combining short fragments selected from known protein structures on the basis of local sequence similarity. These fragments are assembled using a Monte Carlo strategy to generate models that reproduce empirical statistics describing nonlocal protein structure such as overall compactness, hydrophobic burial, and β-strand pairing. By incorporating chemical shift, nuclear Overhauser enhancement, and/or residual dipolar coupling restraints that are insufficient on their own to determine the protein global fold, the RosettaNMR method correctly estimates the global fold of a variety of different proteins, generating models that are that are generally 4 Å or better Cα root-mean-square deviation to the high-resolution experimental structures. Here we review the capabilities of the RosettaNMR approach, describe the underlying methods, and provide practical tips for applying the technique to structure estimation problems.

Introduction

Although protein structure is encoded in protein sequence, predicting protein structure from sequence information alone remains a challenging problem, and accurate high-resolution structural models must be experimentally determined. Structure prediction methods have improved significantly in their ability to predict protein folds using sequence information alone (Venclovas *et al.*, 2003), and the improvement in such methods offers the possibility that structure prediction approaches may be useful for experimental structure determination. Traditional high-resolution structure determination methods have been demonstrated to benefit from the incorporation of knowledge-based potential functions that have traditionally been the realm of structure prediction methods (Clore and Kuszewski, 2002; Kuszewski and Clore, 2000; Kuszewski *et al.*, 1996). Conversely, *de novo* structure prediction methods, supplemented with minimal experimental restraints, can rapidly estimate protein global folds (Bowers *et al.*, 2000; Rohl and Baker, 2002).

Copyright 2005, Elsevier Inc.
All rights reserved.
0076-6879/05 $35.00

Such moderate resolution models of protein structure may be useful directly for functional insight or may be used to accelerate high-resolution structure determination by providing better models early in the structure determination process. Here we describe the combination of the Rosetta structure prediction method with limited experimental nuclear magnetic resonance (NMR) data for rapid estimation of protein folds.

The Rosetta method (Simons *et al.*, 1997, 1999) was originally developed for *de novo* structure prediction. Double-blind assessments of protein structure prediction methods, conducted biannually in the community-wide Critical Assessment of Structure Prediction experiments (Moult *et al.*, 2003), have indicated that the Rosetta algorithm is currently perhaps the most successful method for *de novo* protein structure prediction (Aloy *et al.*, 2003; Bradley *et al.*, 2003; Kinch *et al.*, 2003). Using only primary sequence information, successful Rosetta predictions yield models with typical accuracies of 3–6 Å Cα root-mean-square deviation (RMSD) to the experimentally determined structures for substantial segments of 60 or more residues. In such low to moderate accuracy models of protein structure, the global topology is correctly predicted, the location and arrangement of secondary structure elements are generally correct, and functional residues are frequently clustered to an active site region. The utility of the Rosetta method in providing models leading to functional insight has been demonstrated in a study in which models were predicted for several hundred Pfam (Bateman *et al.*, 2004) families lacking links to known structure (Bonneau *et al.*, 2002). Many of these predictions suggest functional links not accessible by other methods of annotation.

The Rosetta method has been used in combination with experimental restraints derived from NMR to determine protein global folds (Table I), and the combined method is referred to as RosettaNMR. Bowers and associates (2000) described the utilization of chemical shift data and sparse nuclear Overhauser enhancement (NOE) distance restraints in the Rosetta framework to obtain correct global folds (<4 Å Cα RMSD to native) for 8 of 9 proteins tested. Data sets were limited to NOEs between H$^\alpha$ and/or H^N atoms and contained about one restraint per residue. In most cases, less than half of these restraints were long range ($|i-j| > 5$ residues). In the one case in which the model obtained was not within 4 Å of the native structure, no long-range NOEs were available, and the Rosetta-estimated structure differed from the native topology by the arrangement of one helix. In a subsequent study, correct global folds were obtained for 8 of 10 proteins examined using chemical shifts and residual dipolar coupling (RDC) restraints (Rohl and Baker, 2002). For the two proteins for which the correct fold was not obtained, subdomain structures were correctly predicted, but the relative orientation of these subdomains was

TABLE I
PROTEIN STRUCTURES DETERMINED BY ROSETTANMR USING LIMITED RESTRAINTS

Protein	PDB	Length	Class	Chemical shifts[a]	NOE restraints[b]	RDC restraints[c]	RMSD[d]
ISL homeodomain	1bw5	52	α	+	—	249[e]	1.9[f]
Protein G	3gb1	56	α/β	+	52 (52/0)[e]	—	2.6[g]
	1gb1			—	—	300	1.5[f]
Ubiquitin	1d3z	76	α/β	—	10	117	1.4[g]
	1ubq			+	65 (32/33)	68	2.8[f]
	1ubi			—	5	—	1.5[g]
HPr	1poh	85	α/β	+		56	2.1[f]
					86 (49/37)[e]	414[e]	2.3[f]
IM9	1imq	86	α	+		—	1.6[g]
					89 (84/5)	276[e]	5.9[f]
BAF	1ci4	89	α	+		—	3.0[g]
					—	246	2.9[f,h]
Hnrnp KH domain	1khm	89	α/β	—	9	31	2.0[i]
Cyanovirin-N	2ezm	101	β	+	—	327	3.2/2.8[g,i]
Ribosomal L30	1ck2	104	α/β	+	108 (67/41)[e]	503[e]	3.0[f]
					—	—	3.1[g]
Profilin I	1acf	125	α/β	—	123 (74/49)[e]	156[e]	3.0[f]
					—	—	3.1[g]
P14a	1cfe	135	α/β	+	106 (65/41)	—	5.7[g]
GAIP	1cmz	152	α	+	—	291	4.6[f]
CBDN1	1ulo	152	β	+	22 (22/0)	—	9.5[g]
				+	51 (16/35)	—	7.0[g]
				+	160 (19/141)[e]	—	3.9[g]

[a] +, chemical shift assignments were used for fragment selection.
[b] NOE-derived distance constraints. When available, the number of short-range/long-range distance restraints is given in parentheses. Long-range restraints are $|i-j| > 5$.
[c] Residual dipolar coupling restraints.
[d] $C\alpha$ RMSD between the RosettaNMR-determined structure and the reference PDB structure.
[e] Artificially constructed data set.

incorrect in the predicted model as compared to the experimental structure, as a result of the rotation of one subdomain about an axis of the alignment tensor. Because of the degeneracy inherent in RDC restraints, these experimental data cannot distinguish between such symmetry-related models. For one of these proteins in which the predicted model had incorrectly oriented subdomains, cyanovirin-N, the addition of the experimentally detected long-range NOEs between H^N and H^α atoms (Bewley et al., 1998) to the RDC data set allows RosettaNMR to correctly predict the global fold (J. Samayoa and C. A. Rohl, unpublished results).

Combinations of RDC and NOE restraints have been used successfully with RosettaNMR as part of a method for using unassigned NMR data to determine global folds. While Rosetta itself currently can utilize only restraints that are unambiguously assigned, Meiler and Baker (2003) combined Rosetta with a Monte Carlo algorithm for resonance assignment that optimizes the agreement between experimental data and a candidate three-dimensional structure. For each protein sequence, an ensemble of models was generated without reference to the experimental data by the de novo Rosetta fragment assembly protocol, an optimal resonance assignment was generated for each model, and the models were then ranked according to their agreement with the experimental data using these assignments. For nine proteins tested, this procedure gave models between 3 and 6 Å Cα RMSD to the native structure. In four of the nine cases, models of improved accuracy (0.6–1.8 Å Cα RMSD to the native structure) could be obtained by iterative cycles of model generation and assignment. In each cycle, the confidently assigned restraints were used in combination with RosettaNMR to generate a new population of models that was then reassigned and ranked. As more restraints become confidently assigned, the accuracy of models produced by RosettaNMR increases, allowing the further assignment of additional restraints. Examples taken from Meiler and Baker (2003) are included in Table I to illustrate the results obtained with RosettaNMR using combinations of NOE and RDC restraints.

The Rosetta Method

The Rosetta strategy is based on the experimental observation that protein structure is a result of both local and global interactions. The algorithm attempts to mimic the view of folding in which local sequence

[f] Results are taken from Rohl and Baker (2002).
[g] Results are taken from Bowers et al. (2000).
[h] Model structures inconsistent with a dimer were discarded (see Rohl and Baker, 2002).
[i] Results are taken from Meiler and Baker (2003).
[j] Cα RMSD for the N-terminal/C-terminal halves of the molecule independently (see Rohl and Baker, 2002).

preferences bias but do not uniquely define the local structure of a protein. The final native conformation is achieved when fluctuating local structures form favorable nonlocal interactions such as buried hydrophobic residues, paired β-strands, and specific side chain interactions. In the Rosetta *de novo* prediction method, the distribution of structures seen in known protein structures for a particular short sequence is used to approximate the local structural preferences of the chain. This approximation is made by selecting a library of fragments that represents the range of accessible local structures for all short segments of the protein chain from a database of known protein structures. These fragments are then randomly combined using a Monte Carlo simulated annealing search to generate models of the protein. At each step in the search, a position in the chain is randomly selected and a fragment for this position is randomly chosen from the customized library. The backbone torsion angles in the protein chain are replaced with those from the fragment, resulting in a new global configuration of the protein chain. The fitness of individual conformations with respect to nonlocal interactions is evaluated using a scoring function derived from conformational statistics of known protein structures.

Fragment insertion is the basic method of conformation modification in Rosetta. For finer-grained sampling of conformational space once a chain is collapsed to a compact structure, the fragment insertion strategy is modified and combined with continuous torsion angle perturbations to enable effective optimization of energy functions (Rohl *et al.*, 2004a). The Rosetta energy function, described in detail by Rohl and associates (2004a), is similarly adapted to both coarse- or fine-grained sampling. For coarse-grained sampling, a low-resolution description of structure is used. Solvation and electrostatic energies are approximated using observed residue distributions in protein structures. Hydrogen bonding is included using probabilistic descriptions of β-strand pairing geometry and β-sheet patterns. Steric overlap is penalized, but favorable van der Waals interactions are modeled only by rewarding globally compact structures. For finer resolution, a more physically realistic, atomic-level potential function is used that includes an attenuated 6–12 Lennard–Jones potential, an implicit solvation model, a knowledge-based hydrogen bonding potential, and a residue-based pair potential that primarily reflects electrostatics. A knowledge-based torsion potential for backbone dihedral angles is also used when conformation modifications permit introduction of backbone torsion angles not part of the discrete fragment library.

Two alternate representations for side chains are available within Rosetta, and the selection between these representations is made on the basis of the requirements of the energy function in use. For residue-based potential terms, a reduced description is used in which each side chain is represented by a centroid located at the side chain center of mass. For increased detail, atomic coordinates for all side chain atoms, including hydrogens, are utilized. Side chains are generally restricted to discrete conformations as described by a backbone-dependent rotamer library. In simulations using all-atom side chain representations, side chain conformations can be either rapidly optimized by replacing each side chain with the lowest energy rotamer available or subjected to a complete combinatorial optimization using a Monte Carlo simulated annealing search (Kuhlman and Baker, 2000; Rohl et al., 2004a).

Scoring Functions for Experimentally Derived Restraints

NOE Distance Restraints

Several scoring functions for distance restraints, derived either from NOEs or other sources, are available in the Rosetta framework (Table II). The original implementation within Rosetta (Bowers et al., 2000) evaluates distance restraint scores as the sum of upper bounds violations, with the maximum violation per restraint fixed at 10 Å. Subsequently, additional scoring schemes have been incorporated. These schemes differ primarily in the steepness of the gradient of the potential as the extent of bounds violation increases. The Rosetta protocols described below select different default scoring schemes, but within the source code the scoring scheme in use can be selected by calling the *score_set_cst_mode* subroutine. Evaluation of distance restraints within Rosetta can be further tailored by selecting subsets of restraints to evaluate according to the maximal sequence separation

TABLE II
DISTANCE RESTRAINT SCORING SCHEMES

Scheme	Functional form
I	$\sum_{i,j} \min[\max(0, d_{ij} - u_{ij}), 10]$
II	$\sum_{i,j} max(0, d_{ij} - u_{ij})$
III	$\sum_{i,j} \begin{cases} [\max(l_{ij} - d_{ij}, 0, d_{ij} - u_{ij})]^2; & d_{ij} \leq u_{ij} + 0.5\text{Å} \\ d_{ij} - u_{ij} - 0.25\text{Å}; & d_{ij} > u_{ij} + 0.5\text{Å} \end{cases}$
IV	$\sum_{i,j} \max(0, d_{ij}^2 - u_{ij}^2)$

between restrained residues, by calling the *set_noe_stage* subroutine. The relative weight applied to the distance restraint scoring function can be adjusted with the *score_set_cst_weight* subroutine.

The initial published results for combining distance restraints with Rosetta used only restraints between H^N and H^α atoms (Bowers *et al.*, 2000), but restraints between any atom pair can be defined. In the RosettaNMR v1, only reduced side chain representations are used, so distance restraints must be defined in terms of backbone atoms or side chain centroid coordinates. In the current release (RosettaNMR v2.0), distance restraints may also be defined in terms of side chain atoms and/or centroid coordinates. Degenerate restraints are allowed only for H atoms bound to the same heavy atom. These restraints are evaluated using the coordinates of the heavy atom and padding the upper distance bound by 1.3 Å. Centroid-based restraints are evaluated only during portions of the simulation using centroid representations of side chain, and side chain atom distance restraints are evaluated only when full-atom side chain representations are in use. If no centroid-based restraints are defined, then side chain–side chain distance restraints and side chain–backbone restraints are translated to weak (10 Å upper bound) distance restraints on the relevant side chain centroid coordinates for use during the centroid-based portion of simulations.

RDC Restraints

The incorporation of RDC restraints into the Rosetta framework was originally described by Rohl and Baker (2002). Given a set of molecular coordinates, the residual dipolar coupling between atoms m and n, D^{mn}, is given by

$$D^{mn} = D^{mn}_{max} \sum_{ij=\{x,y,z\}} S_{ij} \cos \phi^{mn}_i \cos \phi^{mn}_j \tag{1}$$

where S is the Saupe order matrix and ϕ^{mn}_i is the angle between the mn internuclear vector and the ith axis of the molecular frame. D^{mn}_{max} is given by

$$D^{mn}_{max} = -\left(\frac{\mu_0}{4\pi}\right) \frac{\gamma_m \gamma_n h}{2\pi r^3_{mn}} \tag{2}$$

where γ_m is the gyromagnetic ratio of nucleus m and r_{mn} is the length of the mn internuclear vector. In Rosetta, each molecular conformation for which RDCs are evaluated is treated as a rigid body, and the Saupe order matrix yielding the least-squares fit to the reduced RDCs is determined by singular value decomposition. The scoring function for RDC restraints is the normalized χ^2 between the experimental and calculated reduced RDCs:

$$\chi^2 = \sum \left(\frac{D_{\mathrm{obs}}^{mn} - D_{\mathrm{calc}}^{mn}}{A_{zz}^{mn} D_{\mathrm{max}}^{mn}} \right) \qquad (3)$$

The score is normalized by the principal component of the diagonalized order tensor (A_{zz}^{mn}) to accommodate data from different alignment media, and D_{max}^{mn} is included to normalize for differences in bond lengths and gyromagnetic ratios for different types of couplings. When data from multiple different experimental conditions are used, the χ^2 score is further normalized so that each data set corresponding to a different alignment tensor contributes equally to the final score. Within Rosetta, residual dipolar coupling restraints can be defined for H^N–N, H^α–C^α, C^α–C, H^N–H^α, and C–N backbone bond vectors and for any H^N–H^α atom pair. Bond lengths are fixed at ideal values (1.01 Å for H^N–N, bonds, 1.08 Å for H^α–C^a bonds; heavy atom bond lengths are taken from Engh and Huber, 1991), allowing the factor $1/D_{\mathrm{max}}^{mn}$ to be precalculated for each coupling. For H^α–H^N RDCs, interatomic distances are evaluated for the particular conformation being evaluated. As with the distance restraint score, a default weighting for the RDC χ^2 score is selected by each Rosetta protocol (see below). This weighting can be adjusted in the source code using the *score_set_dpl_weight* subroutine.

Since the original description of the incorporation of RDC restraints into Rosetta, an alternate approach for utilizing this data via projection angle restraints has also been implemented (Meiler *et al.*, 2000; J. Meiler, unpublished results). RDC data sets are automatically translated into projection angle restraints within Rosetta and conformations are scored with respect to these restraints, as described by Meiler and associates (2000). The dipolar projection angle score is not currently incorporated in the energy function optimized during the fragment assembly or refinement protocols described here (see below), but this score is reported for diagnostic use.

Chemical Shift Restraints

Chemical shift assignments for the N, C^α, C, C^β, and H^α nuclei are converted to restraints on ϕ, ψ backbone torsion angles using a modification of the TALOS algorithm (Bowers *et al.*, 2000). For each residue, the TALOS algorithm is used to select the top 10 most likely ϕ, ψ pairs as ranked by the similarity scores, S, calculated from the observed chemical shifts and sequence (Cornilescu *et al.*, 1999). This discrete output is converted into mean predicted ϕ and ψ angles (ϕ_{pred} and ψ_{pred}), and estimates of the errors on these values (ϕ_{err} and ψ_{err}) are calculated according to the empirically determined equations:

$$\phi_{err} = 3.4 \left[\left(\frac{S_{ave} - S_{best}}{S_{ave}} \right) \sigma_\phi \sigma_\psi \right]^{0.67} \tag{4}$$

$$\psi_{err} = 3.2 \left[\left(\frac{S_{ave} - S_{best}}{S_{ave}} \right) \sigma_\phi \sigma_\psi \right]^{0.81} \tag{5}$$

where S_{ave} is the average similarity score of the top 10 TALOS-selected ϕ, ψ pairs, S_{max} is the similarity score of the top TALOS-selected ϕ, ψ pair, and σ_ϕ and σ_ψ are the standard deviations calculated for the set of TALOS-selected ϕ, ψ pairs. Agreement between the observed and predicted ϕ, ψ values is calculated assuming that deviations beyond the calculated errors follow a Poisson distribution:

$$\sum_i \frac{\phi_{obs} - \phi_{pred}}{\phi_{err}} + \log(\phi_{err} + \frac{\psi_{obs} - \psi_{pred}}{\psi_{err}} + \log(\psi_{err}) \tag{6}$$

where the sum is over all residues in the window being evaluated and ϕ_{obs} and ψ_{obs} are the ϕ, ψ angles observed at a given position within this window.

Utilization of Restraints for Structure Estimation

Selection of Fragment Libraries

As noted above, local structure preferences in Rosetta are encapsulated primarily in the selection of a customized fragment library that is used for discrete sampling of torsion space. Experimental restraints on local structure, such as chemical shifts and short-range NOEs, are useful primarily in the selection of these customized fragment libraries. Customized fragment libraries are created for each protein sequence using either the Rosetta-Fragments software package or the automated Rosetta server at http://robetta.bakerlab.org (Chivian et al., 2003; Kim et al., 2004). Every overlapping three- and nine-residue sequence window is compared to every three- and nine-residue window in a nonredundant database of protein structures and scored according to agreement with chemical shifts [Eq. (6)], short-range NOEs evaluated according to scheme II (Table II), and RDC data [Eq. (3)]. Fragments with gross violation (RMS violation >2 Å) of NOE upper bound distance restraints are discarded. A composite score, in which the distance restraint score is weighted by a factor of 10 and the RDC χ^2 score is weighted such that the mean score is 20, is used to rank each fragment. The top matches in each sequence window are added to the fragment library in rank order. The final fragment library typically consists of the 200 top-ranked fragments for each window in the query sequence.

To ameliorate errors or uncertainty in the estimation of ϕ,ψ from chemical shifts, as well as possible errors in other experimental data, 25% of the fragments included in the library are selected according to the original method that evaluates only sequence profile similarity and the agreement of known and predicted secondary structures in each window (described in detail by Rohl *et al.*, 2004a). The short-range information encoded in RDC restraints can also be used if sufficient data are available within a sequence window to determine the Saupe order matrix, but caution should be exercised in using sparse RDC data for fragment selection, because the inherent degeneracy of these restraints can result in decreased accuracy of the selected fragments. See Rohl and Baker (2002) for a discussion of the effects of different types of data on the accuracy of fragments selected. In the absence of chemical shift data, fragments selected using sparse NOE and/or RDC data sets are usually of lower quality than those selected using sequence and predicted secondary structure alone.

On the automated server, additional precautions are implemented to prevent improper overreliance on experimental data for fragment selection. Data sets of less than four restraints per residue are discarded by the server, as are RDC data sets with average densities below two couplings per residue. Additionally, if chemical shift data are not available, fragments selected according to agreement with NOE and/or RDC data are supplemented with an equal number of fragments selected using only sequence profile and secondary structure information. Note that these precautions are not implemented in the current source code release version of Rosetta-Fragments (v1.2) to provide the user with maximum flexibility in selecting the data most appropriate for fragment selection. In general, when chemical shifts are not available, users should exercise caution with respect to using RDC and/or NOE restraints for fragment selection.

Fragment Assembly Protocols

Restraints on long-range and/or global structure are utilized in the process of fragment assembly and model refinement (see below). Short-range distance restraints can also be used in these protocols, although large numbers of short-range restraints typically result in significant increases in computational time without improving the quality of structures obtained because this short-range information is already encoded in the customized fragment set. For the same reason, chemical shift restraints are not evaluated as part of fragment assembly or structure refinement.

In the current release of RosettaNMR (v2.0) several protocols for folding with restraints are available. These protocols are based on the fragment assembly protocol used for *de novo* prediction that is described in detail by Rohl and associates (2004a) and briefly here. Fragment assembly progresses in stages, beginning with nine-residue fragments. The early stages of the simulation ($\sim$2000 attempted fragment insertions) aim to accumulate secondary structure in the chain and do not include energy terms that drive the compaction of the chain. Next, condensation of the chain is encouraged by rewarding compact conformations and pairing of β-strands. In this portion of the protocol ($\sim$20,000 attempted fragment insertions), the energy function cycles between rewarding only nonlocal strand pairs and all strand pairs. This periodic down-weighting of local strand pairing terms is utilized to compensate in part for the algorithm's tendency to overpopulate local strand pairs. The final stage of the protocol ($\sim$12,000 attempted fragment insertions) is a short optimization of the complete Rosetta energy function using three-residue fragments. In general, only the top 25 nine-residue fragments at each position are used for assembly, while all 200 three-residue fragments in each window are used for optimization.

When only RDCs are available, the standard *de novo* structure prediction protocol is used with the RDC χ^2 score added to the Rosetta energy function once compacting terms are in use (Table III, protocol I). The RDC χ^2 score is down-weighted 20-fold whenever energy terms rewarding local strand pairing are down-weighted. Because of the degeneracy of RDC restraints, expanded structures with extended unpaired strands can frequently satisfy RDC restraints as well or better than compact structures. The use of the RDC χ^2 in correlation with terms rewarding compact structures and nonlocal strand arrangements is an effort to counteract the tendency of the RDC data to drive structures toward noncompact configurations. This protocol is that described by Rohl and Baker (2002). For the results in that paper, the top scoring decoy structures were further optimized with a refinement protocol to yield the final models (see below).

When NOE distance restraints are available (with or without RDC data), the fragment assembly protocol is modified (Table III, protocol III). Short-range NOEs only ($i-j < 10$) are used at early stages of the simulation before chain condensation occurs, and this stage is extended to 5000 attempted fragment insertions. The condensation phase is increased to 48,000 attempted insertions, and the evaluation of NOEs of increased sequence separation is cycled on and off, progressively utilizing more long-range NOEs throughout this stage. Additionally, the number of fragments that are considered is progressively increased in this stage from 25 at each position to 200. The final optimization phase is increased to 24,000 attempted three-residue fragment moves and includes all NOEs. Scheme I (Table II) is used to evaluate NOEs throughout this protocol. RDC restraints are evaluated as in the protocol described above, once compacting terms in the energy function are in use. This protocol was used by Bowers

TABLE III
ROSETTANMR FRAGMENT ASSEMBLY AND MODEL REFINEMENT PROTOCOLS

Protocol	Command line option	NOE restraints	RDC restraints	Protocol selected
I	-fast	−	+	Fragment assembly, standard *de novo* protocol
II	(none)	−	+	Fragment assembly (protocol I) followed by centroid-based refinement (protocol V)
III	-fast	+	+/−	Modified fragment assembly
IV	(none)	+	+/−	Fragment assembly with conjugate gradient minimization
V	-refine	+/−	+/−	Centroid-based refinement
VI	-relax	+/−	+/−	Addition of side chains and full atom energy function optimization

and associates (2000) in combination with backbone distance restraints to generate a pool of decoy structures. Compact structures with <1 Å violation per restraint were then further optimized with a full atom refinement protocol to yield the final models reported.

An alternate protocol for utilization of NOE data, which has not been described elsewhere, is also available in RosettaNMR (Table III, protocol IV). In brief, this protocol alternates cycles of fragment insertion with conjugate gradient optimization in torsion space while progressively adding NOEs of increasing sequence separation (C. Rohl, unpublished results). NOEs are evaluated according to scoring scheme III (Table II). In some cases this protocol allows correct folds for proteins of more complex topologies to be obtained than possible using the protocol described by Bowers and associates (2000). Such complex topologies are usually achieved, however, at the expense of significantly increased computational time and introduced steric clashes that require extensive refinement to eliminate.

Model Refinement Protocols

Model refinement in Rosetta is accomplished by a combination of modified fragment insertions and gradient-based optimization (Rohl *et al.*, 2004a). Decoy structures generated by fragment assembly in Rosetta usually do not satisfy experimental restraints as well as structures obtained by traditional NMR structure determination methods. Refinement of Rosetta-generated models improves the agreement of models with the experimental data, and such improvement usually corresponds with improved structural accuracy of the global fold and often improves the ability with which accurate models can be identified by the Rosetta score. It must be emphasized, however, that the RosettaNMR program is not designed for high-resolution refinement of models using complete experimental NMR data sets. In most cases, following refinement with these protocols, Rosetta models still do not satisfy even limited experimental data sets as well as the high-resolution, experimentally determined native structures.

The original release of RosettaNMR (v1.2) includes a simple centroid-based refinement protocol that combines Monte Carlo and Monte Carlo-plus-minimization search strategies (Table III, protocol V) (Li and Scheraga, 1987; Rohl et al., 2004a). The protocol proceeds through three stages of simulated annealing in which the extent of minimization is progressively increased, and the relative weighting of the experimental restraints in the total energy is significantly increased in this refinement relative to the fragment assembly protocol. The temperature program, the number of steps and the maximal perturbation of backbone torsion angles in each stage of the protocol, and the relative weight of the experimental data in the energy function can be modified through the use of a text file specifying these parameters. Improved performance over that obtained with the default parameters can frequently be achieved by monitoring the change in energy as the refinement proceeds, adjusting the temperature and number of attempted moves such that the energy decreases throughout the refinement and the simulation terminates once the energy stops decreasing. The default program is that used by Rohl and Baker (2002) in combination with the first fragment assembly protocol described above to obtain global folds for a variety of proteins using only RDC restraints.

In the current RosettaNMR release (v2.0), an additional protocol is available to add full-atom side chains to models and optimize the Rosetta full atom energy function (Table III, protocol VI). In switching from centroid representations of side chains to full atom representations, a slight adjustment of the backbone is required to accommodate the side chain atoms; this adjustment is accomplished in the first half of the protocol. The second half of the protocol is a Monte Carlo-plus-minimization (Li and Scheraga, 1987; Rohl et al., 2004a) optimization of the Rosetta full-atom energy function, including scoring functions for experimental data. It should be noted that optimization of the Rosetta energy function has not been shown to improve the accuracy of protein models generally; this protocol is provided primarily to enable users to add side chains to Rosetta models.

Model Selection

The fragment assembly protocols described above each generally yields a compact structure in which experimental restraints are partially satisfied, local structure is consistent with the preferences encoded in the sequence, and nonlocal interactions are consistent with those seen for native globular proteins. Because Rosetta simulations are short and the energy function is coarse grained and approximate, the fragment assembly process does *not* usually result in a model with the correct global fold. In general, multiple independent simulations are carried from independent random seeds to produce a collection of different structural models, or "decoys", and the model closest to the native structure must be identified from this pool. In the absence of experimental data, identification of the most correct model from the population of decoys is a challenging problem generally accomplished through the use of clustering to identify models in the broadest energy basin and ranking of models according to detailed atomic potential functions (Bonneau *et al.*, 2002; Bradley *et al.*, 2003; Shortle *et al.*, 1998).

Identification of correct models is greatly simplified by the presence of experimental data. Experimental restraints help guide the Monte Carlo search, resulting in a greater population of structures with the correct fold in a decoy population. Satisfaction of experimental restraints provides a stringent criterion for ranking models (Fig. 1). It should be noted, however, that with sparse experimental data sets, the majority of models obtained by fragment assembly do not have the correct fold, and model selection is

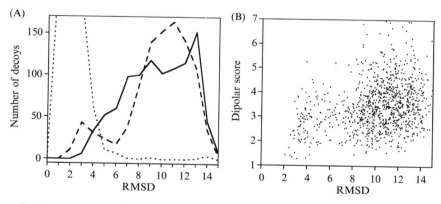

FIG. 1. Decoy populations from Rosetta fragment assembly. (A) Three populations of 1000 Rosetta decoys each were generated by fragment assembly for the protein ubiquitin using either no experimental data (solid line), H^N–N RDC couplings in one alignment medium (68 restraints, dashed line), or H^N–N, C^α–H^α, C–N, and C–H^N couplings in two alignment media (539 restraints, dotted lined). RDC data are taken from Ottiger and Bax (1998). In the absence of experimental data, very few models are closer than 3 Å RMSD to native. With large numbers of restraints, the structure is completely determined by the data, and all decoys have the correct global fold. With sparse restraints, ($\sim$1 coupling per residue), the fold is not completely determined by the data, but the population of near native decoys is significantly increased. (B) In the decoy set generated with sparse RDC restraints (dashed line in A), the accuracy of decoys correlates with the dipolar χ^2 score [Eq. (3)], simplifying the problem of identifying correct models from the decoy population.

an essential part of the RosettaNMR method. In the examples shown in Table I, models were selected by ranking decoys according to the Rosetta score, which combines both agreement with experimental data as well as the Rosetta energy function. In the study by Rohl and Baker (2002), a further requirement that the majority of the top scoring structures share the same global fold was imposed. Bowers and associates (2000) removed structures with poor score components (steric overlap, satisfaction of NOE restraints, and overall compactness) before ranking structures by their composite scores. Whereas the optimal criteria for ranking models likely depends on quality and density of the experimental data available, the composite Rosetta score selects models with favorable protein-like properties that satisfy the experimental restraints. Because agreement with experimental data can be significantly improved by refinement, although at additional computation cost, a useful strategy is to identify the top 5–10% of the fragment assembly structures according to the Rosetta score, apply the refinement protocol to this subset of models, and then rerank the refined models using the composite Rosetta score.

Conclusions

RosettaNMR has been demonstrated to be a useful tool for rapid fold determination from limited experimental methods. The fragment assembly approach of the Rosetta method is well suited to searching the large conformational space that must be sampled in finding the protein global fold. Additionally, because the fragment-assembly search is coarse grained and stochastic, minimal data sets corresponding to complex energy surfaces, such as those associated with degenerate restraints like RDCs, can be used effectively for structure estimation. The optimal combination of NOE and RDC data for use with RosettaNMR has not been systematically investigated, although comparison of results obtained for the same protein with different data sets (Table I) is in some cases illustrative. Generally, the most accurate models are obtained when long-range NOEs are available. In the absence of long-range NOEs, models obtained using only RDC data have the correct topology, but RMSD to native can be high for α-helical proteins as a result of rigid body translations of the helices. With only short-range NOEs, global topology may not be correctly defined.

Since the original description of RosettaNMR, the ability of Rosetta to effectively optimize full atom energy functions has significantly improved. The detailed atomistic energy function and rapid side chain optimization methods used within Rosetta are sufficiently accurate to support protein design, including the design of a protein of novel topology (Chevalier et al., 2002; Dantas et al., 2003; Kuhlman and Baker, 2000; Kuhlman et al., 2003;

Rohl *et al.*, 2004a). Such improvements in the basic Rosetta method open the possibility that incorporation of full atom methods into RosettaNMR could lead to improved model accuracy. Continued development of the Rosetta method is also focused on extending the method to larger proteins with more complex topologies, as well as to adapting the method to incorporate structural homology (Bradley *et al.*, 2003; Chivian *et al.*, 2003; Rohl *et al.*, 2004b), and such improvements will be directly applicable to structure estimation with RosettaNMR.

Supplemental Materials

Licensing information for RosettaNMR and RosettaFragments is available by e-mail from rosettaNMR@rosetta.bakerlab.org or online at http://www.bakerlab.org. The automated Rosetta server can be accessed at http://robetta.bakerlab.org.

References

Aloy, P., Stark, A., Hadley, C., and Russell, R. B. (2003). Predictions without templates: New folds, secondary structure, and contacts in CASP5. *Proteins.* **53**(Suppl. 6), 436–456.

Bateman, A., Coin, L., Durbin, R., Finn, R. D., Hollich, V., Griffiths-Jones, S., Khanna, A., Marshall, M., Moxon, S., Sonnhammer, E. L., Studholme, D. J., Yeats, C., and Eddy, S. R. (2004). The Pfam protein families database. *Nucleic Acids Res.* **32**, D138–D141.

Bewley, C. A., Gustafson, K. R., Boyd, M. R., Covell, D. G., Bax, A., Clore, G. M., and Gronenborn, A. M. (1998). Solution structure of cyanovirin-N, a potent HIV-inactivating protein. *Nat. Struct. Biol.* **5**, 571–578.

Bonneau, R., Strauss, C. E., Rohl, C. A., Chivian, D., Bradley, P., Malmstrom, L., Robertson, T., and Baker, D. (2002). *De novo* prediction of three-dimensional structures for major protein families. *J. Mol. Biol.* **322**, 65–78.

Bowers, P. M., Strauss, C. E. M., and Baker, D. (2000). *De novo* protein structure determination using sparse NMR data. *J. Biomol. NMR* **18**, 311–318.

Bradley, P., Chivian, D., Meiler, J., Misura, K. M., Rohl, C. A., Schief, W. R., Wedemeyer, W. J., Schueler-Furman, O., Murphy, P., Schonbrun, J., Strauss, C. E., and Baker, D. (2003). Rosetta predictions in CASP5: Successes, failures, and prospects for complete automation. *Proteins* **53**(Suppl.6), 457–468.

Chevalier, B. S., Kortemme, T., Chadsey, M. S., Baker, D., Monnat, R. J., and Stoddard, B. L. (2002). Design, activity, and structure of a highly specific artificial endonuclease. *Mol. Cell* **10**, 895–905.

Chivian, D., Kim, D. E., Malmstrom, L., Bradley, P., Robertson, T., Murphy, P., Strauss, C. E., Bonneau, R., Rohl, C. A., and Baker, D. (2003). Automated prediction of CASP-5 structures using the Robetta server. *Proteins* **53**(Suppl. 6), 524–533.

Clore, G. M., and Kuszewski, J. (2002). χ^1 rotamer populations and angles of mobile surface side chains are accurately predicted by a torsion angle database potential of mean force. *J. Am. Chem. Soc.* **124**, 2866–2867.

Cornilescu, G., Delaglio, F., and Bax, A. (1999). Protein backbone angle restraints from searching a database for chemical shift and sequence homology. *J. Biomol. NMR* **13**, 289–302.

Dantas, G., Kuhlman, B., Callender, D., Wong, M., and Baker, D. (2003). A large scale test of computational protein design: Folding and stability of nine completely redesigned globular proteins. *J. Mol. Biol.* **332**, 449–460.

Engh, R., and Huber, R. (1991). Accurate bond and angle parameters for X-ray protein structure refinement. *Acta Crystallogr. A* **47**, 392–400.

Kim, D. E., Chivian, D., and Baker, D. (2004). Protein structure prediction and analysis using the Robetta server. *Nucleic Acids Res.* **32**, W1–W7.

Kinch, L. N., Wrabl, J. O., Krishna, S. S., Majumdar, I., Sadreyev, R. I., Qi, Y., Pei, J., Cheng, H., and Grishin, N. V. (2003). CASP5 assessment of fold recognition target predictions. *Proteins* **53**(Suppl. 6), 395–409.

Kuhlman, B., and Baker, D. (2000). Native protein sequences are close to optimal for their structures. *Proc. Natl. Acad. Sci. USA* **97**, 10383–10388.

Kuhlman, B., Dantas, G., Ireton, G. C., Varani, G., Stoddard, B. L., and Baker, D. (2003). Design of a novel globular protein fold with atomic-level accuracy. *Science* **302**, 1364–1368.

Kuszewski, J., and Clore, G. M. (2000). Source of and solutions to problems in the refinement of protein NMR structures against torsion angle potentials of mean force. *J. Magn. Reson.* **146**, 249–254.

Kuszewski, J., Gronenborn, A. M., and Clore, G. M. (1996). Improving the quality of NMR and crystallographic protein structures by means of a conformational database potential derived from structure databases. *Protein Sci.* **5**, 1067–1080.

Li, Z., and Scheraga, H. A. (1987). Monte Carlo-minimization approach to the multiple-minima problem in protein folding. *Proc. Natl. Acad. Sci. USA* **84**, 6611–6615.

Meiler, J., and Baker, D. (2003). Rapid protein fold determination using unassigned NMR data. *Proc. Natl. Acad. Sci. USA* **100**, 15404–15409.

Meiler, J., Blomberg, N., Nilges, M., and Griesinger, C. (2000). A new approach for applying residual dipolar couplings as restraints in structure elucidation. *J. Biomol. NMR* **16**, 245–252. Erratum in *J. Biomol. NMR* **17**, 185.

Moult, J., Fidelis, K., Zemla, A., and Hubbard, T. (2003). Critical assessment of methods of protein structure prediction (CASP)-round V. *Proteins* **53**(Suppl. 6), 334–339.

Ottiger, M., and Bax, A. (1998). Determination of relative N-H^N, N-C', C-C', and C-H effective bond lengths in a protein by NMR in a dilute liquid crystalline phase. *J. Am. Chem. Soc.* **120**, 12334–12341.

Rohl, C. A., and Baker, D. (2002). *De novo* determination of protein backbone structure from residual dipolar couplings using Rosetta. *J. Am. Chem. Soc.* **124**, 2723–2729.

Rohl, C. A., Chivian, D., Misura, K. M. S., and Baker, D. (2004a). Protein structure prediction using Rosetta. *Methods Enzymol.* **383**, 66–93.

Rohl, C. A., Strauss, C. E. M., Chivian, D., and Baker, D. (2004b). Modeling structurally variable regions in homologous proteins using Rosetta. *Proteins* **55**, 656–677.

Shortle, D., Simons, K. T., and Baker, D. (1998). Clustering of low-energy conformations near the native structures of small proteins. *Proc. Natl. Acad. Sci. USA* **95**, 11158–11162.

Simons, K. T., Kooperberg, C., Huang, E., and Baker, D. (1997). Assembly of protein tertiary structures from fragments with similar local sequences using simulated annealing and Bayesian scoring functions. *J. Mol. Biol.* **268**, 209–225.

Simons, K. T., Ruczinski, I., Kooperberg, C., Fox, B. A., Bystroff, C., and Baker, D. (1999). Improved recognition of native-like protein structures using a combination of sequence-dependent and sequence-independent features of proteins. *Proteins* **34**, 82–95.

Venclovas, C., Zemla, A., Fidelis, K., and Moult, J. (2003). Assessment of progress over the CASP experiments. *Proteins* **53**(Suppl. 6), 585–595.

[10]   Protein Structure Elucidation from Minimal NMR Data: The CLOUDS Approach

By ALEXANDER GRISHAEV and MIGUEL LLINÁS

Abstract

In this chapter we review automated methods of protein NMR data analysis and expand on the assignment-independent CLOUDS approach. As presented, given a set of reliable NOEs it is feasible to derive a spatial H-atom distribution that provides a low-resolution image of the protein structure. In order to generate such a list of unambiguous NOEs, a probabilistic assessment of the NOE identities (in terms of frequency-labeled H-atom sources) was developed on the basis of Bayesian inference. The methodology, encompassing programs SPI and BACUS, provides a list of "clean" NOEs that does not hinge on prior knowledge of sequence-specific resonance assignments or a preliminary structural model. As such, the combined SPI/BACUS approach, intrinsically adaptable to include 13C- and/or 15N-edited experiments, affords a useful tool for the analysis of NMR data irrespective of whether the adopted structure calculation protocol is assignment-dependent.

Introduction

Despite significant recent advances in instrumentation, experimental design, and data analysis, the derivation of macromolecular structures via nuclear magnetic resonance (NMR) remains a slow process, mainly due to the complexity of assigning signals to individual spins. Consequently, much effort is being devoted to the automation of data analysis (Gronwald and Kalbitzer, 2004; Güntert, 2003; Moseley and Montelione, 1999).

Macromolecular structure elucidation via NMR is equivalent to mapping the available spectral information onto specific molecular sites—a case of an "inverse problem." Fortunately, this process is amenable to automation since (1) in general, NMR data are redundant, which helps to cope with the degeneracy of resonance frequencies; (2) the mapping procedure can be broken down into a series of discrete, separate steps at each of which the data analysis is subjected to a well-defined set of logical, programmable rules; (3) the goal of the overall NMR structural elucidation is usually a family of folds of ∼1 Å precision, rather than a rigorously defined structure, which relaxes requirements on the quality of

Copyright 2005, Elsevier Inc.
All rights reserved.
0076-6879/05 $35.00

the mapping; and (4) many aspects of the NMR analysis can be formulated in the language of constrained optimization, a rather well-developed field. A variety of tools created to deal with situations of this kind are adaptable for the specifics of NMR data. Some of the approaches are stochastic, based on Monte Carlo (MC) random walks (Metropolis *et al.*, 1953), simulated annealing (SA) (Kirkpatrick *et al.*, 1983), or genetic algorithms (Bounds, 1987). Others are based on Bayesian statistics (Jaynes, 2003) and constraints propagation (Nadel, 1988). Yet another group of methods, influenced by the spin/vertex and connectivity/edge isomorphism, employ the theorems and tools of graph theory (Balaban, 1976).

In essence, mainstream automation efforts in NMR involve a sequence of steps: signal identification (peak picking), assignment of resonances to the spins, assignment of NOESY cross-peaks to pairs of spins, and finally molecular dynamics (MD)-based structure computation. However, alternatives have been formulated that deviate from the canonical scheme. Their underlying idea is that the NOESY spectrum, by itself, encodes for the topology of the macromolecule. In that sense, these "unconventional" methods, discussed below, aim at a structural interpretation of the NOE data *before* the full resonance assignments are established.

Assignment

Resonances

The overall assignment protocol usually is broken into several discrete steps: *grouping* of resonances into spin systems, *typing* of spin systems by amino acid kind, *linking* of spin systems, and *mapping* of the linked fragments onto the protein's primary structure. Successful automated assignment protocols attempt to address each of these stages. The most robust are reported to achieve ~95% backbone resonance assignments with <5% error rate (Feng *et al.*, 1998; Lukin *et al.*, 1997).

Grouping is relatively straightforward when triple-resonance (^{1}H/^{13}C/^{15}N) data are available, due to the favorable spectral resolution and large ^{1}J(^{1}H,^{13}C) and ^{1}J(^{1}H,^{15}N) coupling constants. Popular methods of grouping searches are (1) along sets of characteristic "root" resonances (Buchler *et al.*, 1997; Lukin *et al.*, 1997; Zimmerman *et al.*, 1997), and (2) via pattern guidance (Croft *et al.*, 1997). The former methods (1) seem to be more stable with respect to peak overlap and missing peaks; however, some of these methods have been formulated only for dealing with backbone resonances down to C^β. The latter method (2) is designed to go further along the spin ladder; however, it is more influenced by degradation of the spectral quality.

Two main approaches are available for spin system typing. One is based on the classification of graphs built from resonance connectivity data, and the other on the amino acid–type dependence of chemical shifts (typically, $^{13}C^{\alpha}$, $^{13}C^{\beta}$). The former, whether based on graph theory and fuzzy math (Xu and Borer, 1994; Xu et al., 1999) or on pattern matching (Buchler et al., 1997; Croft et al., 1997), seems to be more affected by the deterioration of the data quality due to overlap and missing connectivities. The latter (Lukin et al., 1997; Zimmerman et al., 1997), based on Bayesian statistics or related semiquantitative reasoning (Atreya et al., 2000; Meadows et al., 1994), are quite robust; however, all these methods require high-quality triple-resonance connectivity data. In another approach (Gronwald et al., 1998), available structural and NMR information from homologous proteins is used.

Sequential linking of spin system can be based on either J-mediated transfer through peptide bonds or NOESY connectivities. The ambiguity of the former is considerably less than that of the latter. However, these through-peptide-bond linkages have to be established from sets of triple-resonance experiments, which necessitates doubly labeled ($^{13}C/^{15}N/^{1}H$) samples. Residue linking, whether via J- or NOE-connectivities, is usually ambiguous owing to resonance overlap or missing peaks, calling for the usage of global optimization techniques. The approaches that tackle the ambiguity can be divided into stochastic (Rinooy Kan and Timmer, 1987) and deterministic (Floudas, 2000). Both impose additional restrictions, requiring either uniqueness and internal consistency of the sequential linkages or an agreement of linkage and residue typing with primary sequence. In our opinion, recourse to residue typing alone may be un-desirable at these early stages, as the ambiguities of typing, mapping, and linking can lead to misreading the connectivities. Ideally, criteria of internal consistency, partial typing information, and conformity with the primary sequence should be applied simultaneously in the optimal sense.

The reliability of the available best-performing deterministic and sto-chastic methods seems to be similar; however, the former are significantly more efficient from a computational standpoint. In the approach of Buchler et al. (1997), linking is accomplished via MD/SA optimization in "assignments space." In the protocols of Zimmerman et al. (1997) and Lukin et al. (1997), the best—in the sense of their quality and consisten-cy—linkages are established first, gradually decreasing the complexity of the problem until convergence according to some specified criteria. In TATAPRO (Atreya et al., 2000) a somewhat simplistic linking is coupled to simultaneous mapping that aids in decreasing the ambiguities.

Mapping of linked stretches of residues onto the primary structure is yet another case of an optimization problem. Ambiguities, which increase with sequence length, are likely to be unavoidable, since unique matches can, in general, be expected only for the segments of three or more residues with uniquely established amino acid types (Grzesiek and Bax, 1993). At this stage, typing is often ambiguous and some of the linked fragments may be short, so this condition is rarely met in practice.

Similar to linking, mapping can be established via either deterministic or stochastic methods. The performance of both approaches is similar, owing to a considerable amount of information available at this stage. AUTOASSIGN (Zimmerman et al., 1997) provides an example of the deterministic approach, as constraint propagation is used to guide ambiguous placements. In the stochastic protocol by Lukin et al. (1997) and Hitchens et al. (2003), MC/SA simulations in the assignments space are used to optimally match the segments of linked spin systems onto the primary sequence. In MAPPER (Güntert et al., 2000), linked segments are exhaustively mapped to the primary sequence, with branches of the search tree trimmed according to the "most likely to fail" principle. $^{13}C^\alpha$ and $^{13}C^\beta$ chemical shift–based residue typing scores are used to describe the "fitness" of individual fragments. GARANT (Bartels et al., 1997) uses genetic algorithms to overcome ruggedness of the mapping pseudoenergy landscape.

NOEs

Both interactive ("manual") and automated methods of NOESY analysis start by building the lists of distance-related spin pairs within suitable tolerances from each peak's chemical shift coordinates. The tolerances depend on spectral resolution. Because of the chemical shift degeneracy, the average number of assignments per cross-peak is often >10. Thus, in a two-dimensional (2D) NOESY spectrum of a moderately sized protein of ~100 a.a., <10% of cross-peaks can be unambiguously matched to their frequencies at a tolerance of 0.01 ppm and only 2–4% are unambiguous at 0.02 ppm tolerance (Mumenthaler and Braun, 1995). The use of ^{13}C- or ^{15}N-edited three-dimensional (3D) NOESY spectra, when recorded at ~11.75–16.45 T, does not totally alleviate the problem, since the advantage of the heteronuclear editing of one proton frequency is compromised by lower digital resolution of the other (Mumenthaler et al., 1997). Thus, a chemical shift matching tolerance of ~0.02 ppm in 2D NOESY spectra has to be increased to ~0.05 ppm in 3D spectra and to ~0.1 ppm in four-dimensional (4D) spectra. As a result, even 4D NOEs are not unambiguous, with ~60–70% of cross-peaks corresponding to more than one pair of resonances for an ~150 residue protein. In common practice the ambiguities are dealt with by neglecting those assignments incompatible with the calculated folds, as persistent violation of NOEs vis-à-vis computed structures decreases their assignment probabilities. Such an approach is based on the hypothesis that all correct matches are self-consistent in the context of the underlying structure (they support each other when used as structural restraints), and the incorrect assignments are essentially random, hence, mutually inconsistent.

Thus, ambiguous restraints are either incorporated into a molecular dynamics protocol via a cleverly designed cost function, such as in the ARIA method (Nilges, 1993, 1995; Nilges and O'Donoghue, 1998; Nilges et al., 1997), or included simultaneously in self-correcting distance geometry methods NOAH/DIAMOD (Mumenthaler and Braun, 1995; Mumenthaler et al., 1997; Xu et al., 1999). Both procedures rely on the assignment of resonance frequencies, require extensive 3D structure computations, and are variously influenced by the quality of the initial structural models. Moreover, both ARIA and NOAH/DIAMOD hinge on a sufficient fraction of unambiguous NOEs being available to generate a rough molecular fold, which then serves to bootstrap the assignments of the remaining cross-peaks. Several of the recently reported NOE assignment approaches are essentially variants of ARIA that differ either in the use of genetic algorithms (Adler, 2000) or in homology modeling/secondary structure filtering (Duggan et al., 2001). When either ARIA or NOAH/DIAMOD is run with ~2–10% initially unambiguous peaks, the final unique assignment rate is ~60–80%, plus an additional 10% of peaks with multiple assignment origins, with error rates of ~3–5%.

Direct and Unconventional Approaches

"Direct" methods aim at deriving the protein molecular structure (and/or assignments) from NOE data only via the construction of a spatial proton distribution, an intermediate and somewhat fictitious object (Atkinson and Saudek, 1996, 2002; Kraulis, 1994; Malliavin et al., 1992a; Oshiro and Kuntz, 1993). The NOESY/structure mapping, scrambled in the absence of the resonance assignments, is then established by derivation of the protein fold in terms of the proton locations that, in turn, yield assignments. More recently, Malliavin et al. (2001) reported a protocol by which spatial proximity of unassigned residues is predicted from the ^{15}N-edited HSQC–NOESY data only, by evaluation of scalar product-based spectral similarity matching scores. The method was tested on experimental and simulated data of a number of proteins, yielding 58–88% correct sequential predictions, with a higher error rate for nonsequential contacts.

In spite of the fact that the protocol does not solve the residue proximity problem, it demonstrates that a simple NOESY analysis can yield valuable structural information in the absence of resonance assignments. Yet another promising approach exploits graph theory for the identification of regular secondary structure patterns (α-helices and β-sheets) from unassigned data and probabilistic mapping of identified stretches of secondary structure to the sequence (Bailey-Kellogg et al., 2000). It requires a ^{15}N-edited data set (HNHA, HSQC–TOCSY, HSQC–NOESY). Residue typing is done via ^{1}H chemical shift/pattern-based scores, in a manner reminiscent of the MCD strategy of Englander and Wand (1987). Clearly, combination with residual dipolar couplings (Beraud et al., 2002; Valafar and Prestegard, 2003) and inclusion of structure prediction (Meiler and Baker, 2003) should improve a distances-based structure characterization.

A primary difficulty of all these procedures stems from the degeneracy of resonances, which leads to ambiguous NOE identities. Indeed, NOE misassignments can have a profound effect on the quality of all-hydrogen structures, a result of the relatively small total number of constraints per atom (Atkinson and Saudek, 1997). None of the methods referred to above was successful in identifying ambiguous NOEs without assigning resonances first. Another problem is the requirement of high precision and a large number of NOE distance restraints (Oshiro and Kuntz, 1993). The usual method of NOE quantification via "initial rate approximation" would seem to be incapable of producing data of the requisite quality. However, a relaxation matrix (**R**) approach (reviewed by Likić, 1996) should be more rigorous in that it incorporates the complete network of interacting protons, while providing a larger number of distance restraints at higher precision. Such treatment decreases systematic errors due to spin diffusion, thus enabling use of data of higher signal/noise (intermediate mixing times in NOESY). Moreover, under appropriate error analysis (Liu et al., 1995), NOESY data are less prone to be overinterpreted with the **R**-matrix methods.

It should be noted that most of the reported **R**-matrix routines were not intended for unassigned data, their main application being as a powerful structure validation tool (James, 1994). Many of these protocols are hybrid—they generate a fraction of NOESY matrix based on the model structure, a procedure that requires the assignment of resonances (Boelens et al., 1989; Borgias et al., 1990; Zhang et al., 1995). However, some reported **R**-matrix protocols are, in principle, independent of the spectral assignments. Indeed, precise estimates of relaxation matrix elements may be obtained either by enforcing self-consistency of the **R**-matrix (Madrid et al., 1991) or, in principle, by fitting a distance estimate to each cross-peak intensity profile separately (Malliavin et al., 1992b).

The CLOUDS Protocol

As presented above, conventional strategies for automated assignments based on triple-resonance ($^1H/^{13}C/^{15}N$) experiments have obvious merits for expediting protein structure derivation. However, taking into account that the NOESY carries much of the structure information and is one of the most robust among the multidimensional NMR experiments, the question remains as to whether a NOE-centered approach cannot expedite the process.

The goal of the CLOUDS project is to formulate a protocol that generates molecular protein structures from unassigned, unambiguous NOESY data via computation, interpretation, and fitting of the molecular proton density. CLOUDS (Grishaev and Llinás, 2002a) was designed to overcome perceived shortcomings of the unconventional NOE-only structure elucidation protocols. Indeed, as recently as 2002, such an approach was referred to as a "hypothesis" (Atkinson and Saudek, 2002). We have shown that a reliable proton density can be generated from *experimental* NOE data sets of two small protein domains by employing distances computed via MIDGE, an iterative, model-independent **R**-matrix algorithm (Madrid *et al.*, 1991), and addressed the issue of the requisite data quality (Grishaev and Llinás, 2002a,b). It is our contention that for an NOE-only approach to work, a relaxation matrix treatment may well make all the difference. CLOUDS also includes a suite of novel computer programs for interpretation of proton density whose performance has been tested on experimental data (Grishaev and Llinás, 2002b). These procedures are based on Bayesian statistics and constraints propagation. An explicit fitting of the complete identified proton density via a MD/SA optimization also has been developed (Grishaev and Llinás, 2002b). The CLOUDS approach is aimed at protein structure computation from sparse NMR data. As we have shown (Grishaev and Llinás, 2002a), sparseness can be overcome by resorting to "non-NOEs" or "antidistance constraints" (ADCs) (Bruschweiler *et al.*, 1991; De Vlieg *et al.*, 1986). A virtue of CLOUDS is that while computing the structure, as a byproduct, it simultaneously assigns the spectrum.

As formulated, CLOUDS relies on precise and abundant NOE distance restraints calculated via MIDGE analysis of NOESY data. Resonances identified from NOESY spectra are listed blindly, without regard to the locations of the H-atoms in the chemical structure of the molecule. Internuclear distances become input restraints for a gas of protons subjected to an MD/SA procedure that yields a spatial distribution, or *cloud*, of protons. A proton density, or *foc*, is generated by combining a large number of such distributions, and the polypeptide is subsequently fitted to the *foc* via EMBEDS, a restrained molecular dynamics protocol. The method consists of (1) an iterative, model-independent relaxation matrix analysis of the sorted NOESY cross-peaks (program MIDGE), (2) proton density computation (program CLOUDS), and (3) automated docking of the chemical skeleton (i.e., the protein sequence) via programs BAF (BAckbone Finder) and SIF (SIde chain Finder) and then globally, to the *foc*, via EMBEDS. The method was tested on two proteins: the col 2 domain (60 residues) of matrix metalloproteinase-2 (MMP-2) and the kringle 2 domain of plasminogen (83 residues), for which high-quality NMR data are available (Briknarová *et al.*, 1999; Marti *et al.*, 1999). No assignments were used. The solution structure via CLOUDS, although computationally more intensive than via standard schemes, takes considerably less operator time. As such, CLOUDS affords a robust, computationally efficient approach for high-throughput derivation of NMR molecular structures. The resulting structures are similar to the ones obtained through standard protocols based on assigned NOEs.

Components of the complete CLOUDS protocol are as follows:

- **SPI**. The program groups ^{1}H resonances into spin systems. Because its performance is affected by the quality of the input peak list, it requires a degree of human intervention. SPI is flexible and can be adapted to handle additional (triple-resonance, ^{13}C-edited, etc.) types of data.

- **BACUS**. It assesses compatibilities between NOESY and COSY/TOCSY-derived connectivities. It can readily be reformulated to exploit other information available from the SPI analysis such as identified HNs, HAs, HBs, etc. such that it increases both the resolving power and correctness of the algorithm. BACUS generates the input for MIDGE.

- **MIDGE**. It is an iterative relaxation matrix (**R**) approach that obtains interproton distances from NOESY data in terms of self-consistency of τ_c, diagonal elements, leak terms, and treatment of methyl group spectral density. MIDGE is used to analyze NOESY data from several, short and long, mixing times to obtain a single **R** that best fits all experimental data, taking into account the individual error levels.

- **CLOUDS** and **FILTER**. Interproton distances derived from MIDGE and ADCs are input as restraints for gas phase MD/SA runs that lead to spatial distribution of protons, or *clouds*. A superposition of *clouds* yields, after filtering, the *foc*.

- **BAF** and **SIF**. These programs aim at identifying "atomic" *focs* within the polypeptide backbone (BAF) and the side chains (SIF). The only criteria are chemical shifts and *clouds*-based interatomic distances.

- **EMBEDS**. The protocol "best fits" the protein covalent structure to the proton density defined by the *foc*. For this, the standard (XPLOR/

CNS) SA/MD force field was modified by incorporating an ad-hoc term whose potential energy is related to the *foc* density.

Identification of Multidimensional NMR Cross-Peaks via SPI and BACUS

For CLOUDS to achieve its goal, a list of *unambiguous* NOEs must be available. The generation of such input requires two sets of data: a list of resonance frequencies and the mapping of each NOESY cross-peak to the corresponding pair of resonances. Two protocols, SPI and BACUS, based on Bayesian inference, were developed toward this goal. SPI (Grishaev and Llinás, 2002c) produces a list of the resonance frequencies from a set of J- and NOE-correlation experiments, grouped into effective spin systems. BACUS (Grishaev and Llinás, 2004) automatically establishes probabilistic identities of NOESY cross-peaks in terms of resonances listed via SPI. BACUS requires neither assignment of resonances nor an initial structural model. The method exploits the self-consistency of NOESY identities, using J- and NOE-connectivities established via SPI.

As discussed above, all automated methods of NOESY cross-peak assignment are based on the self-consistency of the correct assignments. It may be argued that much of this self-consistency is directly encoded in the graph topology of the NOESY spectrum, similar to extracting the 3D structure from the set of compatible distance restraints. Carrying the analogy further, one could exploit the detection of particular sets of cross-peaks to improve the probabilities of the other matches, analogous to the use of the triangular inequalities for correcting distance estimates. Thus, by taking advantage of connectivities within NOESY spectra, one could, in principle, refine chemical shift–based matching probabilities without the need to cycle through 3D structure calculations.

Even if the 3D molecular structure is unknown, the *local* geometries can be inferred from the experimental J-connectivities, encoded in COSY/TOCSY type spectra. For example, if protons *i* and *j* are proximal, the rest of their J- and NOE-linked spin systems are likely to occur within the neighborhood, which can be used to modify the cross-peak matching probabilities. Such a strategy is the basis of the CANDID program (Hermann *et al.*, 2002) and of the SPI/BACUS approach.

Bayesian Framework for NMR Data Analysis

Bayesian inference provides a suitable mathematical framework for the exploitation of probabilities stemming from independent data sources. In the standard jargon, $P(\mathcal{H}^i | \mathcal{D}_1, \ldots, \mathcal{D}_n)$, the "posterior" probability of hypothesis $\mathcal{H}^i$ conditional on the data $\mathcal{D}_1, \ldots, \mathcal{D}_n$, is estimated from the "prior" probabilities $\mathcal{P}(\mathcal{H}^i)$, as well as the "likelihoods" $\mathcal{P}(\mathcal{D}_k|\mathcal{H}^i)$ of satisfying $\mathcal{D}_k$ conditional on $\mathcal{H}^i$. Bayes' theorem (Jaynes, 2003) states that

$$\mathcal{P}(\mathcal{H}^i|\mathcal{D}_1, \ldots, \mathcal{D}_n) = \frac{\mathcal{P}(\mathcal{H}^i)\prod_{k=1}^{n}\mathcal{P}(\mathcal{D}_k|\mathcal{H}^i)}{\mathcal{P}(\mathcal{D}_1, \ldots, \mathcal{D}_n)} \tag{1}$$

where the normalization factor $\mathcal{P}(\mathcal{D}_1, \ldots, \mathcal{D}_n) = \sum_{j=1}^{m}\mathcal{P}(\mathcal{H}^j)\prod_{k=1}^{n}\mathcal{P}(\mathcal{D}_k|\mathcal{H}^j)$.

It is rewarding to view the standard, interactive NMR assignment procedure as the evaluation of the identities of signals, obtained from a complete set of mutually exclusive identity "hypotheses" $\{\mathcal{H}^1, \ldots, \mathcal{H}^m\}$, by testing the conjecture against a set of available data $\{\mathcal{D}_1, \ldots, \mathcal{D}_n\}$, typically extracted from multidimensional J- and NOE-correlated experiments. Such analysis benefits from prior knowledge that could include, e.g., cross-peak lineshapes, chemical shift patterns, or, if available, data from homologous structures. The hypotheses that best fit both the data and the accumulated knowledge are then selected as the assignments. Thus, such assessments—basically "educated guesses"—can naturally be recast in terms of Bayesian inference.

SPI Identification of Resonances Within Self-Connected Spin Systems

SPI is a semiautomated protocol for the dissection of *unassigned* 2D NMR spectra into amino acid spin systems. In its minimalistic implementation (Grishaev and Llinás, 2002c) SPI analyzes only $^1\text{H}-^1\text{H}$ and $^1\text{H}-^{15}\text{N}$ 2D spectra, and ^{15}N-edited 3D experiments. However, it can readily be adapted to deal with other kinds of multidimensional, isotope-edited data. Spin systems are treated as vectors in the multidimensional space of chemical shifts obtained from a variety of multidimensional NMR experiments. SPI produces a list of resonances grouped into spin systems.

Roots of the 3D spin systems are obtained from the $^1\text{N}-^{15}\text{N}$ HSQC; these are denoted by $(\delta_{\text{HN}}, \delta_{\text{N}})$, where δ_{HN} is the chemical shift of the backbone peptidyl amide ^1H and δ_{N} is that of the attached ^{15}N. Side chain Trp $\text{H}^{\varepsilon 1}/\text{N}^{\varepsilon 1}$ and Asn/Gln $\text{NH}_2^{\delta,\varepsilon}$ groups are identified and excluded from the roots list. All resonances originating from HSQC-TOCSY, HNHA, and HNHB spectra are probabilistically matched to the roots. For a given 3D connectivity $(\delta_{\text{H}}, \delta_{\text{HN}}, \delta_{\text{N}})$ and a spin system i of root $(\delta_{\text{HN},i}, \delta_{\text{N},i})$, the probability of δ_{H} corresponding to i is estimated as

$$P(i|\delta_{HN}, \delta_N) = \frac{G(\delta_{HN}, \delta_{HN,i}; \sigma_{HN}) \times G(\delta_N, \delta_{N,i}; \sigma_N)}{\sum\limits_{j=1}^{M_3} G(\delta_{HN}, \delta_{HN,j}; \sigma_{HN}) \times G(\delta_N, \delta_{N,j}; \sigma_N)} \qquad (2)$$

Here, $G(\delta_1, \delta_2; \sigma) \equiv \exp[-0.5(\delta_1 - \delta_2)^2/\sigma^2]$ and σ_{HN} and σ_N are the chemical shift uncertainties of H^N and ^{15}N resonances in 3D spectra, 0.030 and 0.336 ppm, respectively. Only spin systems for which $|\delta_{HN} - \delta_{HN,i}| < 2\sigma_{HN}$ and $|\delta_N - \delta_{N,i}| < 2\sigma_N$ are taken into account.

To build 2D spin systems, all amide-aliphatic TOCSY connectivities are matched to the H^N/H^α COSY roots. Similar to the 3D case, aromatic and Asn/Gln NH_2 peaks are excluded from COSY and TOCSY peak lists. The initial probabilities are written as functions of H^N frequencies, "H^N-based probabilities." For a peak of coordinates (δ_1, δ_2), the probability is estimated via Eq. (3):

$$P(i|\delta_2, \delta_{HN,i}) = \frac{G(\delta_2, \delta_{HN,i}; \sigma_{2D})}{\sum\limits_{j=1}^{M_2} G(\delta_2, \delta_{HN,j}; \sigma_{2D})} \qquad (3)$$

where $G(\delta_1, \delta_2; \sigma)$ are Gaussian distributions as described for Eq. (2), and σ_{2D} is resonance position uncertainty in the 2D spectra, $\sim$0.015 ppm. For any TOCSY peak, only spin systems for which $|\delta_2 - \delta_{HN,i}| < 2\sigma_{2D}$ were taken into account.

The root-based probabilities obtained from Eq. (2) are used to recalculate spin system membership probabilities by cross-referencing other observed J-connectivities to the H^α chemical shifts, $\delta_{H\alpha,i}$, of the 2D spin systems. For every ambiguous match between a resonance δ_1 and the spin system i with root $(\delta_{HN,i}, \delta_{H\alpha,i})$ obtained as above, the program searches for a COSY/TOCSY peak that is the closest to the coordinates $(\delta_1, \delta_{H\alpha,i})$ within $2\sigma_{2D}$. If such a peak with coordinates $(\delta_{1,i}, \delta_{2,i})$ is found, in the context of Bayesian analysis, the "H^N/H^α-based probabilities" are estimated via

$$P(i|\delta_1, \delta_2, \delta_{HN,i}, \delta_{H\alpha,i}) = \frac{G(\delta_1, \delta_{1,i}; \sigma_{2D}) \cdot G(\delta_{H\alpha,i}, \delta_{2,i}; \sigma_{2D}) \cdot P(i|\delta_2, \delta_{HN,i})}{\sum\limits_{j=1}^{M_2} G(\delta_1, \delta_{1,j}; \sigma_{2D}) \cdot G(\delta_{H\alpha,j}, \delta_{2j}; \sigma_{2D}) \cdot P(j|\delta_2, \delta_{HN,j})} \qquad (4)$$

Otherwise, the probability is set to zero.

Resulting side chain resonances that are uniquely matched to their spin systems can be exploited to refine the remaining matches. Thus, for a resonance at δ_1 ambiguously matched to spin system i, and a spin H with a chemical shift $\delta_{H,i}$, unambiguously matched to i the program searches for the COSY/

TOCSY peak that, within $2\sigma_{2D}$, most closely matches $(\delta_1, \delta_{H,i})$. If such a peak $(\delta'_{1,i}, \delta'_{2,i})$ is found, the "$H^N/H^\alpha/H$-based probabilities" are expressed as

$$
\begin{aligned}
&\mathcal{P}(i|\delta_1, \delta_2, \delta_{HN,i}, \delta_{H\alpha,i}, \delta_{H,i}) = \\
&\frac{\mathcal{G}(\delta_1, \delta'_{1,i}; \sigma_{2D}) \times \mathcal{G}(\delta_{H,i}, \delta'_{2,i}; \sigma_{2D}) \times \mathcal{P}(i|\delta_1, \delta_2, \delta_{HN,i}, \delta_{H\alpha,i})}{\displaystyle\sum_{j=1}^{M_2} \mathcal{G}(\delta_1, \delta'_{1,j}; \sigma_{2D}) \times \mathcal{G}(\delta_{H,i}, \delta'_{2,j}; \sigma_{2D}) \times \mathcal{P}(j|\delta_1, \delta_2, \delta_{HN,j}, \delta_{H\alpha,j})}
\end{aligned}
\tag{5}
$$

Following application of Eq. (5), SPI separately detects and links Gly and Arg spin systems.

The independently identified 2D and 3D spin systems are matched, based on common chemical shifts. Initially, ambiguous matches of the H^α atoms (from HNHA spectrum) are associated with their most probable 3D spin system roots, producing uniquely defined $N/H^N/H^\alpha$ triads. The latter are frequency matched to the roots of the 2D spin systems with Gaussian probabilities where, instead of matching a resonance i to a spin system j, we are matching spin systems j_1 to j_2:

$$
\mathcal{P}_{match}(j_1, j_2) \propto \mathcal{G}(\delta_{HN,j_1}, \delta_{HN,j_1}; \sigma_{HN,3D}) \cdot \mathcal{G}(\delta_{H\alpha,j_2}, \delta_{H\alpha,j_2}; \sigma_{H,3D})
\tag{6}
$$

Here, the match is between a 2D spin system i with root $(\delta_{HN,j_1}, \delta_{H\alpha,j_1})$ and a 3D spin system j with root $(\delta_{N,j_2}, \delta_{HN,j_2}, \delta_{H\alpha,j_2})$; $\sigma_{H,3D}$ is the point size in the indirect (H^{other}) dimension of the 3D spectra. In those cases where the placement of a spin is unambiguous in only one of the uniquely matched systems, the latter are used for intraspin system identification. Ambiguous matches between 2D and 3D spin systems are refined by considering additional spins, etc. Whenever a match is found, the probabilities from Eq. (6) are multiplied by the corresponding Gaussian factor, from the updated common root spin subspace.

The matches are selected in order of decreasing probabilities and 2D and 3D spin systems combined based on consensus resonances present in both 2D and 3D spin systems. Consensus spin systems are then extended into the aliphatic area ($\delta < 6$ ppm). As consensus spin systems grow in size, further resonances are added until convergence.

For proline spin systems, which lack H^N, COSY cross-peaks become roots that the program attempts to combine via *all* COSY/TOCSY connectivities in the aliphatic region. For these, a separate peak list is created as well as a list of the derived fragments with three or more members. Finally, residues Phe, Tyr, Trp, His, Met, Asn, and Gln contain spin subsystems that do not connect each other via standard ^{15}N-edited or $^1H/^1H$ J-correlated spectra. Others, such as Arg, often exhibit two parallel spin systems, one originating from the backbone H^N atom and the other from the nitrogen-bonded H^ε. Such subsystems are also detected and matched.

In the final list, chemical shifts are averaged over the complete set of identically identified cross-peaks.

SPI Performance

SPI mimics the standard interactive analysis of NMR spectra, based on subconscious probabilistic assessments at each step of the analysis. Bayesian inference just quantifies and expedites the decision-making process. SPI also searches for self-consistency when analyzing *observed* J-connectivity data. By cycling the execution and peak list editing steps, the procedure helps to compensate for missing connectivities from the initial input files.

SPI managed to assemble most of the tested (col 2 and kringle 2) spin systems. The analysis of col 2 spectral data proved to be relatively straightforward, yielding 303 unambiguous protons. In the case of kringle 2, the test was more challenging due to the larger molecular size of the protein. A total of 93 amino acid systems remained after combining the 2D and 3D data. Examples of backbone/side chain spin system matching probabilities are shown in Fig. 1. These consensus systems contained 458 protons, against 478 reported via manual analysis (Marti *et al.*, 1999). Most of the differences stem from chemical shift degeneracy or missing (low intensity) cross-peaks. To deal with the first problem, SPI attempts to maximally exploit redundancies within the experimental J-connectivity map. Thus, in the cases of col 2 and kringle 2, ~95% of the previously established (Briknarová *et al.*, 1999; Marti *et al.*, 1999) resonances were uncovered via SPI. The second problem may be overcome by substituting NOEs for the missing J-connectivities, as is often done in the interactive procedure. In the spirit of

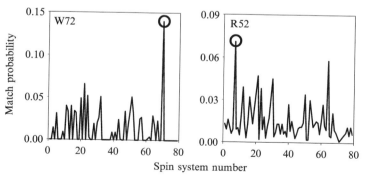

FIG. 1. Examples of SPI matching backbone/side chain spin systems probabilities for kringle 2. Low ambiguity match, W72; high ambiguity match, R52. Correct choices are circled. Modified from Grishaev and Llinás (2002c), with kind permission of Springer Science & Business Media.

the CLOUDS approach, missing H-atoms eventually fall into place, guided by the NOEs of the identified protons in the final structure generation.

Because the basic concept behind SPI is not data specific, the protocol can readily be adapted to incorporate other types (^{13}C-edited, triple-resonance, etc) of J-connectivity data. Thus, the crowding of peaks in a relatively small spectral area (the overlap problem referred to above) and unfavorable excitation profile in the H^{α}/H^{β} region associated with the implemented water suppression scheme can be alleviated by recourse to ^{13}C-edited experiments. Overall, the results obtained on the $H^{N}/H^{\alpha}/H$ resolving power of SPI for 2D spectral data indicate that SPI approaches the intrinsic $^{15}N/^{1}H^{N}$ resolution afforded by 3D experiments.

BACUS: Identification of NOE-Connected Proton Networks

A list of unambiguous NOEs, whether assigned or unassigned, is a prerequisite for any distances-based protocol of NMR protein structure elucidation. CLOUDS is no exception to this rule. All current methods of automated NOESY analysis exploit self-consistency of the correct NOESY assignment vis-à-vis a computed structure. In the BACUS approach, no assumption is made a priori regarding the molecular structure; however, the local environment is considered to be restricted by the previously established spin connectivities.

Like SPI, BACUS is a heuristic approach based on consistency of the entire available experimental data: if protons i and k are known to be proximal, the observation of an NOE attributable to protons j and k should increase the probability of observing an NOE between i and j. In practice, "reporter" protons k are crucial within the BACUS scheme and are afforded by those COSY, TOCSY, or NOESY connected to i (Fig. 2).

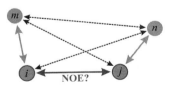

FIG. 2. The BACUS concept. Given knowledge of a set of connectivities between protons m, n, i, and j, what is the probability of observing an NOE, denoted by the horizontal darker double-headed arrow, between protons i and j? Doubly arrowed dashed lines identify NOEs, and light gray double arrows (unlabeled) indicate COSY and/or TOCSY and/or NOESY cross-peaks involving H^m and H^n. The occurrence of some or all of these connectivities "reports" on the H^i and H^j environment and serves to establish the probability for identification of an O_{ij} NOE. Modified from Grishaev and Llinás (2004), with kind permission of Springer Science & Business Media.

The program is designed to obtain unambiguous identities for the NOESY cross-peaks starting from the grid of NMR frequencies, in our case, generated via SPI. Its main goal is the estimation of probabilities for frequency labeling (or "identification") of the NOESY peak O, given the match between its experimental frequency coordinates and a grid of *unassigned* resonance frequencies, *as well as* the state of the current identities for the rest of the NOESY cross-peaks.

Requisite Likelihoods and Priors

Consider a NOESY cross-peak p ambiguously attributed to a pair of protons (i, j), as well as to other such pairs. Let us also consider protons m and n, where $m(n)$ either is $i(j)$ or is connected to $i(j)$ via COSY or TOCSY. The centerpieces of the BACUS formalism are the conditional probabilities $\mathcal{P}(O_{mn}|O_{ij})$ of observing an NOE between protons m and n given observation of an NOE between protons i and j. The $\mathcal{P}(O_{mn}|O_{ij})$ values are derived from the statistical distance probability distributions $\mathcal{P}(r_{mn}|O_{ij})$, where r_{mn} is the distance between protons m and n, and protons i and j are spatially close so as to produce a cross-peak in the NOESY spectrum. Other important components are $\mathcal{P}(V_{mn}|r_{mn})$, the probability distribution of cross-peak volumes V_{mn} conditional on interspin distances r_{mn}, and $\mathcal{P}(O_{mn}|V_{mn})$, the probability of observing a NOESY cross-peak O_{mn} of volume V_{mn}. $\mathcal{P}(O_{mn}|V_{mn})$ encodes for the sensitivity of detection, as affected by experimental noise and line widths. The dependence of V_{mn} on r_{mn} reflects the spin dipolar environment surrounding the pair (m, n) in the presence of spin diffusion. Simple probability propagation enables us to calculate $\mathcal{P}(O_{mn}|O_{ij})$ by integrating out the nuisance variables V_{mn} and r_{mn}:

$$\mathcal{P}(O_{mn}|O_{ij}) = \int\int dV_{mn}dr_{mn}\mathcal{P}(O_{mn}|V_{mn})\mathcal{P}(V_{mn}|r_{mn})\mathcal{P}(r_{mn}|O_{ij}) \quad (7)$$

The $\mathcal{P}(V_{mn}|r_{mn})$ are estimated on the basis of a set of distances extracted from selected protein structures in the Protein Data Bank (PDB) by back-calculating NOESY cross-peak volumes at the experimental mixing times via the relaxation matrix (**R**-matrix) formalism assuming isotropic rigid motion. In practice, $\mathcal{P}(O_{mn}|V_{mn})$ is taken as the probability of the cross-peak maximum being $>3 \times \sigma_{noise}$, where σ_{noise} is the measured root-mean-square (RMS) spectral noise (Fig. 3).

To complete the calculation of $\mathcal{P}(O_{mn}|O_{ij})$ [Eq. (7)], $\mathcal{P}(r_{mn}|O_{ij})$ has to be estimated. The values of this distribution depend on the identities of m and n with respect to i and j, respectively. We have binned these relationships into three classes: spins i or j themselves (class Δ), those COSY connected to i or j (class C), non-COSY (e.g., TOCSY or NOESY) connected to i and

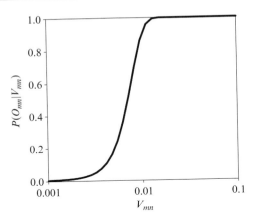

FIG. 3. Probability of observation of a 2D NOESY cross-peak as a function of its volume V_{ij} for the position uncertainty of 0.012 ppm and rms noise level of 5×10^{-5}. Based on standard deviation about the average probability distribution of NOESY peak volumes as a function of distances, back-calculated from ubiquitin (Cornilescu et al., 1998) and BPTI (Berndt et al., 1992) structures via **R**-matrix computation with $\tau_{\mathrm{mix}} = 0.2$ s and $\tau_{\mathrm{c}} = 3.0$ ns. Taken from Grishaev and Llinás (2004), with kind permission of Springer Science & Business Media.

j (class T), etc. The starting point for the derivation is the interproton distance probability distribution $\mathcal{P}(r_{ij})$ in protein structures. This distribution (Fig. 4) was extracted from the structures of six proteins. It is largely independent of protein fold and size and indicates the presence of a series of concentric spheres that, in "liquid-like" fashion, fade into a continuum as the correlations decay with increasing interproton distances. On this basis, we have expressed the $\mathcal{P}(r_{ij})$ as

$$\mathcal{P}(r_{ij}) = \mathcal{P}(r_{ij}|C_{ij})\mathcal{P}(C_{ij}) + \mathcal{P}(r_{ij}|T_{ij})\mathcal{P}(T_{ij}) + \mathcal{P}(r_{ij}|\overline{CT}_{ij})\mathcal{P}(\overline{CT}_{ij}) \qquad (8)$$

where C_{ij}, T_{ij}, and $\overline{CT}_{ij}$ indicate, respectively, COSY, TOCSY, and neither COSY nor TOCSY connected. The conditional distance probability distributions $\mathcal{P}(r_{ij}|C_{ij})$, $\mathcal{P}(r_{ij}|T_{ij})$, and $\mathcal{P}(r_{ij}|\overline{CT}_{ij})$, obtained for the same set of the six protein structures used to generate the $\mathcal{P}(r_{ij})$, are also relatively well structured (Fig. 5).

We also need the probability $\mathcal{P}(r_{ij}|O_{ij}, \overline{CT}_{ij})$ to obtain a distance r_{ij}, given that (i, j) are connected by an observable interresidue NOE. By application of Bayes' theorem [Eq. (1)], we write

$$\mathcal{P}(r_{ij}|O_{ij}, \overline{CT}_{ij}) = \frac{\mathcal{P}(r_{ij}|\overline{CT}_{ij})\mathcal{P}(O_{ij}|r_{ij}, \overline{CT}_{ij})}{\mathcal{P}(O_{ij}|\overline{CT}_{ij})} \qquad (9)$$

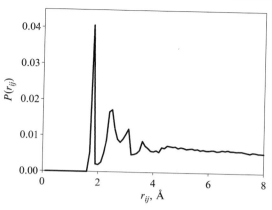

FIG. 4. Interproton distance probability distribution, calculated from the structures of ubiquitin (Cornilescu *et al.*, 1998), BPTI (Berndt *et al.*, 1992), calmodulin (Ikura *et al.*, 1992), crambin (Bonvin *et al.*, 1993; Jelsch *et al.*, 2000), cytochrome *c* (Qi *et al.*, 1996), and human prion protein (James *et al.*, 1997; Zahn *et al.*, 2000). Modified from Grishaev and Llinás (2004), with kind permission of Springer Science & Business Media.

The shape of the distribution is shown in Fig. 6. The probability of observation of an NOE between H_i and H_j is expressed to depend only on the distance r_{ij} by integrating out the nuisance variables. Hence, $\mathcal{P}(O_{ij}|r_{ij}, X_{ij}) = \mathcal{P}(O_{ij}|r_{ij})$, where X refers to any of C, T, or $\overline{CT}$ classes. Therefore, except for a normalization factor, Eq. (9) can be recast as

$$\mathcal{P}(r_{ij}|O_{ij}, \overline{CT}_{ij}) \propto \mathcal{P}(r_{ij}|O_{ij}) \times \mathcal{P}(r_{ij}|\overline{CT}_{ij}) \int dV_{ij}\mathcal{P}(O_{ij}|V_{ij})\mathcal{P}(V_{ij}|r_{ij}) \quad (10)$$

where the integral accounts for $\mathcal{P}(O_{ij}|r_{ij})$.

The probability distributions $\mathcal{P}(r_{im}|W_{im}), \mathcal{P}(r_{jn}|W_{jn})$, and $\mathcal{P}(r_{ij}|O_{ij})$ were input to Monte Carlo simulations to generate random spatial arrangements of protons i, j, m, and n; here W stands for any of the sets of reporter protons defined above. The moves that placed proton pairs (m, n), (i, n), or (j, m) at <1.7 Å were rejected; the resulting probability distributions $\mathcal{P}(r_{mn}|O_{ij})$ were accumulated over $\sim 10^5$ configurations.

As formulated for SPI [Eq. (2)], when matching a detected cross-peak and calculating the probability that it corresponds to a given pair of frequencies, for a pair of (unassigned) resonances i and j of potential NOESY connectivity O_{ij} having measured NOE cross-peak p with chemical shift coordinates $\delta_p \equiv (\delta_{1,p}, \delta_{2,p})$, the prior probabilities of chemical shifts matching, $\mathcal{P}(O_{ij}|\delta_p)$, are estimated as

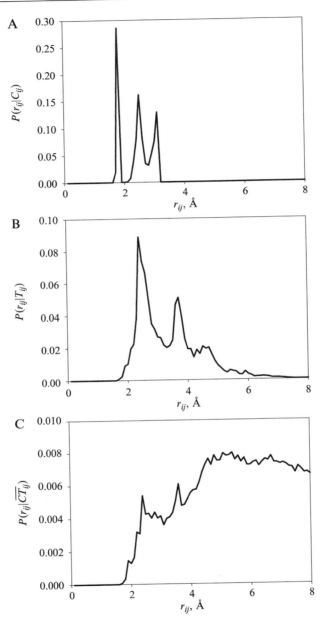

Fig. 5. Distance probability distributions for COSY-connected (A), TOCSY-connected (B), and remaining (C) protons, calculated from the set of protein structures described in the caption to Fig. 4. Modified from Grishaev and Llinás (2004), with kind permission of Springer Science & Business Media.

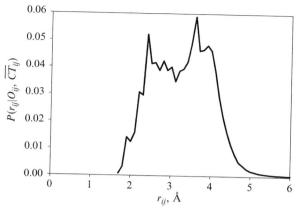

FIG. 6. Distance probability distributions for interresidue NOE-connected protons, calculated according to Eq. (13). Modified from Grishaev and Llinás (2004), with kind permission of Springer Science & Business Media.

$$\mathcal{P}(O_{ij}|\delta_p) = \frac{\mathcal{G}(\delta_{1,p}, \delta_i; \sigma) \times \mathcal{G}(\delta_{2,p}, \delta_j; \sigma)}{\sum\limits_{k,l} \mathcal{G}(\delta_{1,p}, \delta_k; \sigma) \times \mathcal{G}(\delta_{2,p}, \delta_1; \sigma)} \tag{11}$$

where $\mathcal{G}(\delta_{1p}, \delta_i; \sigma) \equiv \exp[-0.5(\delta_{1p} - \delta_i)^2/\sigma^2]$, etc., and σ is the uncertainty of the cross-peak position,

For protons i and j that do not belong to the same residue, the likelihoods $\mathcal{P}(O_R|O_{ij})$ of observing at least one NOESY connectivity O_R "reporter" cross-peak at $q \equiv q(m,n)$ arising from the protons m and n [reporting on i and j, respectively; $q(m,n) \neq p(i,j)$] conditional on (i, j) matching of NOESY cross-peak p, can be formulated as

$$\mathcal{P}(O_R|O_{ij}) \propto \frac{1}{NM} \sum_{m=1}^{M} \sum_{n=1}^{N} \mathcal{P}(O_{mn}|O_{ij}) \times \mathcal{P}(O_{mn}|\delta_q) \tag{12}$$

where $\mathcal{P}(O_{mn}|\delta_q)$ are nonzero priors of matches of the cross-peak q to the resonances m and n [Eq. (11)], and M and N are the numbers of reporter protons for i and j, respectively. Notice that δ_q does not appear within the left term, as it is implied in both O_{ij} and O_R. In the context of Eq. (12), it should be apparent that the O_{mn} connectivities are restricted to $\overline{CT}$ class only, whereas O_{im}, O_{jn} connectivities can arise from C, T, and Δ classes, as previously defined.

For protons i and j that belong to the same spin system, the likelihoods are calculated as probabilities of observing NOESY cross-peaks (i, j), conditioned by the observed C/T connectivities between i and j:

$$P(O_{ij}|X_{ij}) = \int \int dV_{ij}dr_{ij}P(O_{ij}|V_{ij})P(V_{ij}|r_{ij})P(r_{ij}|X_{ij}) \qquad (13)$$

where X_{ij} denotes either C_{ij} or T_{ij} connectivities between i and j. Here, $P(V_{ij}|r_{ij}), P(O_{ij}|V_{ij})$, and $P(r_{ij}|X_{ij})$ are those computed as indicated in the above section. From numerical integration of Eq. (13), $P(O_{ij}|C_{ij}) = 0.9998$ and $P(O_{ij}|T_{ij}) = 0.6907$ were estimated. Because in Eq. (13) the (i, j) pair reports (through $X = C$ or T) on itself, it follows that for (i, j) belonging to the same residue $P(O_{ij}|X_{ij}) \equiv P(O_R|O_{ij})$. The nature and number of connectivities can be generalized to include any type of unambiguous correlation available from homo- or heteronuclear experiments.

NOE Identities

Once the likelihoods and priors are computed, the posterior probabilities for NOE identities can be estimated. Via Bayes' theorem, the "prior" probabilities of NOESY chemical shifts matches $P(O_{ij}|\delta_p)$ can be combined with the likelihoods $P(O_R|O_{ij})$ to obtain the target posterior probabilities $P(O_{ij}|\delta_p, O_R)$, conditional on both chemical shifts and connectivities:

$$P(O_{ij}|\delta_p, O_R) = \frac{P(O_{ij}|\delta_p) \times P(O_R|O_{ij})}{\sum_{k,l} P(O_{kl}|\delta_p) \times P(O_R|O_{kl})} \qquad (14)$$

In our application, before combining with the likelihoods, $P(O_{ij}|\delta_p) < 0.05$ were set to zero and the remaining priors renormalized. Similarly, the priors that led to posteriors <0.05 were discarded and the remaining $P(O_{ij}|\delta_p)$ renormalized. In iterative fashion, a new set of posterior probabilities $P(O_{ij}|\delta_p, O_R)$ was then calculated on the updated priors and the procedure repeated until convergence, i.e., until all posteriors were >0.05. The convergence of BACUS was monitored via the information entropies S_p (Shannon, 1948) of the set of cross-peak prior matches for all peaks p:

$$S_p = -\sum_{k,l} P(O_{kl}) \times \ln[P(O_{kl})] \qquad (15)$$

After convergence of S_p, the program analyzes the obtained matches. Each peak matching odds was defined as the ratio of its largest matching probability to the next largest. Details of the matching criteria have been published (Grishaev and Llinás, 2002c, 2004). The aim is to achieve a satisfactory compromise between the protocol's resolving power and accuracy.

Performance

A total of 1049 cross-peaks were identified in the combined set of 2D NOESY spectra for col 2. The distribution of the entropies of the prior matching probabilities shows that the procedure would have a low chance of success if it were based exclusively on chemical shift matching, as only ~8% of the prior matches are defined uniquely. Posterior matching probabilities converged after 18 iterations in <5 min with a 300-MHz Pentium II processor. Out of 1049 input cross-peaks, 88% were matched with odds better than 2:1. Among these, 0.5% were in disagreement with the reported manual assignments (Briknarová *et al.*, 1999). Only seven of those mismatches were incompatible with the reported family of structures (interproton distances larger than 6 Å).

Figure 7 shows the fraction of NOESY cross-peaks versus the effective number of possible identities for kringle 2, before and after the BACUS procedure. Out of the 1354 input peaks, 1023 (75.6%) were identified uniquely. Of these, 78 (7.6%) were in disagreement with the manual assignments. However, only 24 peaks (2.3% of unique identities) corresponded to proton pairs >6 Å apart in the structure and are likely to represent true errors due to the algorithm. Inconsistencies are most dramatic in terms of distances' violations when the correct assignment hypotheses are excluded from the start. Despite the shortcomings, in most other cases distance differences between the manual and automated assignments are not as large.

The idea of exploiting "reporter" protons, a main feature of BACUS, is conceptually similar to the "network anchoring" approach reported by Hermann *et al.* (2002). This device reduces the impact of chemical shift degeneracy and leads to a significant decrease in ambiguity when analyzing crowded NOESY spectra. When tested against col 2 and kringle 2 data, the protocol compressed the number of possible NOE identities by a factor of 6–9, yielding unambiguous matches for 75–88% of the input cross-peaks (starting from ~1–8% unambiguous) with an ~5% error level. In conclusion, BACUS is robust, does not require resonance assignments, and runs in minutes without cycling through structure calculations. All these features make the combined SPI/BACUS protocol particularly attractive in the context of direct methods such as CLOUDS.

MIDGE

MIDGE is an iterative algorithm that, starting from experimental NOESY data (**A**-matrix), derives an internally consistent estimate for the relaxation matrix (**R**), without any knowledge of the molecular 3D

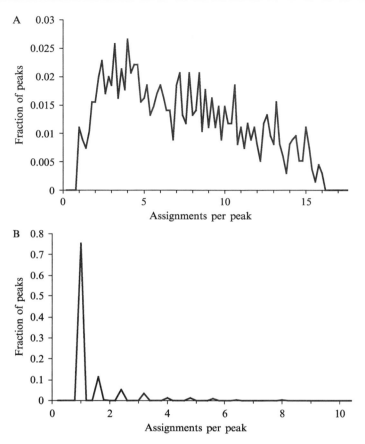

FIG. 7. Distributions of assignments per NOESY peak for kringle 2 before (A) and after (B) BACUS procedure. The numbers of assignment for each peak were calculated as exponents of the corresponding information entropies [Eq. (15)], thus the occurrence of noninteger assignments per peak. Taken from Grishaev and Llinás (2004), with kind permission of Springer Science & Business Media.

structure. The rows and columns in these square matrices span the detected protons. It should be recalled (Macura and Ernst, 1980) that the normalized $\mathbf{A}(\tau_{\mathrm{mix}}) = \exp(-\mathbf{R}\tau_{\mathrm{mix}})$, so that diagonalizing the experimental $\mathbf{A}(\tau_{\mathrm{mix}})$ leads to a diagonal $\mathbf{R}$ through a simple logarithmic transformation. Reversing the diagonalization, the latter is then converted to its common, non-diagonal form. The elements of such $\mathbf{R}$ are the first-order rate constants for the network of dipolar-coupled ^{1}H spins. Each such element is given by $R_{ij} \propto f(\tau_c)d_{ij}^{-6}$, where $f(\tau_c)$ is a known function of the isotropic rotational correlation time τ_c and d_{ij} is the unknown interproton distance (Solomon, 1955).

The essence of MIDGE is that the mostly unresolved diagonal elements of $\mathbf{A}(\tau_{mix})$, crucial for a proper analysis, are iteratively improved based on the observed off-diagonal elements, so that the converged $\mathbf{R}$ is both self-consistent and "optimal" for the set of NOESY data (Madrid et al., 1991). The intensities of unobserved NOESY cross-peaks are set to zero. Measured $\mathbf{A}$ elements are normalized assuming an exponential decay of the sum of the intensities of all peaks originating from a given proton, as a function of τ_{mix}. The diagonal $\mathbf{A}$ intensities can then be set to any value exceeding one, to generate the initial NOESY matrix. The latter is converted into the matrix $\mathbf{R}$ that encompasses the self- (diagonal) and the cross- (off-diagonal) relaxation rates. A relationship between the former and latter is used to obtain an updated estimate for the self-relaxation rates, which replaces the previous values (Grishaev and Llinás, 2002a). Back-transformation then yields an improved estimate for the diagonal elements of $\mathbf{A}$. The process is iterated until convergence of matrices $\mathbf{A}$ and $\mathbf{R}$.

Performance

MIDGE was found stable with respect to the trial set of adjustable τ_c and leak parameters, converging independently of the initial choices of diagonal elements of $\mathbf{A}$. For well-resolved diagonal peaks, the procedure yielded volumes that agreed within ~9% of the measured values. On average, the col 2 and kringle 2 experimental NOESY data (τ_{mix} of 60, 120, and 200 ms for col 2; 60, 90, and 250 ms for kringle 2) provided, respectively, 3.2 and 3.5 restraints per atom (1055 and 1494 restraints, correspondingly). The internuclear distances computed as averages over these mixing times differ by ~5% for col 2 and ~8% for kringle 2.

The self-consistency of the MIDGE protocol alleviates the effects of sparseness in the experimental NOESY matrix. The absence of an a priori structural model—a main advantage of the protocol—permits the derivation of unbiased interproton distances. In practice, numerical instability of the logarithm operation coupled with errors of the NOESY measurement tend to decrease the protocol's performance at longer τ_{mix} values. Despite this limitation, the relaxation matrix approach is preferable to the standard isolated spin pair approximation, being more objective and accurate within its range of applicability.

CLOUDS and FOC: Generation of Unassigned Proton Distributions Without Assignments

The distance restraints from MIDGE were input to an SA/MD protocol in which only nonbonded repulsive terms and NOE-based distance restraints are active. Amide H^N–H^N and H^N–H^α ADCs (Bruschweiler et al., 1991; De Vlieg et al., 1986), fundamental for the protocol, were added during a refinement step. The CNS program (Brünger et al., 1998) was used to carry out the simulation, in which the hydrogen gas was cooled from 2000 K to 10 K for 150 ps. For the on-line CCPN implementation, an ad hoc MD program, Hcloud, written by A. Lemak, was incorporated. A large number of clouds, generated by varying initial random proton coordinates, were superimposed by translation and improper rotation (Kabsch, 1976) to minimize the pairwise RMSDs. Clouds closest to the average, i.e., within 3 standard deviations, were selected for further analysis. Their overlap generates the foc (family of clouds), effectively a proton probability density.

Performance

A total of 1200 SA/MD runs for col 2 produced clouds with restraint violations <0.5 Å. For kringle 2, out of 1100 computed clouds, 5 had a single violation >0.5 Å and were discarded. The obtained clouds were evenly split into mirror image–related subsets, otherwise exhibiting high pairwise similarity. Selected single clouds for col 2 and kringle 2 are shown in Fig. 8. Although the all-H clouds provide rather coarse images of the atomic distribution, the underlying fold architecture is conveyed by the array of amide H^N atoms, from which it is possible to trace rather well-defined segments of the polypeptide backbone.

Totals of 955 col 2 clouds and 1048 kringle 2 clouds remained after similarity-based filtering and were superimposed to generate the molecular focs. Figure 9 shows the backbone (H^N, H^α) atomic focs of col 2 and kringle 2. The overall foc quality is quite satisfactory in that the previously reported NMR backbone structures (Briknarová et al., 1999; Marti et al., 1999) nicely fit the computed hydrogen density. Thus, notwithstanding some poor local—mostly side chain—geometries, the foc provides a realistic description of the molecular fold.

BAF and SIF: Bayesian Identification of foc Densities

To compute the structure, it is necessary to identify the foc, i.e., to achieve the sequence-specific assignment of each of its frequency-labeled atomic components. However, in the CLOUDS strategy, *it is the structure that leads to the assignment* rather than the reverse. The task at hand is

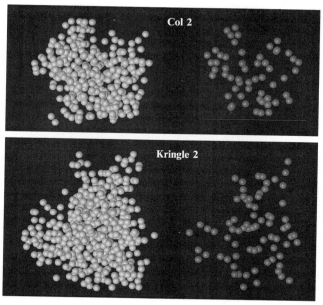

FIG. 8. Individual *clouds* for col 2 (top) and kringle 2 (bottom). Complete *clouds* are shown on the left and H^N-only *clouds* on the right. Modified from Grishaev and Llinás (2002a), copyright 2002, National Academy of Sciences, USA.

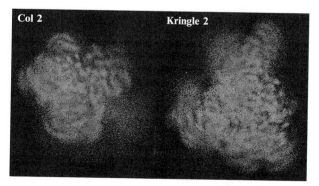

FIG. 9. Proton density distributions (*focs*): col 2 (left) and kringle 2 (right). Individual atomic *focs* are discernible as condensed, pseudospherical density clusters.

somewhat similar to identification of an imperfectly phased electron density in X-ray crystallography. Indeed, our approach to this problem bears some resemblance to the ARP/wARP automated refinement method

(Morris *et al.*, 2002), which iteratively identifies the atomic constituents of the electron density, imposing new rules and improving phases as the procedure progresses. By basing the *foc* identification exclusively on distance and chemical shift information, the protocol neglects covalent structural information from spin system grouping, obtainable from J-coupling-encoded experiments as discussed above for SPI. Our purpose was to explore whether with such a minimal amount of information an identification algorithm would be feasible, a feature useful in the context of a potential interface with, e.g., crystallographic-type analysis of the *foc* as an approach to molecular replacement (Pozharski *et al.*, 2002).

The data pertinent for our $\mathcal{H}^i$ assignment hypotheses for the H_n *foc* atom includes its chemical shift δ_n and/or its spatial position relative to other *foc* atoms. The latter is defined through a number M of interproton distances by reference to the set of atoms of currently established, i.e., updated, identity $H_{\xi j}, 1 \leq j \leq M$. Within this context, Bayes' theorem [Eq. (1)] leads to

$$\mathcal{P}(H^i|\delta_n, r_{\xi_1,n}, \ldots, r_{\xi_M,n}) = \frac{\mathcal{P}(H^i) \times \mathcal{P}(\delta_n|H^i) \prod_{j=1}^{M} \mathcal{P}(r_{\xi_j,n}|H^i)}{\mathcal{P}(\delta_n, r_{\xi_1,n}, \ldots, r_{\xi_M,n})} \qquad (16)$$

where r_{ξ_j}, n is the distance between *foc* atoms $\mathcal{H}_{\xi j}$ and H_n. Our priors are unbiased selection probabilities: e.g., if there are 64 H^N atoms, for $\mathcal{H}^i = H_i^N, \mathcal{P}(\mathcal{H}^i) = 1/64$, while the likelihoods $\mathcal{P}(\delta_n|\mathcal{H}^i)$ and $\mathcal{P}(r_{\xi_j,n}|\mathcal{H}^i)$ are evaluated from available experimental databases. The empirical probability distributions (PDs) for distances between various H atoms were generated from a set of eight reported high-quality protein structures solved via NMR spectroscopy and X-ray crystallography (listed in the caption to Fig. 4), binned to 0.2 Å. The chemical shift PDs built from the BioMagResBank database (Madison, WI) were approximated by Gaussian functions. Distance likelihoods in Eq. (2) were computed as

$$\mathcal{P}(r_{\xi_j,n}|H^i) = \int_{r \in r_{\xi_j,n}} dr \; \mathcal{P}_{\text{foc}}(r) \times \mathcal{P}_{\text{db}}(r|H^i) \qquad (17)$$

where the PDs $\mathcal{P}_{\text{foc}}(r)$ and $\mathcal{P}_{\text{db}}(r|H^i)$ are over the set of individual *clouds* and the structure database, respectively. Only hypotheses yielding posteriors that fall within two orders of magnitude from their highest calculated values were kept for further analysis.

Backbone Finder: BAF

The protocol identifies the string of backbone atoms within the *foc*. For each H_i^N atom, BAF finds the most probable intraresidue H_i^α and sequential $H_{i\pm1}^N$ and $H_{i\pm1}^\alpha$ atoms. Initially, $H_{i\pm1}^N$ are inferred from their distances to

H_i^N, while the H_i^α and $H_{i\pm1}^\alpha$ from both distances to H_i^N and their chemical shifts δ_n. The BAF cycle is iterated until convergence. Glycines, which potentially yield two high-scoring H^α matches separated by distances <2.5 Å, become identified. Likewise unique, Pro H^α atoms can be recognized from both their upstream and downstream sequential amide connections. Because BAF determines the directionality of the amide linkages, the polypeptide C- and N-termini can be identified and the sequential assignment established.

Side Chain Finder: SIF

Once the polypeptide backbone atoms become sequentially assigned, the *foc* identification reduces to best-fitting the sequence-established side chains to the remainder proton density. SIF aims at localizing side chain *foc* H-atoms linked to given H_i^N/H_i^α pairs or, in the case of prolines, H_i^α only. The protocol gradually identifies H atoms along the side chain, with probabilities updated with the current state of knowledge. Similar to BAF, the likelihoods are distance and chemical shift dependent. Initially, the posteriors for the assignment hypotheses are written according to Eq. (16), with $M = 3$, $H_{\xi1} = H_{i+1}^N$ (except after the Pro residue), $H_{\xi2} = H_i^N$ (except Pro), and $H_{\xi3} = H_i^\alpha$. The H^βs are identified first since they tend to be better defined. The H^γ assignments are then derived, and the probabilities of the remaining assignments are modified along the lines described above. The process is continued until no further assignments result.

Following the atomic *foc* identification, SIF evaluates *cis/trans* isomerism of prolylimino bonds. The prior probabilities are database-estimated 0.15 for *cis* and 0.85 for *trans* forms. The likelihoods include the H_{P-1}^α–H_P^α and H_{P-1}^α–H_P^δ distances, where P identifies the Pro residue sequence number.

Performance

Sorting of H^N/H^α pairs via BAF generated 100% correct assignments for both col 2 and kringle 2 *focs*. The three Pro H^α atoms in col 2 were identified from the immediate upstream and downstream H^N/H^α pairs. For kringle 2, five Pro H^α atoms were detected from both ends and the remaining two from one end. For both col 2 and kringle 2, *cis* and *trans* isomers of the X-Pro peptide bonds were all correctly recognized: two *cis* and eight *trans*.

The analysis of side chain hydrogens via SIF, 199 in col 2 and 298 in kringle 2, converged in five iterations. In the case of col 2, 10 differed from the reported manual assignments (Briknarová *et al.*, 1999). The differences involved $H^{\beta1} \leftrightarrow H^{\beta2}$ switches in Phe-4, Pro-14, Phe-17, Pro-18, Cys-29, Arg-34, Arg-39, Asp-48, Lys-51, and Lys-52. Furthermore, in col 2, the

Tyr-26 $H^{\beta 1}$ and the Pro-57 $H^{\gamma *}$ were not identified and remained unassigned. In the case of kringle 2, eight side chain assignments differed from the manual assignments (Marti $et\ al.$, 1999) because of pairwise switches: $H^{\beta 1} \leftrightarrow H^{\beta 2}$ in Phe-41, Lys-47, Tyr-50, Arg-52, Asp-67, and Asn-69; $H^{\gamma *} \leftrightarrow H^{\beta 2}$ in Glu-1; $H^{\varepsilon 21} \leftrightarrow H^{\varepsilon 22}$ in Gln-28; and $H^{\delta 21} \leftrightarrow H^{\delta 22}$ in Asn-43. Incorporation of chemical shift–dependent terms in the expressions for the posteriors averted numerous side chain atom misassignments. The success of BAF and SIF underscores the robustness of its local, "best first" strategy.

EMBEDS: Molecular Structure via Direct Embedding into foc Proton Density

Once the BAF/SIF identification is complete, protein conformation is fitted to the global foc guided by the locations of the identified proton densities. Compared to fitting the X-ray model to the phased experimental electron density, the foc fitting procedure is simplified by the one-to-one correspondence between the atomic sites and the individual proton densities. For this, a total of N_f BAF/SIF-assigned individual atomic $focs$ $\rho^i(\mathbf{r})$ were digitized on a 0.5-Å grid, normalized, and smoothened via convolution with a Gaussian function of width σ, to yield a target density $\rho_c(\mathbf{r}_i)$:

$$\rho_c(\mathbf{r}_i) \propto \int d^3\mathbf{r}' \rho^i(\mathbf{r}') \exp\left[-\frac{(\mathbf{r}_i - \mathbf{r}')^2}{2\sigma^2}\right] \tag{18}$$

A pseudoenergy $E(r)$

$$E(r) \equiv -k_B T \sum_{i=1}^{N_f} \ln\rho_c(\mathbf{r}_i) \tag{19}$$

and its spatial gradient were coded into CNS to describe the agreement between the instantaneous hydrogen coordinates {$\mathbf{r}$} and the target foc densities. MD simulations from randomized initial geometries were carried out for 400 ps at 300 K with σ decreasing from 2 to 0.5 Å. The usual empirical energy terms including bonds, angles, impropers, and repulsive nonbonded interactions were active in addition to the foc-fitting term. Thus, by the end of the simulation, the target density $\rho_c(r)$ approaches the input foc density $\rho(r)$.

Performance

A priori, it is not possible to distinguish a foc proton density from its mirror image, as both are compatible with the imposed distance restraints. The discrimination cannot be obtained during BAF/SIF stages either, since these protocols are distances based as well. However, when we ran EMBEDS on each of the two mirror image–related proton densities, the quality of the fit and the values of the empirical energy terms, as well as

Ramachandran statistics, were found to be consistently poorer for the fit to the mirror image of the correct proton density. This enabled an unambiguous identification of the correct overall chirality. Once the correct *foc* enantiomers were identified, ensembles of 10 structures were fitted to col 2 and kringle 2 proton densities via a full EMBEDS protocol.

In case of col 2, except for the less determined C-terminal region (residues Pro-57–Ala-60), the differences in backbone conformations are <2.0 Å RMSD (heavy atoms). The differences in the C-terminus are likely to arise from a low number of NOE constraints, affected by an enhanced dynamics (backbone order parameters $S^2 = 0.6$–0.2; Briknarová *et al.*, 1999). The col 2 secondary structure elements—two double-stranded antiparallel β-sheets, Phe-19–Phe-21/Thr-24–Tyr-26 and Trp-40–Gly-42/Tyr-53–Phe-55, and an α-helical turn, Tyr-47–Asp-50—are also present in the CLOUDS structures. The robustness of the protocol is also exemplified in the case of the col 2 Gly-33 where, while the *foc* shows two distinct, equally probable, distributions for the H$^\alpha$s (Fig. 10), the algorithm discriminates in favor of the correct geometry.

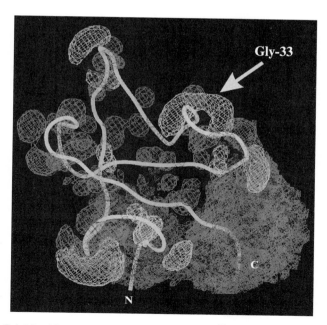

Fig. 10. Col 2 backbone conformation fitted to the H^N and H$^\alpha$ *foc* atomic densities in contour level representation. Contours defining Gly H$^\alpha$s densities are in lighter trace. The Gly-33 H$^{\alpha 2,3}$ density is bimodal. Modified from Grishaev and Llinás (2002b), copyright 2002, National Academy of Sciences, USA.

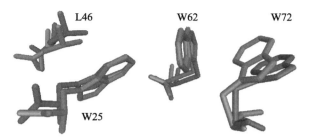

FIG. 11. Hydrophobic core of kringle 2: representative amino acid side chains are shown. CLOUDS (EMBEDS-fitted) structure are depicted in light gray and previously reported structure (Marti *et al.*, 1999) in dark gray. Modified from Grishaev and Llinás (2002b), copyright 2002, National Academy of Sciences, USA.

For kringle 2, the differences between CLOUDS and the reported structures (Marti *et al.*, 1999) are mainly in the orientation of the C-terminal region, Cys-75–Thr-81, and in the conformation of the flexible Phe-41–Asn-45 loop. Again, both regions exhibit relatively low numbers of NOE as well as enhanced backbone mobilities (Marti *et al.*, 1999). In contrast, it is gratifying that the Leu-74–Arg-79 segment of the left-handed 3_1-helix is also obtained by CLOUDS. Furthermore, notwithstanding that the atomic *focs* for the side chain atoms of Trp residues are irregular, the geometries of the hydrophobic cluster neighboring the lysine-binding site (side chains of Trp-25, Leu-46, Trp-62, and Trp-72) are also close to those exhibited by the reported structure (Marti *et al.*, 1999) (Fig. 11).

Conclusions

This chapter describes a direct protocol capable of calculating protein structures using *minimal* experimental NMR data, in a fast and reliable manner. The essence of the method is in the structural interpretation of the NMR spectral data prior to the sequence-specific assignment of resonances. As discussed above, the **R**-matrix analysis via MIDGE affords an objective and rather precise estimation of the inter-proton distances, not necessitating the assignment of resonance frequencies. These distances, when input to a MD/SA optimization procedure, lead to the generation of molecular proton densities (*focs*) that convey fuzzy images of the molecular H-atom spatial distribution. As shown, *focs* can be interpreted in terms of the underlying protein fold. Novel methods aimed at *foc* identification and fitting provide a framework for generating standard NMR-quality protein structures. We also describe a set of automated protocols, aimed at obtaining identities of ambiguous NOESY cross-peaks, without resonance assignments or structural estimates. These novel approaches are both efficient and robust, as demonstrated by their performances on experimental data. By exploiting Bayesian inference at several stages (SPI, BACUS, BAF, SIF), the complete CLOUDS protocol intrinsically differs from the by now well established stochastic schemes of NMR data optimization (e.g., Mumenthaler and Braun, 1995; Nilges, 1995).

The methods outlined in this chapter prove it is possible to elucidate NMR-based macromolecular structures with a sequence of steps different from the canonical resonance assignment → restraints derivation → structure calculation. As demonstrated by the results of the MIDGE/CLOUDS/BAF/SIF/EMBEDS sequence, conformational models can be obtained prior to the resonance assignment and subsequently analyzed to obtain the latter, in turn necessary for routine structure refinement. On the other hand, SPI/BACUS results prove that unambiguous NOESY cross-peak identities can also be established prior to the full sequence-specific resonance assignments. Therefore, the CLOUDS protocol represents a novel paradigm for the calculation of protein structures via NMR. By complementing, conceptually and in practice, important components of the standard methods of structure computation, it broadens the landscape of atomic-resolution approaches to structure elucidation and suggests new direct pathways to rapidly converge from the spectroscopic data to the macromolecular fold.

Program Availability

All components within the CLOUDS suite of programs are available in FORTRAN version from http://www.chem.cmu.edu/groups/Llinas/. User-friendly C/PYTHON versions of programs MIDGE and CLOUDS, including modules FILTER and HClouds, are available from CCPN at www.ccpn.ac.uk within the CcpNmr Analysis software release.

Acknowledgments

The CLOUDS project was developed under NIH funding, Grants HL29409, and GM67964.

References

Adler, M. (2000). Modified genetic algorithm resolves ambiguous NOEs and reduces unsightly NOE violations. *Proteins* **39**, 385–392.
Atkinson, R. A., and Saudek, V. (1996). The direct determination of protein structure from multidimensional NMR spectra without assignment: An evaluation of the concept. *In* "Dynamics and the Problem of Recognition in Biological Macromolecules" (O. Jardetzky and J. F. Lefébre, eds.). Plenum Press, New York.

Atkinson, R. A., and Saudek, V. (1997). Direct fitting of structure and chemical shift to NMR spectra. *J. Chem. Soc. Faraday Trans.* **93**, 3319–3323.

Atkinson, R. A., and Saudek, V. (2002). Hypothesis: The direct determination of protein structure by NMR without assignment. *FEBS Lett.* **510**, 1–4.

Atreya, H. S., Sahu, S. C., Chary, K. V. R., and Govil, G. (2000). A tracked approach for automated NMR assignments in proteins (TATAPRO). *J. Biomol. NMR* **17**, 125–136.

Bailey-Kellogg, C., Widge, A., Kelley, J. J., Berardi, M. J., Buschweller, J. H., and Donald, B. R. (2000). The NOESY Jigsaw: Auotomated protein secondary structure and main-chain assignment form sparse, unassigned NMR data. *J. Comp. Biol.* **7**, 537–558.

Balaban, E. T. (ed.) (1976). "Chemical Applications of Graph Theory." Academic Press, New York.

Bartels, C., Güntert, P., Billeter, M., and Wüthrich, K. (1997). GARANT—a general algorithm for resonance assignment of multidimensional nuclear magnetic resonance spectra. *J. Comput. Chem.* **18**, 139–149.

Beraud, S., Bersch, B., Brutscher, B., Gans, P., and Blackledge, M. (2002). Direct structure determination using residual dipolar couplings: Reaction-site conformation of methionine sulfoxide reductase in solution. *J. Am. Chem. Soc.* **124**, 13709–13715.

Berndt, K. D., Güntert, P., Orbons, L. P., and Wüthrich, K. (1992). Determination of a high-quality nuclear magnetic resonance solution structure of the bovine pancreatic trypsin inhibitor and comparison with three crystal structures. *J. Mol. Biol.* **227**, 757–775.

Boelens, R., Koning, T. M. G., Van der Marel, G. A., Van Boom, J. H., and Kaptein, R. (1989). Determination of biomolecular structures from proton-proton NOEs using a relaxation matrix approach. *J. Magn. Reson.* **82**, 290–308.

Bonvin, A. M., Rullmann, J. A., Lamerichs, R. M., Boelens, R., and Kaptein, R. (1993). "Ensemble" iterative relaxation matrix approach: A new NMR refinement protocol applied to the solution structure of crambin. *Proteins* **15**, 385–390.

Borgias, B. A., and James, T. L. (1990). MARDIGRAS—a procedure for matrix analysis of relaxation for discerning geometry of an aqueous structure. *J. Magn. Reson.* **87**, 475–487.

Bounds, D. G. (1987). New optimization methods from physics and biology. *Nature* **329**, 215–219.

Briknarová, K., Grishaev, A., Bányai, L., Tordai, H., Patthy, L., and Llinás, M. (1999). The second type II module from human matrix metalloproteinase 2: Structure, function and dynamics. *Structure* **7**, 1235–1245.

Brünger, A. T., Adams, P. D., Clore, G. M., Delano, W. L., Gros, P., Grosse-Kunstleve, R. W., Jiang, J. S., Kuszewski, J., Nilges, M., Pannu, N. S., Read, R. J., Rice, L. M., Simonson, T., and Warren, G. L. (1998). Crystallography and NMR system (CNS): A new software system for macromolecular structure determination. *Acta Crystallogr. D.* **54**, 905–921.

Bruschweiler, R., Blackledge, M., and Ernst, R. R. (1991). Multi-conformational peptide dynamics derived from NMR data: A new search algorithm and its application to antamanide. *J. Biomol. NMR* **1**, 3–11.

Buchler, N. E. G., Zuiderweg, E. R. P., Wang, H., and Goldstein, R. A. (1997). Protein heteronuclear NMR assignments using mean-field simulated annealing. *J. Magn. Reson.* **125**, 34–42.

Cornilescu, G., Marquardt, J. L., Ottiger, M., and Bax, A. (1998). Validation of protein structure from anisotropic carbonyl chemical shifts in a dilute liquid crystalline phase. *J. Am. Chem. Soc.* **120**, 6836–6837.

Croft, D., Kemmink, J., Neidig, K. P., and Oshkinat, H. (1997). Tools for the automated assignment of high-resolution, three dimensional protein NMR spectra based on pattern recognition techniques. *J. Biomol. NMR* **10**, 207–219.

De Vlieg, J., Boelens, R., Scheek, R. M., Kaptein, R., and van Gunsteren, W. F. (1986). Restrained molecular dynamics procedure for protein tertiary structure determination from NMR data: A lac repressor headpiece structure based on information on J-coupling and from presence and absence of NOE's. *Isr. J. Chem.* **27,** 181–188.

Duggan, B. M., Legge, G. B., Dyson, H. J., and Wright, P. E. (2001). SANE (Structure Assisted NOE Evaluation): An automated model-based approach for NOE assignment. *J. Biomol. NMR* **19,** 321–329.

Englander, S. W., and Wand, A. J. (1987). Main-chain-directed strategy for the assignment of H1 NMR spectra of proteins. *Biochemistry* **26,** 5953–5958.

Feng, W. Q., Tejero, R., Zimmerman, D. E., Inouye, M., and Montelione, G. T. (1998). Solution NMR structure and backbone dynamics of the major cold-shock protein (CspA) from Escherichia coli: Evidence for conformational dynamics in the single-stranded RNA-binding site. *Biochemistry* **37,** 10881–10896.

Floudas, C. A. (2000). Deterministic global optimization: Theory, methods and applications. *In* "Nonconvex Optimization and Its Applications." Kluwer Academic Publ., Dordrecht.

Grishaev, A., and Llinás, M. (2002a). CLOUDS, a protocol for deriving a molecular proton density via NMR. *Proc. Natl. Acad. Sci. USA* **99,** 6707–6712.

Grishaev, A., and Llinás, M. (2002b). Protein structure elucidation from NMR proton densities. *Proc. Natl. Acad. Sci. USA* **99,** 6713–6718.

Grishaev, A., and Llinás, M. (2002c). Sorting signals from protein NMR spectra: SPI, a Bayesian protocol for uncovering spin systems. *J. Biomol. NMR* **24,** 203–213.

Grishaev, A., and Llinás, M. (2004). BACUS: A Bayesian protocol for the identification of protein NOESY spectra via unassigned spin systems. *J. Biomol. NMR* **28,** 1–10.

Gronwald, W., and Kalbitzer, H. R. (2004). Automated structure determination of proteins by NMR spectroscopy. *Prog. NMR Spectrosc.* **44,** 33–96.

Gronwald, W., Willard, L., Jellard, T., Boyko, R., Rajarathman, K., Wishart, D., Sonnichsen, F., and Sykes, B. D. (1998). CAMRA: Chemical shift based computer aided protein NMR assignments. *J. Biomol. NMR* **12,** 395–405.

Grzesiek, S., and Bax, A. (1993). Amino acid type determination in the sequential assignment rpcedure of uniformly C-13/N-15 enriched proteins. *J. Biomol. NMR* **3,** 185–204.

Güntert, P. (2003). Automated NMR protein structure calculation. *Prog. NMR Spectrosc.* **43,** 105–125.

Güntert, P., Saltzmann, M., Braun, D., and Wüthrich, K. (2000). Sequence-specific NMR assignments of proteins by global fragment mapping with the program MAPPER. *J. Biomol. NMR* **18,** 129–137.

Hermann, T., Guntert, P., and Wuthrich, K. (2002). Protein NMR structure determination with automated NOE assignment using new software CANDID and torsoin angle dynamics algorithm DYANA. *J. Mol. Biol.* **319,** 209–227.

Hitchens, T. K., Lukin, J. A., Zhan, Y. P., Mc Callum, S. A., and Rule, G. S. (2003). MONTE: An automated Monte Carlo based approach to nuclear magnetic resonance assignment of proteins. *J. Biomol. NMR* **25,** 1–9.

Ikura, M., Clore, G. M., Gronenborn, A. M., Zhu, G., Klee, C. B., and Bax, A. (1992). Solution structure of a calmodulin-target peptide complex by multidimensional NMR. *Science* **256,** 632–638.

James, T. L. (1994). Assessment of quality of derived macromolecular structure. *Methods Enzymol.* **239,** 416–439.

James, T. L., Liu, H., Ulyanov, N. B., Farr-Jones, S., Zhang, H., Donne, D. G., Kaneko, K., Groth, D., Mehlhorn, I., Prusine, S. B., and Cohen, F. E. (1997). Solution structure of a 142-residue recombinant prion protein corresponding to the infectious fragment of the scrapie isoform. *Proc. Natl. Acad. Sci. USA* **94,** 10086–10091.

Jaynes, E. T. (2003). "Probability Theory, the Logical Science." Cambridge University Press, Cambridge, UK.

Jelsch, C., Teeter, M. M., Lamzin, V., Pichon-Lesme, V., Blessing, B., and Lecomte, C. (2000). Accurate protein crystallography at ultra-high resolution: Valence-electron distribution in crambin. *Proc. Natl. Acad. Sci. USA* **97**, 3171–3176.

Kabsch, W. (1976). Solution for best rotation to relate 2 sets of vectors. *Acta Crystallogr. A.* **32**, 922–923.

Kirkpatrick, S., Gelatt, C. D., and Vecchi, M. P. (1983). Optimization by simulated annealing. *Science.* **220**, 671–680.

Kraulis, P. J. (1994). Protein 3-dimensional structure determination and sequence-specific assignment of C-13-separated and N-15-separated NOE data: A novel real-space ab-initio approach. *J. Mol. Biol.* **243**, 696–718.

Likić, V. A. (1996). Relaxation analysis of 2D NOE in macromolecules. *Concepts Magn. Reson.* **8**, 223–236.

Liu, H., Spielmann, H. P., Ulyanov, N. B., Wemmer, D. E., and James, T. L. (1995). Interproton distance bounds from 2D NOE intensities: Effect of experimental noise and peak integration errors. *J. Biomol. NMR* **6**, 390–402.

Lukin, J. A., Grove, A. P., Talukdar, S. N., and Ho, C. (1997). Automated probabilistic method for assigning backbone resonances of (^{13}C, ^{15}N)-labeled proteins. *J. Biomol. NMR* **9**, 151–166.

Macura, S., and Ernst, R. R. (1980). Elucidation of cross relaxation in liquids by two-dimensional nmr-spectroscopy. *Mol. Phys.* **41**, 95–117.

Madrid, M., Llinás, E., and Llinás, M. (1991). Model-independent refinement of interproton distances generated from ^{1}H NMR Overhauser intensities. *J. Magn. Reson.* **93**, 329–346.

Malliavin, T. E., Rouh, A., Delsuc, M., and Lallemand, J. Y. (1992a). Approche directe de la détermination de structures moléculaires à partir de l'effect Overhauser nucléaire. *CR Acad. Sci. Paris* **315**(II), 653–659.

Malliavin, T. E., Delsuc, M. A., and Lallemand, J. Y. (1992b). Computation of relaxation matrix elements from incomplete NOESY data sets. *J. Biomol. NMR* **2**, 349–360.

Malliavin, T. E., Barthe, P., and Delsuc, M. A. (2001). FIRE: Predicting the spatial proximity of protein residues form 3D NOESY-HSQC. *Theor. Chem. Acc.* **106**, 91–97.

Marti, D., Schaller, J., and Llinás, M. (1999). Solution structure and dynamics of the plasminogen kringle 2-AMCHA complex: 3_1-Helix in homologous domains. *Biochemistry* **38**, 15741–15755.

Meadows, R. P., Olejniczak, E. T., and Fesik, S. W. (1994). A computer-based protocol for semiautomated assignments and 3D structure determination of proteins. *J. Biomol. NMR* **4**, 79–96.

Meiler, J., and Baker, D. (2003). Rapid protein fold determination using unassigned NMR data. *Proc. Natl. Acad. Sci. USA* **100**, 15404–15409.

Metropolis, N., Rosenbluth, A. W., Rosenbluth, M. N., Teller, A. H., and Teller, E. (1953). Equation of state calculations by fast computing machines. *J. Chem. Phys.* **21**, 1087–1092.

Morris, R. J., Perrakis, A., and Lamzin, V. S. (2002). ARP/wARP model-building algorithms. I. The main chain. *Acta Crystallogr. D.* **58**, 969–975.

Moseley, H., and Montelione, G. T. (1999). Automated analysis of NMR assignments and structures for proteins. *Curr. Opin. Struct. Biol.* **9**, 635–642.

Mumenthaler, C., and Braun, W. (1995). Automated assignment of simulated and experimental NOESY spectra of proteins by feedback filtering and self-correcting distance geometry. *J. Mol. Biol.* **254**, 465–480.

Mumenthaler, C., Guntert, P., Braun, W., and Wuthrich, K. (1997). Automated combined assignment of NOESY spectra and three-dimensional protein structure determination. *J. Biomol. NMR* **10,** 351–362.

Nadel, B. (1988). Tree search and arc consistency in constraint satisfaction algorithms. *In* "Search in Artificial Intelligence" Springer-Verlag, New York.

Nilges, M. (1993). A calculation strategy for the structure determination of symmetric dimers by ¹H NMR. *Proteins* **17,** 297–309.

Nilges, M. (1995). Calculation of protein structures with ambiguous distance restraints. Automated assignment of ambiguous crosspeaks and disulfide connectivities. *J. Mol. Biol.* **245,** 645–660.

Nilges, M., and O'Donoghue, S. I. (1998). Ambiguous NOES and automated NOE assignment. *Progr. NMR Spectr.* **32,** 107–139.

Nilges, M., Macias, M. J., O'Donoghue, S. I., and Oshkinat, H. (1997). Automated NOESY interpretation with ambiguous distance restraints: The refined NMR solution structure of the pleckstrin homology domain from b-spectrin. *J. Mol. Biol.* **269,** 408–422.

Oshiro, C. M., and Kuntz, I. D. (1993). Application of distance geometry to the proton assignment problem. *Biopolymers* **33,** 107–115.

Pozharski, E., Grishaev, A., Greenspan, D., Tulinsky, A., Llinás, M., Petsko, G., and Ringe, D. (2002). The unassigned NMR protein structure (proton clouds) as a model for molecular replacement. *Biophys. J.* **82,** 470A–471A.

Qi, P. X., Beckman, R. A., and Wand, A. J. (1996). Solution structure of horse heart ferricytochrome c and detection of redox-related structural changes by high-resolution 1H NMR. *Biochemistry* **35,** 12275–12286.

Rinooy Kan, A. H. G., and Timmer, G. T. (1987). Stochastic global optimization methods. *Math. Progr.* **39,** 27–78.

Shannon, C. E. (1948). A mathematical theory of communication. *Bell Sys. Tech. J.* **27,** 379–423.

Solomon, I. (1955). Relaxation processes in a system of two spins. *Phys. Rev.* **99,** 559–565.

Valafar, H., and Prestegard, J. H. (2003). Rapid classification of a protein fold family using a statistical analysis of dipolar couplings. *Bioinformatics* **19,** 1549–1555.

Xu, J., and Borer, P. N. (1994). Rigorous deduction theory for assignment of multidimensional NMR spectra using the independent spin coupling network approach. *J. Chem. Inform. Sci.* **34,** 349–356.

Xu, Y., Wu, J., Goresnstein, D., and Braun, W. (1999). Automated 2D NOESY assignment and structure calculation of crambin (S22/I25) with the self-correcting distance geometry based NOAH/DIAMOD programs. *J. Magn. Reson.* **136,** 76–85.

Zahn, R., Liu, A., Luhrs, T., Calzolai, L., Von Schroetter, C., Garcia, F. L., Riek, R., Wider, G., Billeter, M., and Wüthrich, K. (2000). NMR solution structure of the human prion protein. *Proc. Natl. Acad. Sci. USA* **97,** 145–150.

Zhang, Q., Chen, J. Y., Gozansky, E. K., Zhu, F., Jackson, P. L., and Gorenstein, D. G. (1995). A hybrid-hybrid matrix-method for 3D NOE-NOE data analysis. *J. Magn. Reson. B.* **106,** 164–169.

Zimmerman, D. E., Kulikowski, C. A., Huang, Y. P., Feng, W. Q., Tashiro, M., Shimotakahara, S., Chien, C. Y., Powers, R., and Montelione, G. T. (1997). Automated analysis of protein NMR assignments using methods from artificial intelligence. *J. Mol. Biol.* **269,** 592–610.

Section III

Challenging Systems for NMR

[11]   Elucidation of the Protein Folding Landscape by NMR

By H. JANE DYSON and PETER E. WRIGHT

Abstract

NMR is one of the few experimental methods that can provide detailed insights into the structure and dynamics of unfolded and partly folded states of proteins. Mapping the protein folding landscape is of central importance to understanding the mechanism of protein folding. In addition, it is now recognized that many proteins are intrinsically unstructured in their functional states, while partly folded states of several cellular proteins have been implicated in amyloid disease. NMR is uniquely suited to characterize the structures present in the conformational ensemble and probe the dynamics of the polypeptide chain in unfolded and partially folded protein states.

Introduction

The multiplicity of states that are sampled by a protein in the process of folding can be characterized as an "energy landscape" (Bryngelson *et al.*, 1995; Onuchic *et al.*, 1995), where, under conditions that favor the folded state, the bottom of a "folding funnel" represents the single, stable folded three-dimensional structure, and the top of the funnel represents the multiplicity of states of the fully unfolded protein. Depending on the particular protein, there may be intermediate states with more or less structure, represented as subsidiary minima on the side of the funnel, and there may be more than one pathway for a given protein molecule to navigate the landscape to the folded state. A schematic illustration of such a folding landscape is shown in Fig. 1. At the top of the funnel are a few representative structures for the ensemble of the unfolded state. Schematic folding trajectories are shown as the free energy of the individual molecules decreases, and at the same time, the number of native interactions and the number of residue contacts increase. The trajectories converge at the saddle point, which represents the transition state, and folding proceeds to the final state at the bottom of the funnel.

This chapter summarizes nuclear magnetic resonance (NMR) techniques that can be applied to the elucidation of the structural and dynamic characteristics of protein molecules within the entire energy landscape.

METHODS IN ENZYMOLOGY, VOL. 394

Copyright 2005, Elsevier Inc.
All rights reserved.
0076-6879/05 $35.00

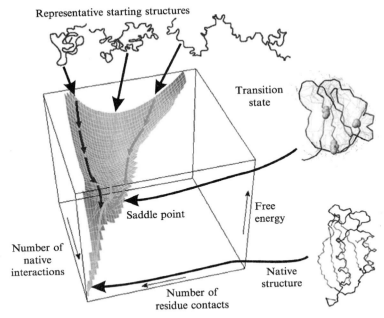

Fig. 1. Schematic energy landscape for protein folding, with a surface derived from a computer simulation of the folding of a highly simplified model of a small protein. The ensemble of denatured conformations is "funneled" to the unique native structure via a saddle point corresponding to the transition state, the barrier that all molecules must cross if they are to fold to the native state. Included in the figure are structure models corresponding to conformational ensembles at different stages of the folding process. The transition state ensemble was calculated from computer simulations constrained by experimental data from mutational studies of acylphosphatase (Vendruscolo et al., 2001). The native structure is shown at the bottom of the surface, and at the top are some members of the ensemble of unfolded species, representing the starting point for folding. Adapted from Dinner et al. (2000) and Dobson (2003), and used with permission. (See color insert.)

Clearly, there is a great deal of information available from high-resolution structure determination techniques such as X-ray crystallography and NMR, for the highly structured states at the bottom of the funnel. Crystallography gives very little information on states that are less than fully ordered: these states comprise a majority of the folding landscape, as illustrated in Fig. 1. In general, mapping the conformational ensemble of unfolded proteins is not possible by X-ray crystallography, because crystals of conformationally disordered molecules are difficult to form, and if formed may not be representative of the structures populated in solution. Spectroscopic methods, especially NMR, are required.

NMR methods for the elucidation of high-resolution three-dimensional structures in solution are relatively well known and will not be addressed here. In this chapter, we survey NMR methods that are applicable to the study of unfolded and partly folded proteins and the types of information that can be obtained. A considerable amount of interest has recently been expressed in the structural characterization of unfolded proteins. A recent issue of *Advances in Protein Chemistry* (volume 62, 2002) was devoted to the study of unfolded proteins. A comprehensive review of the NMR methodology applied to unfolded and partly folded proteins is included in that volume (Dyson and Wright, 2002b), together with other techniques such as Raman optical activity (Barron *et al.*, 2002), fluorescence correlation spectroscopy (Frieden *et al.*, 2002), infrared absorption and vibrational CD (Keiderling and Xu, 2002), and small angle scattering (Millett *et al.*, 2002). Other recent reviews deal with applications of NMR to study the protein folding process (Brockwell *et al.*, 2000; Juneja and Udgaonkar, 2003), and with applications of NMR to systems that fold upon binding to targets (Dyson and Wright, 2002a). A comprehensive recent review targets NMR studies of unfolded and partly folded proteins in many different contexts (Dyson and Wright, 2004). A previous volume (*Methods in Enzymology*, volume 339) contained a review on NMR methods for the elucidation of structure and dynamics in disordered states (Dyson and Wright, 2001). This chapter provides an extension of that earlier work to include recent advances and new NMR methods for the description of the protein folding landscape.

Methods

Information from Chemical Shifts

Early NMR studies of unfolded proteins were hampered by the lack of proton resonance dispersion, a characteristic feature of the spectra of unfolded proteins. However, developments in high-field NMR spectrometers, availability of uniformly and specifically labeled proteins, and the introduction of isotope-edited and triple-resonance pulse sequences have allowed direct characterization of unfolded and partly folded proteins on a residue-specific basis. The dispersion of the ^{13}C and ^{15}N nuclei can be used effectively to make resonance assignments. The resonance frequencies of these nuclei are sensitive to the local amino acid sequence, whereas proton frequencies are more sensitive to secondary structure (Yao *et al.*, 1997). A description of methods for resonance assignments in unfolded and partly folded proteins was given in the earlier review (Dyson and Wright, 2001).

Local propensities for structured conformers within the conformational ensembles of unfolded proteins can be detected with great sensitivity from chemical shift deviations from random coil shifts (also known as "secondary chemical shifts"). Because these deviations are typically small for unfolded proteins, it is important to obtain the most accurate set of random coil chemical shifts, ideally compiled under solution conditions closely approximating those for the protein of interest. Random coil chemical shifts have been compiled under a number of solution conditions, summarized in Table I. For unfolded proteins, the effects of local sequence dependence become significant, particularly for carbonyl shifts. Corrections for local sequence effects can be made using comprehensive tabulations of sequence-dependent corrections to random coil chemical shifts (Schwarzinger et al., 2001; Wang and Jardetzky, 2002).

Variations in secondary chemical shift values provide important insights into the structures populated in the conformational ensemble in incompletely folded proteins. An example of this is shown in Fig. 2, which shows the secondary chemical shifts for $^{13}C^{\alpha}$ and ^{13}CO, corrected for sequence dependence (Schwarzinger et al., 2001), in the NMR spectra of apomyoglobin unfolded at pH 2.3 in the presence and absence of 8 M urea. The secondary chemical shifts in the presence of urea indicate that the polypeptide backbone largely populates the broad β-polyproline II minimum of the Ramachandran plot. When the protein is unfolded at pH 2.3 in the absence of urea, the secondary chemical shifts consistently show a conformational preference for helical backbone conformations in certain areas of the sequence. Secondary chemical shifts, especially when averaged over a sequence of several residues, can be utilized to estimate the percentage of conformers with helical structure in the sequence (Eliezer et al., 2000).

A number of new methods have recently been introduced to decrease the data acquisition time and facilitate the assignment of resonances for both folded and unfolded proteins. New methods for the rapid assignment of backbone resonances include three-dimensional (3D) triple resonance experiments of particular utility for unfolded proteins (Bhavesh et al., 2001). Two-dimensional (2D) zero-quantum and double-quantum $^{1}H-^{15}N$ correlated spectra obtained with band selective Hadamard-type frequency labeling of an additional nuclear spin (^{13}CO, $^{13}C^{\alpha}$, or $^{13}C^{\beta}$) promise fast data acquisition and sufficient resolution for applications to partially folded proteins (Brutscher, 2004).

Information from Coupling Constants

Unfolded polypeptide chains frequently have averaged coupling constants. The differences in observed coupling constant values that occur as a result of small propensities for residual backbone secondary structure are frequently too small to be discerned within the limits of experimental error

TABLE I
RANDOM COIL CHEMICAL SHIFT DETERMINATIONS

Molecule	Nucleus	Solvent	pH	Concentration	Temperature	C.S. Ref	Reference
GGXGG	^{13}C	H_2O	Various	Various	33.5°	CS_2	Keim et al., 1973a
GGXGG	^{13}C	H_2O	Various	Various	33.5°	CS_2	Keim et al., 1973b
GGXGG	^{13}C	H_2O	Various	Various	33.5°	CS_2	Keim et al., 1974
Various	^{13}C	D_2O or DMSO	Various	Various	Various	TMS	Howarth and Lilley, 1978
GGXA	^{13}C	D_2O	Various	100 mM	35°	Dioxane	Richarz and Wüthrich, 1978
GGXA	^{1}H	90% H_2O/10% D_2O or D_2O	Various	50 mM	35°	TSP	Bundi and Wüthrich, 1979
Amino acids	^{13}C	D_2O	7.0	10–20%	40 ± 2°	Dioxane	Surprenant et al., 1980
GGXA	^{1}HN	90% H_2O/10% D_2O	3.0	2 mM	0°, 24°	TSP	Jimenez et al., 1986
GGXA	^{15}N	90% H_2O/10% D_2O, 20 mM acetate	5.0	100 mM	35°	TSP, NH_3	Braun et al., 1994
GGXGG	^{13}C	D_2O, and acetonitrile, TFE mixtures	2.0–3.5	37.5 mg/ml	25°	TSP	Thanabal et al., 1994
GGX(A,P)GG	^{1}H,^{13}C,^{15}N	99.9% D_2O, 1.0 M deuterated urea, 50 mM PO_4, 95% H_2O/5% D_2O	5.0 ± 0.3	20 mM	25°	DSS	Wishart et al., 1995
GGXGG	^{1}H	90% H_2O/10% D_2O H_2O/deuterated trifluoroethanol	5.0 ± 0.1	30 mM	5–55°	TSP, DSS	Merutka et al., 1995
GGXGG	^{1}H	90% H_2O/10% D_2O, 50 mM PO_4, GuHCl	5.0 ± 0.1	5–140 mg/ml	20°	DSS	Plaxco et al., 1997
GGpXGG (phosphorylated Ser, Thr, Tyr)	^{1}H, ^{13}C, ^{15}N, ^{31}P	90% H_2O/10% D_2O	2–9		25°	DSS	Bienkiewicz and Lumb, 1999
GGXGG	^{1}H, ^{13}C, ^{15}N	90% H_2O/10% D_2O, 8 M urea	2.3	20 mM	20°	DSS	Schwarzinger et al., 2000
GGXA	^{1}H	80% H_2O, 20% D_2O, 50 mM PO_4, 1 μM NaN_3, variable pressure	5.00, 5.40	5 mM	32°	DSS	Arnold et al., 2002

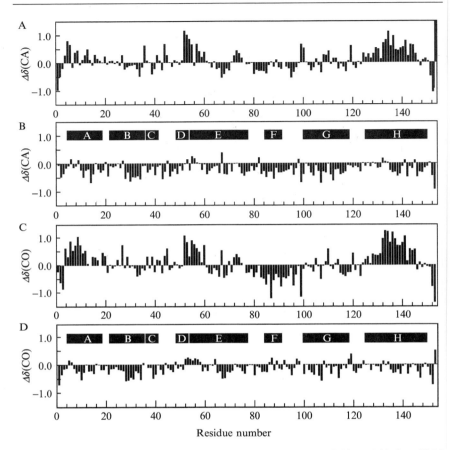

FIG. 2. Sequence-corrected secondary chemical shifts for apomyoglobin unfolded at pH 2.3 (A and C) and at pH 2.3 in 8 M urea (B and D). Secondary shifts for $^{13}C^{\alpha}$ (A and B) and ^{13}CO (C and D). Data from Yao *et al.* (2001) and Schwarzinger *et al.* (2002).

in the measurements. Nevertheless, "random coil" distributions of dihedral angles have also been tabulated (Fiebig *et al.*, 1996; Serrano, 1995; West and Smith, 1998); using these data, estimates have been made of the presence of residual structure from the measurement of coupling constants (West and Smith, 1998).

Dynamic Information from Relaxation Data

NMR provides unique information on the internal motions of polypeptides. Amplitudes and frequencies of backbone and side chain motions on a wide variety of time scales can be obtained using NMR relaxation measurements. The NMR techniques traditionally used for studying polypeptide chain dynamics involve the measurement of T_1, T_2, and heteronuclear NOE for backbone ^{15}N resonances, with analysis of the data using the so-called "model free" approach (Lipari and Szabo, 1982a,b). The model-free formalism is not usually valid for highly unfolded proteins, since the assumption of a single overall correlation time and the temporal deconvolution of internal motions and molecular tumbling is invalid. A distribution of correlation times can be introduced to correct for this problem (Buevich and Baum, 1999; Buevich et al., 2001). Model-free calculations have been used to analyze relaxation measurements on unfolded states (Alexandrescu and Shortle, 1994; Buck et al., 1996), but reduced spectral density mapping (Peng and Wagner, 1992; Farrow et al., 1995) provides a more rigorous methodology for unfolded and partly folded proteins. An example of such an analysis is shown in Fig. 3. Unfolded apomyoglobin at pH 2.3 is highly flexible, as shown by both the measured relaxation parameters (Yao et al., 2001) and the reduced spectral density functions derived from them [Fig. 2 (Yao et al., 2001)]. The spectral densities $J(\omega_N)$ (Fig. 3B) and $J(0.87\omega_H)$ (Fig. 3C) are sensitive only to motions on a picosecond to nanosecond time scale. Sequence-dependent variations in spectral densities indicative of changes in fast time scale motions are most obvious in $J(\omega_N)$, where increased values are observed in several contiguous regions of the backbone. This was ascribed to motional restriction arising from the presence of residual helical structure (consistent with the evidence presented in Fig. 2 for the location of helical backbone dihedral angles). The best correlation found for the trends in these data (Yao et al., 2001) was to the local buried surface area parameter (Rose et al., 1985). In addition, there are increased values of $J(0)$ in certain areas of the sequence. This was interpreted as evidence of transient contacts between opposite ends of the polypeptide (Yao et al., 2001), a conclusion that has been supported by evidence from spin label studies (Lietzow et al., 2002).

The measurement of side chain dynamics has added a new dimension to the overall picture of motion in polypeptides, in both folded and unfolded states. While studies of side chain dynamics in folded proteins are relatively straightforward, using direct ^{13}C relaxation measurements performed at natural abundance (Palmer et al., 1991) or with labels at specific carbon sites (Nicholson et al., 1992), or by measuring ^2H relaxation rates of methyl group deuterons (Millet et al., 2002; Muhandiram et al., 1995; Yang et al., 1998), side chain relaxation experiments are much more challenging for unfolded proteins because of the lack of dispersion of side chain resonances. Kay and co-workers have introduced pulse sequences for

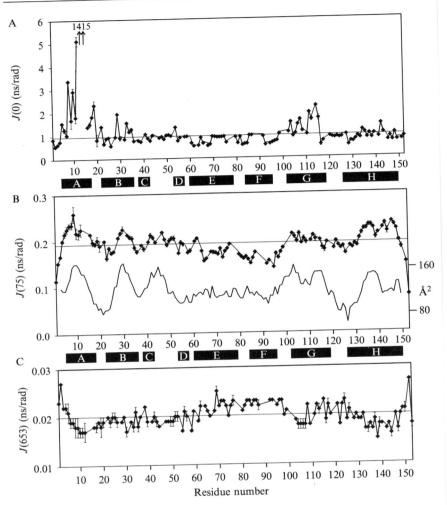

FIG. 3. Calculated values of (A) $J(0)$, (B) $J(75)$, and (C) $J(653)$ for acid-unfolded apomyoglobin, pH 2.3, 25°. (Inset) Average buried surface area, calculated using the values of Rose *et al.* (1985) and averaged over a seven-residue window (right-hand scale). Black bars indicate the positions of helices in the native folded protein. Reproduced from Yao *et al.* (2001), with permission.

probing methyl side chain dynamics in unfolded proteins by transferring methyl ^{2}H relaxation properties to the backbone amide, where they can be detected in the well-dispersed ^{1}H–^{15}N correlation spectrum (Choy et al., 2003; Muhandiram et al., 1997). Measurements of the dynamics of methyl-containing side chains can provide valuable insights into hydrophobic interactions involving aliphatic groups in unfolded states. Insights into the dynamics and interactions of aromatic side chains in unfolded proteins can be obtained from measurement of ^{1}H–^{13}C dipole–dipole cross-correlated relaxation rates (Yang et al., 1999).

A combination of data from backbone ^{15}N and methyl ^{2}H relaxation measurements has recently been used to probe the changes that occur in a fragment of staphylococcal nuclease, termed $\Delta131\Delta$, upon going from a pH of 5 to 3. This fragment is largely unfolded at pH 5, with a significant amount of residual structure, established by backbone relaxation rate measurements and paramagnetic relaxation enhancement of backbone amide proteins (Ackerman and Shortle, 2002; Alexandrescu and Shortle, 1994; Gillespie and Shortle, 1997a,b). A useful index of the stability of this residual structure was provided by the correlation of the amplitudes of motion of the methyl symmetry axis and backbone ^{1}H–^{15}N bond vectors within the same residue (Choy et al., 2003). When the pH is lowered to 3, this residual structure is lost. The amplitudes of the fast motions are not significantly affected, but hydrophobic contacts that stabilize the residual structure at pH 5 are lost at pH 3, giving rise to increased amplitude of motions on a nanosecond time scale (Choy and Kay, 2003).

An important recent innovation for the measurement of microsecond to millisecond time scale motions is spin–spin relaxation rate (R_2) dispersion (Palmer et al., 2001), where exchange contributions to linewidth are directly estimated by changing the strength of the applied radiofrequency field. Information on both backbone (Ishima and Torchia, 2003; Tollinger et al., 2001) and side chain (Mulder et al., 2001, 2002; Skrynnikov et al., 2001) conformational fluctuations in folded proteins can be obtained by this method, which has also found application in the study of unfolded states (Tollinger et al., 2001). Site-specific information on the exchange rates, populations, and chemical shift differences between states in exchange on the microsecond to millisecond time scale is obtained using the relaxation dispersion method. Application of ^{15}N R_2 dispersion measurements to an equilibrium mixture of folded and unfolded states of an SH3 domain yielded folding and unfolding rate constants and revealed intermediate time scale conformational dynamics within a hydrophobic region of the unfolded polypeptide (Tollinger et al., 2001). These measurements have recently been extended to identify low-population folding intermediates in equilibrium with the unfolded and folded states

(Korzhnev *et al.*, 2004) Line shape simulations for exchange-broadened resonances provide a valuable approach for determining folding and un-folding kinetics for fast-folding globular proteins (Huang and Oas, 1995).

Distance Information: NOEs and Spin Labels

High-resolution solution structures of well-folded proteins are deter-mined primarily by the application of distance information in the form of NOEs and dihedral angle information from coupling constants. The addi-tion of information from residual dipolar couplings (Tjandra, 1999; Tjandra and Bax, 1997) provides important long-range information. All of these methods can be applied to unfolded and partly folded proteins, but the information obtained, while unique and difficult to obtain by any other means, is much less specific.

In general, unfolded and partly folded proteins give rise to an extensive array of short and medium-range NOEs. Intraresidue NOEs are generally very strong but are not generally useful in the elucidation of residual structure in an ensemble, since they do not change much between confor-mers with different structures. Sequential NOEs can provide useful infor-mation, particularly on the presence of residual secondary structure. In practice, this means the presence of helical-type conformers, which are indicated by the observation of stronger $d_{NN}(i, i + 1)$ NOEs relative to the $d_{\alpha N}(i, i + 1)$ NOE. It is very difficult to verify the presence of β-structure for two reasons. First, β-structure is inherently long-range; the formation of even a two-stranded β-sheet should be regarded as the for-mation of tertiary structure rather than secondary structure, and implies a level of order that is rarely, if ever, seen in unfolded protein ensembles, though partial β-sheets have been implicated in folding transition states (Vendruscolo *et al.*, 2001). Second, the preferred backbone dihedral angles in unfolded states are in the general area of the large β minimum of the Ramachandran plot, which encompasses both β and polyproline II back-bone conformations. Thus, it may be problematic to distinguish a genuine propensity for β-structure [indicated by sequential $d_{\alpha N}(i, i + 1)$ NOEs] from the conformational ensemble of the unfolded protein [which also give rise to sequential $d_{\alpha N}(i, i + 1)$ NOEs].

The presence of helical or turn-like conformers is confirmed by obser-vation of medium range NOEs, typically $d_{\alpha N}(i, i + 3)$ or $d_{\alpha N}(i, i + 2)$ NOE connectivities, which can frequently be observed for peptides and un-folded proteins. However, detection and assignment of long-range NOEs in unfolded proteins are extremely difficult, despite the availability of pulse sequences designed to detect them (Zhang *et al.*, 1997a,b). No well-authenticated examples of the detection of long-range NOEs in unfolded proteins have yet been reported. It is likely that for most systems so far studied, either the population of the transiently structured forms is too low, or the ensemble containing them is too heterogeneous, for the NOE to be observable. Other NMR evidence for transient long-range interactions, for example, from relaxation data and paramagnetic relaxation enhancement from covalently attached spin labels, is strong in several cases.

Covalently attached paramagnetic nitroxide spin labels cause enhanced relaxation of nuclear spins within a radius of about 15 Å; they were originally used to determine interatomic distances in folded proteins (Kosen et al., 1986; Schmidt and Kuntz, 1984). The method was recently extended to the detection of transient contacts in the conformational ensembles of incompletely folded proteins (Gillespie and Shortle, 1997a,b). The method was reviewed in the previous article in this series (Dyson and Wright, 2001). A spin label reagent, for example, PROXYL (1-oxyl-2,2,5,5-tetramethyl-3-pyrrolidinyl) is attached to a cysteine side chain introduced at a site chosen to minimize influence on the structure of the conformational ensemble. Distance information is obtained by assessing the broadening effect of the spin label on the spectrum, comparing it with a spectrum where the spin label has been reduced to the diamagnetic state. Distance information from the spin label site is provided by differences in the linewidth, relaxation rates, or intensity in these two spectra. Results may be analyzed either qualitatively (Lietzow et al., 2002) or by a more quantitative analysis that gives actual distance ranges (Gillespie and Shortle, 1997b). An example of information from spin label experiments on unfolded and partly folded proteins is shown in Fig. 4. Broadening of resonances in the spectra of spin-labeled apomyoglobin at pH 2.3 provides evidence of interactions within the C-terminal 50 residues, indicating that sequences corresponding to the G- and H-helices of the fully folded protein are frequently in contact under these conditions, consistent with the evidence of residual helical character in the H-helix sequence (Fig. 2). In addition, weaker transient long-range interactions are observed between the N- and C-terminal regions (Fig. 4B), confirming the evidence from $J(0)$ (Fig. 3). Perhaps most significantly, there are many regions of the molecule (for example, the E-helix, Fig. 4A) where there is no evidence of transient long-range interactions. Thus, the NMR information on this unfolded state of a protein rather consistently provides a picture of an ensemble of rapidly interconverting conformers, which nevertheless shows specific areas of residual secondary structure and explicitly native-like transient long-range contacts. These results provide an experimental picture of likely initial steps in the collapse leading to the folding of the protein.

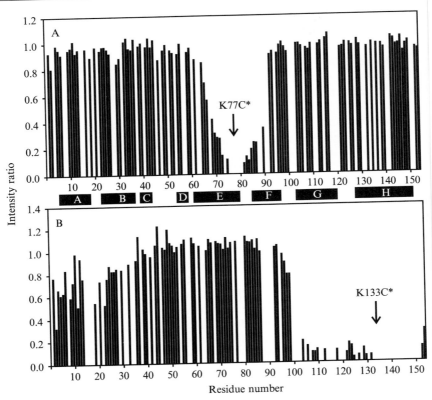

FIG. 4. Paramagnetic enhancement to nuclear spin relaxation for unfolded apomyoglobin at pH 2.3. The histograms show the experimental intensity ratios ($I = I_{para}/I_{diá}$) for each residue with an adequately resolved cross-peak in the ^1H–^{15}N HSQC spectrum of (A) K77C* and (B) K133C*. Adapted from Lietzow et al. (2002), with permission.

Residual Dipolar Couplings and Long-Range Structural Information

Residual dipolar couplings in partially aligned media (Tjandra and Bax, 1997; Tolman et al., 1995) provide extremely useful long-range information on the structure of proteins. Recent reviews have given a detailed description of the experimental methodology and the relationship between residual dipolar couplings and macromolecular structure (Bax, 2003; MacDonald and Lu, 2002; Tolman, 2001). A number of methods have been used to achieve partial alignment of macromolecules in solution: by direct induction in the magnetic field (Prestegard, 1998), by the use of dilute solutions of lipid bicelles (Tjandra and Bax, 1997), using filamentous bacteriophages (Clore *et al.*, 1998), or by incorporation of the sample in stressed polyacrylamide gels (Sass *et al.*, 2000; Tycko *et al.*, 2000). Until recently, dipolar couplings were almost exclusively used for refinement of protein structures determined in solution by NMR, and they have been particularly useful for molecules where the overall topology of the molecule is not well determined. Residual dipolar couplings have been used to provide definitive information on the relative orientation of independently folded protein domains (Al Hashimi *et al.*, 2000; Hudson *et al.*, 2004; Tsui *et al.*, 2000) or the bending of nucleic acid structures (Wu *et al.*, 2003).

Dipolar couplings have the potential to provide valuable information on the overall topology for unfolded and partly folded states, without the need for NOEs or spin labels (Bax, 2003). It is important that the media used to obtain partial alignment do not interact with the unfolded or partly folded proteins of interest, with consequent loss of signal. Stressed polyacrylamide gels have recently been used to study unfolded states (Ackerman and Shortle, 2002; Borges *et al.*, 2004; Ohnishi *et al.*, 2004; Shortle and Ackerman, 2001). The results for an unfolded fragment of staphylococcal nuclease (Ackerman and Shortle, 2002; Shortle and Ackerman, 2001) and for unfolded eglin c (Ohnishi *et al.*, 2004) showed nonzero dipolar couplings even for the most denatured states of the proteins; this was interpreted as evidence for overall native-like topologies for the denatured states of these proteins. An alternative explanation has been given by Annila and co-workers (Louhivuori *et al.*, 2003), who attributed the occurrence of measurable residual dipolar couplings in denatured proteins to the intrinsic properties of random flight polypeptide chains. Nonzero dipolar couplings were also observed for unfolded states of apomyoglobin, at pH 2.3, in the presence and absence of high concentrations of urea (Borges *et al.*, 2004), and for denatured forms of ACBP (Fieber *et al.*, 2004). Representative results for apomyoglobin are shown in Fig. 5. It is noticeable that the couplings for the apomyoglobin in 8 M urea (Fig. 5A) are of uniform sign, compared with those for the acid denatured protein in the absence of urea (Fig. 5B), which show a sign change specifically in areas where residual helical secondary structure has been found (Yao *et al.*, 2001) (Fig. 2). These results have been interpreted (Borges *et al.*, 2004) on the basis of a model in which the unfolded protein behaves as a chain of jointed statistical segments. For a folded protein, the magnitude and sign of the residual dipolar couplings depend on the orientation of each bond vector relative to the alignment tensor of the entire molecule, which reorients as a single entity. For fully unfolded proteins, however, the residual dipolar couplings can be explained by alignment of the statistical segments. Within fully unfolded states, backbone dihedral angles preferentially populate the β- and polyproline II regions of ϕ, ψ space, and the N–H bond

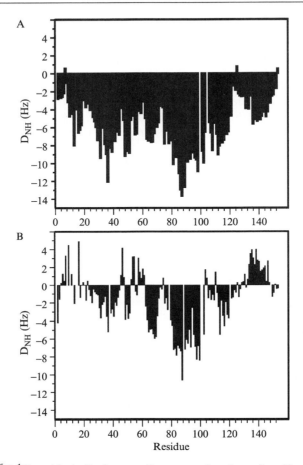

FIG. 5. ^{15}N–^{1}H residual dipolar couplings as a function of residue number for apomyoglobin in 8 M urea, pH 2.3 (A) and for acid unfolded apomyoglobin at pH 2.3 in a stretched polyacrylamide gel (B). Adapted from Borges *et al.* (2004), with permission.

vectors are approximately perpendicular to the long axis of the diffusion tensor for the statistical segment. A small population of helical structure, as seen in the acid-denatured state of apomyoglobin, leads to a change in sign of the residual dipolar couplings in local regions of the polypeptide; the population of helix estimated from the residual dipolar couplings is in excellent agreement with that determined from chemical shifts. Thus, residual dipolar couplings can provide valuable insights into the dynamic conformational propensities of unfolded and partly folded states of proteins.

NMR Under Pressure

Increasing the pressure can perturb the ensemble of conformational states sampled by a protein in solution, due to the presence in the ensemble of conformers with a variety of effective volumes, in rapid equilibrium. A pressure increase shifts the conformational equilibrium in the direction of lower-volume conformers and may stabilize higher energy conformers under normal conditions of pH, temperature, and denaturant concentration; these higher energy states can approximate folding intermediates and unfolded states of the protein. This technique has been used to obtain

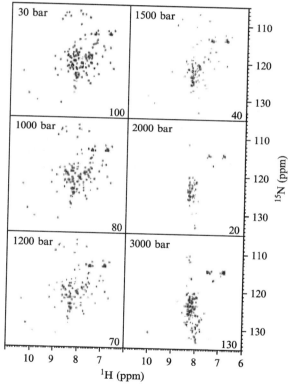

Fig. 6. 750 MHz ^1H–^{15}N HSQC spectra of apomyoglobin at various pressures from 30 to 3000 bar at 35°. The protein solution was prepared in 20 mM MES buffer, 95%^1H_2O/5% ^2H_2O, pH 6.0. Ethanol was added to the sample solution (ethanol/water = 10%/90%) for measurements at 3000 bar to retard the aggregation of the protein. The number in each panel indicates the relative integrated volume of all the cross-peaks normalized to 100 at 30 bar. Reproduced from Kitahara et al. (2002), with permission.

valuable insights into folding intermediates in a number of proteins, including lysozyme (Refaee *et al.*, 2003), α-lactalbumin (Lassalle *et al.*, 2003), and ubiquitin (Kitahara and Akasaka, 2003). A series of NMR spectra of apomyoglobin at increasing pressures from 30 to 3000 bar (Fig. 6) shows changes that corresponded well with the presence of native protein at low pressure, increasing proportions of the molten globule at higher pressures, and the formation of the unfolded protein at the highest pressures. Most recently, high-pressure techniques have been brought to bear on the problem of amyloid and prion protein structure (Kuwata *et al.*, 2002, 2004; Niraula *et al.*, 2002, 2004). Detailed reviews of variable-pressure NMR have recently been published (Akasaka, 2003a,b).

Real-Time NMR and the Folding Process

The combined use of rapid mixing techniques and NMR was used to monitor the folding of α-lactalbumin in real time (Balbach, 2000; Balbach *et al.*, 1995, 1996, 1997). These and similar studies on a wide variety of proteins confirm the highly cooperative nature of the protein folding process, with probes at all sites in the molecule displaying identical folding kinetics. More recently, real-time information on the folding process has been obtained from time-resolved photo-CIDNP (chemically induced dynamic nuclear polarization) NMR, in which laser pulses are used to excite a dye present in the protein solution as it sits in the NMR probe (Mok *et al.*, 2003). Tyrosine, histidine, and tryptophan side chains in the protein are excited, forming short-lived radical pairs, which can be detected by NMR. The folding of several proteins has been explored by this method, including α-lactalbumin (Wirmer *et al.*, 2001), lysozyme (Lyon *et al.*, 1999), and single-tryptophan mutants of HPr (Canet *et al.*, 2003). There are a number of detailed reviews on real-time methods (Dobson and Hore, 1998; Mok *et al.*, 2003; van Nuland *et al.*, 1998).

Conclusions

Continuing innovation is the hallmark of the macromolecular NMR field. As new problems arise, for example, increasing interest in unfolded and partly folded proteins, new techniques or refinements and adjustments of old ones are made to provide the required information. NMR remains unique in its ability to provide information on the entire protein folding landscape, and even, with clever experimentation, on actual processes of folding, assembly, and catalysis. Over the next few years, NMR will undoubtedly continue to play a major role in mapping the energy landscape of protein folding and characterization of the conformational ensemble of intrinsically disordered proteins.

References

Ackerman, M. S., and Shortle, D. (2002). Molecular alignment of denatured states of staphylococcal nuclease with strained polyacrylamide gels and surfactant liquid crystalline phases. *Biochemistry* **41**, 3089–3095.

Akasaka, K. (2003a). Exploring the entire conformational space of proteins by high-pressure NMR. *Pure Appl. Chem.* **75**, 927–936.

Akasaka, K. (2003b). Highly fluctuating protein structures revealed by variable-pressure nuclear magnetic resonance. *Biochemistry* **42**, 10875–10885.

Alexandrescu, A. T., and Shortle, D. (1994). Backbone dynamics of a highly disordered 131 residue fragment of staphylococcal nuclease. *J. Mol. Biol.* **242**, 527–546.

Al Hashimi, H. M., Valafar, H., Terrell, M., Zartler, E. R., Eidsness, M. K., and Prestegard, J. H. (2000). Variation of molecular alignment as a means of resolving orientational ambiguities in protein structures from dipolar couplings. *J. Magn. Reson.* **143**, 402–406.

Arnold, M. R., Kremer, W., Ludemann, H. D., and Kalbitzer, H. R. (2002). 1H-NMR parameters of common amino acid residues measured in aqueous solutions of the linear tetrapeptides Gly-Gly-X-Ala at pressures between 0.1 and 200 MPa. *Biophys. Chem.* **96**, 129–140.

Balbach, J. (2000). Compaction during protein folding studied by real-time NMR diffusion experiments. *J. Am. Chem. Soc.* **122**, 5887–5888.

Balbach, J., Forge, V., van Nuland, N. A. J., Winder, S. L., Hore, P. J., and Dobson, C. M. (1995). Following protein folding in real time using NMR spectroscopy. *Nat. Struct. Biol.* **2**, 865–870.

Balbach, J., Forge, V., Lau, W. S., van Nuland, N. A. J., Brew, K., and Dobson, C. M. (1996). Protein folding monitored at individual residues during a two-dimensional NMR experiment. *Science* **274**, 1161–1163.

Balbach, J., Forge, V., Lau, W. S., Jones, J. A., van Nuland, N. A. J., and Dobson, C. M. (1997). Detection of residue contacts in a protein folding intermediate. *Proc. Natl. Acad. Sci. USA* **94**, 7182–7185.

Barron, L. D., Blanch, E. W., and Hecht, L. (2002). Unfolded proteins studied by raman optical activity. *Adv. Protein Chem.* **62**, 51–90.

Bax, A. (2003). Weak alignment offers new NMR opportunities to study protein structure and dynamics. *Protein Sci.* **12**, 1–16.

Bhavesh, N. S., Panchal, S. C., and Hosur, R. V. (2001). An efficient high-throughput resonance assignment procedure for structural genomics and protein folding research by NMR. *Biochemistry* **40**, 14727–14735.

Bienkiewicz, E. A., and Lumb, K. J. (1999). Random-coil chemical shifts of phosphorylated amino acids. *J. Biomol. NMR* **15**, 203–206.

Borges, R. M., Goto, N. K., Kroon, G., Dyson, H. J., and Wright, P. E. (2004). Structural characterization of unfolded states of apomyoglobin using residual dipolar couplings. *J. Mol. Biol.* **340**, 1131–1142.

Braun, D., Wider, G., and Wüthrich, K. (1994). Sequence-corrected ^{15}N "random coil" chemical shifts. *J. Am. Chem. Soc.* **116**, 8466–8469.

Brockwell, D. J., Smith, D. A., and Radford, S. E. (2000). Protein folding mechanisms: New methods and emerging ideas. *Curr. Opin. Struct. Biol.* **10**, 16–25.

Brutscher, B. (2004). Combined frequency- and time-domain NMR spectroscopy. Application to fast protein resonance assignment. *J. Biomol. NMR* **29**, 57–64.

Bryngelson, J. D., Onuchic, J. N., Socci, N. D., and Wolynes, P. G. (1995). Funnels, pathways, and the energy landscape of protein folding: A synthesis. *Proteins* **21**, 167–195.

Buck, M., Schwalbe, H., and Dobson, C. M. (1996). Main-chain dynamics of a partially folded protein: ^{15}N NMR relaxation measurements of hen egg white lysozyme denatured in trifluoroethanol. *J. Mol. Biol.* **257,** 669–683.

Buevich, A. V., and Baum, J. (1999). Dynamics of unfolded proteins: Incorporation of distributions of correlation times in the model free analysis of NMR relaxation data. *J. Am. Chem. Soc.* **121,** 8671–8672.

Buevich, A. V., Shinde, U. P., Inouye, M., and Baum, J. (2001). Backbone dynamics of the natively unfolded pro-peptide of subtilisin by heteronuclear NMR relaxation studies. *J. Biomol. NMR* **20,** 233–249.

Bundi, A., and Wüthrich, K. (1979). ^{1}H-NMR parameters of the common amino acid residues measured in aqueous solution of the linear tetrapeptides H-Gly-Gly-X-L-Ala-OH. *Biopolymers* **18,** 285–297.

Canet, D., Lyon, C. E., Scheek, R. M., Robillard, G. T., Dobson, C. M., Hore, P. J., and van Nuland, N. A. J. (2003). Rapid formation of non-native contacts during the folding of HPr revealed by real-time photo-CIDNP NMR and stopped-flow fluorescence experiments. *J. Mol. Biol.* **330,** 397–407.

Choy, W. Y., and Kay, L. E. (2003). Probing residual interactions in unfolded protein states using NMR spin relaxation techniques: An application to delta131delta. *J. Am. Chem. Soc.* **125,** 11988–11992.

Choy, W. Y., Shortle, D., and Kay, L. E. (2003). Side chain dynamics in unfolded protein states: An NMR based 2H spin relaxation study of delta131delta. *J. Am. Chem. Soc.* **125,** 1748–1758.

Clore, G. M., Starich, M. R., and Gronenborn, A. M. (1998). Measurement of residual dipolar couplings of macromolecules aligned in the nematic phase of a colloidal suspension of rod-shaped viruses. *J. Am. Chem. Soc.* **120,** 10571–10572.

Dinner, A. R., Sali, A., Smith, L. J., Dobson, C. M., and Karplus, M. (2000). Understanding protein folding via free-energy surfaces from theory and experiment. *Trends Biochem. Sci.* **25,** 331–339.

Dobson, C. M. (2003). Protein folding and misfolding. *Nature* **426,** 884–890.

Dobson, C. M., and Hore, P. J. (1998). Kinetic studies of protein folding using NMR spectroscopy. *Nat. Struct. Biol.* **5,** 504–507.

Dyson, H. J., and Wright, P. E. (2001). Nuclear magnetic resonance methods for elucidation of structure and dynamics of disordered states. *Methods Enzymol.* **339,** 258–270.

Dyson, H. J., and Wright, P. E. (2002a). Coupling of folding and binding for unstructured proteins. *Curr. Opin. Struct. Biol.* **12,** 54–60.

Dyson, H. J., and Wright, P. E. (2002b). Insights into the structure and dynamics of unfolded proteins from nuclear magnetic resonance. *Adv. Protein Chem.* **62,** 311–340.

Dyson, H. J., and Wright, P. E. (2004). Unfolded proteins and protein folding studied by NMR. *Chem. Rev.* **104,** 3607–3622.

Eliezer, D., Chung, J., Dyson, H. J., and Wright, P. E. (2000). Native and non-native structure and dynamics in the pH 4 intermediate of apomyoglobin. *Biochemistry* **39,** 2894–2901.

Farrow, N. A., Zhang, O., Szabo, A., Torchia, D. A., and Kay, L. E. (1995). Spectral density function mapping using ^{15}N relaxation data exclusively. *J. Biomol. NMR* **6,** 153–162.

Fiebig, K. M., Schwalbe, H., Buck, M., Smith, L. J., and Dobson, C. M. (1996). Toward a description of the conformations of denatured states of proteins. Comparison of a random coil model with NMR measurements. *J. Phys. Chem.* **100,** 2661–2666.

Fieber, W., Kristjansdottir, S., and Poulsen, F. M. (2004). Short, long range and transition state interactions in the denatured state of ACBP from residual dipolar couplings. *J. Mol. Biol.* **339,** 1191–1199.

Frieden, C., Chattopadhyay, K., and Elson, E. L. (2002). What fluorescence correlation spectroscopy can tell us about unfolded proteins. *Adv. Protein Chem.* **62,** 91–109.

Gillespie, J. R., and Shortle, D. (1997a). Characterization of long-range structure in the denatured state of staphylococcal nuclease. I. Paramagnetic relaxation enhancement by nitroxide spin labels. *J. Mol. Biol.* **268,** 158–169.

Gillespie, J. R., and Shortle, D. (1997b). Characterization of long-range structure in the denatured state of staphylococcal nuclease. II. Distance restraints from paramagnetic relaxation and calculation of an ensemble of structures. *J. Mol. Biol.* **268,** 170–184.

Howarth, O. W., and Lilley, D. M. J. (1978). Carbon-13-NMR of peptides and proteins. *Prog. NMR Spectros. C* **12,** 1–40.

Huang, G. S., and Oas, T. G. (1995). Submillisecond folding of monomeric lambda repressor. *Proc. Natl. Acad. Sci. USA* **92,** 6878–6882.

Hudson, B. P., Martinez-Yamout, M. A., Dyson, H. J., and Wright, P. E. (2004). Recognition of the mRNA AU-rich element by the zinc finger domain of TIS11d. *Nat. Struct. Mol. Biol.* **11,** 257–264.

Ishima, R., and Torchia, D. A. (2003). Extending the range of amide proton relaxation dispersion experiments in proteins using a constant-time relaxation-compensated CPMG approach. *J. Biomol. NMR* **25,** 243–248.

Jimenez, M. A., Nieto, J. L., Rico, M., Santoro, J., Herranz, J., and Bermejo, F. J. (1986). A study of the NH NMR signals of Gly-Gly-X-Ala tetrapeptides in H_2O at low temperature. *J. Mol. Struct.* **143,** 435–438.

Juneja, J., and Udgaonkar, J. B. (2003). NMR studies of protein folding. *Curr. Sci.* **84,** 157–172.

Keiderling, T. A., and Xu, Q. (2002). Unfolded peptides and proteins studied with infrared absorption and vibrational circular dichroism spectra. *Adv. Protein Chem.* **62,** 111–161.

Keim, P., Vigna, R. A., Marshall, R. C., and Gurd, F. R. N. (1973a). Carbon 13 nuclear magnetic resonance of pentapeptides of glycine containing central residues of aliphatic amino acids. *J. Biol. Chem.* **248,** 6104–6113.

Keim, P., Vigna, R. A., Morrow, J. S., Marshall, R. C., and Gurd, F. R. N. (1973b). Carbon 13 nuclear magnetic resonance of pentapeptides of glycine containing central residues of serine, threonine, aspartic and glutamic acids. *J. Biol. Chem.* **248,** 7811–7818.

Keim, P., Vigna, R. A., Nigen, A. M., Morrow, J. S., and Gurd, F. R. N. (1974). Carbon 13 nuclear magnetic resonance of pentapeptides of glycine containing central residues of methionine, proline, arginine, and lysine. *J. Biol. Chem.* **249,** 4149–4156.

Kitahara, R., and Akasaka, K. (2003). Close identity of a pressure-stabilized intermediate with a kinetic intermediate in protein folding. *Proc. Natl. Acad. Sci. USA* **100,** 3167–3172.

Kitahara, R., Yamada, H., Akasaka, K., and Wright, P. E. (2002). High pressure NMR reveals that apomyoglobin is an equilibrium mixture from the native to the unfolded. *J. Mol. Biol.* **320,** 311–319.

Korzhnev, D. M., Salvatella, X., Vendruscolo, M., Di Nardo, A. A., Davidson, A. R., Dobson, C. M., and Kay, L. E. (2004). Low-populated folding intermediates of Fyn SH3 characterized by relaxation dispersion NMR. *Nature* **430,** 586–590.

Kosen, P. A., Scheek, R. M., Naderi, H., Basus, V. J., Manogaran, S., Schmidt, P. G., Oppenheimer, N. J., and Kuntz, I. D. (1986). Two-dimensional 1H NMR of three spin-labeled derivatives of bovine pancreatic trypsin inhibitor. *Biochemistry* **25,** 2356–2364.

Kuwata, K., Li, H., Yamada, H., Legname, G., Prusiner, S. B., Akasaka, K., and James, T. L. (2002). Locally disordered conformer of the hamster prion protein: A crucial intermediate to PrPSc? *Biochemistry* **41,** 12277–12283.

Kuwata, K., Kamatari, Y. O., Akasaka, K., and James, T. L. (2004). Slow conformational dynamics in the hamster prion protein. *Biochemistry* **43**, 4439–4446.

Lassalle, M. W., Li, H., Yamada, H., Akasaka, K., and Redfield, C. (2003). Pressure-induced unfolding of the molten globule of all-Ala alpha-lactalbumin. *Protein Sci.* **12**, 66–72.

Lietzow, M. A., Jamin, M., Dyson, H. J., and Wright, P. E. (2002). Mapping long-range contacts in a highly unfolded protein. *J. Mol. Biol.* **322**, 655–662.

Lipari, G., and Szabo, A. (1982a). Model-free approach to the interpretation of nuclear magnetic resonance relaxation in macromolecules. 1. Theory and range of validity. *J. Am. Chem. Soc.* **104**, 4546–4559.

Lipari, G., and Szabo, A. (1982b). Model-free approach to the interpretation of nuclear magnetic resonance relaxation in macromolecules. 2. Analysis of experimental results. *J. Am. Chem. Soc.* **104**, 4559–4570.

Louhivuori, M., Paakkonen, K., Fredriksson, K., Permi, P., Lounila, J., and Annila, A. (2003). On the origin of residual dipolar couplings from denatured proteins. *J. Am. Chem. Soc.* **125**, 15647–15650.

Lyon, C. E., Jones, J. A., Redfield, C., Dobson, C. M., and Hore, P. J. (1999). Two-dimensional N-15-H-1 photo-CIDNP as a surface probe of native and partially structured proteins. *J. Am. Chem. Soc.* **121**, 6505–6506.

MacDonald, D., and Lu, P. (2002). Residual dipolar couplings in nucleic acid structure determination. *Curr. Opin. Struct. Biol.* **12**, 337–343.

Merutka, G., Dyson, H. J., and Wright, P. E. (1995). 'Random coil' ^{1}H chemical shifts obtained as a function of temperature and trifluoroethanol concentration for the peptide series GGXGG. *J. Biomol. NMR* **5**, 14–24.

Millet, O., Muhandiram, D. R., Skrynnikov, N. R., and Kay, L. E. (2002). Deuterium spin probes of side-chain dynamics in proteins. 1. Measurement of five relaxation rates per deuteron in (13)c-labeled and fractionally (2)h-enriched proteins in solution. *J. Am. Chem. Soc.* **124**, 6439–6448.

Millett, I. S., Doniach, S., and Plaxco, K. W. (2002). Toward a taxonomy of the denatured state: Small angle scattering studies of unfolded proteins. *Adv. Protein Chem.* **62**, 241–262.

Mok, K. H., Nagashima, T., Day, I. J., Jones, J. A., Jones, C. J., Dobson, C. M., and Hore, P. J. (2003). Rapid sample-mixing technique for transient NMR and photo-CIDNP spectroscopy: Applications to real-time protein folding. *J. Am. Chem. Soc.* **125**, 12484–12492.

Muhandiram, D. R., Yamazaki, T., Sykes, B. D., and Kay, L. E. (1995). Measurement of ^{2}H T_1 and $T_{1\rho}$ relaxation times in uniformly ^{13}C-labeled and fractionally ^{2}H-labeled proteins in solution. *J. Am. Chem. Soc.* **117**, 11536–11544.

Muhandiram, D. R., Johnson, P. E., Yang, D., Zhang, O., McIntosh, L. P., and Kay, L. E. (1997). Specific ^{15}N, NH correlations for residues in ^{15}N, ^{13}C and fractionally deuterated proteins that immediately follow methyl-containing amino acids. *J. Biomol. NMR* **10**, 283–288.

Mulder, F. A., Skrynnikov, N. R., Hon, B., Dahlquist, F. W., and Kay, L. E. (2001). Measurement of slow (micros–ms) time scale dynamics in protein side chains by (15)N relaxation dispersion NMR spectroscopy: Application to Asn and Gln residues in a cavity mutant of T4 lysozyme. *J. Am. Chem. Soc.* **123**, 967–975.

Mulder, F. A., Hon, B., Mittermaier, A., Dahlquist, F. W., and Kay, L. E. (2002). Slow internal dynamics in proteins: Application of NMR relaxation dispersion spectroscopy to methyl groups in a cavity mutant of T4 lysozyme. *J. Am. Chem. Soc.* **124**, 1443–1451.

Nicholson, L. K., Kay, L. E., Baldisseri, D. M., Arango, J., Young, P. E., Bax, A., and Torchia, D. A. (1992). Dynamics of methyl groups in proteins as studied by proton-detected ^{13}C NMR spectroscopy. Application to the leucine residues of staphylococcal nuclease. *Biochemistry* **31**, 5253–5263.

Niraula, T. N., Haraoka, K., Ando, Y., Li, H., Yamada, H., and Akasaka, K. (2002). Decreased thermodynamic stability as a crucial factor for familial amyloidotic polyneuropathy. *J. Mol. Biol.* **320,** 333–342.

Niraula, T. N., Konno, T., Li, H., Yamada, H., Akasaka, K., and Tachibana, H. (2004). Pressure-dissociable reversible assembly of intrinsically denatured lysozyme is a precursor for amyloid fibrils. *Proc. Natl. Acad. Sci. USA* **101,** 4089–4093.

Ohnishi, S., Lee, A. L., Edgell, M. H., and Shortle, D. (2004). Direct demonstration of structural similarity between native and denatured eglin c. *Biochemistry* **43,** 4064–4070.

Onuchic, J. N., Wolynes, P. G., Luthey-Schulten, Z., and Socci, N. D. (1995). Toward an outline of the topography of a realistic protein folding funnel. *Proc. Natl. Acad. Sci. USA* **92,** 3626–3630.

Palmer, A. G., Wright, P. E., and Rance, M. (1991). Measurement of relaxation time constants for methyl groups by proton detected heteronuclear NMR spectroscopy. *Chem. Phys. Lett.* **185,** 41–46.

Palmer, A. G., III, Kroenke, C. D., and Loria, J. P. (2001). Nuclear magnetic resonance methods for quantifying microsecond-to-millisecond motions in biological macromolecules. *Methods Enzymol.* **339,** 204–238.

Peng, J. W., and Wagner, G. (1992). Mapping of spectral density functions using heteronuclear NMR relaxation measurements. *J. Magn. Reson.* **98,** 308–332.

Plaxco, K. W., Morton, C. J., Grimshaw, S. B., Jones, J. A., Pitkeathly, M., Campbell, I. D., and Dobson, C. M. (1997). The effects of guanidine hydrochloride on the 'random coil' conformations and NMR chemical shifts of the peptide series GGXGG. *J. Biomol. NMR* **10,** 221–230.

Prestegard, J. H. (1998). New techniques in structural NMR—anisotropic interactions. *Nat. Struct. Biol.* **5,** 517–522.

Refaee, M., Tezuka, T., Akasaka, K., and Williamson, M. P. (2003). Pressure-dependent changes in the solution structure of hen egg-white lysozyme. *J. Mol. Biol.* **327,** 857–865.

Richarz, R., and Wüthrich, K. (1978). Carbon-13 NMR chemical shifts of the common amino acid residues measured in aqueous solutions of the linear tetrapeptides H-Gly-Gly-X-L-Ala-OH. *Biopolymers* **17,** 2133–2141.

Rose, G. D., Geselowitz, A. R., Lesser, G. J., Lee, R. H., and Zehfus, M. H. (1985). Hydrophobicity of amino acid residues in globular proteins. *Science* **229,** 834–838.

Sass, H. J., Musco, G., Stahl, S. J., Wingfield, P. T., and Grzesiek, S. (2000). Solution NMR of proteins within polyacrylamide gels: Diffusional properties and residual alignment by mechanical stress or embedding of oriented purple membranes. *J. Biomol. NMR* **18,** 303–309.

Schmidt, P. G., and Kuntz, I. D. (1984). Distance measurements in spin-labeled lysozyme. *Biochemistry* **23,** 4261–4266.

Schwarzinger, S., Kroon, G. J. A., Foss, T. R., Wright, P. E., and Dyson, H. J. (2000). Random coil chemical shifts in acidic 8 M urea: Implementation of random coil chemical shift data in NMR View. *J. Biomol. NMR* **18,** 43–48.

Schwarzinger, S., Kroon, G. J. A., Foss, T. R., Chung, J., Wright, P. E., and Dyson, H. J. (2001). Sequence dependent correction of random coil NMR chemical shifts. *J. Am. Chem. Soc.* **123,** 2970–2978.

Schwarzinger, S., Wright, P. E., and Dyson, H. J. (2002). Molecular hinges in protein folding: The urea-denatured state of apomyoglobin. *Biochemistry* **41,** 12681–12686.

Serrano, L. (1995). Comparison between the ϕ distribution of the amino acids in the protein database and NMR data indicates that amino acids have various ϕ propensities in the random coil conformation. *J. Mol. Biol.* **254,** 322–333.

Shortle, D., and Ackerman, M. S. (2001). Persistence of native-like topology in a denatured protein in 8 M urea. *Science* **293,** 487–489.

Skrynnikov, N. R., Mulder, F. A. A., Hon, B., Dahlquist, F. W., and Kay, L. E. (2001). Probing slow time scale dynamics at methyl-containing side chains in proteins by relaxation dispersion NMR measurements: Application to methionine residues in a cavity mutant of T4 lysozyme. *J. Am. Chem. Soc.* **123,** 4556–4566.

Surprenant, H. L., Sarneski, J. E., Key, R. R., Byrd, J. T., and Reilley, C. N. (1980). Carbon-13 NMR studies of amino acids: Chemical shifts, protonation shifts, microscopic protonation behavior. *J. Magn. Reson.* **40,** 231–243.

Thanabal, V., Omecinsky, D. O., Reily, M. D., and Cody, W. L. (1994). The ^{13}C chemical shifts of amino acids in aqueous solution containing organic solvents: Application to secondary structure characterization of peptides in aqueous trifluoroethanol solution. *J. Biomol. NMR* **4,** 47–59.

Tjandra, N. (1999). Establishing a degree of order: Obtaining high-resolution NMR structures from molecular alignment. *Struct. Fold. Des.* **7,** R205–R211.

Tjandra, N., and Bax, A. (1997). Direct measurement of distances and angles in biomolecules by NMR in a dilute liquid crystalline medium. *Science* **278,** 1111–1114.

Tollinger, M., Skrynnikov, N. R., Mulder, F. A., Forman-Kay, J. D., and Kay, L. E. (2001). Slow dynamics in folded and unfolded states of an SH3 domain. *J. Am. Chem. Soc.* **123,** 11341–11352.

Tolman, J. R. (2001). Dipolar couplings as a probe of molecular dynamics and structure in solution. *Curr. Opin. Struct. Biol.* **11,** 532–539.

Tolman, J. R., Flanagan, J. M., Kennedy, M. A., and Prestegard, J. H. (1995). Nuclear magnetic dipole interactions in field-oriented proteins: Information for structure determination in solution. *Proc. Natl. Acad. Sci. USA* **92,** 9279–9283.

Tsui, V., Zhu, L., Huang, T. H., Wright, P. E., and Case, D. A. (2000). Assessment of zinc finger orientations by residual dipolar coupling constants. *J. Biomol. NMR* **16,** 9–21.

Tycko, R., Blanco, F. J., and Ishii, Y. (2000). Alignment of biopolymers in strained gels: A new way to create detectable dipole-dipole couplings in high-resolution biomolecular NMR. *J. Am. Chem. Soc.* **122,** 9340–9341.

van Nuland, N. A. J., Forge, V., Balbach, J., and Dobson, C. M. (1998). Real-time NMR studies of protein folding. *Acc. Chem. Res.* **31,** 773–780.

Vendruscolo, M., Paci, E., Dobson, C. M., and Karplus, M. (2001). Three key residues form a critical contact network in a protein folding transition state. *Nature* **409,** 641–645.

Wang, Y., and Jardetzky, O. (2002). Investigation of the neighboring residue effects on protein chemical shifts. *J. Am. Chem. Soc.* **124,** 14075–14084.

West, N. J., and Smith, L. J. (1998). Side-chains in native and randon coil protein conformations. Analysis of NMR coupling constants and χ_1 torsion angle preferences. *J. Mol. Biol.* **280,** 867–877.

Wirmer, J., Kuhn, J., and Schwalbe, H. (2001). Millisecond time resolved photo-CIDNP NMR reveals a non-native folding intermediate on the ion-induced refolding pathway of bovine alpha-lactalbumin. *Angew. Chem. Int. Ed. Engl.* **40,** 4248–4251.

Wishart, D. S., Bigam, C. G., Holm, A., Hodges, R. S., and Sykes, B. D. (1995). ^1H, ^{13}C, and ^{15}N random coil NMR chemical shifts of the common amino acids: I. Investigations of nearest neighbor effects. *J. Biomol. NMR* **5,** 67–81.

Wu, Z., Delaglio, F., Tjandra, N., Zhurkin, V. B., and Bax, A. (2003). Overall structure and sugar dynamics of a DNA dodecamer from homo- and heteronuclear dipolar couplings and 31P chemical shift anisotropy. *J. Biomol. NMR* **26,** 297–315.

Yang, D. W., Mittermaier, A., Mok, Y. K., and Kay, L. E. (1998). A study of protein side-chain dynamics from new ^2H auto-correlation and ^{13}C cross-correlation NMR experiments: Application to the N-terminal SH3 domain from drk. *J. Mol. Biol.* **276,** 939–954.

Yang, D. W., Mok, Y. K., Muhandiram, D. R., Forman-Kay, J. D., and Kay, L. E. (1999). ^{1}H-^{13}C dipole-dipole cross-correlated spin relaxation as a probe of dynamics in unfolded proteins: Application to the DrkN SH3 domain. *J. Am. Chem. Soc.* **121**, 3555–3556.

Yao, J., Dyson, H. J., and Wright, P. E. (1997). Chemical shift dispersion and secondary structure prediction in unfolded and partly folded proteins. *FEBS Lett.* **419**, 285–289.

Yao, J., Chung, J., Eliezer, D., Wright, P. E., and Dyson, H. J. (2001). NMR structural and dynamic characterization of the acid-unfolded state of apomyoglobin provides insights into the early events in protein folding. *Biochemistry* **40**, 3561–3571.

Zhang, O., Forman-Kay, J. D., Shortle, D., and Kay, L. E. (1997a). Triple-resonance NOESY-based experiments with improved spectral resolution: Applications to structural characterization of unfolded, partially folded and folded proteins. *J. Biomol. NMR* **9**, 181–200.

Zhang, O., Kay, L. E., Shortle, D., and Forman-Kay, J. D. (1997b). Comprehensive NOE characterization of a partly folded large fragment of staphylococcal nuclease Δ131Δ, using NMR methods with improved resolution. *J. Mol. Biol.* **272**, 9–20.

[12] Membrane Protein Preparation for TROSY NMR Screening

By Changlin Tian, Murthy D. Karra,
Charles D. Ellis, Jaison Jacob, Kirill Oxenoid,
Frank Sönnichsen, and Charles R. Sanders

Abstract

The first steps toward undertaking an NMR structural study of a new protein is very often to purify the protein and then to acquire an HSQC or TROSY NMR spectrum, the quality of which is used to assess the feasibility of an NMR-based structural determination. Relatively few integral membrane proteins (IMPs) have been subjected even to this very preliminary stage of NMR analysis. Here, NMR feasibility testing methods are outlined that are tailored for hexahistidine-tagged IMPs that have been expressed in *Escherichia coli*. Generally applicable protocols are presented for expression testing, purification, and NMR sample preparation. A 2D TROSY pulse sequence that has been optimized for use with IMPs is also presented.

Introduction

In recent years, advances in membrane protein expression and spectroscopic technology have led to impressive milestones in the nuclear magnetic resonance (NMR)-based structural biology of this difficult class of proteins (Arora *et al.*, 2001; Chou *et al.*, 2001; Fernandez *et al.*, 2001, 2004; Hwang *et al.*, 2002; Krueger-Koplin *et al.*, 2004; Opella *et al.*, 2002; Oxenoid *et al.*,

Copyright 2005, Elsevier Inc.
All rights reserved.
0076-6879/05 $35.00

2004; Riek *et al.*, 2000). In this chapter, we present protocols found to be useful for determining the feasibility of conducting NMR-based structural studies of hexahistidine-tagged (His$_6$) integral membrane proteins (IMPs). These protocols have been developed and adapted by our laboratory and applied to about 15 different >10-kDa IMPs. Most of these have been primarily helical in secondary structure, with a number also having multiple transmembrane spans. *Escherichia coli*–based expression systems for these proteins have been developed in the laboratories of our collaborators (see acknowledgments) and ourselves, but are not detailed here. Instead, we focus on generic methods for expression testing, subcellular localization of targets, purification into detergent micelles, and NMR sample preparation. We also describe a variant of the two-dimensional (2D)^{1}H/^{15}N TROSY pulse sequence that we have found to be very useful for membrane proteins.

To Deuterate or Not to Deuterate?

In the protocols presented below it is assumed that the final goal is to acquire a ^{1}H/^{15}N TROSY spectrum of a uniformly ^{15}N-labeled target membrane protein to assess the feasibility of conducting more advanced structural studies using NMR. However, as outlined previously (Sanders and Oxenoid, 2000), two questions must first be addressed. First, is it important to uniformly perdeuterate the protein to reduce dipolar relaxation-based line broadening in the TROSY spectrum? Second, is it important to use perdeuterated detergent in the final stage of purification?

In our experience with diacylglycerol kinase (DAGK, a 40-kDa homotrimer with nine transmembrane helices) in 100-kDa detergent complexes, we have not observed any major improvement in the two-dimensional (2D) TROSY spectrum when DAGK is perdeuterated compared to the fully protonated case. However, it should be noted that perdeuteration *is* essential for three-dimensional (3D) experiments involving ^{13}C with directly attached protons. The lack of an imperative to perdeuterate at this stage of feasibility testing using 2D TROSY is fortuitous, since such labeling adds expense and travail to cell culture methods and also requires a (sometimes elaborate) procedure for back-exchanging amide deuterons for protons (cf. Oxenoid *et al.*, 2004).

Through studies of DAGK, it appears that there is also no need to use perdeuterated detergent for the final stage of purification/NMR sample preparation, since no improvement in the quality of the TROSY spectrum results. Moreover, the ^{1}H signal from the high concentration of protonated detergent can be efficiently filtered out provided that an appropriate version of the TROSY sequence is used (see below). For the same reason, use of perdeuterated buffer (i.e., imidazole) is not required.

Which Detergent to Use for NMR?

The protocols given below are based upon the use of either a very harsh detergent (sodium dodecyl sulfate, SDS) or a "just harsh" detergent (dodecyl-N,N-dimethylglycine, "Empigen") to initially solubilize the membrane proteins in E. coli cell extract and/or inclusion bodies. During purification a switch is made to a milder detergent, dodecylphosphocholine (DPC), for final elution and NMR sample preparation. We have found DPC (the unlabeled form is available from Anatrace) to be very effective at maintaining the solubility of a number of IMPs and very often leads to attractive TROSY spectrum for polytopic membrane proteins. DPC is also attractive because it is available in perdeuterated form (Anatrace and Cambridge Isotopes Lab). Nevertheless, even unlabeled DPC is currently very expensive and should be used with care, because it does destabilize some membrane proteins. For example, less stable mutant forms of DAGK are typically observed to be unfolded or molten globule-like in DPC (although the wild-type and stability enhanced forms of the protein remain folded). Fortunately, the protocols presented below are extremely flexible in that other detergents can be substituted for DPC at the final stages of purification, if desired (cf. Czerski et al., 2000; Nagy et al., 2001; Oxenoid et al., 2004; Sanders and Oxenoid, 2000). For instance, we have adapted these protocols so that lysophospholipids can be used as the detergent for NMR samples [Vinogradova et al. (1998), and see the recent paper by Krueger-Koplin et al. (2004) for the merits of using lysolipid detergents]. No matter which is used, it is critical to maintain a detergent concentration at all times that is well in excess of its peculiar critical micelle concentration (CMC). For both DPC and decylmaltoside, we have found that even though the CMC of these detergents is in the range of 2 mM, a minimal concentration of about 10 mM detergent must be present to elute DAGK at the final stage of purification.

Protocols

Overview of Protocol for Testing the Level of Expression of IMPs in E. coli

For screening a number of different expression systems/conditions for a given His$_6$-tagged IMP we typically grow 12-ml cell cultures under a variety of conditions. For convenience in measuring OD$_{600}$s directly in the culture tube, we use 1-in. round, 4-in. tall, flat-bottomed tubes that fit directly into an inexpensive GENESYS 20 spectrophotometer equipped with a 1-in. test tube holder. The OD$_{600}$s measured under these conditions can be normalized to what they would be for the same solution in a 1-cm sample cell.

From 12-ml cultures, 1 ml is removed both prior to and after induction of expression, and cells are harvested and extracted for SDS–polyacrylamide gel electrophoresis (PAGE)/Western blot analysis. The other 10 ml is harvested after induction and used for semipurification of the His$_6$-tagged IMP by metal ion affinity chromatography (MIAC). The protocols for those procedures are presented below and work for IMPs regardless of whether they are expressed in the membrane or in inclusion bodies.

Protocol for Expression Testing Using SDS–PAGE/Western Blot Analysis of Membrane Proteins in Whole Cell Extracts

1. Before induction, measure the OD$_{600}$ and harvest 1 ml of cell culture by centrifuging at full speed for 1–3 min using a desktop centrifuge. Discard the supernatant. Store the cell pellet at $-80°$.
2. At the end of an appropriate induction time, measure the OD$_{600}$ and harvest another 1-ml sample, as well as the remainder of the cell culture ($\sim$10 ml; centrifuge for 7–10 min; this will be used to do the small-scale purification).
3. Resuspend the thawed cell pellets from both the pre- and postinduction samples using 200 μl of lysis buffer [75 mM trizma base, 300 mM NaCl, 0.2 mM ethylenediaminetetraacetic acid (EDTA), 10 μM butylated hydroxytoluene (BHT)] per OD unit in the parent culture. To obtain comparable loadings on SDS–PAGE gels, the cell pellet samples need to be normalized relative to the amount of cell present (based on culture OD$_{600}$ values).
4. Put the tubes containing the cell suspensions on ice and sonicate using a probe sonicator at moderate power for three bursts (5 s on/5 s off).
5. For each sample, combine 18 μl of the lysate with 6 μl of 4× concentrated SDS–PAGE sample buffer. Load 12 μl per lane and run SDS–PAGE. Two gels are required if both Coomassie staining and Western blotting will be conducted. Standard procedures are used for SDS–PAGE staining. However, IMP samples should *not* be boiled before loading on gels (as is common for SDS–PAGE protocols), because IMPs often aggregate irreversibly upon heating in SDS. The procedure used for Western blotting is given below. It should be kept in mind that many expression systems are leaky, such that the protein of interest may be detected both before and after induction.

Expression Testing Using Small-Scale Semipurification of His₆-Tagged Membrane Proteins

1. To analyze expression using small-scale purification, resuspend the cell pellet from the 10-ml sample in 10 ml of lysis buffer plus 200 μl of lysozyme/DNase/RNase (0.010 g of lysozyme, 0.001 g DNase, and 0.001 g RNase in 1.0 ml lysis buffer) plus 50 μl of 100 mM Mg-acetate. Vortex to resuspend, then tumble at room temperature (RT) for 30 min.
2. Add 1.4 ml of 4% SDS to the 10 ml of lysate (to 0.5% SDS). Tumble at RT for 1 hour. At this point the solution should be clear. Then, slowly dilute/mix with buffer to 0.2% SDS.
3. Add ~0.2 ml of packed Ni-NTA resin (QIAGEN "Super-Flow," preequilibrated with 2 bed volumes of buffer A: 40 mM HEPES, 300 mM NaCl, 10 μM BHT, pH 7.5). Tumble at RT for 1 h.
4. Centrifuge at ~4000 rpm for 10–15 min, carefully pour off the supernatant, and then transfer the resin to a 10 × 1-cm column using 3 × 1 ml buffer containing 0.2% SDS plus 25 mM sodium phosphate, pH 7.2. Because so little resin is being used, air pressure may be needed to induce flow through the column.
5. Elute the protein with 3 × 0.3 ml of 0.5% SDS plus 250 mM imidazole, pH 7.8.
6. Combine 30 μl of the eluate with 10 μl of 4 × SDS–PAGE sample buffer, load 5–10 μl per lane, and run SDS–PAGE followed by silver staining. This protocol is not intended to generate pure protein—there is no step for eluting proteins from the resin that are only weakly associated (unlike the standard purification procedure—see below). However, this protocol does provide a many-fold purification of poly-His-tagged proteins relative to simply running SDS–PAGE on extract from lysed cells, such that it is often much easier to identify the band from the protein of interest than when whole cell extracts are run.

Protocol for Western Blotting Analysis of His₆-Tagged Membrane Proteins

The following is based upon Qiagen/Invitrogen protocols described in their product literature. Qiagen's anti-penta-His is used as the primary antibody, and Southern Biotechnology Associate's goat antimouse IgG/AP-conjugate is used as the secondary antibody. The final chromogenic detection of alkaline phosphatase activity is based on use of nitroblue tetrazolium (NBT)/bromochloroindoyl phosphate (BCIP) substrates.

Use flat-bladed tweezers to handle the membrane and a 3 × 5-in. plastic box. Gently agitate on a platform shaker during each wash/incubation step.

Wash steps should be for at least 10 min. A minimum of 30 ml of solution is needed to properly cover the membrane.

1. Run SDS–PAGE and transfer the proteins on the gel to a nitrocellulose membrane according to the instructions that come with the gel apparatus.
2. Wash the membrane twice in tris-buffered saline (TBS) (20 mM trizma, 140 mM NaCl, pH 7.5).
3. Incubate the membrane for 1 h in blocking buffer (3% electrophoresis-grade bovine serum albumin in TBS). If desired, the membrane can be placed in fresh blocking buffer and stored overnight in a closed container at 4°.
4. Wash the membrane twice in TBSTT (0.1% Tween 20 and 0.2% Triton X-100 in TBS).
5. Wash the membrane once in TBS.
6. Incubate the membrane in blocking buffer containing a 1/2000 dilution of penta-His antibody for 1–2 h.
7. Wash the membrane twice in TBSTT.
8. Wash the membrane once in TBS.
9. Incubate the membrane in blocking buffer containing 1/10,000 dilution of goat antimouse IgG/AP-conjugate for 1–2 h.
10. Wash the membrane four times with TBSTT.
11. Wash the membrane once in TBS.
12. Wash the membrane twice with alkaline phosphatase buffer (100 mM diethanolamine, 100 mM NaCl, 5 mM MgCl$_2$, pH 9.5).
13. Add 1 ml of freshly made 10 mg/ml NBT stock solution (dissolve one Sigma 10-mg NBT tablet in 1 ml of water) and 100 μl of freshly made 50 mg/ml BCIP stock solution (dissolve one Sigma 25-mg BCIP tablet in 0.5 ml dimethylformamide) to 30 ml of alkaline phosphatase buffer. Incubate the membrane in the resulting solution for 10–20 min to allow for color development. Waiting longer results in darker signals.
14. Stop the development by washing the membrane twice in distilled water. Air dry the membrane and photograph. Signals will fade over time.

Protocol for Localizing Expressed IMPs in E. coli

The following protocol is used to determine whether the expressed IMP is in the cell membrane or in inclusion bodies.

1. Prepare lysate from 1 ml of cell culture harvested at the end of induction (see steps 3–5 in the first protocol above).

2. Combine 18 μl of the whole-cell lysate with 6 μl of 4 × SDS–PAGE sample buffer.

3. Centrifuge the remaining sample at full speed for 10 min on a desktop centrifuge.

4. Combine 18 μl of the supernatant with 6 μl of 4 × SDS–PAGE sample buffer.

5. Resuspend the cell pellet using 100 μl H_2O. Then combine 18 μl of the pellet suspension with 6 μl of 4 × sample buffer.

6. Load 12 μl of each sample: whole-cell, supernatant, and cell pellet. Run SDS–PAGE and analyze the gel using Coomassie staining or Western blotting. Membrane proteins localized in the cell membrane will appear in the supernatant; those localized in inclusion bodies will appear in the cell pellet sample. An optional extra step is to spin down the membranes from soluble cellular components using ultracentrifugation.

Preparative-Scale Purification of IMPs Expressed into E. coli Membranes

The following protocol is employed for His$_6$-tagged IMPs that are found in the membrane fraction of *E. coli* after expression. This procedure may also be used when the protein is expressed into inclusion bodies, provided that the bodies can be solubilized by the detergent Empigen (which is harsh, but not as harsh as SDS). Empigen is much easier to use in conjunction with MIAC and is preferred over SDS whenever possible.

1. Harvest cells by centrifuging for 15 min at 4° on a preparative centrifuge. Spin at 90% of the maximum allowed rpm for the rotor or for the centrifuge tubes being used (whichever is lower).

2. Discard the supernatant and transfer the cell pellet (which should be very firm and stick to the bottom of the tube) to a preweighed 50-ml polypropylene Falcon tube and weigh the cells. Note that the tube containing the cell pellet can be frozen in liquid N_2 and stored at −80° if desired.

3. Transfer the cell pellet to a sealable bottle and resuspend the cells using 20 ml of lysis buffer (75 mM trizma base, 300 mM NaCl, 0.2 mM EDTA, 10 μM BHT, pH 7.7) for each gram of cells collected.

4. Add phenylmethylsulfonyl fluoride (PMSF; a poisonous protease inhibitor) from a 20-mg/ml stock solution in isopropanol (which can be stored indefinitely in the freezer) to a concentration of 20 mg/100 ml of suspension (1.1 mM).

5. Add lysozyme (0.2 mg/ml), DNase (0.02 mg/ml), and RNase (0.02 mg/ml). Seal the container and incubate by tumbling (not via a stir-bar) at RT for 30 min.

6. Add magnesium acetate from a 500 mM stock solution to 5 mM.

7. Place the container in an ice bath and tip sonicate at moderate power for 5 min (5 s on/5 s off, repetitively).

8. Add dithiothreitol (DTT) to a concentration of 0.5 mM. It is possible to pause at this point by freezing the lysate using liquid N_2 and storing at $-80°$. We have found that MIAC is usually compatible with very modest concentrations of reducing agents.

9. Cool the lysate solution on ice. For every 10-ml portion of lysate (either fresh or thawed after storage in the freezer) add 1 ml of Empigen (Calbiochem) to 3% by volume. Mix thoroughly (but do not stir with a stir-bar) and allow to incubate for about 15 min on ice prior to the addition of nickel resin. Do not freeze the lysed solution. Empigen is a fairly harsh (denaturing) detergent. However, because it is zwitterionic, it is much more compatible with the use of MIAC than is SDS, which must be used very carefully with MIAC.

10. Centrifuge the extracted lysate with a preparative centrifuge and discard any "goop" pellet (this is important).

11. Qiagen "Superflow" Nickel-Agarose resin should be equilibrated by rinsing with buffer A (40 mM HEPES, 300 mM NaCl, 10 μM BHT, pH 7.5). Use 1.2 ml of packed resin for every gram of cells represented by the lysate. Pack the resin into a column and rinse twice with 2 bed volumes of buffer A.

12. Transfer the resin into a tube containing the Empigen-extracted lysed cells (on ice).

13. Tightly close the lid and rotate the tube for 0.5 h (no longer) in the cold room. During this time the detergent-solubilized protein will bind to the nickel resin.

14. Following incubation, isolate the resin by centrifugation.

15. Pour off the supernatant and either freeze the resin in liquid N_2 and store until later use or transfer it to an appropriately sized column. The height of the packed bed should be more than two times the bed diameter, and the total column volume should be about three to five times the bed volume. Wash the resin with 3 × 1 bed volumes of ice-cold buffer A plus 3% Empigen. Freezing protein-on-resin and storing at $-80°$ usually does no harm to the protein of interest, at least when storage is for only a matter of weeks.

16. Turn on the chart recorder and monitor A_{280}.

17. Wash the column with cold wash buffer: buffer A plus 3% Empigen plus 40 mM imidazole, plus fresh 0.2 mM DTT, pH 7.8. Wash until the "junk" peak has finished eluting (as monitored by the chart recorder). The wash buffer contains enough imidazole to knock proteins off the column that have a weak affinity for the nickel ions and do not have the His$_6$ tag.

18. Rinse the column with 8 × 1 bed volumes of cold rinse buffer: 25 mM sodium phosphate, 0.5% DPC, plus fresh 0.2 mM DTT. The purpose of this buffer is to switch from Empigen to DPC.

19. Elute the protein with elution buffer containing 0.5% DPC plus 250 mM imidazole, pH 7.8, plus fresh 0.2 mM DTT. The target protein will elute as a sharp band that can be monitored using the chart recorder. Collect the peak in a tared collection vial. Lower grades of imidazole exhibit considerable background absorbance at 280 nm and will interfere with protein quantitation if based on light absorbance measurements.

20. Weigh the eluted protein solution and measure A_{280} in a quartz cuvette (against elution buffer blank) to determine the concentration and total protein weight.

Preparative-Scale Solubilization and Purification of IMPs Localized in Inclusion Bodies

Some IMP inclusion bodies cannot be solubilized by Empigen, but instead require the use of SDS to achieve solubilization. For example, we have not had success with the Empigen-based procedure with G protein–coupled receptor inclusion bodies and therefore rely upon the following protocol.

1. Cells are lysed exactly as given in steps 1–8 of the preceding protocol.

2. Centrifuge the (thawed) lysate at a high speed on a preparative centrifuge for 15 min at 4° to separate the insoluble inclusion bodies and cell debris from all soluble substances.

3. Discard the supernatant. Resuspend the insolubles using 20 ml of lysis buffer containing fresh 2 mM 2-mercaptoethanol (ME) per gram of original cells.

4. Place the container in an ice bath and tip sonicate at moderate power for 5 min (5 s on/5 s off, repetitively).

5. Repeat steps 2 through 4 twice to wash out any soluble protein that may be bound to the inclusion bodies.

6. Resuspend the insoluble pellet using 20 ml of SDS/urea buffer (20 mM tris, 200 mM NaCl, 8 M urea, 0.2% SDS, fresh 2 mM ME, pH 8.0) per gram of original cells.

7. Rotate container for 1–2 hours at room temperature, periodically using a pipette to disperse any clumps that form.

8. Centrifuge the solution at 18,000 rpm for 20 min at 15° (not at lower temperatures). Transfer the supernatant to a new 50-ml Falcon tube and discard the insoluble material.

9. Pack ~2 ml of Ni-NTA resin per gram of original cells (this assumes a yield of 10 mg of IMP from 5 g of cells) into a column and equilibrate the resin by rinsing with 3 bed volumes of buffer B (20 mM tris, 200 mM NaCl, pH 8.0) followed by 1.5 bed volumes of SDS/urea buffer. It is important that concentrations of SDS higher than 0.2% are not used to avoid interfering with the Ni-NTA association with protein His$_6$ tags.

10. Transfer the equilibrated resin to the solution containing the solubilized inclusion bodies. Tumble at RT for 30 min to allow the poly-His-tagged, solubilized IMPs to bind to the Ni^{2+} resin.

11. Transfer all of the mixture to a column and collect the solution that passes through the column as the unbound fraction for SDS–PAGE analysis.

12. Start monitoring the A_{280} of solutions passing through the column.

13. Rinse the column using SDS/urea buffer until the A_{280} levels off at the baseline value and then repeat using 3 × 1 bed volumes of SDS buffer without urea.

14. Rinse the column using at least 10 × 1 bed volumes of DPC rinse buffer (5 mM tris, 50 mM NaCl, 0.5% DPC, fresh 2 mM ME, pH 8.0) to change the detergent from SDS to DPC. The A_{280} should either stay at or return to the baseline. If a detergent other than DPC is preferred, DPC can be replaced by an appropriate concentration of the detergent of choice at this and subsequent steps. If SDS is going to be the detergent used for elution, step 16 below should be skipped because it will result in partial elution of His$_6$-tagged proteins (SDS significantly reduces the affinity of the tag to MIAC resin).

15. Wash the column using DPC wash buffer (30 mM imidazole in DPC rinse buffer, pH 8.0) until the A_{280} returns to the baseline. This will elute any proteins remaining on the resin that have weak affinity for Ni^{2+} but that do not have a His$_6$ tag.

16. Elute the bound IMP from the column using DPC elution buffer [0.5% DPC, 250 mM imidazole (high grade), fresh 2 mM ME, pH 7.5]. Collect and weigh the eluate.

17. The eluted IMP can be quantified as described at the end of the preceding protocol.

NMR Sample Preparation

1. To a pool of ^{15}N-labeled membrane protein prepared into DPC/imidazole as described above, add D_2O to a concentration of 10% and EDTA to a concentration of 0.5 mM from a 100 mM, pH 7, stock solution. The EDTA will sequester any metal ions that may have eluted with the target protein and will also suppress bacterial growth. Instead of EDTA, a little Chelex resin could be added to the solution and then removed by filtration or centrifugation.

2. Lower the pH to 6.5 using perdeuterated acetic acid and a 20-μl pipettor. DPC does not seem to interfere with pH measurements. If the pH is undershot, raise it to 6.5 using ammonium hydroxide. The use of the weak acid (acetic acid) and weak base (ammonia) allows the avoidance of transient exposure of protein to extremely high or low pH.

3. Concentrate the solution using a centrifugal concentrator. We have had good success with Amicon Ultra, Centricon YM- and PL-series cartridges. Note that most of the detergent is retained with the protein rather than passing through the filter: concentrating the protein means also concentrating the detergent.

4. Transfer the desired volume (normally 550 μl) to a high grade 5-mm NMR tube. Save the rest for non-NMR analysis (it is usually good to have at least 50 μl for this purpose). DPC-based samples should not be frozen, but can be stored at 4°, typically for a number of months.

Acquisition of 2D TROSY NMR Spectra

For routine screening of new membrane proteins, TROSY spectra are usually acquired at 25° and 45°, but spectral quality is almost always higher at the elevated temperature. TROSY yields better-resolved spectra of IMPs than conventional HSQC, even at 600 MHz. TROSY spectra acquired at 750–800 MHz generally yield about 10% more peaks than are observed at 600 MHz.

In our experience, some variants of the 2D TROSY pulse sequence yield much better results than others. Some are ineffective at filtering out the considerable ^{1}H signals from imidazole and detergent. We employ a minimally modified TROSY sequence developed by Weigelt (1998) that has been very effective for IMP samples (Fig. 1). This pulse sequence incorporates sensitivity and gradient enhancement and achieves the selection of the TROSY component with a minimal phase cycle while actively suppressing the unwanted components. It should be noted that,

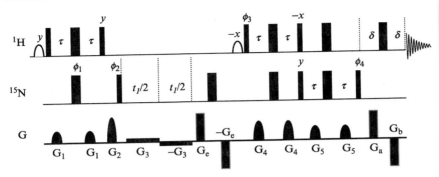

FIG. 1. TROSY [^{1}H, ^{15}N] pulse sequence derived from that of Weigelt (1998) using spin-state selectivity, as well as sensitivity and gradient enhancement. Hard pulses are applied with flip angles of 90° (180°) and the phases of all the pulses are $+x$ except when noted. The carrier frequencies of ^{1}H and ^{15}N are set to 4.7 and 116 ppm, respectively, and pulses are applied with rf field strengths of 21 kHz (^{1}H) and 5.5 kHz (^{15}N). The shaped water-selective 90° pulses have lengths of 1.5 ms and are sinc shaped. P-type signals of the TROSY component are selected using phases of $+y$ and $-x$ for ϕ_3 and ϕ_4, respectively. N-type signals are selected with phases of $-y$ and $+x$ for ϕ_3 and ϕ_4, respectively, along with inversion of gradients G_e. Bilinear gradients G_3 are used to prevent radiation damping during t_1 and the gradients G_a and G_b are used to decode the coherences. All gradients are sine shaped except G_3, G_e, G_a, and G_b, which have rectangular shapes. The length and strengths of the gradient pulses are as follows: G_1 (21 G/cm, 1 ms), G_2 (30 G/cm, 1.2 ms), G_3 (1.8 G/cm), G_e (30 G/cm, 850 μs), G_4 (30 G/cm, 1 ms), G_5 (24 G/cm, 800 μs), G_a (30 G/cm, 100 μs), and G_b (-30 G/cm, 72 μs). Phase cycle: $\phi_1 = x, x, -x, -x$; $\phi_2 = y, -y, -y, y$; receiver $= x, -x, -x, x$. Delays: $\tau = 2.4$ ms, $\delta = 0.25$ ms. Data are collected as 96* × 1024* points with maximum acquisition times of 53 ms (^{15}N) and 124 ms (^{1}H).

as suggested as a possibility by Weigelt, the last ^{1}H 180° pulse of the original sequence was converted from a 3-9-19 selective pulse to a hard pulse and the phases of the other pulses adjusted accordingly, although magic angle gradients are not employed.

Beware of YodA—to the Dark Side It Has Gone!

We conclude on a note of caution regarding a protein that we find contaminates about 20% of the samples prepared according to the protocols given above. YodA is an endogenous *E. coli* protein that sometimes is spontaneously overexpressed and copurifies with His$_6$-tagged membrane proteins expressed in *E. coli*. The protein is the mature form of a periplasmic protein called YodA (or a protein closely related to it). The mature form of this protein has about 190 residues, migrates in the 27-kDa range

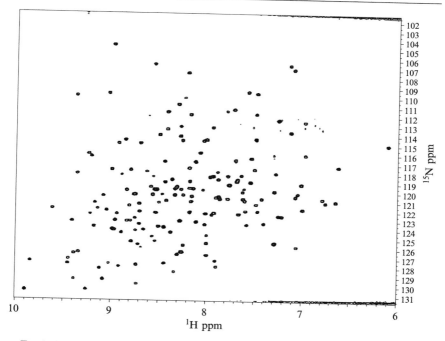

FIG. 2. An 800-MHz TROSY spectrum of the water-soluble *E. coli* YodA protein (U-15N, 1 m*M*) at 25° that was prepared in DPC micelles according to the protocols of this chapter. The spectrum of this protein at 45° is very similar to that shown here.

on SDS–PAGE, and has an N-terminal sequence of HGHHSH—not only will it copurify with poly–His-tagged proteins from MIAC, but it may also be recognized by antibodies that recognize poly-His tags. YodA is a stress response protein that avidly binds soft divalent cations (David *et al.*, 2003). Moreover, it is also known that one of the stress signals that can lead to its massive overexpression is the overexpression of other proteins (David *et al.*, 2003). We have found that YodA often is cooverexpressed with a number of other MTB proteins and occasionally with *E. coli* diacylglycerol kinase. It has a characteristic (and beautiful) NMR spectrum that is shown in Fig. 2. Beware of YodA.

Acknowledgments

This work was supported by U.S. NIH Grants RO1 GM47485 and PO1 GM64676. We thank Philip Gao, Alla Korepanova, and Tim Cross of Florida State University, and Bob Nakamoto and Yelena Peskova of the University of Virginia both for providing us with *E. coli* expression vectors for many different IMPs of *Mycobacterium tuberculosis* and for engaging discussions.

References

Arora, A., Abildgaard, F., Bushweller, J. H., and Tamm, L. K. (2001). Structure of outer membrane protein A transmembrane domain by NMR spectroscopy. *Nat. Struct. Biol.* **8**, 334–338.

Chou, J. J., Gaemers, S., Howder, B., Louis, J. M., and Bax, A. (2001). A simple apparatus for generating stretched polyacrylamide gels, yielding uniform alignment of proteins and detergent micelles. *J. Biomol. NMR* **21**, 377–382.

Czerski, L., and Sanders, C. R. (2000). Functionality of a membrane protein in bicelles. *Anal. Biochem.* **284**, 327–333.

David, G., Blondeau, K., Schiltz, M., Penel, S., and Lewit-Bentley, A. (2003). YodA from *Escherichia coli* is a metal-binding, lipocalin-like protein. *J. Biol. Chem.* **278**, 43728–43735.

Fernandez, C., Adeishvili, K., and Wuthrich, K. (2001). Transverse relaxation-optimized NMR spectroscopy with the outer membrane protein OmpX in dihexanoyl phosphatidyl-choline micelles. *Proc. Natl. Acad. Sci. USA* **98**, 2358–2363.

Fernandez, C., Hilty, C., Wider, G., Guntert, P., and Wuthrich, K. (2004). NMR structure of the integral membrane protein OmpX. *J. Mol. Biol.* **336**, 1211–1221.

Hwang, P. M., Choy, W. Y., Lo, E. I., Chen, L., Forman-Kay, J. D., Raetz, C. R., Prive, G. G., Bishop, R. E., and Kay, L. E. (2002). Solution structure and dynamics of the outer membrane enzyme PagP by NMR. *Proc. Natl. Acad. Sci. USA* **99**, 13560–13565.

Krueger-Koplin, R. D., Sorgen, P. L., Krueger-Koplin, S. T., Rivera-Torres, I. O., Cahill, S. M., Hicks, D. B., Grinius, L., Krulwich, T. A., and Girvin, M. E. (2004). An evaluation of detergents for NMR structural studies of membrane proteins. *J. Biomol. NMR* **28**, 43–57.

Nagy, J. K., Kuhn, H. A., Keyes, M. H., Gray, D. N., Oxenoid, K., and Sanders, C. R. (2001). Use of amphipathic polymers to deliver a membrane protein to lipid bilayers. *FEBS Lett.* **501**, 115–120.

Opella, S. J., Nevzorov, A., Mesleb, M. F., and Marassi, F. M. (2002). Structure determination of membrane proteins by NMR spectroscopy. *Biochem. Cell Biol.* **80**, 597–604.

Oxenoid, K., Kim, H. J., Jacob, J., Sonnichsen, F. D., and Sanders, C. R. (2004). NMR assignments for a helical 40 kDa membrane protein. *J. Am. Chem. Soc.* **126**, 5048–5049.

Riek, R., Pervushin, K., and Wuthrich, K. (2000). TROSY and CRINEPT: NMR with large molecular and supramolecular structures in solution. *Trends Biochem. Sci.* **25**, 462–468.

Sanders, C. R., and Oxenoid, K. (2000). Customizing model membranes and samples for NMR spectroscopic studies of complex membrane proteins. *Biochim. Biophys. Acta* **1508**, 129–145.

Vinogradova, O., Sonnichsen, F., and Sanders, C. R. (1998). On choosing a detergent for solution NMR studies of membrane proteins. *J. Biomol. NMR* **11**, 381–386.

Weigelt, J. (1998). Single scan, sensitivity- and gradient-enhanced TROSY for multidimensional NMR experiments. *J. Am. Chem. Soc.* **120**, 10778–10779.

[13] Solution Structure and Dynamics of Integral Membrane Proteins by NMR: A Case Study Involving the Enzyme PagP

By PETER M. HWANG and LEWIS E. KAY

Abstract

Solution NMR spectroscopy is rapidly becoming an important technique for the study of membrane protein structure and dynamics. NMR experiments on large perdeuterated proteins typically exploit the favorable relaxation properties of backbone amide ^{15}N–^{1}H groups to obtain sequence-specific chemical shift assignments, structural restraints, and a wide range of dynamics information. These methods have proven successful in the study of the outer membrane enzyme, PagP, not only for obtaining the global fold of the protein but also for characterizing in detail the conformational fluctuations that are critical to its activity. NMR methods can also be extended to take advantage of slowly relaxing methyl groups, providing additional probes of structure and dynamics at side chain positions. The current work on PagP demonstrates how solution NMR can provide a unique atomic resolution description of the dynamic processes that are key to the function of many membrane protein systems.

Introduction

Propelled by recent methodological advances, the first detailed studies of membrane proteins by solution state nuclear magnetic resonance (NMR) spectroscopy are beginning to emerge. To date, four multispan structures have been determined including the glycophorin A helical dimer (MacKenzie *et al.*, 1997) and the β-barrels, OmpX (Fernandez *et al.*, 2001), OmpA (Arora *et al.*, 2001), and PagP (Hwang *et al.*, 2002). Significant progress has also been made in characterizing diacylglycerol kinase, as reviewed in this issue. This chapter highlights the techniques used to study PagP and how these can be further extended and applied to membrane proteins in general.

PagP is a 161-residue outer membrane enzyme found in a number of pathogenic Gram-negative bacteria (Hwang *et al.*, 2002) that catalyzes the transfer of palmitate from phospholipids to lipopolysaccharide (LPS) (Bishop *et al.*, 2000), the main lipid component of the outer membrane outer leaflet. As such, LPS forms an important barrier against host

Copyright 2005, Elsevier Inc.
All rights reserved.
0076-6879/05 $35.00

defenses, and PagP activity has been found to be essential for virulence in some microorganisms (Guo *et al.*, 1998; Preston *et al.*, 2003; Robey *et al.*, 2001). The protein is ideal for study by NMR because of its small size, unique enzymatic activity, and biological significance.

Sample Preparation

For all but the smallest of membrane protein systems (<30 kDa total), a ^{2}H, ^{13}C, ^{15}N-enriched sample is required for the multidimensional NMR experiments used to obtain chemical shift assignments. Due to the high cost associated with isotope labeling, a rigorous optimization of protein production and sample preparation is warranted. Fortunately, ^{15}N enrichment alone is sufficient for screening purposes, and we have found that ^{1}H–^{15}N heteronuclear single-quantum correlation (HSQC) spectra recorded on protonated systems with a total molecular weight as large as 100 kDa are still useful for assessing feasibility.

Membrane proteins are notoriously difficult to overexpress. In *Escherichia coli*, however, some membrane proteins can be engineered to aggregate in an unfolded form, so that they do not interfere with cell viability and are not susceptible to degradation. While such strategies can enhance yields by at least an order of magnitude, they do require effective protocols for reconstituting the inactive protein. Very few successful examples of this methodology have been reported for helical membrane proteins (Booth, 2003; Kiefer, 2003), but the approach has become relatively routine for β-barrels (Bannwarth and Schulz, 2003; Buchanan, 1999), a rare class of membrane proteins found only in the outer membranes of Gram-negative bacteria, mitochondria, and chloroplasts. Like other β-barrels, PagP could be made to accumulate in a dense unfolded form (~30 mg/liter of culture) by removing its N-terminal signal sequence (Nielsen *et al.*, 1997) via mutagenesis. Denatured PagP has been successfully reconstituted into DPC, OG/SDS, and CYFOS-7 detergents (Fig. 1), and NMR spectra indicate a similarly folded structure in all three environments, although there are some important differences (see below).

The selection of an appropriate detergent (or lipid) is, of course, key to a successful refolding strategy, but detergents must also be screened for their ability to maintain protein structure and for suitability in NMR studies (Krueger-Koplin *et al.*, 2004). Anionic detergents like sodium dodecyl sulfate (SDS) and perfluorooctonate (PFO) form small micelles, but these totally denature PagP (Fig. 2). On the other hand, solublization in nonionic detergents like octylglucoside (OG) and dodecylmaltoside (DM) tend to produce samples that give very broad NMR signals. Zwitterionic

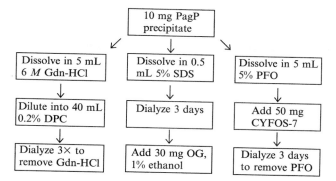

FIG. 1. Flow chart of three alternative folding protocols for PagP. In all cases, the buffer is 50 mM sodium phosphate, pH 6.0.

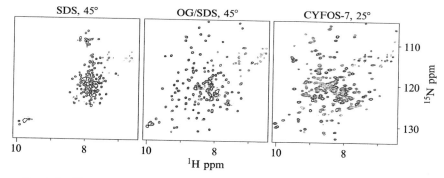

FIG. 2. ^{1}H–^{15}N HSQC spectra of 1 mM ^{2}H, ^{15}N-PagP samples, 50 mM sodium phosphate, pH 6.0, in three different detergent environments: SDS, OG/SDS, and CYFOS-7.

detergents like dodecylphosphocholine (DPC) and dihexanoylphosphatidylcholine (DHPC) provide a good compromise between micelle size and protein stability. For proteins that are particularly difficult to work with, an important advance is the development of lipopeptide detergents (LPDs) (McGregor *et al.*, 2003). LPDs form small micelles, yet they are able to maintain the activity of "fragile" proteins like lactose permease longer than DM. Excellent NMR spectra of PagP in LPD have been obtained, with the rotational correlation time of the complex comparable to that of PagP–DPC and PagP–OG/SDS (McGregor *et al.*, 2003).

Chemical Shift Assignment

Obtaining chemical shift assignments is the first step toward characterizing any system by NMR. In applications to high-molecular-weight proteins where deuteration is critical, the key assignment experiments center around the backbone amide ^{15}N–^1H group. The use of TROSY (Pervushin et al., 1997) is essential in large systems like PagP (50–60 kDa in total), because it minimizes transverse relaxation of amide ^{15}N and ^1H spins leading to substantial improvements in the sensitivity and resolution of out-and-back experiments like the HNCA, HN(CO)CA, HN(CA)CB, HN(COCA)CB, HN(CA)CO, and HNCO (Salzmann et al., 1998; Yang and Kay, 1999). Using these TROSY-modified sequences, >90% of the backbone resonances could be assigned in PagP in both DPC and OG/SDS detergents (Hwang et al., 2002), and the only regions of the protein that could not be assigned were those broadened beyond detection by conformational exchange.

N–H groups are key probes for obtaining atomic level structural and dynamic information, provided that protons can be efficiently incorporated into these sites (in an otherwise deuterated background) and sequence-specific assignments are available. In PagP, incorporating protons at backbone amide positions via solvent exchange is straightforward because the protein is purified from an unfolded state. In addition to protonation at backbone amide sites, it is also possible to incorporate protons into methyl groups of Val, Leu, and Ile ($\delta 1$ only) using $^2H, ^{13}C, ^1H$-methyl-labeled precursors, α-ketoisovalerate, and α-ketobutyrate, in E. coli–based growths (Gardner and Kay, 1998; Goto et al., 1999). The methyl $^{13}C, ^1H$ chemical shifts can be assigned by relaying magnetization to backbone N–H groups using experiments that are specifically optimized for methyl protonated, highly deuterated proteins (Gardner et al., 1996; Hilty et al., 2002); however the sensitivity of these experiments can be quite low for large systems. Recently new experiments for application to high-molecular-weight proteins have been developed that make use of COSY-type relays to correlate methyl groups with backbone amides, aliphatic side chain, or backbone CO carbons, significantly improving sensitivity (Tugarinov and Kay, 2003). For Val and Leu residues, these schemes require an alternate labeling strategy in which one of the methyls in a given residue is $^{13}C^1H_3$ labeled and the other $^{12}C^2H_3$, achieving a linear ^{13}C aliphatic side chain spin system.

Structure Determination

Structural studies of OmpX (Fernandez et al., 2001), OmpA (Arora et al., 2001), and PagP (Hwang et al., 2002) were greatly facilitated by the β-barrel architecture of these proteins, with amide NOEs connecting proximal strands. Combined with hydrogen bond restraints and chemical shift–based dihedral angle restraints (Cornilescu *et al.*, 1999; Wishart and Case, 2001), a reasonably high-resolution global fold (<1.0 Å RMSD for the backbone heavy atoms of the β-barrel) could be obtained for PagP solublized in either DPC (Fig. 3) or in OG/SDS. The overall structure of PagP in both environments is essentially the same, indicating that the type of solubilizing detergent used and the nature of the reconstitution procedure do not strongly perturb the global fold of this enzyme. A recently determined X-ray crystal structure of PagP in LDAO (unpublished observations) is also in close agreement with the NMR-derived structures.

For most soluble proteins, NH–NH NOEs alone would be insufficient for the establishment of a global fold (Gardner *et al.*, 1997), but in β-barrels, inter-strand NOEs between amides are very effective in defining the barrel topology (Fig. 4). However, the positioning of elements outside the PagP β-barrel, including the orientation of loops and the amphipathic α-helix, is poorly defined by NH–NH NOEs alone. The X-ray structure of PagP in LDAO indicates that most of these components are well structured, with the α-helix packing against strands B and C. (A careful analysis of the relative positions of the amphipathic helix and aromatic side chains, based on the crystal structure, has shown that PagP is tilted by 30° relative to the bilayer normal, the first example of a tilted β-barrel membrane protein.) One clear way of improving the quality of NMR-derived

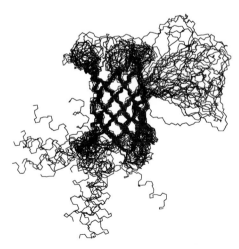

Fig. 3. Superposition of 20 NMR structures of PagP in DPC. Adapted from Hwang *et al.* (2002). Copyright 2002 National Academy of Sciences, USA.

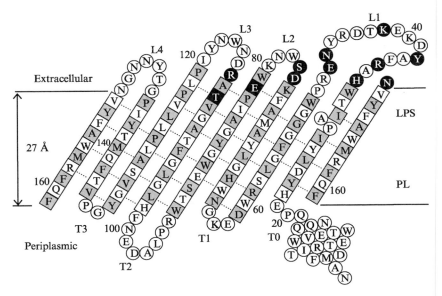

FIG. 4. Topology diagram of PagP. Residues in a β-sheet conformation are represented as squares, and all others as circles. Gray-shaded and nonshaded squares denote those residues with side chains pointing toward the exterior or interior of the β-barrel, respectively. Absolutely conserved polar residues in the extracellular region of the protein (putative active site) are shown in black. Slowly exchanging amides, as detected by ^{1}H–^{15}N HSQC spectra recorded on a D$_2$O-exchanged sample of PagP–OG/SDS, are indicated by dotted lines between hydrogen bonding partners. Adapted from Hwang et al. (2002). Copyright 2002 National Academy of Sciences, USA.

structures is to include NH–CH$_3$ and CH$_3$–CH$_3$ NOEs (Mueller et al., 2000), using the methyl-labeling scheme described above. Methyl NOEs have been shown to improve the accuracy and precision of NMR-derived structures for OmpX (Fernandez et al., 2004), and they would be indispensable for defining helix packing and global folds of α-helical proteins. Another important source of structural information comes from residual dipolar couplings (Prestegard, 1998; Tjandra and Bax, 1997). Unfortunately, most alignment media are incompatible with the detergents used to solubilize membrane proteins (Ma and Opella, 2000). Strain-induced alignment in a gel (Ishii et al., 2001; Sass et al., 2000) has been used to align PagP in DPC micelles. However, the polyacrylamide matrix significantly reduces the effective concentration of the protein, limiting sensitivity. Clearly other alignment strategies must be explored.

Dynamics

Backbone amide ^{15}N and side chain methyl ^{13}C spins are excellent probes of molecular dynamics covering a wide range of time scales. Although this chapter will focus on applications involving backbone amide groups, it is worthwhile to note that methods similar to those for ^{15}N–^{1}H spin systems have been developed to study dynamics using methyl groups. For instance, the relaxation of methyl groups can be optimized using methyl-TROSY (Tugarinov et al., 2003), leading to a CPMG-based multiple quantum relaxation dispersion experiment (Korzhnev et al., 2004) that has been used to characterize conformational exchange in the 82-kDa protein, malate synthase G.

^{15}N T_1, T_2, and heteronuclear ^{1}H-^{15}N NOE measurements are routinely performed on water-soluble proteins (Bruschweiler, 2003), and recently such experiments have also been applied to OmpX (Fernandez et al., 2004), OmpA (Tamm et al., 2003), and PagP (Hwang et al., 2002). Each of those proteins shows a large diversity in the amplitudes of amide bond vector dynamics (on a picosecond to nanosecond time scale). In PagP, T_1 relaxation times are a sensitive indicator of nanosecond motions, identifying three different regions with high mobility on this time scale (Fig. 5). T_2 relaxation times and heteronuclear NOE values show pronounced differences from mean values at the very N-terminus (first few residues prior to the α-helix) and in the large L1 extracellular loop, indicating that these

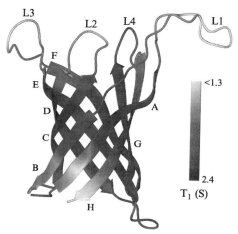

FIG. 5. Ribbon diagram of the PagP β-barrel colored according to measured T_1 times in DPC at 45°. Residues for which T_1 times could not be measured were assigned values midway between flanking residues. (See color insert.)

regions are the most flexible in the protein (picosecond time scale). Notably, these elements were not observed in the X-ray structure of PagP in
LDAO. Many putative active site residues are located in the L1 loop (see
Fig. 4), suggesting that it must rigidify considerably for catalysis to occur. It
is clear that attempts to understand how this enzyme functions must go
significantly beyond static structure determination.

Although the active site of PagP could not be visualized in NMR or X-
ray-derived structures, the crystal structure did establish the presence of a
single inhibitory LDAO detergent molecule buried in the large central
cavity of the barrel. This suggests that the active site of the enzyme is
inside the β-barrel, and this finding is supported by a subsequent mutagenesis study (unpublished observations). Since it is probable that many linear
chain detergents can also fit inside the barrel, inhibiting enzyme activity, we
chose to refold PagP into a new detergent, CYFOS-7, which is similar to
DPC except that it possesses a bulky cyclohexyl ring at the terminus of its
alkyl chain that prevents entry into the barrel. PagP was found to retain
catalytic activity in CYFOS-7 (unpublished observations), unlike all the
other detergents that had been used for previous structural studies. With a
functioning system in hand we now focus on how phospholipid substrate
might diffuse into the barrel and attempt to characterize the dynamic
features of the enzyme that are important for activity.

At 45° the ^{1}H–^{15}N HSQC spectrum of PagP in CYFOS-7 resembles
that of PagP–DPC and PagP–OG/SDS, suggesting a similar structure in all
three environments. However, at 25° the spectrum changes dramatically,
with an approximate doubling of peaks (see Fig. 2). Backbone chemical
shift assignments establish the presence of a second state, with a fractional
population of approximately 0.3. The major "R" form is similar to the
conformers observed in other detergents, but the new "T" state possesses
very distinct chemical shifts (Fig. 6), suggesting a complete restructuring of
the "β-bulge" of strand A and an ordering of the L1 loop. It is not unusual
for membrane proteins to adopt multiple conformations, as can be seen
from the wide range of structures that are crystallized within a single class
of proteins (Chang, 2003; Jiang et al., 2002). The major advantage of using
NMR to probe membrane proteins is that they can be studied in their
functionally active forms and processes involving the interconversion between states can be monitored directly, as in the case of PagP in CYFOS-7
(unpublished observations), described below.

The relative populations of R and T states can be measured directly
from peak volumes in HSQC spectra. However, relaxation losses during
the pulse sequence must be carefully accounted for, since the two conformers have very different relaxation properties. An unenhanced HSQC
pulse scheme was employed, with ^{15}N TROSY component selection

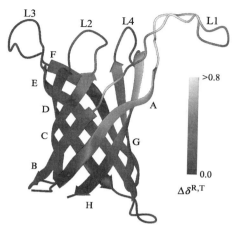

FIG. 6. Ribbon diagram of the PagP β-barrel colored according to the combined chemical shift difference between the R and T states, $\Delta\delta^{R,T} = [(\omega_N\Delta\delta_N)^2 + (\omega_{C\alpha}\Delta\delta_{C\alpha})^2 + (\omega_{CO}\Delta\delta_{CO})^2]^{1/2}, \omega_N = 0.154, \omega_{C\alpha} = 0.276, \omega_{CO} = 0.341$ (Evenas *et al.*, 2001). For the R state, an average of the available chemical shifts in DPC, OG/SDS, and the CYFOS-7 R state was used. If a residue could not be observed in the T state (mainly loop L3) or if the R and T states were identical, $\Delta\delta^{R,T} = 0$. Adapted from Hwang *et al.* (2004). Copyright 2004 National Academy of Sciences, USA. (See color insert.)

achieved using the IPAP procedure (Ottiger *et al.*, 1998; Yang and Nagayama, 1996). In this way relaxation losses could be easily taken into account by considering only the decay of single quantum ^{1}H and ^{15}N magnetization. Equilibrium constants were obtained for the R,T interconversion over a temperature range from 15° to 35°, and from the linear plot of $\ln K_{eq}$ vs. inverse temperature, values of $\Delta H = 10.7$ kcal/mol and $\Delta S = 37.5$ cal/mol/K were calculated for the T-to-R transition.

Kinetic rate constants describing the exchange between R and T states can also be obtained from solution NMR studies. In theory, a NOESY-type experiment could be employed to measure exchange-mediated transfer of proton magnetization (Montelione and Wagner, 1989). However, a better alternative is the N_{ZZ} exchange experiment, in which magnetization is stored as N_Z during the mixing time rather than as H_Z. In this simple 2D experiment, evolution of ^{15}N chemical shift occurs prior to the exchange of magnetization between sites so that cross-peaks at (ω_N^A, ω_H^B) and (ω_N^B, ω_H^A) are obtained, in addition to diagonal peaks at (ω_N^A, ω_H^A) and (ω_N^B, ω_H^B). The original N_{ZZ} exchange experiment (Farrow *et al.*, 1994) was modified to select for the ^{15}N TROSY component using the IPAP approach (Fig. 7). Rate constants, $k_{RT} = 2.8$ s^{-1} and $k_{TR} = 6.5$ s^{-1}, for the R,T interconversion at 25° were extracted from simultaneous fits of the decay/buildup of

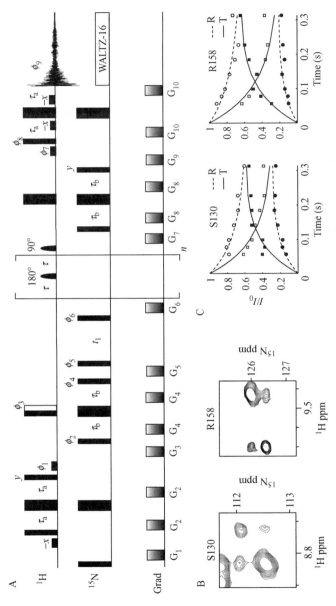

Fig. 7. (A) Pulse sequence of the IPAP-N_{zz} exchange experiment. The ^1H and ^{15}N carriers are positioned at 4.78 (water) and 119 ppm, respectively. All narrow (wide) rectangles denote nonselective 90° (180°) pulses along the x-axis, unless otherwise specified. The shaped (RE-BURP) 180° pulse during the mixing time (in brackets) is selective for the amide region to remove cross-correlation/cross-relaxation effects. The shaped 90° pulse after the mixing time selectively excites the upfield aliphatic region, to eliminate detergent signal that has recovered during the mixing time. The delays used are $\tau_a = 2.3$ ms, $\tau_b = 2.75$ ms, and $\tau = 10$ ms. To record a spectrum in which the ^{15}N doublets are antiphase, the phase cycle is as follows: $\phi_1 = y$; $\phi_2 = 4(x)$, $4(-x)$; $\phi_3 = -x$; $\phi_4 = x$; $\phi_5 = x, -x$; $\phi_6 = 2(-x)$, $2(x)$; $\phi_7 = -x, x$; $\phi_8 = -x, x$; and $\phi_9 = (x, x, -x, -x, -x, -x, x, x)$. For in-phase selection, the phase cycle is

diagonal/cross-peaks as a function of mixing time to a model that accounts for longitudinal ^{15}N T_1 relaxation and a single exchange process (Tollinger *et al.*, 2001).

At temperatures below approximately 35°, both R and T forms of PagP give rise to correlations in HSQC spectra, and it is relatively easy, therefore, to quantify the exchange process using magnetization transfer experiments. At higher temperatures, where only a single set of peaks corresponding to the R state is observed, a different approach must be used. In principle, the kinetics of interconversion, the relative populations of the interconverting states, and chemical shift differences, $\Delta\omega$, between conformers for individual residues can be extracted from CPMG-based relaxation dispersion experiments, as reviewed in detail elsewhere (Palmer *et al.*, 2001). In the case of PagP at 45°, the largest dispersions were well fit individually with similar time constants, suggesting that these profiles report on the same process. All of the dispersion data were subsequently fit to a single global exchange process (Mulder *et al.*, 2001), with $\Delta\omega$ varied for each residue. The fitted $\Delta\omega$ values at 45° correlate well with those measured directly in spectra recorded at 25°, suggesting that the dominant exchange process at this temperature is the R,T interconversion, with rate constants of $k_{RT} = 33$ s^{-1} and $k_{TR} = 298$ s^{-1}.

As described above, positive values of ΔH and ΔS are extracted for the T-to-R transition, consistent with a local unfolding reaction in which the L1 loop and portions of strands A and H (Fig. 6) become more dynamic. The increased dynamics in the R state are quite apparent from a simple ^{1}H–^{15}N HSQC spectrum recorded at 25°, where correlations from the R conformer are significantly broadened relative to the corresponding cross-peaks from the T-state (Fig. 8A). While dispersion curves for residues in the T state can be fit to a slow process matching the T-to-R transition (6.5 s^{-1}), fits of dispersions derived from residues in the R state suggest additional exchange processes that are much faster ($\sim10^3$ s^{-1}) than the R-to-T conversion (2.8 s^{-1}) (Fig. 8B). Thus, a picture emerges in which

---

the same, except that $\phi_1 = -y$, $\phi_3 = x$, and $\phi_4 = y$. Addition or subtraction of the in-phase and antiphase spectra yields the ^{15}N TROSY or anti-TROSY component. Quadrature detection in F_1 is achieved by States-TPPI (Marion *et al.*, 1989) of ϕ_5. The duration and strength of the gradients are as follows: $G_1 = (1$ ms, 5 G/cm); $G_2 = (0.2$ ms, 8 G/cm); $G_3 = (0.5$ ms, 12 G/cm); $G_4 = (0.25$ ms, 10 G/cm); $G_5 = (0.4$ ms, 5 G/cm); $G_6 = (0.3$ ms, 15 G/cm); $G_7 = (0.2$ ms, 7 G/cm); $G_8 = (0.3$ ms, 8 G/cm); $G_9 = (0.25$ ms, 10 G/cm); $G_{10} = (0.3$ ms, 15 G/cm). (B) Region of a correlation map recorded with an IPAP-N$_{zz}$ experiment showing diagonal and cross-peaks for S130 and R158. The largest peak in both panels corresponds to the diagonal peak of the R state. (C) Decay [correlations at (ω_N^A, ω_H^A); open symbols] and buildup [correlations at (ω_N^A, ω_H^B); closed symbols] curves for S130 and R158. The curves are normalized so that the intensities of the diagonal peaks (ω_N^A, ω_H^A) at time = 0 are 1.0 in both R and T states. Adapted from Hwang *et al.* (2004). Copyright 2004 National Academy of Sciences, USA.

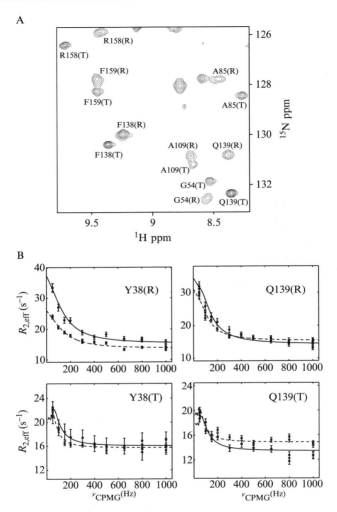

FIG. 8. (A) Selected region of the ^1H–^{15}N TROSY–HSQC spectrum of PagP–CYFOS-7 at 25°. (B) ^{15}N-CPMG relaxation dispersion curves for Y38 and Q139 in both R and T states at 25°, recorded at 800 (upper profile) and 600 MHz. $R_{2,eff}$ is calculated as $-1/T_{CP} \ln(I\nu_{cpmg}/I_o)$, where T_{CP} is a 40 ms constant time CPMG element, and $I\nu_{cpmg}$ and I_o are the intensities of correlations recorded in spectra with and without the T_{CP} period, respectively. T_{CP} is made up of (τ—^{15}N π pulse—τ) elements, where the effective RF field strength, ν_{CPMG}, equals $1/(4\tau)$ (Mulder et al., 2002). Adapted from Hwang et al. (2004). Copyright 2004 National Academy of Sciences, USA.

the enhanced flexibility of the R state facilitates substrate entry into the central cavity, while the formation of a defined conformation of the peripheral L1 loop in the T state may lead to the production of a functional active site.

Conclusion

The significant structural rearrangements that occur during the R, T cycle in PagP demonstrate how a detailed description of dynamics is essential for an understanding of protein function. This is likely to be especially true of membrane proteins, which often depend on large conformational changes to regulate the transmission of solutes or signals across biological membranes. Solution NMR is thus ideally poised to make a significant and unique contribution to the structural biology of this important class of molecules.

References

Arora, A., Abildgaard, F., Bushweller, J. H., and Tamm, L. K. (2001). Structure of outer membrane protein A transmembrane domain by NMR spectroscopy. *Nat. Struct. Biol.* **8,** 334–338.

Bannwarth, M., and Schulz, G. E. (2003). The expression of outer membrane proteins for crystallization. *Biochim. Biophys. Acta* **1610,** 37–45.

Bishop, R. E., Gibbons, H. S., Guina, T., Trent, M. S., Miller, S. I., and Raetz, C. R. (2000). Transfer of palmitate from phospholipids to lipid A in outer membranes of Gram-negative bacteria. *EMBO J.* **19,** 5071–5080.

Booth, P. J. (2003). The trials and tribulations of membrane protein folding *in vitro. Biochim. Biophys. Acta* **1610,** 51–56.

Bruschweiler, R. (2003). New approaches to the dynamic interpretation and prediction of NMR relaxation data from proteins. *Curr. Opin. Struct. Biol.* **13,** 175–183.

Buchanan, S. K. (1999). Beta-barrel proteins from bacterial outer membranes: Structure, function and refolding. *Curr. Opin. Struct. Biol.* **9,** 455–461.

Chang, G. (2003). Structure of MsbA from Vibrio cholera: A multidrug resistance ABC transporter homolog in a closed conformation. *J. Mol. Biol.* **330,** 419–430.

Cornilescu, G., Delaglio, F., and Bax, A. (1999). Protein backbone angle restraints from searching a database for chemical shift and sequence homology. *J. Biomol. NMR* **13,** 289–302.

Evenas, J., Tugarinov, V., Skrynnikov, N. R., Goto, N. K., Muhandiram, R., and Kay, L. E. (2001). Ligand-induced structural changes to maltodextrin-binding protein as studied by solution NMR spectroscopy. *J. Mol. Biol.* **309,** 961–974.

Farrow, N. A., Zhang, O., Forman-Kay, J. D., and Kay, L. E. (1994). A heteronuclear correlation experiment for simultaneous determination of 15N longitudinal decay and chemical exchange rates of systems in slow equilibrium. *J. Biomol. NMR* **4,** 727–734.

Fernandez, C., Adeishvili, K., and Wuthrich, K. (2001). Transverse relaxation-optimized NMR spectroscopy with the outer membrane protein OmpX in dihexanoyl phosphatidylcholine micelles. *Proc. Natl. Acad. Sci. USA* **98,** 2358–2363.

Fernandez, C., Hilty, C., Wider, G., Guntert, P., and Wuthrich, K. (2004). NMR structure of the integral membrane protein OmpX. *J. Mol. Biol.* **336,** 1211–1221.

Gardner, K. H., and Kay, L. E. (1998). The use of 2H, 13C, 15N multidimensional NMR to study the structure and dynamics of proteins. *Annu. Rev. Biophys. Biomol. Struct.* **27,** 357–406.

Gardner, K. H., Konrat, R., Rosen, M. K., and Kay, L. E. (1996). An (H)C(CO)NH-TOCSY pulse scheme for sequential assignment of protonated methyl groups in otherwise deuterated N-15,C-13-labeled proteins. *J. Biomol. NMR* **8,** 351–356.

Gardner, K. H., Rosen, M. K., and Kay, L. E. (1997). Global folds of highly deuterated, methyl-protonated proteins by multidimensional NMR. *Biochemistry* **36,** 1389–1401.

Goto, N. K., Gardner, K. H., Mueller, G. A., Willis, R. C., and Kay, L. E. (1999). A robust and cost-effective method for the production of Val, Leu, Ile (delta 1) methyl-protonated 15N-, 13C-, 2H-labeled proteins. *J. Biomol. NMR* **13,** 369–374.

Guo, L., Lim, K. B., Poduje, C. M., Daniel, M., Gunn, J. S., Hackett, M., and Miller, S. I. (1998). Lipid A acylation and bacterial resistance against vertebrate antimicrobial peptides. *Cell* **95,** 189–198.

Hilty, C., Fernandez, C., Wider, G., and Wuthrich, K. (2002). Side chain NMR assignments in the membrane protein OmpX reconstituted in DHPC micelles. *J. Biomol. NMR* **23,** 289–301.

Hwang, P. M., Choy, W. Y., Lo, E. I., Chen, L., Forman-Kay, J. D., Raetz, C. R., Prive, G. G., Bishop, R. E., and Kay, L. E. (2002). Solution structure and dynamics of the outer membrane enzyme PagP by NMR. *Proc. Natl. Acad. Sci. USA* **99,** 13560–13565.

Hwang, P. M., Bishop, R. E., and Kay, L. E. (2004). The integral membrane enzyme PagP alternates between two dynamically distinct states. *Proc. Natl. Acad. Sci. USA* **101,** 9618–9623.

Ishii, Y., Markus, M. A., and Tycko, R. (2001). Controlling residual dipolar couplings in high-resolution NMR of proteins by strain induced alignment in a gel. *J. Biomol. NMR* **21,** 141–151.

Jiang, Y., Lee, A., Chen, J., Cadene, M., Chait, B. T., and MacKinnon, R. (2002). The open pore conformation of potassium channels. *Nature* **417,** 523–526.

Kiefer, H. (2003). *In vitro* folding of alpha-helical membrane proteins. *Biochim. Biophys. Acta* **1610,** 57–62.

Korzhnev, D. M., Kloiber, K., Kanelis, V., Tugarinov, V., and Kay, L. E. (2004). Probing slow dynamics in high molecular weight proteins by methyl-TROSY NMR spectroscopy: Application to a 723-residue enzyme. *J. Am. Chem. Soc.* **126,** 3964–3973.

Krueger-Koplin, R. D., Sorgen, P. L., Krueger-Koplin, S. T., Rivera-Torres, I. O., Cahill, S. M., Hicks, D. B., Grinius, L., Krulwich, T. A., and Girvin, M. E. (2004). An evaluation of detergents for NMR structural studies of membrane proteins. *J. Biomol. NMR* **28,** 43–57.

Ma, C., and Opella, S. J. (2000). Lanthanide ions bind specifically to an added "EF-hand" and orient a membrane protein in micelles for solution NMR spectroscopy. *J. Magn. Reson.* **146,** 381–384.

MacKenzie, K. R., Prestegard, J. H., and Engelman, D. M. (1997). A transmembrane helix dimer: Structure and implications. *Science* **276,** 131–133.

Marion, D., Ikura, M., Tschudin, R., and Bax, A. (1989). Rapid recording of 2D NMR-spectra without phase cycling—application to the study of hydrogen-exchange in proteins. *J. Magn. Reson.* **85,** 393–399.

McGregor, C. L., Chen, L., Pomroy, N. C., Hwang, P., Go, S., Chakrabartty, A., and Prive, G. G. (2003). Lipopeptide detergents designed for the structural study of membrane proteins. *Nat. Biotechnol.* **21,** 171–176.

Montelione, G. T., and Wagner, G. (1989). 2D chemical-exchange NMR-spectroscopy by proton-detected heteronuclear correlation. *J. Am. Chem. Soc.* **111,** 3096–3098.

Mueller, G. A., Choy, W. Y., Yang, D., Forman-Kay, J. D., Venters, R. A., and Kay, L. E. (2000). Global folds of proteins with low densities of NOEs using residual dipolar couplings: Application to the 370-residue maltodextrin-binding protein. *J. Mol. Biol.* **300,** 197–212.

Mulder, F. A., Mittermaier, A., Hon, B., Dahlquist, F. W., and Kay, L. E. (2001). Studying excited states of proteins by NMR spectroscopy. *Nat. Struct. Biol.* **8,** 932–935.

Mulder, F. A., Hon, B., Mittermaier, A., Dahlquist, F. W., and Kay, L. E. (2002). Slow internal dynamics in proteins: Application of NMR relaxation dispersion spectroscopy to methyl groups in a cavity mutant of T4 lysozyme. *J. Am. Chem. Soc.* **124,** 1443–1451.

Nielsen, H., Engelbrecht, J., Brunak, S., and von Heijne, G. (1997). Identification of prokaryotic and eukaryotic signal peptides and prediction of their cleavage sites. *Protein Eng.* **10,** 1–6.

Ottiger, M., Delaglio, F., and Bax, A. (1998). Measurement of J and dipolar couplings from simplified two-dimensional NMR spectra. *J. Magn. Reson.* **131,** 373–378.

Palmer, A. G., 3rd, Kroenke, C. D., and Loria, J. P. (2001). Nuclear magnetic resonance methods for quantifying microsecond-to-millisecond motions in biological macromolecules. *Methods Enzymol.* **339,** 204–238.

Pervushin, K., Riek, R., Wider, G., and Wuthrich, K. (1997). Attenuated T2 relaxation by mutual cancellation of dipole-dipole coupling and chemical shift anisotropy indicates an avenue to NMR structures of very large biological macromolecules in solution. *Proc. Natl. Acad. Sci. USA* **94,** 12366–12371.

Prestegard, J. H. (1998). New techniques in structural NMR—anisotropic interactions. *Nat. Struct. Biol.* **5**(Suppl.), 517–522.

Preston, A., Maxim, E., Toland, E., Pishko, E. J., Harvill, E. T., Caroff, M., and Maskell, D. J. (2003). Bordetella bronchiseptica PagP is a Bvg-regulated lipid A palmitoyl transferase that is required for persistent colonization of the mouse respiratory tract. *Mol. Microbiol.* **48,** 725–736.

Robey, M., O'Connell, W., and Cianciotto, N. P. (2001). Identification of Legionella pneumophila rcp, a pagP-like gene that confers resistance to cationic antimicrobial peptides and promotes intracellular infection. *Infect. Immun.* **69,** 4276–4286.

Salzmann, M., Pervushin, K., Wider, G., Senn, H., and Wuthrich, K. (1998). TROSY in triple-resonance experiments: New perspectives for sequential NMR assignment of large proteins. *Proc. Natl. Acad. Sci. USA* **95,** 13585–13590.

Sass, H. J., Musco, G., Stahl, S. J., Wingfield, P. T., and Grzesiek, S. (2000). Solution NMR of proteins within polyacrylamide gels: Diffusional properties and residual alignment by mechanical stress or embedding of oriented purple membranes. *J. Biomol. NMR* **18,** 303–309.

Tamm, L. K., Abildgaard, F., Arora, A., Blad, H., and Bushweller, J. H. (2003). Structure, dynamics and function of the outer membrane protein A (OmpA) and influenza hemagglutinin fusion domain in detergent micelles by solution NMR. *FEBS Lett.* **555,** 139–143.

Tjandra, N., and Bax, A. (1997). Direct measurement of distances and angles in biomolecules by NMR in a dilute liquid crystalline medium. *Science* **278,** 1111–1114.

Tollinger, M., Skrynnikov, N. R., Mulder, F. A., Forman-Kay, J. D., and Kay, L. E. (2001). Slow dynamics in folded and unfolded states of an SH3 domain. *J. Am. Chem. Soc.* **123,** 11341–11352.

Tugarinov, V., and Kay, L. E. (2003). Ile, Leu, and Val methyl assignments of the 723-residue malate synthase G using a new labeling strategy and novel NMR methods. *J. Am. Chem. Soc.* **125,** 13868–13878.

Tugarinov, V., Hwang, P. M., Ollerenshaw, J. E., and Kay, L. E. (2003). Cross-correlated relaxation enhanced H-1-C-13 NMR spectroscopy of methyl groups in very high molecular weight proteins and protein complexes. *J. Am. Chem. Soc.* **125,** 10420–10428.

Wishart, D. S., and Case, D. A. (2001). Use of chemical shifts in macromolecular structure determination. *Methods Enzymol.* **338**, 3–34.

Yang, D. W., and Kay, L. E. (1999). TROSY triple-resonance four-dimensional NMR spectroscopy of a 46 ns tumbling protein. *J. Am. Chem. Soc.* **121**, 2571–2575.

Yang, D. W., and Nagayama, K. (1996). A sensitivity-enhanced method for measuring heteronuclear long-range coupling constants from the displacement of signals in two 1D subspectra. *J. Magn. Reson. Ser. A* **118**, 117–121.

[14] NMR Experiments on Aligned Samples of Membrane Proteins

By A. A. De Angelis, D. H. Jones, C. V. Grant, S. H. Park, M. F. Mesleh, and S. J. Opella

Abstract

NMR methods can be used to determine the structures of membrane proteins. Lipids can be chosen so that protein-containing micelles, bicelles, or bilayers are available as samples. All three types of samples can be aligned weakly or strongly, depending on their rotational correlation time. Solution NMR methods can be used with weakly aligned micelle and small bicelle samples. Solid-state NMR methods can be used with mechanically aligned bilayer and magnetically aligned bicelle samples.

Introduction

Membrane proteins are important targets for structure determination. They play key roles in signaling and function as receptors for small molecules and their surrogates that act as drugs. There are two major classes of membrane proteins. The three-dimensional structures of several examples of β-barrel membrane proteins have been determined by solution nuclear magnetic resonance (NMR) spectroscopy (Fernandez *et al.*, 2004; Hwang *et al.*, 2002) and X-ray crystallography (Schulz, 2002). However, the vast majority of membrane proteins are helical (White and Wimley, 1999), and only a small fraction of these have had their structures determined (White, 2004). NMR spectroscopy is emerging as a method capable of determining the atomic-resolution structures of helical membrane proteins in lipid environments that enable them to maintain their native structures and functions.

There are several different NMR approaches to structure determination of helical membrane proteins (Auger, 2000; de Groot, 2000; Drechsler

Copyright 2005, Elsevier Inc.
All rights reserved.
0076-6879/05 $35.00

and Separovic, 2003; Fu and Cross, 1999; Griffin, 1998; Luca *et al.*, 2003; McDowell and Schaefer, 1996; Opella, 1997; Saito *et al.*, 2002; Smith *et al.*, 1996; Thompson, 2002; Watts and Opella, 2000). They differ primarily in the overall rotational correlation time and the extent of alignment of the polypeptides in the samples. Experimental methods that exploit the properties of aligned samples are essential for structure determination of helical membrane proteins. The field is undergoing rapid development, and applications of these methods to membrane proteins have been described in recent articles and reviews (Bertram *et al.*, 2003; Lee *et al.*, 2003; Marassi and Opella, 2003; Mesleh *et al.*, 2003; Nevzorov and Opella, 2003a; Nielsen *et al.*, 2004; Opella and Marassi, 2004; Tian *et al.*, 2003). This chapter presents an integrated view of solution NMR and solid-state NMR methods for structure determination of helical membrane proteins.

Experimental studies of proteins involve the adjustment and optimization of sample conditions, e.g., concentrations, temperature, pH, buffers, and counterions. Notably, in the case of membrane proteins, there are also choices among lipids with various head groups and numbers and lengths of hydrocarbon chains. In the presence of water, lipids and combinations of lipids self-assemble to form protein-containing micelles, bicelles, and bilayers. NMR is a particularly flexible form of spectroscopy, and the availability of samples with these wide ranges of correlation times and degrees of alignment provides opportunities for measuring orientation constraints as input for structure determination.

All of the options represented in Fig. 1 require individual optimization of the sample preparation, instrumentation, and experimental methods for the NMR experiments to yield high-resolution spectra and structural information. Smaller membrane proteins in micelles can be studied using solution NMR methods; however, it is essential to carefully optimize the sample conditions so that the protein-containing micelles reorient as rapidly as possible in solution to obtain well-resolved spectra. The preparation of weakly aligned samples for the measurement of residual dipolar couplings adds further complexity (Chou *et al.*, 2002), since it is very easy to slow the reorientation and degrade the spectral resolution in the course of inducing alignment (Jones and Opella, 2004). It is also possible to use small "isotropic" bicelles instead of micelles for some solution NMR experiments (Chou *et al.*, 2002; Glover *et al.*, 2001; Vold *et al.*, 1997). In contrast, in large bicelles (Ram and Prestegard, 1988; Sanders and Prestegard, 1990; Sanders *et al.*, 1994) and lipid bilayers, which are the most attractive membrane mimetic systems that can be reconstituted from pure polypeptides and lipids for experimental samples, the proteins are immobilized by their environment and can be strongly aligned for solid-state NMR experiments. Because the orientation constraints measured from weakly and

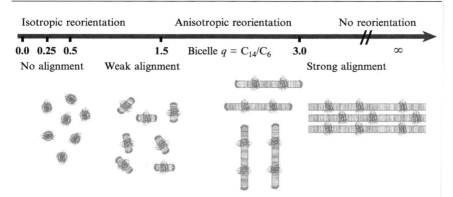

FIG. 1. Membrane-mimetic systems for protein structure determination ordered by correlation times and alignment. From left to right, systems range from isotropic reorientation to immobile with complete alignment. The long-chain to short-chain lipid ratio q is used to define the samples, ranging from micelles ($q = 0$), to small isotropic bicelles ($q \leq 0.5$), to large bicelles ($q = 3$), to bilayers ($q =$ infinity). Sample alignment and correlation times define the transition from solution NMR to solid-state NMR methods.

strongly aligned samples can be interpreted very similarly (Mesleh *et al.*, 2003; Nevzorov and Opella, 2003a), the solution NMR and solid-state NMR approaches to structure determination are complementary.

Protein Expression

NMR experiments place stringent requirements on the purity and homogeneity of samples. The need for isotopic labeling and large amounts of material means that nearly all NMR studies rely on polypeptides prepared by chemical synthesis or expression in bacteria. Other expression systems are under development and will undoubtedly contribute to the preparation of membrane proteins for future studies. Although modern synthetic methods can provide uniquely labeled polypeptides (Koechendorfer *et al.*, 2004), the quest to study larger proteins and the spectroscopic benefits of uniform labeling (Cross *et al.*, 1982) place the emphasis on expression in bacteria.

Many factors are taken into account in optimizing the expression system for a helical membrane protein in *Escherichia coli* (Grisshammer and Tate, 1995). By synthesizing the oligonucleotides, it is possible to optimize the codon usage for *E. coli* regardless of the initial source of the gene. We have found the use of fusion proteins to be essential for the expression of most small and medium size membrane proteins in *E. coli*, whether of prokaryotic or eukaryotic origin; although, in some cases a polyhistidine tag (His-tag) at the C- or N-terminus is sufficient (Tian *et al.*, 2003). Depending on the system and conditions, large amounts of the expressed fusion proteins can be found in the cell membranes, inclusion bodies, or in a few examples, in the cytoplasm. It is often preferable to form large amounts of inclusion bodies, since they appear to protect the polypeptides from proteases and their isolation provides an initial purification step. We have fused the genes of interest to that of the Trp leader sequence (trpΔLE) (Ma *et al.*, 2002; Staley and Kim, 1994), ketosteroid isomerase (KSI) (Kuliopulos and Walsh, 1994; Park *et al.*, 2003), and maltose binding protein (MBP) (Berthold *et al.*, 2003; di Guan *et al.*, 1988). It is often necessary to try several of these systems to obtain reasonable overall yields of protein. Following isolation and purification using standard methods, the polypeptide of interest can be cleaved from the fusion partner with cyanogen bromide or a selective protease. Final purification is generally accomplished with high-performance liquid chromatography (HPLC) or size-exclusion chromatography in the presence of lipids.

Other factors that require optimization to obtain sufficient amounts for economical isotopic labeling include the choice of host strain, growth conditions, and the durations of the pre- and postinduction periods. The use of strains deficient in gene products with known cytoplasmic protease activities, such as Lon, OmpT, DegP, or HtpR, contributes by minimizing the effects of proteolytic degradation (Baker *et al.*, 1984; Strauch and Beckwith, 1988). *E. coli* strain BL21(DE3), which is defective in OmpT and Lon, is widely used. In some cases, mutants of BL21(DE3), such as C41(DE3) and C43(DE3), are used to overcome toxic effects associated with overexpression as inclusion bodies and produce elevated levels of proteins (Miroux and Walker, 1996).

We have described the expression in *E. coli* of a number of full-length and truncated versions of Vpu, an integral membrane protein of HIV-1, using two different systems. The His-tag trpΔLE-Vpu fusion protein cloned into the pMMHa vector (Staley and Kim, 1994) is expressed at levels up to 20% of total cellular protein in *E. coli* strain BL21(DE3) (Ma *et al.*, 2002), while the Vpu-KSI_His-tag fusion protein cloned into pET-31b(+) vector (Novagen) is expressed at even higher levels, up to 40% of total cellular protein in *E. coli* strain C43(DE3) (Park *et al.*, 2003).

The most powerful NMR approaches use protein samples labeled uniformly with ^{2}H, ^{13}C, and ^{15}N in various combinations. It is often useful to have selectively ^{15}N-labeled samples available as well. The most widely used method to obtain enriched proteins has been through the expression of proteins from bacterial cells grown in minimal media with ^{15}N-enriched ammonium chloride as the sole nitrogen source (Cross *et al.*, 1982). This approach can be readily extended to ^{13}C sites through the use of labeled

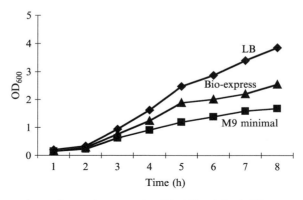

FIG. 2. Comparison of growth curves for LB, M9 minimal, Bio-express (Cambridge Isotope Laboratories) on *E. coli* C43(DE3) transformed with the plasmid vectors carrying the Vpu gene.

glucose as the carbon source and has proven successful in the determination of many protein structures by both solution and solid-state NMR. Nonetheless, there are limitations associated with using minimal media for protein expression. Bacterial growth rates and protein expression can be increased with the use of complex labeled media of biological origin that are commercially available and contain more complex substrates rather than simple sugars and nitrogen salts. The rich media can increase growth rate for the bacteria and increase total protein expression. Figure 2 compares the growth curves for several media on *E. coli* C43(DE3) transformed with the plasmid vectors carrying the Vpu construct. The final yield of pure target protein from LB media, standard M9 minimal media for ^{15}N uniform labeling, Bio-Express rich media for $\sim$20% ^{13}C/$\sim$98% ^{15}N uniform labeling (Cambridge Isotope Laboratories), is 10, 5, and 8 mg/liter of culture, respectively.

Protein Purification

The methods used for the purification of membrane proteins are generally similar to those employed with soluble proteins. However, the characteristic features of helical membrane proteins, namely their high hydrophobicity and tendency to aggregate, make modifications necessary. The isolation and purification of fusion proteins are generally accomplished with a four-step protocol when the trpΔLE, KSI, and MBP fusion systems are used (Ma *et al.*, 2002; Park *et al.*, 2003): (1) isolation of the inclusion bodies containing the fusion protein from the *E. coli* lysate by centrifugation; (2) purification of the fusion protein by affinity (Ni-chelate resin for His-tag, amylose resin for

MBP) chromatography; (3) cleavage of the target protein from its fusion partner using cyanogen bromide or a selective protease; and (4) purification of the target protein by reverse-phase HPLC and/or size-exclusion chromatography in the presence of lipids.

The polyacrylamide gel electrophoresis (PAGE) results shown in Fig. 3 illustrate the purification of the 37-residue hydrophobic polypeptide corresponding to the trans-membrane channel-forming domain of Vpu. To optimize expression for the corresponding KSI-fusion protein, a gene was designed with the codon usage of highly expressed genes in *E. coli* and inserted into the pET-31b(+) vector (Novagen) in BLR(DE3)pLysS *E. coli* cells. The major resolved peak in the HPLC trace in Fig. 3B corresponds to the polypeptide we refer to as Vpu_{2-30+} as verified by mass spectrometry. The fractions containing the polypeptide are collected, the solvent is removed, and the material is stored as a pure powder. The samples for both solution and solid-state NMR spectroscopy are prepared by the addition of lipids and water.

Sample Preparation

Samples of protein-containing micelles are prepared for solution NMR spectroscopy at the earliest possible stage for the purpose of quality control. There is no reason to go further if the protein has the incorrect

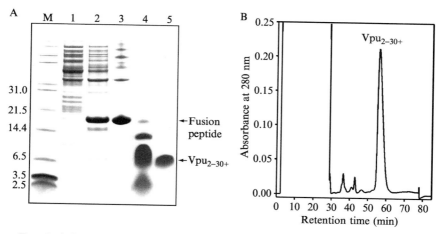

FIG. 3. (A) SDS–PAGE analysis of the purification procedure: lane 1, discarded supernatant fractions; lane 2, inclusion bodies of the KSI-Vpu_{2-30+} fusion peptide; lane 3, purified fusion peptide from Ni-chelate chromatography; lane 4, mixture dissolved in HPLC injection solvent after CNBr cleavage; lane 5, purified Vpu_{2-30+} by RP-HPLC. (B) RP-HPLC purification of Vpu_{2-30+} after CNBr cleavage of the KSI–Vpu_{2-30+} fusion peptide. The major peak at 58 min (approximately 70% of buffer B) represents purified Vpu_{2-30+}.

amino acid sequence, is aggregated, or has some other problem that can be identified in solution NMR spectra. For all membrane-associated poly- peptides, regardless of length or hydrophobicity, samples are prepared for solution NMR with four different micelle-forming lipids [sodium dode- cyl sulfate (SDS), dodecylphosphocholine (DPC), lysomyristoylphospha- tidylglycerol (LMPG), and dihexanoylphosphatidylcholine (DHPC)] (Krueger-Koplin *et al.*, 2004; Vinogradova *et al.*, 1998). Two-dimensional heteronuclear single-quantum correlation (HSQC) spectra of uniformly ^{15}N-labeled samples are highly sensitive monitors of the entire expression, purification, and sample preparation process. Any "doubling" or anoma- lous broadening of resonances is a sign that the polypeptide may be aggregated or improperly folded (McDonnell and Opella, 1993). We do not proceed with structural studies by solution NMR or with sample preparation in bilayers or bicelles for solid-state NMR until these problems are resolved. Samples are prepared by either dissolving the protein in concentrated solutions of micelle-forming lipids and diluting to the de- sired concentration, or by codissolving the protein and lipid in a suitable organic solvent, removing the organic solvent under a stream of nitrogen gas, placing the samples under high vacuum overnight, and then adding the appropriate buffer to solubilize the protein in the micelles that form. Although it may take many trials, it is almost always possible to identify at least one combination that yields well-resolved solution NMR spectra. Acquisition of a well-resolved HSQC spectrum sets the stage for structural studies by solution NMR and provides assurance that it is worthwhile to proceed with the preparation of bilayer and bicelle samples for solid-state NMR experiments.

Weakly Aligned Samples

It is generally difficult to observe and assign interresidue "long-range" NOEs in spectra of helical membrane proteins as distance constraints for structure determination. As a result, conventional approaches to structure determination of membrane proteins in lipid environments have been successful in only a few cases (Almeida and Opella, 1997; MacKenzie *et al.*, 1997). However, measurements of residual dipolar couplings (RDCs) serve to constrain the bond orientations relative to a common alignment frame and provide global structural information. Although RDCs have improved all aspects of the study of globular proteins (Bax, 2003; Bax *et al.*, 2001; de Alba and Tjandra 2002; Gronenborn 2002), they are crucial for determining the structures of helical membrane proteins (Lee *et al.*, 2003; Mesleh *et al.*, 2003).

A prerequisite for the measurement of RDCs is the ability to induce weak alignment in the samples. The most commonly used methods of inducing weak alignment cannot be applied to membrane proteins in the presence of lipids. Filamentous bacteriophages (Clore *et al.*, 1998; Hansen *et al.*, 1998) are unstable in the presence of lipids, and bicelles have a strong affinity for added lipids (Koenig *et al.*, 1999; Sass *et al.*, 1999; Tjandra and Bax 1997). These problems can be overcome by lanthanide-induced alignment (Ma and Opella, 2000; Veglia and Opella, 2000). However, we find that a more general and robust approach to obtaining weak alignment of protein-containing micelles is strain-induced alignment in gel (Chou *et al.*, 2001; Jones and Opella, 2004; Lee *et al.*, 2003; Meier *et al.*, 2002; Mesleh *et al.*, 2003; Sass *et al.*, 2000; Tycko *et al.*, 2000), since the polyacrylamide gel matrix is stable in the presence of lipids over a wide variety of temperatures, ionic strength, and pH (Tycko *et al.*, 2000).

The application of either vertical or radial compression alters the geometry of the pores in the gel, inducing a slight preference in the alignment of the protein-containing micelles. The easiest method for achieving radial compression is with the funnel-like device developed by Bax and colleagues (Chou *et al.*, 2001), although other methods have been developed (Meier *et al.*, 2002; Sass *et al.*, 2000). The interior diameter of the device is slightly larger than that of the NMR sample tube, and the gel is stretched in the z-direction as it is squeezed into the open-ended NMR tube. The bottom of the tube can be conveniently sealed using a plug. A kit consisting of the device, NMR tube, and plug is available from New Era Enterprises. Similarly, the application of vertical compression by applying pressure with a Shigemi plunger to a polyacrylamide gel has been shown to induce weak alignment (Sass *et al.*, 2000; Tycko *et al.*, 2000); in this case the gel is initially prepared with a diameter slightly smaller than that of the NMR tube, and then compressed in the z-direction. In both cases, the magnitude of alignment can be controlled by altering the amount of compression or the composition of the gel (total acrylamide concentration or the cross-link ratio) (Sass *et al.*, 2000; Tycko *et al.*, 2000).

The incorporation of protein-containing micelles into gels is the key step. Typically the protein is allowed to diffuse into the gel (Chou *et al.*, 2001; Ishii *et al.*, 2001; Tycko *et al.*, 2000), and this has been successfully applied to a few small membrane proteins (Chou *et al.*, 2002; Lee *et al.*, 2003; Mesleh *et al.*, 2003; Park *et al.*, 2003). Stretched samples are prepared by soaking a dried gel directly in the device and subsequently squeezing the sample into the NMR tube. The dried gel is cast with the same diameter as the device and the presence of the lipid provides sufficient lubrication to ensure that the gel is not adversely affected during squeezing. Alternatively, compressed samples are generally prepared by casting a gel slightly smaller than the inner diameter of the NMR tube and hydrating the gel Shigemi tube with the plunger held fixed by wrapping with Teflon tape (Ishii *et al.*, 2001; Lee *et al.*, 2003; Mesleh *et al.*, 2003). The height of the plunger is set so that the length of the rehydrated gel is constrained relative to the original size but the total volume remains similar. The ability of a protein to diffuse into a gel is highly dependent on the temperature (Tanaka, 1978), ionic strength (Ricka and Tanaka, 1984), type of lipid, and size of the protein. This approach benefits from simplicity, but not all membrane proteins, especially larger ones, diffuse into gels. Copolymerization and electrophoresis provide alternative mechanisms to incorporate membrane proteins into polyacrylamide.

Copolymerization is a rapid alternative method for incorporating proteins into gels. (Sass *et al.*, 2000; Trempe *et al.*, 2002). We have developed a simple protocol that avoids the large increase in pH that affected earlier applications to membrane proteins (Chou *et al.*, 2001). As previously recognized (Dirksen and Chrambach, 1972), the change in pH results from the quaternary amine N,N,N',N'-tetramethylethylene diamine (TEMED) that stabilizes the ammonium persulfate (APS) free radicals required for polymerization. Degassing the samples under high vacuum eliminates the requirement for TEMED (Chrambach, 1985). However, when samples are prepared in the absence of TEMED, the spectra are often problematic with multiple resonances for each site. We have found that these problems can be avoided by using 0.08% w/v APS and a small amount of TEMED (0.002%) and incubating at 36° for at least 4 h. Alternatively, since riboflavin is superior for polymerization at lower pH (Shi and Jackowski, 1998), samples can be photopolymerized by replacing the APS with 0.04% riboflavin and incubating at room temperature under strong fluorescent light. These samples generally require more time to polymerize (Righetti *et al.*, 1981). Care in degassing should be taken since oxygen is needed to convert riboflavin into the active form but also inhibits free radical formation (Gordon, 1973). The ^1H–^{15}N HSQC spectra of samples obtained by copolymerization using these methods are virtually identical to the ones obtained in the absence of polyacrylamide, although there can be a minor (10%) loss in signal intensity (Sass *et al.*, 2000). Radial compression is easily achieved by copolymerizing the sample directly in the device developed by Bax and co-workers (Chou *et al.*, 2001). Vertical compression of hydrated gels requires a significant amount of force constantly applied on the gel. The device shown in Fig. 4 utilizes a susceptibility matched Shigemi plunger (4f) to compress the gel (4g) in a screw cap NMR tube (4e) using a large nylon screw (4a) that can be rotated to force the plunger down and compress the gel. A small glass rod (4c) is inserted between the nylon screw and Shigemi plunger. This rod has a small brass screw (4d) applying pressure against a

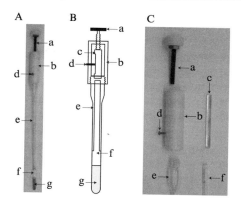

FIG. 4. Apparatus for compressing hydrated polyacrylamide gels. Photograph (A) and schematic drawing (B) of the device. (C) Photograph of the disassembled apparatus. The various components are (a) large nylon screw; (b) nylon cylinder threaded at the top to match the nylon screw and at the bottom to the NMR tube; (c) glass rod notched at one side; (d) small brass screw; (e) screw cap NMR tube; (f) Shigemi plunger; and (g) compressed polyacrylamide gel. The large nylon screw forces the plunger to compress the gel and the glass rod isolates the twisting motion.

notch to prevent twisting. This isolates the screw rotation from the sample, since movement of the plunger can damage the gel. Samples are first copolymerized in a medium wall NMR tube (3.4 mm i.d.), the isotropic values are measured, and the gel is subsequently transferred to the screw cap tube (4.2 mm i.d.) for compression. Samples are compressed until the gel uniformly fills the NMR tube and exhibits strong optical birefringence, and there is a measurable splitting of the deuterium resonance from the solvent HDO. The copolymerization approach enables the J-couplings and chemical shifts to be measured in the same gel environment before and after induction of weak alignment. This eliminates the need to make multiple samples, and it is particularly important for the accurate measurement of alignment-induced chemical shift changes (Lee *et al.*, 2003).

Electrophoresis provides an active mechanism for loading protein-containing micelles into gels. The polyacrylamide gel is prepared in a 3.4-mm-i.d. NMR tube that has a small hole cut at the bottom to provide contact with the anode buffer. After filling the upper and lower reservoirs with running buffer, the protein mixed with loading dye (1% glycerol and 0.0004% xylene cyanol) is carefully layered onto the surface of the gel. Current is applied, electrophoresis proceeds, and the indicator dye allows monitoring of the process. Since the dye and protein migrate differently, the running time must be empirically calibrated using markers with covalently

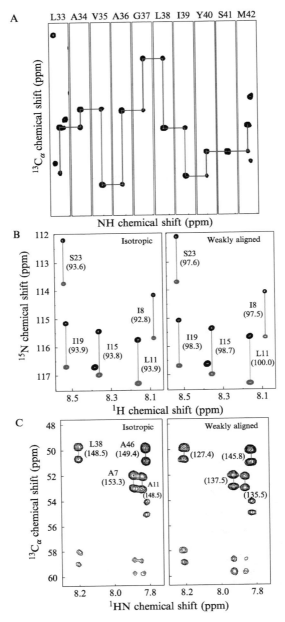

FIG. 5. Examples of solution NMR spectra of membrane proteins in micelles. (A) 3D HNCA. Strips taken parallel to the F1 axis of the HNCA spectrum of Pf1, taken at the (F2, F3) frequencies of the amides of Leu-33–Met-42 of Pf1. (B) ^{1}H–^{15}N IPAP. Expanded region of

attached dyes. The protein runs as a diffuse band, approximately the same height as the loading solution since the gel is poured without a stacking gel. Following electrophoresis, the running buffer is removed, the bottom sealed, and a 50-μl drop of D_2O is added. The sample is incubated overnight to provide sufficient time for the deuterium oxide to diffuse and obtain a lock signal. The ^1H–^{15}N HSQC spectra are virtually identical to those obtained from the protein-containing micelles in solution. Weak alignment is induced using the device shown above in the identical manner as for copolymerized samples.

The same amount of compression scales the alignment by minus one-half in prolate compared to oblate cavities. The increased RDC magnitude of the prolate pores also extends to the ^1H–^1H couplings that increase linewidths (Bax, 2003); consequently, the percentage of gel must be decreased when preparing squeezed samples. In general, it is easier to optimize sample conditions with compressed samples. While the information from these two orientations does not serve to overcome the 4-fold degeneracy associated with helices (Losonczi et al., 1999), it is still useful to confirm assignments and the location of kinks in the helix.

Solution NMR of Weakly Aligned Samples

Heteronuclear triple-resonance experiments that correlate backbone ^1HN, ^{15}N, $^1H_\alpha$, $^{13}C_\alpha$, and ^{13}CO (and side chain $^1H_\beta$ and $^{13}C_\beta$) spins using one-bond and two-bond scalar coupling interactions (Ikura et al., 1990; Kay et al., 1990) are necessary to perform the triple-resonance experiment to assign the backbone resonances and measure the RDCs for membrane proteins with more than 50 residues (Lee et al., 2003; Oxenoid et al., 2004). Figure 5 shows a three-dimensional HNCA strip plot of the membrane-bound form of the Pf1 coat protein in micelles, which correlates the amide ^1H and ^{15}N chemical shifts with the intraresidue and the proceeding residue $^{13}C_\alpha$ shift.

Since RDCs are observed as alterations in the magnitudes of the J-couplings, they are readily measured by determining the differences in the splittings observed in isotropic and weakly aligned samples. Although it is possible to measure the $^1J_{HN}$ coupling by simply removing the proton decoupling pulse during ^{15}N chemical shift evolution in HSQC experiments, the resulting spectral crowding presents problems with the already highly overlapped spectra of helical membrane proteins. Instead, IPAP

---

^1H–^{15}N, an IPAP–HSQC spectra of an isotropic and weakly aligned sample of Vpu$_{2-30+}$. The values for the one-bond ^1H–^{15}N splittings shown in parentheses are in Hz. (C) 3D HNCA-J. A selected $^{13}C_\alpha$–^1H_N slice from the HNCA-J spectra of an isotropic and weakly aligned sample of Pf1. The values for the one-bond $^1H_\alpha$–$^{13}C_\alpha$ splittings shown in parentheses are in Hz.

HSQC experiments that store the in-phase and antiphase components of the ^{15}N magnetization doublet in separate subspectra are used; the frequency difference between the two spectra corresponds to the $^1J_{HN}$ splitting (Ottiger et al., 1998). The experiment can be modified to suppress the natural abundance of ^{15}N signal from the NH_2 groups of the polyacrylamide (Ishii et al., 2001), which can be problematic. We currently use a sensitivity-enhanced version of the two-dimensional IPAP experiment (Ding and Gronenborn, 2003). To measure the $^1J_{C_\alpha-H_\alpha}$ RDCs, we performed the three-dimensional HNCA-J experiments (Montelione and Wagner, 1989; Seip et al., 1994) in both isotropic and weakly aligned samples of Pf1 (Fig. 5). In these experiments, we used 20% randomly ^{13}C- and uniformly ^{15}N-labeled sample to avoid complications from homonuclear carbon–carbon couplings to the measurement of $^1J_{C_\alpha-H_\alpha}$ RDCs. Figure 5 shows examples of solution NMR spectra used for measurement of $^1J_{HN}$ and of $^1J_{C_\alpha-H_\alpha}$ RDCs in weakly aligned samples of Vpu$_{2-30+}$ and Pf1 coat proteins in weakly aligned micelle samples.

Deuterium/hydrogen fractionation experiments (H/D experiments) are routinely applied to membrane proteins in micelles. The magnitudes of the H/D fractionation factors are indicators of the structure and topology of membrane proteins, and they are used to identify residues in trans-membrane helices as well as to determine the polarity of amphipatic helices (Czerski et al., 2000; Veglia et al., 2002). Residue-specific fractionation factors are obtained by fitting the normalized peak intensities in the ^{15}N HSQC spectrum as a function of the mole fraction of ^1H_2O.

Strongly Aligned Samples

Membrane proteins in mechanically aligned bilayers and magnetically aligned bicelles can be studied by solid-state NMR spectroscopy. They are immobilized on NMR timescales and aligned to an extent that rivals that observed for peptides in single crystals. Although proteins in bicelles are affected by the "wobble" motions of the disks, characterized by the order parameter of the system (Czerski and Sanders, 2000; Sanders et al., 1994), there is no evidence of additional protein backbone dynamics in either bilayer or bicelle samples.

It is generally easier to prepare uniaxially aligned samples of membrane proteins in bilayers than it is to crystallize them; nonetheless, it is still a demanding task to prepare the well-aligned samples of membrane proteins that yield well-resolved solid-state NMR spectra. There are two methods for mechanically aligning lipid bilayers on glass plates (Ketchem et al., 1997; Opella et al., 2001): deposition from organic solvents followed by evaporation and lipid hydration, and fusion of unilamellar reconstituted lipid vesicles with the glass surfaces. In both cases, there are numerous factors that affect the alignment of membrane proteins, including the type of lipids, the molar ratio of lipid mixtures, the molar ratio of protein to lipid, and the hydration level of the sample. Aligned lipid bilayers are mixtures of phosphatidylcholine (PC) and phosphatidylglycerol (PG) and/or phosphatidylethanolamine (PE), which are the most abundant phospholipids in the cell membrane. The optimal molar ratio of lipids varies. For example, palmitoyloleoylphosphatidylcholine/palmitoyloleoylphosphatidylglycerol (POPC/POPG) lipids in a molar ratio of 80:20 gave the best results for samples of fd coat protein (Marassi *et al.*, 1997), while dioleoylglycerophosphocholine/dioleoylglycerophosphoglycerol (DOPC/DOPG) lipids in a molar ratio of 90:10 gave the best results for samples of the transmembrane domain of Vpu (Park *et al.*, 2003). The optimal molar ratio of proteins to lipids is typically between 1:100 and 1:200. Following deposition of the lipid/protein solution on the glass slides (no. 00, Marienfeld, Germany), the residual organic solvent is removed under vacuum. The glass slides are stacked and placed in a sealed chamber containing a saturated solution of ammonium phosphate or potassium sulfate solution, pH 7.0, at 42°, which provides a 95% relative humidity atmosphere. Oriented bilayers form after equilibrating the sample in this chamber for more than 12 h. Before inserting them into the square coil of the NMR probe, the stacked glass plates are placed inside a thin film of polyethylene, which is heat-sealed to maintain sample hydration during the experiments. Local heating from radiofrequency irradiation may affect the hydration of bilayer samples, although they are typically stable for several weeks if care is taken.

Bicelles are well-suited for solid-state NMR studies of membrane-associated proteins (Sanders *et al.*, 1994). They are aggregates composed of long-chain phospholipids, e.g., dimyristoylphosphatidylcholine (DMPC), and either short-chain phospholipids or surfactants, e.g., DHPC, in aqueous solution. The long-chain lipids are organized into planar bilayers, with the short-chain lipids arranged in a rim surrounding the bilayer edges. Bicellar solutions are lyotropic liquid crystalline solutions, and at temperatures above the transition temperature of the long-chain phospholipid they form a nematic phase that aligns in magnetic fields. Significantly, enzymes have been shown to retain their biological activity in bicelles (Sanders and Landis, 1995). Membrane proteins inserted in bicelles are aligned in the magnetic field along with the lipid assembly, yielding a sample that can be studied by solid-state NMR (Howard and Opella, 1996). Bicelles are characterized by the long-chain lipid:short-chain lipid molar ratio q. For $q > 2.5$, and total lipid concentrations between 5% and 40% (w/w), typical DMPC/DHPC bicelles form a stable aligned phase with the bilayer normal perpendicular to the direction of the applied magnetic field ("perpendicular" or "unflipped" bicelles) (Sanders and Schwonek, 1992). The direction of the alignment of the bilayer normal can be "flipped" by 90° by addition of lanthanide ions; "parallel" bicelles have the same alignment as bilayers on glass plates (Prosser *et al.*, 1996). Replacement of the carboxyl-ester bonds of DMPC and DHPC with ether linkages (Woese and Wolfe, 1985) prevents acid- or base-catalyzed hydrolysis of the lipids (Cavagnero *et al.*, 1999; Ottiger and Bax, 1999); therefore bicelles composed of ditetradecylphosphatidylcholine (14-O-PC) and dihexylphosphatidylcholine (6-O-PC) (Avanti Polar lipids) in water are well-suited for protein structure determination by solid-state NMR spectroscopy.

Samples of membrane proteins in phospholipid bicelles are prepared by dissolving purified, isotopically labeled polypeptides in a 6-O-PC aqueous solution, and then adding this solution to a dispersion of 14-O-PC in water (De Angelis *et al.*, 2004; Howard and Opella, 1996). Best results are often obtained by first dissolving the pure, lyophilized polypeptide in organic solvent, typically trifluoroethanol (TFE), or a mixture of chloroform/TFE, and then evaporating the solvent to a thin, transparent protein film, which is placed under high vacuum overnight to remove the organic solvents. An aqueous solution of 6-O-PC is added to the dry protein film. A dispersion of 14-O-PC in H_2O is prepared by adding water to the lipid, followed by extensive vortexing and three freeze-heating cycles (liquid nitrogen/45°). The protein/6-O-PC solution is then added to the 14-O-PC dispersion, previously warmed above 45°. The resulting solution is briefly vortexed and taken through four freeze/heat cycles. The resulting sample for $q = 3.2$, 28% w/v, contains 3–5 mg of protein in a 200 μl volume. A short tube with 5 mm o.d. (New Era Enterprises) is filled with ~160 μl of the bicelle solution. The tube is sealed with a tight-fitting plastic cap and is pierced with a thin syringe to remove excess air from the sample and to create a tight seal. These samples are stable for many months when stored at room temperature.

Probes for Solid-State NMR

In addition to high-quality samples, the other key technology for successful solid-state NMR studies of membrane proteins is the NMR probe. The special demands placed on the NMR probes are a consequence of the characteristics of both the NMR samples being studied and the experiments being performed. First, the unusual sample geometry must be considered for the case of mechanically oriented samples that consist of protein in lipid bilayers immobilized on thin glass plates. Second, the radiofrequency properties of the sample must be taken into account. The samples contain a significant amount of water; therefore the samples impact the tuning frequency of the probe. Finally, the probe must be capable of handling high-power irradiation on both ^1H and ^{15}N channels for several milliseconds during dipolar evolution and decoupling periods. Power handling can be problematic in probes designed for standard bore (52-mm) magnets, where the crowding of circuit components often leads to arcing.

An advantage of magnetically aligned bicelle samples is that conventional solenoid coil geometry can be used. The sample coil is oriented perpendicular to the long axis of the probe and thus to the external magnetic field provided by the NMR magnet, as shown in Figs. 6 and 7. The probe circuit must be designed to minimize the detrimental effects of the high-dielectric samples. This is accomplished by aiming for the highest possible quality factor (Q) in the components and their assembly and by utilizing a circuit that has a very large tuning range (>30 MHz on the ^{1}H channel) (Fukushima and Roeder, 1981). For the mechanically aligned bilayer samples, the coil must accommodate a stack of thin glass slides that supports the formation of uniform lipid bilayers. Typically, a coil size of 12 mm × 12 mm × X mm is employed; X depends on the ^{1}H frequency, and is typically 2.6 mm at ^{1}H frequencies greater than 600 MHz and can be as large as 3–5 mm at the lower ^{1}H frequencies. These coils are referred to as "flat" coils because they resemble a flattened solenoid coil. The choice of coil geometry is made for practical sample preparation reasons, and it has been found that the larger, square surface area of the 11 mm × 11 mm glass slides is compatible with the preparation of well-aligned samples.

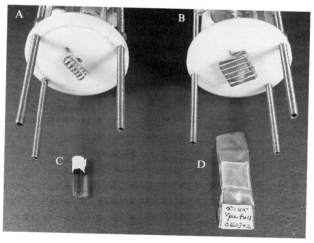

FIG. 6. The sample coils of two configurations of 50-mm solid-state NMR probes are shown with (A) a 5-mm solenoid coil and (B) a 12-mm × 12-mm × 2.6-mm flat coil. A 5-mm NMR tube containing a bicelle sample is shown in (C), while a bilayer sample consisting of 16 total 11-mm × 11-mm glass slides stacked and then sealed in plastic to retain hydration is shown in (D).

A   B

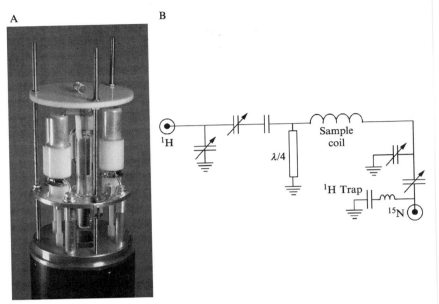

FIG. 7. (A) Double resonance 5-mm solenoid coil probe head shown in profile. (B) The probe circuit shown in electrical schematic form. The arrows indicate the variable trim capacitors of the ^{1}H and ^{15}N tuning networks. The sample coil, ^{1}H 1/4 and ^{1}H series LC trap are all labeled.

Great care must be taken during the construction of the probe to ensure that homogeneous radiofrequency magnetic fields (B_1) can be achieved, particularly with the large lower-symmetry flat coils.

The probes described above utilize a circuit (Fig. 7) similar to those of previously reported designs (Cross *et al.*, 1976; Doty *et al.*, 1981; Jiang *et al.*, 1987; Kan *et al.*, 1980). However, details of the implementation, especially the mechanical properties, have been carefully optimized and contribute to the performance of these probes, especially at relatively high (700–900 MHz) resonance frequencies. This is illustrated with a double-resonance probe tuned to 750 MHz in a standard bore magnet. A range of B_1 nutation frequencies and power depositions is presented in Table I under realistic experimental conditions and represents B_1 fields that can be generated for 20-ms irradiations.

Solid-State NMR of Strongly Aligned Samples

The first step in solid-state NMR studies is to verify the alignment of the bilayer lipids and the protein. Acquisition of a one-dimensional ^{31}P spectrum or of a ^{2}H spectrum after adding small amounts of deuterated

TABLE I
STATIC SOLID-STATE NMR PROBE PERFORMANCE[a]

Probe coil geometry	Channel	Crystalline sample: nutation frequency (kHz) power deposition (W)	Biological sample: nutation frequency (kHz) power deposition (W)
750-MHz solenoid coil	^{1}H	85–90 kHz 100–125 W	65–70 kHz 100–125 W
	^{15}N	70–75 kHz 600–700 W	70–75 kHz 600–700 W
750-MHz flat coil	^{1}H	50–55 KHz 75–100 W	50–55 kHz 100–125 W
	^{15}N	50–55 kHz 700–800 W	50–55 kHz 700–800 W

[a] 750 MHz/54-mm bore.

lipids to the system enables the bilayer orientation to be monitored. The one-dimensional ^{1}H–^{15}N cross-polarization (CP) experiment (Pines *et al.*, 1973) is carried out on ^{15}N-labeled proteins to verify protein alignment. Any unoriented protein is readily recognized by the observation of characteristic "powder pattern" intensity in the spectrum. Resolved resonances in well-aligned bilayer samples display linewidths 2–5 ppm, similar to those observed for peptide single crystals. The cross-polarization experiments are optimized by the choice of contact-time and ^{1}H-carrier frequency and compensated for mismatch by using CP-MOIST (Levitt *et al.*, 1986). Composite-pulse decoupling by SPINAL-16 (Fung *et al.*, 2000) has proven particularly effective for heteronuclear ^{1}H decoupling during acquisition. Each ^{1}H–^{15}N bond in a well-aligned protein contributes a single, sharp line in the spectrum. In the case of complete alignment on glass plates, the one-dimensional spectrum for a trans-membrane helix spans the frequency region between 150 and 225 ppm. By contrast, in the case of bicelles oriented perpendicular to the magnetic field, this range is reversed with respect to the isotropic value of 120 ppm for amide sites, and then scaled by a factor of 0.5 (Nevzorov *et al.*, 2004b). The structural information is obtained from ^{1}H–^{15}N dipolar coupling/^{15}N chemical shift-separated local field spectra (Waugh, 1976) of uniformly ^{15}N-labeled proteins (Marassi and Opella, 2000). The PISEMA (polarization inversion spin exchange at the magic angle) (Wu *et al.*, 1994) and the more recent SAMMY (Nevzorov and Opella, 2003b) experiments were designed to remove ^{1}H–^{1}H couplings during ^{1}H–^{15}N dipolar evolution, either by Lee–Golburg decoupling (Lee and Goldburg, 1963), or by refocusing unwanted couplings with a "magic sandwich" (Rhim *et al.*, 1970). Both of these experiments can be used to measure ^{1}H–^{15}N dipolar couplings in bilayers or in bicelles. SAMMY, which was recently developed to compensate for ^{1}H chemical shift offset dependence, is generally easier to set up than PISEMA and is less demanding of the probe. However, PISEMA is capable of yielding spectra with narrower dipolar linewidths due to its more favorable scaling factor, and it has been analyzed more extensively (Gan, 2000; Ramamoorthy *et al.*, 1999). Figure 8 shows the PISEMA spectra of the uniformly ^{15}N-labeled transmembrane domain of Vpu in a completely aligned bilayer sample (Park *et al.*, 2003) and in "unflipped" bicelles (De Angelis *et al.*, 2004), both displaying the characteristic "wheel-like" patterns obtained from α-helical segments. In these spectra, each amide site contributes a single correlation resonance that is characterized by unique ^{15}N chemical shifts and ^{1}H–^{15}N dipolar coupling frequencies. The differences between the spectra obtained in bilayers and bicelles reflect the 90° difference in orientation of the peptide planes with respect to the static magnetic field in the two different types of samples. Spectral assignments are verified through comparisons with spectra from selectively ^{15}N-labeled samples.

Calculation of Three-Dimensional Structures from Orientation Constraints

On the path toward structure determination, the secondary structure and topology of membrane proteins can be described by the patterns of resonances in PISEMA spectra. The angular constraints derived from the orientation-dependent frequencies of the resonances provide reliable and precise structural information. The independent measurements of ^{1}H–^{15}N heteronuclear dipolar coupling and ^{15}N chemical shifts frequencies for each backbone amide site in a protein relative to a single reference frame, e.g., the magnetic field axis for completely aligned samples, mean that the errors do not propagate in a cumulative manner. This enables the combination of experimental constraints from individual residues and well-established covalent geometry of proteins (bond lengths, dihedral angles, planarity of peptide linkages) to be used as the basis for protein structure determination with atomic resolution.

PISA (polarity index slant angle) wheels (Marassi and Opella, 2000; Wang *et al.*, 2000) and Dipolar Waves (Kovacs *et al.*, 2000; Mesleh and Opella, 2003) involve the analysis of the periodic patterns observed in spectral data. While PISA wheels were first developed to interpret PISEMA spectra of completely aligned proteins in bilayers, these two methods are applicable to results from both solid-state NMR of strongly aligned samples (bilayers, bicelles) and solution NMR data of weakly aligned samples (micelles, small "isotropic" bicelles) (Lee *et al.*, 2003). Although applicable to any type of regular secondary structure, including β-sheet, this approach

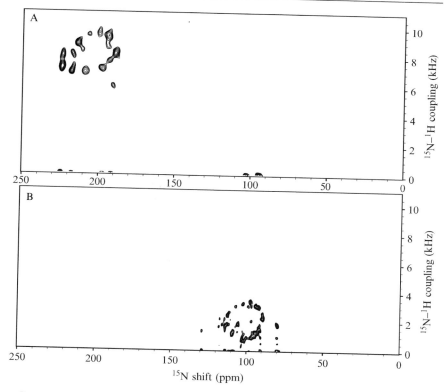

Fig. 8. Two-dimensional ^1H–^{15}N dipolar coupling/^{15}N chemical shift PISEMA spectra of the uniformly ^{15}N-labeled trans-membrane domain of Vpu at 700 MHz: (A) in DOPC/DOPG bilayers on glass plates, mechanically aligned; (B) in large "unflipped" bicelles, magnetically aligned, with $q = [14\text{-}O\text{-}PC]/[6\text{-}O\text{-}PC] = 3.2$. The differences in chemical shifts and dipolar couplings going from bilayers to "unflipped" bicelles reflect the relative $90°$ orientation of the proteins in the two types of samples.

has been developed primarily for helical proteins. The characteristic "wheel-like" patterns of ^1H–^{15}N heteronuclear dipolar couplings versus ^{15}N chemical shifts observed in spectra of completely aligned samples reflect helical wheel projections of residues in both trans-membrane and in-plane helices. In Dipolar Waves, a periodic function is obtained by using a simple distribution on a cone assuming a constant angle between the N–H bond and the helix axis. The magnitudes of the chemical shifts (Kovacs *et al.*, 2000) and of the dipolar couplings (Mesleh and Opella, 2003) can be plotted as a function of residue number, generating sinusoidal waves with a period of 3.6 for an α-helix. This is illustrated in Fig. 9 for an ideal α-helix at different angles in bilayers.

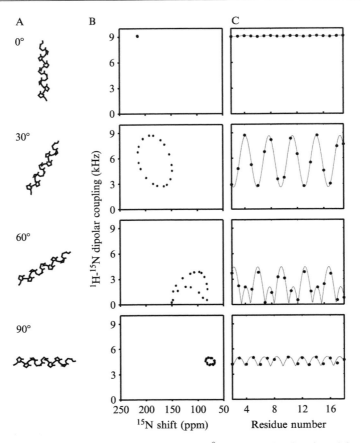

Fig. 9. An ideal α-helix is tilted at 0, 30, 60, and 90° relative to the direction of the applied magnetic field (A). Simulated PISA wheels for the various tilt angles (B). Simulated Dipolar Waves with the same tilt angles (C).

Inspection of the "wheel-like" patterns in the PISEMA spectrum allows an immediate evaluation of the tilt of the α-helical domains, before having completed a full spectral assignment (Fig. 10). Dipolar Waves provide independent validation of the geometry and regularity of α-helices. Comparisons of experimentally measured dipolar couplings, modeling studies, and bioinformatics have shown that the helices found in proteins typically satisfy this ideal approximation quite well (Kim and Cross, 2002; Walther and Cohen, 1999). Deviations from ideality have a pronounced effect on the appearance of PISA wheels and Dipolar Waves. In general, however, the 100° rotations between adjacent residues in the sequence preserve

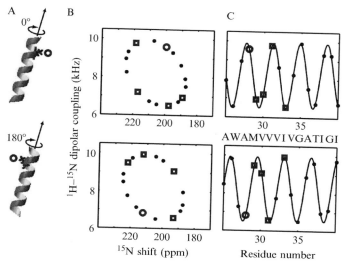

FIG. 10. Examples of mapping of structure onto spectra. ^{15}N-PISEMA spectra are simulated for an α-helix tilted 20° relative to the magnetic field axis (A) are shown with their corresponding PISA wheel patterns (B) for two different rotation angles about the long axis of the helix. Rotation of the helix by 180° results in a rotation of the position of a single resonance (circle) by 180° in the PISA wheels. Multiple sites can also be labeled and their patterns mapped directly from the positions of these residues on the helical wheel. This is shown for the dipolar couplings as a function of residue number in (C).

the general wheel-like pattern of resonances from helical residues. The identification and assignment of resonances from helices in aligned proteins in the solid state represent the first stage in the structure calculation process and yield the tilts and rotations of helices. Without the influence of chemical shift variability, Dipolar Waves are an even more reliable indicator of molecular structure with fits to sine waves that are better than the experimental errors (Mesleh and Opella, 2003). This is shown in Fig. 11 with the PISEMA spectra of the trans-membrane domain of Vpu for a uniformly ^{15}N-labeled and four selectively ^{15}N-labeled samples of the protein. The spectra are superimposed on the ideal PISA wheel calculated for an 18-residue α-helix with uniform dihedral angles. The sequential assignment of one or the identification of a few residues by type and their position in the PISA Wheel (or the phase of the Dipolar Wave) directly determines the rotational orientation of the α-helix about its long axis. The rest of the spectrum can be assigned based on the expected pattern of assignments in the helical wheel, the so-called "shotgun" approach (Marassi and Opella, 2003). Figure 12 illustrates the analysis of the orientationally dependent

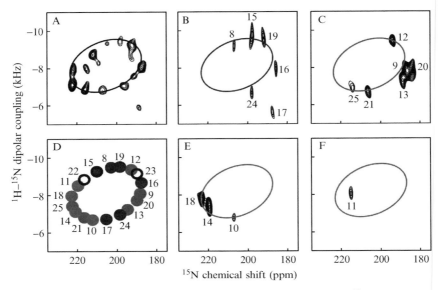

Fig. 11. Shotgun assignment: Two-dimensional ^1H–^{15}N dipolar coupling/^{15}N chemical shift correlation PISEMA spectrum of uniformly and selectively ^{15}N-labeled Vpu$_{2-30+}$ peptide in lipid bilayers superimposed on the ideal PISA wheel as calculated for an 18-residue α-helix, with uniform dihedral angles ($\phi = -57°$; $\psi = -47°$) and the tilt angle of 18°: (A) uniform; (B) isoleucine; (C) valine; (D) ideal PISA wheel; (E) alanine; (F) leucine. The assignments of the amide resonances are indicated by residue numbers.

frequencies measured from the spectrum of uniformly ^{15}N-labeled trans-membrane Vpu, and the assignments derived from the spectra of the selectively labeled samples from Fig. 11. For data obtained in solution as well as in the solid-state, identification of the helices is made by applying a sliding window function to identify regions with a distinct 3.6-residue periodicity. In practice, this limits the number of degrees of freedom. In this example, the kink is evidenced by the discontinuity of the Dipolar Wave in data sets obtained in micelles as well as bilayers.

Initial determinations of protein structure by solid-state NMR relied on the frequencies of independently assigned resonances, using graphic representations of the geometric relationships and restrictions derived from the orientationally dependent spectral parameters measured in the solid-state NMR experiments. Other approaches include Direct Calculation (Bertram *et al.*, 2000; Opella *et al.*, 1987) and Structural Fitting (Nevzorov and Opella, 2003a), which, with complete data sets, give the same results for

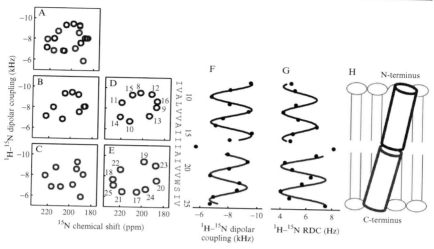

FIG. 12. (A–C) Representations of the experimental PISEMA spectrum of Vpu in aligned bilayers. (A) All residues in the trans-membrane helix. (B) Residues 8–16. (C) Residues 17–25. (D and E) Ideal PISA wheels. (F and G) Dipolar Waves fit to the experimental plots of dipolar couplings as a function of residue number. (F) Dipolar couplings from a completely aligned sample. (G) Residual dipolar couplings from a weakly aligned sample. (H) Tube representation of the trans-membrane helix in lipid bilayers.

helices, and are capable of determining complex irregular structures from these data. With partial data sets or incomplete assignments, they provide complementary views that can be checked against the results obtained by the analysis of PISA wheels and Dipolar Waves. It is possible to perform a direct mathematical analysis of the NMR frequencies in terms of orientation constraints. In the most general approach to structure determination of strongly aligned samples, two or more frequencies for each residue of a protein are needed to determine the orientation of a peptide plane. The orientation of each plane is related to the common axis defined by the direction of the applied magnetic field; therefore all the planes can be assembled into a complete protein structure. The various steps in the process can be separated so that errors and uncertainties can be evaluated. We calculated the structures of the M2 ion-channel peptide (Opella et al., 1999) and fd coat protein (Marassi and Opella, 2003) in bilayers with this approach. The chemical shift and dipolar coupling frequencies from the PISEMA spectra provide the sole input for structure determination. While early determinations of protein structure by solid-state NMR utilized the frequencies of independently assigned resonance, the "shotgun" NMR

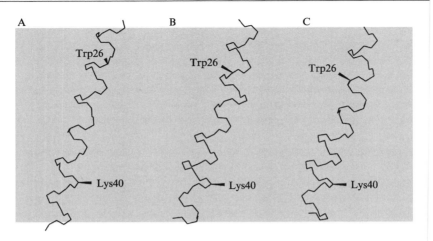

FIG. 13. Structures of the trans-membrane domain of the membrane-bound form of the fd coat protein obtained by various methods. (A) Ideal helical residues fit to dipolar waves. (B) Average structure from structural fitting. (C) Unique structure from direct calculation.

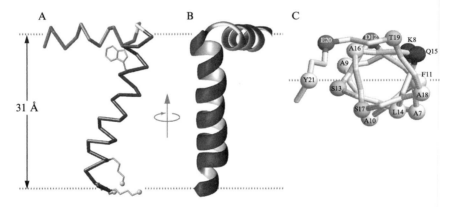

FIG. 14. Three-dimensional views of the fd coat protein in lipid bilayers calculated from the assigned PISEMA spectrum. All of the features of the protein structure are found, including the kink in the hydrophobic trans-membrane helix (A), the orientation of the in-plane amphipathic helix (B), and the conformations of the residues in the loop connecting the two helices (C).

approach relies solely on the spectra from one uniformly and several selectively ^{15}N-labeled samples and on the symmetry properties of PISA wheels to enable the simultaneous sequential assignment of resonances and the measurement of orientationally dependent frequency constraint.

"Structural fitting" is a recently developed approach that is complementary to the direct calculation of protein structures from the orientation-dependent frequencies (Nevzorov and Opella, 2003a). Structure determination is accomplished by fitting the spectrum to a model structure assuming a constant peptide plane geometry. When the resonances are assigned, the Ramachandran angles Φ and Ψ are considered the only degrees of freedom. The whole structure is assembled by the sequential walking from one residue to the next, calculating the angles Φ and Ψ directly from two orientation-dependent frequencies for these residues. The algorithm is potentially applicable in an "assignment-free" manner to spectra of uniformly ^{15}N-labeled proteins; however, the precision of the structural fitting is improved by the addition of assignment information, such as the identification of resonances by residue type from spectra of selectively labeled proteins. The method can incorporate the sources of error in solid-state NMR data, for example, incorrect measurement of the frequencies or residue-to-residue variability of the magnitudes and orientations in the molecular frame of the principal components of the chemical shift tensor. We have determined the structures of fd (Zeri *et al.*, 2003) and Pf1 (Thiriot *et al.*, 2004) bacteriophages and of the Vpu channel-forming trans-membrane domain (Park *et al.*, 2003) by structural fitting.

Figure 13 compares the structures of the trans-membrane domain of the membrane-bound form of fd coat protein in bilayers determined with three different methods of data analysis. All three methods show the presence of a helix kink near residue 39 and a similar overall helix tilt. The complete structure of the fd coat protein in bilayers obtained using direct calculation of peptide planes with the "shotgun" NMR approach to resonance assignment is shown in Fig. 14. It consists of three views, showing the relative position of the trans-membrane and amphipathic domains and a top view of the trans-membrane helix. The 16-Å-long in-plane helix is amphipatic and rests on the membrane surface, while the 35-Å-long trans-membrane helix crosses the membrane, with a change in helix tilt that accommodates the thickness of the phospholipid bilayer of ~31 Å.

Acknowledgments

This research was supported by Grants EB002169, GM064676, and GM066978, the Biomedical Technology Resource for NMR Molecular Imaging of Proteins (EB002031), and a postdoctoral fellowship GM65833 to A.D. from the National Institutes of Health. D.J. was supported by postdoctoral fellowship 70561-RF from the American Foundation of AIDS Research.

References

Almeida, F. C., and Opella, S. J. (1997). fd coat protein structure in membrane environments: Structural dynamics of the loop between the hydrophobic trans-membrane helix and the amphipathic in-plane helix. *J. Mol. Biol.* **270,** 481–495.

Auger, M. (2000). Biological membrane structure by solid-state NMR. *Curr. Issues Mol. Biol.* **2,** 119–124.

Baker, T. A., Grossman, A. D., and Gross, C. A. (1984). A gene regulating the heat shock response in *Escherichia coli* also affects proteolysis. *Proc. Natl. Acad. Sci. USA* **81,** 6779–6783.

Bax, A. (2003). Weak alignment offers new NMR opportunities to study protein structure and dynamics. *Protein Sci.* **12,** 1–16.

Bax, A., Kontaxis, G., and Tjandra, N. (2001). Dipolar couplings in macromolecular structure determination. *Methods Enzymol.* **339,** 127–174.

Berthold, D. A., Stenmark, P., and Nordlund, P. (2003). Screening for functional expression and overexpression of a family of diiron-containing interfacial membrane proteins using the univector recombination system. *Protein Sci.* **12,** 124–134.

Bertram, R., Quine, J. R., Chapman, M. S., and Cross, T. A. (2000). Atomic refinement using orientational restraints from solid-state NMR. *J. Magn. Reson.* **147,** 9–16.

Bertram, R., Asbury, T., Fabiola, F., Quine, J. R., Cross, T. A., and Chapman, M. S. (2003). Atomic refinement with correlated solid-state NMR restraints. *J. Magn. Reson.* **163,** 300–309.

Cavagnero, S., Dyson, H. J., and Wright, P. E. (1999). Improved low pH bicelle system for orienting macromolecules over a wide temperature range. *J. Biomol. NMR* **13,** 387–391.

Chou, J. J., Gaemers, S., Howder, B., Louis, J. M., and Bax, A. (2001). A simple apparatus for generating stretched polyacrylamide gels, yielding uniform alignment of proteins and detergent micelles. *J. Biomol. NMR* **21,** 377–382.

Chou, J. J., Kaufman, J. D., Stahl, S. J., Wingfield, P. T., and Bax, A. (2002). Micelle-induced curvature in a water-insoluble HIV-1 Env peptide revealed by NMR dipolar coupling measurement in stretched polyacrylamide gel. *J. Am. Chem. Soc.* **124,** 2450–2451.

Chrambach, A. (1985). "Advanced Methods in the Biological Sciences the Practice of Quantitative Gel Electrophoresis." VCH Publishers, Deerfield Beach, FL.

Clore, G. M., Starich, M. R., and Gronenborn, A. M. (1998). Measurement of residual dipolar couplings of macromolecules aligned in the nematic phase of a colloidal suspension of rod-shaped viruses. *J. Am. Chem. Soc.* **120,** 10571–10572.

Cross, T. A., DiVerdi, J. A., and Opella, S. J. (1982). Strategy for nitrogen NMR of biopolymers. *J. Am. Chem. Soc.* **104,** 1759–1761.

Cross, V. R., Hester, R. K., and Waugh, J. S. (1976). Single coil probe with transmission-line tuning for nuclear magnetic double resonance. *Rev. Sci. Instrum.* **47,** 1486–1488.

Czerski, L., and Sanders, C. R. (2000). Functionality of a membrane protein in bicelles. *Anal. Biochem.* **284,** 327–333.

Czerski, L., Vinogradova, O., and Sanders, C. R. (2000). NMR-Based amide hydrogen-deuterium exchange measurements for complex membrane proteins: Development and critical evaluation. *J. Magn. Reson.* **142,** 111–119.

de Alba, E., and Tjandra, N. (2002). NMR dipolar couplings for the structure determination of biopolymers in solution. *Prog. NMR Spectrosc.* **40,** 175–197.

De Angelis, A. A., Nevzorov, A. A., Park, S. H., Howell, S. C., Mrse, A. A., and Opella, S. J. (2004). High-resolution NMR spectroscopy of membrane proteins in aligned bicelles. *J. Am. Chem. Soc.* **126,** 1540–1541.

de Groot, H. J. (2000). Solid-state NMR spectroscopy applied to membrane proteins. *Curr. Opin. Struct. Biol.* **10,** 593–600.

di Guan, C., Li, P., Riggs, P. D., and Inouye, H. (1988). Vectors that facilitate the expression and purification of foreign peptides in Escherichia coli by fusion to maltose-binding protein. *Gene* **67**, 21–30.

Ding, K. Y., and Gronenborn, A. M. (2003). Sensitivity-enhanced 2D IPAP, TROSY-anti-TROSY, and E.COSY experiments: Alternatives for measuring dipolar N-15-H-1(N) couplings. *J. Magn. Reson.* **163**, 208–214.

Dirksen, M. L., and Chrambach, A. (1972). Studies on the redox state in poly acrylamide gels. *Separation Sci.* **7**, 747–772.

Doty, F. D., Inners, R. R., and Ellis, P. D. (1981). A multinuclear double-tuned probe for applications with solids or liquids utilizing lumped tuning elements. *J. Magn. Reson.* **43**, 399–416.

Drechsler, A., and Separovic, F. (2003). Solid-state NMR structure determination. *IUBMB Life* **55**, 515–523.

Fernandez, C., Hilty, C., Wider, G., Guntert, P., and Wuthrich, K. (2004). NMR structure of the integral membrane protein OmpX. *J. Mol. Biol.* **336**, 1211–1221.

Fu, R., and Cross, T. A. (1999). Solid-state nuclear magnetic resonance investigation of protein and polypeptide structure. *Annu. Rev. Biophys. Biomol. Struct.* **28**, 235–268.

Fukushima, E., and Roeder, S. B. W. (1981). *In* "Experimental Pulse NMR a Nuts and Bolts Approach" (E. Fukushima and S. B. W. Roeder, eds.), pp. 407–416. Addison-Wesley Publishing Company, Reading, MA.

Fung, B. M., Khitrin, A. K., and Ermolaev, K. (2000). An improved broadband decoupling sequence for liquid crystals and solids. *J. Magn. Reson.* **142**, 97–101.

Gan, Z. (2000). Spin dynamics of polarization inversion spin exchange at the magic angle in multiple spin systems. *J. Magn. Reson.* **143**, 136–143.

Glover, K. J., Whiles, J. A., Wu, G., Yu, N., Deems, R., Struppe, J. O., Stark, R. E., Komives, E. A., and Vold, R. R. (2001). Structural evaluation of phospholipid bicelles for solution-state studies of membrane-associated biomolecules. *Biophys. J.* **81**, 2163–2171.

Gordon, A. H. (1973). "Electrophoresis of Proteins in Poly Acrylamide and Starch Gels." Elsevier/North-Holland Biomedical Press, Amsterdam.

Griffin, R. G. (1998). Dipolar recoupling in MAS spectra of biological solids. *Nat. Struct. Biol.* **5**, 508–512.

Grisshammer, R., and Tate, C. G. (1995). Overexpression of integral membrane proteins for structural studies. *Q. Rev. Biophys.* **28**, 315–422.

Gronenborn, A. M. (2002). The importance of being ordered: Improving NMR structures using residual dipolar couplings. *C. R. Biol.* **325**, 957–966.

Hansen, M. R., Mueller, L., and Pardi, A. (1998). Tunable alignment of macromolecules by filamentous phage yields dipolar coupling interactions. *Nat. Struct. Biol.* **5**, 1065–1074.

Howard, K. P., and Opella, S. J. (1996). High-resolution solid-state NMR spectra of integral membrane proteins reconstituted into magnetically oriented phospholipid bilayers. *J. Magn. Reson.* **B112**, 91–94.

Hwang, P. M., Choy, W. Y., Lo, E. I., Chen, L., Forman-Kay, J. D., Raetz, C. R., Prive, G. G., Bishop, R. E., and Kay, L. E. (2002). Solution structure and dynamics of the outer membrane enzyme PagP by NMR. *Proc. Natl. Acad. Sci. USA* **99**, 13560–13565.

Ikura, M., Kay, L. E., and Bax, A. (1990). A novel approach for sequential assignment of 1H, 13C, and 15N spectra of proteins: Heteronuclear triple-resonance three-dimensional NMR spectroscopy. Application to calmodulin. *Biochemistry* **29**, 4659–4667.

Ishii, Y., Markus, M. A., and Tycko, R. (2001). Controlling residual dipolar couplings in high-resolution NMR of proteins by strain induced alignment in a gel. *J. Biomol. NMR* **21**, 141–151.

Jiang, Y. J., Pugmire, R. J., and Grant, D. M. (1987). An efficient double-tuned $^{13}C/^1H$ probe circuit for CP/MAS NMR and its importance in linewidths. *J. Magn. Reson.* **71,** 485–494.

Jones, D. H., and Opella, S. J. (2004). Weak alignment of membrane proteins in stressed polyacrylamide gels. *J. Magn. Reson.* **171,** 258–269.

Kan, S., Fan, M., and Courtieu, J. (1980). A single-coil triple resonance probe for NMR experiments. *Rev. Sci. Instrum.* **51,** 887–890.

Kay, L. E., Ikura, M., Tschudin, R., and Bax, A. (1990). Three-dimensional triple-resonance NMR spectroscopy of isotopically enriched proteins. *J. Magn. Reson.* **89,** 496–514.

Ketchem, R., Roux, B., and Cross, T. (1997). High-resolution polypeptide structure in a lamellar phase lipid environment from solid state NMR derived orientational constraints. *Structure* **5,** 1655–1669.

Kim, S., and Cross, T. A. (2002). Uniformity, ideality, and hydrogen bonds in transmembrane alpha-helices. *Biophys. J.* **83,** 2084–2095.

Koechendorfer, G. G., Jones, D. H., Lee, S., Oblatt-Montal, M., Opella, S. J., and Montal, M. (2004). Functional characterization and NMR spectroscopy of full-length Vpu from HIV-1 prepared by total chemical synthesis. *J. Am. Chem. Soc.* **8,** 2439–2446.

Koenig, B. W., Hu, J. S., Ottiger, M., Bose, S., Hendler, R. W., and Bax, A. (1999). NMR measurement of dipolar couplings in proteins aligned by transient binding to purple membrane fragments. *J. Am. Chem. Soc.* **121,** 1385–1386.

Kovacs, F. A., Denny, J. K., Song, Z., Quine, J. R., and Cross, T. A. (2000). Helix tilt of the M2 transmembrane peptide from influenza A virus: An intrinsic property. *J. Mol. Biol.* **295,** 117–125.

Krueger-Koplin, R. D., Sorgen, P. L., Krueger-Koplin, S. T., Rivera-Torres, I. O., Cahill, S. M., Hicks, D. B., Grinius, L., Krulwich, T. A., and Girvin, M. E. (2004). An evaluation of detergents for NMR structural studies of membrane proteins. *J. Biomol. NMR* **28,** 43–57.

Kuliopulos, A., and Walsh, C. T. (1994). Production, purification, and cleavage of tandem repeats of recombinant peptides. *J. Am. Chem. Soc.* **116,** 4599–4607.

Lee, M., and Goldburg, W. I. (1963). Nuclear magnetic resonance line narrowing by a rotating rf field. *Phys. Rev. Lett.* **11,** 255–258.

Lee, S., Mesleh, M. F., and Opella, S. J. (2003). Structure and dynamics of a membrane protein in micelles from three solution NMR experiments. *J. Biomol. NMR* **26,** 327–334.

Levitt, M. H., Suter, D., and Ernst, R. R. (1986). Spin dynamics and thermodynamics in solid-state NMR cross polarization. *J. Chem. Phys.* **84,** 4243–4255.

Losonczi, J. A., Andrec, M., Fischer, M. W. F., and Prestegard, J. H. (1999). Order matrix analysis of residual dipolar couplings using singular value decomposition. *J. Magn. Reson.* **138,** 334–342.

Luca, S., Heise, H., and Baldus, M. (2003). High-resolution solid-state NMR applied to polypeptides and membrane proteins. *Acc. Chem. Res.* **36,** 858–865.

Ma, C., and Opella, S. J. (2000). Lanthanide ions bind specifically to an added "EF-hand" and orient a membrane protein in micelles for solution NMR spectroscopy. *J. Magn. Reson.* **146,** 381–384.

Ma, C., Marassi, F. M., Jones, D. H., Straus, S. K., Bour, S., Strebel, K., Schubert, U., Oblatt-Montal, M., Montal, M., and Opella, S. J. (2002). Expression, purification, and activities of full-length and truncated versions of the integral membrane protein Vpu from HIV-1. *Protein Sci.* **11,** 546–557.

MacKenzie, K. R., Prestegard, J. H., and Engelman, D. M. (1997). A transmembrane helix dimer: Structure and implications. *Science* **276,** 131–133.

Marassi, F. M., and Opella, S. J. (2000). A solid-state NMR index of helical membrane protein structure and topology. *J. Magn. Reson.* **144,** 150–155.

Marassi, F. M., and Opella, S. J. (2003). Simultaneous assignment and structure determination of a membrane protein from NMR orientational restraints. *Protein Sci.* **12**, 403–411.

Marassi, F. M., Ramamoorthy, A., and Opella, S. J. (1997). Complete resolution of the solid-state NMR spectrum of a uniformly 15N-labeled membrane protein in phospholipid bilayers. *Proc. Natl. Acad. Sci. USA* **94**, 8551–8556.

McDonnell, P. A., and Opella, S. J. (1993). Effect of detergent concentration on multi-dimensional solution NMR spectra of membrane proteins in micelles. *J. Magn. Reson.* **B102**, 120–125.

McDowell, L. M., and Schaefer, J. (1996). High-resolution NMR of biological solids. *Curr. Opin. Struct. Biol.* **6**, 624–629.

Meier, S., Haussinger, D., and Grzesiek, S. (2002). Charged acrylamide copolymer gels as media for weak alignment. *J. Biomol. NMR* **24**, 351–356.

Mesleh, M. F., and Opella, S. J. (2003). Dipolar waves as NMR maps of helices in proteins. *J. Magn. Reson.* **163**, 288–299.

Mesleh, M. F., Lee, S., Veglia, G., Thiriot, D. S., Marassi, F. M., and Opella, S. J. (2003). Dipolar waves map the structure and topology of helices in membrane proteins. *J. Am. Chem. Soc.* **125**, 8928–8935.

Miroux, B., and Walker, J. E. (1996). Over-production of proteins in Escherichia coli: Mutant hosts that allow synthesis of some membrane proteins and globular proteins at high levels. *J. Mol. Biol.* **260**, 289–298.

Montelione, G., and Wagner, G. (1989). Accurate measurements of homonuclear HN-H.alpha. coupling constants in polypeptides using heteronuclear 2D NMR experiments. *J. Am. Chem. Soc.* **111**, 5474–5475.

Nevzorov, A. A., and Opella, S. J. (2003a). Structural fitting of PISEMA spectra of aligned proteins. *J. Magn. Reson.* **160**, 33–39.

Nevzorov, A. A., and Opella, S. J. (2003b). A "magic sandwich" pulse sequence with reduced offset dependence for high-resolution separated local field spectroscopy. *J. Magn. Reson.* **164**, 182–186.

Nevzorov, A. A., Mesleh, M. F., and Opella, S. J. (2004a). Structure determination of aligned samples of membrane proteins by NMR spectroscopy. *Magn. Reson. Chem.* **42**, 162–171.

Nevzorov, A. A., De Angelis, A. A., Park, S. H., and Opella, S. J. (2004b). "Uniaxial Motional Averaging of the Chemical Shift Anisotropy in Magnetically Oriented Protein-Containing Bicelles." Marcel Dekker, New York.

Nielsen, N. C., Malmendal, A., and Vosegaard, T. (2004). Techniques and applications of NMR to membrane proteins. *Mol. Membr. Biol.* **21**, 129–141.

Opella, S. J. (1997). NMR and membrane proteins. *Nat. Struct. Biol.* **4**, 845–848.

Opella, S. J., and Marassi, F. M. (2004). Structure determination of membrane proteins by NMR spectroscopy. *Chem. Rev.* **104**, 3587–3606.

Opella, S. J., Stewart, P. L., and Valentine, K. G. (1987). Protein structure by solid-state NMR spectroscopy. *Q. Rev. Biophys.* **19**, 7–49.

Opella, S. J., Marassi, F. M., Gesell, J. J., Valente, A. P., Kim, Y., Oblatt-Montal, M., and Montal, M. (1999). Structures of the M2 channel-lining segments from nicotinic acetylcholine and NMDA receptors by NMR spectroscopy. *Nat. Struct. Biol.* **6**, 374–379.

Opella, S. J., Ma, C., and Marassi, F. M. (2001). Nuclear magnetic resonance of membrane-associated peptides and proteins. *Methods Enzymol.* **339**, 285–313.

Ottiger, M., and Bax, A. (1999). Bicelle-based liquid crystals for NMR-measurement of dipolar couplings at acidic and basic pH values. *J. Biomol. NMR* **13**, 187–191.

Ottiger, M., Delaglio, F., and Bax, A. (1998). Measurement of J and dipolar couplings from simplified two-dimensional NMR spectra. *J. Magn. Reson.* **131**, 373–378.

Oxenoid, K., Kim, H. J., Jacob, J., Sonnichsen, F. D., and Sanders, C. R. (2004). NMR assignments for a helical 40 kDa membrane protein. *J. Am. Chem. Soc.* **126**, 5048–5049.

Park, S. H., Mrse, A. A., Nevzorov, A. A., Mesleh, M. F., Oblatt-Montal, M., Montal, M., and Opella, S. J. (2003). Three-dimensional structure of the channel-forming trans-membrane domain of virus protein "u" (Vpu) from HIV-1. *J. Mol. Biol.* **333**, 409–424.

Pines, A., Gibby, M. G., and Waugh, J. S. (1973). Proton-enhanced NMR of dilute spins in solids. *J. Chem. Phys.* **59**, 569–590.

Prosser, R. S., Hunt, S. A., DiNatale, J. A., and Vold, R. R. (1996). Magnetically aligned membrane model systems with positive order parameter: Switching the sign of S_{zzz} with paramagnetic ions. *J. Am. Chem. Soc.* **118**, 269–270.

Ram, P., and Prestegard, J. H. (1988). Magnetic field induced ordering of bile salt/ phospholipid micelles: New media for NMR structural investigations. *Biochim. Biophys. Acta* **940**, 289–294.

Ramamoorthy, A., Wu, C. H., and Opella, S. J. (1999). Experimental aspects of multidimensional solid-state NMR correlation spectroscopy. *J. Magn. Reson.* **140**, 131–140.

Rhim, W. K., Pines, A., and Waugh, J. S. (1970). Violation of the spin-temperature hypothesis. *Phys. Rev. Lett.* **25**, 218–220.

Ricka, J., and Tanaka, T. (1984). Swelling of ionic gels: Quantitative performance of the Donnan theory. *Macromolecules* **17**, 2916–2921.

Righetti, P. G., Gelfi, C., and Bosisio, A. B. (1981). Polymerization kinetics of polyacrylamide gels. 3. Effect of catalysts. *Electrophoresis* **2**, 291–295.

Saito, H., Tuzi., Tanio, M., and Naito, A. (2002). Dynamic aspects of membrane proteins and membrane-associated peptides as revealed by ^{13}C NMR: Lessons from bacteriorhodopsin as an intact protein. *Annu. Rep. NMR Spectrosc.* **47**, 39–108.

Sanders, C. R., 2nd, and Landis, G. C. (1995). Reconstitution of membrane proteins into lipidrich bilayered mixed micelles for NMR studies. *Biochemistry* **34**, 4030–4040.

Sanders, C. R., 2nd, and Prestegard, J. H. (1990). Magnetically orientable phospholipid bilayers containing small amounts of a bile salt analogue, CHAPSO. *Biophys. J.* **58**, 447–460.

Sanders, C. R., 2nd, and Schwonek, J. P. (1992). Characterization of magnetically orientable bilayers in mixtures of dihexanoylphosphatidylcholine and dimyristoylphosphatidylcholine by solid-state NMR. *Biochemistry* **31**, 8898–8905.

Sanders, C. R., Harre, B. J., Howard, K. P., and Prestegard, J. H. (1994). Magneticallyoriented phospholipids micelles as a tool for the study of membrane associated molecules. *Prog. NMR Spectrosc.* **26**, 421–444.

Sass, J., Cordier, F., Hoffmann, A., Cousin, A., Omichinski, J. G., Lowen, H., and Grzesiek, S. (1999). Purple membrane induced alignment of biological macromolecules in the magnetic field. *J. Am. Chem. Soc.* **121**, 2047–2055.

Sass, H. J., Musco, G., Stahl, S. J., Wingfield, P. T., and Grzesiek, S. (2000). Solution NMR of proteins within polyacrylamide gels: Diffusional properties and residual alignment by mechanical stress or embedding of oriented purple membranes. *J. Biomol. NMR* **18**, 303–309.

Schulz, G. E. (2002). The structure of bacterial outer membrane proteins. *Biochim. Biophys. Acta* **1565**, 308–317.

Seip, S., Balbach, J., and Kessler, H. (1994). Determination of backbone conformation of isotopically enriched proteins based on coupling constants. *J. Magn. Reson. Ser. B* **104**, 172–179.

Shi, Q., and Jackowski, G. (1998). One-dimensional polyacrylamide gel electrophoresis. *In* "Gel Electrophoresis of Proteins: A Practical Approach" (B. D. Hanes, ed.), pp. 1–21. Oxford University Press, Oxford, UK.

Smith, S. O., Aschheim, K., and Groesbeek, M. (1996). Magic angle spinning NMR spectroscopy of membrane proteins. *Q. Rev. Biophys.* **29,** 395–449.

Staley, J. P., and Kim, P. S. (1994). Formation of a native-like subdomain in a partially folded intermediate of bovine pancreatic trypsin inhibitor. *Protein Sci.* **3,** 1822–1832.

Strauch, K. L., and Beckwith, J. (1988). An *Escherichia coli* mutation preventing degradation of abnormal periplasmic proteins. *Proc. Natl. Acad. Sci. USA* **85,** 1576–1580.

Tanaka, T. (1978). Dynamics of critical concentration fluctuations in gels. *Phys. Rev. A* **17,** 763–766.

Thiriot, D. S., Nevzorov, A. A., Zagyanskiy, L., Wu, C. H., and Opella, S. J. (2004). Structure of the coat protein in Pf1 bacteriophage determined by solid-state NMR spectroscopy. *J. Mol. Biol.* **341,** 869–879.

Thompson, L. K. (2002). Solid-state NMR studies of the structure and mechanisms of proteins. *Curr. Opin. Struct. Biol.* **12,** 661–669.

Tian, C., Gao, P. F., Pinto, L. H., Lamb, R. A., and Cross, T. A. (2003). Initial structural and dynamic characterization of the M2 protein transmembrane and amphipathic helices in lipid bilayers. *Protein Sci.* **12,** 2597–2605.

Tjandra, N., and Bax, A. (1997). Direct measurement of distances and angles in biomolecules by NMR in a dilute liquid crystalline medium. *Science* **278,** 1111–1114.

Trempe, J. F., Morin, F. G., Xia, Z. C., Marchessault, R. H., and Gehring, K. (2002). Characterization of polyacrylamide-stabilized Pf1 phage liquid crystals for protein NMR spectroscopy. *J. Biomol. NMR* **22,** 83–87.

Tycko, R., Blanco, F. J., and Ishii, Y. (2000). Alignment of biopolymers in strained gels: A new way to create detectable dipole-dipole couplings in high-resolution biomolecular NMR. *J. Am. Chem. Soc.* **122,** 9340–9341.

Veglia, G., and Opella, S. J. (2000). Lanthanide ion binding to adventitious sites aligns membrane proteins in micelles for solution NMR spectroscopy. *J. Am. Chem. Soc.* **122,** 11733–11734.

Veglia, G., Zeri, A. C., Ma, C., and Opella, S. J. (2002). Deuterium/hydrogen exchange factors measured by solution nuclear magnetic resonance spectroscopy as indicators of the structure and topology of membrane proteins. *Biophys. J.* **82,** 2176–2183.

Vinogradova, O., Sonnichsen, F., and Sanders, C. R., 2nd (1998). On choosing a detergent for solution NMR studies of membrane proteins. *J. Biomol. NMR* **11,** 381–386.

Vold, R. R., Prosser, R. S., and Deese, A. J. (1997). Isotropic solutions of phospholipid bicelles: A new membrane mimetic for high-resolution NMR studies of polypeptides. *J. Biomol. NMR* **9,** 329–335.

Walther, D., and Cohen, F. E. (1999). Conformational attractors on the Ramachandran map. *Acta Crystallogr. D Biol. Crystallogr.* **55,** 506–517.

Wang, J., Denny, J., Tian, C., Kim, S., Mo, Y., Kovacs, F., Song, Z., Nishimura, K., Gan, Z., Fu, R., Quine, J. R., and Cross, T. A. (2000). Imaging membrane protein helical wheels. *J. Magn. Reson.* **144,** 162–167.

Watts, A., and Opella, S. J. (2000). Membranes studied by NMR spectroscopy. *In* "Encyclopedia of Spectroscopy and Spectrometery" (G. Tranter, J. Holmes, and J. Lindon, eds.), pp. 1281–1291. Academic Press, New York.

Waugh, J. S. (1976). Uncoupling of local field spectra in nuclear magnetic resonance: Determination of atomic positions in solids. *Proc. Natl. Acad. Sci. USA* **73,** 1394–1397.

White, S. H. (2004). The progress of membrane protein structure determination. *Protein Sci.* **13,** 1948–1949.

White, S. H., and Wimley, W. C. (1999). Membrane protein folding and stability: Physical principles. *Annu. Rev. Biophys. Biomol. Struct.* **28,** 319–365.

Woese, C. R., and Wolfe, R. S. (1985). "The Bacteria," Vol. VIII. Academic Press, New York.
Wu, C. H., Ramamoorthy, A., and Opella, S. J. (1994). High resolution heteronuclear dipolar solid-state NMR spectroscopy. *J. Magn. Reson.* **A109**, 270–272.
Zeri, A. C., Mesleh, M. F., Nevzorov, A. A., and Opella, S. J. (2003). Structure of the coat protein in fd filamentous bacteriophage particles determined by solid-state NMR spectroscopy. *Proc. Natl. Acad. Sci. USA* **100**, 6458–6463.

[15] NMR Techniques Used with Very Large Biological Macromolecules in Solution

By GERHARD WIDER

Abstract

Methods for the characterization of very large biological macromolecules by NMR in solution are presented. For studies of molecular structures with molecular weights beyond 100,000 Da transverse relaxation in common multidimensional NMR experiments becomes a limiting factor. Novel techniques optimize transverse relaxation based on cross-correlated relaxation between dipole–dipole interactions and chemical shift anisotropy (CSA), and include transverse relaxation-optimized spectroscopy (TROSY), cross-correlated relaxation-enhanced polarization transfer (CRINEPT), and cross-correlated relaxation-induced polarization transfer (CRIPT). In combination with various biochemical isotope-labeling techniques these experimental schemes make possible studies of biological macromolecules with molecular masses of up to 1,000,000 Da by NMR in solution. The physical basis and the implementation of the experiments are discussed.

Introduction

The foundations of successful applications of nuclear magnetic resonance (NMR) spectroscopy are high-quality spectra recorded with good sensitivity and spectral resolution. With increasing molecular weight these basic requirements are harder to achieve. The scarcity of NMR studies with biological macromolecules above 30 kDa molecular weight reflects the increasing challenge and costs involved. During the past years, considerable effort has been devoted to extend applications of NMR in solution to larger molecular systems (Pervushin *et al.*, 1997; Riek *et al.*, 1999; Tugarinov and Kay, 2003; Tugarinov *et al.*, 2003; Wider and Wüthrich, 1999; Wüthrich and Wider, 2003). The availability of such NMR techniques is of considerable interest, for example, for structural investigations of larger proteins that

Copyright 2005, Elsevier Inc.
All rights reserved.
0076-6879/05 $35.00

resist crystallization, or for studies of intermolecular interactions in solution involving large molecules and supramolecular assemblies including structure–activity relationships (SAR) by NMR (Shuker et al., 1996).

When studying large molecules and macromolecular assemblies in solution by conventional NMR methods, usually a number of problems arise: (1) extensive signal overlap due to the high complexity of the spectra, (2) low solubility of the solute limits the sensitivity, and (3) low sensitivity and line broadening due to rapid transverse spin relaxation. Spectral overlap can be reduced with a variety of techniques that result in a simplification of spectra: e.g., fractional deuteration (Gardner and Kay, 1998; Lian and Middleton, 2001), amino acid selective labeling (Kainosho, 1997), and segmental labeling (Xu et al., 1999; Yamazaki et al., 1998; Yu, 1999). The limitations caused by low solubility and transverse relaxation pose severe technical challenges. However, advances have been achieved with both NMR hardware and novel NMR techniques. The recent development of NMR cryogenic probes improved the sensitivity of NMR measurements by a factor of 3, and novel experimental approaches, in particular TROSY (transverse relaxation-optimized spectroscopy) (Pervushin et al., 1997), CRINEPT (cross-correlated relaxation-enhanced polarization transfer) (Riek et al., 1999) and CRIPT (cross-correlated relaxation-induced polarization transfer) (Brüschweiler and Ernst, 1992; Dalvit, 1992; Riek et al., 1999), extended the size limit for observation of NMR signals in solution severalfold. These experimental techniques use spectroscopic means to reduce transverse relaxation, making possible studies of molecular systems with molecular weights up to 1000 kDa (Fiaux et al., 2002). The introduction of the novel techniques opens a wide range of new applications for NMR in solution, in particular in the newly emerging field of structural and functional genomics. In the following sections these new techniques for observation of high-quality NMR spectra with molecular sizes beyond 100 kDa are discussed.

Optimizing Transverse Relaxation

Transverse relaxation has a large impact on the quality of NMR spectra. With increasing molecular weight, transverse relaxation becomes larger. Consequently, the resonance lines become broader and the sensitivity decreases. A few years ago TROSY was introduced (Pervushin et al., 1997) and in the meantime has found widespread applications (Fernández and Wider, 2003), since better spectra are readily obtained when working with molecular sizes above 15–20 kDa. The TROSY technique can reduce the effective relaxation of the measured signal during the pulse sequence and during the data acquisition resulting in increased spectral resolution and improved effective sensitivity. The detailed understanding of the

TROSY principle requires relaxation theory based on quantum mechanical principles (e.g., Goldman, 1984; Pervushin, 1997, 2000). Here, a simplified approach based on classic physics shall advance a more intuitive understanding.

In the following TROSY is discussed using the important example of a ^{15}N and a ^{1}H nucleus, such as in ^{15}N–^{1}H amide groups of proteins or in nucleic acid bases. Both nuclei have a spin-$\frac{1}{2}$, which can classically be described by a magnetic dipole having two orientations: up and down. In the ^{1}H-NMR spectrum a doublet is observed originating from the protons attached to ^{15}N nuclei with spin up and spin down, respectively. These doublets are routinely collapsed into a single, centrally located line by decoupling (Bax and Grzesiek, 1993; Wider, 1998), with the expectation of obtaining a simplified spectrum and improved sensitivity. However, in the spectrum of a large protein the individual multiplet components have different linewidths (Pervushin et al., 1997). Decoupling mixes the different relaxation rates that deteriorate the averaged signal for large molecules studied at high magnetic fields. In TROSY no decoupling is applied, and only the narrowest, most slowly relaxing line of each multiplet is retained. TROSY disregards half of the potential signal of an amide moiety. This loss is compensated in large molecules by the slower relaxation of the selected resonance line.

Why do the individual components of the ^{1}H–^{15}N doublet have different relaxation rates for large molecules? The answer requires an analysis of the magnetic fields created by different interactions. These fields become time dependent (and cause relaxation) due to stochastic rotational motions of the molecule containing the amide group. Fortunately, for large proteins only the z-component of the stochastic magnetic fields contributes significantly to transverse relaxation ("dephasing" of the signals), because the molecules rotate too slowly to create the necessary frequency components at the transition frequencies. In the context of relaxation theory, the z-components correspond to terms with spectral density at zero frequency. In the amide moiety two processes lead to significant relaxation: dipole–dipole (DD) and chemical shift anisotropy (CSA) interactions. Individually, the two relaxation processes do not cause differential relaxation either for the two-proton or for the two-nitrogen transitions. The DD and CSA interactions are present simultaneously and their magnetic fields add. This addition leads to larger and smaller z-components, since the dipolar field (but not CSA) has an opposite sign depending on the state of the attached nucleus. Consequently, the relaxation rates of the two transitions are potentially different. However, this static picture presents just a snapshot that can influence relaxation only if the stochastic magnetic fields created by DD interactions and CSA have the same dependence on the rotational motion of the molecules (Goldman, 1984; Pervushin et al., 1997).

The field created by the nuclear dipole depends on $(3 \cos^2 \theta - 1)$, where θ is the angle between the main magnetic field and the direction of the H–N bond in an amide moiety, all other factors being constant. For the magnetic field caused by CSA, the same angular dependence is obtained when an axially symmetric CSA tensor with its main axis along the H–N bond is assumed (Pervushin *et al.*, 1997). As a result, the differential stochastic magnetic fields are maintained irrespective of the molecular tumbling, which leads to different relaxation rates for individual transitions. With increasing strength of the main magnetic field, CSA relaxation increases whereas DD relaxation stays constant. The optimal TROSY effect can thus be obtained by choosing the appropriate field strength. Interestingly, in an amide moiety the TROSY effect for both the ^{1}H and the ^{15}N nuclei has an optimum at about the same magnetic field strength, which is approximately 23.5 T, corresponding to a proton resonance frequency of 1000 MHz (Pervushin *et al.*, 1997). In experiments with ^{1}H and ^{15}N nuclei, the slower relaxing component for both nuclei is selected in a relaxation-optimized experiment.

The fact that amide groups are strategically located in the polypeptide backbone of proteins and in the bases of nucleotides of DNA and RNA molecules makes them a prime target in the optimization of many NMR experiments with biological macromolecules. In practice, complete cancellation of relaxation cannot be expected. In particular, the principal tensor axis of the CSA tensor and the ^{15}N–^{1}H bond is not colinear, and the properties of the CSA tensor vary from residue to residue (Fushman and Cowburn, 1999). In addition, there is some residual transverse relaxation that cannot be influenced by TROSY, especially dipole–dipole couplings with "remote" protons, i.e., all protons outside of the ^{15}N–^{1}H group. Therefore, an optimal TROSY effect is obtained with uniformly deuterated proteins, where remote couplings are limited to nearby amide protons.

TROSY is not limited to amide moieties in biological macromolecules; important applications use CH-groups in aromatic rings (Brutscher *et al.*, 1998; Pervushin *et al.*, 1998), where large signal enhancement could be obtained. More recently, a TROSY experiment for the carbon nuclei in methyl groups was described (Tugarinov *et al.*, 2003). In a $[^{13}$C,^{1}H]-HMQC (heteronuclear multiple-quantum correlation) experiment, the methyl carbon is in a multiple quantum state with one proton, which eliminates their mutual dipolar interactions, and for half of the carbon nuclei the other two attached protons have opposite spin states. Thus, their dipolar fields at the carbon nuclei cancel each other in large molecules.

Optimizing Polarization Transfer

In nearly all multidimensional NMR experiments for studies of biological macromolecules in solution, magnetization is transferred from one type of nucleus to another via scalar spin–spin couplings using INEPT (insensitive nuclei enhanced by polarization transfer) pulse sequences (Morris and Freeman, 1979). With increasing molecular weight, transverse relaxation during the INEPT transfer period becomes a limiting factor that severely compromises sensitivity. On the other hand the performance of the previously little-used CRIPT improves for larger molecules (Brüschweiler and Ernst, 1992). The novel technique CRINEPT (Riek et al., 1999) combines CRIPT (Dalvit, 1992) and INEPT. For even larger molecular weights, polarization transfer using only CRIPT will become an alternative to CRINEPT. CRIPT and CRINEPT make use of the same cross-correlated relaxation, which is the basis for TROSY. Thus for amide groups the optimal magnetic field strength for these polarization transfer mechanisms corresponds to 1 GHz proton resonance frequency.

The INEPT and CRIPT transfer elements are compared in Fig. 1. In INEPT, antiphase magnetization is created via J-coupling during the period τ_1. In CRIPT, the effect of the J-coupling is refocused by omitting the 180° nitrogen pulse in the transfer period τ_2, and antiphase magnetization is obtained solely via cross-correlated relaxation. Ideally the fast relaxing doublet component has completely relaxed during τ_2, whereas the slowly relaxing component has hardly been reduced. This situation is indicated in Fig. 1 by the stick patterns that visualize the doublet appearance at the beginning and at the end of the transfer period at time points a and b, respectively. The asymmetric doublet at the end of the CRIPT transfer period τ_2 can be decomposed into a sum of an in-phase and an antiphase doublet with half the intensity of the original resonance line. In a physical sense this view represents another basis set describing the same situation. However, this basis visualizes that via cross-correlated relaxation at most half the magnetization arrives in an antiphase state and is the basis for the polarization transfer. The same result is obtained with a rigorous quantum mechanical calculation (Brüschweiler and Ernst, 1992). Similar to the situation with TROSY, for large proteins half the magnetization with optimized relaxation may easily surpass the magnetization obtained with averaged relaxation during the transfer period in an INEPT sequence.

Figure 2 shows three experimental schemes that make use of CRINEPT and CRIPT magnetization transfer elements: two-dimensional (2D) [^{15}N, ^{1}H]-CRINEPT-TROSY, 2D [^{15}N,^{1}H]-CRINEPT-HMQC-[^{1}H]-TROSY, and 2D [^{15}N,^{1}H]-CRIPT-TROSY (Riek et al., 1999). Note that all three schemes use water flip-back pulses that keep the water polarization close to its equilibrium value during the whole experiment. This is a requirement when working with very large proteins (see next section). The 2D [^{15}N,^{1}H]-CRINEPT-TROSY experiment (Fig. 2A) consists of three time periods τ used for polarization transfer and for the selection of the fastest and the

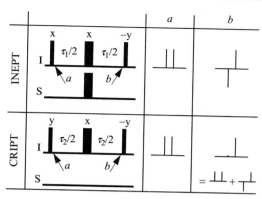

FIG. 1. Representation of two different magnetization transfer elements between nuclei I and S. The pulse sequence segments for an INEPT and a CRIPT transfer are shown along with stick patterns characterizing the doublet signal of spin I at the beginning and at the end of the transfer period at time points a and b, respectively (assuming no relaxation in INEPT and ideal cross-correlated relaxation in CRIPT). The narrow and wide vertical bars indicate 90° and 180° pulses that are applied at the resonance frequency of the I and S spins, respectively; the phase is x unless indicated otherwise above the pulse bar; note the difference in phase of the first pulse in the two-pulse sequence fragments. In INEPT the in-phase doublet evolves into an antiphase signal under J coupling. In CRIPT the evolution under J coupling is refocused and one component of the doublet relaxes much faster; the remaining pattern can alternatively be viewed as a sum of an in-phase and of an antiphase signal with each having half the intensity. The time period τ_1 depends on the scalar coupling between the two nuclei I and S. J_{IS}: $\tau_1 = 1/(2\ J_{IS})$; the time period τ_2 depends on the molecular size and has to be estimated from build-up measurements (Fig. 4); τ_2 can be as short as 1 ms for extremely large molecules.

most slowly relaxing multiplet components, which is in contrast to TROSY, where only the narrowest component is selected. In practice, the additional component relaxes so fast that it is often not detectable. In CRINEPT transverse relaxation optimization is active during all three time periods τ and during the evolution and acquisition periods, t_1 and t_2; in other words, the experiment is fully optimized for relaxation. This implementation of CRINEPT has the disadvantage that the combination of INEPT- and CRIPT-type magnetization transfer causes chemical shift evolution of the proton signals during the first period τ, which cannot be refocused; consequently, half the signal is lost.

The 2D [^{15}N,^{1}H]-CRINEPT-HMQC-[^{1}H]-TROSY (Fig. 2B) is an alternative to the 2D [^{15}N,^{1}H]-CRINEPT-TROSY. The conventional [^{15}N,^{1}H]-HMQC experiment (Müller, 1979) contains two CRINEPT-type transfers. To make the sequence useful for large molecules it needs a few important changes. The transfer periods have to be optimized for CRINEPT transfer; no heteronuclear decoupling during acquisition must be applied, and water

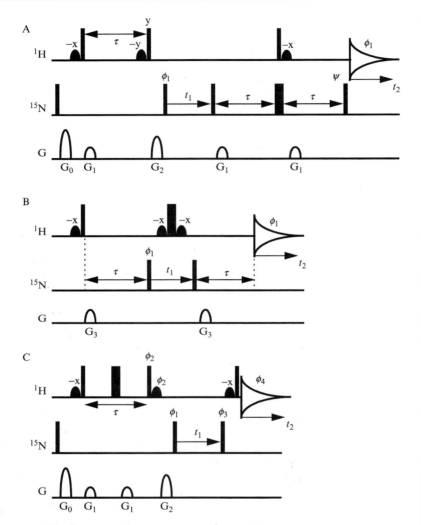

FIG. 2. [^{15}N, ^{1}H] correlation experiments using cross-correlated relaxation for magnetization transfer. (A) 2D [^{15}N,^{1}H]-CRINEPT-TROSY, (B) 2D [^{15}N,^{1}H]-CRINEPT-HMQC-[^{1}H]-TROSY, and (C) 2D [^{15}N,^{1}H]-CRIPT-TROSY. The narrow and wide vertical bars indicate 90° and 180° pulses, respectively, which are applied at the proton (^{1}H) and nitrogen (^{15}N) frequencies with phase x unless indicated otherwise above the pulse bar. Bell shapes on the ^{1}H line represent selective 90° pulses on the water resonance (duration ~1 ms), which keep the water magnetization along the positive z-axis during the whole experiment. On the line marked G, open bell shapes indicate pulsed magnetic field gradients applied along the z-axis with the following approximate durations and strengths: G_0, 1 ms, 30 G/cm; G_1, 300 μs, 20 G/cm; G_2, 400 μs, 40 G/cm; G_3, 800 μs, 10 G/cm. In (A) and (C) the magnetization of

flip-back pulses have to be introduced (see next section). Compared to [^{15}N,^{1}H]-CRINEPT-TROSY, there are only two periods τ and there is no signal loss due to proton chemical shift evolution before detection of the signal. However, there is no TROSY during the evolution period t_1 where the spins are in a multiple quantum state, which results in faster relaxation. In general, the experiment will have higher overall sensitivity but lower resolution in the indirect dimension than the [^{15}N,^{1}H]-CRI-NEPT-HMQC-[^{1}H]-TROSY. The sequence does not contain any selection of multiplet components and the resonances are split in the proton dimension due to scalar coupling. Again, in practice the broader component is often not visible in the spectra due to its fast relaxation.

For molecular structures with rotational correlation times longer than 120 ns, i.e., molecular weights larger than 300 kDa, the 2D [^{15}N,^{1}H]-CRIPT-TROSY experiment finds application (Fig. 2C). The scheme is fully optimized for transverse relaxation and uses cross-correlated relaxation for the polarization transfer and TROSY during the evolution and acquisition periods. It is the shortest scheme with only one transfer period τ, however, at the cost of no multiplet peak selection and in principle, all four components of a ^{15}N–^{1}H correlation peak are present in the spectrum. In general, only the narrowest peak will be visible in the spectrum, and the three other components will be suppressed by their fast transverse relaxation. In contrast to the CRINEPT schemes, CRIPT suppresses signals from more mobile regions of large proteins and from smaller molecules when the transfer delay is set for detection of very large molecular weights, since the optimal transfer time τ is inversely proportional to the rotational correlation time of the amide moiety.

Water Polarization

In NMR spectra of very large proteins a surprisingly strong dependence of the proton signal intensities on the polarization of the solvent water is observed. The dependence applies over the whole spectral range and not only to some amide ^{1}H resonances for which it is well established that magnetization/saturation transfer can occur from water via amide–proton

---

the ^{15}N spins is suppressed at the beginning of the sequence by a 90° pulse on ^{15}N followed by G$_0$. The phase cycles for the experiments are $\phi_1 = (x, -x)$, $\phi_2 = (x, x, -x, -x)$, $\phi_3 = (x, x, x, x, -x, -x, -x, -x)$, and $\phi_4 = (-x, x, x, -x, x, -x, -x, x)$. Quadrature detection in the ^{15}N dimension is achieved using States-TPPI (Marion et al., 1989) applied to the phase ϕ_1. In (A) two free induction decays are recorded for each t_1 increment, with $\psi = \{x, x\}$ and $\psi = \{-x, -x\}$, respectively, and are added with a 90° phase shift in both dimensions (Riek et al., 1999). The duration of the magnetization transfer period τ is typically between 1.5 and 6 ms and decreases with molecular size.

exchange (Bax and Grzesiek, 1993). Moreover, for large compact proteins it is expected that only a minor fraction of the amide protons exchanges at an appreciable rate. In Fig. 3 spectra obtained with different water suppression schemes are compared for a very large protein with 800 kDa molecular weight and a small protein of 6.5 kDa. Only the aliphatic region of the ^{1}H spectrum is shown, excluding the influence of direct effects of amide exchange. The three schemes maintain or temporarily destroy the equilibrium water polarization (Fig. 3). For the small protein, three identical spectra are obtained (Fig. 3C), whereas large differences in intensity are observed in the spectra of the large protein (Fig. 3B), with presaturation producing the least and water flip-back the most signal intensity. But even saturation of the water resonance immediately before acquisition has a

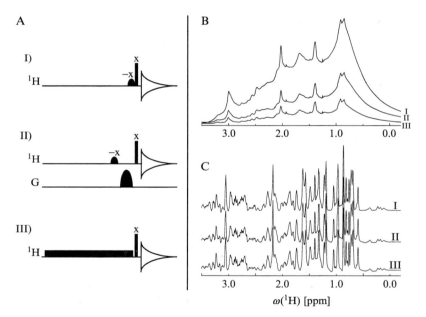

FIG. 3. Aliphatic regions of 1D ^{1}H spectra obtained with experiments using different water suppression techniques. (A) Pulse sequences used: (I) flip-back of water magnetization by a selective 90° pulse at the water frequency with the opposite phase to the 90° broadband excitation pulse directly following; (II) a strong magnetic field gradient pulse is applied between the two pulses described in I; (III) the water resonance is saturated by continuous irradiation of a weak radiofrequency field (presaturation). (B) ^{1}H spectra of GroEL from *E. coli* (molecular weight 800 kDa) and (C) bovine pancreatic trypsin inhibitor (BPTI; 6.5 kDa) measured with the three schemes shown in (A). In the spectra of GroEL, large intensity changes are observed whereas the spectra of BPTI are identical.

large impact. These effects are observed for protonated as well as for perdeuterated proteins studied in H_2O solutions (Hohwy et al., 2004).

The strong coupling between the water protons and the protons in large globular proteins has an additional important consequence. The effective longitudinal relaxation times T_1 of the protein protons strongly depend on the polarization of the water protons. The experiments that conserve the water polarization using water flip-back pulses show the shortest effective T_1 values (Hohwy et al., 2004). The relaxation is much shorter than expected from simple models of relaxation that predict a dramatic increase of T_1 for longer rotational correlation times. The short effective proton longitudinal relaxation times imply that optimal sensitivity is obtained with shorter waiting periods between successive scans than typically used for small proteins, e.g., an interscan period around 0.5 s was found optimal in experiments with the 800-kDa chaperonin GroEL when carefully adjusted water flip-back pulses were used.

The experimental findings were recently explained by a model that predicted and quantified the saturation transfer and the drastic reduction of the longitudinal relaxation times observed (Hohwy et al., 2004). The model shows that fast exchanging hydroxyl groups of the amino acid side chains and internal water molecules are the dominating mechanisms of interaction between water protons and protons in the protein. The hydroxyl groups usually have exchange rates larger than $\sim 250 \text{ s}^{-1}$ for all pH values (see Fig. 2.3 in Wüthrich, 1986). These hydroxyl groups are omnipresent in proteins and, for a large fraction of all protons in a protein, hydroxyl groups are found within a distance of 0.5 nm, permitting efficient cross-relaxation. For the case of the 110-kDa protein DHNA (7,8-dihydroneopterin aldolase from Staphylococcus aureus), a clear correlation was found between the distance to a hydroxyl proton or an interior water molecule and the strength of interaction between nonexchanging protein protons and water protons: this finding confirms the assumption made in the model (Hohwy et al., 2004).

Water flip-back pulses can never completely restore equilibrium magnetization of water protons. As a result, shorter T_1 relaxation of water would increase its steady-state magnetization and thus improve the quality of the NMR spectra of large proteins. Two conflicting effects have to be optimized: the T_1 of water must be reduced without increasing the transverse relaxation of the protein protons. Recent experiments in our laboratory have shown that this requirement can be fulfilled using chelated paramagnetic ions (Hiller et al., 2004). Such relaxation agents permit access of water molecules to the paramagnetic ion but keep the protein at a large distance, which minimizes relaxation of protein resonances. The T_1 of water can easily be reduced by a factor of 20 and still transverse relaxation of protons in large proteins is increased by only a few percent. Moreover, it was found that by a proper choice of the paramagnetic additive, full water equilibrium magnetization can be maintained without the use of any water flip-back pulses. This finding is important because a careful adjustment of the flip-back pulses can be tedious, and some experiments, e.g., CRIPT, often show a large residual water resonance compromising spectral quality, which can be markedly improved by temporarily saturating the water resonance.

Experimental Procedures

When performing NMR experiments with very large biological macromolecules some basic aspects require special attention. Even though the resonance lines of the protein are very broad a good homogeneity of the main magnetic field is crucial for an optimal performance of the water flip-back pulses, which depend on the quality of the line shape of the water resonance. Often better shimming is required than for small molecules where residual water can be removed by techniques such as WATER-GATE (Piotto et al., 1992). Such sequences can, in general, not be applied with very large molecules because they increase the length of the pulse sequence, which must be kept to an absolute minimum to counteract relaxation.

A further aspect concerns the determination of the length of the radio-frequency pulses. For nuclei other than protons, the pulses can usually be taken from measurements with another sample if the receiver coil can be tuned to the same performance with the sample of interest. For protons this procedure is often not precise enough and the durations of the proton pulses have to be determined using the sample with the large protein. Due to the strong influence of the polarization of the water protons on the protons of the protein, the length of a selective water flip-back pulse and the wideband excitation pulse would have to be determined simultaneously, which is not easily possible. For this reason, the 360° pulse for the water resonance is determined, which again requires good field homogeneity and possibly a susceptibility matched sample tube that eliminates pick-up of signal outside the central region of the receiver coil. Such sample tubes benefit all measurements with very large proteins since better suppression of residual water signals can be obtained.

The importance of the water polarization requires a particularly careful adjustment of the water flip-back pulses. The pulses chosen should not be too selective, which is usually not a limitation since the proteins are deuterated and the water resonance is at an ample distance to the nearest proton resonance of interest (amide protons). A pulse length of about 1 ms has proven to be adequate at magnetic field strengths corresponding to 800 or 900 MHz proton resonance frequency. Two selective pulses have to be determined (usually they are chosen with the same duration but different powers): one flipping water magnetization from the z-axis into the transverse plane and one flipping it from the transverse plane back to the z-axis. The two pulses differ due to the different influence of the radiation damping effect on the water signal, which counteracts excitation of the water resonance but supports flipping it back to the positive z-axis. The appropriate selective pulses can then be introduced into the pulse sequence of interest.

Before starting the experiment, the phase and power of the selective pulses at the water resonance frequency have to be fine tuned at their specific position in the pulse sequence. Good water suppression has to be obtained, and at the same time maximal water polarization must be maintained. As a final test of the performance of the experiment, the quality of the water flip-back is tested. Applying a very small excitation pulse at the water frequency permits determination of the maximal water magnetization without disturbing it. The same small pulse is then inserted just before the acquisition in the pulse sequence of interest, and the intensity of the water resonance is monitored for the first few evolution time increments (Hohwy et al., 2004). The intensity will decrease from the maximal value to a steady-state value that reaches typically between 65 and 90% of the equilibrium value depending on the type of experiment and the number of water flip-back pulses. With this monitoring system, misadjustment can easily be detected.

Before obtaining spectra of a new protein, the polarization transfer periods have to be estimated. The values can be obtained from the molecular weight (Riek et al., 2002) or measured using the first free induction decay (FID) of the multidimensional experiment with different transfer periods. For an estimate, the amide regions in the corresponding 1D ^{1}H spectra are integrated and plotted against the polarization transfer delay τ (Fig. 2). Figure 4 shows the results of such build-up measurements with the 2D [^{15}N, ^{1}H]-CRIPT-TROSY sequence (Fig. 2C) applied with the 110-kDa protein DHNA and the 800-kDa GroEL, respectively.

For initial experiments, 2D [^{15}N, ^{1}H]-CRINEPT-HMQC-[^{1}H]-TROSY is often best suited due to its relatively high sensitivity and the combination of INEPT and CRIPT magnetization transfers. However, the resolution is limited in the ^{15}N dimension resulting in broad lines. The evaluation of this first spectrum will then indicate whether CRINEPT or CRIPT experiments are required for the NMR studies. The merits of the different pulse sequences shown in Fig. 2 were previously discussed extensively (Riek et al., 2002). Even with relaxation compensation, the resonances are very broad

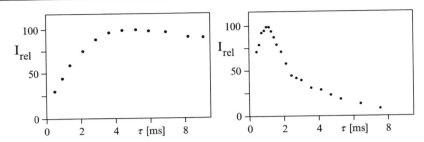

FIG. 4. CRIPT build-up curves for the 110-kDa uniformly $^{15}N,^2H$-labeled 7,8-dihydroneopterin aldolase (DHNA) from *Staphylococcus aureus* (left) and the 800-kDa $[^{15}N,^2H]$-GroEL from *E. coli* (right). A series of 1D ^1H spectra was measured using the 2D $[^{15}N,^1H]$-CRIPT experiment (Fig. 2C) with evolution period $t_1 = 0$. The intensity, I_{rel}, is the integral over the resonances in the low-field amide spectral region. The maxima of the two build-up curves were set independently to a value of 100. The optimal CRIPT transfer period for DHNA is about 5 ms and for GroEL it is about 1.5 ms.

in CRIPT and CRINEPT spectra, reminiscent of the broad lines in absolute value spectra, requiring stronger window functions (Wider *et al.*, 1984). Often, the data have to be transformed with different window functions to make different features of the spectra visible. Correspondingly, special consideration has to be given to the choice of the window function used in the processing of the data as described in detail by Riek *et al.* (2002). As an example of a heteronuclear correlation spectrum of a very large protein and of the quality of spectra that can be obtained, Fig. 5 presents a 2D $[^{15}N, ^1H]$-CRIPT-TROSY spectrum of the perdeuterated 800-kDa tetradecameric chaperonin GroEL from *Escherichia coli*. The experimental details are given in the figure caption. A 2D $[^{15}N,^1H]$-TROSY spectrum of GroEL contains very few resonances from more mobile regions in the protein (Riek *et al.*, 2002) and does not show the typical pattern of a folded protein as observed in the spectrum in Fig. 5.

Perspective

With novel NMR techniques correlation spectra of biological macromolecules up to 1 MDa can be acquired in solution. These methods were already successfully applied to study large proteins (Fernández and Wider, 2003; Fiaux *et al.*, 2002; McElroy *et al.*, 2002; Rudiger *et al.*, 1999; Salzmann *et al.*, 2000; Tugarinov *et al.*, 2002). With increasing molecular weight of the proteins under investigation, the number of polarization transfer steps and the overall duration of the experimental schemes must be reduced.

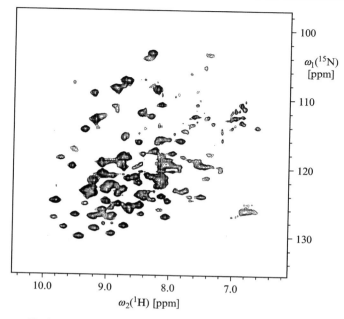

FIG. 5. 2D [^{15}N,^{1}H]-CRIPT-TROSY spectrum of a 1.5 mM sample of perdeuterated and ^{15}N-labeled 800-kDa tetradecameric GroEL containing 20 mM KCl and 25 mM phosphate buffer at pH 6.2, $T = 35°$, measured at 750 MHz using the experimental scheme shown in Fig. 2C. The CRIPT transfer period was 1.6 ms, the maximal evolution time 16 ms, the interscan delay 300 ms, the acquisition time 95 ms, and the total measuring time 12 h; 72 and 1024 complex points were measured in the evolution and in the acquisition periods, respectively; before Fourier transformation the data were multiplied with a sine bell shifted by 20° in the indirect dimension and an exponential function in the direct dimension, zero-filling to 256 times 2048 data points, was applied (Wider *et al.*, 1984). The resulting spectrum was baseline corrected in both dimensions.

Measurements of magnetization transfer efficiencies in the 110-kDa protein DHNA (Braun *et al.*, 2003) indicate that TROSY-type triple resonance experiments (Salzmann *et al.*, 1998, 1999; Yang and Kay, 1999) or experiments with relayed magnetization transfer steps will become increasingly difficult with the current technology when applied to proteins with molecular weights beyond 200–300 kDa (Chung and Kroon, 2003). Triple resonance experiments are usually the basis for sequence-specific resonance assignments, which are the foundation for a detailed analysis of the spectra. When these experiments fail, alternative methods must be developed. A promising approach seems to be the use of experiments based on nuclear Overhauser enhancements (NOEs). Initial studies with a 500-kDa protein complex support this idea and permitted the assignment of resonances (R. Horst, unpublished observations).

The amide resonances provide a wealth of information with which to characterize a protein. Still, it may be desirable to obtain information on amino acid side chains. A promising approach seems to be the protonation of some methyl groups in an otherwise perdeuterated protein. This procedure has been shown to be very successful in obtaining more information in large molecules (Fernández et al., 2004; Mueller et al., 2000) and seems to provide signals in very slowly tumbling molecules (Kreishman-Deitrick et al., 2003). Again, the assignment of the methyl group resonances may have to rely on NOEs, since through-bond correlation experiments (Hilty et al., 2002; Tugarinov and Kay, 2003) most likely will not be sensitive enough with very large molecules. In the future, further applications of TROSY, CRINEPT, and CRIPT will answer questions related to the structure and function of supramolecular structures such as membrane proteins solubilized in micelles or lipid vesicles, proteins attached to nucleic acid fragments, or oligomeric proteins. The methodology will be supported by new isotope labeling techniques for proteins and nucleic acids and further advances in NMR technology.

Acknowledgments

I thank Dr. Reto Horst for discussions on all aspects of NMR spectroscopy with very large biological macromolecules, Sebastian Hiller for providing the data for Fig. 3, and Prof. Kurt Wüthrich for his continuous support and interest in my research.

References

Bax, A., and Grzesiek, S. (1993). Methodological advances in protein NMR. *Acc. Chem. Res.* **26,** 131–138.

Braun, D., Wüthrich, K., and Wider, G. (2003). Dissection of heteronuclear NMR experiments for studies of magnetization transfer efficiencies. *J. Magn. Reson.* **165,** 89–94.

Brüschweiler, R., and Ernst, R. R. (1992). Molecular dynamics monitored by cross-correlated cross relaxation of spins quantized along orthogonal axes. *J. Chem. Phys.* **96,** 1758–1766.

Brutscher, B., Boisbouvier, J., Pardi, A., Marion, D., and Simorre, J. P. (1998). Improved sensitivity and resolution in ^{1}H–^{13}C NMR experiments of RNA. *J. Am. Chem. Soc.* **120,** 11845–11851.

Chung, J., and Kroon, G. (2003). ^{1}H, ^{15}N, ^{13}C-triple resonance NMR of very large systems at 900 MHz. *J. Magn. Reson.* **163,** 360–368.

Dalvit, C. (1992). ^{1}H to ^{15}N polarization transfer via ^{1}H CSA–^{1}H-^{15}N dipole dipole cross correlation. *J. Magn. Reson.* **97,** 645–650.

Fernández, C., and Wider, G. (2003). TROSY in NMR studies of the structure and function of large biological macromolecules. *Curr. Opin. Struct. Biol.* **13,** 570–580.

Fernández, C., Hilty, C., Wider, G., Güntert, P., and Wüthrich, K. (2004). NMR structure of the integral membrane protein OmpX. *J. Mol. Biol.* **336,** 1211–1221.

Fiaux, J., Bertelsen, E. B., Horwich, A. L., and Wüthrich, K. (2002). NMR analysis of a 900K GroEL–GroES complex. *Nature* **418,** 207–211.

Fushman, D., and Cowburn, D. (1999). The effect of noncolinearity of ^{15}N-^{1}H dipolar and ^{15}N CSA tensors and rotational anisotropy on ^{15}N relaxation, CSA/dipolar cross correlation, and TROSY. *J. Biomol. NMR* **13,** 139–147.

Gardner, K. H., and Kay, L. E. (1998). The use of ^{2}H, ^{13}C, ^{15}N multidimensional NMR to study the structure and dynamics of proteins. *Annu. Rev. Biophys. Biomol. Struct.* **27,** 357–406.

Goldman, M. (1984). Interference effects in the relaxation of a pair of unlike spin-1/2 nuclei. *J. Magn. Reson.* **60,** 437–452.

Hiller, S., Wider, G., Etezady-Esfarjani, T., Horst, R., and Wüthrich, K. (2004). Managing the solvent water polarization to obtain improved NMR spectra of large molecular structures. *J. Biomol. NMR.* Submitted.

Hilty, C., Fernández, C., Wider, G., and Wüthrich, K. (2002). Side chain NMR. assignments in the membrane protein OmpX reconstituted in DHPC micelles. *J. Biomol. NMR* **23,** 289–301.

Hohwy, M., Braun, D., and Wider, G. (2004). Characterization of the influence of protein-water interactions on the NMR spectra of large proteins. *J. Biomol. NMR.* Submitted.

Kainosho, M. (1997). Isotope labelling of macromolecules for structural determinations. *Nat. Struct. Biol.* **4,** 858–861.

Kreishman-Deitrick, M., Egile, C., Hoyt, D. W., Ford, J. J., Li, R., and Rosen, M. K. (2003). NMR analysis of methyl groups at 100–500 kDa: Model systems and Arp2/3 complex. *Biochemistry* **42,** 8579–8586.

Lian, L. Y., and Middleton, D. A. (2001). Labelling approaches for protein structural studies by solution-state and solid-state NMR. *Prog. NMR Spectrosc.* **39,** 171–190.

Marion, D., Ikura, M., Tschudin, R., and Bax, A. (1989). Rapid recording of 2D NMR spectra without phase cycling. Application to the study of hydrogen exchange in proteins. *J. Magn. Reson.* **85,** 393–399.

McElroy, C., Manfredo, A., Wendt, A., Gollnick, P., and Foster, M. (2002). TROSY-NMR studies of the 91 kDa TRAP protein reveal allosteric control of a gene regulatory protein by ligand-altered flexibility. *J. Mol. Biol.* **323,** 463–473.

Morris, G. A., and Freeman, R. (1979). Enhancement of NMR signals by polarization transfer. *J. Am. Chem. Soc.* **101,** 760–762.

Mueller, G. A., Choy, W. Y., Yang, D., Forman-Kay, J. D., Venters, R. A., and Kay, L. E. (2000). Global folds of proteins with low densities of NOEs using residual dipolar couplings: Application to the 370-residue maltodextrin-binding protein. *J. Mol. Biol.* **300,** 197–212.

Müller, L. (1979). Sensitivity enhanced detection of weak nuclei using heteronuclear multiple quantum coherence. *J. Am. Chem. Soc.* **101,** 4481–4484.

Pervushin, K. (2000). Impact of transverse relaxation optimized spectroscopy (TROSY) on NMR as a technique in structural biology. *Q. Rev. Biophys.* **33,** 161–197.

Pervushin, K., Riek, R., Wider, G., and Wüthrich, K. (1997). Attenuated T$_2$ relaxation by mutual cancellation of dipole-dipole coupling and chemical shift anisotropy indicates an avenue to NMR structures of very large biological macromolecules in solution. *Proc. Natl. Acad. Sci. USA* **94,** 12366–12371.

Pervushin, K., Riek, R., Wider, G., and Wüthrich, K. (1998). Transverse relaxation-optimized spectroscopy (TROSY) for NMR studies of aromatic spin systems in ^{13}C-labeled proteins. *J. Am. Chem. Soc.* **120,** 6394–6400.

Piotto, M., Saudek, V., and Sklenar, V. (1992). Gradient-tailored excitation for single quantum NMR spectroscopy of aqueous solutions. *J. Biomol. NMR* **2,** 661–665.

Riek, R., Wider, G., Pervushin, K., and Wüthrich, K. (1999). Polarization transfer by cross-correlated relaxation in solution NMR with very large molecules. *Proc. Natl. Acad. Sci. USA* **96**, 4918–4923.

Riek, R., Fiaux, J., Bertelsen, E. B., Horwich, A. L., and Wüthrich, K. (2002). Solution NMR techniques for large molecular and supramolecular structures. *J. Am. Chem. Soc.* **124**, 12144–12153.

Rudiger, S., Freund, S. M. V., Veprintsev, D. B., and Fersht, A. R. (1999). CRINEPT-TROSY NMR reveals p53 core domain bound in an unfolded form to the chaperone Hsp90. *Proc. Natl. Acad. Sci. USA* **99**, 11085–11090.

Salzmann, M., Pervushin, K. G., Wider, G., Senn, H., and Wüthrich, K. (1998). TROSY in triple-resonance experiments: New perspectives for sequential NMR assignment of large proteins. *Proc. Natl. Acad. Sci. USA* **95**, 13585–13590.

Salzmann, M., Wider, G., Pervushin, K., Senn, H., and Wüthrich, K. (1999). TROSY-type triple-resonance experiments for sequential assignments of large proteins. *J. Am. Chem. Soc.* **121**, 844–848.

Salzmann, M., Pervushin, K., Wider, G., Senn, H., and Wüthrich, K. (2000). NMR assignment and secondary structure determination of an octameric 110 kDa protein using TROSY in triple resonance experiments. *J. Am. Chem. Soc.* **122**, 7543–7548.

Shuker, S. B., Hajduk, P. J., Meadows, R. P., and Fesik, S. W. (1996). Discovering high-affinity ligands for proteins: SAR by NMR. *Science* **274**, 1531–1534.

Tugarinov, V., and Kay, L. E. (2003). Ile, Leu, and Val methyl assignments of the 723-residue malate synthase G using a new labeling strategy and novel NMR methods. *J. Am. Chem. Soc.* **125**, 13868–13878.

Tugarinov, V., Muhandiram, R., Ayed, A., and Kay, L. E. (2002). Four-dimensional NMR spectroscopy of a 723-residue protein: Chemical shift assignments and secondary structure of malate synthase G. *J. Am. Chem. Soc.* **124**, 10025–10035.

Tugarinov, V., Hwang, P. M., Ollerenshaw, J. E., and Kay, L. E. (2003). Cross-correlated relaxation enhanced ^{1}H-^{13}C NMR spectroscopy of methyl groups in very high molecular weight proteins and protein complexes. *J. Am. Chem. Soc.* **125**, 10420–10428.

Wider, G. (1998). Technical aspects of NMR spectroscopy with biological macromolecules and studies of hydration in solution. *Prog. NMR Spectrosc.* **32**, 193–275.

Wider, G., and Wüthrich, K. (1999). NMR spectroscopy of large molecules and multimolecular assemblies in solution. *Curr. Opin. Struct. Biol.* **9**, 594–601.

Wider, G., Macura, S., Kunar, A., Ernst, R. R., and Wüthrich, K. (1984). Homonuclear two-dimensional ^{1}H NMR of proteins. Experimental procedures. *J. Magn. Reson.* **56**, 207–234.

Wüthrich, K. (1986). "NMR of Proteins and Nucleic Acids." Wiley, New York.

Wüthrich, K., and Wider, G. (2003). Transverse relaxation-optimized NMR spectroscopy with biomacromolecular structures in solution. *Magn. Reson. Chem.* **41**, S80–S88.

Xu, R., Ayers, B., Cowburn, D., and Muir, T. W. (1999). Chemical ligation of folded recombinant proteins: Segmental isotopic labeling of domains for NMR studies. *Proc. Natl. Acad. Sci. USA* **96**, 388–393.

Yamazaki, T., Otomo, T., Oda, N., Kyogoku, Y., Uegaki, K., Ito, N., Ishino, Y., and Nakamura, H. (1998). Segmental isotope labeling for protein NMR using peptide splicing. *J. Am. Chem. Soc.* **120**, 5591–5592.

Yang, D., and Kay, L. E. (1999). TROSY triple-resonance four-dimensional NMR spectroscopy of a 46 ns tumbling protein. *J. Am. Chem. Soc.* **121**, 2571–2575.

Yu, H. T. (1999). Extending the size limit of protein nuclear magnetic resonance. *Proc. Natl. Acad. Sci. USA* **96**, 332–334.

[16] Structure Determination of Large Biological RNAs

By PETER J. LUKAVSKY and JOSEPH D. PUGLISI

Abstract

Complex RNA structures regulate many biological processes but are often too large for structure determination by nuclear magnetic resonance (NMR) methods. We determined the solution structure of domain II of the hepatitis C viral internal ribosome entry site (HCV IRES), a 25-kDa RNA, using a novel NMR approach. Conventional short-range, distance, and torsion angle NMR restraints were combined with long-range, angular restraints derived from residual dipolar couplings (RDCs) to improve both the local and global precision of the structure. This powerful approach should be generally applicable to the NMR structure determination of large, modular RNAs.

Introduction

Nuclear magnetic resonance (NMR) spectroscopy has become a powerful tool for high-resolution structure determination of RNA oligonucleotides up to 15 kDa (Lukavsky and Puglisi, 2001; Lynch *et al.*, 2000; Varani and Tinoco, 1991; Varani *et al.*, 1996). As with proteins, RNA NMR structure determination requires uniform isotopic labeling with ^{13}C and ^{15}N to achieve complete resonance assignment using multidimensional double- and triple-resonance NMR experiments. Unambiguous resonance assignments form the basis for extraction of structural information from nuclear Overhauser effect spectroscopy (NOESY)–type experiments that yield local distance information between protons separated by less than 5 Å and from NMR experiments that measure torsion angles along the backbone. A maximum achievable number of local restraints is then used in a simulated annealing protocol followed by restrained molecular dynamics (MD) calculations to generate ensembles of structures that satisfy the restraints.

The precision of the structural ensemble depends on both the number and "quality" of the restraints. Long-range NOEs, between residues that are far apart in sequence or link regions of secondary structure, add more to the global structural precision compared to intraresidual NOEs (Allain and Varani, 1997). In general, global precision is more difficult to achieve for RNA molecules, since they often form extended structures, which yield

Copyright 2005, Elsevier Inc.
All rights reserved.
0076-6879/05 $35.00

only a limited number of long-range restraints compared to globular, compactly folded proteins. The number of restraints for RNAs is also often smaller than in proteins of similar molecular weight. Whereas proteins are composed of 20 different amino acids with an average molecular weight of 130 Da, RNA is composed of only four different nucleosides with an average molecular weight of 340 Da. The molecular weight difference per residue results in a lower density of protons/Dalton for RNA (1/3) compared to proteins, and therefore less structural restraints per residue compared to proteins. The difference of four RNA residues versus 20 protein residues is also reflected in less favorable chemical shift dispersion for all RNA nuclei. This makes unambiguous resonance assignments more difficult in RNA compared to proteins, where most NMR experiments gain resolution by the favorable backbone amide nitrogen chemical shift dispersion of about 30 parts per million (ppm). NMR experiments that make use of the 100% natural abundance of ^{31}P in an RNA backbone suffer from severe ^{31}P spectral overlap, small backbone two- and three-bond couplings (3–10 Hz) (Marino et al., 1999), and unfavorable chemical shift anisotropy parameters of the ^{31}P nuclei, which result in short transverse relaxation times. The severe resonance overlap of ribose nuclei and small heteronuclear couplings makes it more difficult to extract backbone torsion angle restraints from NMR experiments for RNAs.

These NMR spectroscopic shortcomings of RNA molecules become even more severe for larger RNAs. Most biological RNAs are far larger than 15 kDa, which is the size at which the methods described above work effectively. RNA molecules often form more extended structures than proteins of similar molecular weight, resulting in slower overall tumbling times and therefore shorter transverse relaxation times. The concomitant increase in linewidth exacerbates the severity of spectral overlap for all RNA nuclei and eventually makes unambiguous assignments impossible. Many biologically interesting RNA molecules have therefore not been accessible to a high-resolution NMR structure determination. Instead, large biological RNAs have been reduced to collections of smaller, thermodynamically stable subdomains, such as helices and loops, taken out of their larger structural context. In favorable cases (Allain et al., 1996, 2000; Battiste et al., 1996; Fourmy et al., 1996), these subdomains maintained the conformation adopted in the larger RNA and helped to overcome the size and resonance overlap problem. High-resolution structures obtained from subdomains can be used to model the larger RNA molecule (Butcher et al., 1999; Cai and Tinoco, 1996). Without additional structural information from the larger RNA molecule, which would define the orientation of subdomains within their larger structural context, defining the structure of the large RNA is fraught with problems. Lack of interdomain NOE restraints (Skrynnikov *et al.*, 2000) in multidomain proteins causes a similar problem of domain orientation (Skrynnikov *et al.*, 2000).

The application of residual dipolar couplings (RDC) to biomolecular NMR supplements local torsion and NOE restraints. RDCs yield orientational restraints that improve the global precision of NMR structures (Bax *et al.*, 2001). Application of RDC-derived restraints to the structure determination of small protein and RNA molecules significantly improves both global and local precision of the structural ensembles. In addition, global structures of multidomain proteins can also be defined (Skrynnikov *et al.*, 2000). Similarly, RDC-derived restraints should also benefit the structure determination of large RNAs, where NMR structures of subdomains can be solved to high resolution, but no structural information is available to define their relative orientation in the context of the larger RNA molecule. Here, we describe an NMR approach that uses RDC-derived angular restraints to improve local structures in RNA and to define the overall shape of the RNA molecule. We present the application of this method to the high-resolution structure determination of the hepatitis C viral (HCV) internal ribosome entry site (IRES) domain II RNA, a 25-kDa RNA (Lukavsky *et al.*, 2003).

Design and Validation of RNA Oligonucleotides for Structural Studies

RNA oligonucleotides for structural studies are designed based on a correctly predicted RNA secondary structure. Although the initial secondary structure of the HCV IRES RNA had been determined carefully using a combination of comparative sequence analysis of related pestiviral 5'-untranslated regions (5'-UTR), thermodynamic modeling, and enzymatic footprinting (Brown *et al.*, 1992), the secondary structure of its domain II had to be revised several times (Honda *et al.*, 1999; Zhao and Wimmer, 2001) (Fig. 1). NMR spectroscopy is a powerful tool to study RNA secondary structure, since the imino proton "fingerprint" region of homonuclear two-dimensional (2D) NOESY spectra between 10 and 15 ppm provides information not only about the type of base pair formed, but also their sequential neighbors (Heus and Pardin, 1991). NMR spectroscopy on an oligonucleotide comprising the entire HCV IRES RNA domain II allowed us to confirm the correct secondary structure (Fig. 1B). Figure 1C shows the region of imino–imino NOE cross-peaks of a 2D S-NOESY spectrum (Smallcombe, 1993), which helped to establish the secondary structure of domain II. A strong imino–imino cross-peak arising from two base-paired uracil residues (U64 and U103) and an NOE to an adjacent G102–C65 Watson–Crick base pair as well as an A-form-like imino–ribose NOE

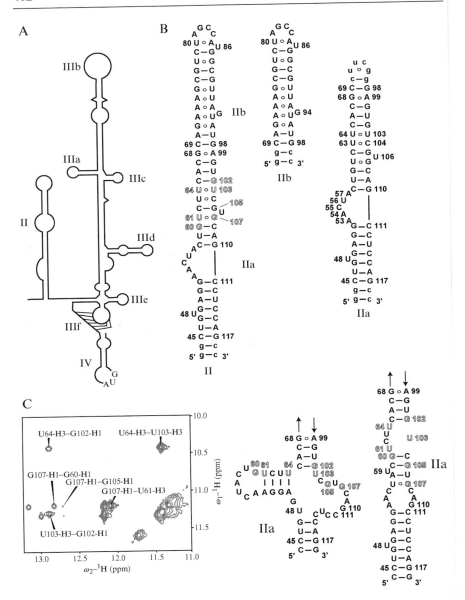

FIG. 1. Secondary structure of HCV IRES RNA and HCV IRES domain II RNAs. (A) Schematic representation of the secondary structure of HCV IRES RNA and its domain organization (Brown *et al.*, 1992). (B) RNA NMR constructs of domain II and subdomains IIa and IIb used for NMR structure determination (Lukavsky *et al.*, 2003). Numbering according

pattern (Heus and Pardi, 1991) (not shown) indicated the formation of a continuous helix below domain IIb rather than the predicted formation of a pyrimidine-rich internal loop or a three-way junction (Fig. 1C). In addition, the G107–U61 base pair showed two NOEs to adjacent G–C Watson–Crick base pairs, which was consistent only with the latest predicted secondary structure shown in Fig. 1B (Zhao and Wimmer, 2001).

Based on these initial NMR studies of domain II, smaller RNA oligonucleotides that correspond to subdomains could be correctly designed. RNA oligonucleotides IIb [34 nucleotides (nt)] and IIa (55 nt) comprising nt 69–98 and nt 45–69 and 98–117 of domain II, respectively, were designed for high-resolution NMR structure determination. Both domain IIb and IIa also contained two additional G–C base pairs analogous to domain II (Fig. 1B), and the apical end of domain IIa was capped by an additional C–G base pair and a UUCG-tetraloop to aid resonance assignments and to serve as a nucleation site for proper folding (Cheong et al., 1990). The addition of stabilizing G–C pairs and capping of helices by tetraloops are standard approaches to RNA oligonucleotide design and provide a stable context for secondary-structure formation.

Chemical shifts were used to confirm that tertiary structures within domain II are accurately represented by the isolated, smaller subdomains IIa and IIb. Chemical shifts are a sensitive measure of the local chemical environment of a nucleus especially in RNA (Cromsigt et al., 2001; Furtig et al., 2004), and the chemical shifts of aromatic base protons are mainly affected by the shielding effects generated by the ring currents of the 5'-neighboring base (Cromsigt et al., 2001). For domain II, aromatic ^{1}H, ^{13}C, and ^{15}N chemical shifts (II versus IIa or IIb) were compared. Similar chemical shifts in the different RNA oligonucleotides were reflected in low root-mean-square deviations (RMSD) for chemical shifts of aromatic C–H groups [0.099 (^{13}C) and 0.074 (^{1}H), respectively] and N–H groups [0.055 (^{15}N) and 0.028 (^{1}H), respectively]. Larger RMSDs were observed only for the C69–G98 Watson–Crick base pair, since the 5'-neighboring nucleotides differed in all three constructs, but close inspection of NOE patterns and cross-peak intensities in S-NOESY spectra indicated that the same structure was formed (data not shown). These data demonstrate that subdomains IIa and IIb adopt the same conformation within the context of the entire domain II (Lukavsky et al., 2003).

---

to Brown et al., (1992). Additional nts are shown in small letters (see text). (C) Region of imino–imino NOE cross-peaks of a 2D S-NOESY spectrum of HCV IRES domain II, which confirmed the secondary structure of domain II shown in (B). Outlined letters show nts, which yield NOEs consistent with the correct secondary structure of domain II in (B) (Zhao and Wimmer, 2001) but inconsistent with two previously predicted secondary structures (Brown et al., 1992; Honda et al., 1999).

Use of Segmental Isotope Labeling for NMR Studies of Large RNAs

Segmental isotope labeling of RNA reduces the complexity of NMR spectra and therefore allows more detailed NMR studies of subdomains of very large RNAs (Kim *et al.*, 2002; Xu *et al.*, 1996). Segmental labeling can also be used to confirm the validity of NMR data acquired on model oligonucleotides. Isotopic labeling of domain II in the context of the entire 100-kDa HCV IRES RNA was used to address the question of whether domain II forms an individually folded domain within the HCV IRES.

The most cost-efficient method of choice for the preparation of a segmentally labeled RNA oligonucleotide is T4 RNA ligase-catalyzed joining of a 3' "donor" RNA having a 5'-terminal monophosphate and a 5' "acceptor" RNA terminating in a 3'-hydroxyl. The product of such a ligation has a standard 3',5'-phosphodiester linkage (Romaniuk and Uhlenbeck, 1983). The efficiency of ligation depends on the structure of the 5' and 3' ends, and it is efficient only for RNAs that have a single-stranded region as the site of ligation. To avoid intramolecular ligation and the formation of by-products, both sides of the 5'-fragment should be dephosphorylated and both sides of the 3'-fragment should be phosphory-lated. This can be easily achieved using hammerhead ribozymes to engineer the RNA 5' and 3' ends of the donor and acceptor in a way such that only the desired product can be formed (Kim *et al.*, 2002).

Based on the secondary structure of domain II (Fig. 1B), which was confirmed by NMR methods, only two single-stranded regions emerged as potential candidates for optimum ligation conditions, namely the single-stranded region in domain IIa (A53–A57) and the apical hairpin loop (U80–U86). Since both sites would allow only 20% or 50%, respectively, of domain II to be ^{15}N labeled in the context of the HCV IRES, we chose nucleotides C104 and G105 as the point of ligation. While this region is base paired in the context of domain II, individually transcribed parts comprising nt 40–104 and nt 105–354, respectively, should be single strand-ed at their ligation site. Both the 5' and 3'-RNA fragments were transcribed *in vitro* using transcription with ^{15}N-labeled nucleotide triphosphates for the 5'-fragment and unlabeled nucleotide triphosphates for the 3'-frag-ment. The purified RNA fragments were then ligated using T4 RNA ligase at a maximum efficiency of 50%, yielding a 0.1 mM NMR sample of segmentally ^{15}N-labeled HCV IRES RNA (Kim *et al.*, 2002). Chemical shifts of imino ^{1}H and ^{15}N resonances observed for domain II alone and in the context of the 100-kDa HCV IRES were almost identical, confirming that domain II forms an independently folded subdomain in the intact IRES (Fig. 2).

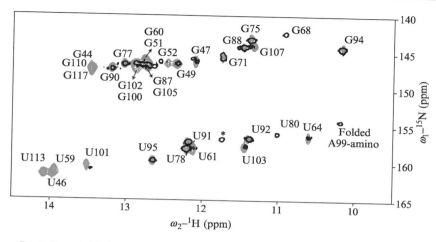

FIG. 2. Domain II alone adopts the same conformation in the context of the 100-kDa HCV IRES. Comparison of imino H–N correlations of isolated $^{13}C,^{15}N$-labeled domain II (25-kDa) and segmentally (nt 40–104) ^{15}N-labeled IRES (100-kDa) RNA (Kim *et al.*, 2002). Domain II alone is in gray and domain II in the context of the IRES is black. Domain II residues, which are not ^{15}N labeled in the segmentally labeled sample, are shown in italic. Resonances of U46, U59, and U61 (segmentally labeled sample) and resonances of G68 and U80 (domain II sample) are observed at lower thresholds of the spectra (data not shown). An unassigned resonance is marked by an asterisk. Reprinted with permission from Lukavsky *et al.* (2003).

Measurement of RDCs

In proteins, RDCs are usually obtained for amide backbone one-bond ^{1}H–^{15}N couplings and side-chain one-bond ^{1}H–^{13}C couplings (Tjandra *et al.*, 1997). This set can be supplemented by additional one-bond ^{15}N–$^{13}C'$, $^{13}C^{\alpha}$–$^{13}C'$, two-bond $^{13}C'$–H^{N}, and three-bond $^{13}C^{\alpha}$–H^{N} RDCs to provide a high density of RDC-derived restraints (Skrynnikov *et al.*, 2000). In RNA, on the other hand, one-bond ^{1}H–^{15}N couplings from imino protons usually yield only a limited number of restraints and backbone three-bond ^{1}H–^{31}P and four-bond ^{13}C–^{31}P couplings are small and cannot be accurately measured. Orientational restraints for RNA structure determination are therefore commonly derived from RDCs measured from one-bond ^{1}H–^{13}C interactions supplemented by a few derived from ^{1}H–^{15}N interactions (Hansen *et al.*, 1998). Bacteriophage Pf1 is the ideal liquid-crystalline medium for RDC measurements of RNA, because negatively charged RNA oligonucleotides are aligned by steric interactions with the negatively charged phage particles rather than by direct binding to the phage, which would increase the overall tumbling time, and, correspondingly, the linewidths of RNA resonances (Hansen *et al.*, 1998). One-bond coupling constants in RNA are usually measured in isotropic and aligned media (upon addition of up to 25 mg/ml of Pf1 phage (Hansen *et al.*, 1998)), and the RDC value can then be extracted from the difference between the value (in Hertz) of the coupling constant obtained in aligned and isotropic media.

Several NMR methods have been developed for the accurate measurement of one-bond couplings. Initially, one-bond ^{1}H–^{15}N and ^{1}H–^{13}C couplings were measured by simply recording 2D heteronuclear single-quantum correlation (HSQC) spectra without decoupling during the t_1 evolution period. Since this method doubles the number of resonances in already crowded RNA 2D HSQC spectra, only a limited number of coupling constants could be measured. More recently, one-bond couplings are measured using either IPAP-HSQC (in-phase and antiphase) experiments, which yield either the upfield or downfield component of the F_1 ^{15}N–^{1}H doublet (Ottiger *et al.*, 1998) or sets of J_{CH}-modulated 2D HSQC spectra (Tjandra and Bax, 1997). Both methods alleviate the problem of spectral overlap, and the latter has been applied to the study of the Sarcin–Ricin loop (Warren and Moore, 2001) and the hammerhead ribozyme RNA (Bondensgaard *et al.*, 2002). For larger RNAs, like the HCV IRES domain II, 2D IPAP HSQC and 2D t_1-coupled HSQC cannot be applied, since the upfield component of the F_1 ^{15}N–^{1}H or ^{13}C–^{1}H doublet is too broad to allow accurate measurement of the coupling constant. Figure 3 shows an adenine C2–H2 cross-peak from a spectrum acquired with a 2D t_1- and t_2-coupled ^{1}H–^{13}C HSQC experiment on the 77-nt RNA comprising domain II (see Fig. 1B). Only the downfield components in the ^{13}C dimension yield sufficiently sharp resonances, while the upfield components of the multiplet exhibit severe broadening (Fig. 3). A similar situation is encountered for the ^{1}H–^{15}N multiplet of imino resonances (data not shown). In addition, ^{1}H–^{13}C correlation experiments require a constant time (CT) frequency-editing period to eliminate one-bond ^{13}C–^{13}C couplings during t_1 evolution (17 ms for base and 25 ms for sugar carbons). In large RNAs, where short transverse relaxation times are encountered, these long CT periods can lead to significant losses in signal intensity and thereby make measurement lengthy or even impossible.

These problems can be overcome by using TROSY-based methods, which significantly reduce signal loss and allow selection of the sharper, downfield ^{13}C or ^{15}N components of the ^{1}H–^{13}C and ^{1}H–^{15}N multiplets. For the one-bond ^{1}H–^{15}N coupling measurements of HCV IRES domains II, IIa, and IIb, the spin-state selective, gradient- and sensitivity-enhanced 2D ^{1}H–^{15}N TROSY by Weigelt (1998) or the original 2D WATERGATE ^{1}H–^{15}N TROSY (Pervushin *et al.*, 1997) were therefore used. One-bond

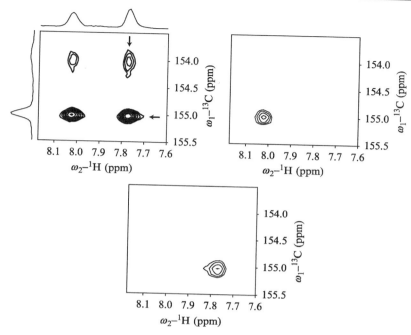

FIG. 3. RDC-measurement using ¹H–¹³C TROSY with CT evolution. Schematic representation of an adenine C2–H2 multiplet from a 2D t_1- and t_2-coupled ¹H–¹³C HSQC experiment of the 77-nt RNA comprising domain II. The 2D ¹H–¹³C ct-TROSY selects either the downfield or the upfield component of the downfield ¹³C–¹H doublet.

¹H–¹³C couplings were measured using a ¹³C version of the Weigelt-TROSY (Weigelt, 1998), which included a CT evolution period (see Fig. 3). The latter ¹H–¹³C TROSY version was preferred over a previously published version (Brutscher *et al.*, 1998) due to its excellent H₂O suppression. All measurements were performed in the same buffer (10 m*M* sodium phosphate buffer at pH 6.40, 100 m*M* sodium chloride, and 5 m*M* magnesium chloride with 4% D₂O) and repeated at least twice in the absence or presence of 8 (II) or 10 (IIa and IIb) mg/ml Pf1 phage at 800 MHz field strength. To ensure complete alignment at such high salt versus low Pf1 phage concentrations (Zweckstetter and Bax, 2001), the samples were prealigned on top of the 800-MHz magnet overnight. The spinner with the sample tube was simply inserted into a piece of cardboard with a hole in the middle to prevent the sample from falling into the magnet and was placed on top of the sample inlet tube. The uncertainty in the measurement was estimated based on the quotient of linewidth and signal-to-noise ratio

(Bax *et al.*, 2001) to be 2.5 Hz for $^{13}C-^1H$ and 1.25 Hz for $^{15}N-^1H$ couplings.

The RDC values were calculated as the difference of the measured coupling (Hertz) in aligned and isotropic media (for $^1H-^{15}N$ RDCs, the negative gyromagnetic ratio has to be taken into account). RDCs, whose standard deviation from the average of at least two individual RDC values exceeded the uncertainty of the measurement (see above), were not used for subsequent structure calculations. Large errors in some RDCs originated from partial overlap or water suppression artifacts near the residual water line. In addition, RDCs from three dynamic uracil residues (U48, U56, and U106) were also eliminated. A final set of 136 (IIa), 105 (IIb), or 60 (II) RDCs was used for the structure refinement of subdomains (IIa and IIb) and domain II.

Use of RDCs in Structure Calculation

To determine the structure of domain II by NMR, high-resolution restraints from the subdomains IIa and IIb were combined with RDC data of the subdomains and the full domain II. Assignment procedures for domains IIa and IIb followed previously published protocols using RNA-Pack NMR pulse sequences on VARIAN NMR spectrometers (Lukavsky and Puglisi, 2001). Complete resonance assignments of domains IIa and IIb yielded 1146 NOE and 419 scalar coupling restraints for domain IIa, 831 NOE and 253 scalar coupling restraints for domain IIb, and a total of 1744 NOE and 523 scalar coupling restraints for domain II, respectively. Random starting structures were generated and subjected to a simulated annealing protocol excluding RDC restraints, followed by restrained molecular dynamics simulations (Lukavsky *et al.*, 2003). The atomic RMSDs for the 20 best IIb and 29 best IIa structures were 2.43 and 4.91 Å, respectively; however, the entire domain II was poorly defined with an RMSD of 7.48 Å for the best 20 structures, since no RDC-derived restraints were used to define the global conformation of domain II.

Calculating structures without RDC-derived restraints was necessary for two reasons. First, NOE restraints tend to favor convergence through cooperative contributions to the folding energy during simulated annealing, thereby creating a funnel-like energy landscape, whereas angular restraints tend to compete with one another and, thus, are not usable during the initial in-silico folding of the structures (Bax *et al.*, 2001). Second, initial structures calculated with RDCs can be used to determine the magnitude of the axial (D_a) and rhombic (R) components of the alignment tensor to extract orientational restraints for the subsequent NMR structure refinement from measured RDCs. In addition to one model-free approach, several model-based prediction methods have been proposed to determine an initial alignment tensor for the refinement of NMR structures (see below), and initial structures calculated without RDCs can be used for this purpose.

Prediction of the Alignment Tensor

Several methods for prediction of the magnitude of the D_a and R components of the alignment tensor have been tested on RNA molecules (McCallum and Pardi, 2003; Warren and Moore, 2001). These include laborious grid search approaches (Clore et al., 1998a), prediction of the alignment tensor from the three-dimensional shape of the biomolecule (Zweckstetter and Bax, 2000), structure-independent prediction from the histogram of the RDCs (Clore et al., 1998b), and singular value decomposition (SVD), which determines D_a and R values consistent with a given RDC data set and structure using a fitting procedure (Losonczi et al., 1999).

Analysis of the histogram of the RDC data was used as a model-free approach for the determination of an initial alignment tensor for NMR refinement of the subdomains IIa and IIb as well as domain II (Table I). All histograms displayed the characteristic powder pattern with RDC values ranging from −44.5 to +37.0 for II, −23.8 to +22.6 for IIa, and −23.8 to +19 for IIb, respectively. In domain IIb, RDCs from aromatic carbon–proton pairs, which are almost perpendicular to the helix axis, clustered in the right-hand side indicative of a roughly helical shape of domain IIb. For domains II and IIa, on the other hand, negative RDCs for aromatic carbon–proton pairs in the lower stem indicated a bend within domain IIa. Analysis of the structural ensembles of domains IIa, IIb, and II calculated without RDCs using the SVD method yielded very similar D_a and R values (Table I). Independent fits performed on helical regions of the subdomains IIa and IIb as well as subdomains IIa and IIb in the context of domain II also confirmed that these parts align as a single species (Losonczi and Prestegard, 1998).

Two sets of structure calculations with RDC restraints were performed on each subdomain using the two individually determined alignment tensors for domains IIa and IIb. Implementation of RDC restraints in the CNS package (Brunger et al., 1998) requires definition of a tetraatomic pseudomolecule OXYZ, which represents the alignment tensor (Bax et al., 2001). The pseudomolecule was added to the simulated annealing structures calculated without RDCs at a fixed position in space away from the RNA and the refinement procedure including RDC restraints was performed as follows: (1) 500 steps of restrained energy minimization; (2) restrained molecular dynamics (rMD) at 2000 K, while increasing the torsion angle

TABLE I
SUMMARY OF ALIGNMENT TENSOR MAGNITUDES FOR DOMAINS IIa, IIb, AND II

Method[a]	D_a(Hz)	R
Singular value decomposition (SVD) method		
Domain IIa	−9.44	0.40
Upper helix of IIa (58–69, 98–110)	−11.26	0.56
Lower helix of IIa (43–52, 111–119)	−8.92	0.62
Average	−9.88 ± 1.23	0.53 ± 0.11
Domain IIb	−11.97	0.15
No hairpin loop of IIb (67–79, 87–100)	−12.30	0.19
Average	−12.13 ± 0.23	0.17 ± 0.03
Domain II	−19.86	0.44
Subdomain IIa of II (43–69, 98–119)	−23.07	0.42
Subdomain IIb of II (70–97)	−21.06	0.54
Average	−21.52 ± 1.38	0.43 ± 0.09
Histogram of the observed RDCs		
Domain IIa	−11.90	0.60
Domain IIb	−11.90	0.40
Domain II	−22.25	0.44
Final tensor after iterative refinement		
Domain IIa	−12.93	0.45
Domain IIb	−15.28	0.17
Domain II	−26.40	0.29

[a] See text for details.

force constant; (3) rMD at 2000 K while increasing the RDC force constant keeping the torsion angle force constant low and (4) rMD at 2000 K while keeping the RDC force constant at 0.25 kcal/Hz^2 and increasing the torsion angle force constant; (5) rMD while cooling to 300 K; and finally (6) 5000 cycles of energy minimization, which included a Lennard–Jones potential, but no electrostatic terms. The ensemble of structures that had the lowest average total energy and final RDC restraint violation energy was then used to determine a new alignment tensor by SVD fitting of observed RDCs to each ensemble member. The average values for D_a and R obtained from the fitting were then used in another round of structure calculations to improve the alignment tensor iteratively until the average value for the rhombic component was unchanged and met acceptance criteria ($\Delta R < \pm 0.015$) (Fig. 4), yielding final alignment tensors for IIa and IIb (Table I). Figure 4 illustrates the iterative improvement of the alignment tensor for domain IIa. The initial D_a and R values determined by the SVD method yielded the ensemble with the lowest final RDC restraint violation and total energy and even an unrefined alignment

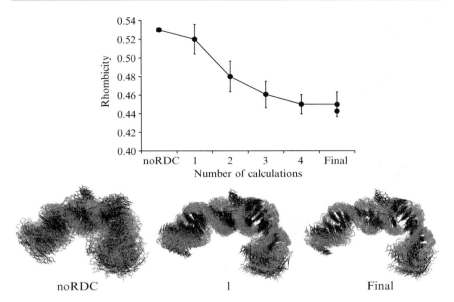

FIG. 4. Iterative refinement of the alignment tensor of domain IIa. The x-axis displays the round of calculations and the y-axis displays the average rhombicity of the ensemble after each round of calculations with standard deviations. The final round of calculations left the rhombic component of the alignment tensor unchanged and the standard deviation met acceptance criteria ($\Delta R < \pm 0.015$), yielding final alignment tensors for IIa (Table I). The three structural ensembles displayed represent the initial ensemble calculated without RDCs, the first ensemble calculated with RDCs, and the final ensemble calculated with RDCs.

tensor already significantly improved the definition of the overall structure of domain IIa (RMSD of 4.91 Å without RDCs versus 2.98 Å after the first round of calculations). The final stage was reached after five rounds of calculations and the final ensemble of domain IIa structures had an RMSD of 2.34 Å. The structure of domain IIb was refined in the same manner improving the RMSD from 2.43 to 1.35 Å.

Refinement of subdomains IIa and IIb in the context of the entire domain II was then performed using final alignment tensors and RDCs from the domain IIa and IIb oligonucleotides (see Fig. 1B). The protocol for structural calculations of domain II was performed in the same two stages as described for domains IIa and IIb and used a total of 1744 distance restraints, including 21 NN hydrogen bond distance restraints, 261 RDC restraints (from domains IIa, IIb, and II), and 523 dihedral restraints. For each alignment tensor (final alignment tensors of IIa and IIb, and the initial alignment tensor of II), a separate pseudomolecule

OXYZ was added to the accepted simulated annealing structures of domain II at a fixed position in space away from the RNA. The refinement procedure including all RDC restraints was then performed very similarly; the only exception was that RDCs were introduced stepwise, first for IIb and II, then for IIa and II, and finally for all RDCs to improve convergence. After the first round of calculations, the individual ensembles were analyzed to improve iteratively the alignment tensor for domain II RDCs without changing the alignment tensors for domains IIa and IIb until the average value for the rhombic component was unchanged and met the same acceptance criteria ($\Delta R < \pm0.015$) as for the individual subdomains IIa and IIb. The final ensemble of domain II structures (Fig. 5B) after five rounds of calculations showed both locally improved subdomains (1.15 Å for IIb and 1.62 Å for IIa) and a well-defined global shape of domain II (2.18 Å) compared to the ensembles obtained without RDCs (Fig. 5A) (RMSDs of 2.43 Å for IIb, 4.38 Å for IIa, and 7.48 Å for II). This

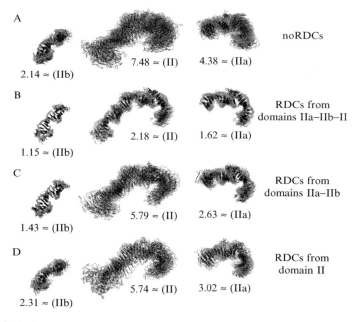

Fig. 5. Final ensembles of domain II structures calculated with different sets of RDCs and corresponding local superpositions of subdomains IIa and IIb. (A) Final ensemble of structures calculated without RDCs. (B) Final ensemble of structures calculated with RDCs from domains IIa, IIb, and II. (C) Final ensemble of structures calculated only with RDCs from domains IIa and IIb. (D) Final ensemble of structures calculated only with RDCs from domain II.

improvement could be achieved only with the combination of RDCs from the subdomains with the global RDCs from domain II. Omitting the 60 RDCs from domain II yielded well-defined local domains (1.43 Å for IIb and 2.63 Å for IIa), while the overall shape of domain II improved only slightly (5.79 Å) as shown in Fig. 5C. Similarly, using only 60 RDCs from domain II also gave only a slight improvement of the overall definition (5.74 Å), since subdomains were less well-defined without RDCs (2.31 Å for IIb and 3.02 Å for IIa) as shown in Fig. 5D.

A limited number of RDC restraints was sufficient to define the global fold of domain II, consistent with prior data (Mollova *et al.*, 2000). For domain II, only 60 additional RDCs, which is less than one RDC restraint per residue, were sufficient to define the global shape of domain II. Mollova *et al.* (2000) demonstrated that only 27 RDCs were sufficient to determine the global structure of tRNA, but only if the local structures of the helical stems is known. In the case of a *de novo* RNA NMR structure determination, where local and global structures are unknown, such well-defined subdomain structures can be obtained using an RDC-based refinement procedure. Only smaller subdomains of the larger RNA allow extraction of a sufficiently large number of RDCs to obtain well-defined structures, since NMR spectra are less crowded compared to larger RNAs. A small number of global RDCs is then sufficient to define orientation of these well-defined substructures and thereby the overall conformation of a large RNA.

Conclusions

We have outlined a powerful and general approach to structure determination of modular large RNAs using NMR spectroscopy. Dissection of a large RNA into subdomains allows extraction of the maximum number of conventional NOE- and scalar coupling-derived restraints, as well as a much larger number of RDCs. This leads to determination of high-quality, well-defined subdomain structures, which provide the basis for a global refinement of the larger RNA using RDCs. In the future, this approach should allow NMR structure determination of even larger RNAs, where several smaller RNA subdomains could be oriented within the larger RNA using segmental labeling techniques to determine a sufficient number of global RDC restraints. These approaches will further increase the power of NMR to probe the biological functions of RNA.

References

Allain, F. H., and Varani, G. (1997). How accurately and precisely can RNA structure be determined by NMR? *J. Mol. Biol.* **267**, 338–351.

Allain, F. H., Gubser, C. C., Howe, P. W., Nagai, K., Neuhaus, D., and Varani, G. (1996). Specificity of ribonucleoprotein interaction determined by RNA folding during complex formulation. *Nature* **380,** 646–650.

Allain, F. H., Bouvet, P., Dieckmann, T., and Feigon, J. (2000). Molecular basis of sequence-specific recognition of pre-ribosomal RNA by nucleolin. *EMBO J.* **19,** 6870–6881.

Battiste, J. L., Mao, H., Rao, N. S., Tan, R., Muhandiram, D. R., Kay, L. E., Frankel, A. D., and Williamson, J. R. (1996). Alpha helix-RNA major groove recognition in an HIV-1 rev peptide-RRE RNA complex. *Science* **273,** 1547–1551.

Bax, A., Kontaxis, G., and Tjandra, N. (2001). Dipolar couplings in macromolecular structure determination. *Methods Enzymol.* **339,** 127–174.

Bondensgaard, K., Mollova, E. T., and Pardi, A. (2002). The global conformation of the hammerhead ribozyme determined using residual dipolar couplings. *Biochemistry* **41,** 11532–11542.

Brown, E. A., Zhang, H., Ping, L. H., and Lemon, S. M. (1992). Secondary structure of the 5′ nontranslated regions of hepatitis C virus and pestivirus genomic RNAs. *Nucleic Acids Res.* **20,** 5041–5045.

Brunger, A. T., Adams, P. D., Clore, G. M., DeLano, W. L., Gros, P., Grosse-Kunstleve, R. W., Jiang, J. S., Kuszewski, J., Nilges, M., Pannu, N. S., Read, R. J., Rice, L. M., Simonson, T., and Warren, G. L. (1998). Crystallography & NMR system: A new software suite for macromolecular structure determination. *Acta Crystallogr. D* **54,** 905–921.

Brutscher, B., Boisbouvier, J., Kupce, E., Tisne, C., Dardel, F., Marion, D., and Simorre, J. P. (1998). Improved sensitivity and resolution in 1H-13C NMR experiments of RNA. *J. Am. Chem. Soc.* **120,** 11845.

Butcher, S. E., Allain, F. H., and Feigon, J. (1999). Solution structure of the loop B domain from the hairpin ribozyme. *Nat. Struct. Biol.* **6,** 212–216.

Cai, Z., and Tinoco, I., Jr. (1996). Solution structure of loop A from the hairpin ribozyme from tobacco ringspot virus satellite. *Biochemistry* **35,** 6026–6036.

Cheong, C., Varani, G., and Tinoco, I., Jr. (1990). Solution structure of an unusually stable RNA hairpin, 5′GGAC(UUCG)GUCC. *Nature* **346,** 680–682.

Clore, G. M., Gronenborn, A. M., and Tjandra, N. (1998a). Direct structure refinement against residual dipolar couplings in the presence of rhombicity of unknown magnitude. *J. Magn. Reson.* **131,** 159–162.

Clore, G. M., Gronenborn, A. M., and Bax, A. (1998b). A robust method for determining the magnitude of the fully asymmetric alignment tensor of oriented macromolecules in the absence of structural information. *J. Magn. Reson.* **133,** 216–221.

Cromsigt, J. A., Hilbers, C. W., and Wijmenga, S. S. (2001). Prediction of proton chemical shifts in RNA. Their use in structure refinement and validation. *J. Biomol. NMR* **21**(1), 11–29.

Fourmy, D., Recht, M. I., Blanchard, S. C., and Puglisi, J. D. (1996). Structure of the A site of *Escherichia coli* 16S ribosomal RNA complexed with an aminoglycoside antibiotic. *Science* **274,** 1367–1371.

Furtig, B., Richter, C., Bermel, W., and Schwalbe, H. (2004). New NMR experiments for RNA nucleobase resonance assignment and chemical shift analysis of an RNA UUCG tetraloop. *J. Biomol. NMR* **28,** 69–79.

Hansen, M. R., Mueller, L., and Pardi, A. (1998). Tunable alignment of macromolecules by filamentous phage yields dipolar coupling interactions. *Nat. Struct. Biol.* **5,** 1065–1074.

Heus, H. A., and Pardi, A. (1991). Novel H-1 Nuclear magnetic resonance assignment procedure for RNA duplexes. *J. Am. Chem. Soc.* **113,** 4360–4361.

Honda, M., Beard, M. R., Ping, L. H., and Lemon, S. M. (1999). A phylogenetically conserved stem-loop structure at the 5' border of the internal ribosome entry site of hepatitis C virus is required for cap-independent viral translation. *J. Virol.* **73**, 1165–1174.

Kim, I., Lukavsky, P. J., and Puglisi, J. D. (2002). NMR study of 100 kDa HCV IRES RNA using segmental isotope labeling. *J. Am. Chem. Soc.* **124**, 9338–9339.

Losonczi, J. A., and Prestegard, J. H. (1998). Nuclear magnetic resonance characterization of the myristoylated, N-terminal fragment of ADP-ribosylation factor 1 in a magnetically oriented membrane array. *Biochemistry* **37**, 706–716.

Losonczi, J. A., Andrec, M., Fischer, M. W., and Prestegard, J. H. (1999). Order matrix analysis of residual dipolar couplings using singular value decomposition. *J. Magn. Reson.* **138**(2), 334–342.

Lukavsky, P. J., and Puglisi, J. D. (2001). RNAPack: An integrated NMR approach to RNA structure determination. *Methods* **25**, 316–332.

Lukavsky, P. J., Kim, I., Otto, G. A., and Puglisi, J. D. (2003). Structure of HCV IRES domain II determined by NMR. *Nat. Struct. Biol.* **10**, 1033–1038.

Lynch, S. R., Recht, M. I., and Puglisi, J. D. (2000). Biochemical and nuclear magnetic resonance studies of aminoglycoside-RNA complexes. *Methods Enzymol.* **317**, 240–261.

Marino, J. P., Schwalbe, H., and Griesinger, C. (1999). *J*-coupling restraints in RNA structure determination. *Acc. Chem. Res.* **32**, 614–623.

McCallum, S. A., and Pardi, A. (2003). Refined solution structure of the iron-responsive element RNA using residual dipolar couplings. *J. Mol. Biol.* **326**, 1037–1050.

Mollova, E., Hansen, M. R., and Pardi, A. (2000). Global structure of RNA determined with residual dipolar couplings. *J. Am. Chem. Soc.* **122**, 11561–11562.

Ottiger, M., Delaglio, F., and Bax, A. (1998). Measurement of *J* and dipolar couplings from simplified two-dimensional NMR spectra. *J. Magn. Reson.* **131**, 373–378.

Pervushin, K., Riek, R., Wider, G., and Wuthrich, K. (1997). Attenuated T2 relaxation by mutual cancellation of dipole-dipole coupling and chemical shift anisotropy indicates an avenue to NMR structures of very large biological macromolecules in solution. *Proc. Natl. Acad. Sci. USA* **94**, 12366–12371.

Romaniuk, P. J., and Uhlenbeck, O. C. (1983). Joining of RNA molecules with RNA ligase. *Methods Enzymol.* **100**, 52–59.

Skrynnikov, N. R., Goto, N. K., Yang, D., Choy, W. Y., Tolman, J. R., Mueller, G. A., and Kay, L. E. (2000). Orienting domains in proteins using dipolar couplings measured by liquid-state NMR: Differences in solution and crystal forms of maltodextrin binding protein loaded with beta-cyclodextrin. *J. Mol. Biol.* **295**, 1265–1273.

Smallcombe, S. H. (1993). Solvent suppression with symmetrically-shifted pulses. *J. Am. Chem. Soc.* **115**, 4776–4785.

Tjandra, N., and Bax, A. (1997). Measurement of dipolar contributions to $^1J_{CH}$ splittings from magnetic-field dependence of *J* modulation in two-dimensional NMR spectra. *J. Magn. Reson.* **124**, 512–515.

Tjandra, N., Omichinski, J. G., Gronenborn, A. M., Clore, G. M., and Bax, A. (1997). Use of dipolar ^1H-^{15}N and ^1H-^{13}C couplings in the structure determination of magnetically oriented macromolecules in solution. *Nat. Struct. Biol.* **4**, 732–738.

Varani, G., and Tinoco, I., Jr. (1991). RNA structure and NMR spectroscopy. *Q. Rev. Biophys.* **24**, 479–532.

Varani, G., Aboul-Ela, F., and Allain, F. H. (1996). NMR investigation of RNA structure. *Prog. Nucl. Magn. Reson. Spectrosc.* **29**, 51–127.

Warren, J. J., and Moore, P. B. (2001). Application of dipolar coupling data to the refinement of the solution structure of the sarcin-ricin loop RNA. *J. Biomol. NMR* **20**, 311–323.

Weigelt, J. (1998). Single scan, sensitivity- and gradient-enhanced TROSY for multidimensional NMR experiments. *J. Am. Chem. Soc.* **120,** 10778–10779.

Xu, J., Lapham, J., and Crothers, D. M. (1996). Determining RNA solution structure by segmental isotopic labeling and NMR: Application to Caenorhabditis elegans spliced leader RNA 1. *Proc. Natl. Acad. Sci. USA* **93,** 44–48.

Zhao, W. D., and Wimmer, E. (2001). Genetic analysis of a poliovirus/hepatitis C virus chimera: New structure for domain II of the internal ribosomal entry site of hepatitis C virus. *J. Virol.* **75,** 3719–3730.

Zweckstetter, M., and Bax, A. (2000). Prediction of sterically induced alignment in a dilute liquid crystalline phase: Aid to protein structure determination by NMR. *J. Am. Chem. Soc.* **122,** 3791–3792.

Zweckstetter, M., and Bax, A. (2001). Characterization of molecular alignment in aqueous suspensions of Pf1 bacteriophage. *J. Biomol. NMR* **20,** 365–377.

Section IV

Macromolecular Dynamics

[17] Hydrodynamic Models and Computational Methods for NMR Relaxation

By J. García de la Torre, P. Bernadó, and M. Pons

Abstract

Interpretation of NMR relaxation data of macromolecules is based on the analysis of their dynamic behavior in solution. For quasirigid molecules, in addition to a minor, separable contribution from local mobility, the main contribution corresponds to the overall rotational diffusion of the complete molecule. Therefore, theoretical descriptions and computational methodologies for hydrodynamic calculations, which yield the full, anisotropic rotational diffusion tensor of rigid molecules, are extremely helpful in the analysis of NMR relaxation. Recent approaches allow realistic predictions of the rotational diffusion tensor from structures at atomic detail. This enables measured relaxation rates and structural models to be compared. Such a comparison (1) provides an independent test of the structural model, (2) provides a framework for the interpretation of local motion, even for highly anisotropic systems, (3) provides a simple method for the detection of additional sources of relaxation, such as chemical exchange, and (4) provides a sensitive method for the detection of nonspecific aggregation or oligomer formation. Although hydrodynamic calculations usually assume a rigid structure, Brownian dynamics simulations extend their range of applications to flexible multidomain structures. Hydrodynamic applications are not restricted to globular proteins. Small DNA fragments, which could be otherwise considered cylindrical objects, can also be treated with atomic detail using the same methodology used for proteins.

Introduction

Biomolecular structures are intrinsically dynamic entities, and there is an increasing consensus that dynamics is often directly related to function. Together with a widespread accessibility of isotopically labeled samples, this fact has led to an increased interest in the study of biomolecular dynamics, especially through nuclear magnetic resonance (NMR) relaxation methods.

Backbone mobility is now routinely measured using ^{15}N NMR relaxation that probes the reorientation of N–H bonds. Additional information

Copyright 2005, Elsevier Inc.
All rights reserved.
0076-6879/05 $35.00

can be obtained from ^{13}C relaxation, with a focus on carbonyl groups, although its use is limited by the higher costs of ^{13}C labeling and possible complications due to homonuclear coupling in uniformly labeled samples. Protein side chains have a much richer dynamics that is presently the object of great interest, but it is outside of the scope of this chapter. Heteronuclear NMR relaxation of backbone nuclei is usually interpreted in a model-free framework (Lipari and Szabo, 1982), whose central assumption is the absence of correlation between global and local motions, allowing their separation. Extensions of the model-free approach consider additional, uncorrelated, local fast motions or extra relaxation sources derived from chemical exchange. Experimental relaxation rates of individual nuclei are fitted to the model with the lowest number of statistically significant parameters, and the results are analyzed in terms of extracted parameters (Palmer et al., 1996). Local motions are captured by order parameters, related to the amplitude of fluctuations, and characteristic time constants. Global motion, universally present in fluid phases, is accounted for by either a single correlation time, implying isotropic reorientation, or up to five correlation times for a fully anisotropic system. The assumption of axial symmetry may be remarkably misleading because, among other reasons, a careful description of global motion is required for the correct interpretation of local mobility.

Individual nuclei belonging to bonds with different orientations with respect to the principal axes of the rotational diffusion tensor relax at different rates. A structural model of the macromolecule is required to deconvolute anisotropic reorientation from other sources (e.g., local mobility) of variability in relaxation rates within a molecule. If the structure of the macromolecule is known and it can be effectively considered a rigid body, an alternative approach for interpreting NMR relaxation data is to compute the rotational diffusion tensor from the structure by applying hydrodynamic theory. A comparison of calculated and experimental relaxation rates will provide an independent confirmation of the model structure and the rigid body assumption and a framework for the interpretation of local variability in relaxation rates. In this chapter we shall discuss recent advances in the application of hydrodynamic theory to predict relaxation rates and some applications to the detection of local motion in rigid structures, characterization of protein oligomers, and detection of flexibility in multidomain structures. A recent review describes additional approaches to the interpretation and prediction of relaxation data (Brüschweiler, 2003). Background information on NMR relaxation has been extensively reviewed. (Fischer et al., 1998; Korzhnev et al., 2001).

Theory and Methods

Rotational Brownian motion (RBM) of a rigid particle is determined by the rotational diffusion tensor, $\mathbf{D}_r$, a symmetric 3×3 matrix whose six components depend on the size and shape of the particle and are proportional to the ratio of absolute temperature to solvent viscosity, T/η_0. Diagonalization of $\mathbf{D}_r$ provides three eigenvalues, D_i, where $i = 1, 2$, or 3, and its principal axes are defined by the eigenvectors $\mathbf{u}_i$. For the simplest case of a spherical particle, $\mathbf{D}_r$ is isotropic and all D_is are identical; conversely, for an arbitrary particle, the D_i values are different and depend sensitively on the overall shape and therefore contain valuable structural information. The time course of observable properties related to RBM of a rigid body is determined by five relaxation time, τ_k, $k = 1, \ldots, 5$, directly derived from D_i values. For an isotropic system, a single correlation time is enough to describe its motion. For an arbitrary system, it is sometimes useful to define the (harmonic) mean or correlation time, τ_c, whose reciprocal is given by

$$1/\tau_c = (1/5) \sum (1/\tau_k) = 2(D_1 + D_2 + D_3)$$

The Brownian reorientation of a particle-fixed vector $\mathbf{v}$, is usually represented by the time-dependent function $<P_2[\cos \beta(t)]>$, where $\beta(t)$ is the angle subtended by an initial orientation of the vector and its orientation after time t has elapsed, $P_2[x] = (3x^2 - 1)/2$, and $<\ldots>$ means an average over the choices of the initial instant. RBM theory predicts that this function decays from 1 to 0 in the form of a sum of exponentials, $\sum a_k \exp(-t/\tau_k)$, where the time constants are the τ_ks and the amplitudes a_k depend on the vector $\mathbf{v}$ being considered and are given by the coordinates of $\mathbf{v}$ relative to the $\mathbf{u}_i$s.

In NMR relaxation, RBM is detected in the frequency domain, and the pertinent, frequency-dependent function is the spectral density, $J_0(\omega)$, which is the Fourier transform of $<P_2[\cos \beta(t)]>$, given by

$$J_0(\omega) = \sum a_k \tau_k / (1 + \omega^2 \tau_k^2)$$

In heteronuclear NMR relaxation, the RBM of the macromolecular structure is measured mainly by the relaxation rates R1 and R2, and NOE values observed for a series of $\mathbf{v}$ vectors, which are along the directions of X–H (X = N or C) bonds, usually one per each residue. These relaxation properties are determined by the values adopted by the $J_0(\omega)$ function for five particular frequencies that are combinations of the Larmor frequencies of H and X: $\omega = 0, \omega_X, \omega_H, \omega_X - \omega_H$, and $\omega_X + \omega_H$. Global motion is sensed differently by bond vectors with different orientations; therefore, relaxation measurements integrate details of the external, overall shape of the molecule with information about its internal structure. Fast internal motions provide additional contributions to relaxation that are not accounted for in a rigid body approximation. Fortunately, R1 and R2 rates are affected similarly by internal motions, and the ratio R2/R1 provides a good approximation to the relaxation properties of the rigid body model (Kay *et al.*, 1989).

Even fairly rigid macromolecules, such as globular proteins, present some amount of internal mobility that contributes, along with overall tumbling, to the RBM of the residues. For most of them, the contribution of internal motion to R1 and R2 nearly cancels out in the ratio R2/R1 and is typically ignored in experimental results and theoretical calculations. For the sake of brevity, we have summarized the basic concepts of RBM and NMR relaxation of rotationally anisotropic molecules (Abragam, 1961; Woessner, 1962); the reader can learn more physical and mathematical details in the original references. The equations employed in the calculations are compiled in García de la Torre *et al.* (2000a).

The core of the prediction of NMR relaxation consists of the calculation of $\mathbf{D_r}$. In general, the calculation of hydrodynamic properties of rigid particles can be done employing bead models (García de la Torre and Bloomfield, 1981) as implemented in the HYDRO program (García de la Torre *et al.*, 1994). This possibility suggests a route applicable to macromolecules, described with chemical detail, at the level of residues or even individual atoms. Figure 1A displays an example in which all the nonhydrogen atoms of a globular protein are represented by identical

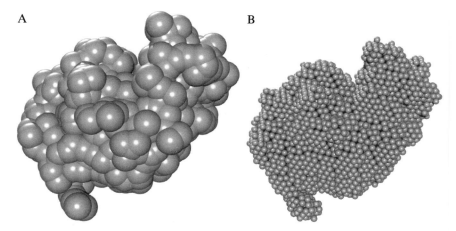

A   B

FIG. 1. (A) Primary hydrodynamic model for lysozyme. Each nonhydrogen atom is represented by a bead of radius $a = 3.2$ Å. (B) Shell model for lysozyme. The exposed surface of the primary hydrodynamic model is represented by a shell of minibeads with radius $\sigma = 0.7$ Å.

spherical beads of radius, a. To match the actual volume of the protein in solution, including immobilized hydration water, the radius of the beads must be appreciably larger than the atomic radius, and this produces an extensive overlap between beads. This is the primary hydrodynamic model (PHM) of the protein, whose properties are sought. The theory implemented in HYDRO was devised for nonoverlapping beads and is not applicable to the PHM. García de la Torre et al. (2000b) proposed a procedure, based on the so-called bead-shell modeling strategy (Bloomfield et al., 1967), in which the exposed (but not the interior) surface of the overlapping spheres in the PHM is represented by a shell of touching but not overlapping "minibeads" of radius σ (see an example in Fig. 1B). Hydrodynamic properties of this shell model are calculated using the standard bead-model treatment for various values of σ, and the results are extrapolated to the limit of $\sigma = 0$, at which the shells represent exactly the hydrodynamic surface. This strategy was shown to yield quite accurate predictions of several hydrodynamic properties (including correlation times) for a number of globular proteins from their atomic structure, using a value of $a \approx 3.0$–3.3 Å (Bernadó et al., 2002; García de la Torre, 2001; García de la Torre et al., 2000b). This radius is concordant with the van der Waals size of protein atoms increased by a contribution from hydration (García de la Torre, 2001) and can be regarded as a standard value for ab initio predictions.

The program HYDRONMR (García de la Torre et al., 2000b) implements this methodology to calculate rigid body heteronuclear NMR relaxation rates from a known three-dimensional structure. From the atomic coordinates, $\mathbf{D}_r$ is obtained by the shell-model procedure, and the X–H vectors are determined and referred to the eigenaxes of $\mathbf{D}_r$. Then, the procedure above described is employed to determine R1, R2, and NOE. The procedure is repeated for several values of a, treated as an adjustable parameter until the results fit the R2/R1 ratio (a more elaborated procedure for this fitting will be described below). However, it was found that the relative deviation of the series of (R2/R1)$_j$ ratios for the series of residues from its mean value, expressed as

$$\nabla_j = [(\text{R2/R1})_j - \langle\text{R2/R1}\rangle]/\langle\text{R2/R1}\rangle$$

practically does not vary with a, and this serves as a direct, parameter-free test of agreement with experimental data.

Rigid Anisotropic Proteins

As an example of a straightforward application of HYDRONMR, we present a calculation for the outer surface protein (OspA, PDB file 1osp). The strongly anisotropic structure of OspA and the arrangement of its

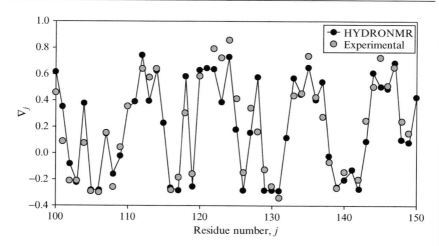

Fig. 2. Values of ∇_j for a series of consecutive residues (100–150) in Osp A.

secondary structure elements result in a remarkable periodicity of R2/R1 ratios along the sequence. This feature is accurately predicted by HYDRONMR, confirming that the residue variability of R2/R1 values is the result of anisotropic motion. Experimental relaxation data (Pawley *et al.*, 2002) are compared to the calculated results in Fig. 2, which plots ∇_j for a series of successive residues. We remark that the calculated values are practically independent, not only of radius *a* but also of other physical and instrumental conditions. We see how the calculated values correlate rather well with the experimental ones, and the trend of the ∇_j values along the polypeptide chain is predicted with notable accuracy. The experimental value of the correlation time of OSP at 318 K is 13.3 ns. Taking the "standard" value $a = 3.3$ Å, the calculated correlation time at 318 K is 14.3 ns, and the experimental value is fitted for $a = 2.9$ Å.

A detailed comparison of individual experimental and calculated R2/R1 values provides further insight into local deviations from the rigid body model. N-terminal residues are highly mobile and calculations overestimate R2/R1. The same situation is observed for a few internal residues (e.g., L98 and G163) suggesting local fast motion at these sites (see Fig. 3). Conversely, for residues G218, I224, and E239 the experimental R2/R1 value is higher than expected considering only dipole–dipole and chemical shift anisotropy (CSA) modulation due to molecular tumbling. Most likely, the additional relaxation mechanism is chemical exchange.

The effects of anisotropic motion and the possible complications derived from the presence of chemical exchange are well known (Tjandra

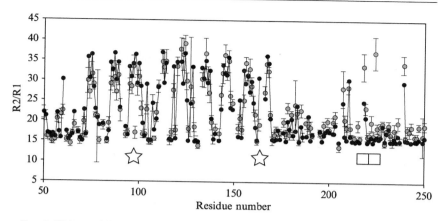

Fig. 3. Values of R2/R1 for a larger series of consecutive residues (50–250) in OspA, showing residues with high local mobility (stars) and exchange (squares).

et al., 1995), and there are experimental approaches to separate the two effects (Kroenke *et al.*, 1998), but hydrodynamic calculations offer a straightforward approach when a three-dimensional (3D) structure is available. Osborne and Wright (2001) used hydrodynamic calculations based on a PHM model to identify residues that are affected only by rotational diffusion.

A complete analysis of relaxation data requires the comparison of calculated and experimental values of R2/R1, in which the radius of the atomic elements can be treated as an adjustable parameter to optimize the agreement. An analysis of a number of proteins has shown that the optimized value of *a* can be used as a parameter to check the consistency of the 3D coordinates, the experimentally determined relaxation rates, and the assumption of a nonaggregating rigid-body molecule that is central to the HYDRONMR strategy. Additionally, relative deviations of individual R2/R1 values from the mean can be analyzed as a parameter-free test of the agreement with experimental data and can be used to identify sites of internal mobility or subjected to other sources of relaxation, such as chemical exchange. A detailed protocol, including filters to eliminate individual residues that are not representative of the global motion, has been described (Bernadó *et al.*, 2002) and is implemented in program HYDRORELAX.

Figure 4 shows the distribution of *a* values for a set of proteins in which three situations are observed. First, for rigid, nonaggregating proteins, HYDRONMR reproduces the experimental relaxation rates using *a* values

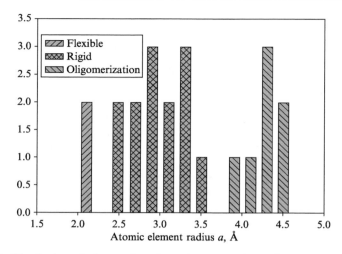

FIG. 4. Distribution of the *a* values for a large set of proteins. The central region corresponds to essentially rigid, nonaggregating structures. Cases with higher and lower *a* are attributed, respectively, to oligomerization and overall flexibility.

clustered around 3.3 Å. The procedures indicated in this section provide an adequate data analysis in this case. Second, proteins known or suspected to be involved in oligomerization processes require substantially larger values of *a*. Finally, flexible proteins show an apparent smaller molecular volume that results in low values of *a*. In the following sections, we describe further applications of hydrodynamic modeling to proteins belonging to the two last cases.

Protein Oligomers

Reversible formation of high-molecular-weight oligomers, even if present in small quantities, results in a substantial increase in the effective correlation time that is equivalent to an increased apparent molecular volume, and attempts to fit the observed relaxation rates require atomic elements with radii much larger than 3.3 Å. If the structure of the oligomers is known, the ability to predict their relaxation rates quantitatively allows the use of relaxation measurements as a function of concentration as a method for the determination of small oligomerization constants. Conversely, a good agreement between measured and computed relaxation rates could be taken as sensitive proof for nonaggregation. Finally, in favorable cases, concentration-dependent relaxation rates can be fitted to an oligomerization model to extract some characteristics of the high-molecular-weight species, in spite of the fact that the observed NMR spectrum corresponds to the monomeric form.

In a recent example, the structure of a tetramer of bovine protein tyrosine phosphatase (BPTP) could be characterized by comparing the observed concentration dependence of relaxation rates and the known structures of a BPTP monomer and its dimer. The ability to calculate the expected rates for the known species allowed the detection of a previously unknown tetramer that explained the observed deviations from the best fit. Interestingly, the same analysis identified a set of residues consistent with a tetramerization interface providing independent proof of the tetramer model. (Bernadó et al., 2003).

Enhanced relaxation rates in high-molecular-weight species form the basis of widely used methods to detect weakly binding ligands to macromolecules. The ability to predict the expected relaxation rates of macromolecules of known structure opens the way to sensitive and structurally informative methods to characterize weak protein–protein complexes.

Flexible Multidomain Proteins

Large-scale interdomain motion may substantially modify the shape of a protein, thus having an appreciable influence in solution properties (García de la Torre et al., 2003). In particular, rotational diffusion is not well described by a single time-independent tensor (Brüschweiler et al., 1995). Baber et al. (2001) suggested the use of an extension of the model-free approach in which all residues belonging to the same domain would have a common order parameter and correlation time describing the amplitude and time scale of interdomain motion. This treatment of relaxation data is valid only when slow internal rearrangements and global tumbling are uncoupled. However, since the dimensions of individual domains are usually comparable, the time scales of interdomain and global motion will be similar and the two motions are likely to be strongly correlated. To separate them, additional information arising from molecular dynamic simulations is required.

Brüschweiler's group has developed a theoretical frame (RED) to combine NMR relaxation data and a molecular dynamic trajectory to study protein dynamics by uncoupling active motional modes (Prompers and Brüschweiler, 2002). Our group has recently applied this frame to a multi-domain protein using a Brownian dynamics trajectory of a simplified model of protein PIN1 (Bernadó et al., 2004). Brownian dynamics simulations of simple systems allow much longer simulations (μs) than molecular dynamics (ns). This is a requirement to correctly describe motions with correlation times of several nanoseconds. Simplified models that realistically capture the rigid body motion of individual models can be produced by replacing each domain by a small cluster of spheres that collectively has the same hydrodynamic properties of the isolated domain as modeled using HYDRONMR.

Nucleic Acids

Although NMR relaxation measurements and interpretations based on rigid-body hydrodynamics are more frequently applied to globular proteins, it is noteworthy that such methodologies can also be applied to small nucleic acids. Thus, oligonucleotides can be hydrodynamically regarded as short rods, and indeed NMR relaxation of these molecules has sometimes been interpreted with the help of the hydrodynamic coefficients for cylindrical particles derived by García de la Torre and co-workers (Tirado and García de la Torre, 1980; Tirado et al., 1984). A recent example has been provided by Boisbouvier et al. (2003). Actually, the hydrodynamic description of such pieces of nucleic acids can be done with atomic detail, using HYDRONMR, as described by Fernandes et al. (2002). This makes it possible to investigate fine details of DNA structure and large-scale dynamics by NMR relaxation.

Computer Programs

HYDRONMR and related computer programs can be downloaded from http://leonardo.fcu.um.es/macromol.

Acknowledgments

Financial support has been provided by grants from Spanish MCyT BQU2003-04517 (J.G.T.) and BIO2001-3115 (M.P.). P.B. is the recipient of an EMBO fellowship.

References

Abragam, A. (1961). "The Principles of Nuclear Magnetism." Clarendon Press, Oxford.
Baber, J., Szabo, A., and Tjandra, N. (2001). Analysis of slow interdomain motion of macromolecules using NMR relaxation data. *J. Am. Chem. Soc.* **123**, 3953–3959.
Bernadó, P., García de la Torre, J., and Pons, M. (2002). Interpretation of ^{15}N NMR relaxation data of globular proteins using hydrodynamic calculations with HYDRONMR. *J. Biomol. NMR* **23**, 139–150.
Bernadó, P., Akerud, T., García de la Torre, J., Akke, M., and Pons, M. (2003). NMR relaxation evidence for tetramers of low molecular weight protein. *J. Am. Chem. Soc.* **125**, 916–923.

Bernadó, P., Fernandes, M. X., Jacobs, D., Fiebig, K., García de la Torre, J., and Pons, M. (2004). Interpretation of NMR relaxation properties of Pin1, a two-domain protein, based on Brownian dynamics simulation. *J. Biomol. NMR* **29**, 21–35.

Bloomfield, V. A., Dalton, W. O., and van Holde, K. E. (1967). Frictional coefficient of multisubunit structures. I. Theory. *Biopolymers* **5**, 135–148.

Boisbouvier, J., Wu, Z., Ono, A., Kainosho, M., and Bax, A. (2003). Rotational diffusion tensor of nucleic acids from ^{13}C NMR relaxation. *J. Biomol. NMR* **27**, 133–142.

Brüschweiler, R. (2003). New approaches to the dynamic interpretation and prediction of NMR relaxation data from proteins. *Curr. Opin. Struct. Biol.* **13**, 175–183.

Brüschweiler, R., Liao, X. B., and Wright, P. E. (1995). Long-range motional restrictions in a multidomain zinc-finger protein from anisotropic tumbling. *Science* **268**, 886–889.

Fernandes, M. X., Ortega, A., López Martínez, M. C., and García de la Torre, J. (2002). Calculation of hydrodynamic properties of small nucleic acids from their atomic structure. *Nucleic Acids Res.* **29**, 3362–3376.

Fischer, M. W. F., Majumdar, A., and Zuiderweg, E. R. P. (1998). Protein NMR relaxation: Theory, applications and outlook. *Prog. NMR Spectrosc.* **33**, 207–272.

García de la Torre, J. (2001). Hydration from hydrodynamics: General considerations and applications of bead modelling to globular proteins. *Biophys. Chem.* **93**, 159–170.

García de la Torre, J., and Bloomfield, V. A. (1981). Hydrodynamic properties of complex, rigid, biological macromolecules. Theory and applications. *Q. Rev. Biophys.* **14**, 81–139.

García de la Torre, J., Navarro, S., López Martínez, M. C., Díaz, F. G., and Lopez Cascales, J. J. (1994). HYDRO: A computer software for the prediction of hydrodynamic properties of macromolecules. *Biophys. J.* **67**, 361–372.

García de la Torre, J., Huertas, M. L., and Carrasco, B. (2000a). HYDRONMR: Prediction of NMR relaxation of globular proteins from atomic-level structures and hydrodynamic calculations. *J. Magn. Reson.* **147**, 138–146.

García de la Torre, J., Huertas, M. L., and Carrasco, B. (2000b). Calculation of hydrodynamic properties of globular proteins from their atomic-level structure. *Biophys. J.* **78**, 719–730.

García de la Torre, J., Pérez Sánchez, H. E., Ortega, A., Hernández Cifre, J. G., Fernández, M. X., and López Martínez, M. C. (2003). Calculation of solution properties of flexible macromolecules. Methods and applications. *Eur. Biophys. J.* **32**, 477–486.

Kay, L., Torchia, D., and Bax, A. (1989). Backbone dynamics of proteins as studied by 15N inverse detected heteronuclear NMR spectroscopy: Application to staphylococcal nuclease. *Biochemistry* **28**, 8972–8979.

Korzhnev, D. M., Billeter, M. A., Arseniev, A. S., and Orekhov, V. Y. (2001). NMR studies of brownian tumbling and internal motions in proteins. *Prog. Nucl. Magn. Reson. Spectrosc.* **38**, 197–266.

Kroenke, C. D., Loria, J. P., Lee, L. K., Rance, M., and Palmer, A. G., III (1998). Longitudinal and transverse 1H-15N dipolar/15N chemical shift anisotropy relaxation interference: Unambiguous determination of rotational diffusion tensors and chemical exchange effects in biological macromolecules. *J. Am. Chem. Soc.* **120**, 7905–7915.

Lipari, G., and Szabo, A. (1982). Model-free approach to the interpretation of NMR relaxation in macromolecules. 1. Theory and range of validity. *J. Am. Chem. Soc.* **104**, 4546–4559.

Osborne, M. J., and Wright, P. E. (2001). Anisotropic rotational diffusion in model-free analysis for a ternary DHFR complex. *J. Biomol. NMR* **19**, 209–230.

Palmer, A. G., III, Williams, J., and McDermott, A. (1996). Nuclear magnetic resonance of biopolymer dynamics. *J. Phys. Chem.* **100**, 13293–13310.

Pawley, N. H., Koide, S., and Nicholson, L. K. (2002). Backbone dynamics and thermodynamics of Borrelia outer surface protein A. *J. Mol. Biol.* **324**, 991–1002.

Prompers, J. J., and Brüschweiler, R. (2002). General framework for studying the dynamics of folded and nonfolded proteins by NMR relaxation spectroscopy and MD simulation. *J. Am. Chem. Soc.* **124,** 4522–4534.

Tirado, M. M., and García de la Torre, J. (1980). Rotational dynamics of rigid, symmetric top macromolecules. Application to circular cylinders. *J. Chem. Phys.* **73,** 1986–1993.

Tirado, M. M., López Martínez, C., and García de la Torre, J. (1984). Comparison of theories for translational and rotational diffusion coefficients of rodlike macromolecules. Application to short DNA fragments. *J. Chem. Phys.* **81,** 2047–2052.

Tjandra, N., Feller, S. E., Pastor, R. W., and Bax, A. (1995). Rotational diffusion anisotropy of human ubiquitin from 15N NMR relaxation. *J. Am. Chem. Soc.* **117,** 12562–12566.

Woessner, D. E. (1962). Nuclear spin relaxation in ellipsoids undergoing rotational Brownian motion. *J. Chem. Phys.* **37,** 647–654.

[18] Solution NMR Spin Relaxation Methods for Characterizing Chemical Exchange in High-Molecular-Weight Systems

By Arthur G. Palmer, III, Michael J. Grey, and Chunyu Wang

Abstract

Transverse relaxation optimized NMR spectroscopy (TROSY) techniques for ^1H–^{15}N backbone amide moieties and for $^{13}CH_3$ methyl groups have permitted the development of Hahn spin echo and Carr–Purcell–Meiboom–Gill (CPMG) experiments for characterizing chemical exchange kinetic phenomena on microsecond–millisecond time scales in proteins with molecular masses >50 kDa. This chapter surveys the theoretical bases for TROSY in spin systems subject to chemical exchange linebroadening, the experimental methods that have been developed to quantitatively characterize chemical exchange in large proteins, and the emerging applications to triose phosphate isomerase, hemoglobin, and malate synthase G, with molecular masses ranging from 54 to 82 kDa.

Introduction

The ultimate goal of structural biology is to understand the relationship between protein structure and biological function. In this context, protein structure is usefully conceived not as a single conformational entity but rather as a dynamic ensemble of conformational states that is populated at equilibrium according to the Boltzmann distribution. With this in mind, under a set of "native" conditions, the lowest energy, or ground state, conformation is the most probable state. However, any given molecule in

Copyright 2005, Elsevier Inc.
All rights reserved.
0076-6879/05 $35.00

the ensemble will undergo activated transitions to additional conformational substates with lifetimes and stabilities dictated by the energy landscape (Frauenfelder *et al.*, 1991). These excursions to higher free energy states can result in conformations that are relevant for biological function (Fersht, 1999; Kern and Zuiderweg, 2003; Kraut *et al.*, 2003; McCammon and Harvey, 1987; Wand, 2001). In these instances, studying these "excited state" conformations, which may be populated by less than a few percent at equilibrium under native conditions, is difficult by conventional techniques in structural biology.

Nuclear magnetic resonance (NMR) spin relaxation is uniquely able to characterize dynamic processes that occur over a wide range of time scales with high spatial resolution in biological macromolecules (Palmer *et al.*, 1996). Stochastic processes that result in the time-dependent modulation of the dipole–dipole (DD), chemical shift anisotropy (CSA), quadrupole, and isotropic chemical shift Hamiltonians provide the principal physical basis of spin relaxation for diamagnetic proteins in solution (Abragam, 1961). The magnitudes of the relaxation rate constants for particular elements of the nuclear spin density operator depend on both the variance of the local magnetic fields induced by the time-varying Hamiltonians and the correlation time for the underlying dynamic process. Stochastic fluctuations occurring on the picosecond to nanosecond time scale result in spin relaxation through modulation of the DD, CSA, and quadrupole interactions, while stochastic processes occurring on the microsecond to millisecond time scale result in spin relaxation through modulation of isotropic chemical shifts. The latter case is called *chemical exchange* and provides a powerful probe for characterizing the kinetics, thermodynamics, and mechanisms of functionally relevant conformational changes in proteins.

Over the past decade, several experimental schemes have been devised to quantitatively measure the contributions of chemical exchange to transverse relaxation in biological macromolecules (Akke and Palmer, 1996; Ishima *et al.*, 1998; Loria *et al.*, 1999a,b; Massi *et al.*, 2004; Mulder *et al.*, 2001b, 2002; Skrynnikov *et al.*, 2001; Szyperski *et al.*, 1993; Wang *et al.*, 2001). Most of these methods are derived from the Hahn echo (Rance and Byrd, 1983), Carr–Purcell–Meiboom–Gill (CPMG) (Carr and Purcell, 1954; Meiboom and Gill, 1958), and $R_{1\rho}$ (Deverell *et al.*, 1970) techniques. In these approaches, the application of radiofrequency (rf) fields is used to modulate the effect of chemical exchange on transverse relaxation, and the effective field dependence of relaxation rate constants, termed *relaxation dispersion*, is analyzed to obtain information on the chemical exchange kinetics (Palmer *et al.*, 2001). A growing number of examples have been described in which exchange broadening and relaxation dispersion have been used to characterize the role of microsecond to millisecond dynamics in molecular recognition (Mulder *et al.*, 2001a; Volkman *et al.*, 2001; Wang *et al.*, 2004), folding and unfolding (Hill *et al.*, 2000; Tollinger *et al.*, 2001; Vugmeyster *et al.*, 2000), and enzymatic catalysis (Butterwick *et al.*, 2004; Eisenmesser *et al.*, 2002). In most of these instances, the requisite spin relaxation techniques have been developed on and applied to small to moderately sized macromolecular systems (i.e., with molecular masses <30 kDa). However, from a biological perspective, many cellular processes involve proteins and protein–ligand, including protein–protein and protein–nucleic acid, complexes of significantly larger molecular mass.

Considerable experimental and theoretical advances made over the past few years have extended the molecular size barrier imposed on solution NMR spectroscopy. Most notable in this regard has been the development of transverse relaxation optimized spectroscopy (TROSY) for minimizing rapid transverse relaxation normally associated with slow rotational diffusion in high-molecular-weight systems (Pervushin *et al.*, 1997, 1998). TROSY can be incorporated into many solution NMR techniques for spin relaxation measurements by replacing the conventional heteronuclear single-quantum correlation (HSQC) indirect evolution and acquisition periods (Kempf *et al.*, 2003; Zhu *et al.*, 2000). This simple expedient extends the range of application to molecular systems of moderate size; however, the modified experiments still measure the relaxation rate constants of the conventional Bloch magnetization components. The large transverse relaxation rate constants of these elements of the density operator ultimately limit the sensitivity of conventional relaxation dispersion experiments for identifying and characterizing chemical exchange in systems beyond 50 kDa. The focus of this chapter is to highlight recent theoretical and experimental advances that enable chemical exchange in high-molecular-weight systems to be characterized by NMR spin relaxation techniques.

Chemical Exchange as a Mechanism of Transverse Relaxation

This chapter focuses on experimental methods for characterizing chemical exchange in large biological macromolecules; consequently, only a limited overview of the theoretical aspects of chemical exchange will be presented to provide a framework for illustrating aspects of exchange-induced relaxation that are important when considering high-molecular-weight systems. More formal theoretical treatments of spin relaxation in chemically exchanging systems have been presented recently (Abergel and Palmer, 2003; Bain, 2003; Desvaux and Berthault, 1999; Palmer, 2004).

For illustrative purposes, only the simplest case of two-site exchange is considered. In this situation, an isolated nuclear spin, or group of spins, exchanges between magnetically distinct sites as described by the kinetic mechanism:

$$A_1 \underset{k_{21}}{\overset{k_{12}}{\rightleftarrows}} A_2 \tag{1}$$

The lifetimes of the exchanging sites are governed by the first-order rate constant k_{ij} for the transition from site i to site j, and the relative equilibrium population distribution is determined by the Boltzman relationship $p_2/p_1 = k_{12}/k_{21} = \exp(-\Delta G/RT)$, where ΔG is the free energy difference between the two sites and $p_1 + p_2 = 1$ and $p_1 > p_2$. As a consequence of the Boltzmann relationship, relatively small differences in free energy between the two sites result in highly skewed population distributions such that $p_1 \gg p_2$. Recent theoretical developments (Abergel and Palmer, 2003; Trott and Palmer, 2004) and experimental applications (Grey et al., 2003; Korzhnev et al., 2004c; Tolkatchev et al., 2003; Tollinger et al., 2001) have considered multisite exchange mechanisms.

Free Precession Transverse Relaxation

Chemical exchange provides an additional mechanism for the transverse relaxation of single-quantum (SQ) and multiple-quantum (MQ) coherences evolving under free precession, that is, in the absence of rf fields. For a system of N spin-1/2 nuclei, the signed coherence, B_i, is written as $B_i = T_1^{k_1} T_2^{k_2} \dots T_N^{k_N}$ in which $k_1 + k_2 + \dots k_N = m_i$ is the coherence order, $k_j = \{-1, 0, +1\}$, $T_j^{\pm 1} \equiv I_j^{\pm}$ are shift operators, and T_j^0 is a longitudinal spin operator, I_{zj}, or the identity operator, E_j. The change in resonance frequency associated with a change from site A_1 to site A_2 is given by

$$\Delta\omega_i = \omega_{2i} - \omega_{1i} = \sum_{j=1}^{N} k_j(\Omega_{2j} - \Omega_{1j}) \tag{2}$$

in which ω_{li} is the resonance frequency for coherence B_i in site l and Ω_{lj} is the resonance frequency for coherence I_j^+ in site l. This expression assumes that differences in scalar coupling constants for the spins of interest between sites A_1 and A_2 are negligible sources of exchange broadening; this is normally a valid assumption for intramolecular conformational fluctuations in macromolecules. The evolution of the coherence B_i during a free-precession period, t, is described by

$$<B_i>(t) = e^{(i\omega_{1i} - R_{2i}^0)t} \langle \mathbf{s} \,|\exp[\mathbf{L}t]\mathbf{s}\rangle <B_i>(0) \tag{3}$$

in which $<B_i>(t) = \text{Trace}\{B_i\sigma(t)\}$, $\sigma(t)$ is the density operator, $|\mathbf{s}\rangle = [p_1^{1/2} p_2^{1/2}]^T$ is a column vector, $\mathbf{L}$ is the symmetrized evolution matrix given by

$$\mathbf{L} = \begin{bmatrix} -k_{12} & (k_{12}k_{21})^{1/2} \\ (k_{12}k_{21})^{1/2} & i\Delta\omega_i - k_{21} \end{bmatrix} \tag{4}$$

and nonselective excitation has been assumed. The relaxation rate constant from sources other than chemical exchange, including DD, CSA, and quadrupole mechanisms, is denoted by R_{2i}^0 and is assumed to be identical for sites A_1 and A_2.

The contribution to the transverse relaxation rate constant of the dominant component of the expectation value of the coherence B_i due to the chemical exchange process is given by (Wang and Palmer, 2002; Woessner, 1961)

$$R_{ex(i)} = R_{2i} - R_{2i}^0 = \frac{k_{ex}}{2} - \frac{1}{\sqrt{8}}$$
$$\left\{ k_{ex}^2 - \Delta\omega_i^2 + \left[(k_{ex}^2 + \Delta\omega_i^2)^2 - 16p_1p_2\Delta\omega_i^2 k_{ex}^2 \right]^{1/2} \right\}^{1/2} \tag{5}$$

in which $R_{2(i)}$ is the total relaxation rate constant from all sources and $k_{ex} = k_{12} + k_{21}$. In the limit of fast exchange, Eq. (5) simplifies to

$$R_{ex(i)} = p_1p_2\Delta\omega_i^2/k_{ex} \tag{6}$$

Other approximations of Eq. (5) have been derived that are valid for exchange on all chemical shift time scales provided that $p_1 \gg p_2$ (Abergel and Palmer, 2004; Skrynnikov et al., 2002; Swift and Connick, 1962).

Transverse Relaxation in the Presence of an Applied RF Field

For any given set of exchange parameters, the exchange contribution to the transverse relaxation rate constant can be modified by the application of an effective rf field in the form of a spin-echo sequence in CPMG experiments and or a spin-lock rf field in $R_{1\rho}$ experiments (Palmer, 2004). In this chapter, only CPMG experiments are considered because, except for a TROSY-detected $R_{1\rho}$ experiment described by Kempf et al. (2003), substantially more progress has been made in adapting CPMG experiments for macromolecules with masses >50 kDa.

In a generalized CPMG experiment, the spin echo sequence is described by

$$\left\{ \left(\frac{\tau_{cp}}{2} - \mathbf{U}_1 - \frac{\tau_{cp}}{2} \right)_n - \mathbf{U}_2 - \left(\frac{\tau_{cp}}{2} - \mathbf{U}_1 - \frac{\tau_{cp}}{2} \right)_n \right\}_m \tag{7}$$

in which τ_{cp} is the spin-echo time period, $\mathbf{U}_1$ and $\mathbf{U}_2$ are pulse sequence elements, and m and n are integers (Korzhnev et al., 2004b). The elements

U_1 and U_2 may be simple $180°$ pulses or periods of pulse-interrupted free precession. Evolution through the CPMG sequence can always be calculated numerically; however, analytical expressions have been derived for particular cases. For example, the conventional CPMG sequence is obtained if $U_1 B_i U_1^{-1} = B_i^\dagger$, $U_2 = E$, and $m = 1$. In this case, evolution is described by (Allerhand and Thiele, 1966)

$$<B_i> (2n\tau_{cp}) = e^{-R_{2i}^0 2n\tau_{cp}} \langle s| \{ \exp[L\tau_{cp}/2] \exp[L^*\tau_{cp}] \exp[L\tau_{cp}/2] \}^n |s\rangle$$
$$<B_i> (0)$$

$$= e^{-R_{2i}^0 2n\tau_{cp}} \langle s| \exp[L\tau_{cp}/2] \{ \exp[L^*\tau_{cp}] \exp[L\tau_{cp}] \}^n$$
$$\exp[-L\tau_{cp}/2] |s\rangle <B_i> (0)$$

(8)

Under conditions for which the evolution of magnetization is dominated by a single exponential decay (Allerhand and Thiele, 1966; Carver and Richards, 1972; Jen, 1978),

$$<B_i> (2n\tau_{cp}) = e^{-\{R_{2i}^0 + R_{ex(i)}(1/\tau_{cp})\}2n\tau_{cp}} \langle s| \exp[L\tau_{cp}/2] |I_1\rangle$$
$$\langle I_1| \exp[-L\tau_{cp}/2] |s\rangle <B_i> (0)$$

(9)

in which $|I_1\rangle$ is the eigenvector corresponding to the largest (least negative) eigenvalue of $\exp[L^*\tau_{cp}] \exp[L\tau_{cp}]$, the exchange contribution to the transverse relaxation rate constant during the CPMG period is given by

$$R_{ex(i)}\left(1/\tau_{cp}\right) = \frac{1}{2}\left(k_{ex} - \frac{1}{\tau_{cp}} \cosh^{-1}[D_+ \cosh(\eta_+) - D_- \cos(\eta_-)] \right)$$ (10)

$$D_\pm = \frac{1}{2}\left[\pm 1 + \frac{\psi + 2\Delta\omega_i^2}{(\psi^2 + \zeta^2)^{1/2}} \right]$$ (11)

$$\eta_\pm = \frac{\tau_{cp}}{\sqrt{2}} \left[\pm\psi + (\psi^2 + \zeta^2)^{1/2} \right]^{1/2}$$ (12)

$\psi = k_{ex}^2 - \Delta\omega_i^2$ and $\zeta = -2\Delta\omega_i k_{ex}(p_1 - p_2)$. Using second-order perturbation theory in the limit of fast exchange, Eq. (10) reduces to (Allerhand and Thiele, 1966; Luz and Meiboom, 1963)

$$R_{ex(i)}(1/\tau_{cp}) = \frac{p_1 p_2 \Delta\omega^2}{k_{ex}} \left[1 - \frac{2\tanh(\tau_{cp}k_{ex}/2)}{\tau_{cp}k_{ex}} \right]$$ (13)

The functional dependence of $R_{ex(i)}(1/\tau_{cp})$ on the effective field strength is illustrated by the calculated relaxation dispersion profiles shown in Fig. 1. Relaxation dispersion curves are analyzed by numerically fitting the theoretical expressions for $R_{ex(i)}(1/\tau_{cp})$ to experimental relaxation rate

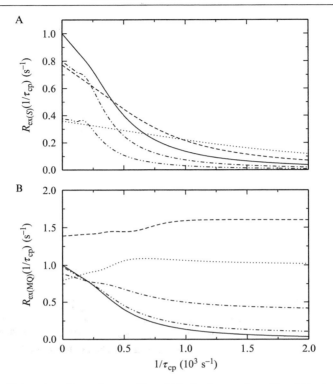

FIG. 1. Relaxation dispersion profiles. (A) Single-quantum $R_{ex(S)}(1/\tau_{cp})$ relaxation dispersion profiles were calculated using Eq. (10) with $\Delta\omega_S = 1000$ s^{-1}, $p_1 = 0.95$, and k_{ex} = 5000 ($\cdots$), 2000 (– – –), 1000 (——), 500 (– · –), and 200 (– ·· –) s^{-1}. Dispersion curves are scaled relative to $R_{ex(S)}(1/\tau_{cp} \to 0)$ for $k_{ex} = \Delta\omega_S = 1000$ s^{-1}. (B) Multiple quantum $R_{ex(MQ)}(1/\tau_{cp})$ relaxation dispersion profiles calculated using Eq. (14) for $k_{ex} = 1000$ s^{-1}, $p_1 = 0.95$, $\Delta\omega_S = 1000$ s^{-1}, and $\Delta\omega_I = 0$ (——), 200 (– ·· –), 500 (– · –), 1000 ($\cdots$), and 2000 (– – –) s^{-1}. Scaling is the same as in (A).

constants measured at various pulsing rates, $1/\tau_{cp}$, to determine the exchange parameters k_{ex}, $\Delta\omega_i$, and $p_1 p_2$.

If the above conditions for $\mathbf{U}_1$ and $\mathbf{U}_2$ do not hold, then the expressions for $R_{ex(i)}(1/\tau_{cp})$ must be modified accordingly. Kay and co-workers have analyzed the generalized CPMG sequence applied to heteronuclear MQ coherences of the form $I^{\pm}S^{\pm}$ with the conditions $m = 1$, $\mathbf{U}_1$ is a selective inversion of the S operator and $\mathbf{U}_2$ is a selective inversion of the I operator (Korzhnev *et al.*, 2004b). In this case, for example, if $B_i = I^+S^+$, then $\mathbf{U}_1 B_i \mathbf{U}_1^{-1} = I^+S^-$, rather than $B_i^{\dagger} = I^-S^-$. Thus, evolution through the CPMG sequence will depend on the evolution of both double-quantum

(DQ) and zero-quantum (ZQ) coherences. The resulting analytical approximation to $R_{ex(MQ)}(1/\tau_{cp})$ is

$$R_{ex(MQ)}(1/\tau_{cp}) = \frac{1}{2}\text{Re}\left\{ k_{ex} - \frac{1}{\tau_{cp}}\cosh^{-1}[D_+\cosh(\eta_+) - D_-\cos(\eta_-)]\right\} \quad (14)$$

in which $D_\pm$ and $\eta_\pm$ are given by Eqs. (11) and (12) with ψ and ζ modified to

$$\psi = k_{ex}^2 - \frac{1}{2}(\Delta\omega_{ZQ}^2 + \Delta\omega_{DQ}^2) + 2i\Delta\omega_I k_{ex}(p_1 - p_2) \quad (15)$$

$$\zeta = -2\Delta\omega_S\{i\Delta\omega_I + k_{ex}(p_1 - p_2)\} \quad (16)$$

The averaging between ZQ and DQ coherences is made more explicit by noting that $\Delta\omega_S = (\Delta\omega_{DQ} + \Delta\omega_{ZQ})/2$ and $\Delta\omega_I = (\Delta\omega_{DQ} - \Delta\omega_{ZQ})/2$. As shown in Fig. 1, Eq. (14) displays some unexpected differences compared with Eq. (10); in particular, $R_{ex(MQ)}(1/\tau_{cp})$ may become an increasing function of $1/\tau_{cp}$.

Conventionally, the transverse relaxation rate constant, $R_{2(i)} = R_{2(i)}^0 + R_{ex(i)}(1/\tau_{cp})$, is measured experimentally by recording $<B_i>(2n\tau_{cp})$ as a function of n (or of $T = 2n\tau_{cp}$ for fixed τ_{cp}) and fitting the resulting time series to a monoexponential decay function. The relaxation dispersion profile is obtained by repeating the measurement for a different value of τ_{cp}. In some experimental situations (vide infra), the relaxation dispersion curve is obtained by measuring $<B_i>(T)$ while varying n and τ_{cp} such that $T = 2n\tau_{cp}$ is constant. If $<B_i>(0)$ is measured in a second experiment, then an effective transverse relaxation rate constant is defined by

$$R_{eff}(1/\tau_{cp}) = -\frac{1}{T}\ln\left[\frac{<B_i>(T)}{<B_i>(0)}\right] \quad (17)$$

This protocol is called "constant relaxation time" NMR spectroscopy (Akke and Palmer, 1996) and is particularly useful in applications in which relaxation from other sources, such as DD interactions with remote spins, is multiexponential (Mulder *et al.*, 2002). For the conventional CPMG sequence described by Eq. (9),

$$R_{eff}(1/\tau_{cp}) = R_{2i}^0 + R_{ex(i)}(1/\tau_{cp}) - \frac{1}{T}\ln[\langle s|\exp[\mathbf{L}\tau_{cp}/2]|\mathbf{I}_1\rangle\langle\mathbf{I}_1|\exp[-\mathbf{L}\tau_{cp}/2]|s\rangle] \quad (18)$$

The last term in this expression is an explicit function of τ_{cp} and in principle must be included in the analysis of the relaxation dispersion profile. For other variants of the generalized CPMG sequence, Kay and co-workers have shown that Eq. (18) is modified to

$$R_{\text{eff}}(1/\tau_{\text{cp}}) = R_{2i}^0 + R_{\text{ex}(i)}(1/\tau_{\text{cp}}) - \frac{1}{T}\ln Q \qquad (19)$$

and have provided algebraic forms for Q suitable for incorporation into fitting algorithms (Korzhnev et al., 2004a,b).

Sensitivity to Chemical Exchange Is Limited by the Magnitude of R_{2i}

The contribution of chemical exchange to the transverse relaxation of both SQ and MQ coherences depends on the exchange parameters k_{ex}, $\Delta\omega_i$, and p_1p_2. To illustrate the dependence of the absolute magnitude of $R_{\text{ex}(i)}$ on the exchange parameters, Eq. (5) was used to generate contour plots of $R_{\text{ex}(i)}$ as a function of $\Delta\omega_S$ for a SQ S-spin coherence and k_{ex} for minor site populations (p_2) of 1, 5, and 10% (Fig. 2A–C). The range of k_{ex} encompasses a subset of the time scales accessible to experimental characterization, and the range of $\Delta\omega_S$ corresponds to possible chemical shift changes of 8.4 and 3.4 ppm that may be expected for backbone amide ^{15}N and side chain methyl ^{13}C spins, respectively, at $B_0 = 14.1$ T. The magnitude and range of $R_{\text{ex}(i)}$ scale dramatically with a change in the minor state population from 1 to 10%. In the case of $p_2 = 1\%$, the maximum value of $R_{\text{ex}(i)}$ in the illustrated region of $\{\Delta\omega_S, k_{\text{ex}}\}$ space is on the order of 15 s^{-1}, while for $p_2 = 10\%$, the range of $R_{\text{ex}(i)}$ spans more than two orders of magnitude. The magnitude of $R_{\text{ex(MQ)}}$ depends on the relative values of $\Delta\omega_I$ and $\Delta\omega_S$ (Fig. 2D), and in particular instances can be more sensitive than the relaxation of SQ coherence to microsecond to millisecond dynamic processes (Fig. 2E) (Wang and Palmer, 2002).

Characterizing chemical exchange in high-molecular-weight systems depends critically on the relative contribution of the exchange process, $R_{\text{ex}(i)}$, to the total transverse relaxation rate constant, R_{2i}. For an intramolecular exchange process, $R_{\text{ex}(i)}$ is independent of molecular size; therefore, the sensitivity of any NMR spin relaxation experiment to chemical exchange will ultimately be limited by the relaxation rate constant for other mechanisms, R_{2i}^0. For example, SQ transverse ^{15}N coherence in a backbone amide group has R_{2S}^0 on the order of 70 s^{-1} in a protein with $\tau_c = 40$ ns (corresponding approximately to a molecular mass of 100 kDa at 300 K) in a static magnetic field of 18.8 T. As a result, when the minor state population in a two-site exchange model is less than a few percent, $R_{\text{ex}(i)}$ is expected to make only a fractional contribution (roughly one-sixth, at most, for $p_2 = 0.01$) to the observed R_{2S} for a protein of this size. In addition, as molecular size increases, conventional schemes for measuring R_{2i} will be sensitive to chemical exchange processes within an increasingly

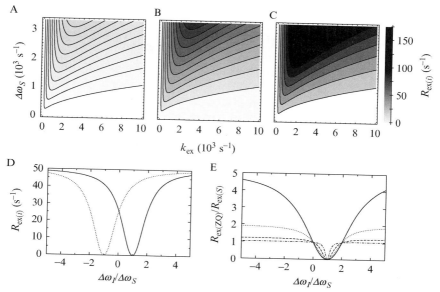

Fig. 2. Parameter dependence of chemical exchange broadening. $R_{ex(S)}$ contour plots were calculated as a function of $\Delta\omega$ and k_{ex} for $p_1 = 0.99$, 0.95, and 0.90 (A–C, respectively) using Eq. (5). The minimum contour levels are 1.1, 5.9, and 12.0 s^{-1} for (A)–(C), respectively, with a maximum of $R_{ex(S)} = 174.2$ s^{-1} occurring for $p_1 = 0.9$ and $k_{ex} = \Delta\omega_S = 3300$ s^{-1}. (D) $R_{ex(ZQ)}$ (——) and $R_{ex(DQ)}$ (·····) calculated as a function of $\Delta\omega_I/\Delta\omega_S$ for $k_{ex} = \Delta\omega_S = 1000$ s^{-1} and $p_1 = 0.95$ using Eq. (5). (E) Relative exchange contributions to ZQ and SQ coherences calculated using Eq. (5) as a function of $\Delta\omega_I/\Delta\omega_S$ for $p_1 = 0.95$ and $k_{ex} = 2000$ (——), 1000 (·····), 500 (– – –), and 200 (– · –) s^{-1}. For a comparison of the exchange contribution to DQ coherence relative to SQ coherence, the plot is simply reflected about the y-axis at $\Delta\omega_I/\Delta\omega_S = 0$.

narrower range of $\{\Delta\omega, k_{ex}, p_1 p_2\}$ parameter space that yield large values of $R_{ex(i)}$. Therefore, identification and quantification of chemical exchange processes in high-molecular-weight systems ultimately rely on experimental schemes that minimize relaxation by other mechanisms.

Mechanisms of Transverse Relaxation Optimization

Transverse relaxation optimization (TROSY) was introduced as an experimental method for enhancing sensitivity and resolution of multidimensional NMR spectroscopy by Wüthrich and coworkers (Pervushin *et al.*, 1997; See also Chapter 15 by Wider in this volume). TROSY is based on differential line broadening of individual multiplet components that arises through relaxation interference, or cross-correlation between relaxation mechanisms, in scalar-coupled spin systems (Mackor and MacLean, 1966, 1967). Transverse relaxation optimization in ^{1}H–^{15}N and ^{13}CH$_3$ spin systems has proven particularly advantageous in characterizing chemical exchange processes in high-molecular-weight systems.

As shown by Eq. (2), chemical exchange broadening of a coherence B_i is not dependent on the longitudinal spin components of the coherence, assuming, as noted above, that site-to-site variations in scalar coupling constants can be neglected. For example, the coherences $S^-, 2S^-I_z$, $S^-I^\alpha = S^-(E/2 + I_z)$, and $S^-I^\beta = S^-(E/2 - I_z)$ all have the same chemical exchange contribution to their respective transverse relaxation rate constants; however, R^0_{2i} for these different coherences can be dramatically different. Thus, given a subset of coherences with coherence order m_i, chemical exchange is optimally characterized by using experiments designed to selectively measure transverse relaxation of the particular coherence element in the set with the smallest value of R^0_{2i}.

For large biological molecules in the spin diffusion limit, transverse relaxation is dominated by the spectral density function at zero frequency, and other contributions can be neglected to a first approximation (Cavanagh et al., 1996). In this regime, $R^0_{2(i)}$ is given by

$$R^0_{2(i)} = -\langle B_i | B_i \rangle^{-1} \left\langle B_i | \int_0^\infty \left[\overline{\mathcal{H}(t), [\mathcal{H}(t+\tau), B_i]} \right] d\tau \right\rangle \qquad (20)$$

in which angle brackets indicate the trace operation and the overbar indicates ensemble averaging. $\mathcal{H}(t)$ includes only the terms of the stochastic Hamiltonian that have eigenfrequencies of zero; these terms are proportional to longitudinal operators (CSA), longitudinal two-spin operators (heteronuclear DD interactions), or spherical tensor operators formed from linear combinations of longitudinal two-spin and ZQ operators (homonuclear DD interactions). Coherences containing longitudinal or transverse ^{1}H spin operators are subject to DD relaxation with remote ^{1}H spins; consequently, deuteration of remote sites normally is necessary to obtain the maximum advantage of TROSY techniques.

^{1}H–^{15}N DD/CSA Relaxation Interference

TROSY was first developed for the backbone ^{1}H–^{15}N amide moiety in proteins (Pervushin et al., 1997). An isolated IS spin system consisting of two spin-1/2 nuclei is subject to relaxation mechanisms arising from DD interactions between the I and S spins, CSA of spin-S, and CSA of spin-I. Assuming that the CSA tensors are axially symmetric, the stochastic Hamiltonian is given by

$$\mathcal{H}(t) = d_{IS}Y_2^0[\theta_{IS}(t)]2I_zS_z + c_SY_2^0[\theta_S(t)]\frac{2}{\sqrt{3}}S_z + c_IY_2^0[\theta_I(t)]\frac{2}{\sqrt{3}}I_z \quad (21)$$

in which $d_{jk} = -\mu_0 h\gamma_j\gamma_k/(8\pi^2 r_{jk}^3)$; $c_k = 3^{-1/2}\omega_k\Delta\sigma_k$; μ_0 is the permeability of free space; h is Planck's constant; γ_k, ω_k, and $\Delta\sigma_k$ are the gyromagnetic ratio, Larmor frequency, and CSA of spin-k, respectively; r_{jk} is the vector between the jth and kth nuclei; $Y_2^0[\theta] = P_2[\cos\theta] = (3\cos^2\theta - 1)/2$ is a modified spherical harmonic function; $\theta_{jk}(t)$ is the orientation of r_{jk} in the laboratory reference frame; and $\theta_k(t)$ is the orientation of the symmetry axis of the CSA tensor of spin-k in the laboratory reference frame.

The relaxation rate constant for in-phase SQ S^- coherence, as measured in an experiment designed to suppress cross-correlation between I–S dipolar and S CSA interactions (Kay $et\ al.$, 1992b; Palmer $et\ al.$, 1992), is calculated using Eq. (20) as

$$\begin{aligned}
R_{2S}^0 &= -\langle S^-|S^-\rangle^{-1}\left\langle S^-\Big|\int_0^\infty \overline{[\mathcal{H}(t),[\mathcal{H}(t+\tau),S^-]]}d\tau\right\rangle \\
&= -d_{IS}^2\langle S^-|S^-\rangle^{-1}\langle S^-|[2I_zS_z,[2I_zS_z,S^-]]\rangle \\
&\quad \int_0^\infty \overline{Y_2^0[\theta_{IS}(t)]Y_2^0[\theta_{IS}(t+\tau)]}d\tau \\
&\quad -\frac{4}{3}c_S^2\langle S^-|S^-\rangle^{-1}\langle S^-|[S_z,[S_z,S^{-1}]]\rangle\int_0^\infty \overline{Y_2^0[\theta_S(t)]Y_2^0[\theta_{IS}(t+\tau)]}d\tau \\
&= [d_{IS}^2/2 + 2c_S^2/3]J(0)
\end{aligned}$$

$$(22)$$

in which $J(\omega)$, has the form

$$\begin{aligned}
J(0) &= \int_{-\infty}^\infty \overline{Y_2^0[\theta(t)]Y_2^0[\theta(t+\tau)]}d\tau \\
&= \frac{2}{5}\tau_c
\end{aligned} \quad (23)$$

for a rigid spherical molecule with an isotropic rotational diffusion correlation time τ_c.

^{1}H–^{15}N DD/^{15}N CSA TROSY results from the differential relaxation of the two resonances of the scalar-coupled ^{15}N doublet. These resonances correspond to the $S\ (=^{15}$N$)$ spin coherences in which the attached $I(=^1$H$)$ spin is in the α or β state, represented by the operators $I^\alpha S^-$ and $I^\beta S^-$, respectively. To facilitate the calculations of the relaxation rate constants for these coherences, the stochastic Hamiltonian of Eq. (21) is expressed as

$$\begin{aligned}
\mathcal{H}(t) &= (d_{IS}Y_2^0[\theta_{IS}(t)] + \frac{2}{\sqrt{3}}c_SY_2^0[\theta_S(t)])I^\alpha S_z - (d_{IS}Y_2^0[\theta_{IS}(t)] \\
&\quad -\frac{2}{\sqrt{3}}c_SY_2^0[\theta_S(t)])I^\beta S_z + \frac{1}{\sqrt{3}}c_IY_2^0[\theta_I(t)](I^\alpha - I^\beta)
\end{aligned}$$

$$(24)$$

using the identities $S_z = (I^\alpha + I^\beta)S_z$, $2I_zS_z = (I^\alpha - I^\beta)S_z$, and $I_z = (I^\alpha - I^\beta)/2$.

Substituting Eq. (24) into Eq. (20) gives the transverse relaxation rate constant of the $I^\beta S^-$ operator:

$$
\begin{aligned}
R_{2\beta}^0 &= -\langle I^\beta S^- | I^\beta S^- \rangle^{-1} \left\langle I^\beta S^- \left| \int_0^\infty \left[\mathcal{H}(t), [\mathcal{H}(t+\tau), I^\beta S^-] \right] d\tau \right. \right\rangle \\
&= d_{IS}^2 \int_0^\infty \overline{Y_2^0[\theta_{IS}(t)]Y_2^0[\theta_{IS}(t+\tau)]} d\tau + \frac{4}{3}c_S^2 \int_0^\infty \overline{Y_2^0[\theta_S(t)]Y_2^0[\theta_S(t+\tau)]} d\tau \\
&\quad - \frac{4}{\sqrt{3}}d_{IS}c_S \int_0^\infty \overline{Y_2^0[\theta_{IS}(t)]Y_2^0[\theta_S(t+\tau)]} d\tau \\
&= d_{IS}^2 \int_0^\infty \overline{Y_2^0[\theta_{IS}(t)]Y_2^0[\theta_{IS}(t+\tau)]} d\tau + \frac{4}{3}c_S^2 \int_0^\infty \overline{Y_2^0[\theta_S(t)]Y_2^0[\theta_S(t+\tau)]} d\tau \\
&\quad - \frac{4}{\sqrt{3}}d_{IS}c_S P_2[\cos\beta] \int_0^\infty \overline{Y_2^0[\theta_{IS}(t)]Y_2^0[\theta_{IS}(t+\tau)]} d\tau \\
&= \left(\frac{1}{2}d_{IS}^2 + \frac{2}{3}c_S^2 - \frac{2}{\sqrt{3}}d_{IS}c_S P_2[\cos\beta] \right) J(0) \\
&= R_{2S}^0 - n_{xy}
\end{aligned}
$$

(25)

in which β is the angle between the IS bond vector and the symmetry axis of the S spin CSA tensor, and η_{xy} is the relaxation rate constant for interference between $^1H-^{15}N$ DD and ^{15}N CSA interactions:

$$
\eta_{xy} = \frac{2}{\sqrt{3}}d_{IS}c_S P_2[\cos\beta]J(0) \tag{26}
$$

A similar calculation for the $I^\alpha S^-$ coherence yields $R_{2\alpha}^0 = R_{2S}^0 + n_{xy}$. A complete calculation, including interactions with remote I spins, shows that (Goldman, 1984)

$$
R_{2\beta}^0 = R_{2S}^0 - \eta_{xy} + \frac{1}{2}R_{1I} \tag{27}
$$

$$
R_{2\alpha}^0 = R_{2S}^0 + \eta_{xy} + \frac{1}{2}R_{1I} \tag{28}
$$

in which R_{1I} is the longitudinal relaxation rate constant of the I spin arising from interactions with remote spins. When d_{IS} and c_S have the same (opposite) signs, the relaxation rate constant of the $I^\beta S^-$ ($I^\alpha S^-$) coherence is smaller than the relaxation rate constants of the $I^\alpha S^-$ ($I^\beta S^-$) operator provided that β is not close to the magic angle, 54.7°, for which $P_2[\cos\beta] = 0$.

The larger CSA contribution to relaxation at high static magnetic field strengths results in the partial cancellation of the auto- and cross-correlated terms (Pervushin *et al.*, 1997). As shown in Fig. 3, the narrow component of the ^{15}N doublet relaxes as much as about eight times more slowly (neglecting chemical exchange and dipolar interactions with remote protons) than the SQ in-phase coherence [given by Eq. (22)] for $\tau_c = 40$ ns. The maximum reduction in transverse relaxation is obtained by deuterating remote ^{1}H sites to minimize R_{1I}. Thus, any exchange contribution to transverse relaxation due to ^{15}N chemical shift changes makes a significantly larger fractional contribution to the relaxation rate of the narrow doublet component in a spin-state selective experiment compared to the relaxation rate of

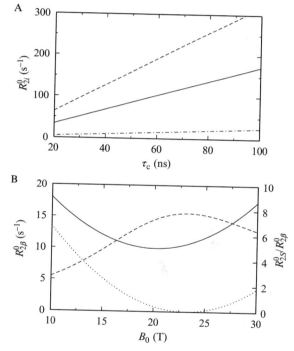

FIG. 3. Line narrowing in ^{1}H–^{15}N DD/^{15}N CSA TROSY. (A) Relaxation rate constants R_{2S}^0 (——), $R_{2\alpha}^0$ (–––), and $R_{2\beta}^0$ (– · –) were calculated using Eqs. (22) and (25) as a function of τ_c for a spherical molecule rotating isotropically. Calculations were performed using $r_{IS} = 1.02$ Å, $\Delta\sigma_N = -172$ ppm, $\beta = 17°$, and $B_0 = 18.8$ T; dipolar interactions with remote protons and spectral densities evaluated at high frequency were neglected. (B) The static magnetic field dependence of $R_{2\beta}^0$ was calculated using Eq. (25) (left axis) for $\beta = 0°$ (· · · · ·) and 17° (——) and a rotational correlation time $\tau_c = 40$ ns. (Right axis) The ratio $R_{2S}^0/R_{2\beta}^0$ (–––) was calculated as a function of static magnetic field strength for $\tau_c = 40$ ns, $r_{IS} = 1.02$ Å, $\Delta\sigma_N = -172$ ppm, and $\beta = 17°$.

in-phase SQ coherence observed in conventional experiments, *especially* in high-molecular-weight systems.

Similar conclusions hold for relaxation of $^1H^N$ SQ coherence associated with the ^{15}N α and β spin states due to cross-correlated ^1H–^{15}N DD and $^1H^N$ CSA interactions (Goldman, 1984; Pervushin *et al.*, 1997). In this case, $S = {}^1H^N$ and $I = {}^{15}N$. R_{1I} is a negligible contribution to the relaxation rate constant, but deuteration of remote ^1H sites reduces external ^1H–^1H DD contributions to R^0_{2S} that have not been included in the above analysis and eliminates ^1H–^1H scalar coupling interactions.

^1H CSA/^{15}N CSA Relaxation Interference

In addition to transverse relaxation optimization of SQ coherence in a ^1H–^{15}N spin system, ZQ coherence, $I^{\pm}S^{\mp}$, intrinsically relaxes slower than DQ coherence, $I^{\pm}S^{\pm}$, due to cross-correlation between ^1H and ^{15}N CSA interactions (Konrat and Sterk, 1993; Kumar and Kumar, 1996; Norwood *et al.*, 1999). Using the above formalism, the transverse relaxation rates of DQ and ZQ coherences are given by the sum and difference, respectively, of auto- and cross-correlated relaxation mechanisms:

$$R^0_{2ZQ} = \frac{2}{3}\{c_I^2 + c_S^2 - 2c_I c_S P_2[\cos(\chi)]\}J(0) \tag{29}$$

$$R^0_{2DQ} = \frac{2}{3}\{c_I^2 + c_S^2 + 2c_I c_S P_2[\cos(\chi)]\}J(0) \tag{30}$$

in which χ is the angle between the principal axes of the ^1H and ^{15}N CSA tensors. Thus, if the sign of the CSA coupling is the same (opposite) for the I and S spins, then ZQ (DQ) coherence has a lower relaxation rate constant, provided that χ is not close to the magic angle.

As shown in Fig. 4, ZQ coherence relaxes ~3–15 times more slowly (neglecting chemical exchange and dipolar interactions with remote protons) than OQ coherence for $\tau_c = 40$ ns. The maximum reduction in transverse relaxation is obtained by deuterating remote ^1H sites to minimize R_{2I} arising from remote dipolar interactions, which is not included in the above equations. The significant increase in $R_{2(DQ)}(1/\tau_{cp} \to \infty)$ with static magnetic field strength may limit the applicability of DQ relaxation dispersion for characterizing chemical exchange in large proteins. Relaxation dispersion of ZQ coherences, on the other hand, is more attractive for this purpose, because the partial cancellation of ^1H and ^{15}N CSA interactions (Pervushin *et al.*, 1999) results in smaller $R_{2(ZQ)}(1/\tau_{cp} \to \infty)$ relaxation rate constants and increased sensitivity to the effects of chemical exchange on transverse relaxation. However, in

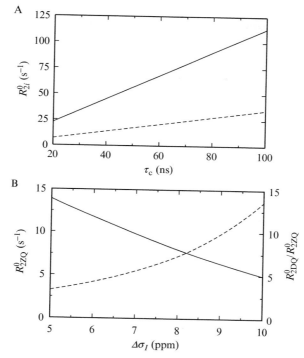

FIG. 4. Line narrowing in ^{1}H CSA/^{15}N CSA TROSY. (A) Relaxation rate constants R^0_{2DQ} (——) and R^0_{2ZQ} (- - -) were calculated using Eqs. (29) and (30) as a function of τ_c for a spherical molecule rotating isotropically. Calculations were performed using $B_0 = 18.8$ T, $\Delta\sigma_I = 5$ ppm, $\Delta\sigma_S = -172$ ppm, and $\chi = 0°$; dipolar interactions with remote spins and spectral densities evaluated at high frequency were neglected. (B) The zero-quantum relaxation rate constant (left axis, ——) and the ratio of double-quantum to zero-quantum relaxation rate constants (right axis, - - -) were calculated for $\tau_c = 40$ ns as a function of $\Delta\sigma_I$ [other parameters were the same as in (A)].

some circumstances, $\Delta\omega_{ZQ} = \Delta\omega_S - \Delta\omega_I$ may approach zero, and the ZQ relaxation dispersion will be relatively insensitive to chemical exchange.

Dipole–Dipole Relaxation Interference in ^{13}CH$_3$ Spin Systems

Kay and co-workers recently demonstrated that transverse relaxation optimization can be obtained for MQ coherences in methyl spin systems (Tugarinov *et al.*, 2003; See also Chapter 13 by Hwang and Kay in this volume). Relaxation in ^{13}CH$_3$ spin systems is more complex than in ^{1}H–^{15}N pairs due to the larger number of spin state transitions associated with the methyl group (10 ^{1}H transitions and 8 ^{13}C transitions) (Kay and Bull, 1992). In an isolated methyl group in the slow tumbling regime, relaxation is dominated by intramethyl group DD interactions; the effects of CSA relaxation are small and not considered here (Vold and Vold, 1978; Werbelow and Grant, 1977). The stochastic Hamiltonian is given by

$$
\begin{aligned}
\mathcal{H}(t) = {} & d_{IS}Y_2^0[\theta_{1S}(t)]2I_{1z}S_z + d_{IS}Y_2^0[\theta_{2S}(t)]2I_{2z}S_z \\
& + d_{IS}Y_2^0[\theta_{3S}(t)]2I_{3z}S_z + d_{II}Y_2^0[\theta_{12}(t)]A_{12} \\
& + d_{II}Y_2^0[\theta_{13}(t)]A_{13} + d_{II}Y_2^0[\theta_{23}(t)]A_{23}
\end{aligned}
\tag{31}
$$

in which $\theta_{jS}(t)$ refers to the orientation of the bond vector between the ^{13}C nucleus and the jth ^{1}H nucleus, $\theta_{jk}(t)$ refers to the orientation of the internuclear vector between the jth and kth ^{1}H nuclei, $S = {}^{13}$C, $I_j = {}^1$H (for the jth methyl ^{1}H spin), and $A_{jk} = 6^{-1/2}[3I_{jz}I_{kz} - I_j \cdot I_k]$. In the limit of fast rotation about the methyl symmetry axis, the above Hamiltonian can be expressed as

$$
\begin{aligned}
\mathcal{H}(t) = {} & d_{IS}P_2[\cos\beta]Y_2^0[\theta_{\text{sym}}(t)](I_{1z}S_z + I_{2z}S_z + I_{3z}S_z) \\
& + d_{II}P_2\left[\cos\frac{\pi}{2}\right]Y_2^0[\theta_{\text{sym}}(t)](A_{12} + A_{13} + A_{23})
\end{aligned}
\tag{32}
$$

in which $\beta = 70.5°$ is the angle between the ^{13}C–H bond vectors and the symmetry axis of the ^{13}CH$_3$ methyl group and $\theta_{\text{sym}}(t)$ is the orientation of the symmetry axis in the laboratory reference frame. The transverse relaxation rate constant for in-phase SQ S^- magnetization (ignoring cross-correlation effects) is calculated using Eq. (20) as

$$
R_{2S}^0 = (3d_{IS}^2/2)P_2^2[\cos\beta]J(0)
\tag{33}
$$

Remarkably, the ZQ and DQ coherences

$$
I^\alpha I^\beta I^\pm S^\pm = (I_2^\alpha I_3^\beta + I_2^\beta I_3^\alpha)I_1^\pm S^\pm + (I_1^\alpha I_3^\beta + I_1^\beta I_3^\alpha)I_2^\pm S^\pm + (I_1^\alpha I_2^\beta + I_1^\beta I_2^\alpha)I_3^\pm S^\pm
\tag{34}
$$

$$
I^\alpha I^\beta I^\pm S^\pm = (I_2^\alpha I_3^\beta + I_2^\beta I_3^\alpha)I_1^\pm S^\pm + (I_1^\alpha I_3^\beta + I_1^\beta I_3^\alpha)I_2^\pm S^\pm + (I_1^\alpha I_2^\beta + I_1^\beta I_2^\alpha)I_3^\pm S^\pm
\tag{35}
$$

yield

$$
\begin{aligned}
R_{2ZQ}^0 = R_{2DQ}^0 & = -\langle I^\alpha I^\beta I^\pm S^\pm | I^\alpha I^\beta I^\pm S^\pm \rangle^{-1} \left\langle I^\alpha I^\beta I^\pm S^\pm | \int_0^\infty \left[\mathcal{H}(t), [\mathcal{H}(t+\tau), I^\alpha I^\beta I^\pm S^\pm]\right] d\tau \right\rangle \\
& = 0
\end{aligned}
\tag{36}
$$

Thus, in the limit of slow overall rotational diffusion and fast methyl rotation, these ZQ and DQ coherences do not relax from intramethyl dipole–dipole interactions. In contrast, Eq. (33) predicts $R_{2S}^0 \approx 60 \ s^{-1}$ for a protein with $\tau_c = 40$ ns. Including external sources of relaxation, $R_{2ZQ} \approx R_{2DQ} \approx 8 \ s^{-1}$ for $\tau_c = 40$ ns. A complete analysis shows that the ZQ coherence is less sensitive to dipole–dipole cross-correlations with remote spins (Tugarinov *et al.*, 2004).

Unlike *IS* DD/*S* CSA TROSY (vide supra), intramethyl DD TROSY is independent of the static magnetic field, so long as the field strength is sufficiently strong that the spin diffusion limit is satisfied $[J(0) \gg J(\omega)]$. The significantly reduced rate of transverse relaxation of MQ coherences due to dipolar cross-correlations allows methyl groups to serve as sensitive probes of microsecond to millisecond dynamics in high-molecular-weight systems.

Quantitative Identification of Spins Affected by Chemical Exchange

This section will focus on experimental techniques that are designed for the rapid identification of spins affected by chemical exchange in high-molecular-weight systems. Although first identifying the presence of chemical exchange–induced relaxation under a given set of experimental conditions may be viewed solely as a prerequisite to performing a complete relaxation dispersion analysis, considerable information potentially can be obtained from quantitative identification of exchange in lieu of a more comprehensive analysis.

Except in situations in which exchange is slow on the chemical shift time scale and p_2 is large enough to generate an observable resonance signal from the minor state, the presence of chemical exchange must be identified by a faster rate of transverse relaxation of the dominant resonance signal than expected from CSA and dipolar interactions. Ideally, the most sensitive approach would be to measure the free precession transverse relaxation rate constant, R_{2i} (of a transverse relaxation optimized coherence or spin state), as well as R_{2i}^0, and determine $R_{ex(i)}$ as the difference between the two rate constants. However, the measurement of free precession relaxation rates is difficult; instead, considerable effort has been made to approximate free precession relaxation using Hahn spin echo, based pulse sequences (Wang and Palmer, 2003). In these experiments, the spin echo sequence $T/2 - 180° - T/2$ produces only a very weak effective field, on the order of $1/T = 25–50 \ s^{-1}$, so that suppression of exchange-induced relaxation is minimized and $R_{ex(i)}^{HE} \approx R_{ex(i)}$, where $R_{ex(i)}^{HE}$ is the exchange contribution to relaxation during the Hahn spin echo. In contrast, a conventional CPMG experiment with $\tau_{cp} = 1$ ms can result in $R_2(1/\tau_{cp} = 1000 \ s^{-1}) \ll R_{ex(i)}$, depending on the chemical shift time scale and k_{ex} for the exchange process, which makes detection of exchange-broadened resonances more difficult.

^{15}N Single Quantum Chemical Exchange Broadening

For backbone amide ^1H–^{15}N spin pairs, Palmer and co-workers designed a set of pulse sequences for measuring the exchange contribution to the transverse relaxation rate of the narrow component of the ^{15}N doublet (Wang *et al.*, 2003). The experimental scheme shown in Fig. 5 involves the measurement of three relaxation rate constants in order to estimate $R_{ex(S)}^{HE} \approx R_{ex(S)}$. The pulse sequence in Fig. 5A is used to measure the relaxation rates of the narrow ($R_{2\alpha}$) and broad ($R_{2\beta}$) components of the ^{15}N doublet depending on whether the ^1H composite pulse immediately following the Hahn spin echo period produces a net 0° or 180° rotation, respectively. The two components of the ^{15}N doublet relax independently during the spin echo of duration T if $T = n/J_{HN}$, where n is an integer and

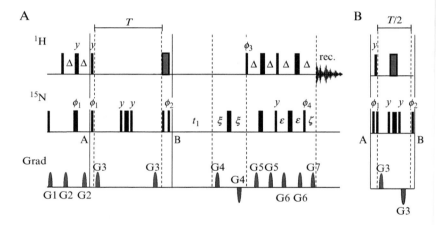

FIG. 5. ^1H–^{15}N DD/^{15}N CSA TROSY Hahn echo pulse sequence for measuring ^{15}N $R_{ex(S)}$ (Wang *et al.*, 2003). Pulse sequences for measuring $R_{2\alpha}$, $R_{2\beta}$, and R_{zz} to detect chemical exchange in large proteins (Wang *et al.*, 2003). The sequence shown in (A) detects relaxation of the narrow doublet component during the Hahn echo period T when the gray pulse element is $(90°_x 90°_y 90°_{-y} 90°_{-x})$ and detects relaxation of the broad doublet component when this element is $(90°_x 90°_y 90°_y 90°_x)$. Relaxation of longitudinal two-spin order is detected if the elements between points A and B in (A) are replaced with pulse elements in (B); the gray element is $(90°_x 90°_y 90°_y 90°_x)$. The block of four 90° pulses is applied at the center of the amide ^1H region to minimize off-resonance effects. Other delays are $\Delta = 1/2J_{NH}$, $\xi > G4$, $\zeta > G7$, $\varepsilon = \Delta - z/2$. All pulses have x phase unless otherwise stated. Phase cycles are $\phi_1 = x, -x$; $\phi_2 = x, x, -x, -x$; $\phi_3 = y$; $\phi_4 = x$; and receiver phase $= -x, x, x, -x$. Gradients G4 and G7 are used for coherence selection; other gradients are for artifact suppression. Echo–antiecho quadrature detection is achieved by inverting ϕ_3, ϕ_4, and the sign of gradient G4; axial peaks are shifted by inverting ϕ_1, and the receiver phase. Echo and antiecho data sets are processed by the Rance–Kay approach (Kay *et al.*, 1992a; Palmer *et al.*, 1991).

J_{HN} is the one-bond ^1H–^{15}N scalar coupling constant (Palmer *et al.*, 1992). In the third experiment, the relaxation rate of longitudinal two-spin order, R_{zz}, is measured by replacing the portion of the sequence between A and B in Fig. 5A with the element in Fig. 5B. In all three experiments, TROSY selection is utilized during the t_1 chemical shift evolution and t_2 detection periods for improved resolution and sensitivity in the two-dimensional correlation spectra.

In the first two experiments, the transverse relaxation rates of the two lines of the ^{15}N doublet are given by Eqs. (27) and (28), and the ratio of the two signal intensities at time T, $I^\alpha(T)$ and $I^\beta(T)$, is used to determine the cross-correlated relaxation rate constant η_{xy}:

$$\eta_{xy} = -\frac{1}{2T}\ln\left[\frac{I^\alpha(T)}{I^\beta(T)}\right] \qquad (37)$$

The relaxation rate constant for longitudinal two-spin order, $R_{zz} \approx R_{1H} + R_{1N}$ and R_{1N} is the ^{15}N longitudinal relaxation rate constant, is used to remove the ^1H R_{1H} contribution in Eq. (27). An apparent relaxation rate constant is determined from

$$R_2^\beta - \frac{1}{2}R_{zz} = \frac{1}{T}\ln\left[\frac{I^\beta(T)}{I_{zz}(T/2)}\right] \approx R_{2S}^0 - \frac{1}{2}R_{1N} - \eta_{xy} + R_{ex(S)}^{HE} \qquad (38)$$

For exchange-broadened spins, R_{2S}^0 is determined from the ratio $\kappa = R_{2S}^0/\eta_{xy}$, which is independent of chemical exchange and fast internal dynamics in high-molecular-weight systems. The ratio can be estimated either experimentally as $\kappa = \langle 1 + (R_{2\beta} - R_{zz}/2)/\eta_{xy}\rangle$, where the angle brackets represent a trimmed mean over all nonexchanging residues, or from theoretical calculations. Equation (27) can be rearranged to provide an estimate of the exchange contribution during the Hahn spin echo as

$$\begin{aligned} R_{ex(S)}^{HE} &= R_2^\beta - \frac{1}{2}(R_{zz} - R_{1N}) - R_{2S}^0 + \eta_{xy} \\ &= R_2^\beta - \frac{1}{2}(R_{zz} - R_{1N}) - \eta_{xy}(\kappa - 1) \end{aligned} \qquad (39)$$

R_{1N} makes a negligible contribution in high-molecular-weight systems and can be ignored.

To illustrate the method, ^{15}N $R_{ex(S)}$ relaxation rate constants for glycerol-3-phosphate (G3P) bound [85%-^2H, U–^{15}N] triosephosphate isomerase (TIM), a symmetric dimer with a molecular weight of 54 kDa, are shown in Fig. 6. For nonexchanging spins, the average ^{15}N $R_{2\alpha}$ is on the order of 6 s^{-1}, compared to $R_{2S}^0 \sim 37$ s^{-1} at a temperature of 298 K and a static magnetic field strength of 18.8 T. Thus, relaxation of the narrow

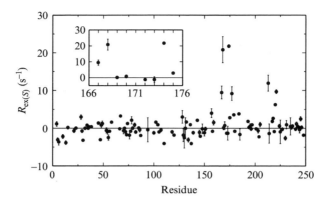

FIG. 6. Chemical exchange in G3P bound [85%-D, U–^{15}N] TIM detected using the pulse sequences in Fig. 5. The data were collected at 800 MHz using a Bruker DRX800 NMR spectrometer equipped with a triple-resonance cryoprobe. The relaxation delay $T = 3/J_{NH} = 21.4$ ms, corresponding to $I^{\alpha}(T)/I^{\beta}(T) = 0.3$ for most residues. Four duplicates of each spectrum were recorded to estimate experimental uncertainties. κ was determined by the trimmed mean of $(R_{2\beta} - R_{zz}/2)/\eta_{xy}$ or by theoretical calculation assuming $r_{NH} = 1.02$ Å, ^{15}N CSA $= -172$ ppm, and the angle between the ^{15}N CSA principal axis and the NH bond is 19°. Both methods gave $\kappa = 1.2$. Reprinted with permission from Wang *et al.* (2003). Copyright (2003) American Chemical Society.

component of the ^{15}N doublet potentially is a factor of six times more sensitivity to chemical exchange than relaxation of in-phase SQ magnetization observed when DD/CSA cross-correlation is suppressed. The distribution of $R_{ex(S)}$ values about 0 s^{-1} in Fig. 6 arises from variation in the magnitude and orientation of the ^{15}N CSA tensor, which is implicitly assumed to be uniform throughout the protein by using an average value of κ in Eq. (39). Such an assumption may be unwarranted at higher static magnetic field strengths where the CSA relaxation mechanism is more significant, and the site-specific variation in the ^{15}N CSA will lower the sensitivity of the experiment to chemical exchange. In Fig. 6, exchange-induced relaxation can clearly be identified for ^{15}N spins of residues at the N- and C-termini of loop 6 in TIM (residues V167–L174, see inset), which is hypothesized to act as a molecular hinge for gating access to the active site.

Although delineating a mechanistic interpretation of the exchange process from a single measurement of $R_{ex(S)}$ is difficult due to the dependence on k_{ex}, $\Delta\omega$, and p_1p_2 in Eq. (5), measurements such as those in Fig. 6 provide considerable useful information. First, the static magnetic field dependence of $R_{ex(S)}$ can be used to estimate the chemical shift time scale for the exchange process through the relationship (Millet *et al.*, 2000)

$$\frac{\delta R_{\text{ex}(S)}}{R_{\text{ex}(S)}} = \alpha \frac{\delta B_0}{B_0} \tag{40}$$

where the scaling parameter α is defined as

$$\alpha = \frac{d\left(\ln R_{\text{ex}(S)}\right)}{d(\ln \Delta\omega)} \tag{41}$$

The chemical shift time scale is given by $0 \leq \alpha < 1, \alpha \approx 1, 1 < \alpha \leq 2$ for slow, intermediate, and fast exchange, respectively. Estimates of α can be obtained experimentally by measuring $R_{\text{ex}(S)}$ at two or more static magnetic field strengths and approximating Eq. (41) as

$$\alpha = \left(\frac{B_{02} + B_{01}}{B_{02} - B_{01}}\right)\left(\frac{R_{\text{ex}2} - R_{\text{ex}1}}{R_{\text{ex}2} + R_{\text{ex}1}}\right) \tag{42}$$

where $R_{\text{ex}j}$ is $R_{\text{ex}(S)}$ measured at static magnetic field strength B_{0j}. With the chemical shift time scale known, simplified expressions, such as Eqs. (6) and (13) in the limit of fast exchange, can be used in subsequent analyses. Second, the temperature dependence of $R_{\text{ex}(i)}$ provides an estimate of the chemical shift time scale (Grey *et al.*, 2003; Mandel *et al.*, 1996) and apparent activation energy for the chemical exchange process (Butterwick *et al.*, 2004; Evenas *et al.*, 2001; Mandel *et al.*, 1996). Assuming an Arrhenius relationship for $R_{\text{ex}(i)}$;

$$\frac{d\ln(R_{\text{ex}(i)})}{d(1/RT)} = \left(\Delta H_{\text{f}}^{\ddagger} - p_1 \Delta H\right)$$

$$\frac{(k_{\text{ex}}^4 - \Delta\omega_S^4)(k_{\text{ex}}^2 - \Delta\omega_S^2)}{(k_{\text{ex}}^4 - \Delta\omega_S^4)(k_{\text{ex}}^2 - \Delta\omega_S^2) + 2p_1 p_2 k_{\text{ex}}^2 \Delta\omega_S^2 (5k_{\text{ex}}^2 - 3\Delta\omega_S^2)}$$

$$-(p_1 - p_2)\Delta H \frac{(k_{\text{ex}}^2 + \Delta\omega_S^2)^2}{(k_{\text{ex}}^2 + \Delta\omega_S^2)^2 - p_1 p_2 k_{\text{ex}}^2 \Delta\omega_S^2 (5k_{\text{ex}}^2 + \Delta\omega_S^2)}$$

$$= \Delta H_{\text{f}}^{\ddagger} + (1 - 3p_1)\Delta H$$

where $\Delta H_{\text{f}}^{\ddagger}$ is the transition state enthalpy of the forward reaction, ΔH is the difference in enthalpy between the two states, R is the universal gas constant, and T is the temperature. The last line results from assuming $k_{\text{ex}} \gg \Delta\omega_S$. Third, if the site populations and chemical shift changes can be determined a priori, such as applications based on pH-dependent processes or ligand-binding events, correlation of R_{ex} with the measured $\Delta\omega_S$ for multiple residues in the limit of fast exchange can be used to estimate k_{ex} for the exchange process.

^1H–^{15}N Multiple Quantum Exchange Broadening

Chemical exchange provides an additional mechanism of transverse relaxation for ZQ and DQ coherence that depends on the signs and magnitudes of the chemical shift changes for the two spins comprising the MQ coherence, as shown by Eq. (2) (Pervushin, 2001; Rance, 1988). Based on Eqs. (29) and (30), the differential relaxation of MQ coherences, $\Delta R_{MQ} = R_{DQ} - R_{ZQ}$, arises solely from cross-correlation effects, including cross-correlated chemical shift modulation (i.e., chemical exchange):

$$\Delta R_{MQ} = \frac{4}{3} c_I c_S P_2[\cos(\chi)] J(0) + R_{ex(DQ)} - R_{ex(ZQ)} \qquad (43)$$

The original experiment for measuring ΔR_{MQ} in ^1H–^{15}N spin systems was based on conventional HSQC methods (Kloiber and Konrat, 2000). Pervushin and co-workers developed a constant-time MQ pulse sequence for measuring ΔR_{MQ} in ^1H–^{15}N spin systems that may be advantageous for high-molecular-weight systems (Pervushin, 2001; Pervushin et al., 1999). In this scheme, shown in Fig. 7, evolution and relaxation of ZQ and DQ

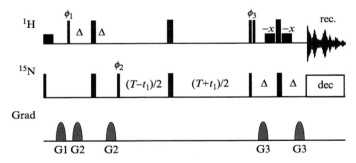

FIG. 7. Pulse sequence for the MQ Hahn echo experiment (Pervushin, 2001). Narrow and wide bars depict 90° and 180° pulses, respectively. Short solid bars are water-selective 90° pulses. All pulses are x-phase unless otherwise indicated. The delay $\Delta = 1/(4J_{NH})$. The Hahn echo period is T. Decoupling during acquisition is performed using GARP (Shaka et al., 1985). Four free induction decays (FIDs) are acquired for each t_1 point. The phase cycle for the first FID is $\phi_1 = x, -x$; $\phi_2 = x, x, -x, -x$; $\phi_3 = x$; and receiver $= x, -x, -x, x$. The phase cycle for the second FID is $\phi_1 = x, -x$; $\phi_2 = x, x, -x, -x$; $\phi_3 = -x$; and receiver $= x, -x, -x, x$. The third and fourth FIDs are obtained using the phase cycles for the first two FIDs except that $\phi_2 = -y, -y, y, y$. The gradients are used to suppress unwanted coherences and pulse imperfections and for solvent suppression using the WATERGATE approach (Piotto et al., 1992). ZQ echo and antiecho data sets are obtained as $FID_1 + FID_3$ and $FID_2 - FID_4$, respectively. DQ echo and antiecho data sets are obtained as $FID_1 - FID_3$ and $FID_2 + FID_4$, respectively. Echo and antiecho data sets are processed by the Rance–Kay approach (Kay et al., 1992a; Palmer et al., 1991).

coherences are recorded during a constant time period. Following the chemical shift evolution and relaxation periods, the selected MQ coherence is transferred back to the ^{1}H spin for detection. Although only one-half of the initial magnetization, which originates as a superposition of ZQ and DQ coherences, is recorded, the sensitivity-enhanced quadrature detection scheme requires only one INEPT element to transfer the selected ZQ or DQ coherence back to detectable ^{1}H magnetization. This reduces the overall length of the pulse sequence compared to a ^{1}H–^{15}N DD/^{15}N CSA TROSY scheme. The ratio of signal intensities in the DQ and ZQ correlation spectra is used to estimate the differential relaxation rate constant:

$$\Delta R_{\mathrm{MQ}} = -\frac{1}{T}\ln\left[\frac{I^{\mathrm{DQ}}(T)}{I^{\mathrm{ZQ}}(T)}\right] \qquad (44)$$

Measurement of ΔR_{MQ} can be used qualitatively in conjunction with measurements of $R_{\mathrm{ex}}^{\mathrm{HE}}$ to identify chemical exchange processes that modulate both ^{1}H and ^{15}N chemical shifts. However, a more quantitative use of ΔR_{MQ} for estimating the chemical shift time scale requires separating the exchange contribution from the other cross-correlated relaxation mechanisms. Lundstroem and Akke (2004) recently presented a detailed quantitative analysis of the static magnetic field dependence of ΔR_{MQ} relaxation rate constants in calmodulin. In the limit of fast exchange, ΔR_{MQ} is linearly dependent on B_0^2 so that field-independent DD cross-correlation terms can be separated from the field-dependent cross-correlated CSA and chemical exchange contributions. In the absence of an independent measurement of η_{cc} (Tessari and Vuister, 2000), an estimate of an average value of $\kappa' = d\eta_{\mathrm{cc}}/dB_0^2$ is derived from the field dependence of η_{cc} for nonexchanging residues and used to estimate $\Delta R_{\mathrm{ex(MQ)}} = R_{\mathrm{ex(DQ)}} - R_{\mathrm{ex(ZQ)}}$. Similar assumptions are inherent to this approximation as described for the ratio κ (vide supra). Once the pure exchange contribution is estimated, the static magnetic field dependence is used to estimate the chemical shift time scale (Lundstroem and Akke, 2004; Wang and Palmer, 2002). In the limit of fast exchange, the ratio $\Delta R_{\mathrm{ex(MQ)}}/R_{\mathrm{ex}(S)}$ gives an estimate of the ratio of chemical shift changes, $\Delta\omega_I/\Delta\omega_S$, using Eq. (6), and provides potentially important structural information (Ishima et al., 1999).

Characterizing Chemical Exchange by CPMG Relaxation Dispersion

Experimental schemes that employ CPMG refocusing periods to modulate the contribution of chemical exchange to the transverse relaxation of SQ and MQ coherences involving ^{1}H–^{15}N and ^{13}CH$_3$ spin systems allow more detailed quantification of exchange kinetic parameters than is possible using Hahn echo experiments alone. Relaxation during CPMG spin echoes is sensitive to chemical exchange processes with $k_{ex} \leq 1/\tau_{cp,min}$, where $\tau_{cp,min}$ corresponds to the smallest value of τ_{cp} that can be utilized experimentally. The effects of sample heating that occur under fast pulsing conditions typically limit the minimum value to $\tau_{cp} > 0.1–1.0$ ms. In most instances, the CPMG experiments described herein are sensitive to exchange processes with $k_{ex} < 10^4$ s^{-1}. Dynamic processes at least an order of magnitude faster can be characterized through the use of off-resonance spin-lock fields in $R_{1\rho}$ experiments (Palmer et al., 2001). In addition, the range of time scales accessible to experimental characterization is extended if $R_{2(i)}^0$ can be measured independently. In this case, the constraint $R_{ex(S)}(1/\tau_{cp} \rightarrow \infty) = 0$ facilitates analysis of the relaxation dispersion data (Grey et al., 2003; Vugmeyster et al., 2000; Wang et al., 2001). Finally, relaxation dispersion data recorded at two or more static magnetic field strengths are required to establish the chemical shift time scale for the exchange process and also enhances the reliability of fitted exchange parameters (Millet et al., 2000).

^{15}N SQ Relaxation Dispersion

The original TROSY–CPMG pulse sequence for measuring ^{15}N $R_2(1/\tau_{cp})$ transverse relaxation rate constants is shown in Fig. 8 (Loria et al., 1999a). At the beginning of the first CPMG spin echo (point a), the magnetization present is a linear combination of $I^{\alpha}S_y$ and $I^{\beta}S_y$ coherences. These coherences relax for a time period $T/2 = 2n\tau_{cp}$ during the first CPMG spin echo period. The S^{3}CT pulse sequence element (Sorensen et al., 1997) during the U period serves to invert one of the doublet components, and relaxation proceeds through the second CPMG spin echo period for an additional $T/2 = 2n\tau_{cp}$. The U period effectively suppresses cross-relaxation between the doublet components; thus, the $I^{\alpha}S_y$ and $I^{\beta}S_y$ coherences relax according to Eqs. (27) and (28) for a time $T = 4n\tau_{cp}$ where n is an integer. The ^{1}H–^{15}N DD/^{15}N CSA TROSY selection pulse sequence elements following the CPMG spin echo periods permit the measurement of exchange line broadening for the narrow component of the ^{15}N doublet.

The utility of this experiment was originally demonstrated on TIM at a static magnetic field strength of 14.1 T and a temperature of 298 K. The average ^{15}N transverse relaxation rate constant for $\tau_{cp} = 0.8$ ms was $R_2(1/\tau_{cp}) = 16.1 \pm 0.3$ s^{-1} using the TROSY–CPMG sequence compared to 40 ± 2 s^{-1} measured with the relaxation-compensated CPMG experiment (Loria et al., 1999b). In an important application, the TROSY–CPMG experiment was used to record ^{15}N relaxation dispersion profiles for tryptophan side chains ^{15}N$^{\varepsilon 1}$ spins in different states of hemoglobin,

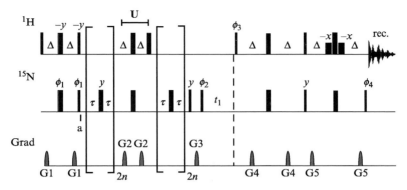

FIG. 8. Pulse sequence for the ^{15}N SQ TROSY–CPMG experiment (Loria *et al.*, 1999a). Narrow and wide bars depict 90° and 180° pulses, respectively. Short solid bars are water-selective 90° pulses. All pulses are *x*-phase unless otherwise indicated. The delays are $\tau = \tau_{cp}/2$ and $\Delta = 1/(4J_{NH})$. The relaxation period is $T = 4n\tau_{cp}$. Quadrature detection was obtained by recording echo and antiecho FIDs for each t_1 point. The phase cycle for the first FID is $\phi_1 = 4(x),\, 4(-x);\, \phi_2 = -y,\, y,\, x,\, -x;\, \phi_3 = y;\, \phi_4 = x$ and receiver $= x,\, -x,\, y,\, -y,\, -x,\, x,\, y,\, -y$. The phase cycle for the second FID is $\phi_1 = 4(x),\, 4(-x);\, \phi_2 = -y,\, y,\, x,\, -x;\, \phi_3 = -y;\, \phi_4 = -x$ and receiver $= -x,\, x,\, -y,\, y,\, x,\, -x,\, y,\, -y$. The gradients are used to suppress unwanted coherences and pulse imperfections, and for solvent suppression using the WATERGATE approach (Piotto *et al.*, 1992). Echo and antiecho data sets are processed by the Rance–Kay approach (Kay *et al.*, 1992a; Palmer *et al.*, 1991).

with a total molecular mass for the $\alpha_2\beta_2$ tetramer of 64.5 kDa (Yuan *et al.*, 2002). Illustrative dispersion curves are shown in Fig. 9. Slow conformational exchange for Trp-37, which resides at the interface between the α_1 and β_2 subunits, under certain experimental conditions, is proposed to arise in response to conformational changes associated with the $T \rightleftharpoons R$ allosteric transition that accompanies formation of the ligated state in hemoglobin.

$^{13}CH_3$ MQ Relaxation Dispersion

As described above, the ^{13}CH$_3$ ZQ and DQ coherences represented by Eqs. (34) and (35) do not relax under the influence of the intramethyl DD interactions. This TROSY line narrowing has been used by Kay and co-workers to develop CPMG methods to characterize chemical exchange broadening in large proteins (Korzhnev *et al.*, 2004a). The pulse sequence for this experiment is shown in Fig. 10 and is based on an HMQC pulse sequence that records the average evolution of ZQ and DQ coherence during t_1. In principle, slightly better resolution at a cost of slightly reduced sensitivity can be obtained by recording only the ZQ component after the t_1 evolution period (Tugarinov *et al.*, 2004).

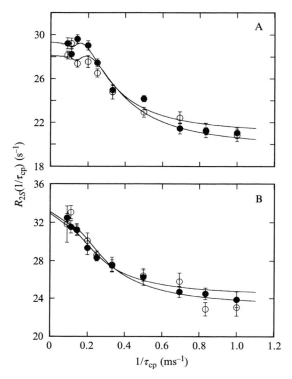

FIG. 9. ^{15}N SQ TROSY–CPMG relaxation dispersion profiles for β37 Trp ^{15}N^{el} of CO-hemoglobin in the presence of inositol hexaphosphate at (A) 34° and (B) 29°. Data were recorded at (●) 11.7 T and (○) 14.1 T. Dispersion profiles were fit using Eq. (10). Reprinted from Yuan *et al.* (2002), with permission from Elsevier.

In this experiment, the elements between points a and b constitute a filter to select only the desired ZQ and DQ coherences. The ZQ and DQ operators in Eqs. (34) and (35) contain longitudinal ^{1}H components; consequently, repetitive pulsing on the ^{1}H spins during the CPMG spin echo sequence is not possible due to the cumulative effects of pulse imperfections. Therefore, only a single ^{1}H 180° pulse is used to refocus ^{1}H chemical shift evolution. As a result, the relaxation decay represents an average over the decay of the ZQ and DQ operators, as described in detail elsewhere (Korzhnev *et al.*, 2004a). Therefore, Eq. (14), rather than Eq. (10), is used to analyze the relaxation dispersion curves. Furthermore, relaxation may be multiexponential due to interactions with remote spins;

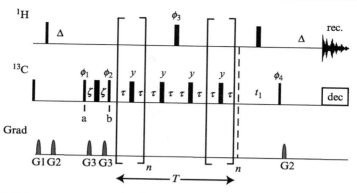

FIG. 10. Pulse sequence for the ^{1}H–^{13}C MQ TROSY–CPMG experiment (Korzhnev *et al.*, 2004a). Narrow and wide bars depict 90° and 180° pulses, respectively. All pulses are x-phase unless otherwise indicated. The delays are $\tau = \tau_{cp}/2, \zeta = 1/(8J_{CH})$, and $\Delta = 1/(2J_{CH})$. The constant relaxation period is $T = 2(n + 1)\tau_{cp}$ and n is an odd integer. A reference spectrum for $T = 0$ is recored by omitting all pulse sequence elements within T except for the pulse ϕ_3. Decoupling during acquisition is performed using WALTZ-16 (Shaka *et al.*, 1983). The phase cycle is $\phi_1 = x, -x; \phi_2 = y, y, -y, -y; \phi_3 = 4(x), 4(-x); \phi_4 = -x$; and receiver $= 2(x, -x), 2(-x, x)$. Quadrature detection is obtained by States-TPPI phase cycling of ϕ_4 (Marion *et al.*, 1989). The gradients are used to suppress unwanted coherences and pulse imperfections.

accordingly, relaxation rate constants are measured using the constant relaxation time approach as illustrated by Eq. (17). ZQ and DQ coherences are sensitive to relaxation by external ^{1}H spins; accordingly, optimal results for Leu and Val amino acids require preparation of samples using a [^{13}CH$_3$, ^{12}CD$_3$]-isovalerate precursor to ensure that only one of the geminal methyl groups is ^{13}C labeled and protonated and the other is ^{12}C natural abundance and deuterated (Korzhnev *et al.*, 2004a).

The ^{13}CH$_3$ TROSY–CPMG experiment has been applied to a highly deuterated, methyl-protonated sample of malate synthase G, a monomeric protein with a molecular mass of 82 kDa. The relaxation dispersion curves for selected Ile amino acid residues are illustrated in Fig. 11. Limiting values of $R_{eff}(1/\tau_{cp} \to \infty) < 25$ s^{-1} are obtained using the TROSY approach for malate synthase G. Surprisingly, ~20% (46 of 231) of the CH$_3$ groups investigated exhibit detectable chemical exchange broadening. Exchange broadening of both side chain CH$_3$ and backbone NH moieties is observed for the $\alpha 2$–βk loop that connects the α/β and barrel domains of malate synthase G.

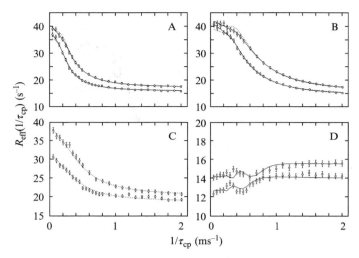

FIG. 11. ^{1}H–^{13}C MQ TROSY–CPMG relaxation dispersion profiles for Ile (A) 242, (B) 260, (C) 424, and (D) 42 of malate synthase G. Data were recorded at 14.1 T (lower curve) and 18.8 T (upper curve) and a temperature of 37°. Dispersion profiles were fit using Eqs. (14) and (19) assuming (dashed lines) $\Delta\omega_I = 0$ and (solid lines) $\Delta\omega_I \neq 0$ ($I = {}^1$H). Solid curves are shown only if a statistically significant improvement was obtained by optimizing $\Delta\omega_I$. Reprinted with permission from Korzhnev *et al.* (2004a). Copyright (2004) American Chemical Society.

$^1H^N$ SQ and $^1H^N$–^{15}N MQ Relaxation Dispersion

TROSY-based CPMG experimental methods have been proposed for recording relaxation dispersion data for ^{1}H^N SQ coherence (Ishima and Torchia, 2003; Ishima *et al.*, 1998; Orekhov *et al.*, 2004) and for ^{1}H^N–^{15}N MQ coherence (Dittmer and Bodenhausen, 2004; Orekhov *et al.*, 2004). These pulse sequences have utilized both ^{1}H–^{15}N DD/^{15}N CSA TROSY (Ishima and Torchia, 2003; Ishima *et al.*, 1998) and ^{1}H CSA/^{15}N CSA TROSY techniques (Orekhov *et al.*, 2004; Pervushin, 2001). To date, ^{1}H^N SQ and ^{1}H^N–^{15}N MQ relaxation dispersion measurements have not been applied to proteins with molecular masses >50 kDa; however, these experiments may prove to be useful methods for characterizing chemical exchange in large proteins that are perdeuterated to reduce these dipolar and scalar interactions with remote spins.

An example of a pulse sequence for a ^{1}H^N SQ CPMG experiment is shown in Fig. 12 (Orekhov *et al.*, 2004). This sequence is based on the CSA/CSA ZQ TROSY technique (Pervushin *et al.*, 1999), also illustrated by the MQ Hahn echo pulse sequence shown in Fig. 7. The pulse sequences of

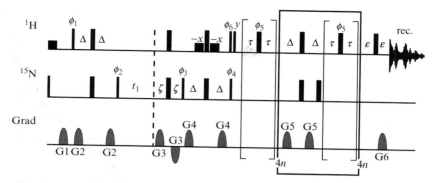

FIG. 12. Pulse sequences for the ZQ TROSY $^1H^N$ SQ CPMG experiment (Orekhov *et al.*, 2004). The pulse sequence elements in the box can be included to improve the accuracy of results for long values of τ_{cp} (Ishima and Torchia, 2003) in a fashion similar to that for the ^{15}N SQ TROSY–CPMG experiment shown in Fig. 8. Narrow and wide bars depict 90° and 180° pulses, respectively. Short solid bars are water-selective 90° pulses. All pulses are x-phase unless otherwise indicated. The delays are $\tau = \tau_{cp}/2$ and $\Delta = 1/(4J_{NH})$. The delays ζ and ε are long enough to encompass the gradients G3 and G5, respectively. The relaxation period is $T = 4n\tau_{cp}$ or $8n\tau_{cp}$ if the boxed sequence elements are included. A reference spectrum is recorded for $T = 0$ in which the bracketed CPMG elements are removed. The phase cycle is $\phi_1 = 2(x, -x)\ 2(y,-y)$; $\phi_2 = 2(-y),\ 2(y),\ 2(-x),\ 2(x)$; $\phi_3 = 4(x),\ 4(-x)$; $\phi_4 = 4(y),\ 4(-y)$; $\phi_6 = y$; and receiver $= x, -x, -x, x, -x, x, x, -x$. The second FID acquired for quadrature detection has G3 inverted and $\phi_6 = -y$. The CPMG pulse phases ϕ_5 are applied to consecutive pulses during the spin echo train using the XY16 scheme $\{x, y, x, y, y, x, y, x, -x, -y, -x, -y, -y, -x, -y, -x\}$ in minimum blocks of four pulses. Gradients G3 and G6 are used for gradient coherence selection. Other gradients are used to suppress unwanted coherences and pulse imperfections and for solvent suppression using the WATERGATE approach (Piotto *et al.*, 1992). Echo and antiecho data sets are processed by the Rance–Kay approach (Kay *et al.*, 1992a; Palmer *et al.*, 1991).

Figs. 7 and 12 are easily modified for $^1H^N$–^{15}N MQ relaxation measurements (Orekhov *et al.*, 2004). In this approach, following evolution of MQ coherence during t_1, only ZQ coherence is transferred back to the transverse relaxation-optimized ^{1}H doublet via a single INEPT transfer step. The spin-state selective $^1H^N$ transverse relaxation is then measured during a constant relaxation time CPMG period. The relaxation compensation approach, shown as the boxed pulse sequence elements, may be used to improve the accuracy of the relaxation rate constants for large values of τ_{cp} (Ishima and Torchia, 2003). The potential utility of this method for characterizing exchange in larger proteins can be illustrated by observing (1) the overall reduction of $R_{eff}(1/\tau_{cp} \to \infty)$ compared to non-spin-state-selective experiments (Ishima and Torchia, 2003), and (2) the improved cancellation of ^{1}H CSA and ^{1}H–^{15}N DD interactions that reduce $R_{eff}(1/\tau_{cp} \to \infty)$ at higher static magnetic field strengths.

Conclusion

The extensive developments and burgeoning applications of NMR spin relaxation methods for characterizing chemical exchange in proteins with molecular masses <30 kDa have spurred efforts to extend these experiments to larger proteins and macromolecules with masses >50 kDa. TROSY techniques for ^1H–^{15}N backbone amide moieties and for $^{13}CH_3$ methyl groups have been particularly powerful in the development of novel Hahn spin echo and CPMG experiments. To date these experiments have been applied to proteins with molecular masses ranging from 54 to 82 kDa. These applications illustrate the power of NMR spectroscopy for investigations of dynamic phenomena in proteins, including folding, molecular recognition, allosterism, and enzyme catalysis.

Acknowledgments

Grants from the National Institutes of Health GM59273 (A.G.P.), GM08281 (M.J.G.), and DK07328 and GM068311 (C.W.) are gratefully acknowledged.

References

Abergel, D., and Palmer, G. (2003). On the use of the stochastic Liouville equation in nuclear magnetic resonance. Application to $R_{1\rho}$ relaxation in the presence of exchange. *Concepts Magn. Reson.* **19A,** 134–148.

Abergel, D., and Palmer, A. G. (2004). Approximate solutions of the Bloch-McConnell equations for two-site chemical exchange. *Chem. Phys. Chem.* **5,** 787–793.

Abragam, A. (1961). "Principles of Nuclear Magnetism." Oxford University Press, New York.

Akke, M., and Palmer, A. G. (1996). Monitoring macromolecular motions on microsecond to millisecond time scales by $R_{1\rho}$-R_1 constant relaxation time NMR spectroscopy. *J. Am. Chem. Soc.* **118,** 911–912.

Allerhand, A., and Thiele, E. (1966). Analysis of Carr-Purcell spin-echo nuclear magnetic resonance experiments on multiple-spin systems. II. The effect of chemical exchange. *J. Chem. Phys.* **45,** 902–916.

Bain, A. D. (2003). Chemical exchange in NMR. *Prog. Nucl. Magn. Reson. Spectrosc.* **43,** 63–103.

Butterwick, J. A., Patrick Loria, J., Astrof, N. S., Kroenke, C. D., Cole, R., Rance, M., and Palmer, A. G. (2004). Multiple time scale backbone dynamics of homologous thermophilic and mesophilic ribonuclease HI enzymes. *J. Mol. Biol.* **339,** 855–871.

Carr, H. Y., and Purcell, E. M. (1954). Effects of diffusion on free precession in nuclear magnetic resonance experiments. *Phys. Rev.* **94,** 630–638.

Carver, J. P., and Richards, R. E. (1972). General two-site solution for the chemical exchange produced dependence of T_2 upon the Carr-Purcell pulse separation. *J. Magn. Reson.* **6,** 89–105.

Cavanagh, J., Fairbrother, W. J., Palmer, A. G., and Skelton, N. J. (1996). "Protein NMR Spectroscopy: Principles and Practice." Academic Press, San Diego, CA.

Desvaux, H., and Berthault, P. (1999). Study of dynamic processes in liquids using off-resonance rf irradiation. *Prog. Nucl. Magn. Reson. Spectrosc.* **35**, 295–340.

Deverell, C., Morgan, R. E., and Strange, J. H. (1970). Chemical exchange by nuclear magnetic relaxation in the rotating frame. *Mol. Phys.* **18**, 553–559.

Dittmer, J., and Bodenhausen, G. (2004). Evidence for slow motion in proteins by multiple refocusing of heteronuclear nitrogen/proton multiple quantum coherences in NMR. *J. Am. Chem. Soc.* **126**, 1314–1315.

Eisenmesser, E. Z., Bosco, D. A., Akke, M., and Kern, D. (2002). Enzyme dynamics during catalysis. *Science* **295**, 1520–1523.

Evenas, J., Malmendal, A., and Akke, M. (2001). Dynamics of the transition between open and closed conformations in a calmodulin C-terminal domain mutant. *Structure* **9**, 185–195.

Fersht, A. (1999). "Structure and Mechanism of Protein Structure: A Guide to Enzyme Catalysis and Protein Folding." W. H. Freeman, New York.

Frauenfelder, H., Sligar, S. G., and Wolynes, P. G. (1991). The energy landscapes and motions of proteins. *Science* **254**, 1598–1603.

Goldman, M. (1984). Interference effects in the relaxation of a pair of unlike spin-1/2 nuclei. *J. Magn. Reson.* **60**, 437–452.

Grey, M. J., Wang, C., and Palmer, A. G. (2003). Disulfide bond isomerization in basic pancreatic trypsin inhibitor: Multisite chemical exchange quantified by CPMG relaxation dispersion and chemical shift modeling. *J. Am. Chem. Soc.* **125**, 14324–14335.

Hill, R. B., Bracken, C., DeGrado, W. F., and Palmer, A. G. (2000). Molecular motions and protein folding: Characterization of the backbone dynamics and folding equilibrium of α_2D using ^{13}C NMR spin relaxation. *J. Am. Chem. Soc.* **122**, 11610–11619.

Ishima, R., and Torchia, D. A. (2003). Extending the range of amide proton relaxation dispersion experiments in proteins using a constant-time relaxation-compensated CPMG approach. *J. Biomol. NMR* **25**, 243–248.

Ishima, R., Wingfield, P. T., Stahl, S. J., Kaufman, J. D., and Torchia, D. A. (1998). Using amide ^{1}H and ^{15}N transverse relaxation to detect millisecond time-scale motions in perdeuterated proteins: Application to HIV-1 protease. *J. Am. Chem. Soc.* **120**, 10534–10542.

Ishima, R., Freedberg, D. I., Wang, Y.-X., Louis, J. M., and Torchia, D. A. (1999). Flap opening and dimer-interface flexibility in the free and inhibitor-bound HIV protease, and their implications for function. *Structure* **7**, 1047–1055.

Jen, J. (1978). Chemical exchange and NMR T_2 relaxation—the multisite case. *J. Magn. Reson.* **30**, 111–128.

Kay, L. E., and Bull, T. E. (1992). Heteronuclear transverse relaxation in AMX, AX2, and AX3 spin systems. *J. Magn. Reson.* **99**, 615–622.

Kay, L. E., Keifer, P., and Saarinen, T. (1992a). Pure absorption gradient enhanced heteronuclear single quantum correlation spectroscopy with improved sensitivity. *J. Am. Chem. Soc.* **114**, 10663–10665.

Kay, L. E., Nicholson, L. K., Delaglio, F., Bax, A., and Torchia, D. A. (1992b). Pulse sequences for removal of the effects of cross correlation between dipolar and chemical-shift anisotropy relaxation mechanisms on the measurement of heteronuclear T_1 and T_2 values in proteins. *J. Magn. Reson.* **97**, 359–375.

Kempf, J. G., Jung, J.-Y., Sampson, N. S., and Loria, J. P. (2003). Off-resonance TROSY $(R_{1\rho}-R_1)$ for quantitation of fast exchange processes in large proteins. *J. Am. Chem. Soc.* **125**, 12064–12065.

Kern, D., and Zuiderweg, E. R. P. (2003). The role of dynamics in allosteric regulation. *Curr. Opin. Struct. Biol.* **13**, 748–757.

Kloiber, K., and Konrat, R. (2000). Differential multiple-quantum relaxation arising from cross-correlated time-modulation of isotropic chemical shifts. *J. Biomol. NMR* **18**, 33–42.

Konrat, R., and Sterk, H. (1993). Cross-correlation effects in the transverse relaxation of multiple-quantum transitions of heteronuclear spin systems. *Chem. Phys. Lett.* **203**, 75–80.

Korzhnev, D. M., Kloiber, K., Kanelis, V., Tugarinov, V., and Kay, L. E. (2004a). Probing slow dynamics in high molecular weight proteins by methyl-TROSY NMR spectroscopy: Application to a 723-residue enzyme. *J. Am. Chem. Soc.* **126**, 3964–3973.

Korzhnev, D. M., Kloiber, K., and Kay, L. E. (2004b). Multiple-quantum relaxation dispersion NMR spectroscopy probing millisecond time-scale dynamics in proteins: Theory and application. *J. Am. Chem. Soc.* **126**, 7320–7329.

Korzhnev, D. M., Salvatella, X., Vendruscolo, M., di Nardo, A. A., Davidson, A. R., Dobson, C. M., and Kay, L. E. (2004c). Low-populated folding intermediates of Fyn SH3 characterized by relaxation dispersion NMR. *Nature* **430**, 586–590.

Kraut, D. A., Carroll, K. S., and Herschlag, D. (2003). Challenges in enzyme mechanism and energetics. *Annu. Rev. Biochem.* **72**, 517–571.

Kumar, P., and Kumar, A. (1996). Effect of remote cross correlations on transverse relaxation. *J. Magn. Reson. Ser. A* **119**, 29–37.

Loria, J. P., Rance, M., and Palmer, A. G. (1999a). A TROSY CPMG sequence for characterizing chemical exchange in large proteins. *J. Biomol. NMR* **15**, 151–155.

Loria, J. P., Rance, M., and Palmer, A. G. (1999b). A relaxation-compensated Carr-Purcell-Meiboom-Gill sequence for characterizing chemical exchange by NMR spectroscopy. *J. Am. Chem. Soc.* **121**, 2331–2332.

Lundstroem, P., and Akke, M. (2004). Quantitative analysis of conformational exchange contributions to ^{1}H–^{15}N multiple-quantum relaxation using field-dependent measurements. Time scale and structural characterization of exchange in a calmodulin C-terminal domain mutant. *J. Am. Chem. Soc.* **126**, 928–935.

Luz, Z., and Meiboom, S. (1963). Nuclear magnetic resonance (N.M.R.) study of the protolysis of trimethylammonium ion in aqueous solution. Order of the reaction with respect to solvent. *J. Chem. Phys.* **39**, 366–370.

Mackor, E. L., and MacLean, C. (1966). Sign of J_{HF} in $CHFCl_2$. *J. Chem. Phys.* **44**, 64–69.

Mackor, E. L., and MacLean, C. (1967). Relaxation processes in systems of two non-identical spins. *Prog. Nucl. Magn. Reson. Spectrosc.* **3**, 129–157.

Mandel, A. M., Akke, M., and Palmer, A. G. (1996). Dynamics of ribonuclease H: Temperature dependence of motions on multiple time scales. *Biochemistry* **35**, 16009–16023.

Marion, D., Ikura, M., Tschudin, R., and Bax, A. (1989). Rapid recording of 2D NMR spectra without phase cycling. Application to the study of hydrogen exchange in proteins. *J. Magn. Reson.* **85**, 393–399.

Massi, F., Johnson, E., Wang, C., Rance, M., and Palmer, A. G. (2004). NMR $R_{1\rho}$ rotating-frame relaxation with weak radio frequency fields. *J. Am. Chem. Soc.* **126**, 2247–2256.

McCammon, J. A., and Harvey, S. C. (1987). "Dynamics of Proteins and Nucleic Acids." Cambridge University Press, Cambridge, UK.

Meiboom, S., and Gill, D. (1958). Modified spin-echo method for measuring nuclear relaxation times. *Rev. Sci. Instrum.* **29**, 688–691.

Millet, O., Loria, J. P., Kroenke, C. D., Pons, M., and Palmer, A. G. (2000). The static magnetic field dependence of chemical exchange linebroadening defines the NMR chemical shift time scale. *J. Am. Chem. Soc.* **122**, 2867–2877.

Mulder, F. A. A., Mittermaier, A., Hon, B., Dahlquist, F. W., and Kay, L. E. (2001a). Studying excited states of proteins by NMR spectroscopy. *Nat. Struct. Biol.* **8**, 932–935.

Mulder, F. A. A., Skrynnikov, N. R., Hon, B., Dahlquist, F. W., and Kay, L. E. (2001b). Measurement of slow (μs-ms) time scale dynamics in protein side chains by ^{15}N relaxation dispersion NMR spectroscopy: Application to Asn and Gln residues in a cavity mutant of T4 lysozyme. *J. Am. Chem. Soc.* **123**, 967–975.

Mulder, F. A. A., Hon, B., Mittermaier, A., Dahlquist, F. W., and Kay, L. E. (2002). Slow internal dynamics in proteins: Application of NMR relaxation dispersion spectroscopy to methyl groups in a cavity mutant of T4 lysozyme. *J. Am. Chem. Soc.* **124**, 1443–1451.

Norwood, T. J., Tillett, M. L., and Lian, L.-Y. (1999). Influence of cross-correlation between the chemical shift anisotropies of pairs of nuclei on multiple-quantum relaxation rates in macromolecules. *Chem. Phys. Lett.* **300**, 429–434.

Orekhov, V. Y., Korzhnev, D. M., and Kay, L. E. (2004). Double- and zero-quantum NMR relaxation dispersion experiments sampling millisecond time scale dynamics in proteins. *J. Am. Chem. Soc.* **126**, 1886–1891.

Palmer, A. G. (2004). NMR characterization of the dynamics of biomacromolecules. *Acc. Chem. Res.* **104**, 3623–3640.

Palmer, A. G., Cavanagh, J., Wright, P. E., and Rance, M. (1991). Sensitivity improvement in proton-detected two-dimensional heteronuclear correlation NMR spectroscopy. *J. Magn. Reson.* **93**, 151–170.

Palmer, A. G., Skelton, N. J., Chazin, W. J., Wright, P. E., and Rance, M. (1992). Suppression of the effects of cross correlation between dipolar and anisotropic chemical-shift relaxation mechanisms in the measurement of spin-spin relaxation rates. *Mol. Phys.* **75**, 699–711.

Palmer, A. G., Williams, J., and McDermott, A. (1996). Nuclear magnetic resonance studies of biopolymer dynamics. *J. Phys. Chem.* **100**, 13293–13310.

Palmer, A. G., Kroenke, C. D., and Loria, J. P. (2001). Nuclear magnetic resonance methods for quantifying microsecond-to-millisecond motions in biological macromolecules. *Methods Enzymol.* **339**, 204–238.

Pervushin, K. (2001). The use of TROSY for detection and suppression of conformational exchange NMR line broadening in biological macromolecules. *J. Biomol. NMR* **20**, 275–285.

Pervushin, K., Riek, R., Wider, G., and Wuthrich, K. (1997). Attenuated T$_2$ relaxation by mutual cancellation of dipole-dipole coupling and chemical shift anisotropy indicates an avenue to NMR structures of very large biological macromolecules in solution. *Proc. Natl. Acad. Sci. USA* **94**, 12366–12371.

Pervushin, K., Riek, R., Wider, G., and Wuthrich, K. (1998). Transverse relaxation-optimized spectroscopy (TROSY) for NMR studies of aromatic spin systems in ^{13}C-labeled proteins. *J. Am. Chem. Soc.* **120**, 6394–6400.

Pervushin, K. V., Wider, G., Riek, R., and Wuthrich, K. (1999). The 3D NOESY-[1H,15N,1H]-ZQ-TROSY NMR experiment with diagonal peak suppression. *Proc. Natl. Acad. Sci. USA* **96**, 9607–9612.

Piotto, M., Saudek, V., and Sklenár, V. (1992). Gradient-tailored excitation for single-quantum NMR spectroscopy of aqueous solutions. *J. Biomol. NMR* **2**, 661–665.

Rance, M. (1988). Intramolecular exchange effects in multiple quantum spectroscopy. *J. Am. Chem. Soc.* **110**, 1973–1974.

Rance, M., and Byrd, R. A. (1983). Obtaining high fidelity spin 1/2 powder spectra in anisotropic media: Phase cycled Hahn-echo spectroscopy. *J. Magn. Reson.* **54**, 221–240.

Shaka, A. J., Keeler, J., Frenkiel, T., and Freeman, R. (1983). An improved sequence for broadband decoupling: WALTZ-16. *J. Magn. Reson.* **52**, 334–338.

Shaka, A. J., Barker, P. B., and Freeman, R. (1985). Computer-optimized decoupling scheme for wideband applications and low-level operation. *J. Magn. Reson.* **64**, 547–552.

Skrynnikov, N. R., Mulder, F. A. A., Hon, B., Dahlquist, F. W., and Kay, L. E. (2001). Probing slow time scale dynamics at methyl-containing side chains in proteins by relaxation dispersion NMR measurements: Application to methionine residues in a cavity mutant of T4 lysozyme. *J. Am. Chem. Soc.* **123,** 4556–4566.

Skrynnikov, N. R., Dahlquist, F. W., and Kay, L. E. (2002). Reconstructing NMR spectra of "invisible" excited protein states using HSQC and HMQC experiments. *J. Am. Chem. Soc.* **124,** 12352–12360.

Sorensen, M. D., Meissner, A., and Sorensen, O. W. (1997). Spin-state-selective coherence transfer via intermediate states of two-spin coherence in IS spin systems: Application to E.COSY-type measurement of J coupling constants. *J. Biomol. NMR* **10,** 181–186.

Swift, T. J., and Connick, R. E. (1962). NMR (nuclear magnetic resonance)-relaxation mechanisms of O^{17} in aqueous solutions of paramagnetic cations and the lifetime of water molecules in the first coordination sphere. *J. Chem. Phys.* **37,** 307–320.

Szyperski, T., Luginbuehl, P., Otting, G., Guentert, P., and Wuethrich, K. (1993). Protein dynamics studied by rotating frame nitrogen-15 spin relaxation times. *J. Biomol. NMR* **3,** 151–164.

Tessari, M., and Vuister, G. W. (2000). A novel experiment for the quantitative measurement of $CSA(^1H^N)/CSA(^{15}N)$ cross-correlated relaxation in ^{15}N-labeled proteins. *J. Biomol. NMR* **16,** 171–174.

Tolkatchev, D., Xu, P., and Ni, F. (2003). Probing the kinetic landscape of transient peptide-protein interactions by use of peptide ^{15}N NMR relaxation dispersion spectroscopy: Binding of an antithrombin peptide to human prothrombin. *J. Am. Chem. Soc.* **125,** 12432–12442.

Tollinger, M., Skrynnikov, N. R., Mulder, F. A. A., Forman-Kay, J. D., and Kay, L. E. (2001). Slow dynamics in folded and unfolded states of an SH3 domain. *J. Am. Chem. Soc.* **123,** 11341–11352.

Trott, O., and Palmer, A. G. (2004). Theoretical study of $R_{1\rho}$ rotating-frame and R_2 free-precession relaxation in the presence of n-site chemical exchange. *J. Magn. Reson.* **170,** 104–112.

Tugarinov, V., Hwang, P. M., Ollerenshaw, J. E., and Kay, L. E. (2003). Cross-correlated relaxation enhanced 1H-13C NMR spectroscopy of methyl groups in very high molecular weight proteins and protein complexes. *J. Am. Chem. Soc.* **125,** 10420–10428.

Tugarinov, V., Sprangers, R., and Kay, L. E. (2004). Line narrowing in methyl-TROSY using zero-quantum ^1H–^{13}C NMR spectroscopy. *J. Am. Chem. Soc.* **126,** 4921–4925.

Vold, R. L., and Vold, R. R. (1978). Nuclear magnetic relaxation in coupled spin systems. *Prog. Nucl. Magn. Reson. Spectrosc.* **12,** 79–133.

Volkman, B. F., Lipson, D., Wemmer, D. E., and Kern, D. (2001). Two-state allosteric behavior in a single-domain signaling protein. *Science* **291,** 2429–2433.

Vugmeyster, L., Kroenke, C. D., Picart, F., Palmer, A. G., and Raleigh, D. P. (2000). ^{15}N $R_{1\rho}$ measurements allow the determination of ultrafast protein folding rates. *J. Am. Chem. Soc.* **122,** 5387–5388.

Wand, A. J. (2001). Dynamic activation of protein function: A view emerging from NMR spectroscopy. *Nat. Struct. Biol.* **8,** 926–931.

Wang, C., and Palmer, A. G. (2002). Differential multiple quantum relaxation caused by chemical exchange outside the fast exchange limit. *J. Biomol. NMR* **24,** 263–268.

Wang, C., and Palmer, A. G. (2003). Solution NMR methods for quantitative identification of chemical exchange in 15N-labeled proteins. *Magn. Reson. Chem.* **41,** 866–876.

Wang, C., Grey, M. J., and Palmer, A. G. (2001). CPMG sequences with enhanced sensitivity to chemical exchange. *J. Biomol. NMR* **21,** 361–366.

Wang, C., Rance, M., and Palmer, A. G. (2003). Mapping chemical exchange in proteins with MW > 50 kD. *J. Am. Chem. Soc.* **125**, 8968–8969.

Wang, C., Karpowich, N., Hunt, J. F., Rance, M., and Palmer, A. G. (2004). Dynamics of ATP-binding cassette contribute to allosteric control, nucleotide binding and energy transduction in ABC transporters. *J. Mol. Biol.* **342**, 525–537.

Werbelow, L. G., and Grant, D. M. (1977). Intramolecular dipolar relaxation in multispin systems. *Adv. Magn. Reson.* **9**, 189–299.

Woessner, D. E. (1961). Nuclear transfer effects in nuclear magnetic resonance (NMR) pulse experiments. *J. Chem. Phys.* **35**, 41–48.

Yuan, Y., Simplaceanu, V., Lukin, J. A., and Ho, C. (2002). NMR investigation of the dynamics of tryptophan side-chains in hemoglobins. *J. Biomol. NMR* **321**, 863–878.

Zhu, G., Xia, Y., Nicholson, L. K., and Sze, K. H. (2000). Protein dynamics measurements by TROSY-based NMR experiments. *J. Magn. Reson.* **143**, 423–426.

[19] Isotropic Reorientational Eigenmode Dynamics Complements NMR Relaxation Measurements for RNA

By SCOTT A. SHOWALTER and KATHLEEN B. HALL

Abstract

^{13}C NMR relaxation measurements alone are often not sufficient to describe the motions of small RNA molecules in solution. In the case where the global tumbling time of the RNA is on the same time scale as its internal motions, standard Lipari–Szabo analysis becomes inadequate, and other methods must be used to describe the dynamics. Here, molecular dynamics simulations of the iron-responsive element (IRE) RNA hairpin are analyzed using isotropic reorientational eigenmode dynamics (iRED) to provide a picture of the motions of the RNA. The results show that indeed there is no separability of global and internal motions, and thus the order parameters determined from experimental data cannot be quantitatively accurate. iRED analysis also identifies correlated motions, providing a new picture of the dynamics of the IRE loop.

Introduction

RNA sequences that bind proteins, small molecules, or other RNAs are often displayed in the loop of a hairpin, as a bulge within a duplex region, or as single strands in the context of a larger structured RNA. It is axiomatic that these sequences will experience a structural change upon interaction with their target molecule; their inherent structural flexibility allows them to adopt their bound conformation. Therefore, any molecular

Copyright 2005, Elsevier Inc.
All rights reserved.
0076-6879/05 $35.00

description of these RNA molecules must include their dynamics, both the time scale and amplitude of their motions, and a physical description of the conformational space they sample.

The study of RNA by nuclear magnetic resonance (NMR) is attractive because it offers both structural and dynamic information. These studies are facilitated by the isolation of small structural elements from larger molecules in the expectation that these small elements have autonomous structures. Examples include the famous UUCG tetraloop (Allain and Varani, 1995), the HIV TAR element (Brodsky and Williamson, 1997), the iron-responsive element (IRE) (Addess *et al.*, 1997; Laing and Hall, 1996), and the U6 snRNA intramolecular stem loop (ISL) (Huppler *et al.*, 2002). While the UUCG tetraloop is a highly constrained structure, with little opportunity for dynamic motion, the other RNAs are capable of conformational changes and so have intrinsic flexibility; IRE (Hall and Tang, 1998) and TAR (Dayie *et al.*, 2002; King *et al.*, 1995) dynamics have been studied by ^{13}C NMR relaxation methods.

NMR spin relaxation is the most powerful method available to experimentally measure dynamics of RNA molecules, but there are several unique problems associated with these measurements as applied to RNA. Motions of the sugar-phosphate backbone are difficult to measure because experimentally tractable probes are sparse where they exist. The phosphate moiety is not a good source of dynamic information because the ^{31}P resonances are often overlapped and chemical shift anisotropy (CSA) dominates the relaxation process. In the riboses, the C1' position is relatively simple to specifically label with ^{13}C to introduce an isolated spin pair, and the contribution of CSA to relaxation at this site is small ($\Delta\sigma = 25$ ppm), leaving relaxation at this site dominated by the magnetic dipole–dipole interaction between the C1' and H1'.

Even when isotopic enrichment to generate isolated spin pairs is relatively easy, complex motion can make the interpretation of dynamics at a given site difficult at best. For this reason, the backbone amide ^{15}N–^{1}H spin pair of uniformly ^{15}N-enriched protein has been a reliable source of dynamic information; the protein backbone has only two torsional degrees of freedom and a description of the dynamics sampled by the relatively rigid peptide plane results in an adequate description of backbone dynamics for most purposes. In contrast, each nucleoside in RNA has seven degrees of freedom that describe its conformational space, making detailed description of the motion experienced by the sugar phosphate backbone challenging to develop from limited relaxation probes.

Analogous to measuring protein side chain dynamics, RNA base mobility could be measured by ^{13}C or ^{15}N relaxation. However, the use of imino or amino ^{15}N nuclei as probes suffers from the rapid exchange of their attached protons with water: the inclusion of a term to describe the rate of that chemical exchange adds considerable uncertainty to the dynamics data. Despite this complication, ^{15}N imino relaxation measurements have been made of a UUCG tetraloop RNA hairpin (Akke et al., 1997) and the HIV-2 TAR RNA (Dayie et al., 2002). The ^{13}C–^{1}H spin pairs on the bases are potentially less complicated to study, since the protons are nonexchangeable. The difficulty here is that only in purine bases are there isolated ^{13}C–^{1}H spin pairs; those on the pyrimidines will suffer from dipolar cross-relaxation between the C5 and C6 carbons (although constant time experiments can reduce this impediment by effectively decoupling the carbons). The magnitude of the CSA of the purine ^{13}C nuclei also becomes an issue for relaxation at higher fields.

Although ^{13}C relaxation measurements can access all structural elements of an RNA molecule, the probes are nevertheless sparse. Unlike the situation in globular proteins where there might be 100+ amide nitrogens that provide a basis against which outliers can be identified, NMR relaxation experiments that use only purine carbons in a small RNA molecule could yield as many examples of "outliers" as there are of "core." Measurement of both ribose and base relaxation can help increase the number of probes, but the motions of the sugar-phosphate backbone and the bases are not guaranteed to be correlated, making a unified interpretation of the data collected difficult without supporting information from other sources.

The most common formalism used for the analysis of NMR relaxation data is that of Lipari and Szabo (1982a), which provides a model-free parameterization of the Bloch–Wangsness–Redfield spectral density function [$J(\omega)$] (Bloch, 1956; Redfield, 1957; Wangsness and Bloch, 1953). This function is the cosine transform of the correlation function [$C(t)$] describing the temporal decay of the lattice functions of the spin interactions contributing to relaxation. In the Lipari–Szabo description, the correlation functions and spectral densities that describe R_1, R_2, and heteronuclear NOE relaxation are written in terms of the overall tumbling time of the molecule (τ_m); order parameters (S^2), which are proportional to the amplitude of local motion at each site; the time scale of the motions associated with the order parameters (τ_e); and in some cases an exchange contribution to R_2, describing motion on a time scale much slower than global tumbling (typically microsecond to millisecond). This model has been extended by Clore et al. (1990) to account for the possibility of local motion on multiple distinct time scales. Further extension of the model to account for anisotropic overall tumbling is also possible.

One assumption that is made in the construction of the Lipari–Szabo formalism is that the correlation function $C(t)$ governing relaxation can be separated into the overall (global) molecular correlation function [$C_o(t)$]

and a correlation function for internal motion $[C_I(t)]$, such that $C(t) = C_o(t)C_I(t)$. This condition can be met when the time scales of the overall and internal processes differ by at least an order of magnitude. Small RNAs frequently have global tumbling correlation times (τ_m) from 2 to 4 ns and can have internal (local) dynamic motions on time scales from picoseconds to milliseconds, with ns motions prominent in some systems. When the time scale of local motions matches that of global tumbling, the Lipari–Szabo formalism is not strictly valid and interpretation of order parameters may lead to significant errors if the separability assumption cannot be independently justified.

Spectral density mapping (Peng and Wagner, 1992) is another method of describing NMR relaxation data, and it has been applied to the HIV TAR RNA hairpin (Dayie et al., 2002). Use of the reduced form of the spectral density for ^{13}C relaxation is not appropriate, since the ^1H and ^{13}C frequencies are too close for the approximation $J(\omega_H \pm \omega_C) \approx J(\omega_H)$, leaving the investigator forced to use the more experimentally problematic full form of the spectral density. Even so, this method is attractive because its basic conclusions do not require the assumption that overall and internal motions are separable. While low values of $J(0)$ and high values of $J(\omega_H)$ derived from spectral density mapping can suggest regions of flexibility, this model is not able to provide a more detailed physical description of the motion sampled. Thus, even when spectral density mapping is employed, the investigator must often resort to use of the Lipari–Szabo formalism for more detailed analysis.

When a small ^{13}C-labeled RNA molecule is studied and exhibits local motions on a wide range of time scales that may render the separability assumption of the Lipari–Szabo formalism invalid, other methods must be used to supplement NMR relaxation measurements to describe the dynamics of the system. Molecular dynamics (MD) trajectories offer a powerful complement to experimental NMR relaxation measurements because they can yield the same site-specific dynamic parameters and can be used to test the assumptions made in the interpretation of experimental data. In addition, they provide a physical model of motion.

We describe here a combined NMR/MD approach featuring application of the isotropic reorientational eigenmode dynamics (iRED) formalism developed by Brüschweiler (Prompers and Brüschweiler, 2001, 2002). This approach allows the investigator to test the validity of the separability assumption, while simultaneously acquiring detailed physical information about the motions experienced by each RNA site and determining the extent of any dynamic correlation between these sites. The iRED method has been described in detail elsewhere and its applicability to the study of both folded and unfolded proteins demonstrated. The protocol proposed here demonstrates the applicability of this formalism to the investigation of RNA systems; data describing the human IRE RNA hairpin are presented as an example. NMR relaxation was measured using ^{13}C-labeled IRE RNAs as described previously. MD simulations were run in AMBER 7.0 using the AMBER94 force field. The results indicate that overall and internal motion are strongly coupled in this system and, when combined with experimental measurements, provide a better description of the system than NMR relaxation data alone.

Lipari–Szabo Analysis of NMR Relaxation

The IRE RNA hairpin is a phylogenetically conserved sequence that comprises part of the binding site for the IRE binding protein. The NMR structure of the six nucleotide loop showed that the first (C6) and fifth (G10) bases formed hydrogen bonding interactions, but the other bases were either stacked (A7) or unconstrained (G8, U9, and C11) (Fig. 1A; numbering system for our hairpin construct). The ribose of C6 is *C3'-endo*, while that of A7 also samples *C2'-endo* [the IRE structure of Addess *et al.* (1997) shows it to be *C3'-endo*]. Addess *et al.* (1997) measured the $^3J_{H1',H2'}$ of the loop riboses and concluded that G8 and U9 were *C2'-endo*, while G10 and C11 riboses interconvert between the two puckers. The repuckering of the sugars and the absence of physical restraints on four of six bases in the loop led to the description of this structure as "floppy"; clearly its constituents underwent significant dynamic motion.

^{13}C NMR relaxation measurements of the IRE RNA required five samples, each containing a single selectively labeled RNA: ^{13}C1'-adenosine, ^{13}C1'-guanosine, ^{13}C1'-cytosine, u-^{13}C-adenosine, and u-^{13}C-guanosine RNAs. The chemical shift overlap of the ^{13}C1' ribose carbons and protons and the purine C8 carbons precluded the use of a single sample. NMR T_1, $T_{1\rho}$, and NOE experiments were adapted from those of Yamazaki *et al.* (1994) for use with RNA. Data were collected at 20°, where the structure was solved, and at 37°, the physiological temperature that is approximately 30° below the melting temperature in this buffer (30 mM NaCl, 10 mM sodium phosphate, pH 6). We anticipated that the loop structure would be very flexible at the higher temperature, since the C6:G10 hydrogen bonded imino proton resonances became very weak.

Relaxation data were acquired at 500 MHz and fit using the Lipari–Szabo formalism (Hall and Tang, 1998). One result of the use of separate samples for the $T_{1\rho}$ experiments was that the carrier could be placed close enough to the ^{13}C resonance frequencies of the RNA that there was effectively no offset; the data could be treated as T_2 parameters for analysis. The resulting order parameters for the purines and the riboses are

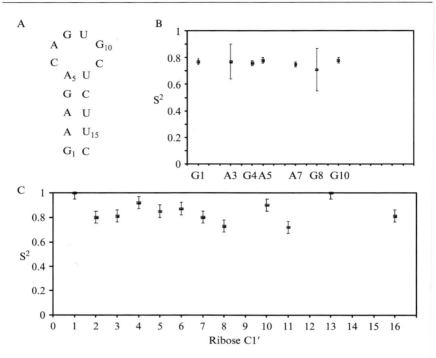

FIG. 1. Experimental characterization of IRE RNA dynamics. (A) Sequence and secondary structure of the IRE RNA hairpin. (B) Purine S^2 derived from Lipari–Szabo analysis of relaxation data collected at 20°. (C) Ribose C1′ S^2 derived from Lipari–Szabo analysis of relaxation data collected at 20°. Error bars in (B) and (C) represent uncertainty estimated through the fitting procedure.

shown in Fig. 1B for 20°; it is apparent that S^2 for many of the riboses are too high. Given the good quality of the data, sources of error that would lead to this result include locally complex dynamics spanning multiple time scales or a breakdown of the assumption that overall correlation times were at least an order of magnitude larger than internal correlation times.

The IRE hairpin used in the NMR experiments is small enough that it tumbles isotropically, with a rotational correlation time of 3.4 ns at 20° and 2.4 ns at 37°. If the unconstrained bases in the IRE loop could rotate about the glycosidic bond, or if the backbone phosphodiester backbone was in motion, then it was plausible to consider that the bases could be moving on time scales similar to the overall tumbling time of the IRE. The Lipari–Szabo analysis would not rigorously apply under these conditions, and the values of order parameters would be inaccurate.

Testing Separability Using Isotropic Reorientational
  Eigenmode Dynamics

Overview

For many small RNAs, the assumption that the internal and global motions experienced by sites of interest are separable is likely to be invalid. Unfortunately, experimentally confirming the applicability of the separability assumption is challenging at best for most of these systems. Biochemical assays such as fluorescence anisotropy measurements can suggest the existence of internal motions on the same time scale as global tumbling, but this does not necessarily invalidate the separation assumption. Even if internal motion occurs on the same time scale as global tumbling, it is possible that the motions will remain separable so long as the global diffusion tensor is not perturbed by the motion.

One of the most robust methods for probing the validity of the separability assumption is application of the iRED formalism to an MD trajectory (Prompers and Brüschweiler, 2002). For the purpose of testing the separability assumption, it is important to compute trajectories with durations several times as long as the global tumbling time of the system. For most small RNAs, this means computing a trajectory on the order of 10–50 ns in length. It is preferable to save snapshots as frequently as storage space will allow, achieving better averaging and good statistics over as wide a range of time scales as possible. In isotropic NMR solutions, all orientations of the RNA will be sampled equally, a situation not likely achieved in an MD trajectory of finite length. The anisotropic reorientation distribution sampled during the simulation can be rendered isotropic mathematically during postprocessing and should not prevent reliable testing of the separability assumption so long as the simulation is sufficiently long.

The nuclear spin relaxation of spins-1/2 measured by NMR spectroscopy is brought about by stochastic temporal fluctuations in the lattice portions of the magnetic dipole–dipole and CSA Hamiltonians, which can be expressed in terms of the rank 2 spherical harmonics $Y_{2M}(\theta, \phi)$ (Abragam, 1961). For isolated $^{13}C-^{1}H$ spin pairs in RNA, the dipole–dipole interaction can be represented to a good approximation by a tensor with a principal axis parallel to the bond vector. It is convenient to assume that the principal axis of the CSA tensor is colinear with that of the dipole–dipole tensor, although the validity of this assumption is not as well established for the spin pairs studied in RNA as it is for the study of protein backbone amides.

Over the course of an MD trajectory, the principal axis directions $\Omega_j(t) = [\theta_j(t), \phi_j(t), j = 1, \ldots, n]$ of the interaction tensors for n sites are stored and used to compute a covariance matrix of the $Y_{2M}(\theta,\phi)$. After isotropic averaging, an $n \times n$ real symmetric matrix is produced with elements of the form (Prompers and Brüschweiler, 2002)

$$M_{ij} = \overline{P_2\big(\cos(\Omega_i - \Omega_j)\big)} \tag{1}$$

where $P_2(x) = (3x^2 - 1)/2$ is the second Legendre polynomial, $\Omega_i - \Omega_j$ is the angle between spin interactions i and j in a given snapshot, and the bar indicates averaging over all snapshots of the trajectory. These matrix elements are not equivalent to the generalized order parameter S^2, because global reorientation is not subtracted and the angle $\Omega_i - \Omega_j$ is calculated between two sites in a single snapshot, rather than a single site in two different snapshots.

The covariance matrix can be diagonalized by solving the eigenvalue problem $\mathbf{M}|m> = \lambda_m|m> (m = 1,\ldots,n)$. The resulting normalized reorientational eigenvectors $|m>$ contain information about which spin interactions reorient in concert under the influence of each motional mode, and the eigenvalues λ_m represent the amplitude of the observed motion. It is the properties of the set of eigenvectors $|m>$ and their associated eigenvalues that will provide the most useful information in testing the separability assumption.

Separable Motion Will Manifest as a Gap in Collectivity vs. Amplitude Plots

The collectivity of an eigenvector is defined by the parameter κ, which is roughly proportional to the percent of sites significantly reoriented by the motion that vector represents (Brüschweiler, 1995):

$$\kappa_m = \frac{1}{N}\exp\left\{ -\sum_{n=1}^{N} ||m>_n|^2\log||m>_n|^2\right\} \tag{2}$$

where $|m>_n$ is the nth component of eigenvector $|m>$ and N is the number of spin interactions. For an internally rigid molecule, the matrix $\mathbf{M}$ will have at most five nonzero eigenvalues, and the collectivities of their associated eigenvectors should be very high ($\kappa > 0.50$). Additional nonzero eigenvalues reflect the internal motion of the system, which may or may not be separable from global reorientation. For a typical macromolecular system with complex internal motions, eigenvalues equal to zero are seldom observed, although many of the lowest eigenvalues are less than 0.05. Internal reorientations tend to be much lower in amplitude than global reorientations, and therefore the internal modes tend to have much smaller eigenvalues. Separability is characterized by a large gap in both eigenvalue and collectivity between the five largest amplitude (global) modes and the $n - 5$ internal modes.

The most straightforward visualization of the validity of the separability assumption is obtained by plotting κ vs. λ_k for each of the eigenvectors of a trajectory. If separability is well satisfied, a distribution of points similar to the characteristic pattern shown in Fig. 2A will be observed. Violation of the separability assumption will result in the absence of both the λ and κ gaps and an overall more linear relationship between the two parameters. This is demonstrated for the IRE in Fig. 2B.

Some caution must be exercised when interpreting these plots in isolation from the rest of the data provided by iRED analysis. When the trajectory length is significantly less than the global tumbling time of

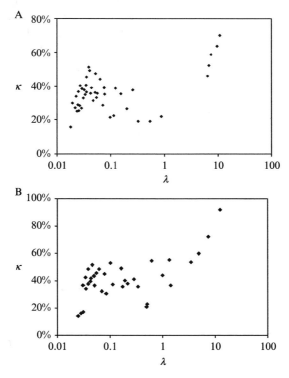

FIG. 2. The distribution of the collectivity parameter κ vs. eigenvalue λ from iRED analysis. (A) Example distribution for a molecule with separable internal and global motion. Plot generated from a simulation of a U1 snRNA stemloop 2 construct bound to human U1A-RBD1. (B) Distribution for the IRE, which suggests that global and internal motion are not separable.

the RNA, the system may not have evolved long enough for all relevant motions to be well sampled. A κ vs. λ_m distribution with no gap between the global and internal modes may be observed in this case, even in systems that display a robust gap when the simulation is extended past the global tumbling time. If a gap is observed, however, the validity of the separability assumption is strongly supported; extending simulations does not tend to remove gaps once they develop. The conclusion that separability is violated is strongest when the shape of the κ vs. λ_m distribution and the properties of the correlation functions describing motion along each eigenvector are interpreted in parallel.

Additional Information from Correlation Functions and Correlation Times

Although the eigenvectors of $\mathbf{M}$ contain detailed information about the dynamic correlations between sites, they contain no information about the time scale on which the observed motion occurs. This information can be extracted by constructing correlation functions describing the decay of motion along each eigenvector over the course of the trajectory:

$$C_M(t) = \sum_{l=-2}^{2} < a_{m,l}^*(\tau + t) \cdot a_{m,l}(\tau) >_\tau \tag{3}$$

where averaging is done over the snapshots of the simulation, and the $a_{m,l}(t)$ are constructed from the instantaneous projection of the snapshots onto the eigenvectors (Prompers and Brüschweiler, 2002). Assuming that these correlation functions decay monoexponentially, a lifetime τ_M associated with each motional mode can be established (Lipari and Szabo, 1982b). In theory, the five eigenvectors with the largest eigenvalues should have τ_M approximately equal to the global tumbling time from Lipari–Szabo analysis. In practice, the longest lifetimes sampled in MD trajectories fall well short of experimental tumbling times.

Even though the decay times of the five largest amplitude modes are likely to be shorter than the experimental global tumbling time, these eigenmodes should have lifetimes at least as long as those of the internal modes if separability is satisfied. In the case of the IRE, 11 eigenmodes, 30% of the total, have lifetimes within an order of magnitude of the experimentally determined global tumbling time (Fig. 3). This situation could arise if (1) much of the internal motion of the system truly does decay on the same time scale as the global motion or (2) the longest lifetime correlation functions were unconverged due to insufficient sampling by a trajectory that was too short, yielding correlation times that are best thought of as lower limits.

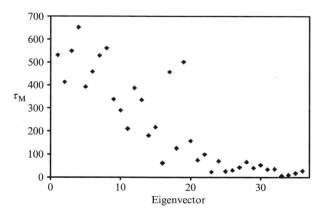

FIG. 3. Characteristic lifetime in picoseconds (τ_M) associated with each eigenvector $|m>$ of the matix **M** generated from iRED analysis of the IRE simulation. τ_M is the lifetime associated with the reorientational mode represented by $|m>$ and therefore cannot be associated with a single residue [unless $\kappa = 1/N$, see Eq. (2)].

Analysis of the shape of the $C_m(t)$ can assist in determining the convergence of the trajectory. Well-behaved correlation functions will asymptotically decay stably to zero, as seen in Fig. 4A, and a predominance of such functions was observed for the IRE, suggesting a well-converged simulation. While a few correlation functions are expected to display a less ideal lineshape, a large number of poorly converged correlation functions indicates a problem with the trajectory or its length. Especially for the eigenvectors with the largest eigenvalues, it is possible that the associated correlation functions will never reach an asymptote (Fig. 4B). Correlation times extracted from these functions represent lower bounds on the true correlation time, at best, and should be interpreted cautiously. A far worse situation is shown in Fig. 4C, where the correlation function has decayed to an apparently stable value significantly greater than zero. Modes that display this residual anisotropy can arise regardless of whether the separability assumption is valid and, so long as they are scarce, should not prevent testing this assumption. If more than 10% of the correlation functions show this behavior, however, the simulation should be extended to improve sampling.

For the IRE RNA, all but a few of the computed correlation functions decayed to an asymptote of zero. Notably, 11 eigenmodes displayed correlation times within an order of magnitude of the experimental global tumbling time. Taken together with the collectivity results in Fig. 2, analysis of the correlation functions and correlation times indicates that

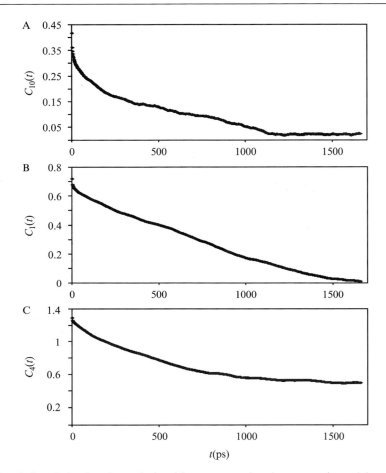

FIG. 4. Correlation functions calculated for representative eigenvectors $|m>$ of the matrix $\mathbf{M}$ generated from iRED analysis of the IRE simulation. (A) $C_{10}(t)$ decays to a stable asymptote within roundoff error of zero (e_{10} is shown in Fig. 5). (B) $C_1(t)$ fails to reach an asymptote suggesting a long lifetime ($\tau_M > 1$ ns) for this mode. (C) $C_4(t)$ asymptotically decays to a value $\gg 0$ suggesting poor sampling of this mode.

separability is not satisfied in the IRE RNA and that discussing "global" vs. "internal" motion is invalid for this system. This result means that the Lipari–Szabo formalism provides a qualitative description of the properties of the IRE RNA and an alternate description of the experimental relaxation data is needed.

Prediction of Correlated Motion

Even for molecules where the assumption of separable motion is reasonable, iRED analysis can offer a useful complement to NMR relaxation experiments, which cannot provide information about correlations between motions as they are currently performed and analyzed. For RNAs, many of which are known to undergo collective conformational changes upon interacting with ligands, understanding the degree of correlated motion and how these motions spatially map onto the average structure may be critical for a full understanding of function. The eigenvectors of the matrix **M** provide this information by directly reporting the correlation networks that couple the various sites, when the correlation function associated with the mode is well behaved. For example, eigenvector 10 from the IRE simulation (Fig. 5) displays correlations between the bases of A7, U9, and C11, three mobile sites spanning the loop of the hairpin. The correlation function for eigenvector 10 is shown in Fig. 4A.

Lastly, while the Lipari–Szabo formalism is a very robust tool for interpreting NMR relaxation data, it is inherently unable to provide a physical description of the motion sampled by the sites (hence the descriptor "model free"). If the separability assumption has been validated, then either iRED or the similar reorientational eigenmode dynamics (RED) analysis could be used to predict order parameters for each spin pair (Prompers and Brüschweiler, 2001). When the simulation trajectory and

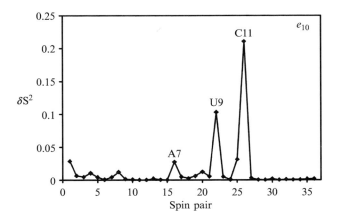

FIG. 5. e_{10} from iRED analysis of the IRE trajectory shown as the principal order parameter components $\delta S_j^2 = \lambda_m ||m>_j|^2$ for each spin pair. This mode represents correlated reorientation of the ribose of A7 and the pyrimidine rings of U9 and C11, three sites that span the flexible loop of the hairpin.

experimental data yield similar order parameters, the trajectory can be used to test the validity of several motional models, such as the 1D Gaussian axial fluctuation model (Brüschweiler and Wright, 1994).

Handling Nonseparable Motion

If iRED analysis of an MD trajectory suggests that the separability assumption cannot be applied to a system, as was the case for the IRE RNA, the experimental data will have to be analyzed using something other than the Lipari–Szabo formalism. The issue of how to do this has received attention from several groups in the recent literature, and two analytical approaches have emerged that seem to offer possible solutions.

It is possible to derive a model for the spectral density function without invoking the separability assumption, which has a functional form identical to the Lipari–Szabo spectral density function, and which contains parameters analogous to S^2 and τ_e (Vugmeyster et al., 2003). One advantage of this rederivation is that it is still a macroscopic description of the system; no motional models must be assumed in its application. The main difference between the new nondecoupled parameterization and the original formalism lies in the interpretation of the experimental S^2 and τ_e values. Vugmeyster et al. (2003) have shown that when separability is invalid, experimental order parameters tend to be systematically overestimated and the extent of overestimation can be predicted using their model.

Another useful formalism is based on the slowly relaxing local structure (SRLS) model (Polimeno and Freed, 1993, 1995) and has been applied to the study of macromolecular systems that violate separability (Tugarinov et al., 2001, 2002). This model employs a microscopic description of the tensors that couple local motion to global tumbling and can provide a very detailed description of the dynamics of these systems. Application of the SRLS formalism requires globally fitting relaxation data acquired at no less than two magnetic field strengths in order to obtain a unique solution. Although this requirement presents a serious experimental complication, the results can be worthwhile, especially when the results of iRED analysis are already available.

Acknowledgments

We would like to thank Dr. Rafael Brüschweiler for numerous insightful discussions and Dr. Nathan Baker for collaboration on the IRE RNA simulations. S.A.S. is the recipient of an NSF predoctoral fellowship. This work was supported by the NIH.

References

Abragam, A. (1961). "Principles of Nuclear Magnetism." Clarendon Press, Oxford, UK.

Addess, K. J., Basilion, J. P., Klausner, R. D., Rouault, T. A., and Pardi, A. (1997). Structure and dynamics of the iron-responsive element RNA: Implications for binding of the RNA by iron regulatory binding proteins. *J. Mol. Biol.* **274,** 72–83.

Akke, M., Fiala, R., Patel, D., and Palmer, A. G., III (1997). Base dynamics of a UUCG tetraloop RNA hairpin characterized by ^{15}N spin relaxation: Correlations with structure and stability. *RNA* **3,** 702–709.

Allain, F. H.-T., and Varani, G. (1995). Structure of the P1 helix from Group I self-splicing introns. *J. Mol. Biol.* **250,** 333–353.

Bloch, F. (1956). Dynamical theory of nuclear induction. II. *Phys. Rev.* **102,** 104–135.

Brodsky, A. S., and Williamson, J. R. (1997). Structure of HIV-2 TAR argininamide complex. *J. Mol. Biol.* **265,** 624–639.

Brüschweiler, R. (1995). Collective protein dynamics and nuclear spin relaxation. *J. Chem. Phys.* **102,** 3396–3403.

Brüschweiler, R., and Wright, P. E. (1994). NMR order parameters of biomolecules: A new analytical representation and application to the gaussian axial fluctuation model. *J. Am. Chem. Soc.* **116,** 8426–8427.

Clore, G. M., Szabo, A., Bax, A., Kay, L. E., Driscoll, P. C., and Gronenborn, A. M. (1990). Deviations from the simple two-parameter model-free approach to the interpretation of nitrogen-15 nuclear magnetic relaxation of proteins. *J. Am. Chem. Soc.* **112,** 4989–4991.

Dayie, K. T., Brodsky, A. S., and Williamson, J. R. (2002). Base flexibility in HIV-2 TAR RNA mapped by solution ^{15}N, ^{13}C NMR relaxation. *J. Mol. Biol.* **317,** 263–278.

Hall, K. B., and Tang, C. (1998). ^{13}C relaxation and dynamics of the purine bases in the iron responsive element RNA hairpin. *Biochemistry* **37,** 9323–9332.

Huppler, A., Nikstad, L. J., Allman, A. M., Brow, D. A., and Butcher, S. E. (2002). Metal binding and base ionization in the U6 RNA intramolecular stem-loop structure. *Nat. Struct. Biol.* **9,** 431–435.

King, G. C., Xi, Z., Michnicka, M. J., and Akratos, C. (1995). ^{13}C relaxation in an RNA hairpin. *In* "Stable Isotope Applications in Biomolecular Structure and Mechanisms," pp. 145–154. Los Alamos National Laboratory, Los Alamos, NM.

Laing, L. G., and Hall, K. B. (1996). A model of the iron responsive element RNA hairpin loop structure determined from NMR and thermodynamic data. *Biochemistry* **35,** 13586–13596.

Lipari, G., and Szabo, A. (1982a). Model-free approach to the interpretation of nuclear magnetic resonance relaxation in macromolecules. 1. Theory and range of validity. *J. Am. Chem. Soc.* **104,** 4546–4559.

Lipari, G., and Szabo, A. (1982b). Model-free approach to the interpretation of nuclear magnetic resonance relaxation in macromolecules. 2. Analysis of experimental results. *J. Am. Chem. Soc.* **104,** 4559–4570.

Peng, J. W., and Wagner, G. (1992). Mapping of spectral density functions using heteronuclear NMR relaxation measurements. *J. Magn. Reson.* **98,** 308–332.

Polimeno, A., and Freed, J. H. (1993). A many-body stochastic approach to rotational motions in liquids. *Adv. Chem. Phys.* **83,** 89–210.

Polimeno, A., and Freed, J. H. (1995). Slow motional ESR in complex fluids: The slowly relaxing local structure model of solvent cage effects. *J. Phys. Chem.* **99,** 10995–11006.

Prompers, J. J., and Brüschweiler, R. (2001). Reorientational eigenmode dynamics: A combined MD/NMR relaxation analysis method for flexible parts in globular proteins. *J. Am. Chem. Soc.* **123,** 7305–7313.

Prompers, J. J., and Brüschweiler, R. (2002). General framework for studying the dynamics of folded and nonfolded proteins by NMR relaxation spectroscopy and MD simulation. *J. Am. Chem. Soc.* **124,** 4522–4534.

Redfield, A. G. (1957). On the theory of relaxation processes. *IBM J. Res. Dev.* **1,** 19–31.

Tugarinov, V., Liang, Z., Shapiro, Y. E., Freed, J. H., and Meirovitch, E. (2001). A structural mode-coupling approach to ^{15}N NMR relaxation in proteins. *J. Am. Chem. Soc.* **123,** 3055–3063.

Tugarinov, V., Shapiro, Y. E., Liang, Z., Freed, J. H., and Meirovitch, E. (2002). A novel view of domain flexibility in *E. coli* adenlyate kinase based on structural mode-coupling ^{15}N NMR relaxation. *J. Mol. Biol.* **315,** 155–170.

Vugmeyster, L., Raleigh, D. P., Palmer, A. G., III, and Vugmeister, B. E. (2003). Beyond the decoupling approximation in the model free approach for the interpretation of NMR relaxation of macromolecules in solution. *J. Am. Chem. Soc.* **125,** 8400–8404.

Wangsness, R. K., and Bloch, F. (1953). The dynamical theory of nuclear induction. *Phys. Rev.* **89,** 728–739.

Yamazaki, T., Muhandiram, R., and Kay, L. E. (1994). NMR experiments for the measurement of carbon relaxation properties in highly enriched uniformly ^{13}C, ^{15}N-labeled proteins: Application to [13]C$_\alpha$ carbons. *J. Am. Chem. Soc.* **116,** 8266–8278.

Section V

Macromolecular Complexes

[20] NMR Techniques for Identifying the Interface of a Larger Protein–Protein Complex: Cross-Saturation and Transferred Cross-Saturation Experiments

By Ichio Shimada

Abstract

NMR provides detailed structural information for protein complexes with molecular weights up to 30 kDa. However, it is difficult to obtain such information on larger proteins using NMR. To identify the interface of a complex with a molecular weight of over 50 kDa, chemical shift perturbation or hydrogen–deuterium (H–D) exchange experiments have been frequently used. The binding sites determined by these methods are quite similar, but not identical, to the contact surface identified by X-ray crystallography. The difference in the binding sites can be explained by the fact that the chemical shift and H–D exchange rates are affected by various factors, such as changes in the microenvironment and subtle conformational changes induced by the binding. Therefore, an alternative NMR strategy is required to identify the interaction site in large protein–protein complexes. The cross-saturation experiment is an NMR measurement for precise identification of the interface of larger protein complexes. This method extensively utilizes deuteration for proteins and the cross-saturation phenomenon along with TROSY detection. In this chapter, the principle of the cross-saturation experiment will be illustrated and then the extended version of the method, transferred cross-saturation, and its applications to larger protein complexes will be demonstrated.

Introduction

The identification of protein complex interfaces provides insights into the various biological systems, such as signal transduction, immune systems, and cellular recognition. Moreover, the recent expansion of structural genomics, in which the three-dimensional structures of proteins from a variety of organisms, including humans, are determined in a genome-wide fashion, increases the significance of investigations of protein–protein interactions. In this context, several attempts have been made to investigate protein–protein interactions using nuclear magnetic resonance (NMR)

Copyright 2005, Elsevier Inc.
All rights reserved.
0076-6879/05 $35.00

(Matsuo *et al.*, 1999; Mayer and Meyer, 2001; Miura *et al.*, 1999; Rajesh *et al.*, 1999; Walters *et al.*, 1999).

In particular, NMR methods that use chemical shift perturbation (Foster *et al.*, 1998) and/or hydrogen–deuterium (H–D) exchange (Paterson *et al.*, 1990) for the main-chain amides of the complex have frequently been used to identify the interfaces of larger protein–protein complexes with molecular weights over 50 kDa in solution. However, these methods are not necessarily perfect, so we have developed the cross-saturation method to enhance the analysis of the interface in large complexes (Takahashi *et al.*, 2000).

Chemical Shift Perturbation and H–D Exchange Studies for the System of Immunoglobulin G and Protein A

We have investigated the interaction sites between immunoglobulin G (IgG) and an Ig-binding protein, protein A (Gouda *et al.*, 1992, 1998; Torigoe *et al.*, 1990). Protein A, which is a cell wall component of *Staphylococcus aureus*, specifically binds to the Fc portion of IgG (Langone, 1982). The extracellular part of protein A contains five tandem, highly homologous Fc-binding domains, designated as E, D, A, B, and C. In previous studies, we have shown that the B domain of protein A (FB) in solution is composed of a bundle of three α-helices, and the residues responsible for the binding to the Fc fragment were identified based upon the combined data of the H–D exchange experiments and ^{1}H–^{15}N correlation spectroscopy, using FB uniformly labeled with ^{15}N (Gouda *et al.*, 1992, 1998; Torigoe *et al.*, 1990).

The three-dimensional structure of FB bound to the Fc fragment has also been determined by an X-ray study (Deisenhofer, 1981). It is of great interest that the residues identified using NMR are similar, but not identical, to those of the contact surface determined by the X-ray study. Some of the residues that showed significant changes in their chemical shifts and H–D exchange rates upon binding to the Fc fragment were not in the contact area determined by X-ray crystallography. The contradiction between NMR and X-ray crystallography may be due to the fact that chemical shifts and the H–D exchange rates are affected by various factors, such as changes in the microenvironment and subtle conformational changes induced by Fc binding. Therefore, an alternative NMR strategy is obviously required for identification of the interaction site in larger protein–protein complexes.

Principle of the Cross-Saturation Method

Figure 1 schematically illustrates the principle of the cross-saturation (CS) method. Protein I, with residues to be identified as interface residues and uniformly labeled with ^{2}H and ^{15}N, forms a complex with a

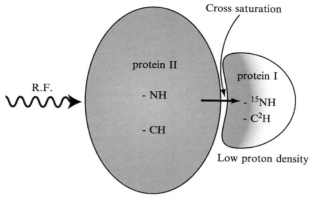

FIG. 1. Principle of the cross-saturation experiment. The protein with residues to be identified in the complex interface (protein I) is uniformly labeled with ^2H and ^{15}N. The saturation caused by the irradiation of the nonlabeled protein (protein II) is transferred to the doubly labeled molecule and is limited to the interface.

non-labeled target protein (protein II). Accordingly, the complex is composed of molecules with low and high proton densities.

In the case of larger proteins, if the aliphatic proton resonances are irradiated nonselectively using an RF field, in addition to the aliphatic resonances, the aromatic and amide resonances in the target protein are quickly saturated. This phenomenon is known as the spin diffusion effect (Akasaka, 1981; Endo and Arata, 1985; Ito et al., 1987; Kalk and Berendesen, 1976). Although the protein uniformly labeled with ^2H and ^{15}N is not directly affected by the RF field, it is expected that the saturation can be transferred from the target molecule (protein II) to the doubly labeled molecule (protein I) through the interface of the complex by cross-relaxation. If the proton density of the doubly labeled molecule is sufficiently low, the saturation transferred to the doubly labeled molecule is limited to the interface. Therefore, we can identify the residues at the interface of protein I by observing reductions of peak intensities in the $^1H-^{15}N$ heteronuclear single-quantum correlation (HSQC) spectra measured by TROSY coherence transfer (Pervushin et al., 1997, 1998; Salzmann et al., 1998).

Application of the Method to the FB–Fc Complex

Figure 2 is the pulse scheme used in the present study. It consists of an alternative band selective WURST-2 saturation scheme (Kupce and Wagner, 1995), followed by a water flip-back TROSY–HSQC experiment

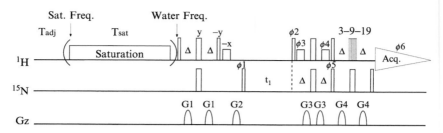

FIG. 2. Pulse scheme for the cross-saturation experiment. Pulses are applied along the x direction, if not otherwise shown. Narrow bars and wide bars indicate 90° pulses and 180° pulses, respectively. The line marked Gz indicates the durations and the amplitudes of the sine-shaped, pulsed magnetic field gradient applied along the z-axis: $G_1 = 600$ μs, 7.5 G cm^{-1}; $G_2 = 1000$ μs, 10 G cm^{-1}; $G_3 = 600$ μs, 14.5 G cm^{-1}; $G_4 = 600$ μs, 20 G cm^{-1}. The delay, Δ, is 2.25 ms. The following phase cycling scheme was used: $\phi_1 = \{y, -y, x, -x\}$; $\phi_2 = \{-y\}$; $\phi_3 = \{-y\}$; $\phi_4 = \{-x\}$; $\phi_5 = \{-y\}$; ϕ_6 (receiver) = $\{y, -y\}$. In the ^{15}N (t_1) dimension, a phase-sensitive spectrum is obtained by recording a second free induction decay (FID) for each increment of t_1, with $\phi_1 = \{y, -y, -x, x\}$; $\phi_2 = \{y\}$; $\phi_3 = \{y\}$; $\phi_4 = \{-x\}$; $\phi_5 = \{y\}$; ϕ_6 (receiver) = $\{-x, x\}$. To measure a spectrum with RF irradiation of an aliphatic region, a band-selective WURST-2 saturation scheme is applied during the T_{sat} period prior to the TROSY–HSQC scheme. In this case, the carrier frequency is switched to the irradiation frequency before the WURST-2 saturation period (T_{sat}) and back to the water frequency just before the TROSY–HSQC scheme. The additional relaxation delay, T_{adj}, can be set to an appropriate value for obtaining a sufficient signal-to-noise ratio.

(Pervushin *et al.*, 1997, 1998). We employed a broadband decoupling scheme for the saturation of the aliphatic proton resonances.

The present method was applied to ^2H, ^{15}N-labeled FB complexed with a nonlabeled Fc fragment of a human myeloma protein IgG(κ), Ike-N. By use of the saturation scheme, the ^1H-NMR spectrum observed for the Fc fragment revealed that saturation caused by irradiation of the aliphatic resonances efficiently spreads to the aromatic and amide resonances via spin diffusion (Fig. 3). Furthermore, the saturation scheme used here is highly selective for the aliphatic proton region, and the leakage of the irradiation to the water resonance was negligible. Therefore, irradiation of the FB–Fc complex does not affect the intensities of the FB amide protons with rapid hydrogen exchange rates (Fig. 4).

Figure 5 shows the ^1H–^{15}N TROSY–HSQC spectra observed for the complex of the doubly labeled FB and the Fc fragment in 10% H_2O/ 90% ^2H_2O, without and with the irradiation, respectively. The effect of the irradiation with a saturation time of 1.2 s on the FB molecule in the complex is clearly observed in Fig. 5B. The intensities of some cross-peaks are obviously reduced by the irradiation. This indicates that the saturation in the Fc fragment of the complex is transferred to the bound FB through

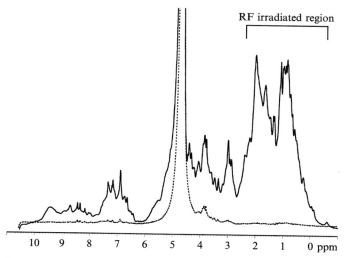

FIG. 3. ^{1}H-NMR spectra of the Fc fragment. The spectra were recorded with irradiation (dotted line) and without irradiation (solid line). The band-selective WURST-2 saturation scheme was used for irradiation. The irradiation area is shown in the figure.

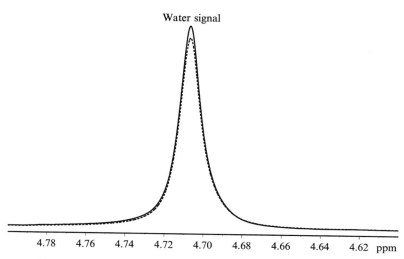

FIG. 4. ^{1}H-NMR signals originating from water in the FB–Fc sample. The spectra were recorded with irradiation (dotted line) and without irradiation (solid line). The band-selective WURST-2 saturation scheme was used for the irradiation. The irradiation area is the same as in Fig. 3.

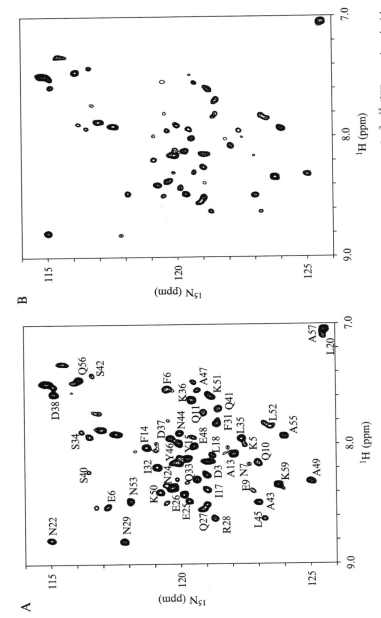

FIG. 5. The results of the cross-saturation experiment. $^1H-^{15}N$ TROSY–HSQC spectra observed for $[U-^2H,^{15}N]$FB complexed with the Fc fragment under the condition of 10% $H_2O/90\%$ ^2H_2O without (A) and with (B) irradiation. The spectra were measured with a reasonable protein concentration (0.5–1.0 mM) and within a reasonable measuring time (20 h).

the interface. It should be noted that the concentration of H_2O in the sample solution used in the present experiment is quite low.

The efficiency of the suppression of spin diffusion in the FB molecule depends on the concentration of H_2O in the sample solution. The FB molecule in the complex takes the conformation of a bundle of three α-helices. Therefore, under the condition of 90% H_2O/10% ^2H_2O, which is conventionally used for NMR measurements, the amide proton of the deuterated FB at the ith position is spatially close, within 4 Å, to those at the $i-1$ and $i+1$th positions, leading to a strong dipole–dipole interaction between the intramolecular amide protons (Fig. 6). However, with the complex in 10% H_2O/90% ^2H_2O, the amide proton of the deuterated FB exists almost in isolation from the other amide protons; therefore, the effect of spin diffusion on the bound FB is effectively suppressed.

Figure 7 shows the intensity ratios of the cross-peaks observed with irradiation to those without irradiation. Under the condition of 10% H_2O/90% ^2H_2O, small intensity ratios are observed for the residues in helix I (Gln-10–His-19) and helix II (Glu-25–Asp-37) (Fig. 7A). Interestingly, the ratios observed for the region of helices I and II show smaller values at every third or fourth residue (Gln-11, Tyr-15, and Leu-18 in helix I and Asn-29, Ile-32, and Lys-36 in helix II), suggesting that one side of each of

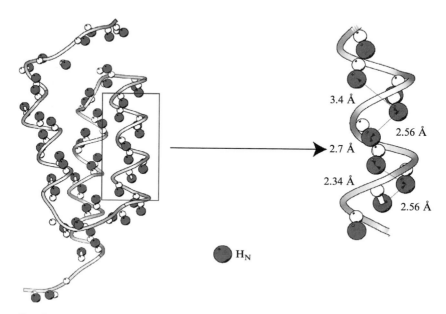

FIG. 6. The ribbon presentation of FB. Amide protons are shown as balls in the figure.

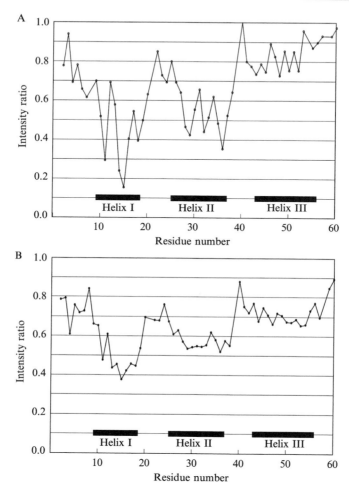

FIG. 7. Plots of cross-peak intensity ratios in the cross-saturation experiments. Intensity ratios of cross-peaks originating from the backbone amide groups with irradiation to those cross-peaks without irradiation were measured under the conditions of 10% H_2O/90% ^2H_2O (A) and 90% H_2O/10% ^2H_2O (B). The ratios for Thr-1, Pro-21, Pro-39, and Pro-58 were not available. The intensity ratio for Arg-28 was calculated based on the side chain εNH signal.

helices I and II is responsible for binding to the Fc fragment. These results are in agreement with the X-ray study.

We carried out the same experiment under the solvent condition of 90% H_2O/10% ^2H_2O. The profile of the plots (Fig. 7B) for the helical region no longer exhibits any distinctive pattern due to spin diffusion. Therefore, we conclude that a specific intermolecular CS effect can be observed if the concentration of H_2O in the solvent is sufficiently low and the perdeuterated sample can be prepared.

Comparison of the Binding Sites of FB to Fc Determined by Cross-Saturation, X-Ray Crystallography, Chemical Shift Perturbation, and H–D Exchange Experiments

Figure 8 compares the binding sites determined by X-ray crystallography and by NMR methods including the CS method described in the present study. In Fig. 8A, the residues that are listed as contact residues of the Fc fragment by the X-ray study are mapped on the structure of FB (Deisenhofer, 1981; Gouda et al., 1992). Figure 8B and C shows the binding sites identified by chemical shift perturbation data and by changes of the H–D exchange rates for the amide groups of the backbone of FB upon binding to the Fc fragment, respectively (Gouda et al., 1998). As shown in Fig. 8B and C, the NMR data indicate that helices I and II of FB are primarily responsible for the binding to the Fc fragment. However, it is of interest that some residues that were revealed to be on the contact surface by the X-ray study are little affected upon Fc binding. Furthermore, the changes in the chemical shifts and the H–D exchange rates induced by Fc binding range from the surface to the interior of the FB molecule.

Based upon our preliminary calculations using a relaxation matrix, it is suggested that the FB residues within about 7 Å of the interface possess intensity ratio values less than 0.5. Therefore, in Fig. 8D, residues with a value less than 0.5 (colored red) are contact residues. The contact residues thus identified significantly overlap those determined by the X-ray crystallographic study. Consequently, we conclude that the CS method makes it possible to determine the interface of the FB–Fc complex more precisely than NMR methods used previously.

Separation of Direct Intermolecular Contacts from the Effects Due to Ligand-Induced Conformational Changes

In this section, it will be shown that a combination of CS and chemical shift perturbation experiments can be used for separation of direct intermolecular contacts from the effects of ligand-induced conformational changes (Takeda et al., 2003).

CD44 is the main cell surface receptor for hyaluronic acid (HA) and contains a functional HA-binding domain (HABD) composed of a Link module with N- and C-terminal extensions (flanking regions) (Aruffo et al.,

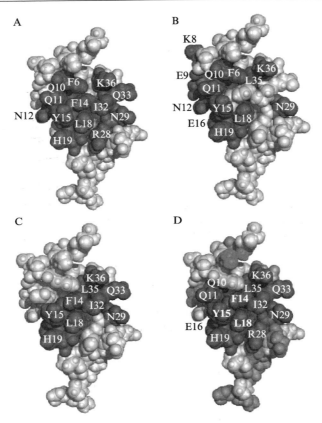

FIG. 8. Comparison of the binding sites of FB to the Fc fragment. (A) The X-ray study, (B) the chemical shift perturbation, (C) the H–D exchange experiments, and (D) the cross-saturation experiment. Residues with accessible surface areas that are covered upon binding of the Fc fragment (Phe-6, Gln-10, Gln-11, Asn-12, Phe-14, Tyr-15, Leu-18, His-19, Arg-28, Asn-29, Ile-32, Gln-33, and Lys-36) (Deisenhofer, 1981) are colored red in (A). Residues showing absolute values of chemical shift difference ($[(\Delta H_N)^2 + (\Delta N \times 0.15)^2]^{1/2}$) of more than 0.2 ppm (Phe-6, Lys-8, Glu-9, Gln-10, Gln-11, Asn-12, Ala-13, Tyr-15, Glu-16, Leu-18, His-19, Asn-29, Gly-30, Leu-35, Lys-36, and Ser-40) are colored red in (B). Of these residues, Ala-13, Gly-30, and Ser-40 are buried in the molecule. On the basis of the BioMagResBank database at http:// www.bmrb.wisc.edu, the scale factor of 0.15 used for normalization of the magnitude of ^{15}N chemical shift changes (in ppm units) was derived from the average (over all residue types but Pro) standard deviations for the backbone ^1H_N and ^{15}N chemical shifts. Residues showing protection factors larger than 10 upon binding of the Fc fragment (Phe-14, Tyr-15, Glu-16, Ile-17, Leu-18, His-19, Asn-29, Ile-32, Gln-33, Leu-35, Lys-36, Leu-46, Ala-49, and Lys-50) are colored red in (C) (Gouda et al., 1998). Of these residues, Glu-16, Ile-17, Leu-46, Ala-49, and Lys-50 are buried in the molecule. Residues with intensity ratios less than 0.5 (Gln-10, Gln-11, Phe-14, Tyr-15, Glu-16, Leu-18, His-19, Arg-28, Asn-29, Ile-32, Gln-33, Leu-35, and Lys-36), 0.5 to 0.6, 0.6 to 0.7, 0.7 to 0.8, and larger than 0.8 are colored red, orange, yellow, green, and blue, respectively, in (D). Molecular graphics images were produced using the MidasPlus program from the Computer Graphics Laboratory, University of California, San Francisco (supported by NIH RR-01081) (Ferrin et al., 1988; Huang et al., 1991). (See color insert.)

1990; Lesley *et al.*, 1993; Naot *et al.*, 1997). Recently, the three-dimensional structure of HABD was determined by X-ray; however, the structure of the complex between HABD and HA is not available (Teriete *et al.*, 2004).

To identify the residues of CD44 HABD that contact HA, CS experiments were performed. HA with an average molecular mass of 6.9 kDa (termed $HA_{\sim 34}$) was selected as a saturation-donating partner from several HA oligomers. RF irradiation applied to the complex resulted in selective intensity losses for the CD44 HABD resonances by the CS phenomena. The affected residues are distributed on both the consensus fold and additional structural elements comprising the flanking regions (Fig. 9A).

The weighted-averaged ^1H and ^{15}N chemical shift changes between the $HA_{\sim 34}$ bound and unbound states of CD44 HABD are plotted in Fig. 9B. Interestingly, these residues are mostly localized on the C-terminal extension and the α_1-helix, and they are generally inconsistent with the contact residues. Significant chemical shift changes upon HA binding are observed in the $^{13}C\alpha$ resonances of the residues in this region, supporting the idea that the conformational changes upon HA binding occur in the C-terminal extension and the α_1 helix.

Chemical shift perturbation is a widely used NMR method to map protein interfaces. However, in this study, chemical shift perturbation would not be an appropriate method to identify the ligand-binding site of CD44 HABD because of the small chemical shift changes at the contact site and the presence of significant conformational changes in the flanking regions. Several chemical shift perturbation experiments for protein–carbohydrate complexes have been reported (Asensio *et al.*, 1998, 2000a,b; Espinosa *et al.*, 2000; Kahmann *et al.*, 2000). A common feature of these experiments is that the chemical shift perturbations of the protons on the putative ligand-binding sites are at most 0.35 ppm. This feature may reflect the fact that the common molecular scaffold of carbohydrates has no aromatic group that would create magnetic anisotropies. On the other hand, the efficiency of CS does not depend on the presence of an aromatic group in the ligand molecules. Therefore, the CS method is especially helpful to determine interfaces of proteins with carbohydrates that have no source of ring current shifts.

A combination of the CS and chemical shift perturbation methods provides a useful means to separate the direct intermolecular contacts from the effects due to ligand-induced conformational changes.

Transferred Cross-Saturation Measurements

As described in the previous section, we could determine the contact residues on FB in the complex of FB and the Fc fragment with high accuracy. However, to determine the contact residues, the CS method is

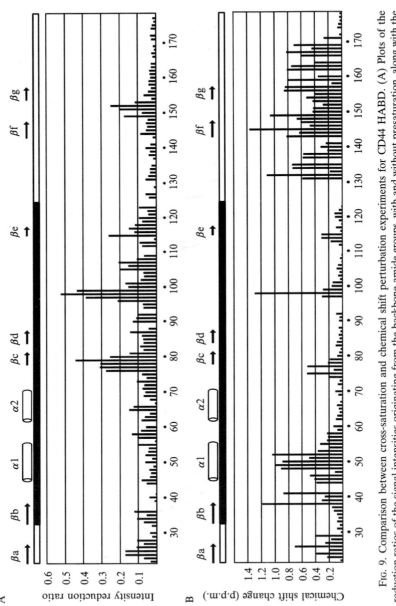

Fig. 9. Comparison between cross-saturation and chemical shift perturbation experiments for CD44 HABD. (A) Plots of the reduction ratios of the signal intensities originating from the backbone amide groups, with and without presaturation, along with the determined secondary structure. (B) Plots of the weighted averaged ^{1}H and ^{15}N chemical shift changes calculated with the function $\Delta\delta = [(\Delta\delta_{H_N}^2 + 0.25\,\Delta\delta_{^{15}N}^2)/2]$.

limited, in that it is difficult to apply to protein complexes with a molecular weight over 150 kDa or with weak binding affinity since the resonances originating from the complexes should be directly observed in the method. To overcome these limitations, we developed another version of the CS measurement, termed transferred cross-saturation (TCS) measurement, in which the NMR resonances originating from a free protein under the fast exchange process between the free and bound states on the NMR time scale are utilized (Nakanishi *et al.*, 2002).

Under conditions of an excess amount of protein I relative to protein II and a fast exchange rate between the free and bound states of protein I, it is expected that the CS effect that occurs in the bound state of protein I would be efficiently observed in the free state of protein I due to the long T_1 relaxation in the deuterated protein I, as it works well in the transferred NOE experiments (Clore and Gronenborn, 1982, 1983) (Fig. 10).

To investigate whether this TCS method works well, we used an intact mouse IgG with a molecular weight of 150 kDa as the target protein and FB as the ligand protein. The molecular weight of the complex of the intact IgG and FB is 164 kDa. The complex used in the present research is suitable for evaluation of the TCS method, since the binding site on FB for the Fc fragment is available and thus can be compared with the binding site identified by the TCS method. Moreover, the intact mouse IgG_1 used in the present study is known to possess a lower affinity to protein A than the human IgG_1 used in the previous study. Therefore, fast exchange between the free and bound states of FB is expected to occur, so the CS effect should be transferred to the free state of FB with high efficiency.

$^1H–^{15}N$ shift correlation spectra were observed for uniformly ^{15}N-labeled FB in the free state and uniformly $^2H–^{15}N$-labeled FB in the

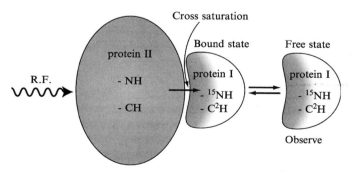

FIG. 10. Principle of the transferred cross-saturation method between a receptor protein (protein II) and a ligand protein (protein I).

presence of the intact IgG at a ratio of 10:1 FB:intact IgG. The chemical shifts of the cross-peaks originating from the backbone amide protons of FB in the presence of intact IgG are almost identical to those from free FB and different from those from bound FB. Therefore, under the present conditions, only cross-peaks originating from free FB were detected on the $^1H–^{15}N$ shift correlation spectrum.

For the FB-intact IgG complex, the TCS experiment was carried out. Figure 11 shows the effect of saturation time on the cross-peak intensity ratios by changing the irradiation time from 0.6 to 2.6 s. The residues with intensity ratios of less than 0.5 were Gln-11, Phe-14, Tyr-15, and Leu-18 in helix I and Arg-28 in helix II. Figure 12 shows the mapping of the affected residues on the structure of FB. Compared with the previous mapping of the contact residues in the FB–Fc complex (Fig. 8D), an almost identical binding surface on FB was obtained. Residues Phe-14, Tyr-15, and Leu-18 are mainly responsible for Fc binding (Gouda *et al.*, 1998). The part of the binding site on FB that is composed of Asn-29, Ile-32, Gln-33, Leu-35, and Lys-36 in helix II exhibited larger intensity ratios than those obtained from the previous study. The differences in the intensity ratios can be explained by the fact that the profile of the surface on the Fc portions where helix II of FB binds differs between the mouse and human IgGs.

In the present study, we identified the contact residues in the FB-intact IgG complex, with a molecular weight of 164 kDa. Since the resonances

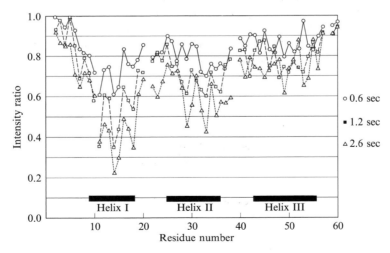

FIG. 11. Effect of the saturation time on the intensity ratios of the cross-peaks originating from the backbone amide groups. Some residues are selectively affected by changing the irradiation time from 0.6 to 2.6 s.

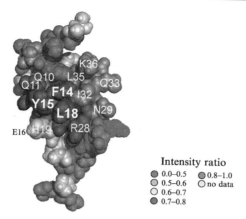

Intensity ratio
- 0.0–0.5   0.8–1.0
- 0.5–0.6   no data
- 0.6–0.7
- 0.7–0.8

FIG. 12. Binding sites of the the FB–IgG complex determined by the transferred cross-saturation experiments. Residues with intensity ratios less than 0.5, 0.5–0.6, 0.6–0.7, 0.7–0.8, and larger than 0.8 are colored red, orange, yellow, green, and blue, respectively. Molecular graphics images were produced using MidasPlus (University of California, San Francisco) (Ferrin *et al.*, 1988; Huang *et al.*, 1991). (See color insert.)

originating from the ligand protein in the free state are used to identify the contact residues in the complex, the TCS method would be easy to apply to much larger protein–protein complexes, such as those between ligand proteins and membrane proteins in bilayer lipids, and also those between ligand proteins and cells. In the following sections, two applications of TCS experiments are shown.

Application of TCS to Membrane Protein Systems

K^+ channels play a crucial role in regulating membrane potential, signal transduction, and various physiological events. A K^+ channel derived from *Streptomyces lividans*, KcsA, functions as a homotetramer with a molecular weight of 70.4 kDa (Cortes and Perozo, 1997; Cuello *et al.*, 1998; Heginbotham *et al.*, 1997). Recently, the three-dimensional structure of the KcsA K^+ channel has been determined (Doyle *et al.*, 1998; Zhou *et al.*, 2001). Structural information on the interaction mode between ion channels and pore-blocking toxins is crucial for gaining deeper insights into the structure and function of the channels and also for designing drugs affecting nervous system function. However, the nature of the complex of the solubilized ion channels and the pore-blocking toxins, with its slow tumbling motion in solution leading to broad NMR signals from the complex, has hindered detailed NMR analyses of the interaction between the ion channels and the pore-blocking toxins. In this section we will examine the interaction between the KcsA K$^+$ channel and AgTx2, a pore blocker of K$^+$ channels, by using TCS (Takeuchi *et al.*, 2003).

We carried out the TCS experiments as shown in Fig. 13. Comparing spectra with and without irradiation, we calculated ratios of peak intensities. The affected residues formed a contiguous surface on the structure of AgTx2 (Fig. 14B, left). In contrast, no residues are affected on the back of the molecule.

The model of the complex between AgTx2 and the KcsA K$^+$ channel was built by using molecular dynamics-simulated annealing to satisfy the CS experiment (Fig. 15) (Matsuda *et al.*, 2004). In the complex model, the site of AgTx2 fits quite well with the cleft on the extracellular vestibule of the KcsA K$^+$ channel. At the interface of the complex, Arg-24 and Arg-31 of AgTx2 are in close proximity to Arg-64 of the KcsA K$^+$ channel. Based upon the model, we successfully explained specifities of K$^+$ channels for the pore blockers (Takeuchi *et al.*, 2003).

Application of TCS to Huge Inhomogeneous Systems

von Willebrand factor (vWF) is a plasma protein that mediates platelet adhesion at damaged sites of vessel walls by linking the subendothelial collagen with platelet receptors, glycoproteins Ib/IX/V complex (Ruggeri,

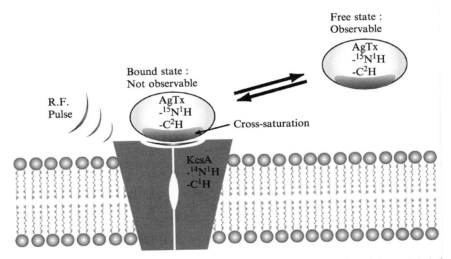

FIG. 13. Schematic representation of the TCS experiment. The saturation of the nonlabeled KcsA K$^+$ channel caused by the irradiation is transferred to the free state of AgTx2, which is labeled with ^{2}H and ^{15}N.

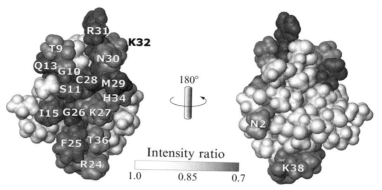

FIG. 14. Mapping of the residues affected by the irradiation in the TCS experiments. The correlation between intensity ratio and black and white tone is shown in the figure. The left and right figures are 180° rotations about the vertical axis relative to each other.

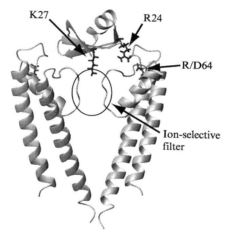

FIG. 15. The model of the AgTx2-KcsA K^+ channel complex. The model was built by a simulated annealing calculation, as described in the text. The three pairs of energetically coupled positions, Arg-24, Lys-27, and Arg-31 on AgTx, and Asp-64 in one subunit, the ion selective filter, and Asp-64 in another subunit of the other side on KcsA, are indicated by blue sticks and red surfaces, respectively.

1997). The A3 domain (amino acids 920–1111 of vWF) contains the major binding site for collagen types I and III (Cruz et al., 1995; Lankhof et al., 1996). The crystal studies of the A3 domain demonstrated that it also assumed the same fold as the α2-I domain but lacked the MIDAS motif, suggested to be of importance in collagen binding (Bienkowska *et al.*, 1997; Huizinga *et al.*, 1997). Therefore, this finding sparked interest in the collagen-binding site of the A3 domain. In fact, site-directed mutations introduced on the top surface of the A3 domain, which correspond to the binding site of the α2-I domain, caused no loss of collagen-binding activity (Bienkowska *et al.*, 1997; van der Plas *et al.*, 2000).

Collagen, which is characterized by its long triple-helical structure (polyproline II-like helices) formed by three tightly interwoven polypeptide chains, is the major component of extracellular matrices. Under physiological conditions at neutral pH, each triple-helix self-assembles into huge fibrillar supramolecules (Kadler *et al.*, 1996) (Fig. 16). Crystallization of the supramolecular complex between fibrillar collagen and the A3 domain may be difficult, so a collagen-like peptide was used for the X-ray study. It is also difficult to investigate the binding site on the A3 domain in the complex using traditional NMR strategies. To investigate the collagen-recognition mechanism of the A3 domain, we applied the TCS method to the complex with intact collagen (Nishida *et al.*, 2003). TCS experiments were performed under conditions of an excess amount of the labeled A3 domain relative to the unlabeled collagen. Based on the spectra in the presence of collagen, with and without irradiation, we calculated the peak intensity ratios. The residues affected by irradiation are distributed on the same surface of the A3 domain (Fig. 17). Therefore, we conclude that collagen binds to the A3 domain on the "front" surface, transversely across helix 3-strand C-helix 4.

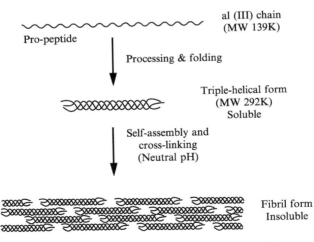

al (III) chain
(MW 139K)

Pro-peptide

Processing & folding

Triple-helical form
(MW 292K)
Soluble

Self-assembly and
cross-linking
(Neutral pH)

Fibril form
Insoluble

FIG. 16. Forms of collagen under various solution conditions.

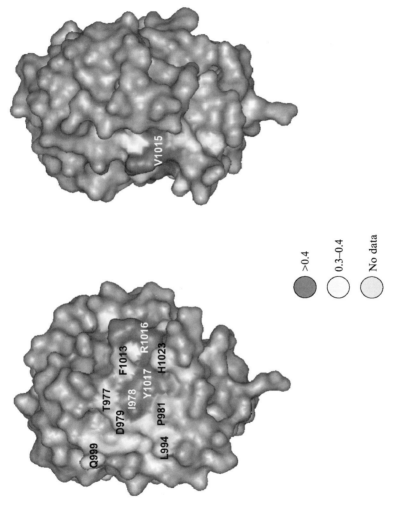

FIG. 17. The surface reproduction of the binding trenches of the A3 domain. The left and right panels are 90° rotations about the vertical axis, relative to each other (PDB code, 1AO3) (Bienkowska et al., 1997). The residues with signal intensity ratios greater than 0.4 and within the range from 0.3 to 0.4 are colored red and yellow, respectively (PDB code, 1AO3) (Bienkowska et al., 1997). Pro-981, which lacks cross-saturation data, due to the absence of the amide proton, is colored cyan. (See color insert.)

To determine whether the A3 domain could bind to the collagen triple helix in the fibrillar form without any steric hindrance, we made a docking model of the A3 domain and collagen microfibril, based upon a model of collagen microfibril (PDB code, 4CLG) (Chen *et al.*, 1991). In the model, binding of the A3 domain was free of significant bad contacts with the fibrillar collagen.

Comparing the model of the A3 domain and the collagen peptides (PDB code, 2CLG) to the X-ray structure of the α2-I domain complexed with the collagen-mimetic peptide, we conclude that the triple-helical collagen is bound at the "front" surface of the A3 domain (Fig. 18A), whereas the α2-I domain recognizes collagen at its "top" face (Fig. 18B). It is interesting that the A3 domain and the α2-I domain share the identical fold and the same collagen binding activity but have different interaction

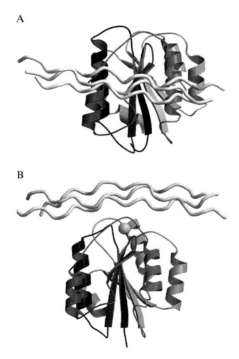

A

B

FIG. 18. Comparison of the collagen-binding sites between the A3 domain and the α2-I domain. (A) Docking model of the A3 domain and the collagen peptides. (B) Ribbon diagram of the α2-I domain complexed with the collagen-mimetic peptide (PDB code 1DZI) (Emsley *et al.*, 2000). This figure was made with the MOLSCRIPT (Kraulis, 1991) and RASTER3D (Merrit and Bacon, 1997) software.

modes. Structural genomics initiatives will undoubtably produce many more protein structures in the near future. To utilize the results of these initiatives, investigations focused on the interactions of proteins are important.

Conclusion

In identifying the interface of a larger protein–protein complex, the CS and TCS methods are obviously superior to traditional NMR methods that use chemical shift perturbation and H–D exchange experiments. This is due to the fact that the CS method extracts more direct information on through-space interaction between the two molecules. The present methods should in general be applicable to larger protein complex systems, such as extracellular matrices, tissues, or living cells under biologically relevant conditions, with TROSY detection accommodated to other complex systems.

Acknowledgments

I am grateful to Drs. Hideo Takahashi, Hiroaki Terasawa, Masanori Osawa, Masayoshi Sakakura, Tamiji Nakanishi, Koh Takeuchi, Noritaka Nishida, and Mitsuhiro Takeda. This work was supported by a grant from the Japan New Energy and Industrial Technology Development Organization (NEDO).

References

Akasaka, K. (1981). Longitudinal relaxation of protons under cross saturation and spin diffusion. *J. Magn. Res.* **45,** 337–343.

Aruffo, A., Stamenkovic, I., Melnick, M., Underhill, C. B., and Seed, B. (1990). CD44 is the principal cell surface receptor for hyaluronate. *Cell* **61,** 1303–1313.

Asensio, J. L., Canada, F. J., Bruix, M., Gonzalez, C., Khiar, N., Rodriguez-Romero, A., and Jimenez-Barbero, J. (1998). NMR investigations of protein-carbohydrate interactions: Refined three-dimensional structure of the complex between hevein and methyl β-chitobioside. *Glycobiology* **8,** 569–577.

Asensio, J. L., Canada, F. J., Siebert, H. C., Laynez, J., Poveda, A., Nieto, P. M., Soedjanaamadja, U. M., Gabius, H. J., and Jimenez-Barbero, J. (2000a). Structural basis for chitin recognition by defense proteins: GlcNAc residues are bound in a multivalent fashion by extended binding sites in hevein domains. *Chem. Biol.* **7,** 529–543.

Asensio, J. L., Siebert, H. C., von Der Lieth, C. W., Laynez, J., Bruix, M., Soedjanaamadja, U. M., Beintema, J. J., Canada, F. J., Gabius, H. J., and Jimenez-Barbero, J. (2000b). NMR investigations of protein-carbohydrate interactions: Studies on the relevance of Trp/Tyr variations in lectin binding sites as deduced from titration microcalorimetry and NMR studies on hevein domains. Determination of the NMR structure of the complex between pseudohevein and N,N',N''-triacetylchitotriose. *Proteins* **40,** 218–236.

Bienkowska, J., Cruz, M., Atiemo, A., Handin, R., and Liddington, R. (1997). The von Willebrand factor A3 domain does not contain a metal ion-dependent adhesion site motif. *J. Biol. Chem.* **272,** 25162–25167.

Chen, J. M., Kung, C. E., Feairheller, S. H., and Brown, E. M. (1991). An energetic evaluation of a "Smith" collagen microfibril model. *J. Protein. Chem.* **10,** 535–552.

Clore, G. M., and Gronenborn, M. A. (1982). Theory and application of the transferred nuclear Overhauser effect to the study of the conformations of small ligands bound to proteins. *J. Magn. Res.* **48,** 402–417.

Clore, G. M., and Gronenborn, M. A. (1983). Theory of the time dependent transferred nuclear Overhauser effect: Applications to structural analysis of ligand-protein complexes in solution. *J. Magn. Res.* **53,** 423–442.

Cortes, D. M., and Perozo, E. (1997). Structural dynamics of the Streptomyces lividans K$^+$ channel (SKC1): Oligomeric stoichiometry and stability. *Biochemistry* **36,** 10343–10352.

Cruz, M. A., Yuan, H., Lee, J. R., Wise, R. J., and Handin, R. I. (1995). Interaction of the von Willebrand factor (vWF) with collagen. Localization of the primary collagen-binding site by analysis of recombinant vWF a domain polypeptides. *J. Biol. Chem.* **270,** 10822–10827.

Cuello, L. G., Romero, J. G., Cortes, D. M., and Perozo, E. (1998). pH-dependent gating in the Streptomyces lividans K$^+$ channel. *Biochemistry* **37,** 3229–3236.

Deisenhofer, J. (1981). Crystallographic refinement and atomic models of a human Fc fragment and its complex with fragment B of protein A from Staphylococcus aureus at 2.9- and 2.8-Å resolution. *Biochemistry* **20,** 2361–2370.

Doyle, D. A., Morais Cabral, J., Pfuetzner, R. A., Kuo, A., Gulbis, J. M., Cohen, S. L., Chait, B. T., and MacKinnon, R. (1998). The structure of the potassium channel: Molecular basis of K$^+$ conduction and selectivity. *Science* **280,** 69–77.

Emsley, J., Knight, C. G., Farndale, R. W., Barnes, M. J., and Liddington, R. C. (2000). Structural basis of collagen recognition by integrin $\alpha 2\beta 1$. *Cell* **101,** 47–56.

Endo, S., and Arata, Y. (1985). Proton nuclear magnetic resonance study of human immunoglobulins G1 and their proteolytic fragments: Structure of the hinge region and effects of a hinge region deletion on internal flexibility. *Biochemistry* **24,** 1561–1568.

Espinosa, J. F., Asensio, J. L., Garcia, J. L., Laynez, J., Bruix, M., Wright, C., Siebert, H. C., Gabius, H. J., Canada, F. J., and Jimenez-Barbero, J. (2000). NMR investigations of protein-carbohydrate interactions binding studies and refined three-dimensional solution structure of the complex between the B domain of wheat germ agglutinin and N,N',N''-triacetylchitotriose. *Eur. J. Biochem.* **267,** 3965–3978.

Ferrin, T. E., Huang, C. C., Jarvis, L. E., and Langridge, R. (1988). The MIDAS database system. *J. Mol. Graphics* **6,** 2–62.

Foster, M. P., Wuttke, D. S., Clemens, K. R., Jahnke, W., Radhakrishnan, I., Tennant, L., Reymond, M., Chung, J., and Wright, P. E. (1998). Chemical shift as a probe of molecular interfaces: NMR studies of DNA binding by the three amino-terminal zinc finger domains from transcription factor IIIA. *J. Biomol. NMR* **12,** 51–71.

Gouda, H., Torigoe, H., Saito, A., Sato, M., Arata, Y., and Shimada, I. (1992). Three-dimensional solution structure of the B domain of staphylococcal protein A: Comparisons of the solution and crystal structures. *Biochemistry* **31,** 9665–9672.

Gouda, H., Shiraishi, M., Takahashi, H., Kato, K., Torigoe, H., Arata, Y., and Shimada, I. (1998). NMR study of the interaction between the B domain of staphylococcal protein A and the Fc portion of immunoglobulin G. *Biochemistry* **37,** 129–136.

Heginbotham, L., Odessey, E., and Miller, C. (1997). Tetrameric stoichiometry of a prokaryotic K$^+$ channel. *Biochemistry* **36,** 10335–10342.

Huang, C. C., Pettersen, E. F., Klein, T. E., Ferrin, T. E., and Langridge, R. (1991). Conic: A fast renderer for spacefilling molecules with shadows. *J. Mol. Graphics* **9,** 230–236.

Huizinga, E. G., Martijn van der Plas, R., Kroon, J., Sixma, J. J., and Gros, P. (1997). Crystal structure of the A3 domain of human von Willebrand factor: Implications for collagen binding. *Structure* **5,** 1147–1156.

Ito, W., Nishimura, M., Sakato, N., Fujio, H., and Arata, Y. (1987). A ^{1}H NMR method for the analysis of antigen-antibody interactions: Binding of a peptide fragment of lysozyme to anti-lysozyme monoclonal antibody. *J. Biochem. (Tokyo)* **102,** 643–649.

Kadler, K. E., Holmes, D. F., Trotter, J. A., and Chapman, J. A. (1996). Collagen fibril formation. *Biochem. J.* **316**(Pt. 1), 1–11.

Kahmann, J. D., O'Brien, R., Werner, J. M., Heinegard, D., Ladbury, J. E., Campbell, I. D., and Day, A. J. (2000). Localization and characterization of the hyaluronan-binding site on the link module from human TSG-6. *Struct. Fold. Des.* **8,** 763–774.

Kalk, A., and Berendesen, H. J. C. (1976). Proton magnetic relaxation and spin diffusion in proteins. *J. Magn. Res.* **24,** 343–366.

Kraulis, P. (1991). MOLSCRIPT: A program to produce both detailed and schematic plots of protein structures. *J. Appl. Crystallogr.* **24,** 924–950.

Kupce, E., and Wagner, G. (1995). Wideband homonuclear decoupling in protein spectra. *J. Magn. Res. B* **109,** 329–333.

Langone, J. J. (1982). Protein A of Staphylococcus aureus and related immunoglobulin receptors produced by streptococci and pneumonococci. *Adv. Immunol.* **32,** 157–252.

Lankhof, H., van Hoeij, M., Schiphorst, M. E., Bracke, M., Wu, Y. P., Ijsseldijk, M. J., Vink, T., de Groot, P. G., and Sixma, J. J. (1996). A3 domain is essential for interaction of von Willebrand factor with collagen type III. *Thromb. Haemost.* **75,** 950–958.

Lesley, J., Hyman, R., and Kincade, P. W. (1993). CD44 and its interaction with extracellular matrix. *Adv. Immunol.* **54,** 271–335.

Matsuda, T., Ikegami, T., Nakajima, N., Yamazaki, T., and Nakamura, H. (2004). Model building of a protein-protein complexed structure using saturation transfer and residual dipolar coupling without paired intermolecular NOE. *J. Biomol. NMR* **29,** 325–338.

Matsuo, H., Walters, K. J., Teruya, K., Tanaka, T., Gassner, G. T., Lippard, S. J., Kyogoku, Y., and Wagner, G. (1999). Identification by NMR spectroscopy of residues at contact surfaces in large, slowly exchanging macromolecular complexes. *J. Am. Chem. Soc.* **121,** 9903–9904.

Mayer, M., and Meyer, B. (2001). Group epitope mapping by saturation transfer difference NMR to identify segments of a ligand in direct contact with a protein receptor. *J. Am. Chem. Soc.* **123,** 6108–6117.

Merritt, E. A., and Bacon, D. J. (1997). Raster3D: Photorealistic molecular graphics. *Methods Enzmol.* **277,** 505–524.

Miura, T., Klaus, W., Gsell, B., Miyamoto, C., and Senn, H. (1999). Characterization of the binding interface between ubiquitin and class I human ubiquitin-conjugating enzyme 2b by multidimensional heteronuclear NMR spectroscopy in solution. *J. Mol. Biol.* **290,** 213–228.

Nakanishi, T., Miyazawa, M., Sakakura, M., Terasawa, H., Takahashi, H., and Shimada, I. (2002). Determination of the interface of a large protein complex by transferred cross-saturation measurements. *J. Mol. Biol.* **318,** 245–249.

Naot, D., Sionov, R. V., and Ish-Shalom, D. (1997). CD44: Structure, function, and association with the malignant process. *Adv. Cancer Res.* **71,** 241–319.

Nishida, N., Sumikawa, H., Sakakura, M., Shimba, N., Takahashi, H., Terasawa, H., Suzuki, E. I., and Shimada, I. (2003). Collagen-binding mode of vWF-A3 domain determined by a transferred cross-saturation experiment. *Nat. Struct. Biol.* **10,** 53–58.

Paterson, Y., Englander, S. W., and Roder, H. (1990). An antibody binding site on cytochrome c defined by hydrogen exchange and two-dimensional NMR. *Science* **249,** 755–759.

Pervushin, K., Riek, R., Wider, G., and Wuthrich, K. (1997). Attenuated T2 relaxation by mutual cancellation of dipole-dipole coupling and chemical shift anisotropy indicates an avenue to NMR structures of very large biological macromolecules in solution. *Proc. Natl. Acad. Sci. USA* **94**, 12366–12371.

Pervushin, K. V., Wider, G., and Wuthrich, K. (1998). Single transition-to-single transition polarization transfer (ST2-PT) in [^{15}N, ^{1}H]-TROSY. *J. Biomol. NMR* **12**, 345–348.

Rajesh, S., Sakamoto, T., Iwamoto-Sugai, M., Shibata, T., Kohno, T., and Ito, Y. (1999). Ubiquitin binding interface mapping on yeast ubiquitin hydrolase by NMR chemical shift perturbation. *Biochemistry* **38**, 9242–9253.

Ruggeri, Z. M. (1997). von Willebrand factor. *J. Clin. Invest.* **99**, 559–564.

Salzmann, M., Pervushin, K., Wider, G., Senn, H., and Wuthrich, K. (1998). TROSY in triple-resonance experiments: New perspectives for sequential NMR assignment of large proteins. *Proc. Natl. Acad. Sci. USA* **95**, 13585–13590.

Takahashi, H., Nakanishi, T., Kami, K., Arata, Y., and Shimada, I. (2000). A novel NMR method for determining the interfaces of large protein-protein complexes. *Nat. Struct. Biol.* **7**, 220–223.

Takeda, M., Terasawa, H., Sakakura, M., Yamaguchi, Y., Kajiwara, M., Kawashima, H., Miyasaka, M., and Shimada, I. (2003). Hyaluronan recognition mode of CD44 revealed by cross-saturation and chemical shift perturbation experiments. *J. Biol. Chem.* **278**, 43550–43555.

Takeuchi, K., Yokogawa, M., Matsuda, T., Sugai, M., Kawano, S., Kohno, T., Nakamura, H., Takahashi, H., and Shimada, I. (2003). Structural basis of the KcsA K^{+} channel and agitoxin2 pore-blocking toxin interaction by using the transferred cross-saturation method. *Structure (Camb.)* **11**, 1381–1392.

Teriete, P., Banerji, S., Noble, M., Blundell, C. D., Wright, A. J., Pickford, A. R., Lowe, E., Mahoney, D. J., Tammi, M. I., Kahmann, J. D., Campbell, I. D., Day, A. J., and Jackson, D. G. (2004). Structure of the regulatory hyaluronan binding domain in the inflammatory leukocyte homing receptor CD44. *Mol. Cell* **13**, 483–496.

Torigoe, H., Shimada, I., Saito, A., Sato, M., and Arata, Y. (1990). Sequential ^{1}H NMR assignments and secondary structure of the B domain of staphylococcal protein A: Structural changes between the free B domain in solution and the Fc-bound B domain in crystal. *Biochemistry* **29**, 8787–8793.

van der Plas, R. M., Gomes, L., Marquart, J. A., Vink, T., Meijers, J. C., de Groot, P. G., Sixma, J. J., and Huizinga, E. G. (2000). Binding of von Willebrand factor to collagen type III: Role of specific amino acids in the collagen binding domain of vWF and effects of neighboring domains. *Thromb. Haemost.* **84**, 1005–1011.

Walters, K. J., Gassner, G. T., Lippard, S. J., and Wagner, G. (1999). Structure of the soluble methane monooxygenase regulatory protein B. *Proc. Natl. Acad. Sci. USA* **96**, 7877–7882.

Zhou, Y., Morais-Cabral, J. H., Kaufman, A., and MacKinnon, R. (2001). Chemistry of ion coordination and hydration revealed by a K^{+} channel-Fab complex at 2.0 Å resolution. *Nature* **414**, 43–48.

[21] Enzyme Dynamics During Catalysis Measured by NMR Spectroscopy

By Dorothee Kern, Elan Z. Eisenmesser, and Magnus Wolf-Watz

Abstract

Many biological processes, in particular enzyme catalysis, occur in the microsecond to millisecond time regime. While the chemical events and static structural features of enzyme catalysis have been extensively studied, very little is known about dynamic processes of the enzyme during the catalytic cycle. Dynamic NMR methods such as ZZ-exchange, line-shape analysis, Carr–Purcell–Meiboom–Gill (CPMG), and rotating frame spin-lattice relaxation ($R_{1\rho}$) experiments are powerful in detecting conformational rearrangements with interconversion rates between 0.1 and 10^5 s^{-1}. In this chapter, the first application of these methods to enzymes *during catalysis* is described, in addition to studies on several other enzymes in their free states and in complex with ligands. From the experimental results of all systems, a picture arises in which flexibility in the microsecond to millisecond time regime is intrinsic and likely to be an essential property of the enzyme. Quantitative analysis of dynamics at multiple sites of the enzyme reveal large-scale collective motions. For several enzymes, the frequency of motion is comparable to the overall turnover rate, raising the possibility that conformational rearrangements may be rate limiting for catalysis in these enzymes.

Introduction

Nature has evolved enzymes to speed up chemical reactions with remarkable efficiency and specificity. Classic enzymology based on organic chemistry, followed by structural biology methods, has provided insights into the chemical mechanisms of many enzymes. However, a detailed understanding of the molecular events and the corresponding energy landscape of the entire catalytic process is still in its infancy. In particular, enzyme dynamics and its relationship to catalytic function remain poorly characterized. The most basic principle of enzyme catalysis is the ability of the enzyme to lower the transition state energy, thereby catalyzing the chemical reaction. A wealth of information about the kinetics and thermodynamics of enzyme-catalyzed reactions has been obtained by monitoring

Copyright 2005, Elsevier Inc.
All rights reserved.
0076-6879/05 $35.00

substrate conversion into products. However, much less is known about the kinetics and energetics of conformational changes in the enzyme. Since it is the protein component that alters the transition state energy, enzyme function depends on transitions from ground states to higher energy states of the enzyme and the reactant. Many enzymatic reactions occur on time scales of microseconds to milliseconds, and it has been suggested that dynamics of the enzyme on these time scales might be linked to catalysis (Fersht, 2000). The challenge is (1) to measure conformational dynamics in enzymes during turnover experimentally and (2) to elucidate the role of these motions for enzymatic power.

Several techniques have been used to detect dynamic processes of enzymes during catalysis, such as fluorescence resonance energy transfer, atomic force microscopy, and stopped-flow fluorescence. While those methods report on dynamics of individual sites, or motions of the entire enzyme molecule, nuclear magnetic resonance (NMR) spectroscopy of proteins in solution allows detection of motions at a multitude of specific atomic sites simultaneously (Fischer *et al.*, 1998; Palmer, 1997; Palmer *et al.*, 1996). Protein dynamics over a time scale from picoseconds to days can be monitored: nuclear spin relaxation rates are sensitive to fast (picoseconds to nanoseconds) and slow (microseconds to milliseconds) motions, magnetization transfer, or H/D exchange methods report on movements in the time scale of milliseconds to days. While the principal theory of relaxation has been known for a long time, there have been important developments of experimental methods for measuring and analyzing dynamics by NMR spectroscopy, summarized in recent reviews (Ishima and Torchia, 2000; Jardetzky, 1996; Kay, 1998; Palmer *et al.*, 1996, 2001, and Chapter 18 of this volume).

This chapter will focus on dynamic processes detected in the microsecond to second time scale *during catalysis* using several different NMR relaxation experiments. These motions correspond to transitions between kinetically distinct substates in the protein energy landscape. Evaluation of picosecond to nanosecond dynamics, also referred as the ruggedness of the landscape, although likely important for function, is not covered in this chapter. In an NMR experiment, conformational exchange processes between different magnetic environments in the submillisecond time regime contribute to dephasing of transverse coherence and is consequently manifested in an increase of transverse relaxation rate (R_2). From transverse relaxation experiments, three physical parameters can be extracted: kinetics (rate constant of interconversion, k_{ex}), thermodynamics (populations of the conformations), and indirect structural information (chemical shift difference, $\Delta\omega$) for a two-site exchange process. Chemical shift differences cannot be directly translated into structural changes; they show only that the electronic environment around an atom is different for the two states.

The following basic equations describe the relation between R_2 and the parameters mentioned:

$$R_2 = R_{2o} + R_{ex} \tag{1}$$

$$R_{ex} \sim p_A * p_B * \Delta\omega^2/k_{ex} \tag{2}$$

for interconversion that is fast on the NMR time scale. R_{2o} is the relaxation rate constant in the absence of exchange, R_{ex} is the chemical exchange contribution to R_2, p_A and p_B are the relative populations of the two exchanging species A and B, $\Delta\omega$ is the chemical shift difference between A and B, and k_{ex} is the sum of the forward (k_{AB}) and reverse rate (k_{BA}) constants. For details about the NMR techniques and the more complex equations for intermediate or slow exchange we refer to, for instance, Palmer *et al.* (2001 and Chapter 18, this volume). While relaxation can be measured on any magnetically active nuclei, we will limit our discussion to the most commonly used nuclei in protein dynamics, ^{15}N, ^{13}C, and ^{1}H.

Experimental Setup

In a solution NMR experiment, proteins can perform their functions during data acquisition. Importantly, dynamic processes can be characterized under steady-state conditions. Consequently, the basic experimental setup must fulfill this requirement during the data acquisition time, which is in the order of hours to days for quantitative studies of dynamics, due to the low sensitivity of NMR. Enzymes catalyzing reversible reactions are perfect candidates, since steady-state conditions can in principle be maintained indefinitely by simply adding the substrate(s) to the enzyme. In contrast to a ligand-binding event, enzymatic reactions are more complex and can be comprised of many steps. Any useful interpretation of dynamic NMR experiments performed in active enzymes therefore requires knowledge of which enzyme species are populated and which kinetic processes are detected in addition to classic enzyme kinetic constants. For accurate data analysis, it is favorable to biochemically "tune" the system to exchange between two sites. However, since both substrate and enzyme concentrations can be varied, more complex reactions such as coupled equilibria can be quantified.

Dynamics in Cyclophilin A Catalyzing a Single-Substrate Single-Product Reaction

We have shown recently that protein dynamics can in principle be measured by NMR in enzymes *during catalysis* on the first model system, human cyclophilin A (CypA) (Eisenmesser *et al.*, 2002). CypA catalyzes the reversible *cis/trans* isomerization of prolyl peptide bonds. It was originally identified as the target of the immunosuppressive drug cyclosporin A and has been shown to be essential for efficient HIV-1 replication. Despite these important medical roles, the true natural function of CypA is unknown.

In order to characterize motions in CypA during catalysis, changes in R_2 as a function of substrate concentration were measured. Two different substrates were used, a model tetrapeptide Suc-Ale-Phe-Pro-Phe-4NA and the N-terminal domain of the HIV-1 capsid protein (CA^N). In the presence of substrate, at least 3 different enzyme forms are populated, free enzyme and CypA bound to the *cis* and *trans* substrate isomers (Scheme 1). The following paragraph will describe experimental approaches to separate motions due to the binding event from those that coincide with isomerization.

Two approaches were used to identify motions in CypA during catalysis, measurements of changes in R_2 as a function of substrate concentration and relaxation dispersion experiments of CypA in its free and substrate-saturated state. Two different substrates were used, a model tetrapeptide Suc-Ale-Phe-Pro-Phe-4NA and the N-terminal domain of the HIV-1 capsid protein (CA^N). In the presence of substrate, at least three different enzyme forms are populated, free enzyme and CypA bound to the *cis* and *trans* substrate isomers (Scheme 1). Experimental approaches to separate motions due to the binding event from those that coincide with isomerization will be described below.

^{15}N Transverse Relaxation Rates of CypA as a Function of Substrate Concentration

Addition of a peptide substrate to CypA initiates turnover. Consequently changes in R_2 due to the action of the enzyme can readily be measured. Indeed, an increase of R_2 during turnover could be detected for several amides, indicative of conformational exchange. To dissect R_{ex} contributions from binding and isomerization (steps 2 and 3 versus step 4 in Scheme 1), R_2 was measured as a function of substrate concentration. Evaluation of Scheme 1 in combination with Eq. (2) reveals that the R_{ex} contribution from steps 2 and 3 will increase with addition of substrate until a maximum is reached when substrate concentration is in the range of the K_d and then decrease because of the changes of populations, in this case, the ratio between free and substrate-bound enzyme. Complete saturation, which would abolish R_{ex} due to binding, could not be reached at 25° due to the low affinity and solubility of the substrate ($K_d = 1.19$ mM). The numerical solution for the dependence of R_{ex} as a function of substrate is given in Eisenmesser *et al.* (2002). Note that k_{ex} is now the sum of the pseudo-first-order on-rate constant that changes with substrate concentration and a first-order off-rate constant. In contrast, the R_{ex} contribution from step 4 should gradually increase until substrate saturation in reached,

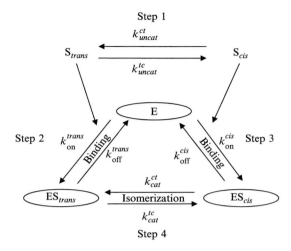

SCHEME 1. Kinetic model of CypA catalysis. E is the free enzyme, S is the free substrate, and ES_{trans} and ES_{cis} are the two Michaelis–Menten complexes with the substrate in *trans* and *cis*, respectively. k_{off} is the off-rate, k_{on} the on-rate, and k_{ex} the rate constant for chemical exchange ($k_{ex} = k_{cat}^{ct} + k_{cat}^{tc}$). Superscripts *cis* and *trans* identify the *cis* and *trans* isomer, respectively. While the enzyme is described by a three-site exchange, the minimal kinetic model for the substrate is a four-site exchange.

because the populations and k_{ex} between the two interconverting species, the two ES complexes, do not change; however, more ES complexes are formed with increasing substrate. Figure 1 shows a representative residue for each of these scenarios. In addition to a qualitative characterization of residues that are influenced by binding or the catalytic step, kinetic and thermodynamic parameters can be obtained from fits of the experimental data to a two-site or three-site exchange model (Fig. 1) (Eisenmesser *et al.*, 2002). Since only a small number of residues could be quantitatively analyzed with the three-site model, the rates for motions due to isomerization could be estimated, but not exactly determined (Eisenmesser *et al.*, 2002).

How do the dynamics measured on the enzyme compare to the rates of substrate interconversion? To answer this question, the fitted rates of enzyme dynamics were compared to the microscopic rate constants of substrate turnover. The latter rates could be determined by NMR line-shape analysis of the substrate resonances in the presence of various catalytic amounts of CypA (Kern *et al.*, 1995), while a classic enzymatic assay (Fischer *et al.*, 1984) provides only the macroscopic kinetic constants k_{cat} and K_m that are composed of several microscopic steps. Interestingly, the rates of enzyme dynamics are similar to the microscopic rates of substrate turnover.

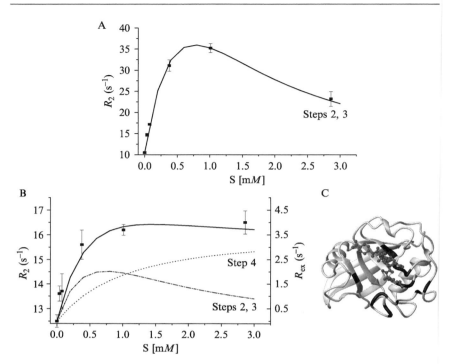

FIG. 1. Exchange dynamics in CypA during turnover quantified by measurements of the dependence of R_{ex} as a function of total substrate concentration. Transverse relaxation rates (R_2) were measured by standard methods using a fixed CPMG field strength of 400 Hz. (A) Experimental data were fit according to a two-site model including contributions from binding only for Lys-82. (B) The continuous increase of R_{ex} with increasing substrate concentration for the active site residue Arg-55 is indicative of significant contributions from steps 2, 3, and 4 in Scheme 1, binding and isomerization, respectively. The R_{ex} contribution from binding is shown with dashed/dotted lines, and the R_{ex} contribution from isomerization is displayed in dotted lines. The fit to the full three-site model comprising both processes is shown as a solid line. The left y-axis corresponds to the experimentally measured R_2 values, and the right y-axis corresponds to the R_{ex} contributions for steps 2, 3, and 4. (C) Dynamic hotspots in CypA are displayed in black during catalysis of the peptide Suc-Ala-Phe-Pro-Phe-4-NA. The peptide is shown in ball and stick representation in the *cis*-bound form according to the X-ray structure (Zhao and Ke, 1996) (adapted from Eisenmesser, 2002).

Catalysis of the HIV-1 Capsid Protein (CA^N) by CypA

This section will emphasize the application of dynamic NMR techniques to characterize CypA catalysis on the biologically relevant substrate CA^N. CypA is packaged into HIV virions via interaction with the capsid protein in a 12:1 ratio of capsid to CypA. While it is known that CypA is essential for efficient HIV-1 replication (Braaten and Luban, 2001; Franke et al., 1994), the mechanism is not understood.

The co-crystal structure of CA^N and CypA showed the Gly-89–Pro-90 peptide bond exclusively in the *trans* conformation, which led to the general conception that CA^N is not being catalyzed by CypA (Gamble et al., 1996). Summers and co-workers (Gitti et al., 1996) identified slow *cis/trans* interconversion of this peptide bond in free CA^N resulting in two distinct peaks for the *cis* and *trans* isomer (Fig. 2A).

To address the question of CypA catalysis, two-dimensional ^{15}N heteronuclear NMR (ZZ) exchange spectroscopy was performed on CA^N in the absence and presence of catalytic amounts (12:1) of CypA (Fig. 2) (Bosco et al., 2002). NMR exchange spectroscopy identifies conformational exchange processes that occur with rates of chemical exchange between 0.1 and 100 s^{-1} (Farrow et al., 1994). In the absence of CypA, *cis/trans* isomerization is slow ($k_{ex} < 0.1$ s^{-1}). Therefore exchange peaks are not observed in the two-dimensional (2D) exchange spectrum (Fig. 2A). Addition of catalytic amounts of CypA accelerates *cis/trans* isomerization of the Gly-89–Pro-90 peptide bond, resulting in exchange peaks (Fig. 2B and C). The presence of exchange peaks provides direct experimental evidence that CypA catalyzes Gly-89–Pro-90 *cis/trans* isomerization in CA^N. The catalytic efficiency of CypA catalysis for the substrate CA^N was determined by fitting the intensity of the exchange and auto peaks as a function of mixing time (Bosco et al., 2002). This work showed that the prolyl isomerase CypA can catalyze *cis/trans* isomerization in folded protein substrates, suggesting a possible biological role of its catalytic activity.

Dynamics in Adenylate Kinase Catalyzing a Two-Substrate Two-Product Reaction

A unique approach to address the role of protein dynamics for enzyme function is to study a thermophilic/mesophilic enzyme pair with respect to their catalytic activity and dynamics. Proteins from thermophilic organisms are more stable in order to survive high temperatures without denaturation (Jaenicke and Bohm, 1998). Based on structural similarity, which is especially striking in enzyme active sites, and based on Arrhenius theory, one would expect that a thermophilic enzyme catalyzes a reaction with similar rates to its mesophilic counterpart at the same temperature. However, the thermophilic enzyme is generally much less active compared to the mesophilic enzyme at the same temperature and, in fact, the absolute activities are comparable at the growth optima of their respective host organisms (D'Amico et al., 2003; Zavodszky et al., 1998). An intriguing hypothesis is that the cost of increased stability is decreased flexibility, which results in

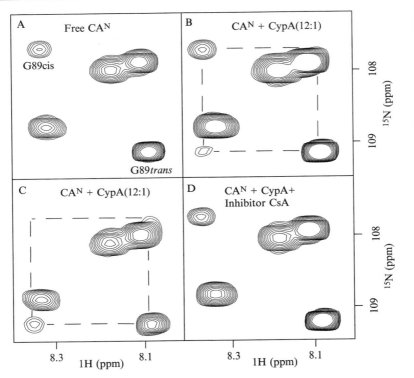

FIG. 2. Detection of *cis/trans* isomerization in HIV-1 CA^N catalyzed by CypA. Distinct *cis* and *trans* peaks in a ZZ-exchange spectrum (Farrow *et al.*, 1994) for the amide signal of Gly-89 preceding Pro-90 are indicative of slow isomerization (A). The appearance of exchange peaks connecting the *cis* and *trans* peak (visualized by dotted lines) upon addition of a catalytic amount of CypA indicates catalysis of *cis/trans* isomerization (B). The buildup of the exchange peak with increasing mixing time from 55 ms (B) to 550 ms (C) was used to determine the exact rate of catalyzed isomerization (Bosco *et al.*, 2002). (D) Isomerase activity was inhibited by CsA. Adapted from Bosco *et al.* (2002).

reduced catalytic power. In other words, a folded and active enzyme is a fine balanced state between stability (to maintain a defined folded state) and flexibility (for function).

Structure, Stability, and Activity of Mesophilic and Hyperthermophilic Adenylate Kinase (MesoAdk and ThermoAdk)

In order to understand this "marriage," (i) structure, (ii) stability, (iii) activity, and (iv) dynamics of a mesophilic and hyperthermophilic adenylate kinase (mesoAdk and thermoAdk) were compared. This enzyme catalyzes the reversible interconversion of AMP and ATP into two ADP molecules (Scheme 2A). Binding of AMP and ATP to the two lids is followed by a large conformational rearrangement, a closure by up to 30 Å for several amino acid residues in the ATP lid (Fig. 3B, C). This rearrangement enables the chemical step, the phosphotransfer reaction (Vonrhein et al., 1995). It has been recently shown that (i) the structures of mesoAdk and thermoAdk are very similar, (ii) the stability of thermoAdk is strongly increased, and (iii) catalytic turnover numbers are about tenfold lower for thermoAdk compared to mesoAdk at 20° (Wolf-Watz et al., 2004).

Dynamics of MesoAdk and ThermoAdk During Catalysis at 20°

Recently developed relaxation dispersion experiments allow a sensitive and quantitative analysis of motions with rates between 1 and $10^4 \, s^{-1}$, since R_{ex}, but not R_{2o}, is modulated by the applied Carr-Purcell-Meiboom-Gill (CPMG) field strength during the relaxation delay (Mulder et al., 2001; Palmer et al., 2001). ^{15}N relaxation dispersion experiments were acquired in a constant time manner (Mulder et al., 2001) taking the different field strengths in an interleaved fashion, which reduces time-dependent uncertainties due to magnet or sample variation.

In contrast to CypA, Adk catalysis is of a complex nature (a random bi-bi mechanism has been proposed; see Scheme 2A). In order to extract useful information from relaxation measurements during turnover, the system must be biochemically tuned. ^{15}N relaxation dispersion experiments performed on Adk under complete substrate saturating conditions exclude contributions from substrate binding to the CPMG field dependence of R_{2obs}. Interestingly, the relaxation data revealed that the opening of the lids and not the actual phospho-transfer is the rate-limiting step. All amides with exchange contributions could be globally fit with common opening and closing rates (Fig. 3). A much smaller opening rate results in an equilibrium far shifted to the closed state. Strikingly, the rate of opening is identical to the k_{cat} determined with a spectroscopic assay measuring ADP production for both enzymes. Moreover, the opening is about tenfold slower in thermoAdk compared to mesoAdk at 20° and the sole reason for the reduced overall turnover of the thermophilic enzyme (Wolf-Watz et al., 2004).

Experimental Support for Rate-Limiting Lid-Opening Event

What is the experimental evidence for the proposed interpretation of relaxation dispersion experiments measured during catalysis? First, residues showing dispersion are distributed throughout the protein and can be rationalized based on the conformational changes via closure of substrate

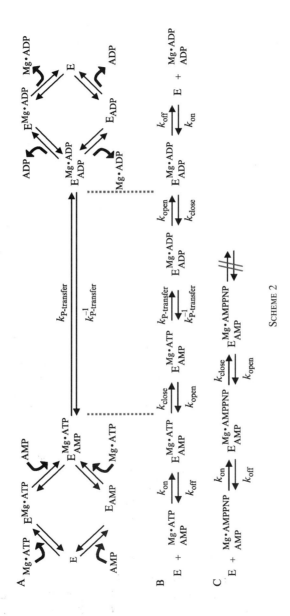

SCHEME 2

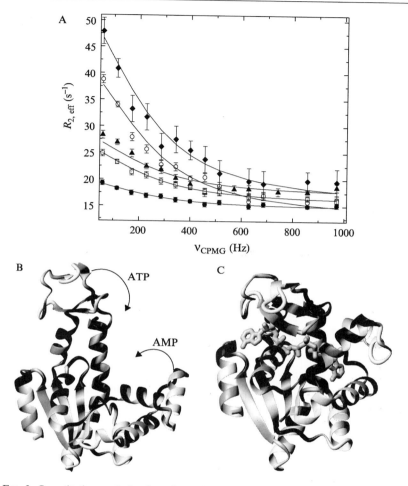

FIG. 3. Quantitative analysis of conformational exchange dynamics in mesoAdk during catalysis. (A) ^{15}N relaxation dispersion curves for individual amides, measured at 20° at saturating concentrations of ATP, ADP, AMP and Mg, are shown with experimental uncertainties estimated as previously described (Mulder *et al.*, 2002). Data were globally fit assuming a two-site exchange model (Carver and Richards, 1972) yielding closing and opening rates of 1370 ± 110 s^{-1} and 286 ± 85 s^{-1}, respectively (compare to a k$_{cat}$ of 263 ± 30 s^{-1}). All residues with conformational exchange are shown in black on the X-ray structures of *E. coli* Adk in the open (B) (Muller *et al.*, 1996) and closed (C) (Muller and Schulz, 1992) state with the inhibitor Ap$_5$A bound in the middle of the protein (C) (adapted from Wolf-Watz *et al.*, 2004).

lids. Second, replacement of ATP with the nonconvertible substrate analog AMPPNP yields a ^{1}H, ^{15}N-HSQC spectrum and relaxation data that are extremely similar to the spectrum acquired during turnover conditions. This result rules out the phopsho-transfer as the cause for the detected dispersions. Third, many chemical shifts during turnover are similar to the chemical shifts of the enzyme in complex with a tight-binding inhibitor that represents the closed state. Fourth, calculated chemical shift differences from the dispersion data are comparable to the shift differences measured for the protein in the open and closed state (Wolf-Watz et al., 2004).

The modified kinetic model including the lid opening and closure (Scheme 2B, C) was buttressed by two independent NMR titration experiments monitoring either amide chemical shifts as a function of substrate concentration (Fig. 4A) or ^{31}P substrate chemical shifts as a function of enzyme concentration (Fig. 4E). For both markers, the slow opening rate in thermoAdk at 20° gave rise to two distinct resonances at intermediate ligand concentrations (Fig. 4). The lineshapes could be fit according to Scheme 2 using the physical parameters obtained from the relaxation dispersion experiments described above (Fig. 4B, F). In sharp contrast, only one average resonance is observed through the entire titrations for mesoAdk (Fig. 4D), indicative of a faster opening rate. The same spectroscopic behavior is seen for thermoAdk at 70° (Fig. 4C). Apparently, the opening rate at this elevated temperature is now faster. Quantification of the opening rate by ^{15}N relaxation dispersion experiments is prohibited by a closing rate that is too fast to be detectable by CPMG dispersion experiments. In summary, the application of relaxation dispersion experiments and line-shape analysis revealed that catalytic power of Adk is limited by a dynamic process, the lid opening, and not by chemistry.

Microsecond Motions in Other Enzymes

In addition to the above-described measurements of protein dynamics *during* catalysis, experiments monitoring microsecond to millisecond motions have been performed on a number of enzymes in their free and ligand-bound state. This section will summarize results from only a few of those studies. However, we want to stress that the general conclusions are based on a larger number of publications in the area of protein dynamics measured by NMR.

In triosephosphate isomerase (TIM), one of the best physicochemical characterized enzymes, movement of the active-site loop was studied by one-dimensional solution NMR using ^{19}F of a fluorinated Trp or ^{31}P of the substrate analogue glycerol 3-phopshate as marker (Rozovsky et al., 2001) and by ^{2}H solid-state NMR of the same deuterium-labeled Trp in the active

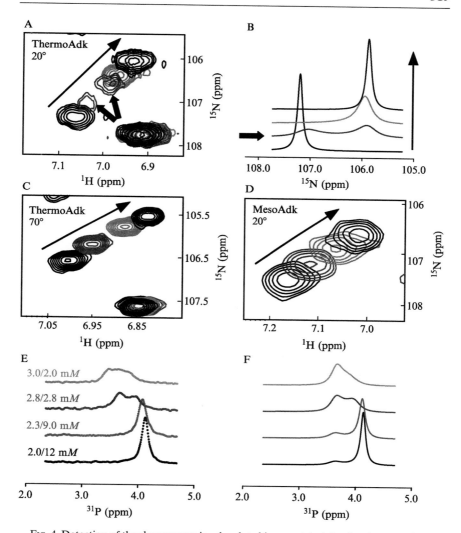

FIG. 4. Detection of the slow process in adenylate kinase catalysis by titration experiments. Chemical shift changes of thermoAdk (A, C) and mesoAdk (D) are shown with increasing substrate concentration indicated by thin arrows. Note that for thermoAdk at 20° at intermediate substrate concentrations, two peaks are observed (marked with bold arrows in (A)) indicative of a slow exchange process. (B) Lineshape simulation of the titration data from (A) according to Scheme 2 using the density matrix formalism (Binsch, 1969) with the same kinetic parameters as determined from the relaxation experiments in Fig. 3. Experimental (E) and theoretical (F) ^{31}P lineshapes of AMP at different concentrations of protein/ligands (AMP/AMPPNP/Mg) as indicated were fit with similar parameters to those in (B) (adapted from Wolf-Watz et al., 2004).

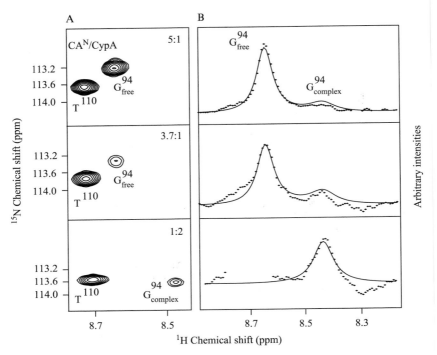

Fig. 5. Quantification of binding kinetics between CypA and CA^N. (A) ^{1}H–^{15}N TROSY HSQC spectra are shown for a titration of CA^N with CypA using CA^N/CypA ratios of 5:1, 3.7:1, and 1:2. (B) ^{1}H slices for Gly-94 were fitted with binding parameters $k_{on} = 3.4 \times 10^6\ M^{-1}\ s^{-1}$, $k_{off} = 45\ s^{-1}$, and $K_d = 13\ \mu M$, where the simulated K_d agrees well with the previously reported K_d of 16 μM determined by Biacore. In the 1:2 ratio several points in the ^{1}H slice had to be removed due to overlap from Thr-110.

site loop (Williams and McDermott, 1995). Both studies clearly reveal that ligand binding is followed by loop opening/closing in the microsecond time regime. While the populations determined and rates of the opening and closing processes are semiquantitative so far and somewhat different in both studies, the results clearly demonstrate that the time scale of this motion is in the same order as the turnover number.

The enzyme dihydrofolate reductase (DHFR) uses NADPH to reduce 5,6-dihydrofolate to 5,6,7,8-tetrahydrofolate. DHFR has been studied extensively by enzyme kinetics, crystallography, NMR, and computation. A complete kinetic scheme for this two-substrate two-product reaction is available (Fierke *et al.*, 1987), and loop and domain movements have been studied both by crystallography (Sawaya and Kraut, 1997) and NMR (Falzone *et al.*, 1994; Osborne *et al.*, 2001, 2003; Schnell *et al.*, 2004). An emerging picture arises in which the conformational transitions between an open, closed, and occluded conformation of the Met-20 loop play a crucial role for catalysis. Wright and co-workers have studied the dynamics of this loop by (1) 2D NOESY experiments for free DHFR, revealing exchange rates of about 35 s^{-1} (Falzone *et al.*, 1994), (2) ^{15}N backbone dynamics using the model-free approach (Osborne *et al.*, 2001), and (3) chemical shift analysis of different complexes with substrate analogues (Osborne *et al.*, 2003), and (4) relaxation dispersion experiments (Schnell *et al.*, 2004). The data indicate that the active site loop interconverts between the closed and occluded conformations.

^{15}N and ^{1}H backbone relaxation rates measured by $R_{1\rho}$ and CPMG experiments in HIV-1 protease identified microsecond motions in the flaps and the β-sheet intermonomer interface in the free enzyme that are suppressed by a tight-binding inhibitor (Ishima *et al.*, 1999). Since this flap is closing upon substrate or inhibitor binding, a link between the dynamics and function is apparent.

As pointed out earlier, transverse relaxation experiments provide rates of conformational transitions but no information about the amplitude and direction of movement. If the chemical shifts of the interconverting states can be determined, they can offer some information about the nature of the conformational transition. Another elegant approach to gain structural insight into rearrangements on the microsecond time regime has been performed on the tetrameric carbonmonoxy-hemoglobin (HbCO) using ^{15}N–^{1}H residual dipolar couplings (RDC) in weakly oriented samples (for details about this method, see Prestegard *et al.*, Chapter 7, this volume and Lukin *et al.*, 2003). The RDC data in combination with ^{15}N relaxation experiments show that HbCO does not reside in one conformation but rather rapidly interconverts between the crystallographically observed R and R2 states.

CPMG dispersion experiments performed on free ribonuclease A (RNase A) identified 28 backbone amides with conformational exchange (Cole and Loria, 2002); many of them are in fact located in the active site. Similar rates of about 1600 s^{-1} at 298 K suggest that a collective process is detected. For this enzyme, the turnover numbers span a wide range depending on the substrate (from 5 to 3000 s^{-1}) (delCardayre *et al.*, 1995; Tarragona-Fiol *et al.*, 1993; Witzel and Barnard, 1962). Notably, the highest turnover number is comparable to the rates of conformational exchange. This correlation was pointed out by Cole and Loria (2002) and interpreted as conformational motion should consequently not be limiting for catalysis. However, this result more likely leads to the possibility that rates of conformational rearrangements represent the upper limit for catalysis of good substrates.

Similar qualitative results were seen in backbone dynamics studies on ribonuclease binase (Wang *et al.*, 2001), α-lytic protease (Davis and Agard, 1998), and glycerol-3-phosphate cytidyltransferase (Stevens *et al.*, 2001), in which microsecond motions were detected in active sites and allosteric sites that may have implications for catalysis.

Conclusions

Application of a series of dynamic NMR methods that detect conformational exchange on the microsecond to millisecond time scale in enzymes exhibits the following features: (1) dynamics can be measured at multiple sites simultaneously, (2) thermodynamic (populations of exchanging species), kinetic (rate constants of interconversion), and structural information (chemical shifts, RDCs) can be obtained for the dynamic processes, (3) dynamics can be studied in enzymes during catalysis, and (4) enzyme dynamics corresponding to different microscopic steps of the enzymatic cycle can be separated. The comparison of enzyme dynamics measured during turnover and dynamics data on a number of other resting enzymes to the turnover numbers suggests that in some fast enzymes ($k_{cat} >$ 100 s^{-1}) nature has evolved to the point where the catalytic power might be limited by the frequency of conformational rearrangements (Wolf-Watz *et al.*, 2004). The application of CPMG, $R_{1\rho}$, and other methods to a larger number of enzymes during catalysis is clearly needed to test this hypothesis.

Acknowledgments

This work was supported by NIH Grants GM67963 and GM62117 to D.K. and a postdoctoral fellowship from the Swedish Research Council to M.W.W. Part of the NMR studies were carried out at the NHMFL at Florida with support from the NSF.

References

Binsch, G. (1969). A unified theory of exchange effects on nuclear magnetic resonance line shapes. *J. Am. Chem. Soc.* **91**, 1304–1309.

Bosco, D. A., Eisenmesser, E. Z., Pochapsky, S., Sundquist, W. I., and Kern, D. (2002). Catalysis of cis/trans isomerization in native HIV-1 capsid by human cyclophilin A. *Proc. Natl. Acad. Sci. USA* **99**, 5247–5252.

Braaten, D., and Luban, J. (2001). Cyclophilin A regulates HIV-1 infectivity, as demonstrated by gene targeting in human T cells. *EMBO J.* **20**, 1300–1309.

Carver, J. P., and Richards, R. E. (1972). A general two-site solution for the chemical exchange produced dependence of T2 upon the Carr-Purcell pulse separation. *J. Magn. Reson.* **6**, 89–105.

Cole, R., and Loria, J. P. (2002). Evidence for flexibility in the function of ribonuclease A. *Biochemistry* **41**, 6072–6081.

D'Amico, S., *et al.* (2003). Activity-stability relationships in extremophilic enzymes. *J. Biol. Chem.* **278**, 7891–7896.

Davis, J. H., and Agard, D. A. (1998). Relationship between enzyme specificity and the backbone dynamics of free and inhibited alpha-lytic protease. *Biochemistry* **37**, 7696–7707.

delCardayre, S. B., Ribo, M., Yokel, E. M., Quirk, D. J., Rutter, W. J., and Raines, R. T. (1995). Engineering ribonuclease A: Production, purification and characterization of wild-type enzyme and mutants at Gln11. *Protein. Eng.* **8,** 261–273.

Eisenmesser, E. Z., Bosco, D. A., Akke, M., and Kern, D. (2002). Enzyme dynamics during catalysis. *Science* **295,** 1520–1523.

Falzone, C. J., Wright, P. E., and Benkovic, S. J. (1994). Dynamics of a flexible loop in dihydrofolate reductase from *Escherichia coli* and its implication for catalysis. *Biochemistry* **33,** 439–442.

Farrow, N. A., Zhang, O., Forman-Kay, J. D., and Kay, L. E. (1994). A heteronuclear correlation experiment for simultaneous determination of 15N longitudinal decay and chemical exchange rates of systems in slow equilibrium. *J. Biomol. NMR* **4,** 727–734.

Fersht, A. (2000). "Structure and Mechanism in Protein Science." W. H. Freeman, New York.

Fierke, C. A., Johnson, K. A., and Benkovic, S. J. (1987). Construction and evaluation of the kinetic scheme associated with dihydrofolate reductase from *Escherichia coli. Biochemistry* **26,** 4085–4092.

Fischer, G., Bang, H., and Mech, C. (1984). Determination of enzymztic catalysis for the cis-trans-isomerization of peptide binding in proline-containing peptides. *Biomed. Biochim. Acta* **43,** 1101–1111.

Fischer, M. W. F., Majumdar, A., and Zuiderweg, E. R. P. (1998). Protein NMR relaxation: Theory, applications and outlook. *Prog. NMR Spectrosc.* **33,** 207–272.

Franke, E. K., Yuan, H. E., and Luban, J. (1994). Specific incorporation of cyclophilin A into HIV-1 virions. *Nature* **372,** 359–362.

Gamble, T. R., Vajdos, F. F., Yoo, S., Worthylake, D. K., Houseweart, M., Sundquist, W. I., and Hill, C. P. (1996). Crystal structure of human cyclophilin A bound to the amino-terminal domain of HIV-1 capsid. *Cell* **87,** 1285–1294.

Gitti, R. K., Lee, B. M., Walker, J., Summers, M. F., Yoo, S., and Sundquist, W. I. (1996). Structure of the amino-terminal core domain of the HIV-1 capsid protein. *Science* **273,** 231–235.

Ishima, R., and Torchia, D. A. (2000). Protein dynamics from NMR. *Nat. Struct. Biol.* **7,** 740–743.

Ishima, R., Freedberg, D. I., Wang, Y. X., Louis, J. M., and Torchia, D. A. (1999). Flap opening and dimer-interface flexibility in the free and inhibitor-bound HIV protease, and their implications for function. *Struct. Fold. Des.* **7,** 1047–1055.

Jaenicke, R., and Bohm, G. (1998). The stability of proteins in extreme environments. *Curr. Opin. Struct. Biol.* **8,** 738–748.

Jardetzky, O. (1996). Protein dynamics and conformational transitions in allosteric proteins. *Prog. Biophys. Mol. Biol.* **65,** 171–219.

Kay, L. E. (1998). Protein dynamics from NMR. *Nat. Struct. Biol.* **5**(Suppl.), 513–517.

Kern, D., Kern, G., Scherer, G., Fischer, G., and Drakenberg, T. (1995). Kinetic analysis of cyclophilin-catalyzed prolyl cis/trans isomerization by dynamic NMR spectroscopy. *Biochemistry* **34,** 13594–13602.

Lukin, J. A., Kontaxis, G., Simplaceanu, V., Yuan, Y., Bax, A., and Ho, C. (2003). Quaternary structure of hemoglobin in solution. *Proc. Natl. Acad. Sci. USA* **100,** 517–520.

Mulder, F. A., Mittermaier, A., Hon, B., Dahlquist, F. W., and Kay, L. E. (2001). Studying excited states of proteins by NMR spectroscopy. *Nat. Struct. Biol.* **8,** 932–935.

Mulder, F. A., Hon, B., Mittermaier, A., Dahlquist, F. W., and Kay, L. E. (2002). Slow internal dynamics in proteins: Application of NMR relaxation dispersion spectroscopy to methyl groups in a cavity mutant of T4 lysozyme. *J. Am. Chem. Soc.* **124,** 1443–1451.

Muller, C. W., and Schulz, G. E. (1992). Structure of the complex between adenylate kinase from *Escherichia coli* and the inhibitor Ap5A refined at 1.9 A resolution. A model for a catalytic transition state. *J. Mol. Biol.* **1,** 159–177.

Muller, C. W., *et al.* (1996). Adenylate kinase motions during catalysis: An energetic counterweight balancing substrate binding. *Structure* **4**, 147–156.

Osborne, M. J., Schnell, J., Benkovic, S. J., Dyson, H. J., and Wright, P. E. (2001). Backbone dynamics in dihydrofolate reductase complexes: Role of loop flexibility in the catalytic mechanism. *Biochemistry* **40**, 9846–9859.

Osborne, M. J., Venkitakrishnan, R. P., Dyson, H. J., and Wright, P. E. (2003). Diagnostic chemical shift markers for loop conformation and substrate and cofactor binding in dihydrofolate reductase complexes. *Protein Sci.* **12**, 2230–2238.

Palmer, A. G., 3rd (1997). Probing molecular motion by NMR. *Curr. Opin. Struct. Biol.* **7**, 732–737.

Palmer, A. G., Williams, J., and McDermott, A. (1996). Nuclear magnetic resonance studies of biopolymer dynamics. *J. Phys. Chem.* **100**, 13293–13310.

Palmer, A. G., 3rd, Kroenke, C. D., and Loria, J. P. (2001). Nuclear magnetic resonance methods for quantifying microsecond-to-millisecond motions in biological macromolecules. *Methods Enzymol.* **339**, 204–238.

Petsko, G. A. (2001). Structural basis of thermostability in hyperthermophilic proteins, or "there's more than one way to skin a cat". *Methods Enzymol.* **334**, 469–478.

Rozovsky, S., Jogl, G., Tong, L., and McDermott, A. E. (2001). Solution-state NMR investigations of triosephosphate isomerase active site loop motion: Ligand release in relation to active site loop dynamics. *J. Mol. Biol.* **310**, 271–280.

Sawaya, M. R., and Kraut, J. (1997). Loop and subdomain movements in the mechanism of *Escherichia coli* dihydrofolate reductase: Crystallographic evidence. *Biochemistry* **36**, 586–603.

Schnell, J. R., Dyson, H. J., and Wright, P. E. (2004). Structure, dynamics, and catalytic function of dihydrofolate reductase. *Annu. Rev. Biophys. Biomol. Struct.* **33**, 119–140.

Stevens, S. Y., Sanker, S., Kent, C., and Zuiderweg, E. R. (2001). Delineation of the allosteric mechanism of a cytidyltransferase exhibiting negative cooperativity. *Nat. Struct. Biol.* **8**, 947–952.

Tarragona-Fiol, A., Eggelte, H. J., Harbron, S., Sanchez, E., Taylorson, C. J., Ward, J. M., and Rabin, B. R. (1993). Identification by site-direced mutagenesis of aminoacids in the B2 subsite of bovine pancreatic ribonuclease A. *Protein Eng.* **6**, 901–906.

Vonrhein, C., Schlauderer, G. J., and Schulz, G. E. (1995). Movie of the structural changes during a catalytic cycle of nucleoside monophosphate kinases. *Structure* **3**, 483–490.

Wang, L., Pang, Y., Holder, T., Brender, J. R., Kurochkin, A. V., and Zuiderweg, E. R. (2001). Functional dynamics in the active site of the ribonuclease binase. *Proc. Natl. Acad. Sci. USA* **98**, 7684–7689.

Williams, J. C., and McDermott, A. E. (1995). Dynamics of the flexible loop of triosephosphate isomerase: The loop motion is not ligand gated. *Biochemistry* **34**, 8309–8319.

Witzel, H., and Barnard, E. A. (1962). Mechanism and binding sites in the ribonuclease reaction. II. Kinetic studies on the first step of the reaction. *Biochem. Biophys. Res. Commun.* **17**, 295–299.

Wolf-Watz, M., *et al.* (2004). Linkage between dynamics and catalysis in a thermophilic-mesophilic enzyme pair. *Nat. Struct. Mol. Biol.* **11**, 945–949.

Zavodszky, P., *et al.* (1998). Adjustment of conformational flexibility is a key event in the thermal adaptation of proteins. *Proc. Natl. Acad. Sci. USA* **95**, 7406–7411.

Zhao, Y., and Ke, H. (1996). Mechanistic implication of crystal structures of the cyclophilin-dipeptide complexes. *Biochemistry* **35**, 7362–7368.

[22]   Structure Determination of Protein/RNA Complexes by NMR

By Haihong Wu, L. David Finger, and Juli Feigon

Abstract

Structure determination of protein/RNA complexes in solution provides unique insights into factors that are involved in protein/RNA recognition. Here, we review the methodology used in our laboratory to overcome the challenges of protein/RNA structure determination by nuclear magnetic resonance (NMR). We use as two examples complexes recently solved in our laboratory, the nucleolin RBD12/b2NRE and Rnt1p dsRBD/snR47h complexes. Topics covered are protein and RNA preparation, complex formation, identification of the protein/RNA interface, protein and RNA resonance assignment, intermolecular NOE assignment, and structure calculation and analysis.

Introduction

A complete understanding of RNA biology requires study of proteins that interact with RNA, because most RNAs are stably bound to or transiently associated with one or more proteins *in vivo* (Hall, 2002; Perez-Canadillas and Varani, 2001). Several classes of protein domains that mediate RNA binding have been identified [e.g., the RBD (also known as RRM or RNP-motif) (Birney *et al.*, 1993), dsRBD (Saunders and Barber, 2003), KH (Nagai, 1996), La (Alfano *et al.*, 2004), zinc finger (Berg, 2003), and CAT (Yang *et al.*, 2002) domains] and their structures determined by both X-ray crystallography and nuclear magnetic resonance (NMR). Of greatest interest to those studying RNA-binding proteins is how these proteins select their target sequence. Although structure determination of the isolated protein domains by NMR is straightforward, solving the solution structures of protein/RNA complexes remains extremely challenging. To date (May 2004), only 13 protein/RNA complexes have been determined by NMR and deposited in the Protein Data Bank (Table I). While the number of crystal structures significantly exceeds this, NMR has some unique advantages over crystallography and can provide aditional insight into the factors involved in recognition of RNA by proteins. For example, formation of protein/RNA complexes almost always involves conformational changes in the protein, the RNA, or both (Leulliot

Copyright 2005, Elsevier Inc.
All rights reserved.
0076-6879/05 $35.00

TABLE I

NMR-Derived Protein/RNA Complexes in the Protein Data Bank (PDB)[a,c]

PDB ID	Type of RNA-binding domain	No. of aa	No. of nt	Total MW (kDa)	Year deposited	Reference
1AUD	RBD (RNP-type)	101	30	22	1997	Allain et al. (1997)
1A1T	Zinc finger	55	20	14	1997	De Guzman et al. (1998)
1CN9[b]	rRNA protein	104	33	22	1999	Mao et al. (1999)
1D6K	rRNA protein	94	37	23	1999	Stoldt et al. (1999)
1DZ5	RBD (RNP-type)	101 × 2	22 × 2	38	2000	Varani et al. (2000)
1EKZ	dsRBD	76	30	19	2000	Ramos et al. (2000a)
1F6U	Zinc finger	55	19	13	2000	Amarasinghe et al. (2000)
1FJE	RBDs (RNP-type)	175	22	27	2000	Allain et al. (2000a)
1K1G	KH	131	11	18	2001	Liu et al. (2001)
1L1C	CAT	56 × 2	29	22	2002	Yang et al. (2002)
1RGO	Zinc fingers	70	9	11	2003	Hudson et al. (2004)
1RKJ	RBDs (RNP-type)	175	21	27	2003	Johansson et al. (2004)
1T4L	dsRBD	88	32	20	2004	Wu et al. (2004)

[a] A search for protein/RNA complexes in the PDB will give approximately 30 complexes. We have not included in the list peptide/RNA complexes ≤ 36 amino acids and those that are duplicates (i.e., minimized average structures).
[b] Related PDB IDs; 1CK5, 1CK8, 1CN8.
[c] As of May 1, 2004.

and Varani, 2001; Williamson, 2000). Because isolated protein or RNA domains may be partially unstructured, they may be difficult to crystallize in biologically relevant conformations, while NMR methods can be used to identify both structured and unstructured regions. Thus, the structures of the free components, as well as the complex, are sometimes more amenable to studies by NMR. NMR methods can also provide information on the dynamics as well as the binding affinity of the complex. Here we review methods currently being used in our laboratory for preparation of protein/ RNA complexes suitable for NMR studies, monitoring complex formation, determination of the protein–RNA interface, assignment of the RNA in the complex, identification of intermolecular NOEs, and structure calculation. We use as examples two protein/RNA complexes whose structures were determined in our laboratory, containing the two most common RNA-binding modules, the RBD and the dsRBD (Fig. 1). The complex of two tandem RBDs separated by a 12-amino-acid linker of nucleolin with a 21-nt RNA is the largest nonsymmetrical protein/RNA complex determined by NMR reported to date (Johansson *et al.*, 2004) and thus illustrates challenges associated with larger complexes. The dsRBD of *Saccharomyces*

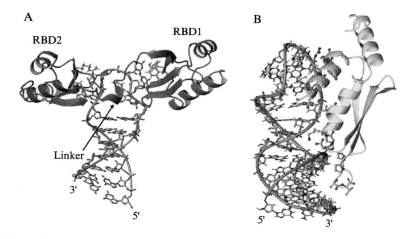

Fig. 1. Lowest energy structures of the (A) nucleolin RBD12/b2NRE (Johansson *et al.*, 2004) and (B) Rnt1p dsRBD/snR47h complexes (Wu *et al.*, 2004). The proteins are shown as a ribbon and the RNAs as sticks. The nucleolin RBD12/b2NRE complex contains the first two RBDs plus linker residues of hamster nucleolin (175 aa) and a 21-nucleotide RNA hairpin derived from mouse pre-rRNA (Ghisolfi-Nieto *et al.*, 1996). The two RBDs ($\beta\alpha\beta\beta\alpha\beta$-fold) and linker interact with the conserved loop consensus UCCCGA. The Rnt1p dsRBD/snR47h complex contains the C-terminal dsRBD of yeast Rnt1p (88 aa) and a 32-nucleotide RNA derived from the 5′-hairpin of the small nucleolar box C/D RNA (snoRNA) snR47 precursor (Chanfreau *et al.*, 2000). The dsRBD ($\alpha\beta\beta\beta\alpha$-fold + additional C-terminal α) recognizes the conserved fold of the AGNN tetraloop.

cerevisiae RNase III binds structure specifically to double-stranded RNA capped by an AGNN tetraloop and thus illustrates challenges associated with RNA-binding proteins that do not recognize a specific sequence, but instead interact primarily with the RNA backbone (Wu *et al.*, 2004).

Sample Preparation

Protein Preparation

Like studies of proteins alone, milligram quantities of unlabeled and isotopically enriched protein are necessary for solution studies of protein/RNA complexes. Procedures for the efficient preparation of ^{15}N-, ^{13}C,^{15}N-, and ^{2}H,^{13}C,^{15}N-labeled proteins have been previously described (Gardner and Kay, 1998; Marley *et al.*, 2001; Muchmore *et al.*, 1989). Since isotopic enrichment can become prohibitively expensive in NMR studies, protein expression in *Escherichia coli* should be optimized for maximum yield (Makrides, 1996). For many eukaryotic RNA-binding proteins, the presence of "rare-codons" in a given gene (http://www.lifesci.ucsb. edu/~maduro/codonusage/usage2.0c.htm) is a limiting factor in protein expression yields (Kane, 1995; Makrides, 1996). In many cases, the efficacy of eukaryotic protein labeling and production is greatly increased by use of *E. coli* BL21 (DE3) codon-plus strains (RIL, RP, and Rosetta), which are commercially available from Stratagene and Novagen.

Since the protein will eventually be complexed with RNA, it is of utmost importance to remove any contaminating RNases from the protein. RNase-free protein can be easily obtained using a three-step, native purification protocol that consists of (1) affinity chromatography [immobilized metal affinity chromatography (IMAC), GST-fusion, MBP-fusion, etc.] (Makrides, 1996), (2) ion-exchange chromatography, and (3) size-exclusion chromatography. In most cases, the protein is sufficiently pure for NMR studies after the first affinity chromatography step, but the remaining two steps are necessary to ensure complete removal of trace RNase activity. Because protein purification is most often done with native protein, nucleic acid contaminants from lysates are sometimes carried by the RNA-binding protein throughout all three purification steps. Proteins exhibiting this problem are treated with 0.1% polyethyleneimine (PEI) at high ionic strength (1 *M* NaCl) to remove nucleic acid contaminants (Burgess, 1991). Note that addition of this step also requires ammonium sulfate precipitation of the protein to remove excess PEI (Burgess, 1991) and subsequent dialysis before affinity chromatography. In addition, removal of the affinity tags from the recombinant proteins is often necessary or desirable and can easily be accomplished using the protease appropriate for the engineered cleavage site. We have noticed that some proteins like nucleolin RBD12, which was originally expressed with (His)$_6$-tag with a thrombin cleavage site (Allain *et al.*, 2000a,b), have secondary thrombin cleavage sites within protein constructs. This type of problem usually requires that the protein be expressed from a vector having an engineered protease site that uses more specific proteases like enterokinase (LaVallie *et al.*, 1993) or tobacco etch virus (TEV) protease (Polayes *et al.*, 1994). Enterokinase can sometimes be prohibitively expensive; therefore, we prefer to use TEV protease because it can be produced in the laboratory (Lucast *et al.*, 2001).

RNA Preparation

At the beginning of protein/RNA complex structural studies in our laboratory, RNAs are prepared with no isotopic labeling in order to try different methods of complex formation and to optimize solution conditions for the complex. Unlabeled RNA for studies of complexes can be prepared either chemically (Scaringe, 2001) (Dharmacon, Inc. Lafayette, CO) or enzymatically using phage RNA polymerases [T7 (Milligan and Uhlenbeck, 1989) or SP6 (Stump and Hall, 1993)]. For structure determination of the RNA in the protein/RNA complex, uniformly ^{13}C,^{15}N-labeled and ^{13}C,^{15}N-base-type-specific-labeled RNA samples are necessary and can be prepared enzymatically using the appropriate ^{13}C,^{15}N-enriched nucleotide triphosphates (NTPs) (Dieckmann and Feigon, 1997), which can obtained from commercial sources (VLI Research Inc., Cambridge Isotopes) or prepared in house by well-established procedures (Batey *et al.*, 1995; Nikonowicz *et al.*, 1992). Regardless of the method used to synthesize the RNA (i.e., chemical or enzymatic), purification from failure sequences and/or nontemplated addition byproducts is necessary and, in our experience, is most efficiently accomplished by denaturing polyacrylamide gel electrophoresis (PAGE) (7 M urea) with subsequent electroelution and anion-exchange chromatography for the removal of acrylamide polymers (Feigon *et al.*, 2001). Other laboratories have shown that RNA purification can also be accomplished by high-performance liquid chromatography (HPLC) using anion-exchange (Anderson *et al.*, 1996) or reverse-phase chromatography (Murray *et al.*, 1994). In addition, RNAs can be prepared enzymatically with a designed 3'-end sequence and purified using a *trans*-hammerhead ribozyme cleavage with subsequent HPLC purification by anion-exchange chromatography (Shields *et al.*, 1999). Once purified, the RNA should be exchanged into the appropriate buffer using dialysis or a desalting column. Optimally, the buffer used for the RNA will be the same as that used for the protein. However, in the case of the nucleolin RBD12/b2NRE complex, a hairpin-dimer equilibrium was observed for free b2NRE at the higher salt concentration used for the protein. Therefore, a simple RNA folding protocol was utilized to ensure that when forming the protein/RNA complex, only the hairpin conformation of b2NRE was present (Finger *et al.*, 2003; Johansson *et al.*, 2004).

Complex Formation

The first step in NMR studies of protein/RNA complexes is finding optimal solution conditions for the complex. This generally means acquiring one-dimensional (1D) and two-dimensional (2D) NMR spectra of the complex in different pH and salt conditions, as is usually also done for the free proteins and RNAs. In our experience, the success of NMR studies of protein/RNA complexes is sometimes dependent on the method in which the complex is made. For instance, the Rnt1p dsRBD/snR47h complex precipitates when the RNA is titrated into the protein, while titrating protein into RNA does not cause precipitation. Some key aspects to the success of any study of protein/RNA complexes are first knowing the stoichiometry of the complex (i.e., 1:1 or 2:1 for the protein:RNA ratio) and having accurate concentrations of the protein and RNA stocks from which the complex is to be made. The stoichiometry of a complex is usually known before NMR studies through basic biochemical experiments (e.g., gel shift assays). Concentrations of the protein and RNA stocks are determined using the molar extinction coefficients calculated using the biopolymer calculator (http://paris.chem.yale.edu/extinct.html). Because these extinction coefficients are calculated assuming no secondary or tertiary structure, the RNA concentration is determined under conditions where it is single stranded (i.e., no added salt and/or high temperature). We have found that this method is quite reliable for determining the correct protein:RNA ratio. Complex formation is also routinely followed by NMR, as discussed below.

Following Complex Formation via NMR and Chemical
  Shift Mapping

The easiest way to follow complex formation via NMR is to add unlabeled RNA to a ^{15}N-labeled protein, while monitoring changes in the chemical shifts of the amide signals using ^{1}H–^{15}N-heteronuclear single-quantum correlations (HSQCs) as a function of RNA concentration. For complexes in slow exchange, a subset of the amide resonances will split into two peaks, corresponding to free and bound. Complete disappearance of the "free" peaks indicates stoichiometric binding of the protein to the RNA binding site(s). For complexes in fast exchange, the subset of resonances will just shift as a function of added RNA. Initial titrations should be continued until the peaks stop shifting. In optimal cases, this will happen at the appropriate protein:RNA ratio. If a large excess of RNA is needed to observe the maximum chemical shift changes, then the complex may be binding with too low of an affinity to observe intermolecular NOEs. Acquisition of ^{1}H–^{15}N-HSQCs as a function of RNA concentration not only ensures that the correct protein:RNA ratio is reached, but also allows chemical shift changes on the protein to be mapped to globally define the protein–RNA interface (Allen *et al.*, 2001). To do this without large changes in volume and buffer conditions, we usually dissolve a substoichiometric aliquot of lyophilized RNA in the protein solution. Once dissolved, the solution is returned to the NMR tube and a 2D ^{1}H–^{15}N-HSQC spectrum is acquired. Substoichiometric aliquots of RNA are then added in the same manner, and ^{1}H–^{15}N-HSQC spectra are taken consecutively until changes in the spectra are no longer observed. This was the method of choice for the nucleolin RBD12/b2NRE complex (Johansson *et al.*, 2004). Note, the RNA could also be added as a concentrated solution, but there were complications in that method because b2NRE displayed a hairpin-dimer equilibrium at high RNA concentrations and above 5 mM monovalent salt (Finger *et al.*, 2003). In this case, the substoichiometric aliquots of b2NRE were diluted in water, folded by heating and annealing, flash-frozen, and then lyophilized. This procedure ensured that the hairpin conformation of b2NRE was presented to the protein during titrations (Johansson *et al.*, 2004). In the case of the Rnt1p dsRBD/snR47h complex, a similar titration would work only if the concentrations of the protein did not exceed 0.5 mM. Above this concentration, titrations done this way resulted in precipitation of the Rnt1p dsRBD/snR47h complex. Although titration data at low concentration of protein/RNA complex are useful for chemical shift perturbation mapping of the protein, higher concentration samples needed to obtain 2D and three-dimensional (3D) spectra for assignment were made by titrating 0.1 mM Rnt1p dsRBD samples into 0.1 mM snR47h RNA samples to a stoichiometric equivalence followed by concentration of the complex sample to ~1.0 mM by ultrafiltration (Amicon). Interestingly, this method of complex formation avoided the precipitation that was observed when a 1 mM RNA sample was titrated with Rnt1p dsRBD.

Complex formation should also be followed by observing changes in the RNA resonances as a function of added protein to ensure that the protein is binding a specific site on the RNA. In this case, ^{1}H–^{13}C-HSQC or ^{1}H–^{13}C-heteronuclear multiple-quantum correlation (HMQC) spectra of uniformly ^{13}C,^{15}N-labeled (or ^{13}C,^{15}N-base-type-specific-labeled) RNA are acquired to monitor chemical shift changes of the aromatic and anomeric proton and carbon resonances as a function of protein concentration.

This also allows initial identification of the interaction surface of the RNA. For instance, mapping the changes in chemical shift between the nucleolin RBD12/sNRE (Allain *et al.*, 2000a) and RBD12/b2NRE complexes allowed us to determine that the consensus sequences were structurally similar in the two complexes even without the nucleolin RBD12/b2NRE structure (Finger *et al.*, 2003). It is sometimes possible to follow complex formation on the RNA as a function of protein concentration by observing changes in the imino proton spectra; however, it is frequently the case that imino protons in stem regions do not display large changes in chemical shift upon complex formation and resonances from iminos at the protein–RNA interface are often line broadened.

Although chemical shift perturbation analysis is useful to identify the protein–RNA interface, such analysis must be made with some caution, since chemical shift changes can be observed even if the residue is remote from the protein/RNA interface due to allosteric changes in the protein and/or RNA or to small changes in solution conditions (Foster *et al.*, 1998; Ramos *et al.*, 2000b). Recently, cross-saturation methods for mapping protein–RNA interaction surfaces have been introduced (Lane *et al.*, 2001; Ramos *et al.*, 2000b). In this method, resonance "windows" that are specific to one of the components of the protein–RNA complex (i.e., H1′, H5, and imino resonances of the RNA) are irradiated, and saturation transfer is detected on protein resonances (i.e., amides) (Ramos *et al.*, 2000b). Although saturation transfer can better localize the protein–RNA interface, this method can lead to an underestimate of the extent of the protein–RNA interface. In conclusion, a combination of both methods could lead to a more accurate prediction of the protein–RNA interface in the absence of structures.

Assignment Methodology

Assignment of Protein Resonances in the Complex

Since 1990, a large number of multidimensional heteronuclear triple-resonance experiments have been developed to correlate intraresidue and/or sequential spin systems of uniformly $^{13}C,^{15}N$-labeled proteins using one-bond and two-bond scalar coupling interactions. Using a combination of these experiments, $^1H^N$ and ^{15}N resonances are correlated with intraresidue and sequential spin systems, which, coupled with characteristic chemical shifts of the $^{13}C^\alpha$ and $^{13}C^\beta$ resonances, leads to unambiguous sequential assignment of the protein (Sattler *et al.*, 1999). Protein assignment methods will not be reviewed here. For smaller protein–RNA complexes, like the Rnt1p dsRBD–snR47h complex (20 kDa), protein assignments can be made with just $^{13}C,^{15}N$ labeling. However, for the nucleolin RBD12–b2NRE complex (28 kDa), obtaining protein backbone and side chain resonance assignments using the triple resonance methods mentioned above requires that the protein, in addition to being uniformly $^{13}C,^{15}N$ labeled, be uniformly and partially deuterated (100 and 70%) as well. Protein assignment experiments of partially and perdeuterated samples require slight modifications of existing protein correlation pulse programs to incorporate deuterium decoupling during chemical shift evolutions (Gardner and Kay, 1998; Sattler et al., 1999). In addition to deuteration, assignment of the protein in larger protein–RNA complexes usually benefits from acquiring some spectra on high-field NMR spectrometers (800/900 MHz) and application of transverse relaxation optimized spectroscopy (TROSY), which makes use of the favorable relaxation properties of the single-transition spin state for resolution and sensitivity enhancement (Pervushin, 2000).

Assignment of RNA Resonances in the Complex

NMR spectra of RNAs in protein–RNA complexes are more difficult to assign than the protein due to the lack of unambiguous correlation-based assignment methodology, poor chemical shift dispersion in both proton and carbon dimensions, and unfavorable relaxation properties. The assignment of RNAs in complex with proteins is basically an extension of the methods reviewed for the study of RNA alone (Dieckmann and Feigon, 1997; Furtig et al., 2003; Wijmenga and van Buuren, 1998). In our two examples, assignment of the RNA relied heavily on traditional NOE-based approaches (Wüthrich, 1986). One useful strategy for assigning the sequential base-to-sugar connectivities of RNAs in complexes is to make use of free RNA studies. For instance, chemical shift mapping of the b2NRE in complex with nucleolin RBD12 showed that the bottom six base pairs of the eight base-pair stem change very little upon protein binding. Therefore, the assignment of the free RNA stem facilitated assignment of the RNA in the complex. Although a 3D-NOESY-HMQC on a $^{13}C,^{15}N$-labeled RNA in complex with unlabeled protein could be used to assign the RNA in the complex, we have had more success using ^{13}C-F1fF2f NOESY experiments (Peterson et al., 2004) with samples consisting of $^{13}C,^{15}N$-labeled nucleolin RBD12 in complex with unlabeled b2NRE. This approach allows analysis of a 2D-NOESY of the protein–RNA complex that has all protein resonances removed, leaving only the RNA cross-peaks, which can be assigned by traditional methods (Wüthrich, 1986) used for RNA or DNA alone (Fig. 2). Once a sequential assignment of H1[9] and H6/H8 is established, various through-bond experiments on complexes

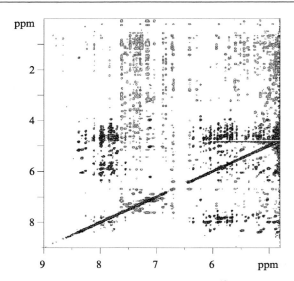

FIG. 2. Overlay of 750-MHz 2D NOESY (gray) and 2D ^{13}C-F1fF2f NOESY (Peterson *et al.*, 2004) (black) spectra of the nucleolin RBD12/b2NRE complex at 310 K used for assignment of RNA resonances. Since the nucleolin RBD12 protein is ^{13}C,^{15}N labeled, the protein resonances are "filtered out," leaving only the RNA intramolecular NOEs in the 2D ^{13}C-F1fF2f NOESY spectrum. The mixing time (τ_m) in both experiments was set to 150 ms.

containing ^{13}C,^{15}N-labeled RNA are used to assign all other protons (Furtig *et al.*, 2003; Wijmenga and van Buuren, 1998). Through-bond experiments that are routinely used in our laboratory include HCN-type experiments to correlate intranucleotide H1'–H6/H8, and HCCH-TOCSY and HCCH-COSY (Dieckmann and Feigon, 1997) to correlate the ribose spin systems (Table II). Often, the use of heteronuclear chemical shifts in the third dimension cannot resolve all assignment ambiguities. In these cases, complexes with ^{13}C,^{15}N-base-type-specific labeled RNAs are used (Dieckmann and Feigon, 1997). Another isotopic labeling technique that utilizes perdeuterated nucleotides was recently developed and used to assign the resonances of a 101-nucleotide RNA (D'Souza *et al.*, 2004). This new assignment strategy could potentially assist in the structure determination of large protein–RNA complexes as well. In addition, nucleotide analog substitution can be used to overcome severe spectral overlap characteristic of repeated RNA sequences (i.e., poly-N) and problems associated with sequences that cannot be assigned using NOE-based strategies due to the lack of secondary structure. This methodology, which has been previously applied for a protein–DNA complex (Mitton-Fry *et al.*, 2004),

TABLE II
EXPERIMENTS USED TO ASSIGN RNA RESONANCES IN PROTEIN/RNA COMPLEXES

| Experiment | Label on | | Use | Comments |
	Protein	RNA[a]		
NOESY (H$_2$O)			Assignment of imino sequential walk	Useful only when the RNA has base pairs
^{1}H–^{15}N-HMQC (H$_2$O)		X	Confirmation and assignment of exchangeable protons	HMQC used due to better relaxation properties of MQ
F1fF2f-^{13}C-NOESY (D$_2$O)	X		Assignment of the base–sugar sequential	Peterson et al. (2004)
^{1}H–^{13}C-HMQC (D$_2$O)		X	Confirmation and assignment of nonexchangeable protons	HMQC used due to better relaxation properties of MQ
2D or 3D HCN (D$_2$O)		X	Resolves assignment ambiguities in RNA	TROSY version (Riek et al., 2001)
AH2 HCCH-TOCSY (D$_2$O)		X	Unambiguously correlates adenine H2 and H8 protons	TROSY version (Simon et al., 2001)
2D or 3D HCCH-COSY (D$_2$O)		X	Unambiguously correlates H2' to H1'	
3D HCCH-TOCSY (D$_2$O)		X	Unambiguously correlates the entire sugar spin system	May require specific deuteration (Tolbert and Williamson, 1997) and use of a C(CC)H TOCSY (Dayie et al., 1998)
^{1}H–^{13}C NOESY HMQC (D$_2$O)		X	Identification of intra- and intermolecular NOEs	HMQC used due to better relaxation properties of MQ

[a] Labeled RNA samples can be uniformly ^{13}C,^{15}N labeled or ^{13}C, ^{15}N-U, C, G, or A labeled only.

assumes that the substitution affects only the local structure and has little effect on the global conformation of the complex. The observed chemical shift changes are due to alterations in the local chemical environment caused by the nucleotide analog substitution. This assignment strategy has recently been used to determine the structure of the zinc finger domain of TIS11d and the AU-rich element of its mRNA substrate (Hudson et al., 2004).

Assignment of Intermolecular NOEs

In protein–RNA complexes, the structure of the protein–RNA-binding interface is of the greatest interest and is mainly defined in NMR structures by intermolecular NOEs. Intermolecular NOEs can be distinguished from intramolecular NOEs in 2D and 3D NOESY experiments that employ X-filters and X/half-filters between the labeled and unlabeled components of complexes (Breeze, 2000). Due to the short T2 relaxation time of both protons and carbons in large protein–RNA complexes, the 2D NOESY experiments with X-filters and X/half-filters have a great advantage over 3D NOESY experiments in detecting intermolecular NOEs and are easier to interpret (Table III). We have optimized a suite of four filtered and/or edited NOESY experiments to detect intermolecular NOEs in large RNAs or protein–RNA complexes (Peterson *et al.*, 2004). In both the nucleolin RBD12/b2NRE and Rnt1p dsRBD/snR47h complexes, intermolecular NOEs between aromatic and anomeric protons and protein side chains

TABLE III
EXPERIMENTS USED TO ASSIGN INTERMOLECULAR NOEs

Experiment	Sample with label			Use	Comments
	Protein	RNA	A, G, U, or C only		
NOESY (D_2O)				Assignment of intra- and intermolecular NOEs	
3D ^{13}C-NOESY HMQC (D_2O)	X			Assignment of intra- and intermolecular NOEs	HMQC used due to better relaxation properties of MQ
3D ^{13}C-NOESY HMQC (D_2O)		X	X	Assignment of intra- and intermolecular NOEs	HMQC used due to better relaxation properties of MQ
3D ^{15}N-HSQC NOESY (H_2O)	X			Assignment of intra- and intermolecular NOEs	
^{13}C-F2f NOESY (D_2O)	X			Identification of intermolecular NOEs	Peterson *et al.* (2004)
^{13}C-F1fF2f NOESY (D_2O)		X	X	Identification of intermolecular NOEs	Peterson *et al.* (2004)

were easily identified by comparison of ^{13}C-F2f NOESY and ^{13}C-F1fF2f NOESY spectra because these regions have relatively good dispersion. The intermolecular NOEs identified in the above manner were largely absent in 3D NOESY-^{1}H–^{13}C-HMQC spectra of both complexes. In the case of the Rnt1p dsRBD/snR47h complex, many intermolecular NOEs were between the RNA phosphodiester backbone and the protein side chains, which are difficult to assign because of spectral overlap. The use of ^{13}C,^{15}N-base-type-specific-labeled RNA samples was necessary to resolve much of the ambiguity in the assignments. For instance, a ^{13}C-F1fF2e NOESY spectrum recorded on a ^{13}C,^{15}N-U-labeled snR47h/unlabeled Rnt1p dsRBD complex shows NOEs only between protons of ^{13}C,^{15}N-labeled uridines and those of unlabeled nucleotides and the protein. Thus, intermolecular NOE cross-peaks, which are overlapped in the ^{13}C-F2f NOESY spectrum (gray peaks), can be assigned to uridine protons and protein side chain protons (Fig. 3). This helps to narrow the choices of assignments and, in most cases, leads to unambiguous assignments.

Structure Calculations

Structure determination of proteins and nucleic acids by NMR requires quantification of NMR parameters that are sensitive to the conformation of the molecule, such as dipolar relaxation, scalar coupling, and chemical shifts (Guntert, 1998). Among them, NOEs are most important, because integration of NOESY cross-peaks is the source of the distance restraints used to calculate the structure. The attainable precision of distance constraints is limited by the signal-to-noise ratio, the peak overlap, and the unknown effects of molecular mobility, especially at the molecular interface (Guntert, 1998). Therefore, we generally utilize NOE-derived distance restraints that are categorized into strong, medium, and weak categories with the van der Waals radius used as the lower limit. Protein dihedral angle restraints, which are commonly used in conjunction with distance restraints in structure calculations, can be determined by measuring scalar couplings (Cavanagh et al., 1996) or inferred using ^{13}C chemical shifts (Oldfield, 1995). Dihedral angle constraints for the RNA component of the complex can also be obtained by measurement of scalar couplings (Marino et al., 1999).

Our laboratory has used simulated annealing protocols in Xplor (Brünger, 1992) with molecular dynamics approaches (Guntert, 1998) in structure calculation of both the nucleolin RDB12/b2NRE and the Rnt1p dsRBD/snR47h complexes. In our experience, the efficiency of thesimulated annealing protocols in searching for minimization of the pseudoenergy potentials varies with respect to the nature of the protein–RNA complexes.

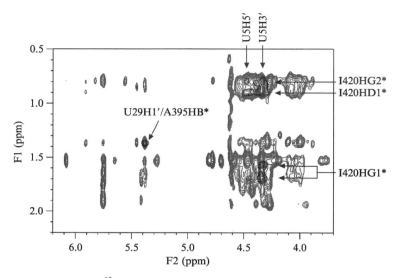

FIG. 3. Overlay of 2D ^{13}C-filtered/edited NOESY spectra of Rnt1p dsRBD/snR47h used for assignment of intermolecular NOEs. The gray cross-peaks are from a 2D ^{13}C-F2f NOESY (Peterson *et al.*, 2004) acquired on a ^{13}C,^{15}N–labeled Rnt1p dsRBD/unlabeled snR47h sample in D$_2$O. The black cross-peaks are from a 2D ^{13}C-F1fF2e NOESY (Peterson *et al.*, 2004) acquired on unlabeled Rnt1p dsRBD/U-selectively ^{13}C,^{15}N-labeled snR47h sample in D$_2$O. Both spectra were acquired on a Bruker 600-MHz spectrometer at 30°. Note, the gray peaks are the intermolecular NOE cross-peaks between protein side chain protons and RNA ribose protons, and the black peaks are intermolecular NOE peaks between uridine protons and the protein side chain protons. The assignments of the cross-peaks in the ^{13}C-F1fF2e NOESY spectrum are indicated.

Thus, different strategies are used for the structure calculation of each protein–RNA complex. Starting from extended templates with randomized dihedral angles, the protein and the RNA molecules of both complexes were separated by 70 Å and folded simultaneously through a simulated annealing protocol (from 2000 K to 100 K). For the dsRBD/snR47h complex, this step resulted in converged structures of the complex but with a high number (30–40) of NOE violations (>0.2 Å) even among the lowest energy structures. A second refinement (from 1000 K to 100 K) step was necessary to find an energy minimum with less than one NOE violation on average per structure (>0.2 Å) and no dihedral angle violations among the 15 lowest energy structures (Fig. 4A). On the other hand, the nucleolin RBD12/b2NRE complex, which has a more intricate protein–RNA interface than the dsRBD/snR47h complex, was commonly trapped in several conformations far from the energy minimum, thereby resulting in a low convergence rate. To increase convergence, the protein and RNA were

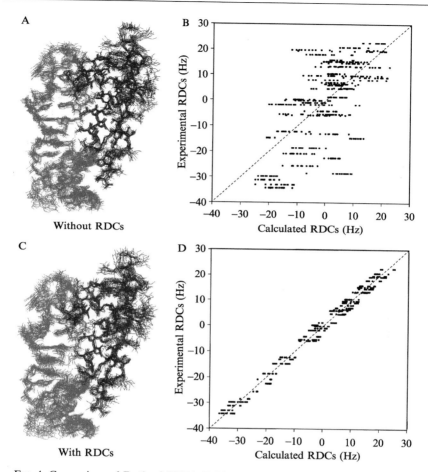

FIG. 4. Comparison of Rnt1p dsRBD/snR47h complex structures and dipolar couplings before (A and B) and after (C and D) refinement with 43 residual dipolar couplings. The 15 lowest energy structures of the Rnt1p dsRBD/snR47h complex before (A) and after (C) refinement with RDCs are superimposed with RMSD of 0.96 ± 0.10 Å and 1.04 ± 0.14 Å, respectively. The correlation between the experimental dipolar couplings and the calculated dipolar couplings from Rnt1p dsRBD/snR47h complex structures is given before (B) and after (D) refinement with RDCs.

separated again by 70 Å in space with random orientation and subjected to additional rounds of folding, similar to the procedure described for the U1A/PIE complex (Allen *et al.*, 2001).

Another strategy to determine protein–RNA complex structures is to calculate protein and RNA structures individually and then dock (Halperin *et al.*, 2002) the folded structures together using restrained rigid-body molecular dynamics (Clore, 2000), as demonstrated in the structure calculations of the Dead Ringer–DNA complex (Iwahara *et al.*, 2002). We have tested this approach on the two example protein–RNA complexes, with success for the Rnt1p dsRBD/snR47h complex and failure for the nucleolin RBD12/b2NRE complex (unpublished data). Failure of the latter probably reflects the fact that the structure of the b2NRE loop is not defined in the absence of intermolecular NOEs (i.e., the loop is not a preformed structure on which to dock). A new high ambiguity driven docking approach (HADDOCK) (Dominguez *et al.*, 2003) incorporates biochemical and/or biophysical information as ambiguous interaction restraints (AIRs), which are derived from chemical shift perturbation data, mutagenesis data, intermolecular NOEs, and RDCs, in the docking of protein structures. Its robustness in structure determination of protein–RNA complexes remains to be evaluated.

Application of Residual Dipolar Couplings (RDCs)

Although traditional NMR restraints like NOEs and scalar couplings are able to define the local structure of the protein and RNA components and binding interface of protein–RNA complexes very well, the accuracy and precision of large protein–RNA complex structures usually suffer from a lack of global restraints (de Alba and Tjandra, 2002; Prestegard *et al.*, 2000). Recently, these problems have been overcome by the incorporation of residual dipolar couplings (RDCs), which provide information about the global structure by reporting the orientation of bond vectors with respect to an alignment tensor of the entire molecule in the external magnetic field (Furtig *et al.*, 2003) into structure calculations. We have utilized the global structural information provided by RDCs in the structure determination of the two protein–RNA complexes used as examples in this chapter. In both cases, the RDCs were measured for protein $^1J_{HN}$ in dilute aqueous liquid crystalline media [i.e., C12E6/hexanol (Ruckert and Otting, 2000)] using t_1-coupled HSQCs. These RDCs were incorporated in the last stage of the structure refinement of nucleolin RBD12/b2NRE and Rnt1p dsRBD/snR47h complexes. The axial and rhombic components of the alignment tensor were determined using a grid search procedure (Clore *et al.*, 1998) during the simulated annealing refinement. For the nucleolin RDB12/b2NRE complex, the incorporation of 80 protein RDCs greatly improved the accuracy and precision of the complex structure, especially the orientation of two protein domains, RBD1 and RBD2, with respect to each other. For the Rnt1p dsRBD/snR47h complex, the incorporation of 43 protein RDCs did not improve the precision of the structures. The overall RMSD values of the 15 lowest energy structures are 0.96 ± 0.10 Å

(Fig. 4A) and 1.04 ± 0.14 Å (Fig. 4C) before and after the refinement against RDCs, respectively. However, the agreement between measured and predicted values of RDCs, which is a measure of the accuracy of the structures, was improved after refinement with RDCs (Fig. 4B and D). This demonstrates that application of RDCs in structure calculations can improve the accuracy of the global fold of the structure by providing long-range structural information and are highly complementary to the short-range NOE-derived distance restraints. In the structure refinement of the U1A/PIE complex (Allen et al., 2001; Bayer et al., 1999), a sample of ^{15}N-labeled U1A/^{13}C,^{15}N-labeled RNA complex in H_2O was used to measure $^1J_{HN}$ of protein amides and $^1J_{HN}$ and $^1J_{HC}$ of the RNA. This approach avoids the possible complication of differences in alignment tensors arising from the use of multiple samples. Another problem for large protein–RNA complexes is that spectral overlap for the RNA can preclude the measurement of a significant number of RDCs, thereby limiting their use. When measuring one-bond couplings of the nucleolin RBD12/b2NRE complex (28 kDa) from J-coupled spectra, we noticed that one of the components of the cross-peak doublets has a much broader linewidth than the other one, a well–documented phenomenon that is due to differences in relaxation (Kontaxis et al., 2000). To determine RDCs of large molecules and complexes more accurately and to increase the number of RDCs that can be measured, TROSY-based experiments have been developed (Evenas et al., 2001; Lerche et al., 1999; Luy and Marino, 2003; Yang et al., 1999).

Finally, we note that the program "ENTANGLE" (Allers and Shamoo, 2001) is useful for facilitating systematic analysis of protein–RNA interactions. ENTANGLE is a JAVA-based program that is very easy to use. The program uses available structures of the protein–RNA complex in their PDB format and searches for appropriate hydrogen bonding, electrostatic, hydrophobic, and van der Waals interactions between protein and RNA.

Acknowledgments

This work was supported by NIH Grants GM37254 and GM48123, and NSF Grant MCB980872 to J.F.

References

Alfano, C., Sanfelice, D., Babon, J., Kelly, G., Jacks, A., Curry, S., and Conte, M. R. (2004). Structural analysis of cooperative RNA binding by the La motif and central RRM domain of human La protein. Nat. Struct. Mol. Biol. 11, 323–329.
Allain, F. H. T., Howe, P. W. A., Neuhaus, D., and Varani, G. (1997). Structural basis of the RNA-binding specificity of human U1A protein. EMBO J. 16, 5764–5774.

Allain, F. H.-T., Bouvet, P., Dieckmann, T., and Feigon, J. (2000a). Molecular basis of sequence-specific recognition of pre-ribosomal RNA by nucleolin. *EMBO J.* **19,** 6870–6881.

Allain, F. H.-T., Gilbert, D. E., Bouvet, P., and Feigon, J. (2000b). Solution structure of the two N-terminal RNA-binding domains of nucleolin and NMR study of the interaction with its RNA target. *J. Mol. Biol.* **303,** 227–241.

Allen, M., Varani, L., and Varani, G. (2001). Nuclear magnetic resonance methods to study structure and dynamics of RNA-protein complexes. *Methods Enzymol.* **339,** 357–376.

Allers, J., and Shamoo, Y. (2001). Structure-based analysis of protein-RNA interactions using the program ENTANGLE. *J. Mol. Biol.* **311,** 75–86.

Amarasinghe, G. K., De Guzman, R. N., Turner, R. B., Chancellor, K. J., Wu, Z. R., and Summers, M. F. (2000). NMR structure of the HIV-1 nucleocapsid protein bound to stem-loop SL2 of the psi-RNA packaging signal. *J. Mol. Biol.* **301,** 491–511.

Anderson, A. C., Scaringe, S. A., Earp, B. E., and Frederick, C. A. (1996). *RNA* **2,** 110–117.

Batey, R. T., Battiste, J. L., and Williamson, J. R. (1995). HPLC purification of RNA for crystallography and NMR. *Methods Enzymol.* **261,** 300–322.

Bayer, P., Varani, L., and Varani, G. (1999). Refinement of the structure of protein-RNA complexes by residual dipolar coupling analysis. *J. Biomol. NMR* **14,** 149–155.

Berg, J. M. (2003). Fingering nucleic acids: The RNA did it. *Nat. Struct. Biol.* **10,** 986–987.

Birney, E., Kumar, S., and Krainer, A. R. (1993). Analysis of the RNA-recognition motif and RS and RGG domains: Conservation in metazoan pre-mRNA splicing factors. *Nucleic Acids Res.* **21,** 5803–5816.

Breeze, A. L. (2000). Isotope-filtered NMR methods for the study of biomolecular structure and interactions. *Prog. Nucl. Magn. Reson. Spectrosc.* **36,** 323–372.

Brünger, A. T. (1992). "X-PLOR (Version 3.1) Manual." Yale University Press, New Haven, CT.

Burgess, R. R. (1991). Use of polyethyleneimine in purification of DNA-binding proteins. *Methods Enzymol.* **208,** 3–10.

Cavanagh, J., Fairbrother, W. J., Palmer, A. G., III, and Skelton, N. J. (1996). "Protein NMR Spectroscopy: Principles and Practice." Academic Press, San Diego, CA.

Chanfreau, C., Buckle, M., and Jacquier, A. (2000). Recognition of a conserved class of RNA tetraloops by *Saccharomyces cerevisiae* RNase III. *Proc. Natl. Acad. Sci. USA* **97,** 3142–3147.

Clore, G. M. (2000). Accurate and rapid docking of protein-protein complexes on the basis of intermolecular nuclear overhauser enhancement data and dipolar couplings by rigid body minimization. *Proc. Natl. Acad. Sci. USA* **97,** 9021–9025.

Clore, G. M., Gronenborn, A. M., and Tjandra, N. (1998). Direct structure refinement against residual dipolar couplings in the presence of rhombicity of unknown magnitude. *J. Magn. Reson.* **131,** 159–162.

Dayie, K. T., Tolbert, T. J., and Williamson, J. R. (1998). 3D C(CC)H TOCSY experiment for assigning protons and carbons in uniformly 13C- and selectively 2H-labeled RNA. *J. Magn. Reson.* **130,** 97–101.

de Alba, E., and Tjandra, N. (2002). NMR dipolar couplings for the structure determination of biopolymers in solution. *Prog. Nucl. Magn. Reson. Spectrosc.* **40,** 175–197.

De Guzman, R. N., Wu, Z. R., Stalling, C. C., Pappalardo, L., Borer, P. N., and Summers, M. F. (1998). Structure of the HIV-1 nucleocapsid protein bound to the SL3 psi-RNA recognition element. *Science* **279,** 384–388.

Dieckmann, T., and Feigon, J. (1997). Assignment methodology for larger RNA oligonucleotides: Application to an ATP-binding RNA aptamer. *J. Biomol. NMR* **9,** 259–272.

Dominguez, C., Boelens, R., and Bonvin, A. M. J. J. (2003). HADDOCK: A protein-protein docking approach based on biochemical or biophysical information. *J. Am. Chem. Soc.* **125,** 1731–1737.

D'Souza, V., Dey, A., Habib, D., and Summers, M. F. (2004). NMR structure of the 101-nucleotide core encapsidation signal of the Moloney murine leukemia virus. *J. Mol. Biol.* **337,** 427–442.

Evenas, J., Mittermaier, A., Yang, D., and Kay, L. E. (2001). Measurement of (13)C(alpha)-(13)C(beta) dipolar couplings in (15)N,(13)C,(2)H-labeled proteins: Application to domain orientation in maltose binding protein. *J. Am. Chem. Soc.* **123,** 2858–2864.

Feigon, J., Butcher, S. E., Finger, L. D., and Hud, N. V. (2001). Solution nuclear magnetic resonance probing of cation binding sites on nucleic acids. *Methods Enzymol.* **338,** 400–420.

Finger, L. D., Trantirek, L., Johansson, C., and Feigon, J. (2003). Solution structures of stem-loop RNAs that bind to the two N-terminal RNA-binding domains of nucleolin. *Nucleic Acids Res.* **31,** 6461–6472.

Foster, M. P., Wuttke, D. S., Clemens, K. R., Jahnke, W., Radhakrishnan, I., Tennant, L., Reymond, M., Chung, J., and Wright, P. E. (1998). Chemical shift as a probe of molecular interfaces: NMR studies of DNA binding by the three amino-terminal zinc finger domains from transcription factor IIIA. *J. Biomol. NMR* **12,** 51–71.

Furtig, B., Richter, C., Wohnert, J., and Schwalbe, H. (2003). NMR spectroscopy of RNA. *Chembiochem.* **4,** 936–962.

Gardner, K. H., and Kay, L. E. (1998). The use of 2H, 13C, 15N multidimensional NMR to study the structure and dynamics of proteins. *Annu. Rev. Biophys. Biomol. Struct.* **27,** 357–406.

Ghisolfi-Nieto, L., Joseph, G., Puvion-Dutilleul, F., Amalric, F., and Bouvet, P. (1996). Nucleolin is a sequence-specific RNA-binding protein: Characterization of targets on pre-ribosomal RNA. *J. Mol. Biol.* **260,** 34–53.

Guntert, P. (1998). Structure calculation of biological macromolecules from NMR Q. *Rev. Biophys.* **31,** 145–237.

Hall, K. B. (2002). RNA-protein interactions. *Curr. Opin. Struct. Biol.* **12,** 283–288.

Halperin, I., Ma, B. Y., Wolfson, H., and Nussinov, R. (2002). Principles of docking: An overview of search algorithms and a guide to scoring functions. *Proteins* **47,** 409–443.

Hudson, B. P., Martinez-Yamout, M. A., Dyson, H. J., and Wright, P. E. (2004). Recognition of the mRNA AU-rich element by the zinc finger domain of TIS11d. *Nat. Struct. Mol. Biol.* **11,** 257–264.

Iwahara, J., Iwahara, M., Daughdrill, G. W., Ford, J., and Clubb, R. T. (2002). The structure of the Dead ringer-DNA complex reveals how AT-rich interaction domains (ARIDs) recognize DNA. *EMBO J.* **21,** 1197–1209.

Johansson, C., Finger, L. D., Trantirek, L., Mueller, T. D., Kim, S., Laird-Offringa, I. A., and Feigon, J. (2004). Solution structure of the complex formed by the two N-terminal RNA-binding domains of nucleolin and a pre-rRNA target. *J. Mol. Biol.* **337,** 799–816.

Kane, J. F. (1995). Effects of rare codon clusters on high-level expression of heterologous proteins in *Escherichia coli*. *Curr. Opin. Biotechnol.* **6,** 494–500.

Kontaxis, G., Clore, G. M., and Bax, A. (2000). Evaluation of cross-correlation effects and measurement of one-bond couplings in proteins with short transverse relaxation times. *J. Magn. Reson.* **143,** 184–196.

Lane, A. N., Kelly, G., Ramos, A., and Frenkiel, T. A. (2001). Determining binding sites in protein-nucleic acid complexes by cross-saturation. *J. Biomol. NMR* **21,** 127–139.

LaVallie, E. R., Rehemtulla, A., Racie, L. A., DiBlasio, E. A., Ferenz, C., Grant, K. L., Light, A., and McCoy, J. M. (1993). Cloning and functional expression of a cDNA encoding the catalytic subunit of bovine enterokinase. *J. Biol. Chem.* **268,** 23311–23317.

Lerche, M. H., Meissner, A., Poulsen, F. M., and Sørensen, O. W. (1999). Pulse sequences for measurement of one-bond (15)N-(1)H coupling constants in the protein backbone. *J. Magn. Reson.* **140,** 259–263.

Leulliot, N., and Varani, G. (2001). Current topics in RNA-protein recognition: Control of specificity and biological function through induced fit and conformational capture. *Biochemistry* **40,** 7947–7956.

Liu, Z. H., Luyten, I., Bottomley, M. J., Messias, A. C., Houngninou-Molango, S., Sprangers, R., Zanier, K., Kramer, A., and Sattler, M. (2001). Structural basis for recognition of the intron branch site RNA by splicing factor 1. *Science* **294,** 1098–1102.

Lucast, L. J., Batey, R. T., and Doudna, J. A. (2001). Large-scale purification of a stable form of recombinant tobacco etch virus protease. *Biotechniques* **30,** 544–554.

Luy, B., and Marino, J. P. (2003). JE-TROSY: Combined J- and TROSY-spectroscopy for the measurement of one-bond couplings in macromolecules. *J. Magn. Reson.* **163,** 92–98.

Makrides, S. C. (1996). Strategies for achieving high-level expression of genes in *Escherichia coli. Microbiol. Rev.* **60,** 512–538.

Mao, H., White, S. A., and Williamson, J. R. (1999). A novel loop-loop recognition motif in the yeast ribosomal protein L30 autoregulatory RNA complex. *Nat. Struct. Biol.* **6,** 1139–1147.

Marino, J. P., Schwalbe, H., and Griesinger, C. (1999). J-coupling restraints for structural refinements of RNA. *Acc. Chem. Res.* **32,** 614–623.

Marley, J., Lu, M., and Bracken, C. (2001). A method for efficient isotopic labeling of recombinant proteins. *J. Biomol. NMR* **20,** 71–75.

Milligan, J. F., and Uhlenbeck, O. C. (1989). Synthesis of small RNAs using T7 RNA polymerase. *Methods Enzymol.* **180,** 51–62.

Mitton-Fry, R. M., Anderson, E. M., Theobald, D. L., Glustrom, L. W., and Wuttke, D. S. (2004). Structural basis for telomeric single-stranded DNA recognition by yeast Cdc13. *J. Mol. Biol.* **338,** 241–255.

Muchmore, D. C., McIntosh, L. P., Russell, C. B., Anderson, D. E., and Dahlquist, F. W. (1989). Expression and nitrogen-15 labeling of proteins for proton and nitrogen-15 nuclear magnetic resonance. *Methods Enzymol.* **177,** 44–73.

Murray, J. B., Collier, A. K., and Arnold, J. R. P. (1994). A general purification procedure for chemically synthesized oligoribonucleotides. *Anal. Biochem.* **218,** 177–184.

Nagai, K. (1996). RNA-protein complexes. *Curr. Opin. Struct. Biol.* **6,** 53–61.

Nikonowicz, E. P., Sirr, A., Legault, P., Jucker, F. M., Baer, L. M., and Pardi, A. (1992). Preparation of 13C and 15N labelled RNAs for heteronuclear multi-dimensional NMR studies. *Nucleic Acids Res.* **20,** 4507–4513.

Oldfield, E. (1995). Chemical shifts and three-dimensional protein structures. *J. Biomol. NMR* **5,** 217–225.

Perez-Canadillas, J. M., and Varani, G. (2001). Recent advances in RNA-protein recognition. *Curr. Opin. Struct. Biol.* **11,** 53–58.

Pervushin, K. (2000). Impact of transverse relaxation optimized spectroscopy (TROSY) on NMR as a technique in structural biology. *Q. Rev. Biophys.* **33,** 161–197.

Peterson, R. D., Theimer, C. A., Wu, H. H., and Feigon, J. (2004). New applications of 2D filtered/edited NOESY for assignment and structure elucidation of RNA and RNA-protein complexes. *J. Biomol. NMR* **28,** 59–67.

Polayes, D. A., Goldstein, A., Hughes, A. J., Johnston, S. A., and Dougherty, W. G. (1994). TEV protease, recombinant: A site-specific protease for efficient cleavage of affinity tags from expressed proteins. *Focus* **16,** 2–5.

Prestegard, J. H., Al-Hashimi, H. M., and Tolman, T. R. (2000). NMR structures of biomolecules using field oriented media and residual dipolar couplings. *Q. Rev. Biophys.* **33,** 371–424.

Ramos, A., Grunert, S., Adams, J., Micklem, D. R., Proctor, M. R., Freund, S., Bycroft, M., St. Johnston, D., and Varani, G. (2000a). RNA recognition by a Staufen double-stranded RNA-binding domain. *EMBO J.* **19**, 997–1009.

Ramos, A., Kelly, G., Hollingworth, D., Pastore, A., and Frenkiel, T. (2000b). Mapping the interfaces of protein-nucleic acid complexes using cross-saturation. *J. Am. Chem. Soc.* **122**, 11311–11314.

Riek, R., Pervushin, K., Fernandez, C., Kainosho, M., and Wuthrich, K. (2001). [(13)C,(13)C]- and [(13)C,(1)H]-TROSY in a triple resonance experiment for ribose-base and intrabase correlations in nucleic acids. *J. Am. Chem. Soc.* **123**, 658–664.

Ruckert, M., and Otting, G. (2000). Alignment of biological macromolecules in novel nonionic liquid crystalline media for NMR experiments. *J. Am. Chem. Soc.* **122**, 7793–7797.

Sattler, M., Schleucher, J., and Griesinger, C. (1999). Heteronuclear multidimensional NMR experiments for the structure determination of proteins in solution employing pulsed field gradients. *Prog. Nucl. Magn. Reson. Spectrosc.* **34**, 93–158.

Saunders, L. R., and Barber, G. N. (2003). The dsRNA binding protein family: Critical roles, diverse cellular functions. *FASEB J.* **17**, 961–983.

Scaringe, S. A. (2001). RNA oligonucleotide synthesis via 5′-silyl-2′-orthoester chemistry. *Methods* **23**, 206–217.

Shields, T. P., Mollova, E., Marie, L. S., Hansen, M. R., and Pardi, A. (1999). High-performance liquid chromatography purification of homogenous-length RNA produced by trans cleavage with a hammerhead ribozyme. *RNA* **5**, 1259–1267.

Simon, B., Zanier, K., and Sattler, M. (2001). A TROSY relayed HCCH-COSY experiment for correlating adenine H2/H8 resonances in uniformly 13C-labeled RNA molecules. *J. Mol. Biol.* **20**, 173–176.

Stoldt, M., Wohnert, J., Ohlenschlager, O., Gorlach, M., and Brown, L. R. (1999). The NMR structure of the 5S rRNA E-domain-protein L25 complex shows preformed and induced recognition. *EMBO J.* **18**, 6508–6521.

Stump, W. T., and Hall, K. B. (1993). SP6 RNA polymerase efficiently synthesizes RNA from short double-stranded DNA templates. *Nucl. Acids Res.* **21**, 5480–5484.

Tolbert, T. J., and Williamson, J. R. (1997). Preparation of specifically deuterated and C-13-labeled RNA for NMR studies using enzymatic synthesis. *J. Am. Chem. Soc.* **119**, 12100–12108.

Varani, L., Gunderson, S. I., Mattaj, I. W., Kay, L. E., Neuhaus, D., and Varani, G. (2000). The NMR structure of the 38 kDa U1A protein – PIE RNA complex reveals the basis of cooperativity in regulation of polyadenylation by human U1A protein. *Nat. Struct. Biol.* **7**, 329–335.

Wijmenga, S. S., and van Buuren, B. N. M. (1998). The use of NMR methods for conformational studies of nucleic acids. *Prog. Nucl. Magn. Reson. Spectrosc.* **32**, 287–387.

Williamson, J. R. (2000). Induced fit in RNA-protein recognition. *Nat. Struct. Biol.* **7**, 834–837.

Wu, H., Henras, A., Chanfreau, G., and Feigon, J. (2004). Structural basis for recognition of the AGNN tetraloop RNA fold by the double-stranded RNA-binding domain of Rnt1p RNase III. *Proc. Natl. Acad. Sci. USA* **101**, 8307–8312.

Wüthrich, K. (1986). "NMR of Proteins and Nucleic Acids." John Wiley, New York.

Yang, D., Venters, R. A., Mueller, G. A., Choy, W. Y., and Kay, L. E. (1999). *J. Biomol. NMR* **14**, 333–343.

Yang, Y. S., Declerck, N., Manival, X., Aymerich, S., and Kochoyan, M. (2002). Solution structure of the LicT-RNA antitermination complex: CAT clamping RAT. *EMBO J.* **21**, 1987–1997.

Section VI

Ligand Discovery

[23] Utilization of NMR-Derived Fragment Leads in Drug Design

By JEFFREY R. HUTH, CHAOHONG SUN,
DARYL R. SAUER, and PHILIP J. HAJDUK

Abstract

The advent of large-scale NMR-based screening has enabled new strategies for the design of novel, potent inhibitors of therapeutic targets. In particular, fragment-based strategies, in which molecular portions of the final high-affinity ligand are experimentally identified prior to chemical synthesis, have found widespread utility. This chapter will discuss some of the practical considerations for identifying and utilizing these fragment leads in drug design, with special emphasis on some of the lessons learned from more than a decade of industry experience.

Introduction

Nuclear magnetic resonance (NMR)–based screening has become a powerful complement to traditional high-throughput screening (HTS) technology to identify lead compounds that have high potential for further optimization. Unlike conventional HTS, most NMR-based screening applications focus on the identification of low-molecular-weight, low-affinity compounds (typically referred to as fragments) from which high-affinity drug candidates can be constructed. There have been many excellent reviews on the applications as well as the theoretical and experimental aspects of NMR-based screening (Coles *et al.*, 2003; Hajduk *et al.*, 1999c; Huth and Sun, 2002; Jahnke *et al.*, 2003; Lepre *et al.*, 2002; Meyer and Peters, 2003; Peng *et al.*, 2001; Stockman and Dalvit, 2002; Stockman *et al.*, 2001). In this chapter, we will briefly cover these areas to reacquaint the reader and highlight new developments or considerations. The first section will review the NMR methods for fragment-based screening and discuss their use in core replacement and high-throughput elaboration strategies for lead generation. The second section will focus on the identification and utilization of second-site ligands, with particular emphasis on linker design.

Finding the Hot Spot: Detecting First-Site Ligands

Through an analysis of the energetics of protein–ligand complexes, Kuntz and co-workers (1999) demonstrated that the binding energy of a

Copyright 2005, Elsevier Inc.
All rights reserved.
0076-6879/05 $35.00

ligand for its receptor increases rapidly as the number of nonhydrogen atoms increases to approximately 15, and then plateaus with only marginal free-energy gains with increasing size. This implies that there are only relatively small regions of the protein surface that impart the majority of the free energy of binding with the ligand, while any additional interaction surface can serve to modulate specificity. Such areas on protein surfaces have been coined energetic hot spots, and they have been observed not only in protein–ligand interactions but also in protein–protein interactions (Kortemme and Baker, 2002). As a result, the low-molecular-weight compounds screened by NMR will typically have highest affinity for the hot spots on the protein surface. We tend to denote these initial leads that bind to the energetic focal point of the active site as first-site ligands. Two different detection methods are used in NMR-based screening for the detection of first-site ligands: monitoring target resonances (target based) or monitoring ligand resonances (ligand based).

Target-Based Methods for First-Site Screening

NMR screening based on chemical shift perturbations of the resonances of the protein target is a profoundly reliable method for detecting protein–ligand interactions. In this method, specific ligand binding is detected through monitoring the chemical shift changes of protein signals that occur upon complex formation. Usually the target is either ^{15}N or ^{13}C labeled, and two-dimensional (2D) [^{15}N,^{1}H] or [^{13}C,^{1}H] correlation spectra are collected in the presence or absence of test compounds. Screening is typically performed against mixtures of 10–30 compounds, and the mixtures that produce the largest chemical shift perturbations are deconvoluted to identify the leads. For ligand binding in the fast exchange regime, the ligand's binding affinity can be calculated by monitoring the chemical shift changes as a function of ligand concentration.

Target-based NMR screening has significant advantages over other screening methods. The observation of discrete, stable chemical shift perturbations is the strongest evidence for a specific, well-defined binding event. As a result, target-based NMR screening is essentially immune to false positives that can arise from nonspecific binding. Artifacts from compound aggregation (McGovern et al., 2002) can readily be detected by nonspecific broadening of the protein resonances. In addition, the unique chemical shift perturbation fingerprint induced by ligand binding can be used not only to detect binding but also to identify the ligand-binding site and even to orient the ligand on the receptor protein (McCoy and Wyss, 2000; Medek et al., 1999). Finally, the spectral editing essentially eliminates the ligand signals, enabling the experiments to be performed even at very high ligand concentrations. With the advance of cryogenic probes (Hajduk *et al.*, 1999b) and cost-effective ^{13}C-methyl labeling (Hajduk *et al.*, 2000a), target-based NMR screening utilizing ^{15}N or ^{13}C labeling can now routinely be applied to targets with molecular weights (MW) in excess of 40 kDa. An alternative nucleus that can potentially afford significant gains in sensitivity for target-based screening is fluorine-19. ^{19}F occurs at high natural abundance (100%) and has NMR sensitivity similar to ^{1}H. The large ^{19}F chemical shift range together with essentially zero background signal makes it possible to rapidly collect one-dimensional ^{19}F spectra even with low protein concentrations. Many examples now exist describing the incorporation of ^{19}F into proteins with little or no effect on protein structure, function, or ligand binding (Gerig, 1994), suggesting that such labeled proteins will be valid targets for NMR-based screening. The full potential of utilizing ^{19}F for target-based screens has yet to be explored.

While target-based NMR screening can now be applied to targets as large as 40 kDa, only a fraction of therapeutically relevant targets falls into this MW range. Extensive deuteration and the use of TROSY pulse sequences can significantly increase the MW limit (Fernandez and Wider, 2003). However, the high cost associated with such labeling has limited this application mostly to lead validation and characterization. Another disadvantage of target-based screening is that cost-effective and efficient labeling can currently be achieved only in bacterial systems. While isotopic labeling in insect cell systems has recently been reported (Strauss *et al.*, 2003), such methods are still in their infancy. Thus, for high-MW targets or those that can be produced only in baculovirus or mammalian cells, target-based NMR screening will not be appropriate, and ligand-based NMR methods should be explored.

Ligand-Based Methods for First-Site Screening

A number of ligand-based NMR screening techniques have been developed (Meyer and Peters, 2003; Stockman and Dalvit, 2002). All these methods utilize the large changes in NMR parameters that occur when a small ligand binds to its macromolecular target. Strategies that exploit changes in relaxation rates (Fejzo *et al.*, 1999; Jahnke, 2002), diffusion rates (Hajduk *et al.*, 1997b), nuclear Overhauser effects (Meyer *et al.*, 1997), and saturation transfer phenomena (Chen and Shapiro, 1998; Mayer and Meyer, 1999) have all been described. Since the resonances of the ligand are observed in these experiments, the size, composition, and oligomeric status of the macromolecular target are essentially unlimited.

While powerful, all these ligand-observed methods have a number of intrinsic drawbacks. First, as most of these methods rely on rapid exchange between the free and bound states, high-affinity ligands can be missed (false negatives). Second, ligands with low solubility either will not be observed (false negatives) or can aggregate and confound spectral interpretation (false positives). A further drawback is that it is essentially impossible to distinguish between specific and nonspecific ligand binding using these techniques.

Most of these concerns have been effectively addressed with recently described competitive screening methods (Dalvit *et al.*, 2002). In this method, a probe molecule with good solubility and moderate affinity is selected, and screening is performed through monitoring the ability of test compounds to displace the probe molecule. Ideally, the chemical shifts of the probe molecule should be unique to avoid overlap with the library compounds. ^{13}C- or ^{19}F-labeled probe compounds offer such advantages with less interference from signals from the protein target, test compounds, solvent, and buffer components. Another advantage of using ^{19}F is that the fluorine atom frequently occurs in small organic compounds. This makes it quite possible to obtain a ^{19}F-containing probe compound without recourse to special synthetic effort. One limitation of competitive one-dimensional (1D) methods is that they detect only the ligands that bind to the same site as the probe compound or competitive allosteric sites. In addition, a critical control in these experiments is that the probe molecule can be efficiently competed away by competitor compounds. Unfortunately, on more than half of the targets for which we have attempted competitive 1D screening, the probe molecule could not be competed away by more potent compounds (data not shown). This is even the case for highly soluble, nonaggregating probe molecules such as adenosine. While the exact source of these phenomena is not clear, it most likely arises from nonspecific, low-affinity sites on the protein target, as has been observed for other systems (Murali *et al.*, 1993; Post, 2003).

Utilizing First-Site Ligands in Drug Design

Each NMR-based screening method has its own advantages and disadvantages, and their optimal use will depend on the specific characteristics of a given protein target. However, what unites all of these methods is that they were specifically developed to enable fragment-based strategies in drug design. The methods themselves have been quite successful at their desired objectives: the identification of low-MW leads that bind to hot spots on protein surfaces and that would be difficult or impossible to detect using any other method. The key question is how to utilize these hits in drug design. Three strategies for utilizing NMR-derived fragments in drug design will be described in the following sections: (1) high-throughput core

[23]   ,acement, and (3) fragment linking (SAR by
elaboration, (2) core   will be given to practical considerations in
NMR). Special emp...creens, as many of these issues have not been
performing second-...he literature.
previously addresse...

Elaboration

High-Through...om NMR-based screening are only of low MW and
While th...iewed as small molecule scaffolds that bind to the hot
affinity, th...ocal point, on the protein surface and can be rapidly
spot, or ...1 combinatorial chemistry or high-throughput parallel
elabora...any scaffolds have been utilized in the design of combi-
synth...(Fauchere et al., 1998; Fecik et al., 1998), it is typically
natc...r any scaffold possesses inherent binding affinity for any
un...is significant to note that an NMR-derived lead with a K_D
...l mM represents more than 4 kcal/mol of binding energy.
...ads are ideal for use in target-directed library design. The
...h libraries is significantly enhanced when structural informa-
protein–ligand complex can be obtained either from NMR or
...allography.
...sign of target-directed libraries begins with the identification or
...tion of one or more functionalities in the NMR leads that are
...e to rapid elaboration. As mentioned above, the placement of
...nctionalities on the scaffold is ideally directed by structural infor-
..., although available structure–activity relationships on the small
...ecule scaffold can often guide the design. A wide range of chemistries
...s available for the elaboration of suitable scaffolds. Recent advances in the
application of polymer-supported reagents (Bartolozzi et al., 2003) and
solid-phase extraction (SPE) (Booth and Hodges, 1997, 1999) allow for
the rapid execution of parallel synthesis to conveniently produce analogs in
good yields, high purity, and with minimal postsynthetic processing. As
such, polymer-assisted solution phase (PASP) synthesis is rapidly emerging
as a method of choice for library preparation via parallel synthesis (Hinzen,
2000). Examples of PASP reactions frequently used for library generation
by our high-throughput organic synthesis group are given in Table I. These
include amide formation (Yun, et al., 2002), sulfonamide formation
(Salvino and Dolle, 2003; Salvino et al., 2000), urea formation (Booth and
Hodges, 1997, 1999), biaryl generation (Suzuki couplings) (Colacot et al.,
2002), amine generation via reductive alkylation (Bhattacharyya et al., 2000,
2003), ether formations (Mitsunobu couplings) (Gentles et al., 2002), and
nucleophilic aromatic substitution reactions. Microwave-accelerated
synthesis in combination with polymer-supported reagents and SPE has

TABLE I
COMMON TRANSFORMATIONS UTILIZED TO PRODUCE DRUG ~~~VERY LIBRARIES VIA~~ PARALLEL SYNTHESIS

Chemistry	Chemical mo~~~~
Amide formation	R_1—C(=O)—OH + H_2N—R_2 (also with secondary am~~ R_2)
Sulfonamide formation	R_1—S(=O)(=O)—Cl + H_2N—R_2 ⟶ R_1 (also with secondary amines)
Urea formation	R_1—NCO + H_2N—R_2 ⟶ R_1—N(H)—C(=O)—N (also with secondary amines)
Biaryl formation	⟨1⟩—Br + (HO)₂B—⟨2⟩ ⟶ ⟨1⟩—⟨
Amine formation	R_1—C(=O)—R_2 + H_2N—R_3 ⟶ R_1—C(R_2)(H)—N(H)—R_3 (also with secondary amines)
Mitsunobu	⟨Ph⟩—OH + HO—R ⟶ ⟨Ph⟩—O—R
Nucleophilic aromatic substitution	⟨Het⟩—Br + NH₂—R ⟶ ⟨Het⟩—NH—R (also with secondary amines)

proven to be useful for library preparation, as it allows for the synthesis of analogs in minutes rather than hours, and it can greatly increase the speed and efficiency of compound synthesis (Sauer *et al.*, 2003).

Several factors must be considered when selecting chemistries suitable for library formation. One factor is that the library members should contain biologically suitable functionalities and be lead-like in character.

In addition, the potential availability of large numbers of commercially available monomers that can be utilized for each class of chemistry, to rapidly generate large numbers of molecules around a given scaffold, is also taken into consideration. An examination of the Comprehensive Medicinal Chemistry database reveals that benzene and heterocyclic rings are the most abundant substructures among drug agents and that tertiary amines and amides are the most prevalent functional groups (Ghose *et al.*, 1999). With the large number of commercially available substrates and robust chemical methods that are available for preparing these classes of molecules, it is not surprising that they represent a majority of the compounds currently prepared via the parallel synthesis of libraries used for drug design studies.

We have successfully utilized high-throughput core elaboration on multiple projects at Abbott. One example was in the design of low micromolar inhibitors of the antibacterial target ErmAM starting from a millimolar lead (Hajduk *et al.*, 1999a). A more recent example is in the design of potent and selective protein tyrosine phosphatase 1B (PTP1B) inhibitors (Xin *et al.*, 2003). PTP1B is a critical enzyme in the insulin-signaling pathway that dephosphorylates the insulin receptor and down-regulates insulin. Intriguingly, PTP1B-gene knockout mice not only display enhanced insulin sensitivity and improved glucose tolerance but also exhibit suppression of diet-induced obesity (Ukkola and Santaniemi, 2002). Thus, PTP1B represents an attractive target for type II diabetes and obesity. To identify novel PTP1B inhibitors, NMR fragment–based screening was initiated, and 2,3-dimethylphenyloxalylaminobenzoic acid (**1**) was identified as an active site binder with a K_D value of 100 μM (Fig. 1). Enzymatic kinetic assays confirmed that **1** is a competitive, reversible inhibitor of PTP1B. From the crystal structure of PTP1B complexed to an analog of **1**, a design strategy was implemented to access the second pTyr-binding site on PTP1B to both improve affinity and gain specificity against highly related phosphatases such as T cell PTP (TCPTP). As this second site was ~10 Å away, a linker group was optimized to bridge this space and position functional groups suitable for rapid elaboration into the second site. Structure-based design led to the discovery of **2**, which exhibited a K_i value of 2.5 μM and contained an alkyl carboxylic acid suitable for high-throughput amide bond formation. A library of 76 primary amines and L-amino acids was selected and coupled to the acid core. From this library, compound **3** was obtained, which exhibited an over 30-fold gain in potency against PTP1B ($K_I = 0.08$ μM). Most significantly, this compound exhibited 5-fold selectivity against TCPTP. This represents the first reported series of potent, non-phosphonic-acid-containing inhibitors of PTP1B with specificity against TCPTP.

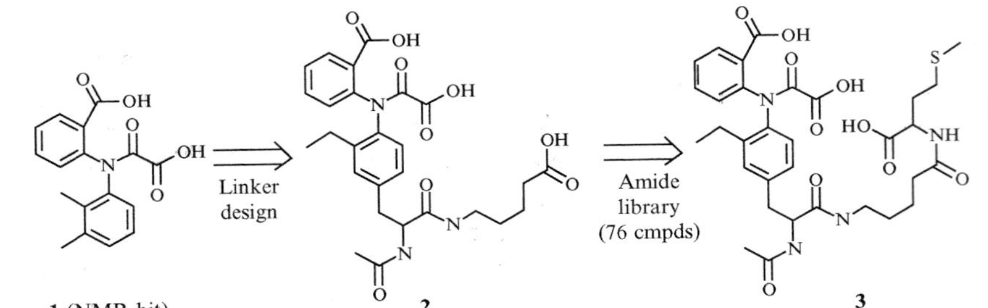

1 (NMR hit)
$K_D = 100 \ \mu M$ (NMR)
$K_I = 93 \ \mu M$

2
$K_I = 2.5 \ \mu M$ (PTP1B)
$K_I = 2.2 \ \mu M$ (TCPTP)

3
$K_I = 0.08 \ \mu M$ (PTP1B)
$K_I = 0.38 \ \mu M$ (TCPTP)

Fig. 1. High-throughput core elaboration strategy for the design of PTP1B inhibitors (Xin et al., 2003). Oxamic acid–containing compounds (e.g., **1**) were discovered by NMR to bind to the catalytic pTyr-binding site. A linker moiety containing a carboxylic acid group was incorporated (**2**), from which a library of 76 compounds was prepared using high-throughput amide bond formation. A number of compounds from this library (e.g., **3**) exhibited nanomolar potency against PTP1B and good selectivity against TCPTP.

Core Replacement

The advent of fully automated, ultrahigh-throughput screening has significantly improved the ability to screen ever-larger compound collections and identify molecules suitable for lead optimization. However, once the optimization process begins, it becomes increasingly difficult to break out of existing structural paradigms without unacceptable losses in potency. Unfortunately, such paradigm shifts are often critical to overcome unanticipated safety, pharmacological, or intellectual property issues. The intrinsic lead-like properties of NMR hits can be used to modify the current lead to address these issues. In a typical NMR-based screen, multiple structurally disparate leads that all bind to the same subsite on a protein target can be identified. Any of these leads can be used in a mix-and-match strategy to replace portions of the existing inhibitor.

One application of the core replacement strategy was in the replacement of the guanidine functionality in urokinase inhibitors with aminobenzimidazoles (Hajduk et al., 2000b). A more recent example is in the design of novel, cell-permeable PTP1B inhibitors. As mentioned earlier, using NMR screening in combination with structure-guided design, novel and potent PTP1B inhibitors have been obtained (Xin et al., 2003). While potent, these compounds are highly charged, which results in lack of cellular permeability. It was further determined that the oxamate moiety was a primary contributor to this lack of permeability and therefore needed to be replaced. To achieve such a goal, focused libraries of monocarboxylic acid–containing heterocycles were screened by NMR. Several isoxazole carboxylates were identified as active site ligands, exemplified by isoxazole **4** with a K_D value of 800 μM (Fig. 2). The fact that none of these compounds could be identified using a pNPP hydrolysis assay highlights the power of NMR screening to robustly detect these weak ligands. The X-ray crystal structure of PTP1B complexed with an analog of compound **4**, together with structural data on compound **5**, led to the design of compound **6** (see Fig. 2). This compound exhibits a K_I value of 6.9 μM against PTP1B and represents a novel class of PTP1B inhibitors with no amide bonds and only a single carboxylic acid. Surprisingly, these compounds displayed good selectivity among PTPases, with more than 30-fold selectivity over TCPTP. Most importantly, compound **6** showed moderate cellular permeability in a Caco-cell membrane assay (Papp > 1 × 10^{-6} cm/s). Consistent with this, compound **6** exhibited dose-dependent cellular activity and reversed PTP1B-dependent dephosphorylation of STAT3 in a COS-7 cell assay (Liu *et al.*, 2003).

One of the key requirements for rapid and successful core replacement is structural information on both the parent compound and the small molecule lead. These data are crucial for the design of new compounds

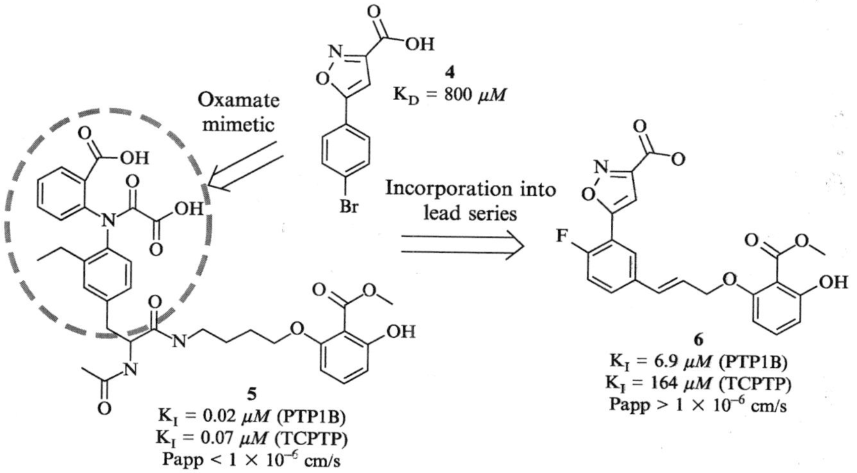

FIG. 2. Core replacement strategy for the design of PTP1B inhibitors (Liu *et al.*, 2003). Isoxazole-containing compounds (e.g., **4**) were discovered by NMR to bind to the same site as the oxamate moiety of the lead series (e.g., **5**). Based on X-ray structural data, the oxamate core was replaced with an isoxazole lead, yielding potent, selective, and cell-permeable inhibitors of PTP1B (e.g., **6**).

that optimally exploit the binding energy of the new core and maintain the position of existing substituents. In the case of the isoxazole replacements for PTP1B as described above, significant conformational changes in the protein were observed between the oxamate-bound (e.g., **5**) and isoxazole-bound (e.g., **4**) complexes. This led to shifts in ligand position and opportunities for linker design that could not have been derived from the chemical structures of the compounds alone. Strikingly, this resulted in a reduction in the linker length from nine atoms for the oxamate to only four atoms for the isoxazoles (e.g., compare compounds **5** and **6** in Fig. 2). Despite this change, X-ray structural analysis confirmed that the methyl-salicylate moieties in both **5** and **6** are maintained in essentially identical positions.

Detecting Second-Site Ligands

As first-site ligands are defined as those fragments that occupy the energetic hot spot within the active site, second-site ligands are defined as those compounds that have affinity for peripheral pockets in the active site *in the presence of a first-site ligand*. The simultaneous binding of two fragments to proximal sites on a protein surface enables one of the more powerful utilizations of fragments in drug design: the linked-fragment approach. In this strategy, high-affinity ligands are constructed by tethering fragments that have been experimentally identified to simultaneously bind near each other on the protein surface. One of the most significant advantages of the linked-fragment approach is that the amount of chemical space that is sampled increases exponentially with the size of the fragment library (Hajduk *et al.*, 1997a). For instance, an NMR screen in which 10,000 compounds are tested for independent binding to two subsites on a protein covers as much chemical space as screening ~100 million single compounds. As will be described in the sections that follow, several NMR methods have been developed to enable NMR-driven linked-fragment approaches to drug design.

Practical Considerations for Second-Site Screens

In principle, two ligands that bind to proximal subsites can be identified from a single NMR screen against a single library. While this can and does occur, it is more often the case that a screen must be repeated in the presence of a ligand identified in an initial screen. There are many reasons for this. First, most compounds have a greater affinity for the energetic hot spot on the protein surface and tend to cluster at a single site—even if they do have some affinity for a neighboring site. Second, the binding of the first ligand can significantly alter the potential energy landscape of the target site and allow the identification of different classes of ligands. The alteration in the binding energies is often favorable (i.e., cooperative binding), due at least in some instances to preformation of the active site or even the potential for ligand–ligand interactions. Therefore, the first step in setting up a second-site screen is the selection of an optimal first-site ligand. Many properties of the first-site ligand should be optimized, including spectral quality, affinity, and size. Theoretically, the first-site ligand can bind weakly, but higher-affinity ligands reduce the false-positive hit rate due to competitive binding. The size of the first-site ligand is also important, as smaller ligands have a reduced risk of occluding the second-site pocket. This may require selecting a structurally related ligand rather than the exact substructure of a lead, as was the case for second-site screening against adenosine kinase (Hajduk *et al.*, 2000c).

Once a suitable first-site ligand has been identified, a library of compounds is screened for probing secondary sites on the drug target. In our experience, a ligand for a peripheral pocket will typically have a K_D value ranging from 5 to 50 mM—which is significantly weaker than what is observed for first-site ligands. Thus, screens must be designed to detect ligands of this affinity. When using target-based NMR screening methods for second-site screening (see below), our preferred library consists of approximately 3500 compounds with an average molecular weight of 150 Da stored as 1 M dimethyl sulfoxide (DMSO) stocks. As most proteins can tolerate up to 5% DMSO, this typically translates to screening mixtures of five compounds at 10 mM each. Under these conditions, a ligand with a K_D value as high as 50 mM can potentially be detected.

Target-Based Methods for Second-Site Screening

Target-based methods for identifying second-site ligands do not differ from those employed for identifying first-site ligands, except for the inclusion of a first-site ligand and perhaps higher concentrations of test compound. Finding those compounds that bind near the first-site ligand requires knowledge of the chemical shift assignments for amides or methyl groups in the peripheral site. In practice, these assignments can be obtained from full NMR assignment of the protein resonances or by using differential chemical shift approaches (Hajduk *et al.*, 2000c; Medek *et al.*, 1999). During the course of the screen, those ligands that bind simultaneously with the first-site ligand must be differentiated from those that compete for binding. In favorable cases, new signals will be significantly perturbed by the second-site ligand that are not affected by the first-site ligand. However, as both ligands bind in the same active site, it is typical that many resonances will be perturbed by both compounds. In such cases, careful titration experiments of each ligand alone and in the presence of the other can be used to unambiguously distinguish between competitive and simultaneous binding. In the case of competitive binding, the apparent binding affinity of one ligand will be reduced in the presence of the other. For simultaneous binding, the measured K_D will be essentially independent of the presence of the other ligand, while for cooperative binding the affinity of one ligand is actually enhanced in the presence of the other. These latter two cases (simultaneous and cooperative binding) indicate the potential for linking.

Ligand-Based Methods for Second-Site Screening

One-dimensional homonuclear methods have also been designed to extend second-site screening to larger proteins and membrane-associated proteins. Using a modification of the SLAPSTIC method (Jahnke *et al.*, 2001), spin labels such as Tempo are chemically attached to the first-site ligand. In a second-site screen in the presence of the modified first-site lead, detection of a neighboring ligand is manifested by enhanced relaxation rates of protons on the second-site ligand (Jahnke *et al.*, 2000). One advantage of this approach is that, in theory, any lead detected will by definition bind simultaneously with and proximal to the first-site ligand. Another advantage is that the binding orientation of the second-site lead can be determined by quantitating the magnitude of the spin label effects. Thus, information for linker design is immediately available. As with any 1D method, false positives can result from nonspecific binding of the spin-labeled first-site ligand to the protein. A second and more severe limitation to this method is the need for a chemically accessible point on the first-site ligand that will allow incorporation of a spin label without reducing the affinity of the lead or blocking access to the second site.

Another 1D method used to detect binding of second-site ligands is termed ILOE for the InterLigand NOE experiment (Li *et al.*, 1999). Similar to the transferred NOE experiment described 20 years ago by Clore and Gronenborn (1983), this method measures the NOE buildup between protons on neighboring ligands when binding for at least one of them is in the fast exchange regime. As with the spin-label methods, ILOE screens have the added advantage of providing orientation information that can be used to design linkers between the two leads. Although the quantitation of distances based on ILOE peak heights is difficult (London, 1999), relative distances between pairs of protons on the two ligands is sufficiently reliable to determine the protons of the two leads that are closest to one another. One disadvantage of this method is lower sensitivity compared to chemical shift and spin label methods. Published examples relied on 0.4- to 2.0-mM protein samples, although these data were collected without a cryoprobe. Another disadvantage that we have discovered is a tendency to overweight NOEs resulting from nonspecific binding. In four second-site screens that we have implemented using this approach, nonspecific binding resulted in false positives for two of the targets—making the ILOE screen unreliable (unpublished data). Similar to the observations described above for competitive 1D screening, protein-dependent ILOEs were observed that could not be competed away by a higher affinity reference ligand. This makes it imperative to validate that the leads derived from ILOE screens are specific ligands that bind to the site of interest.

Utilizing Second-Site Ligands in Drug Design

The linked-fragment strategy has been utilized in two related approaches: (1) the *de novo* construction of high-affinity ligands and (2) the optimization of existing leads by identification of new substituents. The construction of high-affinity ligands from pieces that have been identified via NMR has been described for FKBP (Shuker *et al.*, 1996), stromelysin (Hajduk *et al.*, 1997c, 2002), adenosine kinase (Hajduk *et al.*, 2000c), LFA (Liu *et al.*, 2001), and DHPR (Sem *et al.*, 2004). In all of these cases, screens were performed to identify ligands for peripheral sites that bound simultaneously with a first-site ligand. The strategies for *de novo* construction vs. lead optimization differ only in the sense of the desired end: generation of a novel structural series vs. improved physicochemical properties of an existing series.

We have recently reported on the use of the linked-fragment approach in the search for novel, potent, and selective inhibitors of PTP1B (Szczepankiewicz *et al.*, 2003). As already mentioned above, core elaboration strategies had yielded potent inhibitors of PTP1B by accessing the second pTyr pocket (see Fig. 1). However, this strategy allowed the incorporation of only a small subset of suitably functionalized monomers into the oxalylarylaminobenzoic acid lead. To increase the chances of finding an optimal second-site ligand, a screen was employed to identify ligands that bound specifically to the noncatalytic phosphotyrosine binding site by following the chemical shift perturbations of Met-258 (Liu *et al.*, 2003; Szczepankiewicz *et al.*, 2003). Compounds such as 3-aminosalicylic acid (**7**) and quinoline-2-carboxylic acid (**8**) were found to bind to the second site and, upon linking to a known first-site ligand (**9**), nanomolar inhibitors of PTP1B were produced (e.g., **10** and **11**; see Fig. 3). It is significant to note that compounds such as **7** and **8** could not have been utilized in the core replacement strategy because of incompatible functionality (carboxylic acids were used for amide bond formation).

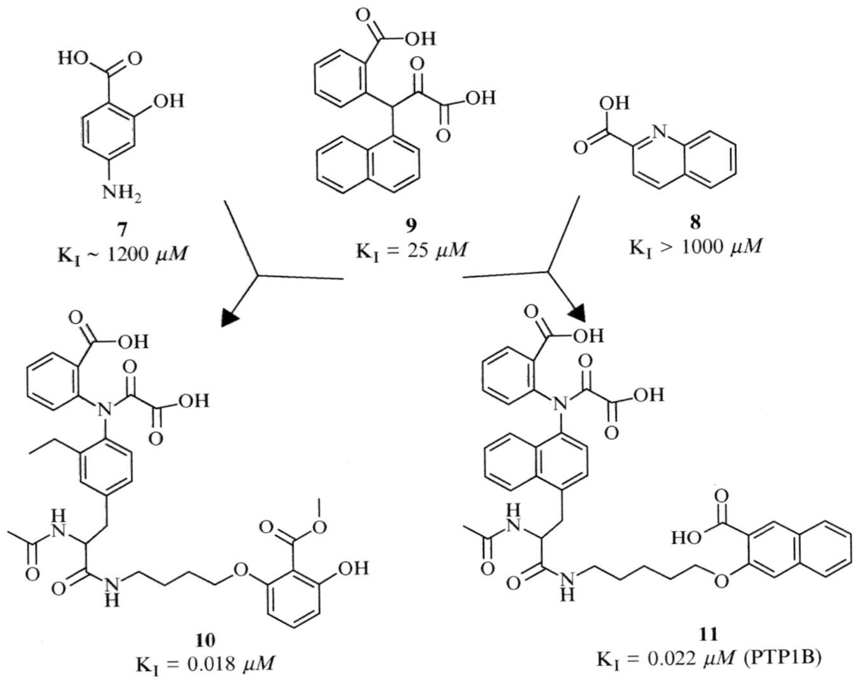

Fig. 3. Linked-fragment strategies for the design of PTP1B inhibitors (Szczepankiewicz *et al.*, 2003). Based on an NMR second-site screen, aminosalicylates (e.g., **7**) and quinaldic acids (e.g., **8**) were found to bind to the noncatalytic pTyr site on PTP1B. When linked to oxamate-containing first-site ligands (e.g., **9**), nanomolar inhibitors of PTP1B were produced (e.g., **10** and **11**).

Linker Design

In all of the linked-fragment examples described in the literature thus far, linking of two leads was done somewhat empirically using structural information from NMR and X-ray studies of ligand complexes. Because very small changes in the orientation of a compound can have dramatic influences on potency, it is not practical to design a single linked compound. Rather, a small selection of compounds is designed and prioritized by ease of chemical synthesis. For FKBP and stromelysin, small numbers of compounds were initially prepared. Very little linker dependence was observed for FKBP, as linkers of three to five atoms were all essentially equipotent (Shuker *et al.*, 1996). The situation was very different for stromelysin, in which a 10- to 300-fold loss in potency was realized when the linker length was changed by even a single atom (Hajduk *et al.*, 1997c). Therefore, great care should be taken in the design and selection of appropriate linkers.

In an attempt to maximize the chances of positioning the two fragments in an optimal manner, linkers can be used that allow a large degree of conformational freedom. Shown in Fig. 4 are the results of simulations for the conformational flexibility of nine three- or four-atom linkers tethering a phenyl to a cyclohexyl ring. On the *y*-axis is plotted the angle formed by the two bond vectors at the end of each linker, and on the *x*-axis is the distance between the carbons of each ring that are linked. More visual representations of the data are shown in Fig. 5, where each line indicates the vector for the bond formed by a carbon in the linked phenyl ring and its attached atom in the linker. Comparison of our results to small molecule X-ray structures indicates that the simulations produce valid conformations (data not shown). Not surprisingly, a purely aliphatic linker exhibits little conformational diversity, while inclusion of oxygen or sulfur in the linker significantly improves the flexibility. For a three-atom linker, the rank order for conformational diversity is hydroxyethyl (CCO) > thioethyl (CCS) > propyl (CCC). The rank order for conformational diversity for four-atom linkers is methylethyl thioether (CCSC) > hydroxypropyl (CCCO) > 2-butenyl (CC=CC) > methylethyl ether (CCOC) > acylsulfonamide (ASULF) ~ butyl (CCCC). The result that the four-atom 2-butenyl linker (CC=CC) actually exhibits greater conformational freedom than a butyl linker (CCCC) may at first seem counterintuitive. However, the removal of two protons from the linker allows a greater variety of local conformations due to lack of steric hindrance, at the cost of rotation about the central bond. Such simulations of simple linker systems are easily performed in most molecular modeling packages and can help prioritize specific types of linkers for a given project.

In addition to simply evaluating conformational flexibility, multiple software packages exist for aiding in actual linker design. For example, LUDI (Bohm, 1995) and ALLEGROW, based on GroMol (Bohacek and McMartin, 1994) and QXP (McMartin and Bohacek, 1997), both contain modules that allow for incorporation of linkers between two proximal molecules. Our experience in designing linkers for multiple projects would suggest that such computational algorithms can be used in a number of ways. First, in the initial design of linker length and geometry, such programs can aid in visualizing a wide variety of choices. In this initial phase, however, it is recommended that simple linkers with the greatest flexibility be pursued to demonstrate the proof-of-concept that potency gains will be realized, even though flexible linkers may impart entropic losses (Jahnke *et al.*, 2003). After linked compounds have been successfully produced,

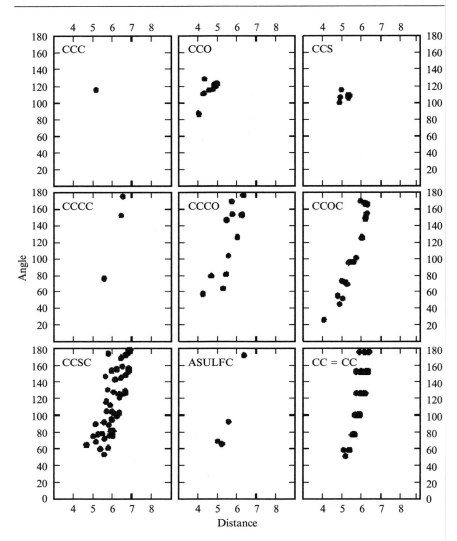

FIG. 4. Plots showing the relative conformational diversity sampled by various linkers between a phenyl and cyclohexyl ring. Low-energy conformations for each tethered system were calculated within Insight (Accelrys). Shown on the *x*-axis is the carbon-to-carbon distance (in Å) between the attached carbons on each ring for all low-energy conformers. Shown on the *y*-axis is the angle between the vectors formed by the attached carbon on each ring and the attached linker atom for all low-energy conformers. Note that each point can represent multiple conformations if the distance and angle are identical.

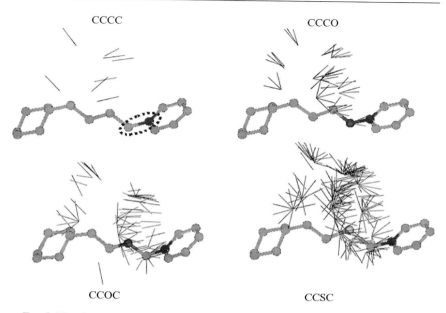

FIG. 5. Visual representations of the low-energy conformations sampled by four different linkers between a phenyl and cyclohexyl ring. Low-energy conformations for each tethered system were calculated within Insight (Accelrys). The lowest energy conformation is shown in ball and stick. The positions of the phenyl-linker bond [denoted by the dashed ellipse for the butyl (CCCC) linker in the upper left panel] for all other low-energy conformations are designated by the thin lines. (See color insert.)

these computational algorithms can then be used to refine and rigidify the linker, preferably guided by structural information on both the linked and unlinked fragments. It is important to note that subtle changes in the linker can often have significant effects. For example, in the initial design of biaryl-hydroxamate inhibitors for stromelysin, a simple hydroxyethyl linker between the hydroxamate (**12**) and biaryl (**13**) moieties was sufficient for high affinity (see Fig. 6) (Hajduk *et al.*, 1997c). This strategy was not as productive, however, in linking the naphthylhydroxamate moiety (**14**) (Hajduk *et al.*, 2002). Simple linkers such as a dihydroxyethyl group yielded potency gains, but the resulting compounds (e.g., **16**) were more than 10-fold weaker than the biarylhydroxamates (e.g., **15**), even though the naphthylhydroxamate moiety binds with 100-fold higher affinity than the acetohydroxamate. To regain affinity, the linker had to be redesigned. Ultimately, a sulfone linker was designed that yielded compounds with potencies less than 100 n*M* (e.g., **17**). These studies suggest that multiple

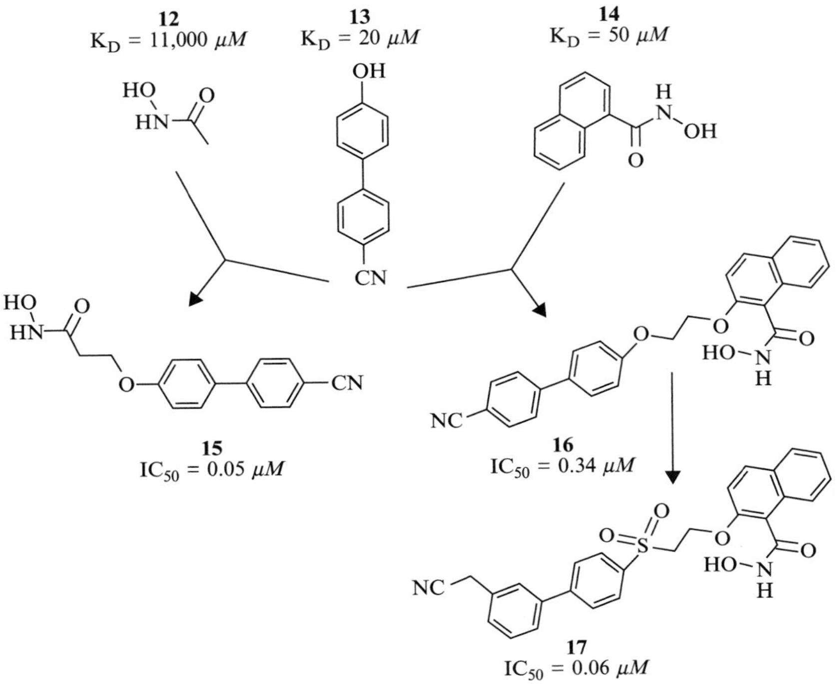

Fig. 6. Summary of the linked-fragment strategies employed in the design of MMP inhibitors (Hajduk *et al.*, 1997c, 2002) In an initial study, acetohydroxamate (**12**) was linked to a biarylphenol (**13**) to yield potent inhibitors (**15**) with a simple hydroxyethyl linker. Linking of the naphthylhydroxamate (**14**) yielded only moderately potent inhibitors (**16**), even though the naphthylhydroxamate exhibited 100-fold more potency than acetohydroxamate. Multiple linkers had to be explored to ultimately produce compounds with sub-100 nM IC_{50} values (e.g., **17**).

approaches can and should be taken to explore various linking strategies and to maximize the probability of producing high-affinity ligands.

Conclusions

In summary, there are a variety of NMR-based screening methods that can be productively utilized to identify fragment leads for use in drug design. The actual design strategy will vary depending on the particular issues at hand. Significantly, these design strategies can be used throughout the entire course of preclinical drug discovery—including initial lead identification, potency optimization, and even modulation of physicochemical properties to aid in candidate selection. As referenced herein, there are now many examples in the literature of NMR-driven, fragment-based discovery programs that have successfully produced high-affinity ligands for their respective targets. Even more significant, however, is the fact that compounds from several programs employing fragment-based design [e.g., matrix metalloproteinase inhibitors (Wada *et al.*, 2002) and inhibitors of the LFA–ICAM interaction (Liu *et al.*, 2001)] have already gone beyond the preclinical setting into the clinic. This is ample validation that the methods described in this chapter can be used to produce ligands not only with high affinity but also with the appropriate drug-like properties for use in therapeutic intervention in humans.

References

Bartolozzi, A., Grogan, M. J., and Kates, S. A. (2003). Advances in the applications of polymer-supported reagents for organic synthesis. *Methods Enzymol.* **369**, 347–366.

Bhattacharyya, S., Fan, L., Vo, L., and Labadie, J. (2000). *Comb. Chem. High Throughput Screen.* **3**, 117–124.

Bhattacharyya, S., Rana, S., Gooding, O. W., and Labadie, J. (2003). *Tetrahedron Lett.* **44**, 4957–4960.

Bohacek, R. S., and McMartin, C. (1994). Multiple highly diverse structures complementary to enzyme binding sites: Results of extensive application of a de novo design method incorporating combinatorial growth. *J. Am. Chem. Soc.* **116**, 5560–5571.

Bohm, H. J. (1995). Site-directed structure generation by fragment-joining. *Perspect. Drug Disc. Des.* **3**, 21–33.

Booth, R. J., and Hodges, J. C. (1997). Polymer-supported quenching reagents for parallel purification. *J. Am. Chem. Soc.* **119**, 4882–4886.

Booth, R. J., and Hodges, J. C. (1999). Solid-supported reagent strategies for rapid purification of combinatorial synthesis products. *Acc. Chem. Res.* **32**, 18–26.

Chen, A., and Shapiro, M. J. (1998). NOE pumping: A novel NMR technique for identification of compounds with binding affinity to macromolecules. *J. Am. Chem. Soc.* **120**, 10258–10259.

Clore, G. M., and Gronenborn, A. M. (1983). Theory and applications of the transferred nuclear Overhauser effect to the study of the conformations of small ligands bound to proteins. *J. Magn. Reson.* **48**, 402–417.

Colacot, T. J., Gore, E. S., and Kuber, A. (2002). High-throughput screening studies of fiber-supported catalysts leading to room-temperature Suzuki coupling. *Organometallics* **21**, 3301–3304.

Coles, M., Heller, M., and Kessler, H. (2003). NMR-based screening technologies. *DDT* **8**, 803–810.

Dalvit, C., Flocco, M., Knapp, S., Mostardini, M., Perego, R., Stockman, B. J., Veronesi, M., and Varasi, M. (2002). High-throughput NMR-based screening with competition binding experiments. *J. Am. Chem. Soc.* **124**, 7702–7709.

Fauchere, J. L., Boutin, J. A., Henlin, J. M., Kucharczyk, N., and Ortuno, J. C. (1998). Combinatorial chemistry for the generation of molecular diversity and the discovery of bioactive leads. *Chemom. Intell. Lab. Sys.* **43**, 43–68.

Fecik, R. A., Frank, K. E., Gentry, E. J., Menon, S. R., Mitscher, L. A., and Telikepalli, H. (1998). The search for orally active medications through combinatorial chemistry. *Med. Res. Rev.* **18,** 149–185.

Fejzo, J., Lepre, C. A., Peng, J. W., Bemis, G. W., Ajay, Murcko, M. A., and Moore, J. M. (1999). The SHAPES strategy: An NMR-based approach for lead generation in drug discovery. *Chem. Biol.* **6,** 755–769.

Fernandez, C., and Wider, G. (2003). TROSY in NMR studies of the structure and function of large biological macromolecules. *Curr. Opin. Struct. Biol.* **13,** 570–580.

Gentles, R. G., Wodka, D., Park, D. C., and Vasudevan, A. (2002). Standardization protocols and optimized precursor sets for the efficient application of automated parallel synthesis to lead optimization: A Mitsunobu example. *J. Comb. Chem.* **4,** 442–456.

Gerig, J. T. (1994). Fluorine NMR of proteins. *Prog. NMR Spectrosc.* **26,** 293–370.

Ghose, A. K., Viswanadhan, V. N., and Wendoloski, J. J. (1999). A knowledge-based approach in designing combinatorial or medicinal chemistry libraries for drug discovery. 1. A qualitative and quantitative characterization of known drug databases. *J. Comb. Chem.* **1,** 55–68.

Hajduk, P. J., Meadows, R. P., and Fesik, S. W. (1997a). Discovering high-affinity ligands for proteins. *Science* **278,** 497–499.

Hajduk, P. J., Olejniczak, E. T., and Fesik, S. W. (1997b). One-dimensional relaxation- and diffusion-edited NMR methods for screening compounds that bind to macromolecules. *J. Am. Chem. Soc.* **119,** 12257–12261.

Hajduk, P. J., Sheppard, G., Nettesheim, D. G., Olejniczak, E. T., Shuker, S. B., Meadows, R. P., Steinman, D. H., Carrera, G. M., Marcotte, P. A., Severin, J., Walter, K., Smith, H., Gubbins, E., Simmer, R., Holzman, T. F., Morgan, D. W., Davidsen, S. K., and Fesik, S. W. (1997c). Discovery of potent nonpeptide inhibitors of stromelysin using SAR by NMR. *J. Am. Chem. Soc.* **119,** 5818–5827.

Hajduk, P. J., Dinges, J., Schkeryantz, J. M., Janowick, D., Kaminski, M., Tufano, M., Augeri, D. J., Petros, A., Nienaber, V., Zhong, P., Hammond, R., Coen, M., Beutel, B., Katz, L., and Fesik, S. W. (1999a). Novel inhibitors of erm methyltransferases from NMR and parallel synthesis. *J. Med. Chem.* **42,** 3852–3859.

Hajduk, P. J., Gerfin, T., Boehlen, J.-M., Häberli, M., Marek, D., and Fesik, S. W. (1999b). High-throughput nuclear magnetic resonance-based screening. *J. Med. Chem.* **42,** 2315–2317.

Hajduk, P. J., Meadows, R. P., and Fesik, S. W. (1999c). NMR-based screening in drug discovery. *Q. Rev. Biophys.* **32,** 211–240.

Hajduk, P. J., Augeri, D. A., Mack, J., Mendoza, R., Yang, J., Betz, S. F., and Fesik, S. W. (2000a). NMR-based screening of proteins containing 13C-labeled methyl groups. *J. Med. Chem.* **122,** 7898–7904.

Hajduk, P. J., Boyd, S., Nettesheim, D., Nienaber, V., Severin, J., Smith, R., Davidson, D., Rockway, T., and Fesik, S. W. (2000b). The identification of novel inhibitors of urokinase using NMR-based screening. *J. Med. Chem.* **43,** 3862–3866.

Hajduk, P. J., Gomtsyan, A., Didomenico, S., Cowart, M., Bayburt, E. K., Solomon, L., Severin, Smith, R., Walter, K., Holzman, T. F., Stewart, A., McGaraughty, S., Jarvis, M. F., Kowaluk, E. S., and Fesik, S. W. (2000c). Design of adenosine kinase inhibitors from the NMR-based screening of fragments. *J. Med. Chem.* **43,** 4781–4786.

Hajduk, P. J., Shuker, S. B., Nettesheim, D. G., Xu, L., Augeri, D. J., Betebenner, D., Craig, R., Albert, D. H., Guo, Y., Meadows, R. P., Michaelides, M., Davidsen, S. K., and Fesik, S. W. (2002). NMR-based optimization of MMP inhibitors with improved bioavailability. *J. Med. Chem.* **45,** 5628–5639.

Hajduk, P. J., Mack, J. C., Olejniczak, E. T., Park, C., Dandliker, P. J., and Beutel, B. A. (2004). SOS-NMR: A saturation transfer NMR-based method for determining the structures of protein-ligand complexes. *J. Am. Chem. Soc.* **126**, 2390–2398.

Hinzen, B. (2000). Polymer-supported reagents: Preparation and use in parallel organic synthesis. *Methods Princ. Med. Chem.* **9**, 209–237.

Huth, J. R., and Sun, C. (2002). Utility of NMR in lead optimization: Fragment-based approaches. *Comb. Chem. High Throughput Screen.* **5**, 631–644.

Jahnke, W. (2002). Spin labels as a tool to identify and characterize protein-ligand interactions by NMR spectroscopy. *Chembiochem.* **3**, 167–173.

Jahnke, W., Perez, L. B., Paris, C. G., Strauss, A., Fendrich, G., and Nalin, C. M. (2000). Second-site NMR screening with a spin-labeled first ligand. *J. Am. Chem. Soc.* **122**, 7394–7395.

Jahnke, W., Rudisser, S., and Zurini, M. (2001). Spin label enhanced NMR screening. *J. Am. Chem. Soc.* **123**, 3149–3150.

Jahnke, W., Florsheimer, A., Blommers, M. J. J., Paris, C. G., Heim, J., Nalin, C. M., and Perez, L. B. (2003). Second-site NMR screening and linker design. *Curr. Top. Med. Chem.* **3**, 69–80.

Kortemme, T., and Baker, D. (2002). A simple physical model for binding energy hot spots in protein-protein complexes. *Proc. Natl. Acad. Sci. USA* **99**, 14116–14121.

Kuntz, I. D., Chen, K., Sharp, K. A., and Kollman, P. A. (1999). The maximal affinity of ligands. *Proc. Natl. Acad. Sci. USA* **96**, 9997–10002.

Lepre, C. A., Peng, J. W., Fejzo, J., Abdul-Manan, N., Pocas, J., Jacobs, M., Xie, X., and Moore, J. M. (2002). Applications of SHAPES screening in drug discovery. *Comb. Chem. High Throughput Screen.* **5**, 583–590.

Li, D., DeRose, E., and London, R. (1999). The inter-ligand Overhauser effect: A powerful new NMR approach for mapping structural relationships of macromolecular ligands. *J. Biomol. NMR* **15**, 71–76.

Liu, G., Huth, J. R., Olejniczak, E. T., Mendoza, R., DeVries, P., Leitza, S., Reilly, E. B., Okasinski, G. F., Fesik, S. W., and von Geldern, T. W. (2001). Novel p-arylthio cinnamides as antagonists of leukocyte function-associated antigen-1/intracellular adhesion molecule-1 interaction. 2. Mechanism of inhibition and structure-based improvement of pharmaceutical properties. *J. Med. Chem.* **44**, 1202–1210.

Liu, G., Xin, Z., Liang, H., Abad-Zapatero, C., Hajduk, P. J., Janowick, D. A., Szczepankiewicz, B. G., Pei, Z., Hutchins, C. W., Ballaron, S. J., Stashko, M. A., Lubben, T. H., Berg, C. E., Rondinone, C. M., Trevillyan, J. M., and Jirousek, M. R. (2003a). Selective protein tyrosine phosphatase 1B inhibitors: Targeting the second phospho-tyrosine binding site with non-carboxylic acid-containing ligands. *J. Med. Chem.* **46**, 3437–3440.

Liu, G., Xin, Z., Pei, Z., Zhao, H., Hajduk, P. J., Abad-Zapatero, C., Hutchins, C. W., Lubben, T. H., Ballaron, S. J., Haasch, D. L., Kaszubska, W., Rondinone, C. M., Trevillyan, J. M., and Jirousek, M. R. (2003b). Fragment screening and assembly: A highly efficient approach to a selective and cell active protein tyrosine phosphatase 1B inhibitor. *J. Med. Chem.* **46**, 4232–4235.

London, R. E. (1999). Theoretical analysis of the inter-ligand Overhauser effect: A new approach for mapping structural relationships of macromolecular ligands. *J. Magn. Reson.* **141**, 301–311.

Mayer, M., and Meyer, B. (1999). Characterization of ligand binding by saturation transfer difference NMR spectroscopy. *Angew. Chem. Int. Ed.* **38**, 1784–1788.

McCoy, M. A., and Wyss, D. F. (2000). Alignment of weakly interacting molecules to protein surfaces using simulations of chemical shift perturbations. *J. Biomol. NMR* **18**, 189–198.

McGovern, S. L., Caselli, E., Grigorieff, N., and Shoichet, B. K. (2002). A common mechanism underlying promiscuous inhibitors from virtual and high-throughput screening. *J. Med. Chem.* **45**, 1712–1722.

McMartin, C., and Bohacek, R. S. (1997). QXP: Powerful, rapid computer algorithms for structure-based design. *J. Comp. Aided Mol. Des.* **11**, 333–344.

Medek, A., Hajduk, P. J., Mack, J., and Fesik, S. W. (1999). The use of differential chemical shifts for determining the binding site location and orientation of protein-bound ligands. *J. Am. Chem. Soc.* **122**, 1241–1242.

Meyer, B., Weimar, T., and Peters, T. (1997). Screening mixtures for biological activity by NMR. *Eur. J. Biochem.* **246**, 705–709.

Meyer, B., and Peters, T. (2003). NMR spectroscopy techniques for screening and identifying ligand binding to receptors. *Angew. Chem. Int. Ed.* **42**, 864–890.

Murali, N., Jarori, G. K., Landy, S. B., and Nageswara Rao, B. D. (1993). Two-dimensional transferred nuclear Overhauser effect spectroscopy (TRNOESY) studies of nucleotide conformations in creatine kinase complexes: Effects due to weakly nonspecific binding. *Biochemistry* **32**, 12941–12948.

Peng, J. W., Lepre, C. A., Fejzo, J., Abdul-Manan, N., and Moore, J. M. (2001). Nuclear magnetic resonance-based approaches for lead generation in drug discovery. *Methods Enzymol.* **338**, 202–230.

Post, C. B. (2003). Exchange-transferred NOE spectroscopy and bound ligand structure determination. *Curr. Opin. Struct. Biol.* **13**, 581–588.

Salvino, J. M., and Dolle, R. E. (2003). The development and application of tetrafluorophenol-activated resins for rapid amine derivatization. *Methods Enzymol.* **369**, 151–163.

Salvino, J. M., Kumar, N. V., Orton, E., Airey, J., Kiesow, T., Crawford, K., Mathew, R., Krolikowski, P., Drew, M., Engers, D., Krolinkowski, D., Herpin, T., Gardyan, M., McGeehan, G., and Labaudiniere, R. (2000). Polymer-supported tetrafluorophenol: A new activated resin for chemical library synthesis. *J. Comb. Chem.* **2**, 691–697.

Sauer, D. R., Kalvin, D., and Phelan, K. M. (2003). Microwave-assisted synthesis utilizing supported reagents: A rapid and efficient acylation procedure. *Organic Lett.* **5**, 4721–4724.

Sem, D. S., Bertolaet, B., Baker, B., Chang, E., Costache, A. D., Coutts, S., Dong, Q., Hansen, M., Hong, V., Huang, X., Jack, R. M., Kho, R., Lang, H., Ma, C.-T., Meininger, D., Pellecchia, M., Pierre, F., Villar, H., and Yu, L. (2004). Systems-based design of bi-ligand inhibitors of oxidoreductases: Filling the chemical proteomic toolbox. *Chem. Biol.* **11**, 185–194.

Shuker, S. B., Hajduk, P. J., Meadows, R. P., and Fesik, S. W. (1996). Discovering high-affinity ligands for proteins: SAR by NMR. *Science* **274**, 1531–1534.

Stockman, B. J., and Dalvit, C. (2002). NMR screening techniques in drug discovery and drug design. *Prog. NMR Spectrosc.* **41**, 187–231.

Stockman, B. J., Farley, K. A., and Angwin, D. T. (2001). Screening of compound libraries for protein binding using flow-injection nuclear magnetic resonance spectroscopy. *Methods Enzymol.* **338**, 231–246.

Strauss, A., Bitsch, F., Cutting, B., Fendrich, G., Graff, P., Liebetanz, J., Zurini, M., and Jahnke, W. (2003). Amino-acid-type selective isotope labeling of proteins expressed in baculovirus-infected insect cells useful for NMR studies. *J. Biomol. NMR* **26**, 367–372.

Szczepankiewicz, B. G., Liu, G., Hajduk, P. J., Abad-Zapatero, C., Pei, Z., Xin, Z., Lubben, T., Trevillyan, J. M., Stashko, M. A., Ballaron, S. J., Liang, H., Huang, F., Hutchins, C. W., Fesik, S. W., and Jirousek, M. R. (2003). Discovery of a potent, selective protein tyrosine phosphatase 1B inhibitor using a linked-fragment strategy. *J. Am. Chem. Soc.* **125**, 4087–4096.

Ukkola, O., and Santaniemi, M. (2002). Protein tyrosine phosphatase 1B: A new target for the treatment of obesity and associated comorbidities. *J. Intern. Med.* **251,** 467–475.

Wada, C. K., Holms, J. H., Curtin, M. L., Dai, Y., Florjancic, A. S., Garland, R. B., Guo, Y., Heyman, H. R., Stacey, J. R., Steinman, D. H., Albert, D. H., Bouska, J. J., Elmore, I. N., Goodfellow, C. L., Marcotte, P. A., Tapang, P., Morgan, D. W., Michaelides, M. R., and Davidsen, S. K. (2002). Phenoxyphenyl sulfone N-formylhydroxylamines (retrohydroxamates) as potent, selective, orally bioavailable matrix metalloproteinase inhibitors. *J. Med. Chem.* **45,** 219–232.

Xin, A., Oost, T. K., Abad-Zapatero, C., Hajduk, P. J., Pei, Z., Szczepankiewicz, B. G., Hutchins, C. W., Ballaron, S. J., Stashko, M. A., Lubben, T., Trevillyan, J. M., Jirousek, M. R., and Liu, G. (2003). Potent, selective inhibitors of protein tyrosine phosphatase 1B. *Biorg. Med. Chem. Lett.* **13,** 1887–1890.

Yun, Y. K., Porco, J. A., Jr., and Labadie, J. (2002). Polymer-assisted parallel solution phase synthesis of substituted benzimidazoles. *Synlett* 739–742.

[24]   Discovery of Ligands by a Combination of Computational and NMR-Based Screening: RNA as an Example Target

By Moriz Mayer and Thomas L. James

Abstract

NMR for screening of knowledge-based focused libraries of compounds provides an efficient, cost-effective method to develop promising drug leads that target functionally important RNA structures. A knowledge-based focused library may be constructed from virtual (i.e., computational) screening of commercial or proprietary databases of available compounds for binding to the three-dimensional structure of a selected RNA target. Alternatively, the library may be constructed from compounds with properties deemed desirable, e.g., molecular moiety commonly found in drugs or known to bind RNA. The library ideally should be composed of small water-soluble, nonpeptide, nonnucleotide organic compounds. Various simple, robust NMR experiments are described that enable experimental screening of such a library for binding to a selected RNA structure. Some of the NMR experiments enable rapid mapping of the interaction site on the RNA to verify that the targeted structure is hit rather than the double helical region or a commonly occurring tetraloop. Other experiments enable elucidation of the ligand's binding moiety. Of course, any compounds thus identified should represent promising scaffolds suitable for easy chemical modification to enhance their pharmaceutical properties for subsequent drug development.

Copyright 2005, Elsevier Inc.
All rights reserved.
0076-6879/05 $35.00

Introduction

As discussed in several recent papers and reviews, nuclear magnetic resonance (NMR) is positioned uniquely in the drug discovery field. NMR experiments enable screening of potential ligands, determination of binding affinity, and acquisition of structural information about any ligands that bind to the receptor (Coles *et al.*, 2003; Meyer and Peters, 2003; Peng *et al.*, 2001; van Dongen *et al.*, 2002; Wyss *et al.*, 2002). The approach presented here relies on combining structure-based *in silico* screening with low-throughput, but high-information-content, NMR-based screening to discover ligands that can bind to RNA. The basic concept and requirements for structure-based screening in general and specifically targeting RNA motifs will be explained briefly. In the second part, some of the NMR methods available to test the putative RNA binders experimentally by NMR are described in more detail.

High-throughput screening (HTS) of large compound libraries, which can be created by combinatorial chemistry, provides a means to find new compounds to bind RNA (Luedtke and Tor, 2003; Mei *et al.*, 1998; Swayze *et al.*, 2002). However, even if a million compounds were made, chemistries available for parallel synthesis, not to mention the starting point for these syntheses, would skew the distribution of compounds available. It has been estimated that the number of potential drug molecules is 10^{10}–10^{50} (Valler and Green, 2000). However, screening more than 10^6 compounds is very expensive, and there is much uncertainty involved in experimental screening of tens of thousands of compounds, which are often of questionable quality (Bajorath, 2002). The expense certainly restricts the HTS approach to industrial settings. Also, the development and validation of a high-throughput screen take time. Since NMR-based screening can detect binding directly, an option is starting the experimental screening on a smaller number of compounds quickly and using information from that in follow-up screens.

Discovery of ligands from *focused* libraries of small molecules by NMR screening is efficient. There are a variety of ways to design such focused libraries. One way is to use a structure-based virtual screen (Filikov *et al.*, 2000). Another possibility is to use a set of surrogate drug-like scaffolds of very low molecular weight selected for their beneficial physicochemical properties (Fejzo *et al.*, 1999; Hajduk *et al.*, 2000). Both of these concepts can be defined as knowledge-based approaches. Ideally, *in silico* screening yields a set of compounds highly enriched toward binding to the specified target, which means fewer compounds need to be tested experimentally (Lyne, 2002; Rudisser and Jahnke, 2002).

Potentially, scaffolds showing binding affinity in the NMR screen can be used as a basis for designing a combinatorial library around the binding moiety. In another approach, the integration of *in silico* and NMR screening has resulted in the design of a fluorescently labeled compound, which was subsequently used in HTS (Hajduk *et al.*, 2002; Moore *et al.*, 2004). On a different note, HTS is often unsuitable for detection of ligands with affinities weaker than ca. 5 μM; so for cases where detection of higher affinity leads is not possible, NMR screening offers a viable alternative. Consequently, NMR screening of libraries designed from a knowledge-based approach is not intended to replace conventional HTS but rather complement it (Hajduk and Burns, 2002).

RNA as a Drug Target

There are good reasons for choosing RNA as a drug target. The number of verified cellular and viral RNA targets is ever increasing; a variety of biochemical processes, including replication, translation, viral transcription, and control of gene expression, is dependent on or controlled by protein–RNA and RNA–RNA interactions (Hermann, 2000). Therefore, small molecule ligands binding to RNA motifs involved in these interactions provide ways to inhibit specific processes by either stabilizing a nonfunctional conformation or competitively blocking the RNA–RNA or RNA–protein interaction (Crowley *et al.*, 2002; Hermann, 2003). For example, molecules binding to active sites of the ribosome have been used as antibacterial agents. Also, inhibiting a single RNA molecule could prevent the production of thousands of proteins.

Because the design of RNA-binding compounds has received far less attention than the design of compounds to bind proteins, it is not surprising that relatively few small molecules with high affinity and specificity toward RNA are currently known. Among these compounds, most fall into a few categories, e.g., aminoglycosides, nucleotide derivatives, and basic amino acids. A promising recent development shows that this may be changing; the first completely new class of antibiotics to be approved by the FDA in 35 years was oxazolidinones, which apparently block ribosomal peptidyl-transferase activity (Colca *et al.*, 2003).

Structure-Based Identification of Compounds Targeting RNA

High-quality structural data for the target are required for efficient receptor-based computational screening. These data can originate either from X-ray crystallography or NMR spectroscopy. However, this obvious prerequisite is also one of the main limitations; the three-dimensional (3D) structure of most potential targets has not been solved yet. New instrumentation and crystallization techniques have increased the number of RNA structures being solved by X-ray, however, and the most prominent example was solving the 30S small ribosomal subunit by Ramakrishnan's group in 2000 (Wimberly *et al.*, 2000).

Choosing the appropriate 3D RNA structure for docking is the first critical decision that needs to be made in an *in silico* screen. The docking of small molecules to an RNA construct is complicated by the fact that many of the interesting RNA motifs from a biochemical and structural point of view are loop and bulge motifs. These regions are typically more flexible than the purely double-stranded sections of the RNA (Davis *et al.*, 2003; Foloppe *et al.*, 2004; Gallego and Varani, 2001). Often RNA binds small molecules by induced fit, which presents both a challenge and an opportunity in the computational screening approach. While such adaptive binding is of great importance for protein-based docking, it is especially critical when docking to an inherently flexible RNA motif. One way to deal with the RNA conformational change induced by the ligand is to identify an ensemble of possible conformers and dock to each of the static conformers (Hermann, 2002; Wei *et al.*, 2004). Another option is to make the docking procedure flexible for both the ligand and the RNA, which has been our approach to a limited extent (Lind *et al.*, 2002). There is a drawback for both techniques: ensemble docking does not allow flexibility during the docking process, and flexible docking samples only local wells, i.e., the flexibility is limited to low-energy conformational transitions. So even when using these techniques, adaptive binding with large conformational rearrangement is not accurately represented. On the other hand, the possibility of docking a small molecule to a nonfunctional RNA conformation will ideally lead to a ligand that locks in or stabilizes this conformation (Murchie *et al.*, 2004). Consequently, in addition to direct competition with protein or RNA-binding sites, small molecules can also exploit allosteric effects. In recent years, it has become evident that an important facet of the conformational flexibility of RNA is that RNA sometimes acts as a molecular switch. Ligand binding preferentially to one conformation in such a switch can severely perturb the switch equilibrium, leading to inhibition of function.

Scoring is the other critical aspect encountered in all docking procedures. Ranking of the docked molecules is done by scoring functions, which use force fields or empirical free energy values to calculate the affinity of the receptor for the ligand. Different energy terms, such as van der Waals, Coulomb, hydrogen bonding, and solvation, may be evaluated during the docking procedure (Stahl and Rarey, 2001). Popular docking programs that allow ligand flexibility and often include the possibility of calculating binding affinities with different scoring functions are DOCK (Kick *et al.*, 1997), GOLD (Jones *et al.*, 1995), ICM (Abagyan *et al.*, 1994), FRED (Stahl and Rarey, 2001), and FlexX (Kramer *et al.*, 1999). In most docking protocols, different docked conformations (poses) of the potential ligand are generated, and the correct pose must be identified according to its calculated interaction energy. The scoring protocols are often intentionally truncated to preserve the speed of the calculation. As a result, absolute binding energies are not accurately calculated, and false positives and negatives will arise (Bissantz *et al.*, 2000).

Docking Procedure

We employ a hierarchical approach for the docking of small molecule libraries (see Lind *et al.*, 2002). This allows us to scan quickly through a library to exclude unfavorable compounds, so we can perform computationally more demanding flexible docking on the more interesting compounds. Starting from the available chemicals directory (ACD), we applied relaxed Lipinski-type filters to exclude ligands too charged, too large, or too flexible for reasonable bioavailability (Lipinski, 2000). We then used the relatively fast algorithm of DOCK 4.0 (Kick *et al.*, 1997) to dock flexible ligands to the rigid structure of the HIV-1 TAR RNA receptor used in this study. In the DOCK scoring function, the electrostatic energy term needed to be halved to accurately redock and score a test set of known RNA–ligand complexes; evidently, the DOCK scoring function available at that time overemphasized electrostatics. The 50,000 highest scoring compounds from DOCK were then reminimized and rescored with the program ICM, which has additional solvation and hydrogen-bonding terms (Abagyan and Totrov, 1994). The scoring function of ICM was also empirically parameterized. This led to a scoring function that emphasized the hydrogen-bonding capability of RNA and reduced the influence of solvation (Lind *et al.*, 2002). Lastly, for the 5000 top scoring compounds from the previous round, we allowed flexibility of both ligand and RNA in a 5 Å radius around the binding site to allow for a certain amount of binding site adaptation. Visual inspection of compounds scoring in the ca. top 0.1% of all docked compounds was the final filter; many compounds were excluded because either they had moved out of the binding pocket or large parts of the ligand made no contact to the target at all. Prior to experimental testing for binding by NMR, the final hurdle for the compounds that passed all the above-mentioned filters and scored high enough to be selected was availability, price, and solubility.

NMR-Based Screening of RNA–Ligand Interactions

Our focus is to use simple one-dimensional (1D) or two-dimensional (2D) screening methods to detect ligand binding to small, unlabeled RNA constructs. We will not discuss 2D NMR ligand-based methods, such as transferred NOE, or methods that are useful only for ligands binding to large targets, such as diffusion edited experiments. For a more detailed description of available receptor- or ligand-based NMR screening methods, see Peng *et al.* (2001) and Hajduk and colleagues (Chapter 23, this volume).

Critical in our view is the assumption that even minimal RNA constructs of naturally occurring motifs can adopt the correct structure, and, therefore, function. This has the advantage that for most RNA constructs, expensive isotopic labeling will not be required for the screening process. For example, paromomycin binds to a model 27-nucleotide RNA construct with the same affinity and protects the same nucleotides from chemical probing as it does upon binding to the ribosome, showing that the chosen construct accurately represents the natural RNA-binding site (Fourmy *et al.*, 1996; Lynch *et al.*, 2000). A comparison of the crystal structure of the 30S subunit and the NMR structure of the 27-mer fragment complexed with paromomycin demonstrated that the structures were quite similar aside from two adenine residues in the bulge, which showed differences most likely originating from crystal packing effects or the freezing of the dynamic bases into a single conformation due to the low temperature employed (Lynch *et al.*, 2003). This implies that target-based screening is easily applicable to RNA because small constructs consisting of less than 40 nucleotides can generally be used. We will also discuss some of the available ligand-based screening methods for the small RNA constructs in the second part of the next section.

RNA Target-Based Screening Methods

Monitoring Imino Proton Chemical Shifts

RNA target-based NMR screening methods can be utilized with resonances that have been assigned. Shown in Fig. 1 is the imino chemical shift region of HIV-1 TAR RNA, where target-based screening is easily implemented (Mayer and James, 2004). The imino protons resonating in this region belong to double-stranded RNA, and only one peak per Watson–Crick base pair is observed; overlap is therefore usually not a big problem. One major drawback is that flexible regions or loops in which no stable hydrogen bonds are formed cannot be monitored directly. However, binding of small molecules is often accompanied with some sort of structural rearrangement of the RNA-binding site, which propagates to neighboring regions with consequent changes in observable imino resonances. So, unless a ligand binds to a very large loop or bulge, line broadening or changes

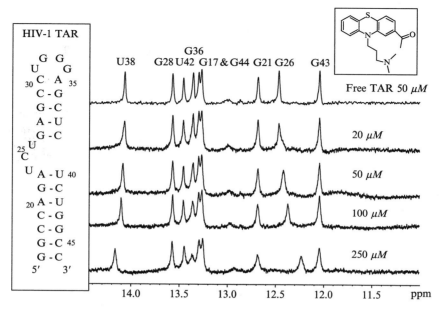

FIG. 1. Imino proton region of NMR spectra of 50 μM HIV-1 TAR RNA with increasing amounts of acetylpromazine (structures shown in insets) added. Resonances G26 and U38 shift the most, while the U40 resonance of the lower stem closing base pair is not observable.

in the chemical shifts will be observed. Another advantage of monitoring imino resonances is that for weak binders in the fast exchange regime, dissociation constants can be determined directly by measuring chemical shift differences with increasing amounts of ligand (Roberts, 2002; Yu et al., 2003). Figure 1 demonstrates how the TAR RNA imino resonances of G26 and U38 change upon addition of increasing amounts of acetylpromazine. A K_d value of 270 μM was calculated by fitting the measured chemical shift differences as the concentration of ligand was increased from 20 to 1000 μM. For high-affinity ligands such as paromomycin that are in slow exchange on the chemical shift scale, the binding stoichiometry can be determined by addition of submolar amounts of ligand (Fourmy et al., 1998; Lynch et al., 2000). For our screening purposes, 50 μM RNA solutions in 90% H_2O were sufficiently concentrated to obtain good imino spectra on a 600-MHz spectrometer with 512 scans or about 35 min acquisition time, so the amount of RNA used is relatively low.

Monitoring Pyrimidine H5–H6 Chemical Shifts

As shown in Fig. 2, the pyrimidine H5–H6 proton region in RNA is usually well dispersed and accessible by 2D TOCSY spectra. The strong vicinal coupling of the H5–H6 protons make this proton pair an ideal candidate, and signal overlap for structured RNA constructs with up to 40 nucleotides is typically not an issue (Hwang *et al.*, 1999; Lind *et al.*, 2002). As long as the binding region is not completely void of pyrimidine residues, monitoring of these peaks by 2D TOCSY or COSY spectra is fairly failsafe. A further advantage is that bulges and loops containing pyrimidines can be monitored directly, so no indirect conclusions (as with imino resonances) need to be made. We were able to collect TOCSY spectra with a good signal-to-noise ratio in about 2–3 h on a 140 μM RNA sample. Of course, if the TOCSY spectrum indicates an RNA–ligand interaction, NOESY spectra can be acquired to define binding more

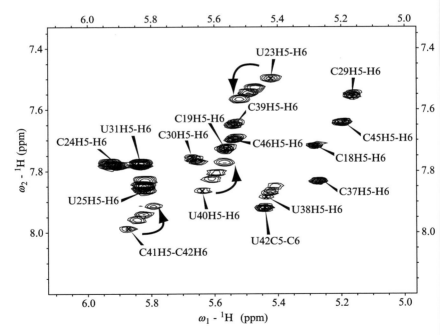

FIG. 2. HIV-1 TAR RNA chemical shift data of the H5–H6 proton spectral region of 150 μM TAR RNA obtained from four 2D TOCSY spectra acquired at 0, 50, 100, and 250 μM acetylpromazine (cf. Fig. 1) concentrations. Resonances U23, U40, and C41 experience the largest shifts as indicated by the arrows.

completely. We believe using NOESY spectra purely as a screening tool is inefficient, since the RNA consumption is too high and transferred NOESY experiments can just as well be substituted by much faster 1D STD or WaterLOGSY experiments (vide infra).

^{19}F NMR Screening

The advantages of using ^{19}F for ligand-based NMR screening have been described in a number of papers recently (Dalvit et al., 2003). Receptor-based ^{19}F NMR screening has also been shown to be effective for smaller proteins (Feeney et al., 1980; Leone et al., 2003) and also seems to be an interesting alternative for RNA constructs by labeling individual bases with ^{19}F nuclei. There would be no need to observe labile imino resonances or to acquire the more time-consuming 2D spectra. With the small RNA constructs used here, simple 1D ^{19}F observed spectra would be sufficient (Chen et al., 1997). 5-Fluorouridine-5'-triphosphate is available commercially, so it can be introduced in a regular in vitro transcription reaction. Site-specific labeling of individual residues is possible via phosphoramidite chemistry (Olejniczak et al., 2002). The large chemical shift dispersion and sensitivity toward ligand-binding events make ^{19}F an ideal noninvasive probe to detect and quantify RNA–ligand interactions. Since ^{19}F chemical shifts are so sensitive to the environment, additional information on binding site characteristics can be extracted readily from solvent-induced shifts, e.g., by changing from D_2O to H_2O (Mirau et al., 1982).

Ligand-Based Screening Methods

We discuss here three simple 1D screening methods to detect ligand binding to small RNA constructs. All initial binding tests were conducted by testing mixtures of three to five compounds, which were later deconvoluted if one or more binding molecules were detected (Lepre, 2001).

Line Broadening

Observation of line broadening of small molecules in the presence of RNA is frequently the easiest and fastest indication that the compound is interacting with the macromolecular receptor (Fejzo et al., 1999). For RNA–ligand mixtures, line broadening is often severe and can pose difficulties when trying to assign or analyze ligand resonances in a complex. Figure 3A shows the regular 1D NMR spectrum of a 2 mM acetylpromazine sample, while Fig. 3B shows the strong line broadening arising from

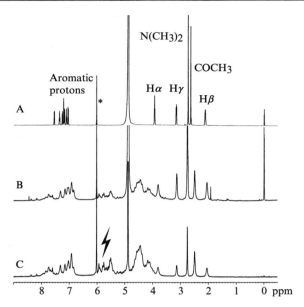

FIG. 3. (A) Reference 1D NMR spectrum of 2 mM acetylpromazine alone. (B) Reference NMR spectrum of acetylpromazine in the presence of 100 μM TAR RNA in D$_2$O buffer at 15°. (C) Saturation transfer difference (STD) spectrum acquired with saturation of the RNA at 5.8 ppm for 2 s. Binding can be detected by the line broadening of the ligand's resonance signals in the presence of RNA and by observation of signals from the ligand in the STD NMR spectrum.

addition of ca. 100 μM TAR RNA. For the small RNA constructs between 6 and 15 kDa typically used for structure determination or screening purposes, changes in the ligand resonance's R_2 relaxation rates due to reduced molecular motion of the small molecule upon binding to the RNA cannot alone be responsible for the extent of line broadening often observed. Indeed, exchange broadening is often seen. With moderately fast to intermediate exchange, a case observed for most weak ligands, the chemical shift difference between free and bound state influences the line shape. This is manifested as a contribution of the exchange process to the observed transverse relaxation rate (and therefore linewidth) of the ligand $R_{2,obs}$ [Eq. (1)] (Marchioro *et al.*, 2002). Large chemical shift differences, $\Delta\delta = \delta_{free} - \delta_{bound}$, in the free and bound state of the ligand can result in large line-broadening effects of the small molecules interacting with the

RNA if the dissociation rate constant k_{off} is not extremely fast (Lian *et al.*, 1994).

$$R_{2,obs} = p_L R_{2,free} + p_{EL} R_{2,bound} + \frac{p_L^2 p_{EL} 4\pi^2 (\delta_{free} - \delta_{bound})^2}{k_{off}} \quad (1)$$

The degree of line broadening for a particular resonance is dependent on the fractions of free and bound ligands, p_L and p_{EL}, respectively, on the transverse relaxation rates in the free and bound state, $R_{2,free}$ and $R_{2,bound}$, and on the exchange rate between the free and the bound state. Therefore, a large excess of ligand results in sharper lines (Peng *et al.*, 2001).

Saturation Transfer Difference (STD) NMR

Screening for binding compounds by saturation transfer difference NMR is less sensitive for the smaller RNA constructs used in our assays compared to the larger proteins used in prior studies (Mayer and Meyer, 1999). This results from the reduced spin diffusion efficiency within RNA, which is due to the lower proton density of RNA versus proteins and to the smaller size (with consequent faster tumbling) of the macromolecule. However, the use of D_2O instead of H_2O has led to an improvement in sensitivity by enhancing spin diffusion in two ways: the viscosity of D_2O is larger than H_2O at the low temperatures used, and the longitudinal relaxation rate R_1 of RNA is reduced because several alternative relaxation sinks are eliminated, allowing for more magnetization build-up in the RNA (Mayer and James, 2002). We are now able to acquire high-quality STD spectra from 25 μM RNA and ligand in ca. 10 to 20-fold excess.

In a list of candidate ligands, ranking of ligands relative to each other is achieved by comparing STD intensities and line-broadening effects of the respective compounds tested at identical concentrations (Mayer and James, 2004). Detailed epitope mapping experiments are also possible under these conditions; a higher ligand excess is beneficial for reducing the ligand resonance's linewidth, which is often severely broadened at low ligand-to-protein ratios (vide supra). The spectral region that will typically be the most accessible for irradiation of the RNA target in the STD experiment and where the majority of potential ligands do not have signals is between 5 and 6 ppm, where the H1' ribose and H5 aromatic protons of the RNA resonate. Figure 3 shows the 1D reference spectra of free acetylpromazine (Fig. 3A) and of acetylpromazine in the presence of 100 μM HIV-1 TAR RNA (Fig. 3B), as well as the STD NMR spectrum (Fig. 3C) of acetylpromazine in the presence of 100 μM HIV-1 TAR RNA.

STD NMR signals of the ligand prove the interaction with the RNA. Differential STD intensities for different protons, as observed between spectra B and C, can be used to map ligand moieties most likely involved in binding.

A recent application of epitope mapping was the accurate positioning of ligands into the binding site from experimental STD NMR data by a complete relaxation matrix approach in which the calculated STD intensities are compared to experimental STD intensities, and the ligand is positioned accordingly with rotatable bonds of both ligand and target optimized to reflect the experimental intensities (Bhunia et al., 2004; Jayalakshmi and Rama Krishna, 2004). This could be used for RNA binding, as well, and would be of great value when combining docking and STD NMR experiments. For such cases, the STD epitope mapping experiments can give an indication of whether a ligand binds in the correct site even in the absence of intermolecular NOEs or SAR-by-NMR data.

Water LOGSY

Another commonly applied screening technique called Water Ligand Observation with Gradient Spectroscopy or WaterLOGSY (Dalvit et al., 2000) also relies on transfer of saturation to ligand resonances. [The basic phenomenon is the same as Shimada describes (Chapter 20, this volume) for studying protein–protein interactions.] Here, the large reservoir of selectively inverted water is used to build up NOE effects on the ligands. Depending on whether the ligand interacts with the receptor, this observed NOE will be positive for nonbinding molecules or negative for binding molecules. Therefore compounds that bind can be directly detected by the inverted sign (Dalvit et al., 2001). A recent study compared the sensitivity of STD NMR and WaterLOGSY experiments on a 160-residue RNA construct (Lepre et al., 2002; Johnson et al., 2003). The conclusion was that the WaterLOGSY experiment gave higher sensitivity. RNA is highly hydrated and contains many exchangeable protons that should be beneficial for this experiment. However, in our experience the WaterLOGSY experiment has not given higher sensitivity for the smaller (20–40 residue) RNA constructs we have been using.

Particularly interesting in the study mentioned above is that a SHAPES-style library was used to discover RNA-binding drugs (Johnson et al., 2003). Although a moderate number of hits was found, there was only one compound that was specific toward the selected RNA target. It may therefore be advantageous to design a SHAPES library geared toward RNA molecules. Also, the acceptable ratio between specificity and affinity needs to be determined for RNA targets, as most ligands, especially highly positively charged ones such as the aminoglycosides, seem to bind nonspecifically at higher ligand concentrations (Griffey et al., 1999).

Conclusions

Development of knowledge-based methods to generate focused libraries of compounds that bind to specific RNA motifs is being fueled by a growing interest in identifying RNA-specific small molecule inhibitors. Indeed, two recent publications used this approach of combining docking with follow-up screens using FRET and NMR-based structural fingerprinting of some of the hits (Davis et al., 2004; Foloppe et al., 2004). A privileged molecule approach has been proposed by our laboratory as well as the group at Vertex Pharmaceuticals (Johnson et al., 2003), but further work needs to be done to generate RNA-specific scaffold libraries. On the NMR side, we have presented a number of simple and robust experiments that allow detection of active compounds and their absolute or relative affinity. NMR as a primary tool for screening small compound libraries for RNA binding was used in our laboratory (Lind et al., 2002; Mayer and James, 2004) and at Abbott (Yu et al., 2003). Indeed, all NMR experiments mentioned here are able to detect the weak binders expected from the focused library approach. In summary, we can expect a wealth of information in the coming years regarding both computational methods and structural insights into what makes small molecules bind to RNA.

Acknowledgments

We thank P. Therese Downing, Christophe Guilbert, and Mark Kelly for valuable discussions or reading of the manuscript. Work from our laboratory described herein was supported by a fellowship to M.M. from the German Academic Exchange Program (DAAD) and by NIH Grant AI46967.

References

Abagyan, R., and Totrov, M. (1994). Biased probability Monte Carlo conformational searches and electrostatic calculations for peptides and proteins. J. Mol. Biol. 235, 983–1002.

Abagyan, R. A., Totrov, M. M., and Kuznetsov, D. N. (1994). ICM—a new method for protein modeling and design—applications to docking and structure prediction from the distorted native conformation. J. Comp. Chem. 15, 488–506.

Bajorath, F. (2002). Integration of virtual and high-throughput screening. Nat. Rev. Drug Discov. 1, 882–894.

Bhunia, A., Jayalakshmi, V., Benie, A. J., Schuster, O., Kelm, S., Rama Krishna, N., and Peters, T. (2004). Saturation transfer difference NMR and computational modeling of a sialoadhesin-sialyl lactose complex. Carbohydr. Res. 339, 259–267.

Bissantz, C., Folkers, G., and Rognan, D. (2000). Protein-based virtual screening of chemical databases. 1. Evaluation of different docking/scoring combinations. *J. Med. Chem.* **43,** 4759–4767.

Chen, Q., Shafer, R. H., and Kuntz, I. D. (1997). Structure-based discovery of ligands targeted to the RNA double helix. *Biochemistry* **36,** 11402–11407.

Colca, J. R., McDonald, W. G., Waldon, D. J., Thomasco, L. M., Gadwood, R. C., Lund, E. T., Cavey, G. S., Mathews, W. R., Adams, L. D., Cecil, E. T., Pearson, J. D., Bock, J. H., Mott, J. E., Shinabarger, D. L., Xiong, L., and Mankin, A. S. (2003). Cross-linking in the living cell locates the site of action of oxazolidinone antibiotics. *J. Biol. Chem.* **278,** 21972–21979.

Coles, M., Heller, M., and Kelssler, H. (2003). NMR-based screening technologies. *Drug Discov. Today* **8,** 803–810.

Crowley, P. B., Rabe, K. S., Worrall, J. A., Canters, G. W., and Ubbink, M. (2002). The ternary complex of cytochrome F and cytochrome C: Identification of a second binding site and competition for plastocyanin binding. *Chembiochem.* **3,** 526–533.

Dalvit, C., Pevarello, P., Tato, M., Veronesi, M., Vulpetti, A., and Sundstrom, M. (2000). Identification of compounds with binding affinity to proteins via magnetization transfer from bulk water. *J. Biomol. NMR* **18,** 65–68.

Dalvit, C., Fogliatto, G., Stewart, A., Veronesi, M., and Stockman, B. (2001). Waterlogsy as a method for primary NMR screening: Practical aspects and range of applicability. *J. Biomol. NMR* **21,** 349–359.

Dalvit, C., Ardini, E., Flocco, M., Fogliatto, G. P., Mongelli, N., and Veronesi, M. (2003). A general NMR method for rapid, efficient, and reliable biochemical screening. *J. Am. Chem. Soc.* **125,** 14620–14625.

Davis, A. M., Teague, S. J., and Kleywegt, G. J. (2003). Application and limitations of X-ray crystallographic data in structure-based ligand and drug design. *Angew. Chem. Int. Ed.* **42,** 2718–2736.

Davis, B., Afshar, M., Varani, G., Murchie, A. I. H., Karn, J., Lentzen, G., Drysdale, M. J., Bower, J., Potter, A. J., Starkey, I. D., Swarbrick, T. M., and Aboul-Ela, F. (2004). Rational design of inhibitors of HIV-1 TAR RNA through the stabilisation of electrostatic "hot spots." *J. Mol. Biol.* **336,** 343–356.

Feeney, J., Roberts, G. C., Thomson, J. W., King, R. W., Griffiths, D. V., and Burgen, A. S. (1980). Proton nuclear magnetic resonance studies of the effects of ligand binding on tryptophan residues of selectively deuterated dihydrofolate reductase from Lactobacillus casei. *Biochemistry* **19,** 2316–2321.

Fejzo, J., Lepre, C. A., Peng, J. W., Bemis, G. W., Ajay, Murcko, M. A., and Moore, J. M. (1999). The shapes strategy: An NMR-based approach for lead generation in drug discovery. *Chem. Biol.* **6,** 755–769.

Filikov, A. V., Mohan, V., Vickers, T. A., Griffey, R. H., Cook, P. D., Abagyan, R. A., and James, T. L. (2000). Identification of ligands for RNA targets via structure-based virtual screening: HIV-1 TAR. *J. Comp. Aided Mol. Des.* **14,** 593–610.

Foloppe, N., Chen, I. J., Davis, B., Hold, A., Morley, D., and Howes, R. (2004). A structure-based strategy to identify new molecular scaffolds targeting the bacterial ribosomal A-site. *Bioorg. Med. Chem.* **12,** 935–947.

Fourmy, D., Recht, M. I., Blanchard, S. C., and Puglisi, J. D. (1996). Structure of the A site of *Escherichia coli* 16S ribosomal RNA complexed with an aminoglycoside antibiotic. *Science* **274,** 1367–1371.

Fourmy, D., Yoshizawa, S., and Puglisi, J. D. (1998). Paromomycin binding induces a local conformational change in the A-site of 16 S rRNA. *J. Mol. Biol.* **277,** 333–345.

Gallego, J., and Varani, G. (2001). Targeting RNA with small-molecule drugs: Therapeutic promise and chemical challenges. *Acc. Chem. Res.* **34,** 836–843.

Griffey, R. H., Hofstadler, S. A., Sannes-Lowery, K. A., Ecker, D. J., and Crooke, S. T. (1999). Determinants of aminoglycoside-binding specificity for rRNA by using mass spectrometry. *Proc. Natl. Acad. Sci. USA* **96**, 10129–10133.

Hajduk, P. J., and Burns, D. J. (2002). Integration of NMR and high-throughput screening. *Comb. Chem. High Throughput Screen.* **5**, 613–621.

Hajduk, P. J., Bures, M., Praestgaard, J., and Fesik, S. W. (2000). Privileged molecules for protein binding identified from NMR-based screening. *J. Med. Chem.* **43**, 3443–3447.

Hajduk, P. J., Betz, S. F., Mack, J., Ruan, X., Towne, D. L., Lerner, C. G., Beutel, B. A., and Fesik, S. W. (2002). A strategy for high-throughput assay development using leads derived from nuclear magnetic resonance-based screening. *J. Biomol. Screen.* **7**, 429–432.

Hermann, T. (2000). Strategies for the design of drugs targeting RNA and RNA–protein complexes. *Angew. Chem. Int. Ed.* **39**, 1891–1905.

Hermann, T. (2002). Rational ligand design for RNA: The role of static structure and conformational flexibility in target recognition. *Biochimie* **84**, 869–875.

Hermann, T. (2003). Chemical and functional diversity of small molecule ligands for RNA. *Biopolymers* **70**, 4–18.

Hwang, S., Tamilarasu, N., Ryan, K., Huq, I., Richter, S., Still, W. C., and Rana, T. M. (1999). Inhibition of gene expression in human cells through small molecule-RNA interactions. *Proc. Natl. Acad. Sci. USA* **96**, 12997–13002.

Jayalakshmi, V., and Rama Krishna, N. (2004). Corcema refinement of the bound ligand conformation within the protein binding pocket in reversibly forming weak complexes using STD-NMR intensities. *J. Magn. Reson.* **168**, 36–45.

Johnson, E. C., Feher, V. A., Peng, J. W., Moore, J. M., and Williamson, J. R. (2003). Application of NMR shapes screening to an RNA target. *J. Am. Chem. Soc.* **125**, 15724–15725.

Jones, G., Willett, P., and Glen, R. C. (1995). Molecular recognition of receptor sites using a genetic algorithm with a description of desolvation. *J. Mol. Biol.* **245**, 43–53.

Kick, E. K., Roe, D. C., Skillman, A. G., Liu, G., Ewing, T. J. A., Sun, Y., Kuntz, I. D., and Ellman, J. A. (1997). Structure-based design and combinatorial chemistry yield low nanomolar inhibitors of cathepsin D. *Chem. Biol.* **4**, 297–307.

Kramer, B., Rarey, M., and Lengauer, T. (1999). Evaluation of the flexx incremental construction algorithm for protein-ligand docking. *Proteins* **37**, 228–241.

Leone, M., Rodriguez-Mias, R. A., and Pellecchia, M. (2003). Selective incorporation of 19F-labeled Trp side chains for NMR-spectroscopy-based ligand-protein interaction studies. *Chembiochem.* **4**, 649–650.

Lepre, C. A. (2001). Library design for NMR-based screening. *Drug Discov. Today* **6**, 133–140.

Lepre, C. A., Peng, J., Fejzo, J., Abdul-Manan, N., Pocas, J., Jacobs, M., Xie, X., and Moore, J. M. (2002). Applications of shapes screening in drug discovery. *Comb. Chem. High Throughput Screen.* **5**, 583–590.

Lian, L. Y., Barsukov, I. L., Sutcliffe, M. J., Sze, K. H., and Roberts, G. C. (1994). Protein–ligand interactions: Exchange processes and determination of ligand conformation and protein–ligand contacts. *Methods Enzymol.* **239**, 657–700.

Lind, K. E., Du, Z., Fujinaga, K., Peterlin, B. M., and James, T. L. (2002). Structure-based computational database screening, *in vitro* assay, and NMR assessment of compounds that target TAR RNA. *Chem. Biol.* **9**, 185–193.

Lipinski, C. A. (2000). Drug-like properties and the causes of poor solubility and poor permeability. *J. Pharmacol. Toxicol. Methods* **44**, 235–249.

Luedtke, N. W., and Tor, Y. (2003). Fluorescence-based methods for evaluating the RNA affinity and specificity of HIV-1 rev-RRE inhibitors. *Biopolymers* **70,** 103–119.

Lynch, S. R., Recht, M. I., and Puglisi, J. D. (2000). Biochemical and nuclear magnetic resonance studies of aminoglycoside-RNA complexes. *Methods Enzymol.* **317,** 240–261.

Lynch, S. R., Gonzalez, R. L., and Puglisi, J. D. (2003). Comparison of X-ray crystal structure of the 30s subunit-antibiotic complex with NMR structure of decoding site oligonucleotide-paromomycin complex. *Structure (Camb.)* **11,** 43–53.

Lyne, P. D. (2002). Structure-based virtual screening: An overview. *Drug Discov. Today* **7,** 1047–1055.

Marchioro, C., Davalli, S., Provera, S., Heller, M., Ross, A., and Senn, H. (2002). *In* "BioNMR in Drug Research" (O. Zerbe, ed.), pp. 321–340. Wiley-VCH, Weinheim, Germany.

Mayer, M., and James, T. L. (2002). Detecting ligand binding to a small RNA target via saturation transfer difference NMR experiments in $D(2)O$ and $H(2)O$. *J. Am. Chem. Soc.* **124,** 13376–13377.

Mayer, M., and James, T. L. (2004). NMR-based characterization of phenothiazines as a RNA binding scaffold. *J. Am. Chem. Soc.* **126,** 4453–4460.

Mayer, M., and Meyer, B. (1999). Characterization of ligand binding by saturation transfer difference NMR spectroscopy. *Angew. Chem. Int. Ed.* **38,** 1784–1788.

Mei, H.-Y., Cui, M., Heldsinger, A., Lemrow, S. M., Loo, J. A., Sannes-Lowery, K. A., Sharmeen, L., and Czarnik, A. W. (1998). Inhibitors of protein-RNA complexation that target the RNA: Specific recognition of human immunodeficiency virus type 1 TAR RNA by small organic molecules. *Biochemistry* **37,** 14204–14212.

Meyer, B., and Peters, T. (2003). NMR Spectroscopy techniques for screening and identifying ligand binding to protein receptors. *Angew. Chem. Int. Ed.* **42,** 864–890.

Mirau, P. A., Shafer, R. H., and James, T. L. (1982). Binding of 5-fluorotryptamine to polynucleotides as a model for protein-nucleic acid interactions: Fluorine-19 nuclear magnetic resonance, absorption, and fluorescence studies. *Biochemistry* **21,** 615–620.

Moore, J., Abdul-Manan, N., Fejzo, J., Jacobs, M., Lepre, C., Peng, J., and Xie, X. (2004). Leveraging structural approaches: Applications of NMR-based screening and X-ray crystallography for inhibitor design. *J. Synchrotron Radiat.* **11,** 97–100.

Murchie, A. I. H., Davis, B., Isel, C., Afshar, M., Drysdale, M. J., Bower, J., Potter, A. J., Starkey, I. D., Swarbrick, T. M., Mirza, S., Prescott, C. D., Vaglio, P., Aboul-ela, F., and Karn, J. (2004). Structure-based drug design targeting an inactive RNA conformation: Exploiting the flexibility of HIV-1 TAR RNA. *J. Mol. Biol.* **336,** 625–638.

Olejniczak, M., Gdaniec, Z., Fischer, A., Grabarkiewicz, T., Bielecki, L., and Adamiak, R. W. (2002). The bulge region of HIV-1 TAR RNA binds metal ions in solution. *Nucleic Acids Res.* **30,** 4241–4249.

Peng, J. W., Lepre, C. A., Fejzo, J., Abdul-Manan, N., and Moore, J. M. (2001). Nuclear magnetic resonance–based approaches for lead generation in drug discovery. *Methods Enzymol.* **338,** 202–230.

Roberts, G. C. (2002). *In* "BioNMR in Drug Research" (O. Zerbe, ed.), pp. 309–319. Wiley-VCH, Weinheim, Germany.

Rudisser, S., and Jahnke, W. (2002). NMR and in silico screening. *Comb. Chem. High Throughput Screen* **5,** 591–603.

Stahl, M., and Rarey, M. (2001). Detailed analysis of scoring functions for virtual screening. *J. Med. Chem.* **44,** 1035–1042.

Swayze, E. E., Jefferson, E. A., Sannes-Lowery, K. A., Blyn, L. B., Risen, L. M., Arakawa, S., Osgood, S. A., Hofstadler, S. A., and Griffey, R. H. (2002). SAR by MS: A ligand based technique for drug lead discovery against structured RNA targets. *J. Med. Chem.* **45,** 3816–3819.

Valler, M. J., and Green, D. (2000). Diversity screening versus focussed screening in drug discovery. *Drug Discov. Today* **5,** 286–293.

van Dongen, M., Weigelt, J., Uppenberg, J., Schultz, J., and Wikstrom, M. (2002). Structure-based screening and design in drug discovery. *Drug Discov. Today* **7,** 471–478.

Wei, B. Q., Weaver, L. H., Ferrari, A. M., Matthews, B. W., and Shoichet, B. K. (2004). Testing a flexible-receptor docking algorithm in a model binding site. *J. Mol. Biol.* **337,** 1161–1182.

Wimberly, B. T., Brodersen, D. E., Clemons, W. M., Morgan-Warren, R. J., Carter, A. P., Vonrhein, C., Hartsch, T., and Ramakrishnan, V. (2000). Structure of the 30S ribosomal subunit. *Nature* **407,** 327–339.

Wyss, D. F., McCoy, M. A., and Senior, M. M. (2002). NMR-based approaches for lead discovery. *Curr. Opin. Drug Discov. Dev.* **5,** 630–647.

Yu, L., Oost, T. K., Schkeryantz, J. M., Yang, J., Janowick, D., and Fesik, S. W. (2003). Discovery of aminoglycoside mimetics by NMR-based screening of *Escherichia coli* A-site RNA. *J. Am. Chem. Soc.* **125,** 4444–4450.

Author Index

A

Abad-Zapatero, C., 555, 556, 557, 561, 562
Abagyan, R., 574, 575, 583
Abdul-Manan, N., 549, 573, 576, 581, 582
Aberg, A., 227
Abergel, D., 432, 433, 434
Abildgaard, F., 116, 321, 335, 338, 341
Aboul-Ela, F., 399, 574, 583
Abragam, A., 422, 431, 471
Abremski, K., 214
Acharya, K. R., 144, 156, 157
Ackerman, M. S., 156, 307, 311
Acton, T. B., 129, 212, 213, 214, 217, 219, 220, 221, 231, 233, 238, 239
Adamiak, R. W., 579
Adams, J., 526
Adams, L. D., 573
Adams, M. D., 3
Adams, M. W. W., 176, 177
Adams, P. D., 124, 283, 409
Addess, K. J., 466, 469
Adeishvili, K., 321, 335, 338
Adler, M., 123, 265
Afshar, M., 574, 583
Agard, D. A., 521
Ahmed, S., 17
Airey, J., 553
Aitio, H., 44, 46
Ajay, 579
Akasaka, K., 313, 314, 485
Akerud, T., 427
Akke, M., 427, 431, 432, 437, 451, 453, 467, 509, 510, 511, 512
Akratos, C., 466
Al-Hashimi, H. M., 44, 62, 63, 175, 180, 202, 311, 537
Alattia, J. R., 11, 232
Albert, D. H., 561, 565, 566, 567
Alexandrescu, A. T., 155, 176, 305, 307
Alfano, C., 525
Allain, F. H., 399, 400, 466, 526, 529, 532
Allard, P., 133

Allen, M., 531, 539, 541
Allerhand, A., 435
Allers, J., 541
Allman, A. M., 466
Almeida, F. C., 356
Almo, S. C., 71, 116, 176
Aloy, P., 245
Altschul, S. F., 218
Amarasinghe, G. K., 526
Amalric, F., 527
Amezcua, C. A., 8, 11
Amor, J. C., 8
An, J. L., 116
Anderson, A. C., 529
Anderson, D. E., 25, 528
Anderson, E. M., 534
Anderson, P. J., 157
Anderson, S., 18, 129, 211, 214, 223, 226, 227
Ando, Y., 314
Andrec, M., 44, 45, 46, 49, 64, 118, 120, 177, 179, 361, 409
Androphy, E. J., 25
Anglister, J., 122
Angwin, D. T., 549
Anklin, C., 103, 117, 118, 119, 121, 133
Annila, A., 44, 46, 176, 311
Appella, E., 127
Aramini, J. M., 89, 90, 95, 129, 131, 176
Arango, J., 305
Arata, Y., 484, 485, 491, 492, 496
Archer, S. J., 71
Ardini, E., 579
Argos, P., 6
Armstrong, C. M., 211
Arnold, J. R. P., 529
Arnold, M. R., 303
Aronzon, D., 86
Arora, A., 321, 335, 338, 341
Arrowsmith, C. H., 84, 85, 102, 116, 212, 214, 231
Arseniev, A. S., 420
Aruffo, A., 491
Asbury, T., 351

Subject Index

A

Aliasing, rapid nuclear magnetic resonance data collection, 82

ARIA, nuclear Overhauser effect spectroscopy analysis, 264–265

Assignment Validation Software, *see* Automated resonance assignment and protein structure elucidation

Automated resonance assignment and protein structure elucidation
Assignment Validation Software suite, 122–123
backbone resonance assignments using AutoAssign, 120–121, 263
GARANT, 264
interspectral registration and quality assessment of peak lists, 119–120
local data organization and archiving, 116
MAPPER, 263–264
minimal constraint approaches to rapid automated fold determination, 131
nuclear Overhauser effect spectroscopy data analysis using AutoStructure
bottom-up data interpretation approach, 123–124
control flow, 124–127
implementation, 129
input data
description, 127
quality control issues, 127–128
testing, 129, 131
organizational challenge, 113
overview of processes, 112–113, 144
peak picking, 118–119
prospects, 135
side chain resonance assignments, 122
spectral processing, 116–118
SPINS integrated platform for automated structure analysis, 132–134
standardized data collection, 113
structure quality assessment tools, 131–132

AutoStructure, *see* Automated resonance assignment and protein structure elucidation

B

BACUS, *see* CLOUDS

BAF, *see* CLOUDS

Bovine protein tyrosine phosphatase, hydrodynamic analysis of nuclear magnetic resonance relaxation, 427

Bovine serum albumin, fold stability analysis using nuclear magnetic resonance, 152–157

BPTP, *see* Bovine protein tyrosine phosphatase

BSA, *see* Bovine serum albumin

C

Calmodulin, fold validation using nuclear magnetic resonance, 159, 164, 166

Carr–Purcell–Meiboom–Gill sequence
chemical exchange characterization, 434–438, 453–459
enzyme catalysis dynamics studies, 515, 521–522

CD44, hyaluronic acid interactions, 491, 492

In-Cell nuclear magnetic resonance
aims and overview, 18–20
comparison with magnetic resonance spectroscopy, 18
conformational change and dynamics studies, 33–35
drug-binding studies, 35–36
labeling
carbon-13, 25–28
nitrogen-15
background signals, 20–22
expression in labeled rich media, 23–24
protein overexpression effects, 22–23

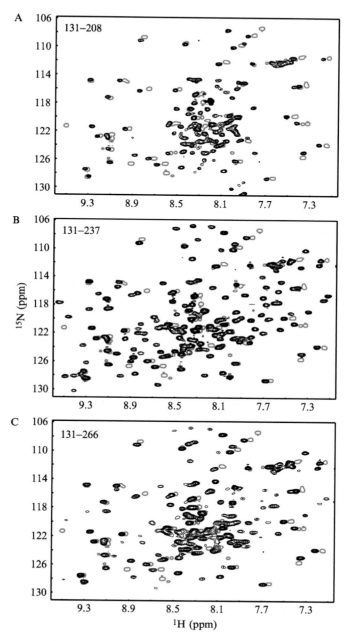

CARD AND GARDNER, CHAPTER 1, FIG. 2. *(continued)*

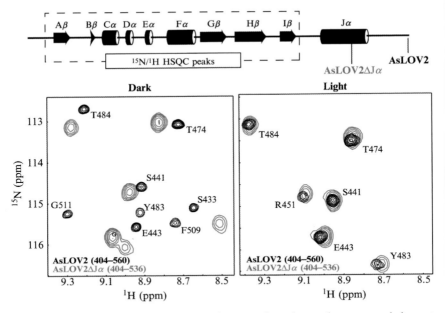

CARD AND GARDNER, CHAPTER 1, FIG. 3. Discovery of novel secondary structural elements by comparison of multiple constructs in the photosensory LOV2 domain of *Avena sativa* phototropin 1 (AsLOV2). Comparisons of $^{15}N/^1H$ HSQC peaks originating from residues within the predicted PAS domain showed significant chemical shift changes among constructs with differing C-termini, as demonstrated in spectra from AsLOV2 (black) and the shorter AsLOV2ΔJα (red). The data strongly suggested that the C-terminal region interacted with the PAS domain in the dark. When these experiments were repeated with blue light illumination, these $^{15}N/^1H$ HSQC peaks became independent of the construct length, providing evidence that light-induced conformational changes led to the displacement of the C-terminal Jα helix. Adapted with permission from Harper *et al.* (2003).

CARD AND GARDNER, CHAPTER 1, FIG. 2. Utility of *E. coli* cell lysate screening for determining domain boundaries. Fragments of human PAS kinase were expressed in *E. coli* BL21(DE3) as C-terminal fusions to GB1 using a T7 RNA polymerase expression vector (Amezcua *et al.*, 2002). Each spectrum was generated from a lysate generated by a 1-liter culture grown in uniformly ^{15}N-labeled M9 media, which was concentrated to 1 ml before acquiring 2D $^{15}N/^1H$ HSQC spectra (40 min, 37°). (A–C) An overlay of spectra from GB1–PAS kinase fusions (black contours; PAS kinase residues are listed on each panel) with a reference from an isolated GB1 sample (red).

A

M1	R2	I3	E4	V5	T6	I7	A8	K9	T10
S11	P12	L13	P14	A15	G16	A17	I18	D19	A20
L21	A22	G23	E24	L25	S26	R27	R28	I29	Q30
Y31	A32	F33	P34	D35	N36	E37	G38	H39	V40
S41	V42	R43	Y44	A45	A46	A47	N48	N49	L50
S51	V52	I53	E54	A55	T56	K57	E58	D59	K60
Q61	R62	I63	S64	E65	I66	L67	Q68	E69	T70
W71	E72	S73	A74	D75	D76	W77	F78	V79	S80

KONTAXIS *ET AL.*, CHAPTER 3, FIG. 3. (*continued*)

B

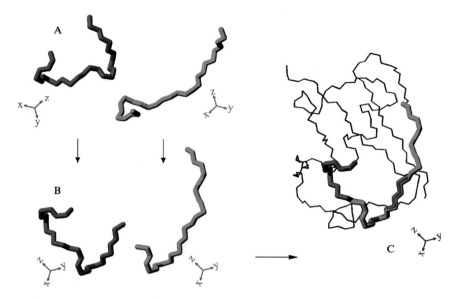

KONTAXIS *ET AL.*, CHAPTER 3, FIG. 5. Pictorial representation of the MFR+ assembly process, illustrated for ubiquitin. (A) Backbone representations of MFR+-derived fragments for residues 13–21 (red, light gray) and 17–25 (green, dark gray) fragments in arbitrary relative orientation, together with their corresponding alignment tensor frames. (B) After rotation such that their respective local alignment tensors have identical orientations. (C) Fragments are translated (with fixed orientation) such that the coordinate RMSD relative to the previously assembled chain is minimized. The figure was generated using the program MOLMOL (Koradi *et al.*, 1996).

KONTAXIS *ET AL.*, CHAPTER 3, FIG. 3. Summary of MFR search results for DinI, (A) before and (B) after refinement of the fragments with respect to dipolar restraints. Each subpanel displays the (ϕ,ψ) backbone angles found for a given residue in the MFR search, with gray regions marking the most occupied region in the database for the particular residue type. Solid lines connect (ϕ,ψ) pairs found for adjacent residues in the same fragment, and "dead ends" such as observed for P12 in (A) correspond to the fragment where this residue is the last in the selected stretch. In (A), blue (dark gray) lines correspond to MFR search results heavily weighted toward dipolar couplings (using parameters of Table II); red (light gray) lines correspond to MFR results when emphasizing chemical shifts. After refinement, using parameters of Table III and convergence criteria of Table IV, the majority of residues display unique (ϕ,ψ) backbone angles (B). The black line connects the (ϕ,ψ) pairs seen in the previously determined NMR structure (PDB entry 1GHH).

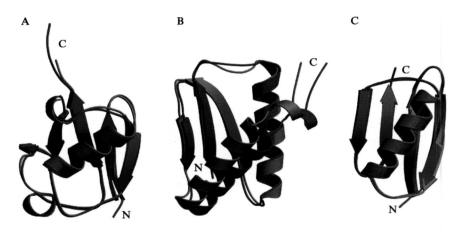

Kontaxis *et al.*, Chapter 3, Fig. 6. Comparison of MFR+-derived structures and previously solved X-ray (A and C) and NMR (B) structures of (A) ubiquitin, (B) DinI, and (C) GB3. Blue (light gray) structures correspond to the PDB reference coordinates; red (dark gray) represents the MFR+-derived ribbon. For ubiquitin (A), the disordered C-terminus (residues 74–76) could not be built by the MFR+ method and is not shown. For DinI (B), the structure could not be assembled uniquely from dipolar couplings acquired in a single medium. Even with data from two media, some ambiguity remains (Fig. 3B) and is responsible for the erroneous lateral displacement of helix α2 by ca. 3.7 Å relative to the reference structure. For clarity, superposition of the MFR+ and reference structure is optimized for residues 2–53, highlighting the displacement of helix 2. Reference structures correspond to PDB entries 1UBQ, 1GHH, and 1IGD. Figures were generated using the programs Molscript (Kraulis, 1991) and Raster 3D (Merritt and Murphy, 1994).

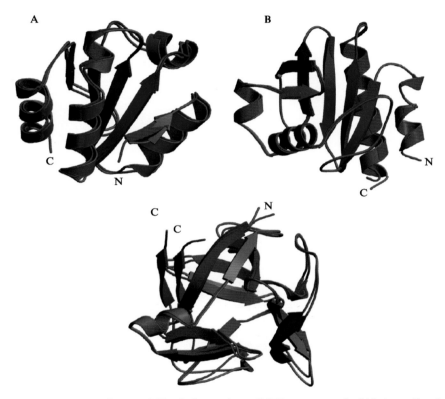

Kᴏɴᴛᴀxɪs *ᴇᴛ ᴀʟ.*, Cʜᴀᴘᴛᴇʀ 3, Fɪɢ. 7. Comparison of MFR+ structures [red (dark gray)] and X-ray reference structures [blue (light gray)] for (A) reduced human thioredoxin (PDB entry 1ERT), (B) profilin (PDB entry 1ACF), and (C) interleukin-1β (PDB entry 4ILB). MFR+ structures were derived from experimental chemical shifts and using two sets of dipolar couplings, simulated for the X-ray structures, for media containing 50 mg/ml bicelles and 15 mg/ml Pf1. Figures were generated using the programs Molscript (Kraulis, 1991) and Raster 3D (Merritt and Murphy, 1994).

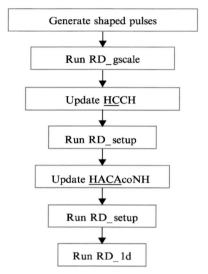

Huang *ET AL.*, Chapter 5, Fig. 2. Flowchart outlining the use of the RDpack. Steps required solely for the *de novo* implementation of RD NMR experiments are shown in red boxes, and steps required for rapid adjustment of parameter sets are displayed in green boxes. First, shaped pulses are generated by use of shell scripts, and the power levels for pulsed field gradients are adjusted to the available hardware configuration (using the macro RD_gscale). Second, the 3D HCCH parameter set is updated by providing proton and carbon high-power pulse widths, power levels, and carrier positions. Execution of the macro "RD_setup" transfers these parameters to the entire suite of RD experiments. Third, the 3D HACAcoNH parameter set is updated by providing nitrogen high-power pulse width, power level, and carrier position. These parameters are then transferred to nitrogen-resolved RD experiments by use of RD_setup. Finally, the macro RD_1d starts the acquisition of the first FID of all 11 parameter sets, while also allowing rapid assessment of the relative sensitivity of the various experiments.

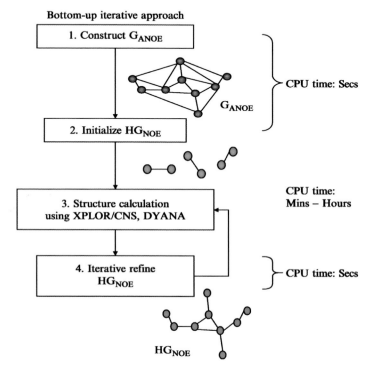

HUANG *ET AL.*, CHAPTER 5, FIG. 3. The control flow of AutoStructure. AutoStructure uses a bottom-up iterative approach. It has four major steps. Step 1 constructs an ambiguous distance network G_{ANOE}, in which all vertices represent protons and proton pairs are connected when their chemical shift values are matched with an NOE peak's chemical shift values within a loose match tolerance. A heuristic HG_{NOE} is initialized from G_{ANOE} at Step 2. After HG_{NOE} is initialized, an initial fold is generated at Step 3. Step 4 iteratively refines HG_{NOE} from the structures generated from Step 3.

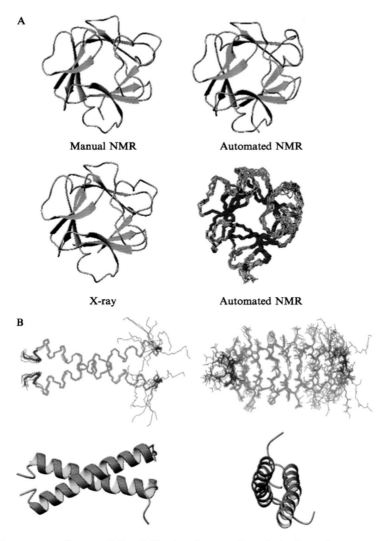

A

Manual NMR   Automated NMR

X-ray   Automated NMR

B

HUANG ET AL., CHAPTER 5, FIG. 4. Results of automatic analysis of protein structures from NMR data. (A) Comparison of backbone structures of human basic fibroblast growth factor (FGF) determined by manual analysis of NMR data (PDB code 1 bld), by automated analysis of the same NMR data using AutoStructure/XPLOR, or by X-ray crystallography (PDB code 1bas). The superposition of 10 NMR structures of human basic FGF computed by AutoStructure with XPLOR is also shown. Backbone conformations are shown only for residues 29 to 155, since the N-terminal polypeptide segment is not well defined in either the automated or manual analyses. For this portion of the structure, the backbone RMSD values within the families of structures determined by AutoStructure are ~0.7 Å, and the backbone RMSD between the AutoStructure and the X-ray crystal structure or manually determined NMR structures is ~0.8 Å. (B) Solution NMR structure of TM1bZip N-terminal segment of human α-tropomyosin determined by AutoStructure with DYANA (Greenfield *et al.*, 2001). The top panels show superpositions of backbone (*left*) and all heavy (*right*) atoms, respectively. Secondary structures are colored in red. The bottom panel shows ribbon diagrams of one representative structure.

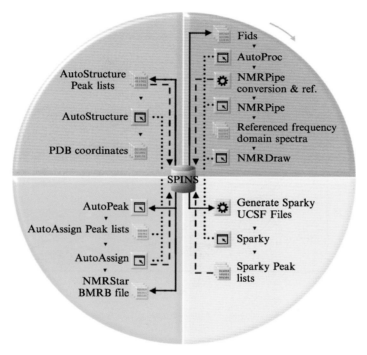

HUANG *ET AL.*, CHAPTER 5, FIG. 5. The integrated SPINS platform for automated analysis of NMR data. This figure depicts the flow of data through the SPINS software from raw FIDs to backbone assignments. (1) The raw FID data are housed in the SPINS database. (2) AutoProc queries the SPINS database for autoreferencing and processing of experimental data using NMRPipe. (3) Sparky software is used for manual peak picking and peak list editing. (4) AutoPeak software is used to validate peak lists as well as prepare AutoAssign input. (5) AutoAssign software is used for automated backbone resonance assignments. The SPINS platform also integrates AutoStructure software for NOESY data analysis, together with DYANA/CNS/XPLOR software for 3D structure generation and software providing estimates of structure quality scores.

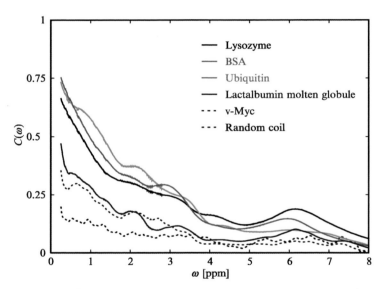

HOFFMANN ET AL., CHAPTER 6, FIG. 2. Autocorrelation functions of protein 1D ^{1}H spectra. The following proteins are shown: lysozyme (black), BSA (red), ubiquitin (green), the molten globule state of α-lactalbumin (blue), the partially folded protein v-Myc (blue, dashed line), and a theoretical random coil polypeptide assuming random coil ^{1}H chemical shifts (black, dashed line). Only energy difference data with $\Delta\omega > 0.25$ ppm are shown (see text).

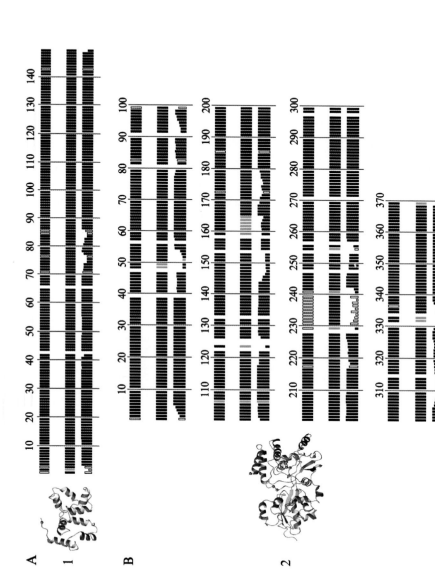

HOFFMANN *ET AL.*, CHAPTER 6, FIG. 5. (*continued*)

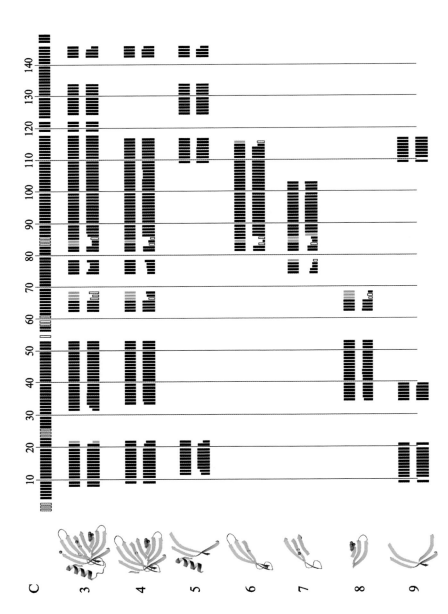

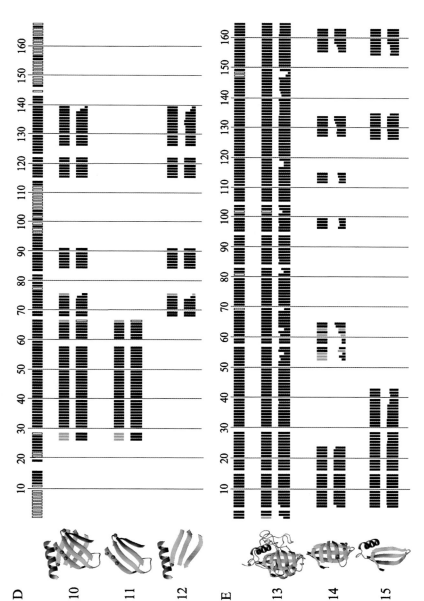

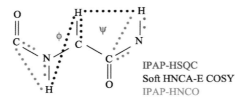

IPAP-HSQC
Soft HNCA-E COSY
IPAP-HNCO

PRESTEGARD *ET AL.*, CHAPTER 7, FIG. 1. Illustration of the nine residual dipolar couplings that can be measured for a dipeptide unit using the low percentage ^{13}C labeling scheme. Data from two consecutive residues are combined to determine the dihedral angles ϕ and ψ around the central C_α carbon.

HOFFMANN *ET AL.*, CHAPTER 6, FIG. 5. Graphic representation of assignment results obtained with our Monte Carlo-based approach for calmodulin (A), maltose-binding protein (B), Q83 (C), Icln (D), and CypD (E). The upper row in (A–E) represents the entire query protein primary sequence with each bar symbolizing one residue. Missing bars indicate residue positions occupied by prolines. NMR-detectable residues are shown as black bars and NMR-unobservable residues as unfilled black bars. Each individual test run is numbered as in Table I. The result of each test run is summarized by two rows of bars. Those query protein residues that are not part of homology segments (which have no structural equivalent in the homology model) as well as prolines were omitted from both rows. Upper row: correct assignment, black; erroneous assignment, gray. Lower row: residue with $C\alpha(i-1)$ chemical shift matching/mismatching with $C\alpha(i)$ chemical shifts of its predecessor, black/gray, respectively. Filled bars represent NMR-observable residues; unfilled bars indicate NMR-unobservable residues. Note that no statement about interresidue $C\alpha$ chemical shift matching/mismatching can be made for residues adjacent (C-terminal) to NMR-unobservable residues. Maximal bar height symbolizes 100% assignment reproducibility after 20 independent assignment cycles. Reduced reproducibility is accordingly indicated by lower bar heights. The ribbon drawings display those parts of the query proteins that are made of the residues shown in the upper and lower rows. The positions of erroneously assigned residues are shown as small spheres. All ribbon drawings were generated with MOLMOL (Koradi *et al.*, 1996).

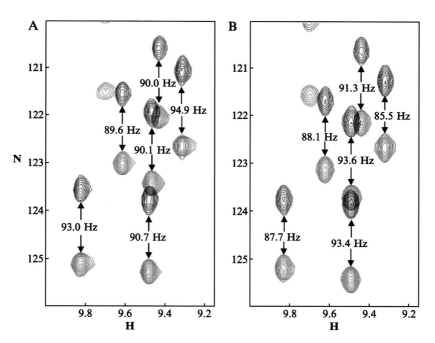

PRESTEGARD *ET AL.*, CHAPTER 7, FIG. 6. IPAP-HSQC of PF0385 under isotropic (A) and aligned (B) conditions. The peaks corresponding to the sum and difference spectra are colored red and black, respectively. ^{15}N–^{1}H splittings are indicated. The isotropic sample contained 1 mM PF0385 in 20 mM sodium phosphate, 15 mM KCl, 90% H_2O. The aligned sample contained 0.5 mM PF0385 in 20 mM sodium phosphate, 85 mM NaCl, 10 mg/ml Pf1 phage, 90% H_2O.

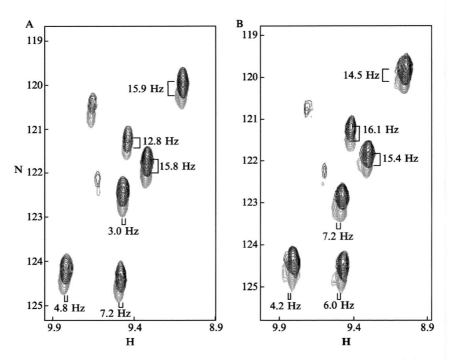

PRESTEGARD *ET AL.*, CHAPTER 7, FIG. 9. IPAP-HNCO of PF0385 under isotropic (A) and aligned (B) conditions. The samples were 1 m*M* PF0385 in 20 m*M* sodium phosphate, 15 m*M* KCl, 90% H_2O for the isotropic and the same buffer, 85 m*M* NaCl, 10 mg/ml Pf1 phage for aligned. The peaks corresponding to the sum and difference spectra are colored red and black, respectively.

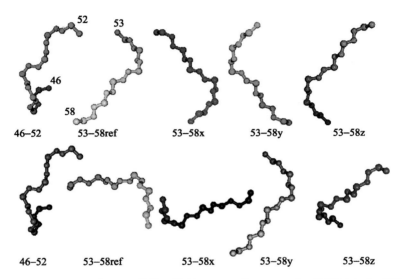

PRESTEGARD *ET AL.*, CHAPTER 7, FIG. 11. Fragment alignment using RDC data for PF0255 from two different media. At the left of each line is the piece of the fragment before proline 53. The remaining four depictions of the piece after proline have been produced by rotating the reference structure by $180°$ about the x, y, and z axes of the principal alignment frame. The structures in the second line have been rotated to overlay the first piece in both lines using the program chimera (Huang *et al.*, 1996).

PRESTEGARD *ET AL.*, CHAPTER 7, FIG. 13. Backbone superimposition of the RDC-based structure for residues 46–58 of PF0255 on the corresponding segment from the X-ray structure: 1RYQ.

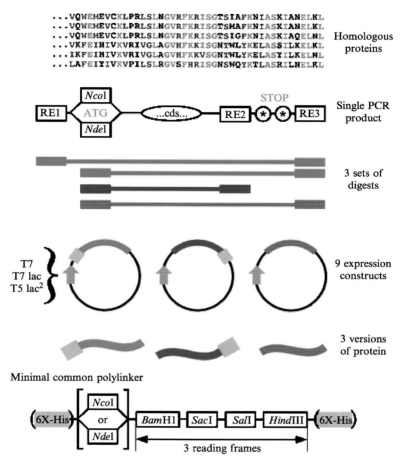

...VQWEMEVCKLPRLSLNGVRFKRISGTSIAFKNIASKIANELKL
...VQWEMEVCKLPRLSLNGVRFKRISGTSMAFKNIASKIANELKL
...VQWEMEVCKLPRLSLNGVRFKRISGTSIGFKNIASKIAQELNL   Homologous
...VKFEIHIVKVRIVGLAGVHFKKISGNTWLYKELASSILKELKL   proteins
...IKFEIHIVKVRIVGLAGVHFKKVSGNTWLYKELASYILKELNL
...LAFEIYIVKVPILSLRGVSFHRISGNSWQYKTLASRILNELKL

Single PCR product

3 sets of digests

9 expression constructs

3 versions of protein

Minimal common polylinker

3 reading frames

ACTON *ET AL.*, CHAPTER 8, FIG. 1. Strategy for multiplexed protein expression. One or more representatives from a family of homologous proteins or protein domains are chosen. For each domain to be expressed, a PCR product is designed to contain either an *Nco*I or an *Nde*I site at the initiator ATG for the coding sequence along with three additional restriction sites (RE1, RE2, and RE3) from the MCP for cloning different versions of the protein into the various expression vectors. RE1 and either *Nco*I or *Nde*I are included in the 5′ PCR primer, while the 3′ primer includes RE2 followed by one or two stop codons and RE3. Using three different combinations of digests, nine different expression variants can be generated from the same PCR primers, differing in the promoter driving expression, the placement of an affinity tag (if any), and the identity of any nonnative amino-acid residues that result from the cloning strategy. The minimum common polylinker found in each of the nine custom expression vectors is shown at the bottom.

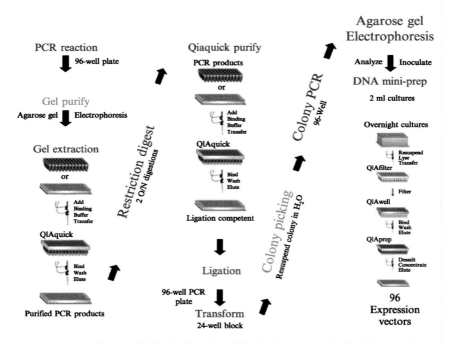

ACTON ET AL., CHAPTER 8, FIG. 3. Biorobot 8000 cloning schematic. Each key step in the cloning strategy is indicated; blue type denotes those steps that are completely automated, and red type indicates those steps that require some manual input. Roughly 1 week of one full-time equivalent is needed to complete all of the cloning steps for 96 target proteins. Several of the procedures are modifications of Qiagen-based protocols, such as the Qiaquick Purification and the DNA Mini-Prep protocols. However, most have been completely created in the Rutgers NESG Protein Production laboratory. A more detailed description of the cloning procedure, as well as the automated protocols, are provided elsewhere (Acton et al., 2005).

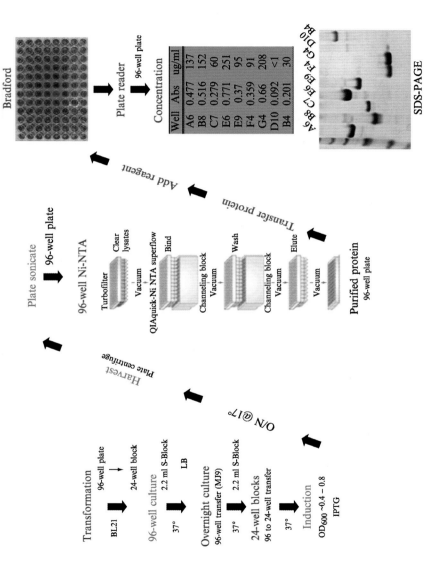

ACTON ET AL., CHAPTER 8, FIG. 4. (continued)

ACTON ET AL., CHAPTER 8, FIG. 4. High-throughput analytical scale protein expression screening using robotic methods. This schematic shows the step-by-step procedure used for small-scale expression screening. Completely automated steps are shown in blue, and red denotes steps that are partially automated. The entire process is conducted in 96-well plates or a corresponding number of 24-well blocks. The right top shows a modified 96-well Bradford assay (Bradford, 1976) with aliquots from the 96-well Ni-NTA purification. The plate configuration is the normal 8 rows by 12 columns. More intense blue wells denote a higher concentration of purified protein and hence constructs that express high levels of soluble proteins. These targets are slated for large-scale production. The relative concentration is calculated by the 96-well plate reader and is reported in spreadsheet format (see the blue box). An SDS–PAGE gel shows the results of the purification and relative agreement with the calculated values.

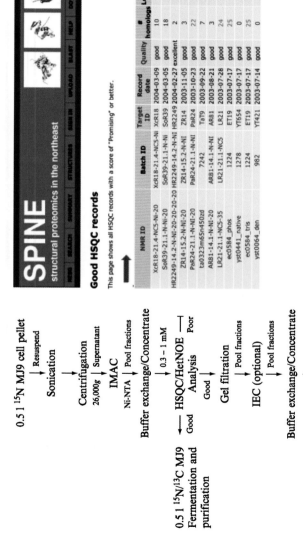

SPINE
structural proteomics in the northeast

HOME | SEARCH | SUPPORT | STRUCTURES | SIGN IN | UPLOAD | BLAST | HELP | DOWNLOAD

Good HSQC records

This page shows all HSQC records with a score of "Promising" or better.

NMR ID	Batch ID	Target ID	Record date	Quality	# homologs	Length	Image
XcR18-21.4-NC5-Ni-20	XcR18-21.4-NC5-Ni	XcR18	2004-03-09	good	10	135	[view]
SoR39-21.1-N-Ni-20	SoR39-21.1-N-Ni	SoR39	2004-03-05	good	18	197	[view]
HR2249-14.2-N-Ni-20-20-20	HR2249-14.2-N-Ni	HR2249	2004-02-27	excellent	2	223	[view]
ZRL4-15.2-N-Ni-20	ZRL4-15.2-N-Ni	ZR14	2003-11-05	good	3	114	[view]
PaR24-21.1-N-Ni-20	PaR24-21.1-N-Ni	PaR24	2003-10-23	good	22	169	[view]
ta0323m65n450zd	7242	TaT9	2003-09-22	good	7	124	[view]
ARB1-14.1-N-Ni-20	ARB1-14.1-N-Ni	ARB1	2003-08-21	good	3	149	[view]
LR21-21.1-NC5-35	LR21-21.1-NC5	LR21	2003-07-28	good	24	116	[view]
ec0584_phos	1224	ET19	2003-07-17	good	25	114	[view]
yst0441_native	1278	YT654	2003-07-17	good	0	141	[view]
ec0584_tris	1224	ET19	2003-07-17	good	25	114	[view]
yst0064_den	982	YT421	2003-07-14	good	0	148	[view]

Flowchart (Left):

0.5 l ^{15}N MJ9 cell pellet
→ Resuspend
Sonication
Centrifugation
26,000g → Supernatant
IMAC
Ni-NTA → Pool fractions
Buffer exchange/Concentrate
0.3 – 1 mM

0.5 l ^{15}N/^{13}C MJ9 Fermentation and purification — Good → HSQC/HetNOE Analysis
Good / Poor

Gel filtration → Pool fractions
IEC (optional) → Pool fractions
Buffer exchange/Concentrate
0.3 – 1 mM
NMR data collection

ACTON ET AL., CHAPTER 8, FIG. 6. (*Left*) Protein purification for NMR screening. Isotope-enriched cell pellets are resuspended in lysis buffer, sonicated, and cleared by centrifugation. Following Ni-NTA (Qiagen) IMAC purification, protein-containing fractions are pooled, concentrated, and exchanged into three buffers, which vary pH among other components. HSQC and HetNOE analysis is performed on these samples. If "good" spectra are obtained, further purification (gel filtration and optionally ion-exchange chromatography) is performed. (*Right*) View of "Good HSQC" Summary Page from the SPINE Database. This page lists those samples that are amenable to structural determination by NMR. Important aspects of the interface include the ability to view an image of the two-dimensional ^{15}N-^{1}H-HSQC for the listed target by selecting "[view]" under the image column. In addition, the number found in the "# homologs" column indicates how many additional protein targets from this Rost cluster family are in the NESG target list; selecting this link provides a list of these homologs and the progress by the consortium on each member of the family.

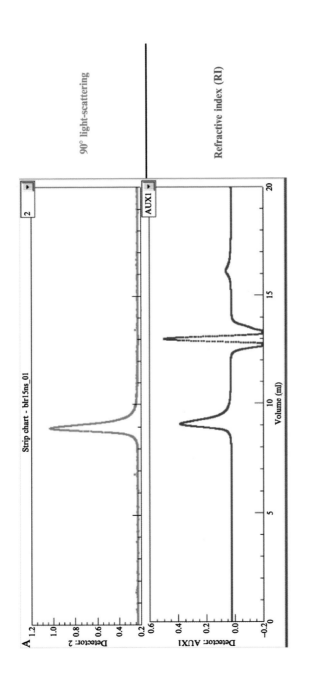

B

Aggregation Screening Record for blr15

- **Recommended Buffers** :
  - No Salt: 10mM Tris, 5mM DTT

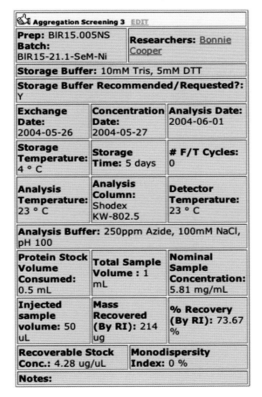

Aggregation Screening 3 EDIT		
Prep: BlR15.005NS Batch: BlR15-21.1-SeM-Ni	**Researchers:** Bonnie Cooper	
Storage Buffer: 10mM Tris, 5mM DTT		
Storage Buffer Recommended/Requested?: Y		
Exchange Date: 2004-05-26	**Concentration Date:** 2004-05-27	**Analysis Date:** 2004-06-01
Storage Temperature: 4 ° C	**Storage Time:** 5 days	**# F/T Cycles:** 0
Analysis Temperature: 23 ° C	**Analysis Column:** Shodex KW-802.5	**Detector Temperature:** 23 ° C
Analysis Buffer: 250ppm Azide, 100mM NaCl, pH 100		
Protein Stock Volume Consumed: 0.5 mL	**Total Sample Volume :** 1 mL	**Nominal Sample Concentration:** 5.81 mg/mL
Injected sample volume: 50 uL	**Mass Recovered (By RI):** 214 ug	**% Recovery (By RI):** 73.67 %
Recoverable Stock Conc.: 4.28 ug/uL	**Monodispersity Index:** 0 %	
Notes:		

ACTON *ET AL.*, CHAPTER 8, FIG. 8. Aggregation screening. A combination of analytical gel filtration, static light scattering, and refractive index detects the volume and mass of each protein species in solution. (A) The static light scattering of NESG target protein BlR15 in a "no salt" buffer is shown in the top chromatogram and indicates a single peak under these buffer conditions. The bottom chromatogram traces the refractive index; the single peak in the lower molecular weight region (corresponding to the single peak in the light scattering) shows that most of the mass injected into the column is contained in this peak, indicating that the majority of the protein in this buffer is monomeric. (B) The Aggregation Screening results for NESG target BlR15 are summarized in this view from SPINE.

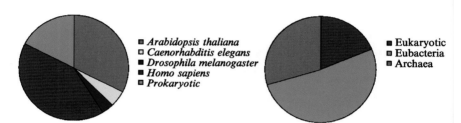

ACTON *ET AL.*, CHAPTER 8, FIG. 9. (*Left*) Phylogenetic distribution of NESG target proteins. Currently, 80% of the NESG targets are from the eukaryotic model organisms *Arabidopsis thaliana, Caenorhabditis elegans, Drosophila melanogaster*, and *Homo sapiens*. As illustrated in the pie chart, the majority of proteins are derived from the human and *Arabidopsis* genomes. The remaining 20% of NESG targets for Rutgers protein production (*S. cerevisiae* efforts are focused in Toronto) are of prokaryotic origin, including proteins from both archaea and eubacteria; see Table I for a complete listing of the eukaryotic organisms and the number of proteins targeted from each of these proteomes. (*Right*) Phylogenetic distribution of NESG protein structures deposited in the PDB. As indicated in the pie chart, ~20% of the NESG structures are of eukaryotic proteins, ~30% are of archaeal origin, and the remaining are structures of proteins from eubacteria. However, the majority of these prokaryotic proteins for which structures have been determined are members of protein domain families that also include eukaryotic members.

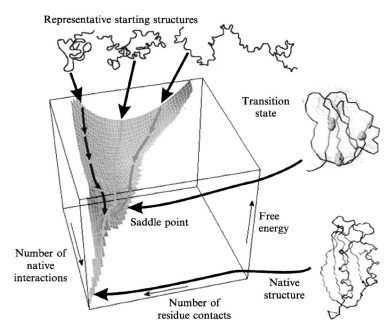

Representative starting structures

Transition state

Free energy

Saddle point

Number of native interactions

Number of residue contacts

Native structure

DYSON AND WRIGHT, CHAPTER 11, FIG. 1. Schematic energy landscape for protein folding, with a surface derived from a computer simulation of the folding of a highly simplified model of a small protein. The ensemble of denatured conformations is "funneled" to the unique native structure via a saddle point corresponding to the transition state, the barrier that all molecules must cross if they are to fold to the native state. Included in the figure are structure models corresponding to conformational ensembles at different stages of the folding process. The transition state ensemble was calculated from computer simulations constrained by experimental data from mutational studies of acylphosphatase (Vendruscolo *et al.*, 2001). The native structure is shown at the bottom of the surface, and at the top are some members of the ensemble of unfolded species, representing the starting point for folding. Adapted from Dinner *et al.* (2000) and Dobson (2003), and used with permission.

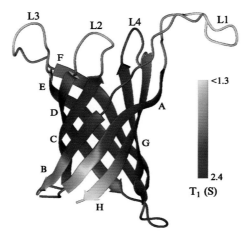

HWANG AND KAY, CHAPTER 13, FIG. 5. Ribbon diagram of the PagP β-barrel colored according to measured T_1 times in DPC at 45°. Residues for which T_1 times could not be measured were assigned values midway between flanking residues.

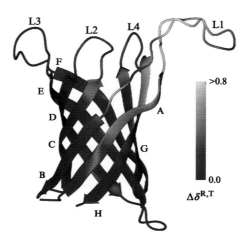

HWANG AND KAY, CHAPTER 13, FIG. 6. Ribbon diagram of the PagP β-barrel colored according to the combined chemical shift difference between the R and T states, $\Delta\delta^{R,T} = [(\omega_N\Delta\delta_N)^2 + (\omega_{C\alpha}\Delta\delta_{C\alpha})^2 + (\omega_{CO}\Delta\delta_{CO})^2]^{1/2}$, $\omega_N = 0.154, \omega_{C\alpha} = 0.276, \omega_{CO} = 0.341$ (Evenas *et al.*, 2001). For the R state, an average of the available chemical shifts in DPC, OG/ SDS, and the CYFOS-7 R state was used. If a residue could not be observed in the T state (mainly loop L3) or if the R and T states were identical, $\Delta\delta^{R,T} = 0$. Adapted from Hwang *et al.* (2004). Copyright 2004 National Academy of Sciences, USA.

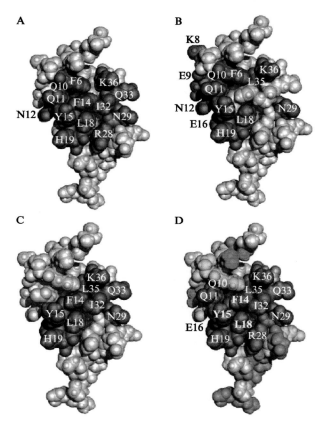

SHIMADA, CHAPTER 20, FIG. 8. Comparison of the binding sites of FB to the Fc fragment. (A) The X-ray study, (B) the chemical shift perturbation, (C) the H–D exchange experiments, and (D) the cross-saturation experiment. Residues with accessible surface areas that are covered upon binding of the Fc fragment (Phe-6, Gln-10, Gln-11, Asn-12, Phe-14, Tyr-15, Leu-18, His-19, Arg-28, Asn-29, Ile-32, Gln-33, and Lys-36) (Deisenhofer, 1981) are colored red in (A). Residues showing absolute values of chemical shift difference ($[(\Delta H_N)^2 + (\Delta N \times 0.15)^2]^{1/2}$) of more than 0.2 ppm (Phe-6, Lys-8, Glu-9, Gln-10, Gln-11, Asn-12, Ala-13, Tyr-15, Glu-16, Leu-18, His-19, Asn-29, Gly-30, Leu-35, Lys-36, and Ser-40) are colored red in (B). Of these residues, Ala-13, Gly-30, and Ser-40 are buried in the molecule. On the basis of the BioMagResBank database at http://www.bmrb.wisc.edu, the scale factor of 0.15 used for normalization of the magnitude of ^{15}N chemical shift changes (in ppm units) was derived from the average (over all residue types but Pro) standard deviations for the backbone $^{1}H_N$ and ^{15}N chemical shifts. Residues showing protection factors larger than 10 upon binding of the Fc fragment (Phe-14, Tyr-15, Glu-16, Ile-17, Leu-18, His-19, Asn-29, Ile-32, Gln-33, Leu-35, Lys-36, Leu-46, Ala-49, and Lys-50) are colored red in (C) (Gouda et al., 1998). Of these residues, Glu-16, Ile-17, Leu-46, Ala-49, and Lys-50 are buried in the molecule. Residues with intensity ratios less than 0.5 (Gln-10, Gln-11, Phe-14, Tyr-15, Glu-16, Leu-18, His-19, Arg-28, Asn-29, Ile-32, Gln-33, Leu-35, and Lys-36), 0.5 to 0.6, 0.6 to 0.7, 0.7 to 0.8, and larger than 0.8 are colored red, orange, yellow, green, and blue, respectively, in (D). Molecular graphics images were produced using the MidasPlus program from the Computer Graphics Laboratory, University of California, San Francisco (supported by NIH RR-01081) (Ferrin et al., 1988; Huang et al., 1991).

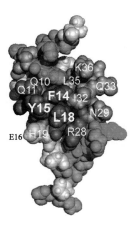

Intensity ratio
○ 0.0–0.5 ○ 0.8–1.0
○ 0.5–0.6 ○ no data
○ 0.6–0.7
○ 0.7–0.8

SHIMADA, CHAPTER 20, FIG. 12. Binding sites of the the FB–IgG complex determined by the transferred cross-saturation experiments. Residues with intensity ratios less than 0.5, 0.5–0.6, 0.6–0.7, 0.7–0.8, and larger than 0.8 are colored red, orange, yellow, green, and blue, respectively. Molecular graphics images were produced using MidasPlus (University of California, San Francisco) (Ferrin *et al.*, 1988; Huang *et al.*, 1991).